WORLDMARK
CHRONOLOGY
of the Nations

WORLDMARK CHRONOLOGY of the Nations

Volume 2 – Americas

Timothy L. Gall and Susan B. Gall, Editors

GALE GROUP

Detroit
San Francisco
London
Boston
Woodbridge, CT

The Gale Group
Editorial
Shelly Dickey, Project Editor
Matthew May, Assistant Editor
With contributions from the Cultures and Customs Team
Rita Runchock, Managing Editor
Production
Mary Beth Trimper, Production Director
Evi Seoud, Production Manager
Wendy Blurton, Senior Buyer
Product Design
Cynthia Baldwin, Production Design Manager
Michelle DiMercurio, Senior Art Director
Graphic Services
Barbara J. Yarrow, Graphic Services Director
Randy Bassett, Image Database Supervisor
Pamela R. Reed, Imaging Coordinator
Permissions
Maria Franklin, Permissions Manager
Margaret Chamberlain, Permissions Specialist

Library of Congress Cataloging-in-Publication Data

Worldmark chronology of the nations / Timothy L. Gall and Susan Bevan Gall, editors.
   p. cm.
Includes bibliographical references and index.
v. 1. Africa -- v. 2. Americas -- v. 3. Asia -- v. 4. Europe.
ISBN 0-7876-0521-2 (set) -- ISBN 0-7876-0522-0 (v. 1) -- ISBN 0-7876-0523-9 (v. 2)
-- ISBN 0-7876-0524-7 (v. 3) -- ISBN 0-7876-0525-5 (v. 4)
1. Geography. I. Gall, Timothy L. II. Gall, Susan B.
G133.W93   1999
910 21--dc21

                            99-044217
                            CIP

Contents

Contributors

Editors: Timothy L. Gall and Susan Bevan Gall
Associate Editors: Daniel M. Lucas, Eleftherios Netos
Editor, Photo Research: Michael Cikraji
Typesetting and Design: Bridgette Nadzam
Graphics: Rebecca Kimble, Hannah Lissauer
Editorial Assistants: Jill Coppola, Rebecca Kimble, Susan Stern, Jennifer Wallace

ADVISORS

JANE L. THOMAS. Librarian, McNeil High School, Round Rock Independent School District, Austin, Texas

WENDI GRANT. Teacher, Geography, McNeil High School Library, Round Rock Independent School District, Austin, Texas

FLO RANKIN. Library Media Specialist, Hoover High School, North Canton, Ohio

MARK KURT. Teacher, Lake Forest High School, Lake Forest, Illinois

KEITH PLATTE. Teacher, Social Studies, Kalamazoo Central High School, Kalamazoo, Michigan

CONTRIBUTORS

OLUFEMI A. AKINOLA, Ph.D. W.E.B. DuBois Institute, Harvard University

LESLIE ASHBAUGH, Ph.D. Department of Sociology, Seattle University

VICTORIA J. BAKER. Department of Anthropology, Eckerd College

IRAJ BASHIRI. Professor of Central Asian Studies, Department of Slavic and Central Asian Languages and Literature, University of Minnesota

THEA BECKER. Researcher/Writer, Cleveland, Ohio

HEATHER BOWEN. Researcher/Writer, Washington, D.C.

GABOR BRACHNA. Researcher/Writer, Cleveland, Ohio

SALVADOR GARCIA CASTANEDA. Department of Spanish and Portuguese, The Ohio State University

ERIK CHING. History Department, Furman University

FRANCESCA COLECCHIA. Modern Languages Department, Duquesne University

LEAH ERMARTH. Worldspace Foundation, Washington, D.C.

JENNIFER FORSTER. Researcher/Writer, Kent, Ohio

ALLEN J. FRANK, Ph.D., Researcher/Writer

DIDIER GONDOLA. History Department, Macalester College, St. Paul, Minnesota

ROBERT GROELSEMA. Ph.D. African Bureau, U.S. Agency for International Development (USAID)

HIMANEE GUPTA. Researcher/Writer, Honolulu, Hawaii

BRUCE HEILMAN. University of Dar es Salaam in Tanzania

ROSE M. KADENDE-KAISER. Director, Women's Studies Program, Mississippi State University

EZEKIEL KALIPENI. Department of Geography, University of Illinois at Urbana-Champaign

RICHARD A. LOBBAN, Jr., Ph.D. Department of Anthropology, Rhode Island College

IGNACIO LOBOS. Journalist, Honolulu, Hawaii

DERYCK O. LODRICK. Visiting Scholar, Center for South Asian Studies, University of California, Berkeley

DANIEL M. LUCAS. Researcher/Writer, Cleveland, Ohio

PATRIZIA C. MCBRIDE. Department of German, Scandinavian, and Dutch, University of Minnesota

MEGAN MENTREK. Researcher/Writer, Cleveland, Ohio

WILLIAM MILES. Department of Political Science, Northeastern University

EDITH T. MIRANTE. Project Maje, Portland, Oregon

CAROL MORTLAND. Crate's Point, The Dalles, Oregon

NYAGA MWANIKI. Department of Anthropology and Sociology, Western Carolina University

ELEFTHERIOS NETOS. Ph.D. Candidate, Kent State University

BRUCE D. ROBERTS, Department of Anthropology and Sociology, University of Southern Mississippi

GAIL ROSEWATER. Researcher/Writer, Cleveland, Ohio

JENNIFER SPENCER. Researcher/Writer, Columbus, Ohio

JEANNE-MARIE STUMPF. Researcher/Writer, Cleveland, Ohio.

CARMEN URDANETA. M.A. Researcher/Writer, Boston, Massachusetts

KIMBERLY VARGO. Researcher/Writer, Cleveland, Ohio

JEFF WASICK. Kent State University.

GERHARD H. WEISS. Department of German, Scandinavian, and Dutch, University of Minnesota

ROSALIE WIEDER. Researcher/Writer, Cleveland, Ohio

JEFFREY WILLIAMS, Ph.D. Cleveland State University

Preface

Worldmark Chronology of the Nations contains entries on 192 countries of the world. Arranged in four volumes—Africa, Americas, Asia, and Europe—*Worldmark Chronology of the Nations* follows the organizations of its sister sets, *Worldmark Encyclopedia of the Nations* and *Worldmark Encyclopedia of Cultures and Daily Life*. Within each volume, entries are arranged alphabetically.

Each volume begins with a general timeline of world history. This timeline provides a history of the world from prehistoric times to the present, lending a context to the entries on individual nations.

The nations profiled are those that exist at the end of the twentieth century. Profiles cover the territory within the modern-day geographic borders. History of earlier nation-states and empires is included in entries for each modern-day nation that has evolved within their region. Emphasis is on social and cultural history, helping the user of *Worldmark Chronology of the Nations* gain an understanding of a country's history beyond its succession of political leaders and military conflicts.

Each entry begins with an overview essay. This essay provides a general introduction to the country, its history, and its people from the beginning of time to the present. Trends in all aspects of a nation's development—politics, the military, literature, art, industry, religion, society, and relations with its neighbors and the other nations of the world—are included.

Following the introduction is a timeline of history, with dated entries describing the key people and events that shaped the nation. Notable people whose achievements helped shape the society and who contributed to world history are profiled. Entries may cover an era, a decade, a range of years, a specific year, a period of months, a specific month, or a specific date. A heading summarizes the event's significance, while the paragraphs that follow provide further details.

A current map accompanies each entry; in addition, over 100 historic maps illustrate the location of earlier empires, kingdoms, and depict transfers of territory between nation-states. Sidebars provide context for events, empires, people, and organizations of significance, such as the Cold War, Ottoman Empire, and League of Nations.

A comprehensive bibliography following the organization of the work (by continent and country), glossary of terms, and comprehensive index appear at the back of each volume.

Over fifty university professors, professional writers, journalists, and expert reviewers contributed to the preparation of this *Worldmark Chronology of the Nations*. Many have carried out historical research in the country about which they wrote. All are skilled researchers with expertise in their chosen area of study.

Acknowledgments

The editors express appreciation to staff of The Gale Group who were involved at various stages of the project: Linda Irvin and Kelle Sisung, who assisted with development of the initial concept of the work; Shelly Dickey, Project Editor, Matthew May, Assistant Editor, who guided the editorial development of the entries with contributions from the Cultures and Customs Team, Rita Runchock, Managing Editor; Cynthia Baldwin and Michelle DiMercurio, who were responsible for the design of the volumes' covers; and Mary Beth Trimper, Evi Seoud, and Wendy Blurton, who supervised the printing and binding process. In addition, the editors express appreciation to William Becker, archivist of the Cleveland State University Photo Archive, for his assistance with photo research. Also helping with images were members of the Gale Group Graphic Services department, Barbara J. Yarrow, Randy Bassett, and Pamela R. Reed; and Permissions department, Maria Franklin and Margaret Chamberlain.

SUGGESTIONS ARE WELCOME: The first edition of a work of this size and scope is an ambitious undertaking. We look forward to receiving suggestions from users on ways to enhance and improve future editions. Please send comments to:

Editors
Worldmark Chronology of the Nations
The Gale Group
27500 Drake Road
Farmington Hills, MI 48331–3535
(248) 699-4253

Timeline of World History

c. 2,600,000 B.C. *Homo Australopithecus* **walks the earth**

Homo Australopithecus, the earliest ancestor of *Homo Sapiens Sapiens* (present-day human beings) lives in sub-Saharan Africa. *Homo australopithecans* are hunter-gatherer nomads who fashion stone tools.

c. 100,000 B.C. **Emergence of** *Homo Sapiens Sapiens* **in Africa**

The earliest ancestors of the modern human species, *Homo Sapiens Sapiens,* emerges in Africa. They have larger brains than their predecessors and are hunter-gatherers who communicate by means of sophisticated language. By 30,000 B.C. they are the only human-like hominid (two-legged primate) left.

c. 8000 B.C. **Agricultural revolution**

Humans begin raising crops and domesticating animals. This results in the greatest change in human lifestyle to this point; humans start settling in fertile river valleys. This change in life patterns leads to the development of the first civilizations.

c. 3200 B.C. **Sumerians establish first civilization**

The first civilization emerges in Sumer, in Mesopotamia, along the Euphrates River. The establishment of civilization brings with it government, religion, urbanization, and a specialized economy (including trade between cities). Significantly, the Sumerians establish a pictographic system of writing called Cuneiform. A writing system allows for record keeping as well as communication, particularly for trading purposes.

c. 2850 B.C. **Old Kingdom established in Egypt**

During the period of the Old Kingdom, the lands along the Nile River in northern Africa are the second region of the world to form a civilization. Arguably the most impressive achievement of the Old Kingdom are the pyramids which are built as the tombs of pharaohs (kings). The Egyptians of the Old Kingdom also establish a calendar based on the annual flooding of the Nile and develop a form of writing known as hieroglyphics which they often write on papyrus scrolls (early paper).

2200 B.C. **Emergence of Chinese civilization**

Chinese civilization emerges along the Yellow River with the establishment of the Xia (Hsia) dynasty. This is the first step in the formation of a Chinese empire. To this day, Chinese civilization is the oldest in existence.

1728–1686 B.C. **Code of Hammurabi**

King Hammurabi of Babylonia writes a law code for his subjects. The laws, which seek to maintain stability in the kingdom, pertain to topics as diverse as marriage, taxes, and business contracts. The Code of Hammurabi is the oldest law code in the world.

c. 1500–1000 B.C. **Emergence of Indian civilization, basis of Hinduism**

Indian civilization emerges along the Indus River valley. An early achievement of the Indus civilization is the compilation of the first *Vedas*, spiritual texts which form the basis of Hinduism.

c. 1200 B.C. **Emergence of Hebrew civilization, Judaism**

The emergence of the Hebrew civilization gives rise to Judaism, a monotheistic religion. Although the Hebrews subsequently succumb to internal division and foreign invasion, their religion is one of the most influential in the world. Both Christianity and Islam are offshoots of Judaism.

c. 1000 B.C. **Emergence of the Kingdom of Kush**

The Kingdom of Kush emerges in present-day Sudan. Centered around the city of Meroe, the Kingdom of Kush is the first sub-Saharan African civilization.

c. 1000 B.C. **Emergence of first American civilizations in Mexico and Peru**

Civilization emerges in the Americas in the areas of present-day Mexico and Peru.

8th century B.C.   Emergence of Classical Greek civilization, *Iliad* and the *Odyssey* compiled

Classical Greek civilization arises out of a dark age. The Greeks organize themselves into city-states both on the Greek mainland as well as on hundreds of the islands that make up the Aegean and Ionian seas. Great seafarers, the Greeks establish colonies and trade throughout the Mediterranean and Black Seas.

Greek civilization is also well-known for its many cultural achievements. Homer's compilation of the *Iliad* and the *Odyssey* are among the oldest works of Western literature. Greek civilization reaches its height during the Athenian Golden Age (see 5th century B.C.).

563–483 B.C.   Life of Buddha

Siddharta Gautama, an Indian prince, seeks eternal happiness as a way of escaping suffering in the world. After a personal odyssey, he adopts the name Buddha (enlightened one) and forms a belief system that becomes known as Buddhism. In subsequent centuries the Buddhist faith spreads throughout much of Asia including Japan, Korea, and China (where it becomes the dominant form of religious belief).

551–479 B.C.   Life of Confucius

Chinese philosopher Confucius advocates human behavior based upon morality and justice. His writings, compiled in the *Analects,* are part of a program of political reform.

509 B.C.   Roman Republic founded

The Roman monarchy is overthrown and replaced by a republican system of government. Under this system, citizens with voting rights (landholding males) elect representatives to legislate for them. The foundation of the republic leads to stronger and more effective government which eventually unites all of the Italian peninsula. The republican system of government breaks down when Rome subsequently expands to encompass all of the Mediterranean world as well as much of southern and western Europe. The Roman Republic becomes the Roman Empire.

480–431 B.C.   Athenian Golden Age

During the Athenian Golden Age, Athens becomes the center of Greek politics and culture. The city's democratic government allows all of its citizens (roughly a third of its population) to participate in government at some point in their lives. Art, philosophy, and theater flourish. The Parthenon is erected, Socrates (469–399) debates pupils and populace, and Euripides (485?–406?) writes plays. The Golden Age ends with the outbreak of the Peloponnesian War (431–404) against Sparta, which ends in an Athenian defeat.

c. 4th century B.C.   Emergence of Mayan civilization

Mayan civilization emerges in the area of present-day Mexico. The Maya are the first civilization to emerge in the Americas and are best-known for their sophisticated mathematics which they apply to the study of astronomy and which enables them to devise a highly accurate calendar.

356–323 B.C.   Life of Alexander the Great

In only thirteen years King Alexander III (the Great) of Macedon creates an empire that extends from Greece to the Indus River. His conquests spread Greek civilization throughout the Middle East and into Asia.

4th–2nd centuries B.C.   Hellenistic Era

In the wake of Alexander's death, his empire fragments as his successors quarrel for the spoils. Nevertheless, a new cosmopolitan culture emerges that combines Greek and Eastern influences.

3rd century B.C.   Erection of the Great Wall of China

In an effort to keep foreign invaders at bay, the Chinese emperor constructs a wall stretching across the northern frontier of China. Known as the Great Wall, this fortification stretches for over a thousand miles (1,620 kilometers) and becomes one of the Seven Wonders of the World.

A.D. 1st century   Axumite Kingdom established

The Axium Kingdom emerges in Ethiopia. This state soon adopts Christianity and becomes a Christian stronghold in sub-Saharan Africa as it has remained to the present.

5 B.C.–A.D. 29   Life of Jesus, Rise of Christianity

In Roman Judea, Jesus of Nazareth claims to be the Jewish messiah (savior). Although he gains a following among those who embrace his message, the Roman authorities view him as an obstacle to their rule and he is sentenced to death. After his crucifixion, his followers claim that he is resurrected. His life is the basis of a new religion, Christianity, that comes to dominate the Roman Empire by the fourth century. It subsequently spreads throughout Europe into Asia, Africa, and, ultimately, the Americas and Australia. It is presently the largest faith in the world with over one billion followers.

5th century   Middle Ages, rise of feudalism

The Middle Ages follow the fall of the Roman Empire in Western Europe. This period is characterized by the establishment of small kingdoms based upon the economic system of feudalism. Under this regime, peasant serfs work for lords who provide protection from invaders. In turn, the lords provide services for a king in return for land.

440–61   Reign of Pope Leo the Great

Under Pope Leo the Great, Rome increases its power over the other four patriarchates (the five original seats of church government: Rome, Constantinople, Antioch, Jerusalem, Alexandria) of Christianity. Subsequently, the pope becomes the leader of the Roman Catholic branch of Christianity.

528   Justinian's Code

The Byzantine (Eastern Roman) emperor Justinian, (483–565) establishes a code of laws for his empire called the *Corpus Juris Civilis*, which forms the foundation of much of European law today.

7th–8th centuries   Rise and spread of Islam

Under the direction of Mohammed (570–632), the Islamic religion emerges among the Arabs. An offshoot of Judaism and Christianity, Mohammed claims that he is the last prophet of God. After Mohammed's death, the Muslims (followers of Islam) create an Islamic empire that stretches from the North African coast and Spain as far east as Persia. Islam later spreads throughout much of sub-Saharan Africa, and central and southeast Asia. It is presently the second-largest religion in the world.

8th century   Emergence of Ghana

The African kingdom of Ghana emerges in the northwest part of Africa and is the first sub-Saharan African civilization. The kingdom establishes trade across the Sahara with Morocco. Although Ghana is an advanced civilization it falls victim to invasion and disintegrates in the eleventh century.

8th–9th centuries   Carolingian Renaissance

The rise of the Carolingian Empire in western Europe, results in a renewed stress on learning. Newly-established monasteries become centers of education and preservation where monks study Greek and Roman classics (as well as the scriptures) while also copying and translating the old texts.

10th century   Rise of Russia

The Varangians establish the first Russian state centered in Novgorod. This later becomes the foundation of the Russian Empire that stretches across Siberia.

1054   East-West split in Christianity

Christianity splits into two branches: Roman Catholicism, centered in Rome; and Eastern Orthodoxy, centered in Constantinople, the capital of the Byzantine Empire. The split is of vital importance over time as the two halves of Europe undergo significantly different historical development.

11th–13th centuries   Crusades foster greater East-West contact

In an effort to oust the Muslims from the Holy Land, Christian warriors from western Europe wage a series of holy wars (crusades) against the Muslim Turks (who control the Middle East). All but the first crusade are military defeats for the Europeans. (Indeed, the Fourth Crusade, rather than attacking the Holy Land, captures Constantinople in 1204.) Nonetheless, they help bring western Europe out of its isolation during the Middle Ages.

1192   Feudal Japan emerges

Feudalism emerges in Japan. Under this system, the emperor becomes a figurehead and political power rests in the hands of the *Shogun*, the chief warlord. Japan remains under this system until the Meiji Restoration in 1868.

c. 1200   Height of Inca civilization

The Inca establish an empire in the Quechua mountains along the Pacific coast in South America. The Inca government is a theocracy in which the emperor holds absolute authority. Among the Inca achievements are the creation of thousands of suspension bridges across rivers and numerous religious sculptures.

12th century   Rise of the Mongol Empire

Under the leadership of Genghis Khan (1162–1227), the Mongols establish an empire that stretches from China, through central Asia, into Europe.

c. 1300   Aztec civilization

The Aztec civilization in present-day Mexico reaches its height in the fourteenth century. The Aztecs have an advanced civilization that includes sophisticated cities and government. Along with the great pyramids and temples they build, they are known for their bloody ritual human sacrifices meant to please their gods.

1347–50   Black Death sweeps through Europe

Brought in from Asia, Bubonic Plague (known as the Black Death) spreads across Europe and kills approximately half its population.

14th–16th centuries   Era of the Renaissance

A period of intellectual and cultural rebirth, the Renaissance, begins in northern Italy and later takes root in northern Europe, particularly Holland. The central characteristic of the Italian Renaissance is the emphasis on classical (ancient Greek and Roman) civilization. Renaissance artists include Leonardo da Vinci (1452–1519), Michaelangelo (1475–1564), and Raphael (1483–1520). Much of their work graces churches and cathedrals in Rome, most notably St. Peter's.

1445   Gutenberg invents printing press

Johann Gutenberg (1400–67) invents the printing press which allows for the widespread publication of books. Mass production of books leads to widespread access to ideas and is instrumental in bringing people closer together.

1453   Constantinople falls to the Ottoman Turks

Constantinople, the capital of the Byzantine Empire, falls to the Ottoman Turks. The Ottomans, a Muslim people, establish an empire that, at its height, includes all of Asia Minor, all of the Balkans and central Europe as far as Vienna, the North African coast, the Fertile Crescent, and the Caucasus. The empire lasts until 1923.

Late 15th century–19th century   African slave trade

The African slave trade begins with the Portuguese. Over the course of four centuries approximately twenty-two million black Africans are uprooted forcibly. The half who survive the voyage are sent primarily to the Americas where they primarily work in agriculture.

1492   Columbus makes first voyage to the Americas

While searching for a shorter route to Asia by sailing west, Christopher Columbus (an Italian in the service of the Spanish) instead reaches the West Indies and the South American coast. However, not recognizing his mistake, he embarks on three return journeys over the next decade. His voyages promote further European exploration that results in the eventual colonization of North and South America.

Early 16th century   Protestant Reformation

A revolt against certain practices of the Catholic Church led by German monk Martin Luther (1483–1546) results in a full-scale revolt against the Papacy in western Europe. Religious wars ensue between the Catholics and Protestants (those who split from Rome) and leave much of western Europe politically and economically devastated.

1500–1700   European exploration and colonization of the Americas

European explorers make a series of voyages to the Americas. Although their early journeys focus on discovery, they soon turn to expansion. Spain and Portugal conquer most of the Americas (although England, France, and the Netherlands stake claims as well) and establish vast empires in the "New World." The Europeans soon begin exploiting their new territories for mineral and agricultural wealth (often through slave labor). Most of the native population dies of disease or warfare while the Europeans begin establishing their territories.

1519–21   Magellan circumnavigates the globe

An expedition led by Portuguese explorer Ferdinand Magellan (1480–1521) sets sail to circumnavigate the globe. Although Magellan and most of his crew die, the survivors return to Portugal in 1521 and prove that the world is round.

1519–25   Spanish conquest of the Aztecs

The Spanish, under Hernan Cortez (1485–1547) conquer the Aztec Empire in Mexico. The rapid conquest of the Aztecs illustrates the advantages held by Europeans over non-Europeans in weapons technology.

1564–1616   Life of William Shakespeare

English playwright and poet William Shakespeare creates his artistic and literary masterpieces. Not only do his works become widely-acclaimed as the best writings in the English language, they receive international recognition, are translated into dozens of languages, and his plays are performed throughout the world.

1581   Beginning of Russian expansion into Siberia

Russian traders begin expanding their activities eastward into Siberia, incorporating the land into the Russian Empire as they travel. This begins one of the greatest expansions in history as the Russian Empire eventually grows to encompass one-sixth of the entire land area of the world.

17th–18th centuries   Age of Reason, Enlightenment

The principles of liberal political thought emerge during the Enlightenment. Writers such as John Locke (1632–1704), Voltaire (1694–1778), and Jean-Jacques Rousseau (1712–78) stress the importance of individual liberty based on private property, religious toleration, and the social contract between a government and its people. These ideas become the key ideology behind the American and French Revolutions and remain an important part of Western political thought through the end of the twentieth century.

1618–48   Thirty Years War

When Catholic Habsburg prince Ferdinand II becomes king of Protestant Bohemia, Protestant and Catholic states in Germany, and eventually throughout central Europe, fight one another for political and religious dominance.

c. 1750   Industrial Revolution

Industrialization begins to develop in England. Characterized by steam-powered manufacturing and mass production, the industrial revolution sets the stage for the modern age. The development of industry soon spreads to western Europe and North America and propels them to the forefront of world economic, political, and military power. With the West dominant throughout the globe, the rest of the world attempts to industrialize as well.

1770   Cook charts the east coast of Australia

English Captain James Cook (1728–79) charts the east coast of Australia while on a South Sea expedition. His discovery paves the way for the eventual European colonization of the island-continent.

1776   Adam Smith writes *The Wealth of Nations*

British economist Adam Smith writes *The Wealth of Nations*. This work stresses that the key to expanding a nation's wealth lies in the establishment of a capitalist economic system characterized by free enterprise, private property, and an absence of government interference in the economy.

1776–83   American War of Independence

In the name of Enlightenment principles of political freedom, Great Britain's thirteen American colonies declare their independence and, by 1783, emerge victorious as the United States of America. The new country comes to dominate the North American continent and by the twentieth century becomes the most powerful nation in the world.

1789   French Revolution

With its goals of *Liberté*, *Egalité*, and *Fraternité* (Liberty, Equality, and Fraternity), revolution breaks out in France. By 1792 it overthrows the monarchy and institutes a republic. Although the revolution ultimately degenerates into the imperial dictatorship of Napoleon Bonaparte (1769–1821) it serves as a model for subsequent liberal nationalist revolutions throughout the world.

Early 1800s   Latin American revolts against Spain

Spain's colonies in Latin America rise in revolt against Spain. By the mid-1820s, the once vast Spanish Empire in South America is reduced to minor island possessions.

1839–42   Opium War

The Chinese forbid the importation of opium into their ports by the the British, who ship it from India. The British wage war on China forcing them to reopen their ports. The Chinese defeat is a severe setback to the forces that oppose Western imperialism.

1848   Revolutions in Europe

Liberal nationalist revolutions sweep across the European continent. In the short-term, the revolutions fail, although their long-term impact is profound as they form the basis of future liberal nationalist political programs that play an important role in European politics in the late nineteenth century.

1848   Karl Marx publishes the *Communist Manifesto*

In the wake of revolutions that sweep across Europe, Karl Marx publishes his *Communist Manifesto*, which advocates the overthrow of capitalism in favor of socialist societies in which all property is communal and workers rule. Although his work has no immediate impact, his ideas form the basis of future Communist revolutions in the twentieth century.

1868   Meiji Restoration

In an effort to combat Western imperialism, Japanese patriots stage a revolution that "restores" the emperor's authority. As a result of this successful revolt, Japan embarks upon a policy of rapid modernization that makes it a leading power by the first decade of the twentieth century. Its defeat of Russia in the Russo-Japanese War (1904–05) is the first time that a non-European power defeats a European power in modern times.

1885   Karl Benz designs first automobile

The German Karl Benz designs the first automobile. Within decades, his invention becomes the primary means of ground transportation in the industrialized world.

1903   Wright brothers make first powered flight

The American Wright brothers make the first powered flight in a light airplane in Kitty Hawk, North Carolina. By the late twentieth century, air travel becomes one of the safest and primary means of long-distance transportation.

1911–49   Chinese Revolution

In 1911, the Manchu dynasty falls and is replaced by a republic led by Sun Zhongshan (Sun Yat-Sen, 1866–1925). The revolution is aimed at modernizing China and throwing off Western imperialism. Sun is succeeded by Jiang Jieshi (Chiang Kai-Shek, 1887–1975). By the late 1920s, however, a Communist movement led by Mao Zedong (Mao Tse-Tung, 1893–1976) challenges Jiang for control and a bloody civil war ensues. By 1949 the Communists are victorious and drive Jiang's forces off the mainland to the island of Formosa (Taiwan).

1914–18   World War I

World War I pits two great alliances against each other: the Triple Alliance (Germany, Austria-Hungary, the Ottoman Empire, and, later, Bulgaria) versus the Triple Entente (Great Britain, France, Russia, later joined by the United States) and their associates. The war results in an Entente victory but only after most of Europe is destroyed. The war kills around nine million people. Political instability follows the war and leads to World War II in 1939.

1917 Russian Revolution

The monarchical tsarist regime falls in Russia. After a brief period in which a provisional government takes over, Russia is governed by the Bolsheviks, who establish the first Communist government in the world. The Communists seek to create a socialist state run by the working class. In practice, however, their one-party regime relies upon force and widespread suppression of human rights to remain in power.

1939–45 World War II

World War II pits the expansionist Axis (Germany, Italy, and Japan) and its associates against the Allies (Great Britain, France, China, the United States, and the Soviet Union) and its partners. By 1945 the Allies are victorious. Their victory, however, comes at great cost as most of Europe and Asia are destroyed. The war kills around sixty million people, including twenty-seven million Soviet citizens and six million European Jews. Jews are singled out for extermination by Nazi Germany in a policy known as the Holocaust.

1947–91 Cold War

The wartime alliance between the United States and the Soviet Union (known as the superpowers) breaks down into a relationship of mutual distrust and hostility. The Cold War is the commonly used name for the prolonged rivalry and tension between the United States and the Soviet Union which lasts from the end of World War II to the break-up of the U.S.S.R. in 1991. The Cold War encompasses the predominantly democratic and capitalist nations of the West, which are allied with the U.S., and the Soviet-dominated nations of eastern Europe, where Communist regimes are imposed by the U.S.S.R. in the late 1940s. Although the United States and the Soviet Union never go to war with each other, the Cold War results in conflicts elsewhere in the world, such as Korea, Vietnam, and Afghanistan, where the superpowers fight wars either by proxy or with their own troops against allies of their superpower rival.

1950s–60s Decolonization

Throughout the world European colonial empires collapse and previously subjugated peoples receive their independence. Freedom from colonial rule presents many challenges to the newly-independent states. Economic underdevelopment, widespread poverty, and tenuous political stability all plague these new countries.

1969 Man walks on the moon

On July 20, 1969, U.S. astronauts Neil Armstrong (b. 1930) and Edwin "Buzz" Aldrin (b. 1930) become the first men to walk on the moon. This marks the first time human beings have ever set foot upon another celestial body. It is only sixty-six years since the Wright brothers' first flight and only eight years since Soviet cosmonaut Yuri Gagarin made the first manned space flight.

1980–Present Information revolution

The increasing importance of computers in daily life—particularly for the transmission of information results in the so-called "Information Revolution."

1989 Collapse of Communism in Eastern Europe

As part of the liberalizing policies of Soviet leader Mikhail Gorbachev (b. 1930), the U.S.S.R. allows its East European satellites to go their own way. As a result, Communist regimes throughout Eastern Europe collapse.

1991 Collapse of the Soviet Union

Gorbachev's liberalization policies lead to a rise in nationalism among the peoples of the Soviet Union. When his reforms prove unable to rescue Communism, the Communist Soviet Union disintegrates into its constituent republics. Russia, the chief republic of the old U.S.S.R. is a mere shadow of its predecessor and is plagued by political and economic instability for the rest of the decade.

1992 European Union formed

The nations of the European Community (Belgium, Denmark, France, Germany, Greece, Ireland, Italy, Luxembourg, Netherlands, Portugal, Spain, and the United Kingdom) form the European Union. More than an economic union (as its predecessor was), the European Union seeks to establish a common currency, defense and foreign policy.

1997: Summer Asian economic crisis

Beginning in Thailand, a financial crisis strikes East Asia. Caused by over-investment and currency devaluation, the crisis brings economic growth in East Asia to a halt and threatens global prosperity. Prior to this economic reverse East Asian economies are the fastest-growing in the world and comprise one-third of the world economy.

Antigua and Barbuda

Introduction

Antigua and Barbuda is a state located in the Leeward Islands, lying southeast of the U.S. Commonwealth of Puerto Rico and north of the French Overseas Department (province) of Guadaloupe. Antigua has an area of 108 square miles (280 square kilometers) and Barbuda's area is 62 square miles (161 square kilometers). Barbuda lies 25 miles (40 kilometers) north of Antigua. The country also includes uninhabited Redonda, located 25 miles (40 kilometers) southwest of Antigua. The total coastline is 95 miles (153 kilometers). Its terrain is partly volcanic and partly coral; Antigua has deeply indented shores lined by reefs and shoals, with many natural harbors and beaches. Boggy Peak in southwestern Antigua is the nation's highest point, at over 1,300 feet (400 meters) above sea level. Antigua's northwestern coast is dotted by many tiny islets; the central area is fertile plain. Barbuda has a large harbor on the west side; Antigua is visible from Barbuda but it is not possible to see Barbuda from Antigua. Redonda is a low lying rocky islet.

The islands' climate is fairly tropical. There are no permanent rivers on either Antigua or Barbuda so the water supply is of great concern. Temperatures average 75°F (24°C) in the winter and 84°F (29°C) in the summer. There are periodic droughts and occasional hurricanes. The country's population is over 65,000, but only about two percent of the population lives on Barbuda. St. John's in Antigua is the capital, and about half the country lives in its surroundings. Most of the population is descended from African slaves brought to the islands in the seventeenth and eighteenth centuries. There are small numbers of persons of European, Arab, and Asian Indian ancestry. Anglicans account for about half the population, and other Protestant groups (Baptist, Methodist, Pentecostal, Seventh-Day Adventist, Moravian, Nazarene) account for most of the rest.

The first inhabitants of Antigua and Barbuda, the Siboney, arrived on the islands some 4,500 years ago, migrating up the Antilles island chain from South America. Arawak settlers arrived in the first century A.D. The Arawak Indians were primarily farmers and fishers. The Caribs were the last Amerindian group to settle the islands before the arrival of Europeans.

The Colonial Era

Columbus arrived in 1493, but the Spanish did not attempt to start a settlement until nearly thirty years later, which they soon abandoned. Early attempts by the British and French to settle on Antigua also failed, partly due to the island's scarce water supply. The first successful European settlement was finally established by the British in the 1630s.

At first, indentured laborers (workers tied to the land for a fixed period before obtaining their freedom) were brought to the island as agricultural workers. Sugarcane eventually became the main industry, and African slaves were imported to cultivate the labor-intensive crop. By the early 1700s, sugarcane was Antigua's only agricultural product. Sugarcane plantation agriculture flourished into the late eighteenth century.

In 1736, a serious coup plot by the slaves was uncovered before being implemented. When the slaves were finally emancipated in 1834, only Antigua and Bermuda disregarded the transitional five-year apprenticeship system practiced in other British territories and immediately granted the slaves total freedom. The emancipated population moved off the plantations and began establishing new villages along the coast.

The sugarcane industry lost its predominance beginning in the 1840s, due to the substitution of sugar beets from Europe. The international price for sugarcane fell throughout the second half of the nineteenth century, and the economy stagnated. At the beginning of the twentieth century, Antigua and most of the sugar-growing British West Indies were impoverished. When construction of the Panama Canal began, many Antiguans and other men throughout the Caribbean emigrated to Panama for work.

During World War II (1939–45), in order to secure the defense of the Western Hemisphere, the United States established a military presence on Antigua that still remains today. The military operations introduced Antiguans to the possibility of opportunities other than agricultural labor. A strike by sugarcane workers pitted Antigua's canefield laborers against the island's dominant sugarcane estate. V.C. (Vere Cornwall) Bird (b.1910), the president of the union, rose to prominence and became Antigua's premier in 1961. Except for a few years in the 1970s, he remained in power until 1994. The sugar industry was completely dismantled in the 1970s, and tourism quickly grew to become Antigua's leading economic activity. Antigua and Barbuda became an independent nation in 1981. After independence was achieved, the new country was upset in the 1980s by a series of scandals linked to Bird and two of his sons (Vere Bird, Jr. and Lester Bird) who also held managing positions in the administration. Nevertheless, Lester Bird was made prime minister in 1994 after his father's resignation.

Timeline

2400 B.C.   First known inhabitants of Antigua and Barbuda

Settlements of the first people to inhabit Antigua and Barbuda date back to 2400 B.C. The earliest settlers migrate up the Antilles island chain from South America and establish small independent communities that gradually become distinct from those in South America.

First century A.D.   Arawak settlers

Settlers from the Orinoco River Basin of South America first arrive on Antigua and Barbuda some 2,000 years ago. Called Arawaks, their communities line the coast of Antigua, and they are skilled artisans, making tools, pottery, and decorated shells. They are primarily farmers and fishermen.

1200s   Carib settlers

Migrating from the Orinoco Basin of South America, the Caribs are the last Amerindian group to settle on the islands of the Caribbean before the advent of European exploration. The Caribs are a fierce people who maintain control by subduing the Arawaks already present on the island. The Caribs are able to displace and replace their enemies through the practice of killing male opponents and assimilating captured women and children into their culture. Caribs migrate to Antigua until about 1700. They expand their control over the Caribbean until the arrival of the Europeans.

1493   Columbus arrives on Antigua

Christopher Columbus (1451–1506) arrives on Antigua and names the island after the church of Santa Maria de la Antigua in Seville, Spain. Columbus introduces sugarcane to the island but commercial production of this crop only begins two centuries later.

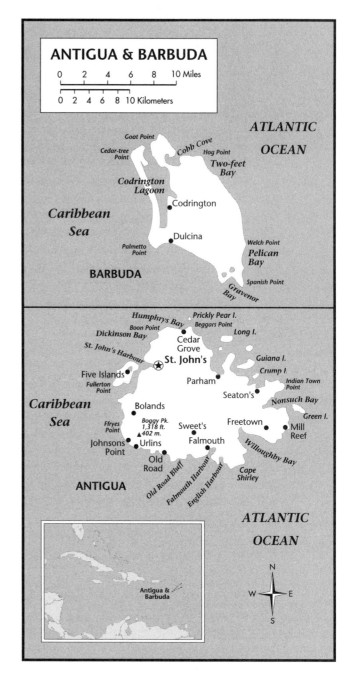

1520   Spanish found settlement on Antigua

Spanish explorers attempt to start a permanent settlement, but the lack of fresh water and attacks by Caribs cause them to leave.

Colonial Development

1627: July 2   Earl of Carlisle given title to several islands in the Lesser Antilles

Several English merchants arrange to make their patron, James Hay, Earl of Carlisle in Scotland, owner of Saint

Vincent and most of the Lesser Antilles. The earl is well-liked by King James I and owes the merchants a lot of money. The king issues a royal patent to the earl, making him "Lord Proprietor of the English Caribbee Islands." The patent covers twenty-one territories, including Antigua.

1628   British settlement attempted on Barbuda

The English start a settlement on Barbuda but the poor soil and raids by Caribs cause them to abandon it.

1629   French found settlement on Antigua

The French, led by Pierre Belain d'Esnambuc (1585–1637), attempt to settle Antigua but soon abandon it.

1632   British found settlement on Antigua

Sir Thomas Warner (d.1649) and a group of English colonists arrive from nearby Saint Kitts to establish a British settlement on Antigua. In its first years, the Caribs repeatedly attack the new settlement but the English settlers are determined to establish a colony where they can grow tobacco and sugarcane.

1640s   Immigrants from Barbados arrive

As sugarcane plantation agriculture develops on Barbados, more and more small plot holders are forced out of business. Some of these poor farmers move to Antigua and Nevis.

1650s   Sugarcane replaces tobacco as main crop

Many planters across the islands of the West Indies are discovering that it is more profitable to grow sugarcane than tobacco, however sugarcane cultivation is much more labor-intensive than tobacco. At first, planters rely on indentured laborers and prisoners-of-war to cultivate the sugarcane, but eventually slaves are imported from Africa. Sugarcane production remains the basis of the island's economy for the next 300 years.

1665   French capture Antigua

During the Second Dutch War, France allies with Holland against England. The French seize plantations in the British West Indies and some British residents flee to the mainland colony of Virginia. On Antigua, the French torch most of the houses and allow their Carib allies to engage in plunder and rape.

1667: July 21   Treaty of Breda

England and the Netherlands sign the Treaty of Breda, ending their hostilities. Antigua formally becomes a British colony.

1678   British establish settlements on Barbuda

The first British settlers establish homes on Barbuda.

1680   Population reaches 4,500

As more and more slaves are imported from Africa, the island's population swells. Over 2,000 African slaves live on the island, and by 1710, almost 13,000 are there.

1674   First large sugar estate established

Sir Christopher Codrington (d. 1698), son of an influential early settler on Barbados, establishes the first large sugar estate in Antigua. He introduces the latest cultivation and processing technology to Antigua.

1685   Codrington leases Barbuda for sugarcane plantations

Sir Codrington leases Barbuda from the British government for a pittance in order to provide a source for slaves and supplies for his sugarcane enterprise. Sir Codrington wants the Caribbean colonies to have representation in London, and proposes the establishment of schools and hospitals on the islands. He later becomes the governor-general of the Leeward Islands.

The Codrington family leases Barbuda until 1870.

Early 1700s   Sugar monoculture established

By the early 1700s, almost all agricultural land on Antigua is devoted to growing sugarcane. Over two dozen sugar mills are in operation.

1710   Governor of Antigua assassinated

Representatives of Antigua's elected government band together and kill the island's governor.

1736   Slave coup plot uncovered

Plantation owners uncover a plot by slaves to overthrow the island's ruling elite. The slaves had planned to stage their rebellion when many of the island's white residents would be attending the king's birthday celebration. The slaves conspired to ambush the festivity and kill the whites in attendance. After the plot is discovered, nearly ninety slaves are executed—some are hanged or tortured to death on the wheel (a medieval torture device that stretches apart a person's limbs), but most are burned alive.

1750   Number of sugar plantations grows

Nearly 200 sugar plantations exist on Antigua, and another 100 will be added within the next fifteen years.

1751   Sugar plantation operations depend on slaves

The inventory of one large plantation, Betty's Hope (see 1984) lists 277 slaves, including 59 women and 39 men, who made up the field gangs; 26 girls and 20 boys, suitable for "light work"; and twenty one "elderly" and 28 "infants," considered unfit for work.

From January until July, the plantation workday begins at dawn. The mill is started by a team of men or oxen who turn the heavy roundhouse until its sails face into the island winds. The windpower will drive the rollers of the mill, crushing the harvested and stripped sugarcane. After field gangs cut and strip the cane into four-foot lengths, it is fed into the mill's rollers. Murky sugar juice dribbles into a channel (usually wooden) running to the boiling house. The leftover pressed stalks—the *bagasse*—are spread on the ground to be dried for fuel.

Craftsmen, a higher class of slave, work in the boiling house. They add lime to the sugar juice to remove impurities. Next, they channel the juice into the large open metal tanks called *taches*. Using the *bagasse* as fuel, the juice is boiled in the taches until most of the water boils off and sugary caramel remains.

The sugar is next transferred to wooden boxes to cool for a few hours. It crystallizes into rough rocks. These rocks of sugar are packed for further drying in wooden barrels called hogsheads which are stored in the curing house for two weeks. The barrels are then sealed and shipped to Europe.

1774   City of St. John's develops

St. John's is developing into Antigua's urban center. As Cyril Hamshere writes in *The British in the Caribbean*: "Janet Schaw, the 'Lady of Quality' whose journal describes a visit to Antigua and St. Kitts in 1774, liked the little town of St. John's that rose from the harbor up a hill slope. Generally neat and pretty, it still bore the scars of the fire in 1769 and the hurricane of 1772. Public buildings were constructed of stone, and the church had an organ. Coming from Scotland Janet remarked on the low houses without chimneys. These were built on the street and friends passing by at meal-times would pop their heads in at the windows for a greeting and a chat." (Hamshere, p. 133).

1784   Nelson stationed on Antigua

Captain Horatio Nelson (1758–1805) sails to Antigua and becomes the head of English Harbour, the naval headquarters on the island. During his stay, Nelson enforces the laws that give British ships a monopoly over trade with the islands of the British West Indies. The Royal Navy supports the sugar industry, which relies on slave labor. Nelson later goes on to become one of England's most famous admirals.

Early 1800s   Introduction of mongooses

European colonists often import one species of plant or animal intentionally or by accident into a region. Sometimes this is done in order to control another species that is seen as a pest. However, the introduction of foreign species often has a serious environmental impact on native ones. In Antigua, the British import mongooses in order to exterminate rats from the canefields. However, the mongooses' devastation on native animal species eventually leads to the extinction of snakes and burrowing owls on the island.

1834: August 1   Slavery abolished

Although the British have prospered through the use of slave labor in the West Indies, the peculiar institution is cruel and highly inefficient. Slaves are unwilling laborers and are forced to work through fear, suffering, and brutality. Slavery also desensitizes and demoralizes the British colonists. Slave owners are compensated for the loss of their slaves by the British government. At the time of emancipation, Antigua has a slave population of over 29,000. The non-slave population only consists of some 4,000 free mulattos and nearly 2,000 whites.

Emancipation covers all of Britain's colonies, but each colony has its own history of slavery, with some more brutal than others. For example, slaves are treated differently on Antigua and Barbuda. Slaves on Antigua are primarily field hands, and they are strictly controlled and supervised. The slaves of Barbuda, however, enjoy a high degree of autonomy. The Barbudan slaves work as skilled laborers (carpenters, weavers, boat builders, and leatherworkers). The Barbudans are also permitted to grow their own food.

Britain has control over most of the Caribbean and the high seas, and is now able to enforce the abolition of the slave trade. At the time of emancipation, many other British territories in the Caribbean implement a five-year apprenticeship system that is designed as a period of transition between slavery and freedom. Only Antigua and Bermuda choose not to have the apprenticeship system and their slaves are entirely freed. Although the former slaves are free, they still continue to work on the plantations for meager wages because that is the only economic activity on the island.

Decline of the Sugarcane Industry

1841   Fire

A fire in St. John's ruins many of the town's buildings.

1840s   Urbanization

Many former slaves move off the estates and establish new villages or take up residence in villages that have abolitionist churches. With the establishment of over two dozen villages, the island takes on an increasingly urban character.

Sugarcane loses its monopoly status in the 1840s, as European farmers begin cultivating sugar beets. Prices for sugarcane fall steadily throughout the second half of the nineteenth century. Planters continue relying on sugarcane, and so the economy stagnates.

1843 Earthquake

An earthquake destroys the cathedral in St. John's and damages other buildings.

1847 Hurricane

A devastating hurricane strikes the island and causes about £100,000 in damages.

1854 Naval dockyard closes

Antigua is the only British territory in the West Indies that has a good harbor. English Harbour serves as the main dockyard in the British West Indies during the height of colonial rule. The naval dockyard, established in 1725, finally closes. The Royal Navy continues to occasionally utilize the dockyard until 1889.

1855 Naval dockyard museum building constructed

A building is erected to present the history of the naval dockyard, known as Nelson's Dockyard, from its beginnings as a center for British navy operations (see 1784).

1860 Barbuda annexed to Antigua

Barbuda is annexed to Antigua as a way of ending administrative problems on Barbuda.

1862 Antigua issues first postage stamps

Antigua issues its first postage stamps, picturing British Queen Victoria.

1869 Redonda annexed to Antigua

Redonda is an isolated small island that lies thirty miles southwest of Antigua. The island is located between Montserrat and Nevis and is actually closer to both than it is to Antigua. However, since Antigua is the administrative headquarters for the Leeward Islands, Redonda is placed under Antigua's control.

1871–1956 Federation of the Leeward Islands

Antigua and Barbuda are placed under the administration of the Federation of the Leeward Islands.

1903–14 Antiguans emigrate to Panama

During the construction of the Panama Canal, thousands of men from across the British West Indies emigrate to work on the project.

1918 Riot in St. John's

Several plantations are torched and a riot spreads into St. John's. The police shoot upon the rioters and several people are killed.

1929: October New York Stock Exchange crashes

Antigua's economy stagnates after a tepid resurgence in the sugar industry during World War I (1914–18). The New York Stock Exchange crashes, driving the world's economy into a depression as commerce withers. Many sugar plantations throughout the Caribbean declare bankruptcy. People resort to subsistence agriculture for a living, barter, and fabricate their own materials throughout the 1930s.

1930 Redonda abandoned

Phosphate mining, conducted on the small island of Redonda, is discontinued, after reaching its height in the late nineteenth century. The island remains uninhabited and is occasionally visited by boaters.

1939 Trade Union Act passes

The Trade Union Act allows workers on Antigua to collectively bargain for higher wages, fewer hours, and better working conditions. The average wage for someone working in the fields is about the same as it was a century ago. During the economic decline of the late nineteenth century, the number of estate owners declined, so that by the time the Trade Union Act passes, most of the island's agricultural land is owned by a single company. The Antigua Trades and Labour Union also forms in 1939.

1941–45 Antigua during World War II

During World War II (1939–45), the United States secures a long-term lease to keep air and naval forces on Antigua and several other British territories in the Caribbean. Many U.S. soldiers and defense contractors inhabit the island. The agreement is made after the United States sends dozens of warships to Britain to aid its defense. The military bases provide the opportunity for Antiguans to learn technical skills. (See also 1956.)

Post-war Development

1951: January Cane workers strike

Vere Cornwall Bird (b.1910), president of the Antigua Trades and Labour Union, issues an ultimatum on behalf of the cane-field workers to Alexander Moody-Stuart, the managing director of Antigua Sugar Estates. Bird confronts Moody-Stuart under a now-famous tamarind tree outside the southeastern town of Bethesda, stating that field workers would continue their strike because they are only paid one shilling per day (the rate has been about the same for nearly a century).

1951 Adult suffrage

Universal adult suffrage to vote is granted, marking the beginning of Britain's forfeiture of colonial control.

1956　Ministerial government begins

Antigua begins a ministerial form of government. The Executive Council now has three ministers in addition to its nominated members. V.C. Bird is appointed minister of trade and production.

1956　United States establishes Antigua Air Station

The United States receives permission from the United Kingdom to maintain the Antigua Air Station, an airbase on Antigua.

1957　First summer carnival

The annual carnival moves from the week after Christmas to the end of July. The carnival is Antigua's premier cultural event of the year, and most events take place in St. John's. There are parading troupes of steel bands, dressed according to a theme, and calypso and beauty contests. The climax of the event is J'Ouvert on the first Monday in August, when thousands of celebrants crowd the streets at 4:00 AM in a frenzy of dancing.

1958–62　Federation of the West Indies

Antigua becomes part of the short-lived Federation of the West Indies, with Sir Grantley Adams of Barbados as its premier. The idea of creating a federation of British states in the Caribbean had first been discussed in 1932 at a conference in Dominica. The federation's members include Grenada, Barbados, Trinidad, Jamaica, and the islands of the Leeward and Windward Antilles. At first, the federation is popular and people see it as a way for the assortment of small islands to gain international recognition. However, as the honeymoon period ends, political differences between the islands intensify; Jamaica wants the federation to be a loose association, while Trinidad wants a strong federal government. Jamaica secedes in September 1961 and Trinidad follows in January 1962. The entire federation dissolves on May 31, 1962.

1961　V.C. Bird becomes head of government

V.C. Bird becomes chief minister after the trade union leaders win control of the legislature. Bird now controls the government and a powerful labor union.

1961　Nelson's Dockyard becomes historic monument

Preservationsists restore the dockyard in English Harbour on Antigua that was headquarters for the British colonial navy under Admiral Horatio Nelson (see 1784). It is designated as an historic monument.

1963　Reforestation program starts

The government begins a reforestation program, linking it with efforts to improve soil and water conservation. Most of the island's natural vegetation had been cleared when the sug-

arcane industry was developing during the late eighteenth century.

Late 1960s　Reservoir opens

A large shallow reservoir opens to collect rainwater for the island's plumbing. Antigua's soil does not hold much water and most rainwater runs off into the sea. The government also builds smaller reservoirs and ponds on the island. Houses on Antigua typically have cisterns that collect rainwater.

1967: February 27　Antigua joins Commonwealth of Nations

After the Federation of the West Indies breaks apart, Antigua becomes an associated state of the Commonwealth of Nations. Britain now has control only over Antigua's defense and foreign policy. Opposition to complete independence comes from the residents of Barbuda, who seek constitutional guarantees for autonomy in land, finances, and local legislative powers.

1967　Sugar industry nationalized

By the mid-1960s, Antigua Syndicate Estates controls the entire sugar industry on Antigua. The company has severe financial problems and so the government buys out the company's 13,000 acres and its mills in order to protect several thousand jobs. The sugar industry continues to decline, however, and by 1972 the last sugar factory on Antigua closes.

1968: February 12　Antigua Workers Union protests

The Antigua Workers Union, led by George Walter, is demanding that the government officially recognize it as a labor union. A crowd of some ten thousand are assembled, and Chief Minister Bird declares a state of martial law and his popularity plummets. The police have to use tear gas to break up the crowd.

1968　Deep water harbor opens

With the opening of Antigua's deep water harbor, cruise ships are now able to visit the island.

1971　George Walter becomes new premier

George Walter is elected and becomes Antigua's new premier. During the early 1970s, Walter's political opposition comes primarily from V.C. Bird, whom he has defeated. Bird is possibly the source of a series of bomb threats in the early 1970s.

1973　Energy crisis

An energy crisis afflicts North America, due to an embargo on oil by the Oil Producing Exporting Countries (OPEC) cartel. The energy problem leads to an economic recession, which affects Antigua's fragile economy. An oil refinery that had opened in the 1960s closes, and the sugar industry folds.

1975 Antigua and Barbuda Development Bank opens

The Antigua and Barbuda Development Bank, wholly owned by the government, begins operating.

The Fiscal Incentives Act of 1975 specifies tax holidays of ten to fifteen years in order to promote foreign investment. The incentives are designed to promote the growth of the island's industrial park near the airport.

1976 Income tax abolished

The income tax, introduced in 1924, is abolished for residents (based on a six-month stay). However, a flat tax is imposed on income from trade or business, and a flat tax on property also goes into effect.

Mid-1970s Rastafarianism comes to Antigua and Barbuda

Rastafarianism started in Jamaica as an Afrocentric movement in response to growing social inequality. The beliefs are based on the teachings of the Jamaican Marcus Garvey (1887–1940) who, in the 1920s, advocated African racial pride and the eventual return of blacks to Africa. He believed that an African king would rise up and rescue Africans from oppression. In 1930, Ras Tafari (1892–1975) of Ethiopia was crowned Emperor Haile Selassie I (r. 1930–36, 1941–74). Garvey's adherents believed the Ethiopian emperor was the king that he had foreseen. Emperor Haile Selassie claimed to be a direct descendent of King David. The Rastafarians reject the European culture imposed by colonists and seek a simpler life. Rastafarians are associated with the youth black power movement that is spreading across the Caribbean. By the late 1970s, Rastafarians are recognized internationally for two of their characteristics: wearing their hair in dreadlocks (long cords of twisted and uncombed hair) and *ganja* (marijuana). The smoking of marijuana is a sacrament in Rastafarianism.

1976 V.C. Bird returns to power

After five years out of power, V.C. Bird's popularity rises again and he becomes Antigua's premier. Bird's two sons (Lester and Vere, Jr.) are also launching political careers.

1977 Space Research Corporation comes to Antigua

Space Research Corporation is a U.S.–Canadian company controlled by Gerald Bull. The company designs precision artillery pieces and plans to secretly relocate part of its operations from Barbados to Antigua. There is public objection to the company after it comes to Antigua because it violates the United Nations's arms trade embargo against South Africa. The U.S. Army funds the company to develop a highly accurate artillery weapon. Gerald Bull is later convicted of illegal arms trafficking and goes to jail for a few months, then is

mysteriously assassinated in Brussels in 1990. After the Persian Gulf War, weapons inspectors from the U.N. discover in 1991 that Iraq possesses artillery pieces produced by the company.

1980s Emigration to North America

About seventeen percent of the country's total population emigrates to North America during the 1980s. At that time, over 12,000 Antiguans and Barbudans legally emigrate to the United States, and over 700 go to Canada.

1980 The rise of tourism

The number of tourists coming to Antigua and Barbuda is about 100,000, up from 65,000 in 1970 and 30,000 in 1960. The number of available hotel rooms exceeds 3,400, up from about 500 in 1960.

The development of tourism results in exclusive tourist establishments, including gated communities and fenced-off resort beaches. Revenue from tourism, however, is mostly spent to purchase imports rather than put into investments. The fate of the Antiguan economy depends upon the dominant U.S. economy.

Independence

1981: November 1 Antigua and Barbuda becomes independent state

With the issues of autonomy for Barbuda still unresolved, Antigua and Barbuda becomes an independent state within the Commonwealth of Nations.

1981: November 11 Antigua and Barbuda joins UN

Antigua and Barbuda joins the United Nations, becoming the organization's 157th member. The following month, the new nation joins the Organization of American States (OAS).

Early 1980s Americans in Antigua

In the years after Antigua and Barbuda becomes an independent nation, hundreds of pregnant Antiguan women travel to the U.S. Virgin Islands to give birth. Since the babies are born on U.S. soil, they are entitled to U.S. citizenship. Several thousand Americans live on Antigua; all but a few hundred are children of Antiguan mothers.

1982 Medical school founded

The University of Health Sciences at St. John's opens a school of medicine.

1982 American fugitive hides on Antigua

Robert Vesco, a wealthy financier from the United States, has been living in Antigua, evading law enforcement officials since the early 1970s, when he was charged with making an illegal donation to the campaign to reelect Richard Nixon president of the United States. He had been in seclusion in the Bahamas but comes to Antigua, where the government permits him to moor his boat for a monthly fee.

There is also suspicion that Vesco and other criminals are involved with the Sovereign Order of New Aragon, a group of foreigners that plans to found an independent principality on Barbuda.

1983: October   U.S.-led invasion of Grenada

Antigua fully supports the U.S.-led invasion of Grenada, and several members of the Antigua and Barbuda Defense Force are sent to Grenada as a token supplement to the U.S. military. A Marxist military government allied with Cuba has usurped power from the government on Grenada.

1984   Reconstruction of Betty's Hope Plantation begins

An important sugar plantation on Antigua, known as Betty's Hope, is designated for preservation. The government of Antigua and Barbuda funds reconstruction and preservation of the site. Island historians design an interpretation center to house artifacts from the 1700s. At Betty's Hope, sugar, molasses, and rum were once manufactured by an estimated 300 African slaves. Reproductions of papers of the Codrington family detailing over two centuries of sugar and slave trade, as well as daily life on the estate, will be on display at Betty's Hope. Originals of these documents are in the British Museum in London, England, and microfilm copies are housed at the Museum of Antigua in St. John's.

1984: January 13   U.S. State Department seeks closer relationship with Antigua

After the United States invasion of Grenada in October 1983, the United States and Antigua enter into a closer relationship, even though the United States acknowledges V.C. Bird's administration as corrupt. The United States maintains two military bases in Antigua and Barbuda: the Antigua Air Station is operated by the U.S. Air Force and there is also a U.S. Navy base on Antigua. The air base tracks satellite launches from Cape Canaveral in Florida and the naval station monitors submarine activity in the Caribbean. During the 1980s, the Antiguan government begins increasing its rent and fees for the bases, and people and businesses begin encroaching onto the lands.

1985: July   Airport scandal

Vere Bird, Jr. serves as minister of aviation and public utilities. Under his watch, renovations and repairs begin on the runway of Antigua's airport. The project is supposed to cost $1 million, but the job ends up costing over $11 million and the work is shoddy. The French company that the government hired to do the job turns out to be a local company controlled by Vere Bird, Jr. He is later investigated in 1987 but not officially accused of any wrongdoing.

1987: May   Unification referendum

The prime ministers of Antigua and Barbuda and six other Organization of Eastern Caribbean States (Dominica, Grenada, Montserrat, Saint Kitts and Nevis, Saint Lucia, Saint Vincent and the Grenadines) members vote to create a single nation, subject to national referenda. The referenda are later defeated and the seven nations remain separate.

Late 1980s   Deep Bay hotel scandal

In the late 1980s, Lester Bird, as chair of the Deep Bay Development Company, authorizes the construction of Antigua's largest hotel. The hotel costs over $90 million to build and is constructed by an Italian contractor that uses its own material and labor. Within a few years, the hotel falls into disrepair, is severely under-utilized, and is valued at only a fraction of what it cost to build.

1988   Waterfront renovations

Heritage Quay, the waterfront district of St. John's, opens with duty-free shops, offices, and apartments. The renovation project costs about $15 million. Many of the duty-free shops are owned by government officials.

1989   Desalination plant opens

Antigua continually has problems with water shortages, because the soil absorbs little rainwater and most of the water drains into the sea. The desalination plant, authorized by Vere Bird, Jr., costs over $70 million and is plagued with operational problems.

1989: April   Gunrunning scandal

A Danish cargo vessel from Israel stops at St. John's and leaves an unmarked container, which is loaded into another vessel and taken to Colombia. The cargo taken to Colombia consists of machine guns and ammunition and is delivered to the Medillín drug cartel, which uses the weapons to conduct assassinations in order to intimidate the Colombian government. However, the guns have been shipped from Israel specifically for Antigua and are not supposed to be sold to anyone else, even though the United States is the sole weapons supplier to the region since the invasion of Grenada in 1983. The weapons are supposedly for the Antigua and Barbuda Defense Force to use for a security training school that is being set up by Israeli military consultants. However, the school trains in terrorism and assassination techniques. The school is authorized by Vere Bird, Jr., son of the Prime Minister. In May 1990, Prime Minister Bird agrees to permit a judicial inquiry of the allegations against his son. However, no formal proceedings are launched against Vere Bird, Jr.

1989: September 10–12   Hurricane Hugo

A powerful hurricane strikes several islands in the Caribbean (St. Kitts, Montserrat, Dominica, Guadeloupe, Antigua, British Virgin Islands, U.S. Virgin Islands, and Puerto Rico). The hurricane continues towards the northwest and eventually strikes the United States mainland. The hurricane causes over two dozen fatalities in the Caribbean.

Early 1990s   Fish consumption highest in Caribbean

Antiguans consume more fish on a per person basis (101 pounds/46 kilograms live weight equivalent) than any other nation or territory in the Caribbean.

1990   The decline of agriculture

Agriculture's contribution to the economy falls to less than fifteen percent, compared to over forty percent in 1960. The decline in the sugar industry leaves sixty percent of the country's 66,000 acres under government control. Crops suffer from droughts and insect pests, and cotton and sugar plantings suffer from soil depletion and the unwillingness of the population to work in the fields.

1991   Lester Bird briefly resigns from government service

Lester Bird and three other ministers resign in protest to government corruption. Within a few months, however, he returns to the government as minister of planning.

1994: March 8   Lester Bird becomes prime minister

Lester Bird, head of the Antigua Labor Party, is appointed prime minister by the governor general after the resignation of his father, Vere Cornwall Bird.

1994: June 30   U.S. embassy closes

The United States has maintained its smallest embassy in Antigua, but many new embassies are opening in Eastern Europe and the former Soviet Union, stretching the budget of the U.S State Department. With the fall of the Soviet Union, the threat of communism spreading in the Caribbean is gone. U.S. affairs in Antigua and Barbuda become the responsibility of the embassy in Barbados.

1995   Betty's Hope Plantation receives UNESCO award

Betty's Hope Plantation, a restored sugar plantation on Antigua (see 1984), receives the United Nations Education, Scientific, and Cultural Organization (UNESCO) World Cultural Project Award in recognition of its preservation of Antigua's history.

1995: September 6   Hurricane Luis

Hurricane Luis strikes Antigua, killing two people. The storm tears off roofs and strips utility poles. Over half the buildings on the island are heavily damaged. Cruise ships and airlines avoid the island immediately afterwards because of the severe damage. By the end of the year, however, recovery is well underway.

Bibliography

Berleant-Schiller, Riva. *Antigua and Barbuda.* Oxford: Clio, 1995.

Coram, Robert. *Caribbean Time Bomb: the United States' Complicity in the Corruption of Antigua.* New York: William Morrow and Company, Inc., 1993.

Dyde, Brian. *Antigua and Barbuda: the Heart of the Caribbean.* London: Macmillan Publishers, 1990.

Hamshere, Cyril. *The British in the Caribbean.* Cambridge, MA: Harvard University Press, 1972.

Henry, Paget. *Peripheral Capitalism and Underdevelopment in Antigua.* New Brunswick, N.J.: Transaction Books, 1985.

Knight, Franklin W. *The Caribbean, the Genesis of a Fragmented Nationalism.* New York: Oxford University Press, 1990.

Kurlansky, Mark. *A Continent of Islands: Searching for the Caribbean Destiny.* Reading, Mass.: Addison-Wesley Publishing Co., 1992.

Lazarus-Black, Mindie. *Legitimate Acts and Illegal Encounters: Law and Society in Barbuda and Antigua.* Washington, D.C.: Smithsonian Institution Press, 1994.

Sletto, Bjorn. "Antigua's Old Mills Turn with New Winds." *Americas,* November–December 1996, vol. 48, no. 6, p. 6+.

Argentina

Introduction

Argentina occupies most of the southern half of South America, taking up nearly a quarter of the entire continent. With an area of over 1 million square miles (2.7 million square kilometers), it is South America's second-largest nation (after Brazil), and the eighth-largest in the world. It is very diverse geographically, with environments ranging from the tropical forests of the north to the frozen Antarctic lands of Tierra del Fuego to the south. The country's major regions include the pampas (plains), home of the famous *gauchos,* or cowboys; the Andes mountains along the country's western edge; the rolling hills and plains of Patagonia in the southeast; and savanna, forests, and lowlands in the north. Argentina has a population of thirty-five million; the capital city of Buenos Aires has an estimated population of about twelve million. Spurred by its livestock industry, Argentina became Latin America's wealthiest nation in the late nineteenth and early twentieth centuries. Since then, however, it has struggled with economic problems and political instability, periods of military rule and, at times, severe restraints on the civil liberties of its own people.

When the first Europeans arrived in present-day Argentina in the sixteenth century, they found no advanced native civilization, such as the Incas of Peru or the Aztecs of Mexico. The indigenous tribes were mostly nomads and hunters. The first Spanish attempt to found a colony in 1536 at the current site of Buenos Aires failed. Instead, the first permanent settlement was established at Santiago del Estero in 1553. (A second settlement was founded at the site of Buenos Aires in 1580.) The territory that later became Argentina was initially made part of the Spanish-controlled Viceroyalty of Peru. Then, in 1776, Argentina became the center of a new colonial entity, the Viceroyalty of the Río Plata, which included present-day Bolivia, Uruguay, and Paraguay and had its capital at Buenos Aires.

Argentina's campaign for independence from Spain lasted from 1810 to 1816, slowed by a significant internal division that overshadowed Argentine politics throughout the nation's history: the rivalry between the inhabitants of Buenos Aires (known as the *porteños,* or port people) and the people living in the country's interior. Argentina and the other provinces of the Viceroyalty of the Río Plata declared independence on July 9, 1816 as a united country. Eventually, all the territories broke away to form nations of their own.

An Independent Nation

Argentina's independence was guaranteed by the military campaigns of José de San Martín, who drove the Spanish from neighboring Chile and Peru between 1816 and 1821.

Although independent, Argentina continued to be torn by factional feuding. In addition to the split between the pampas (plains) and the coast, there was a continuing power struggle between two political parties: the Unitarios, who wanted a strong centralized government, and the Federales, who wanted more autonomy for the individual provinces. At times this rivalry even broke out into armed warfare. Argentina's various groups were finally brought under control and unified—as would occur numerous times in the future—by General Juan Manuel de Rosas (1793–1877), a strong, authoritarian, military leader, who wielded power as governor of Buenos Aires. Rosas also led an armed campaign to drive the Indian population from the pampas region. Rosas held power from 1835 to 1852, ruling without a constitution.

In 1853 a new constitution was approved (with revisions, it has remained in effect to the present). However, it was not accepted by all Argentines, and hostilities broke out once again. Buenos Aires finally approved the constitution, with modifications, in 1862 and became the capital of the Republic of Argentina. From this point on, Argentina made rapid economic progress, becoming a wealthy, modern nation in the late 1800s and early 1900s. The late nineteenth century also ushered in a period of stability and peace, broken only by the nation's participation in the War of the Triple Alliance, between 1865 and 1870, fighting together with Brazil and Uruguay to bring down a Paraguayan dictator. In 1879–80 what remained of the native Indian population was completely banished from the Argentine pampas, either by being exterminated or by being forced to relocate to other areas.

Throughout this period, the government was controlled by conservatives who protected the interests of the wealthy landowners. An important political development of the 1890s was the founding of an opposition party that represented the workers and the middle class, the Unión Civica Radical, commonly known as the Radicals. By the second decade of the twentieth century, the Radicals gained their first political victories, winning control of the government and such reforms as secret ballot elections. Both the Conservatives and Radicals were united in maintaining Argentine neutrality in World War I (1914–18).

Military Rule

After a postwar decade of Radical rule, Argentina's long period of unbroken civilian democracy and democratic rule came to an end with the military coup of 1930 that ousted president and longtime Radical leader Hipólito Yrigoyen from power. The period from 1930 to 1943, general known as the "infamous decade," was characterized by military rule, political corruption, and, among some elements in Argentine society, growing sympathy with the Nazi politics of Adolf Hitler in Germany. However, Argentina remained officially neutral during World War II. It was during the war that the military officer Juan Domingo Perón (1895–1974) rose to prominence in several government positions. Perón, an advocate for the working class and Argentina's single most influential twentieth-century leader, was elected president of the country in 1946.

Perón initially enjoyed great popularity throughout the nation, enhanced by the public's affection for his charismatic wife, Eva "Evita," who played an active political role in his administration. However, by 1949 the Perón presidency took on the trappings of a dictatorship, with the constitution suspended, and censorship imposed on the press. In 1955 Perón was removed from the presidency by a military coup and exiled to Spain. However, his supporters, known as "Peronists," remained an active force in Argentine politics.

In the years between 1955 and 1973, Argentina was mostly ruled by military governments. Those civilian leaders that were elected were eventually deposed by the military. Opposition to the military governments at last grew so strong that free elections were allowed in 1973, and Peronist Héctor Cámpora (1909–80) was chosen as president. Cámpora quickly resigned as president so that Juan Perón, who had returned after seventeen years in exile, could assume the presidency himself. However, the aging leader died the following year, and his third wife, Isabel, took over as president. She was ousted in a coup in 1976 and Argentina entered another extended period of military rule, this one lasting until 1983.

In the late 1970s, Argentina's military rulers carried out what came to be known as the "dirty war," a campaign of repression and terror to silence their political opponents who included radical leftists and Perónists. Thousands of people were killed, and thousands mysteriously "disappeared" and were never seen again. There was a worldwide outcry against the human rights violations committed during this period, and

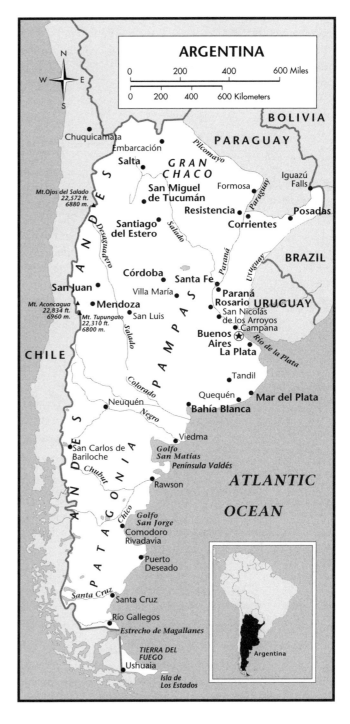

the military officers responsible. Following the return of civilian rule, the officers responsible were eventually brought to trial for crimes against humanity.

Increasingly plagued by economic problems at home, Argentina's leaders invaded the British-held Falkland Islands in April 1982 but suffered humiliation when the British recaptured the islands two months later. Free elections were held in 1983, and Raúl Alfonsín (b. 1927) was elected president. In 1985 five top military officers were tried for their actions during the "dirty war" of the 1970s. Peronist candidate Carlos Menem (b. 1935) was elected president in 1989 and reelected in 1995, as the nation struggled to cope with runaway inflation and other economic problems. Nevertheless, his Perónist party retained its majority in the Senate in 1995 and its plurality in the Chamber of Deputies in 1997.

Timeline

20,000–10,000 B.C. First inhabitants migrate to South America from Asia

The first inhabitants of present-day Argentina are thought to have come from Asia, crossing the Bering Strait near the end of the Ice Age. By A.D. 500, there are some twenty native tribes living in the region. Their civilizations are generally not as advanced as those of the best-known Indian groups in Latin America, such as the Aztecs, Maya, and Incas.

16th century Nomadic tribes inhabit area

By the time of the first European exploration in the sixteenth century, the Indian population in the southern part of South America numbers only about half a million (compared to some 25 million in central Mexico), approximately 300,000 of which live in present-day Argentina. Most belong to nomadic groups and are skilled hunters. The Diaguita, living in the Calchaquí Valley, are the region's largest group and one of its most advanced. They farm with hand-held plows and build irrigation dams, and they have a sophisticated tradition of ceramics and metalworking. The Moliche live in Patagonia to the southeast, and the Ona, Yahgan, and Alikuf live on the island of Terra del Fuego to the south, which they continue to control until 1880.

Spanish Exploration

1502 Malvinas Islands sighted by Amerigo Vespucci

Amerigo Vespucci (1451–1512), an Italian sailing for the Spanish crown, sights the Malvinas (later Falkland) Islands. Later Spanish or English explorers to view the islands include Esteban Gómez (1520), Sarmiento de Gamboa (1580), Thomas Cavendish (1592), John Davis (also in 1592), and Richard Hawkins (1594), as well as a Dutch sailor, Sebald de Weert (1600).

1516 Spanish explorer Juan Díaz di Solís sails down the Río de la Plata

Juan Díaz di Solís (1470?–1516) is the first European to enter the estuary of the Río de la Plata, between present-day Uruguay and Argentina. Believing he has found a passageway between the Atlantic and Pacific oceans, di Solís enters the estuary and lands on the bank of the Uruguay River. The explorer and all but one member of his party are killed by the Charrúa Indians.

1520 Magellan reaches Patagonia

Portuguese explorer Ferdinand Magellan (1480–1521) discovers the strait later named after him and lands on the coast of Patagonia, the region of Argentina which comprises all the area from the Rio Negro to the southern extremity of the continent (about 300,00 square miles, or 777,000 square kilometers).

1526 Cabot explores Río de la Plata

Italian Sebastian Cabot (1475–1557), sailing for Spain, explores the region in 1526, also sailing north to the Paraná and Paraguay rivers. Cabot gives the Río de la Plata (Silver River) its name, believing that there are silver deposits in the region. The mistaken belief in such deposits is also the origin of the name "Argentina," which is derived from the Spanish word for silver. Cabot also founds the first Spanish settlement in the Río de la Plata region, a fort called Sancto Spíritus, which is destroyed by marauding Indians in 1529.

1536: February Pedro de Mendoza reaches site of Buenos Aires

The Spanish explorer Pedro de Mendoza (1487–1537) attempts to found a colony at the present site of Argentina's capital, calling it *Ciudad de Nuestra Señora Santa María de los Buenos Aires* (City of Our Lady Saint Mary of the Fair Winds), later shortened to Buenos Aires. However, Indian attacks and disease eventually discourage the settlers, and they sail upriver to present-day Paraguay, founding the city of Asunción in 1537. The site of Buenos Aires is abandoned.

1553 First permanent settlement in Argentina is founded

Spaniards from Peru establish Argentina's first permanent settlement, Santiago del Estero. Mendoza and San Juan are founded by Spanish settlers from Chile in 1561–62. All together in the sixteenth century, the Spanish found twenty-five cities in what will become Argentina.

Colonial Rule

1580 Juan de Garay refounds Buenos Aires

Departing from Asunción, the present capital of Paraguay, Spaniard Juan de Garay (c. 1527–c. 1583), together with sixty Spanish settlers and several hundred Indians, establish a new settlement at the site where Mendoza has attempted to found his colony. The name "Buenos Aires" is retained, and it becomes a separate province with its own governor in 1617.

From the late sixteenth to the late eighteenth centuries, the Spanish take relatively little interest in the southern portion of South America. Its native population resists control, and it lacks the rich mineral resources of the lands to the north. The territory that is to become Argentina forms part of the vast Viceroyalty of Peru, established in the 1530s.

1580 Arm of the Spanish Inquisition is established in Buenos Aires

The Holy Office of the Spanish Inquisition is instituted in Buenos Aires around 1580. It enhances the already considerable power of the Catholic Church in the Río de la Plata region.

17th and 18th centuries   Indian wars slow expansion

Fighting between the Spanish settlers and Argentina's Indian groups continues sporadically throughout the seventeenth and eighteenth centuries. A fierce raid in 1589 kills cattle and demolishes property in San Miguel del Tucumán. In the mid-seventeenth century, the Spaniards fight the Diaguita, Mocobí, and Chaco tribes. Between 1657 and 1659, an Indian uprising in the Calchaquí Valley is led by a Spanish rebel, Pedro Bohórquez. The Spanish continue fighting the Mocobí into the 1700s. Between 1740 and 1780 they battle to conquer the Chaco, with whom they finally make peace in 1776.

1609 First Jesuit mission established

Jesuit priests found their first mission in northeast Argentina, in the area later known as the province of Misiones. Most of the region's Guaraní tribes are converted to Christianity. However, the missionaries have little success with the nomadic tribes of southern Argentina, who remain hostile and pose a continuing threat to settlers. The Jesuits continue their conversion efforts for the next two centuries.

The first documented performances of sacred music by the Jesuits take place in 1611. Cathedral performances are recorded from 1622.

1611 Ordenanzas call for reducing Indian slave labor; black slavery expands

Recommendations (*ordenanzas*) issued to the Spanish Crown by special investigator Francisco de Alfaro include reducing the number of Indians engaged in slave labor. Subsequently the use of black slaves increases. Black Africans are brought to Buenos Aires first through independent slave traders, then by the French, and, after 1715, by the British South Sea Company. Between 1715 and 1739, the British supply between 8,000 and 9,000 black slaves a year to Buenos Aires.

1613 University of Córdoba is founded

Argentina's oldest university is established in the city of Córdoba.

1630 Miracle of the statue of Luján

A man attempts to drive an ox cart with a statue of the Virgin Mary across the place where the town of Luján will later stand. The cart cannot be moved no matter how much force is used, so a chapel is built to house it. Later, a grand church is erected at the site to receive pilgrims who journey to see the Virgin of Luján.

1630–1680 Colonization slowed by economic depression in Europe

Declining silver production in the interior of the continent, coupled with a depression in Europe slows development of the Río de la Plata region in the seventeenth century, as the supply of silver for export from Bolivia falls and the European demand for it declines as well. There is little population growth in the South American population centers.

1690 First landing on Malvinas/Falkland Islands

After a number of sightings in the sixteenth century, an English captain, John Strong, is the first to land on the Malvinas Islands. Claiming them for England, he names them for an official in the British navy, Viscount Falkland.

1700 Exports of hides flourishes in Buenos Aires

In the eighteenth century, with the spread of wild cattle herds throughout the pampas, or plains, the trade in hides becomes a major source of income for Buenos Aires. In the first quarter of the century, the hide export trade grows to some 75,000 hides a year, exported to clothe and feed the armies of Europe.

1717 Composer Domenico Zipoli arrives in Córdoba

The high point of musical life in colonial Argentina is the migration of celebrated Italian composer and organist Domenico Zipoli (1688–1726) to Córdoba.

1760s–70s Disputes over the Malvinas/Falkland Islands

Both the French and British make attempts to claim territory on the Malvinas. France founds the settlement of Port Louis there in 1764, and the British establish Port Egmont in 1766. Disputes with the British continue into the 1770s.

1776 Viceroyalty of the Río de la Plata established

After gaining control of Spain, the Bourbon dynasty institutes reforms in the administration of its colonies. Argentina is split off from the Viceroyalty of Peru to become part of a new viceroyalty formed by Charles III. Its capital is Buenos Aires, and it also includes what will become Uruguay, Paraguay, and Bolivia. The new monarchy also sets up *intendencias,* local governments that oversee land distribution, taxation, and other economic matters.

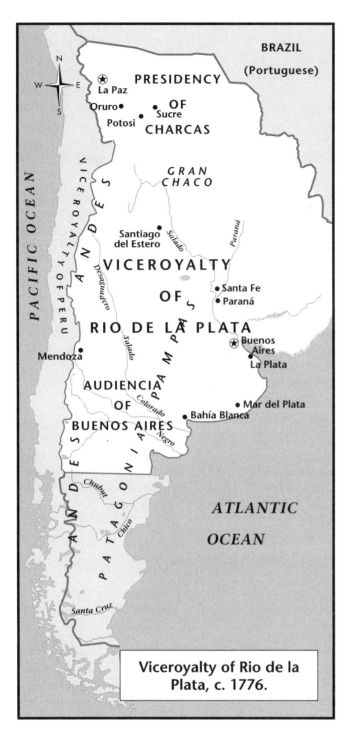

Viceroyalty of Río de la Plata, c. 1776.

1778: February 2   Free trade extended to the Río de la Plata

A new free trade policy opens up South America's Atlantic Coast, beginning a period of prosperity that lasts until 1810. Goods can now be shipped directly from Argentina to Spain. A chamber of commerce, the *consulado*, is formed in Buenos Aires in 1795.

1789   First Argentine play written

The first Argentine theatrical work is *Siripo* by Manuel José de Lavardén. Another early play is *El amor de la estanciera,* an anonymous work written around 1792, and a forerunner of later dramas about the Argentine *gaucho,* or cowboy.

1805   Birth of composer Amancio Alcorta

Alcorta (1805–62) is considered the first Argentine-born composer. His career signifies the coming of age of Argentinian music.

1806   The British invade Buenos Aires

Taking advantage of local dissatisfaction with colonial rule, and of Spain's involvement in the Napoleonic wars in Europe, Britain twice invades Buenos Aires and tries to occupy the city. However, the Argentines do not want to be ruled by the British any more than by Spain. They rally and fight back under the command of Santiago Liniers (1753–1810), and expel the British forces both in 1806 and 1807.

1810   Argentina's National Library is founded

The library has been at its present site since 1902. In 1997 it houses 2.1 million books.

1810: May 25   Buenos Aires ousts Spanish viceroy, chooses own leader

Encouraged by their victory over Britain, the people of Buenos Aires decide to assert their independence from all foreign domination. At a *cabildo abierto* (open town meeting) they set up a governing *junta* (committee) with appointed members. This historic meeting at the Plaza de Mayo is a watershed in the South American campaign for independence from Spain, and May 25 is celebrated in Argentina as a national holiday.

1810: November 7   Battle of Suipacha launches six-year war for independence

Argentinean recruits win their first victory over Spanish forces. The ensuing struggle for independence is complicated by internal divisions between the Argentines in Buenos Aires (the *porteños,* or people who live in the port) and those living in the interior provinces, who retain more loyalty to Spain, especially to Ferdinand, the Spanish king deposed by the French. In addition, the Buenos Aires government itself is unstable, changing hands numerous times over the next few years.

The decisive military victory within Argentina is won by José de San Martín (1778–1850) at Chacabuco in 1817. However, to remove the Spanish military threat permanently, San Martín decides to attack the Spanish strongholds in the neighboring territories of Chile and Peru, both of which are liberated by his forces (Chile in 1817 and Peru in 1821).

1812 Lyrics to Argentina's national anthem are written

Vincente Lopez y Planes writes the words to Argentina's national anthem, *"Himno nacional Argentino."* They are set to music by Blas Parera.

The United Provinces of the Río Plata

1813: January Revolutionary congress meets

An assembly of representatives from different provinces of Argentina meets and passes its own laws, replacing those of its Spanish governors. Among the provisions of these laws are the following: Indian enslavement and physical punishment of prisoners are abolished; Europeans are dismissed from official positions; a national currency and coat of arms are adopted; and titles of nobility are eliminated. An office of supreme director is established for the newly independent United Provinces of the Río de la Plata.

1814 Argentine Post Office is established

1816: July 9 Declaration of independence is drafted

At the Congress of Tucumán, the provinces of the former viceroyalty declare independence as the United Provinces of the Río Plata. However, Bolivia, Uruguay, and Paraguay do not accept Argentinean rule and eventually establish their own independent governments. The territory that remains later becomes the Argentine Republic.

Within Argentina itself, disunity reigns as well. In addition to the division between Buenos Aires and the interior, Argentina is also divided into two political factions. The Unitarios seek a strong central government with its capital in Buenos Aires. The Federales want a looser confederation that allows for more local independence. The new nation has a rapid succession of governments, and hostilities break out between the Unitarios and the Federales. The impasse between the competing factions continues through the early and mid-1820s.

1818 Argentine flag is adopted

The Argentines adopt a national flag. The background is blue and white, the colors worn by troops who battled the British in 1806 and 1807. Superimposed on it is a drawing of a sun with a human face, called the "Sun of May." It stands for freedom and also appears on the Argentine coat of arms.

1820 Argentina lays claim to the Malvinas

The Malvinas/Falkland Islands are claimed officially by the Argentine government.

1822 Native game of *pato* is banned

Pato, a polo-like game in which players on horseback fight over a sack containing a live duck, is officially banned because of violent fighting between players. The game, primarily a working-class pastime developed by *gauchos,* is played by two teams on a field three miles (five kilometers) long. The object is to get the sack past one's opponents and through a net at the opposite end of the playing field. *Pato* makes a comeback during the Rosas regime (1835–52). Eventually the duck is replaced by a ball with six handles, which are used for passing it from one player to another. The National Pato Federation is founded in 1941.

1825–1828 Cisplatine War between Argentina and Brazil

At the end of 1825 hostilities break out between Brazil and Argentina over control of the Banda Oriental (later known as Uruguay). The Portuguese in Brazil have made claims to the region during the colonial period, but the newly independent United Provinces of the Río de la Plata claim it in 1824. Brazil declares war on Argentina on December 1, 1825. After three years of fighting, a peace treaty is negotiated through British mediation, and the independent state of Uruguay is created.

1826 Rivadavia is elected president

Under a new constitution, Bernardino Rivadavia (1780–1845), a Unitario, is elected Argentina's first president. He introduces a series of progressive governmental and social reforms but is unable to gain the support of the rural *caudillos* (local bosses), who fear that the central government poses a threat to their power. Rivadavia is forced to resign the presidency in 1827, after only one year in office, and civil war (1828–29) resumes between the Unitarios and Federales.

1829 Rosas becomes governor of Buenos Aires

Amid the chaos of factional feuding, General Juan Manuel de Rosas (1793–1877), a wealthy rancher and exporter, assumes leadership of the Federales, becoming governor of Buenos Aires. Unlike other leaders in the capital city, Rosas has the support of the *caudillos* in the countryside. Between 1832 and 1835, Rosas resigns the governorship to lead a military campaign against the region's Indian population, driving them further from parts of the Argentine pampas, settled by *criollos* (native-born descendants of European colonists).

1830s Echeverría brings Romanticism to Argentina

After spending five years in Paris, writer Esteban Echeverría (1809–51) returns to his homeland of Argentina and publishes poems influenced by the European Romantic movement in literature. In works such as *La cautiva* (The Captive; 1837), Echeverría introduces the archetypal theme of the pampas into Argentine literature. Echeverría is also active in the political opposition to Argentina's dictator, Juan Manuel de Rosas. His satire *El matadero* (The Slaughterhouse; 1838) uses the slaughterhouse as a symbol of the Rosas regime.

1832–33 British recapture Malvinas/Falkland Islands

The British once again assert their sovereignty over the Malvinas when Captain J. J. Onslow and his crew occupy the islands in late 1832 and 1833. Ignoring protests by Argentina, the British keep their settlement there and also maintain a small naval fleet. The islands serve as a naval base for the Royal Navy, the most powerful fleet in the world.

1834 Birth of José Hernández, author of *Martín Fierro*

José Hernández (1834–86), author of Argentina's most renowned literary work, *Martín Fierro*, lives a life that combines letters and action. Hernández serves in the army, works as a journalist, and founds the newspaper *El Río de la Plata*. In the 1870s, he serves as a legislator and participates in founding the city of La Plata. Concerned with the rights of the gauchos, Hernández also writes *Instrucción del estanciero* (Education of a Rancher).

1835–52 Rosas holds dictatorial power

Resuming the governorship of Buenos Aires after his campaign against the Indians of the pampas, Rosas dominates Argentinean politics from 1835 to 1852. Although he insists on absolute allegiance from local leaders, his government is primarily concerned with national security and does not address education or other civic issues. Rosas rules as a dictator, with no constitution, protecting the interests of ranchers like himself. His opponents are exiled or silenced by secret police.

Rosas twice has to defend his rule against foreign interests allied with the opposition Unitario party—France in 1838 and both France and Great Britain between 1845 and 1848. Naval blockades of Buenos Aires are mounted on both occasions.

In spite of his strong-arm tactics, Rosas is still admired by many Argentines today for unifying their country and protecting it against foreign aggression.

1839 Echeverría founds activist group, Asociación de Mayo

Poet Esteban Echeverría founds the anti-Rosas activist group Asociación de Mayo, whose members include future Argentine political leaders and literary figures. Echeverría's 1937 liberal tract, *Dogma socialista*, is considered a manifesto of this group.

1845 Publication of literary milestone, *Facundo*

This major Latin American prose work is written by Domingo Faustino Sarmiento (1811–88), the educational reformer who will become Argentina's first elected civilian president in 1868. Sarmiento writes *Facundo*, a denunciation of dictator Manuel Rosas, while in political exile in Chile. The full name of the work is *Civilización y barbarie: vida de Juan Facundo Quiroga, y aspecto físico, costumbres, y hábitos de la República Argentina* (Life in the Argentine Republic in the Day of the Tyrants; or, Civilization and Barbarism).

Facundo, which purports to be a biography of a *gaucho* military officer during the Rosas regime, is the first important work to depict the life of the *gauchos,* and the model for the Argentine genre of the *gaucho* novel. *Facundo* is regarded as the major nineteenth-century Argentine prose work, and some have called it the most important work of literature produced in Latin America.

1852: February 3 Rosas overthrown by coalition including former allies

Rosas's refusal to implement a federal constitution leads Justo José de Urquiza (1800–70), governor of the Entre Rís province and a former ally of the dictator, to unseat him. Urquiza is aided by a coalition of Unitarios and sympathetic Federales, with help from Brazil. Rosas's forces are defeated at the battle of Monte Caseros, and the dictator goes into exile in England, where he remains until his death.

1853 New constitution drafted

Under the leadership of Juan Bautista Alberdi (c. 1814–86), a statesman and champion of Argentine democracy, a special convention draws up a constitution creating a confederation of provinces. It establishes a representative democracy but gives the president broad powers, including the right to suspend the constitution and declare a state of siege, taking over the powers of the Congress and the courts. Catholicism is declared the official state religion.

1854 Birth of paleontologist Florentino Ameghino

Ameghino (1854–1911) is one of the world's first paleontologists. Many of the fossils he gathers in Patagonia and on the pampas find a home in Argentina's Natural Science Museum, built in 1877.

1854 Urquiza becomes president

Under the new constitution, Urquiza is installed as president of the confederation.

During his term in office he establishes diplomatic and trade ties with South American and European nations, oversees federal funding for public education, encourages agricultural settlements to promote immigration, and authorizes construction of Argentina's first railroad.

During this period, Buenos Aires refuses to join the confederation, fearing that it will lose political control of its own province and with it, vital customs revenues. The city establishes its own legislature, schools, and bank. Warfare breaks out between Buenos Aires and the provinces of the confederation in 1859. In the 1861 Battle of Pávon, Urquiza's army is defeated by Buenos Aires's forces under the leadership of General Bartolomé Mitre (1821–1906). Mitre takes over the leadership of the country, and Buenos Aires becomes its capital.

1857   First rail track laid

Planning for Argentina's rail system had begun in 1854 with a projected railway between Rosario and Córdoba. It is inaugurated three years later with six miles of track. In 1855 Argentina and Chile had signed an agreement for construction of a railroad across the Andes Mountains. By 1890 Argentina's rail network—a symbol both of economic progress and British investment—has grown to 5,800 miles transporting five million tons of cargo and ten million passengers annually. By the turn of the century, the network of rail lines is the most extensive in South America. Spanning almost 10,000 miles.

1860s   Soccer introduced to Argentina

British sailors bring the game of soccer to Argentina, where it is first adopted by the local British community. By 1891 organized games are being played, and soccer balls, nets, and goal posts are being imported from Europe.

The Argentine Republic

1862   Nation unified under a single constitution and president

Buenos Aires accepts the constitution of 1853 with modifications and becomes the capital of the Argentinean republic. Bartolomé Mitre (1862–68) is the first in a series of intellectual, reformist presidents who usher in a long period of prosperity and political stability. Others include Domingo Faustino Sarmiento (1868–74) and Nicolás Avellaneda (1874–80).

These years are marked by great economic progress. The railroads are expanded, and refrigeration of meat is introduced, making possible large-scale beef exports to Europe. Immigration grows rapidly beginning in the 1860s, bringing in the labor force needed to work the farms and fill the growing number of urban jobs. Nearly six million European immigrants (including large numbers of Italians) arrive in Argentina between 1871 and 1914; the nation's population increases by 150 percent between 1850 and 1880. The population of Buenos Aires grows from 90,000 in 1869 to 670,000 in 1895. By the beginning of the twentieth century, it is the largest city in South America.

1862   Composer Alberto Williams born

Williams (1862–1952) is renowned as Argentina's foremost composer and the founder of the Buenos Aires Conservatory. As a young man, Williams studies music in Buenos Aires, giving piano recitals at the Teatro Colón. Winning a government scholarship for study abroad at the age of twenty, he travels to Paris, where he studies with eminent musicians including composer César Franck. Returning to Argentina, Williams begins a career as a pianist and composer. Starting in 1890, he begins to incorporate Argentine folk motifs in his music, introducing a new nationalistic music to his homeland. His first work in this style is *El rancho abandonado* (The Abandoned Ranch), published in 1890.

Possibly Williams's most lasting contribution to the musical life of his country is his founding in 1893 of the Buenos Aires Conservatory, later renamed the Conservatorio Williams. Serving as its director until 1941, Williams also maintains an international career as a composer and conductor. Williams also establishes a music publishing company, La Quena, that is still in operation today. His compositions include nine symphonies and other works for orchestra, piano pieces, chamber music works, and vocal works.

1865–70   War of the Triple Alliance

The only major interruption in a period of peace and economic progress, the War of the Triple Alliance is fought by Brazil, Argentina and Uruguay against Paraguay's dictator, Francisco Solano López (1836–70) (see Uruguay). The Triple Alliance is victorious, and Paraguay is defeated.

1866   Publication of del Campo's *Fausto*

Estanislao del Campo (1834–86) is best known for his poem *Fausto*, in which a simple gaucho describes the experience of seeing Gounod's opera *Faust* performed in the Teatro Colón. *Fausto* is among Argentina's best loved literary works.

1868–1874   Sarmiento reforms the educational system

Domingo Faustino Sarmiento, Argentina's first civilian president (known as the "schoolteacher president"), presides over a complete reorganization of Argentina's system of public education. His measures to improve educational standards, include consulting with famed U.S. educator Horace Mann, bringing women schoolteachers from the United States to teach in Argentina, and founding five teacher training schools. Argentine literacy grows from twenty-five percent at the beginning of Sarmiento's presidency to twice that figure by the end of the nineteenth century.

1870   Newspaper *La nación* is founded

La nación, today Argentina's leading newspaper, is founded by Bartolomé Mitre. Its literary supplement plays a major role in Argentine cultural life. *La Prensa,* another important daily paper, is established a year earlier by José Carlos Paz.

1870–90 New technology revolutionizes Argentine agriculture

The technological revolution in agriculture begun in the United States spreads to Argentina by the latter part of the nineteenth century. The reaper and thresher are introduced in 1870, barbed wire fencing for cattle breeders in 1876, and steel windmills in 1890.

1872–79 Poet José Hernández publishes gaucho classic, *Martín Fierro*

José Hernández (1834–86) writes *El Gaucho Martín Fierro*, a long narrative poem describing the lives and character of the *gauchos* who live on Argentina's vast pampas. *Martín Fierro,* which is published in two parts (1872, 1879), is among the first works to incorporate the everyday language of the *gauchos*. An epic-length poem in 7,210 verses, it narrates the life of the archetypal *gaucho*, Martín Fierro, who is drafted into the army to fight Indians on the frontier, escapes and lives as an outlaw, and eventually returns to his home. This work, which expresses its author's sociopolitical ideas, wins universal popularity. It is still a best seller in Argentina, and considered the Argentine national epic.

1874 Birth of poet Leopoldo Lugones

Lugones (1874–1951), considered by many to be Argentina's greatest poet, writes about his country's landscape and people.

1875 Birth of novelist Enrique Rodríguez Larreta

Larreta (1875–1961) writes the first internationally acclaimed Latin American novel, *La gloria de Don Ramiro* (The Glory of Don Ramiro), about life in the era of King Philip II of Spain.

1877 Natural Science Museum is founded

The Natural Science Museum in La Plata is home to a world-class fossil collection, much of it from the Argentine pampas and Patagonia. Florentino Ameghino (1854–1911), an early paleontologist, collects many of these specimens.

1877 First Argentine opera performed

Stage productions are mounted as early as 1757 at the Casa de Operas y Comedias. By the nineteenth century, opera is rapidly gaining ground at venues including the Coliseo Provisional, the original Teatro Colón, and the Teatro de la Opera. The first Argentine opera, *La gatta bianca*, is staged at the Teatro de la Opera. Opera becomes one of the nation's most popular types of music. To date, Argentines have written over fifty operas, which have been produced both at home and abroad. The operas of Argentine composer Arturo Beruti (1862–1938) have won international popularity.

1879–80 General Roca defeats the Indians of the pampas

General Julio Roca (1843–1914) drives the remaining Indian population out of the southern plains, including Patagonia. Following this "Conquest of the Desert," the Indians are banished from their settlements and forced to work on local ranches or as domestic servants in Buenos Aires. Thousands who refuse to submit are killed. This action fully opens the region to farming and ranching, even further expanding Argentina's wealth. Some 390,000 square miles (one million square kilometers) become available, most of it to be acquired by a small number of landowners.

1880s Secularization leads to rupture with the Vatican

The establishment of state-run public education in Buenos Aires and the legalization of civil marriage lead to a standoff between the government and the Catholic church. The papal nuncio is expelled from Buenos Aires and diplomatic relations with the Vatican are broken off.

1880s Argentine novelists adopt social realism

Reflecting the growth and urbanization of Buenos Aires, Argentine novelists turn to the conventions of social realism and naturalism to depict life in the increasingly modernized city. A major realist work is *La gran aldea* (The Big Village, 1884) by Lucio Vicente López, which chronicles the growth of commercialism and mourns the passing of traditional values. Eugenio Cambaceres' novels of the 1880s deal realistically with the unsavory aspects of city life in the sometimes shocking naturalistic tradition of French novelists Flaubert and Zola. In the first decade of the twentieth century, this tradition is carried on by Manuel Gálvez (b. 1882).

1880 Roca elected president

Roca serves two terms as president (1880–86 and 1898–1904), with one of the intervening terms served by his brother-in-law, Miguel Juárez Celman. Roca's government and those that follow are conservative and primarily serve the interests of wealthy landowners and other members of Argentina's elite. They preside over a period of great prosperity, as Argentina becomes a leading exporter of meat and grain and one of the world's wealthiest nations. Conservative governments remain in office until 1916.

1880: September 21 Buenos Aires becomes a federal district

Buenos Aires, which has served as capital to both the nation as a whole and to its most powerful province (also called Buenos Aires), is separated from its province and made into a separate district. This move eases tensions with other parts of the country, which resent the capital's disproportionate influence.

1887   Birth of singer Carlos Gardel, "king of the tango"

The legendary Gardel (1887–1935) is widely regarded as the best interpreter of tango music. His records are still popular today.

1889   Peak year for European immigration to Argentina

European immigration reaches a record high. More than 250,000 immigrants arrive at the port of Buenos Aires, many to work as agricultural laborers.

1890s   Working classes represented by new political parties

Amid the reigning conservatism of the late nineteenth century and the domination of the large landowners, two new political parties are formed to represent the middle and working classes. The Unión Cívica Radical, founded in the early 1890s by Leandro N. Alem (c. 1842–96), agitates for fair and honest elections and against the corruption of the Celman government. Much of the party's support comes from Spanish and Italian immigrants. Argentina's Socialist Party—the first in Latin America—is founded in 1896 by Juan B. Justo and is especially popular in Buenos Aires.

1890   Economic recession forces resignation of Celman

Miguel Juárez Celman is forced to resign the presidency when a recession hits the nation. The nation pulls through the crisis during the 1890–92 tenure of President Carlos Pellegrini.

1893: November   Argentine Open polo championship introduced

The Argentine Open premieres in 1893 and is held annually.

1896   National Museum of Fine Arts opens

The opening of the National Museum in Buenos Aires helps promote the work of Argentine artists both at home and abroad by giving the country a major art exhibition center. Today it houses modern works by both Argentine and foreign artists, as well as older works depicting Argentine history.

1897   Argentina's oldest soccer team formed

The Quilmes Athletic Club, Argentina's first soccer team, is established. Others soon follow: Rosario Central (1899), Río de la Plata (1901), Independiente (1904), and Boca Juniors (1905).

1899: August 24   Birth of author Jorge Luis Borges

Essayist, poet, and short story writer Borges (1899–1986) achieves renown as the most internationally acclaimed twentieth-century Argentine writer. Borges is born in Buenos Aires and lives in Switzerland and Spain during his adolescence, before returning to Argentina as a young man. He begins his career writing poetry and essays, then turns to fiction, with *Historia universal de* la infamia (A Universal History of Infamy, 1935) and other works. After 1938, his stories become increasingly fantastical, complex, and enigmatic, challenging ordinary perceptions about identity and reality itself. They are collected in such volumes as *Ficciones* and *The Aleph and Other Stories* (a compilation in English translation). Under the pseudonym Honorio Bustos Demecq, Borges also collaborates with Adolfo Bioy Casares (b. 1914) on a series of detective novels.

In 1955 Borges is appointed director of Argentina's National Library, a largely honorary position, given that the author has become totally blind by this point. He is also professor of English and American Literature at the University of Buenos Aires. In addition to his own literary achievements, Borges is remembered for his profound influence on young writers, in Europe as well as North and South America.

1900s   The tango gains popularity in Argentina

The tango, Argentina's national dance, has its roots in the *milonga,* which has evolved among the working classes in the poor neighborhoods of Buenos Aires and is viewed by the upper classes as too openly sensual. The tango's cultural heritage also includes the Spanish *habanera* and African rhythms brought to the Río de la Plata region by former black slaves. Around the turn of the twentieth century, the tango, which is more polished and refined than the *milonga,* becomes socially acceptable among all groups. By the 1920s, it has caught on in Europe, with the help of the 1921 movie *The Four Horsemen of the Apocalypse,* in which film star Rudolph Valentino plays an Argentine *gaucho* and dances the tango cheek to cheek with his partner.

The principal instrument used in tango music is the *bandeneon,* which combines features of an accordion and a concertina. Violins are commonly played as well. The most famous tango singer is Carlos Gardel (1887–1935). The best-known modern composer of tango music is Astor Piazzolla (1921–92), who writes music for many films and even for Russia's Bolshoi Ballet company.

1902   Arbitration accepted in Beagle Channel dispute

Argentina and Chile agree to international arbitration to settle their twenty-year-old territorial dispute over the Beagle Channel, which divides the Isla Grande de Tierra del Fuego between the two nations.

1908   Teatro Colón opens in Buenos Aires

Construction of the Argentina's major opera house, begun in 1887, is completed. Considered the best concert hall in Latin America, it seats around four thousand, and is one of the world's most elegant opera theaters. Its inauguration begins a distinguished tradition of international opera performances. The Teatro Colón hosts world-famous performers including

Enrico Caruso (1873–1921) and Placido Domingo (b. 1941), both of whom are tenor singers. It is also the home of the country's national symphony orchestra and ballet.

1909   Institute of Physics founded

A physics institute is founded at the University of La Plata, with the backing of the government and aided by the expertise of German physicists. It is directed by Richard Gans, who trains many of Argentina's future physicists.

1911   Prize-winning author Ernesto Sábato born

Ernesto Sábato (b. 1911) is born in Rojas, Argentina. He studies physics at the National University of La Plata, earning a doctoral degree, and later studies at the Curie Laboratory in Paris and the Massachusetts Institute of Technology. Sábato teaches at the National University of La Plata from 1940 to 1945, but is dismissed from his teaching post in 1945 because of his political views. The same year, he publishes a book of observations, *Uno y el universo* (One and the Universe), which becomes a critical and popular success. Sábato's other books include the novels *El túnel* (The Tunnel, 1948) and *Sobre héroes y tumbas* (On Heroes and Tombs, 1961); *Tres aproximaciones a la literatura de nuestro tiempo* (Three Approximations to the Literature of Our Times, 1968), a book of critical essays on the works of Jorge Luis Borges, Alain Robbe-Grillet, and Jean-Paul Sartre; and the novel *Abaddón el exterminado* (Angel of Darkness, 1978).

1912   New election law ushers in period of Radical rule

Against opposition from his own party, Conservative President Roque Sáenz Peña pushes a landmark electoral reform law through Congress. It provides for universal male suffrage and secret ballot elections—a major political goal of the opposition Radical party—and establishes a true multiparty democracy in Argentina, giving the nation its first popularly elected president in 1916.

1914   Argentina remains neutral in World War I

The strict neutrality proclaimed by the Conservative government in office at the start of World War I in 1914 is maintained by the Radicals when they come to power two years later.

1914   Experimental writer Julio Cortázar born

Cortázar (1914–84), born in Brussels, grows up and receives his education in Argentina, where he works as a teacher and translator while beginning his writing career. In 1951 Cortázar moves to Paris and publishes his first book, the short story collection *Bestiario* (Bestiary). Other short story collections include *Final del juego* (End of the Game, 1956) and *Las armas secretas* (Secret Weapons, 1958). As a novelist and short story writer, Cortázar becomes internationally known for his experimental writing techniques.

Cortázar's most highly acclaimed work, *Rayuela* (Hopscotch), published in 1963, is a novel distinguished by its unusual structure. The reader must choose the final arrangement of the different parts, following instructions provided by the author. Many consider it the most important Latin American novel published in the 1960s. A short story by Cortázar serves as the basis for the acclaimed U.S. film, *Blow Up*, (1966). Another work that breaks new ground in narrative structure is *62: Modelo para armar* (62: A Model Kit, 1972). Other books by Cortázar include *Todos los fuegos el fuego* (All Fires the Fire, 1966) and *Libro de Manuel* (Manual for Manual, 1973). Cortázar acquires French citizenship in 1981 but retains his Argentine citizenship as well.

1916   Major female poet, Alfomsina Storni, publishes first volume of poetry

Storni (1892–1938) publishes *La inquietud del rosal*, her first book of verse. Storni is regarded as the first Argentine poet to write from a uniquely female perspective.

1916   Radicals gain power as Yrigoyen is elected

Longtime Radical party leader Hipólito Yrigoyen is elected to the presidency in Argentina's first elections based on universal male suffrage, ending decades of nearly unbroken conservative rule. He is succeeded in 1922 by another Radical, Dr. Marcelo Torcuato de Alvear (1868–1942). Yrigoyen wins reelection in 1928. The Radicals initiate social and educational reforms but suffer from administrative problems, and their policies arouse increasing opposition.

1916   Composer Alberto Ginastera born

Alberto Ginastera (1916–83) wins recognition as Argentina's most internationally acclaimed composer, and one of the best known Latin American composers. Ginastera graduates from the Buenos Aires Conservatory in 1935, winning the gold medal in composition. In 1942 he receives a Guggenheim Fellowship to study in the United States. However, he does not actually go there to study until after World War II. Between 1945 and 1947, Ginastera lives in New York City, attending summer composition classes with Aaron Copland at the Tanglewood Music Festival in Massachusetts. Returning to Argentina in 1948, the composer forms the Argentine branch of the International Society for Contemporary Music (ISCM).

Early in his career he bases his compositions on Latin American folk material. His later works are less nationalistic and more abstract, utilizing twelve-tone systems and other advanced composition techniques, including microtonality, aleatoric (random) components, and *Sprechstimme* (a combination of speech and song). Ginastera's best-known works include the ballet *Estancia* (1941); the *Cantata para América Mágica* (1961); the Piano Concerto No. 1 (1961); and the operas *Don Rodrigo* (1964), *Bomarzo* (1967), and *Beatrix*

Cenci (1971). *Beatrix Cenci* is among the works performed during the inauguration of Washington D.C.'s Kennedy Center for the Performing Arts in 1971.

1919: January   "Tragic week" general strike is violently suppressed

Argentina's burgeoning anarchist movement, which is influential among the labor unions, receives a severe setback when a general strike is suppressed in a week of violence that includes government repression as well as street fighting between anarchists and right-wing militants.

1920s   Poetry journal *Martín Fierro* founded

Established by a group of young authors in Buenos Aires, this journal, named after the famous nineteenth-century narrative poem by José Hernández, aims to reinvigorate Argentina's native literature. It publishes both abstract poetry and poems on folk themes.

1920s   Argentine physicists publish papers on relativity

Inspired by Albert Einstein's 1925 visit to Latin America, physicists Ramón Enrique Gaviola and Enrique Lodel publish studies of relativity.

1926   Novel *Don Segundo Sombra* published

Don Segundo Sombra, by Ricardo Güiraldes (1886–1927), carries the native *gaucho* theme into the twentieth century, portraying the decline of this folk figure. It is Argentina's most famous novel.

1929   Grupo Renovacíon founded to promote modern music

This organization is founded by the three Castro brothers, who are all Argentine composers: Juan José (1895–1968), José María (1892–1964), and Washington (b. 1901).

The "Infamous Decade"

1930   Worldwide depression sparks economic crisis in Argentina

Argentina is hit hard by the U.S. stock market crash in October 1929 and the ensuing global depression. Its export market and investment from abroad decline simultaneously. Agricultural prices, also affected by weather conditions, fall, and there is a decline in imports, as well as rising inflation.

1930: September 6   Irigoyen ousted in military coup

Eighty-year-old president Irigoyen (1850–1933), reelected in 1928, proves unable to take effective measures against the nation's deepening economic crisis, and he is ousted in a military coup led by General José Félix Uriburu (1868–1932), who assumes the role of provisional president until 1932.

1930   Argentina competes in first World Cup soccer tournament

The Argentine team makes it to the finals of the first World Cup contest, held in Uruguay. After defeating, among others, the French and Chilean teams, the Argentines ultimately lose to the Uruguayans, 4–2.

1930–1943   The "infamous decade"

With the president deposed and the radical party torn by internal strife, the Conservatives return to power. In 1932, General Agustín P. Justo (1876–1943) is elected president. Both his government and that of Roberto Ortiz (1886–1942), elected in 1938, are dominated by the Concordancia conservative-radical coalition. Characterized by corruption and election fraud, the entire period from 1930 to 1943 is often referred to as the "infamous decade." However, important economic programs are carried out, including industrial expansion and public works projects. These and major social legislation help the nation weather the worldwide economic depression of the 1930s. Urban unemployment remains at about five percent, and a recovery is under way by 1934.

1931   Soccer becomes a professional sport

Soccer attains "pro" status, with well-attended league games. The Río de la Plata and Boca Juniors become the top two teams, winning supporters nationwide.

1931   Writer Victoria Ocampo founds literary magazine *Sur*

Essayist Victoria Ocampo (1890–1978), Latin America's most celebrated female author, founds the periodical *Sur* in Buenos Aires. Using her own personal wealth, she publishes poems and short stories by major modern European and North American authors in translation and also publishes the works of most major twentieth-century Spanish American writers. *Sur* enjoys international popularity in literary circles, and Ocampo, a friend of British author Virginia Woolf, is credited with furthering feminism and modernism in Argentine literature.

1932   Author Manuel Puig born

Manuel Puig (1932–90) is born in Buenos Aires Province and attends a U.S.-run boarding school in the capital. He becomes a passionate fan of American movies as a youngster, learning English at the age of ten so that he can understand the words. Also a fan of French and Italian film, Puig studies filmmaking in Italy in the 1950s. He begins writing movie screenplays in Argentina in the 1960s and working as an assistant film director. He later continues his writing career in New York City, where he authors novels that incorporate the techniques of film. Puig's works draw criticism in his homeland for the sexual content of some of his works, and for his open homosexuality.

Manuel Puig's novels include *La traicon de Rita Hayworth* (Betrayed by Rita Hayworth, 1968), and *Boquitas pintadas* (Heartbreak Tango, 1969). Puig is best known for his 1976 novel *El beso de la mujer anaña* (The Kiss of the Spider Woman), later adapted both as a movie and as a Broadway musical. In 1986 Puig receives the Curzio Malaparte Award from Italy for his 1983 novel *Sangre de amor correspondido* (Blood of Requited Love).

1933   Roca-Runciman Pact

In response to Britain's 1932 adoption of Imperial Preference trade policies favoring Australia and South Africa, Argentina signs a trade agreement granting preferential treatment to British exports in the Argentine market in return for retaining its access to the British market, especially the market for its meat exports.

1935   Central Bank of the Argentine Republic founded

Argentina's central reserve bank is established.

1936   Conciertos de la Nueva Música founded

This concert series features twentieth-century music, with performances held in the Teatro del Pueblo in Buenos Aires.

1936   Carlos Saavedra Lamas wins Nobel Peace Prize

This Argentine diplomat (1878–1959) wins international recognition for his role in mediating a conflict between Bolivia and Paraguay.

1938   Birth of novelist Luisa Valenzuela

Valenzuela is one of Argentina's leading modern novelists.

1941   National Pato Federation founded

A federation is founded to support the native Argentine game of *pato*. The polo-like game, played on horseback, now uses a ball with handles that is passed between players instead of a live duck in a sack, as was originally the case. Instead of striking the ball with a stick as in polo, the goal is to push it into a large net.

1943   Military ousts Castillo

President Ramón Castillo (1873–1944), elected in 1942, is deposed in a military coup, followed by a brief period of military government, during which army officer Juan Domingo Perón (1895–1974) becomes increasingly prominent and popular. Pro-labor and reformist, Perón holds several top government positions in the early 1940s, including Minister of War and head of the department in charge of labor and social welfare. In 1944 he is named vice president.

1945   *Ulysses* published in Spanish translation

A Spanish translation of James Joyce's (1882–1941) groundbreaking 1922 novel *Ulysses* is published in Buenos Aires, influencing contemporary Latin American authors. The modernist novel's influence can be seen in Leopoldo Marechal's 1948 work *Adán Buenosayres*, whose structure, like that of *Ulysses*, is based on a classic epic work, in this case Dante's *Inferno*. By the 1960s, other techniques pioneered by Joyce—including stream-of-consciousness and elaborate word play—have become popular in Latin American fiction.

The Perón Era

1946: January   Perón becomes president

Juan Perón (1895–1974) is elected president and carries out a nationalistic and labor-oriented program of reforms. Labor reforms instituted under Perón include improved working conditions in factories, accident compensation, female and child labor laws, pension plan legislation, a standard work day, Sunday rest laws, and paid vacations. Perón nationalizes the central bank, foreign-owned public utilities, the railroads, and shipping. He also implements policies to give the nation's workers a greater share of its wealth. In the mid-1930s, wage laborers receive thirty-eight percent of the national income; twenty years later, that figure has risen to forty-six percent. By 1947, Argentina's entire national debt has been paid off.

Perón's second wife, Eva Duarte "Evita" Perón (1919–52), is extremely popular with the Argentine public, especially working-class women. Born into a poor family, she is an aspiring actress who becomes Perón's mistress in the 1930s. She helps him in his rise to power in the early 1940s and becomes his wife in 1945. Eva Perón becomes the champion of the *descamisados* ("shirtless ones"), or urban working-class poor, who regard her with an almost religious devotion. She creates a foundation funded by labor unions, business, and the government, that supports hospitals, schools, and charities. She is also the driving force behind many social reform measures and closely involved in gaining the vote for women. The play *Evita* is based on her life.

1947   Women win the right to vote

1947   Physiologist Houssay wins Nobel Prize for Medicine

Argentine physiologist Bernardo A. Houssay (1887–1971) wins the Nobel Prize for Medicine for his discovery of the role of the pituitary gland in the metabolism of carbohydrates.

1949   Perón suspends constitution

In the 1950s Perón is forced to temper his policies in response to a growing economic crisis. Western Europe, recovering

Juan Perón (1895–1974) and his wife, Eva (1919–52), wave to enthusiastic crowds in Buenos Aires in 1950. A populist, Perón is a strong supporter of labor and the working classes. (AP/Wide World Photos)

from World War II (1939–45), once again becomes an economic competitor, resulting in lower prices for agricultural produce. Export earnings drop thirty percent in 1948, and Argentina no longer has its large postwar foreign exchange reserves. With inflation rising rapidly, Perón implements a two-year freeze on wages, endangering his support by labor. He also modifies his nationalist stance to encourage foreign investment in Argentina, alienating Argentine business interests. In 1954, he signs an agreement giving the U.S. corporation Standard Oil developments rights to the oilfields of Patagonia.

In response to growing opposition to his economic policies, Perón's rule becomes dictatorial. He suspends Argentina's constitution, places restrictions on freedom of speech, and imposes censorship of the press. With his wife's death, Perón loses yet more of his popular support.

1952: July 26   Nation plunges into mourning for death of Eva Perón

Eva Perón dies of cancer at age thirty-three. The death of Eva Perón, whose popularity has surpassed that of her husband, poses a major setback for his political career.

1954   Perón legalizes divorce, loses church support

The Catholic Church, originally supportive of Perón, turns on him in the 1950s as his government begins to infringe on their traditional authority, legalizing divorce and ending religious instruction in the public schools. In addition, the Catholic establishment resents the quasi-religious Perón personality cult, especially the near-worship of Eva Perón and the calls to

have her canonized following her death. The rift becomes even worse after a November 1954 speech in which Perón accuses the Church of antigovernment activities. In May 1955, Perón demands the separation of church and state.

On June 12, 1955 the Church organizes a mass anti-Perón demonstration at a Corpus Christi celebration, which draws over 100,000 people.

1955: June   Military planes bomb presidential palace in aborted coup attempt

The military bombs Peron's residence during a mass rally, killing several hundred Perón supporters. In response, Peronists set fire to churches. Perón inflames his supporters yet further in an August 31 rally, urging them to take up arms against the opposition.

Military Rule

1955: September 16   Military ousts Perón in coup d'état

The armed forces stage a rebellion, forcing Perón from office and into exile in Spain for seventeen years. During these years, his supporters, continue to exert a strong influence on Argentine politics, even though many have been imprisoned and killed by the military following the 1955 coup and their organizations banned. For nearly twenty years, Argentina's government is largely in the hands of the military.

1956   Military government outlaws divorce

Divorce, which has been legalized under Perón, is once again declared illegal by Argentina's military regime. (It becomes legal again in 1987.)

1958–62   Arturo Frondizi serves as president

Frondizi (b. 1908), a Radical, is elected president in February 1958. He works to alleviate the nation's economic problems, launching an anti-inflation program and encouraging foreign investment. He tries to accommodate his Peronist opposition by legalizing their party, which has been banned in the aftermath of the 1955 coup. However, they enjoy such a strong showing in the 1962 legislative elections that the military nullifies the election and overthrows Frondizi himself.

1962–66   Illia presidency

Once again the military allows formation of a civilian government, this time under another Radical party member, Arturo Illia (1900–83). Like his predecessor, Illia encounters opposition by the Peronists and allows them to run in legislative elections (1965), and, as in 1962, they win a majority of the votes. Once again, the military takes over, this time appointing General Juan Carlos Onganía to the presidency.

1962   Lunfardo Academy founded to preserve distinctive folk dialect

The Lunfardo Academy of Buenos Aires is established to uphold the street dialect known as *lunfardo*. Containing words borrowed from Italian, Portuguese, and other languages, *lunfardo* began as a vernacular used by criminals. In the first part of the twentieth century, poets and tango singers, including the famed Carlos Gardel, begin using the colorful street slang. In the 1940s *lunfardo* is officially banned, but not truly abandoned. The Lunfardo Academy is founded by intellectuals and media figures who do not want this distinctive cultural tradition to be lost.

1966–70   Onganía regime

General Onganía imposes repressive measures to consolidate his power and maintain order. He bans political parties, dissolves Congress, and suspends elections. His government seizes control of the University of Buenos Aires and the nation's other universities and ousts all professors whose political views they find objectionable. He institutes an economic recovery program that reduces inflation to around seven percent by 1969. However, the working class fares badly under his policies, which include raising the retirement age, reducing minimum wage, and then freezing wages altogether. By 1968, opposition to Onganía is growing in spite of his authoritarian rule. Labor, leftist, and student opposition find an outlet in the General Confederation of Labor of the Argentines (CGTA), formed in 1968.

1969: May 29   Militant protest is quelled by the army

In May 1969, students and auto workers stage an uprising in the city of Córdoba that comes to be known as the *cordobazo*. In response to cuts in funding of higher education, job cuts, and other effects of Onganía's economic austerity program, student protesters join a labor demonstration organized by the CGTA in the center of the city. The army and the police quell the protest by force, touching off two days of street fighting and rioting that leave at least 100 persons dead or injured. This incident touches off a wave of strikes and protests in other cities and brings students, workers, and business groups together in their opposition to the military regime.

1970   Leloir wins Nobel Prize for Chemistry

Luis Federico Leloir (1906–87) wins Nobel Prize for Chemistry for his dicovery of sugar nucleotides.

1970: February 1   Train wreck kills 142 outside Buenos Aires

An express train on the last leg of a trip from San Miguel de Tucuman to Buenos Aires crashes into a stalled commuter train eighteen miles from the capital, killing 142 people and injuring hundreds more. The express train, traveling at sixty-five miles per hour, is carrying 500 people; the commuter train holds 700 people returning to the capital at the end of the weekend. The crash is blamed both on the mechanical failure of the stalled train and on the human error of the signalman who fails to alert the express that there is a stalled train in its path.

1970: March   Militant Peronists kidnap former president

A group of young Peronists calling themselves the Montoneros kidnap and kill former President Pedro Aramburu (1903–1970), adding to the growing tension between the military government and opposition groups of the Peronists and also of the extreme left wing. Dictator Onganía is forced out of office; he is succeeded by two more generals—Roberto Levingston and Alejandro Lanusse—in the next two years.

The Return of Perón

1973: March   Peronist candidate wins the presidency in free elections

Argentina's military rulers permit free elections to be held but disqualify the exiled Perón from running. Héctor Cámpora (1909–80), a Perón loyalist, is elected president. However, Cámpora soon resigns the presidency so that Perón, who returns to Argentina in June of 1973, can take over the office. As a sign of the divisions plaguing the nation, violence breaks out among the thousands of supporters who turn out to greet Perón at Ezeiza International Airport on June 20, 1973, killing several hundred and injuring over a thousand.

1973: September 23   Juan Perón is elected president

A special election is called, and Perón assumes his old office, winning sixty percent of the vote. His third wife, Isabel Perón (b. 1931), becomes vice president. As president, the 77-year-old Perón faces spiraling inflation, growing guerrilla violence, and even violence between opposing factions among his own followers. The programs of his predecessor, Héctor Cámpora, bring inflation down to thirty percent by late 1973 and increase real wages. Perón then begins a nationalization program similar to those he instituted in the 1940s and 1950s. He also takes steps to destroy Argentina's guerrilla groups, outlawing even the supposedly "Peronist" Monteneros.

1974: July 1   Perón dies; Isabel Perón becomes world's first female president

Perón, whose health is failing, dies during his first year back in office, and his wife, Isabel, becomes president. She is the first woman to serve as head of state in a Latin American country, and the first woman in the world to serve as the president of a nation. She leads a party that was seriously divided between right- and left-wing factions, and her government faces rising political violence and economic disruption. In

November 1974 the military imposes a state of siege and begins the "dirty war" against guerrillas and other alleged subversives that will continue through the 1970s. By mid-1975 Argentina's inflation level is the highest in the world.

1974 Argentina installs Latin America's first nuclear power plant

The Atucha I electric generating plant in Buenos Aires province is the first nuclear power station in Latin America.

1976: March 24 Isabel Perón is ousted in military coup

Isabel Perón is deposed by the military and placed under house arrest. A three-member ruling junta led by General Jorge Rafael Videla (b. 1925) takes control of the government. Videla is later appointed president.

The Return of Military Rule and the "Dirty War"

1976–83 Argentina under military rule

The nation remains under military rule throughout the late 1970s and early 1980s when General Roberto Viola succeeds Videla as president in 1981 and, later the same year, Leopoldo Galtieri (b. 1926) succeeds viola. The primary challenges facing the military government are the nation's economic ills—primarily, spiraling inflation and a large budget deficit—and guerrilla violence.

The government steers the nation toward a free-market economy, reversing Perón's nationalization policies, removing supports for industry, and attempting to crush the country's labor movement (many of whose leaders are among those selected for political persecution). Exports grow, producing a welcome trade surplus by 1976, but the working class suffers as real wages fall, food prices rise, and social welfare programs are cut. In addition, inflation still hovers around 100 percent annually.

The harsh anti-terrorist measures imposed by the military come to be known as the "dirty war." They involve massive human rights violations, as the government targets trade unionists, students, teachers, and other intellectuals in its attempt to stamp out Marxist guerrilla activities and other political violence. The death penalty is reintroduced, and thousands of Argentines are killed. As many as 30,000 "disappear," killed by death squads and buried secretly in mass graves.

1977 Ecological "green belt" is created around Buenos Aires

The Metropolitan Area Ecological Belt State Enterprise places air and water pollution controls, as well as limits on building density, on a 93-mile (150-kilometer) area surrounding the capital.

1977: April 30 Vigil by *las madres* protests state terrorism

The group *las madres de los desparedicos* ("mothers of the disappeared ones") is founded by activist Azunclena De Vincenti. On April 30, seven of *las madres* publicly protest the disappearance of family members in a silent vigil at the Plaza de Mayo in Buenos Aires, carrying placards that ask where their loved ones are. *Las madres* becomes a powerful political symbol of the resistance to human rights violations by the government.

1978 Argentina hosts World Cup competition

Argentina both hosts and wins the World Cup soccer competition. The following year a team of young Argentines win the junior world soccer championship in Japan

1980 Esquivel wins Nobel Prize for protesting government repression

Adolfo Pérez Esquivel (b. 1931), a sculptor and architect, wins the Nobel Peace Prize for criticizing the government's human rights violations and aiding political prisoners.

1980–81 Support for military weakens as economy falters

By 1980 Argentina's trade deficit returns, and foreign investors lose confidence in the nation's economy, leading to several financial crashes. By 1981 the nation's foreign debt has reached $35.6 billion.

1981 Timerman memoir draws international attention

Jacobo Timerman (b. 1923), a Jewish newspaper editor and political dissident, publishes *Prisoner Without a Name, Cell Without a Number*, a memoir of his brutal persecution by the military government. Timerman's story attracts international attention to human rights violations in Argentina.

1981: May 1 Mass protest over government human rights violations

A crowd of over six thousand Argentines commemorates the first vigil by *las madres de los desparecidos*, gathering at the Plaza de Mayo to protest the continuing persecution of those deemed to be political subversives.

1981: December Galtieri leads coup against military leaders

General Leopoldo Galtieri (b. 1926) overthrows the acting military rulers and takes control of the government. He attempts to bring the economy under control by privatization and cuts in government spending.

1982: March 30   Labor group holds mass demonstration

Expressing the growing grievances of the country's workers, the General Confederation of Labor, which has been outlawed by the government, stages a mass rally in the Plaza de Mayo in Buenos Aires. Demonstrators clash with police, and many are arrested. General strikes are held in 1982 and 1983 to protest government corruption and low wages.

1982: April 2   Argentina invades the Falkland Islands

Seeking to consolidate popular support for his government, Galtieri orders an invasion of the sparsely inhabited Falkland Islands (the Malvinas), which the British have held since 1833. The invasion enjoys broad public approval and spurs a surge of nationalism among Argentines. However, the British send troops to the South Atlantic to reclaim the islands.

1982: June 11–12   Pope John Paul II visits Argentina

In an attempt to end the war in the Falklands, John Paul II makes a last-minute trip to Argentina, becoming the first pope to visit the country. Thousands of Argentines attend when he celebrates mass, and his visit fuels a religious revival already under way, especially among young people. As an aftermath of the visit, the Argentine church calls for an end to the state of siege within Argentina as well as an investigation into the "disappearances" of the military regime's political opponents since the inception of the "dirty war" in 1976. Two months after the papal visit, the church issues a document titled "The Path to Reconciliation."

1982: June   Falklands recaptured by the British

On June 14, after more than two months of air, sea, and land battles, 10,000 Argentine troops surrender, and the British retake the Falklands. On June 15, President Galtieri acknowledges defeat. All together, the seventy-two-day war costs some 2,000 British and Argentine lives. Britain continues to maintain 2,500 troops on the islands. Britain and Argentina do not resume full economic and diplomatic ties for another seven years.

Argentina's government is badly discredited by this military failure, and the Argentine economy has steadily worsened. Galtieri resigns from office on June 17 and is replaced by interim leader Reynaldo Bignone (b. 1928). Under Bignone, the government passes a law granting amnesty to military officers who took part in the "dirty war," but the law is annulled by the new civilian government in 1983.

1982: October–December   Anti-government demonstrations multiply

In late 1982 demonstrations are organized by the church, labor unions, and human rights groups. An October protest in the Plaza de Mayo draws 10,000 people. In Buenos Aires, 20,000 protest against city taxes in November. In December, a general strike occurs in which ninety percent of the labor force participates. Also in December, 300,000 people gather in the Plaza de Mayo to protest government repression and call for free elections.

1983: February   Free elections announced

General Bignone, who succeeds Galtieri as president, announces that free elections for a civilian government will take place on October 30, and the new president will be inaugurated in January of 1984.

1983: May   Embalse nuclear plant completed

The 600-mcgawatt Embalse nuclear plant, built jointly by Argentine, Italian, and Canadian firms, is completed after eight years at a cost of $US 1.3 billion.

1983   New peso introduced

The Argentine government introduces a new peso. The existing currency is so valueless that one new peso equals 10,000 old pesos.

Return to Civilian Government

1983: October   Alfonsín elected president

Radical Party leader Raúl Alfonsín (b. 1926) is elected president. It is the first time since 1946 that a non-Peronist candidate has won a free election in Argentina. In addition, Alfonsín's Radical Party wins a number of provincial governorships and control of the legislature. Alfonsín is inaugurated in December 1983. To deal with the nation's $45 billion foreign debt and inflation rate of over 1,000 percent, Alfonsín is forced to accept economic austerity measures imposed by the International Monetary Fund.

1984   Milstein wins Nobel Prize for Medicine

César Milstein (b. 1927) wins Nobel Prize for Medicine for his contributions to the study of antibodies.

1984   Dispute with Chile over the Beagle Channel settled

The nearly one hundred-year-old border dispute between Argentina and Chile over the Beagle Channel is resolved, following Argentine rejection of British arbitration in favor of Chile in 1976 and a subsequent 1978 agreement by the two nations to accept arbitration by the Vatican. The two nations sign the Declaration of Peace and Friendship in Rome, which gives Chile possession of the disputed islands in the channel. Though the Argentinian Congress may have nullified the agreement (leading to further dispte with Chile), eighty percent of the Argentinian people ratify the Declaration in a national referendum.

1984: September   Major human rights report is published

The *Nunca Más* (Never Again) report detailed human rights violations during the "dirty war."

1985   Ernesto Sábato wins Cervantes Prize

Argentine author Ernesto Sábato (b. 1911) receives Spain's Cervantes Prize, the most prestigious literary award in Hispanic literature.

1985   Argentine film wins Academy Award

The Official Story, a film about a woman who discovers that she has adopted the daughter of one of "the disappeared," wins the Academy Award for Best Foreign Film. The movie also garners actress Norma Aleandro a Best Actress nomination.

1985   Officers convicted for human rights violations

Alfonsín's decision to try and convict top military leaders guilty of human rights violations in the "dirty war" of 1976–83 weakens an already strained relationship with the military. Five highly placed military officials are tried for war crimes, including torture and killings, committed during the government crackdown of the late 1970s. General Jorge Vidala, one of the presidents appointed during the period of military rule, is sentenced to life in prison. Another ex-president, Roberto Viola, is sentenced to seventeen years in jail.

1985: May   Government begins literacy program

Opening some three thousand literacy centers throughout the country, the Argentine government launches the National Plan of Functional Literacy and Continuing Education. The eventual goal is to operate 20,000 centers by 1989, teaching literacy skills to 1.2 million people who cannot read at all, and 5.2 million who are functionally illiterate.

1985: June   Alfonsín unveils economic plan

President Alfonsín announces an economic program (the Austral Plan) introducing a new currency, the austral, to replace the peso. He also cuts public spending and freezes wages and prices. State industries are to be privatized and industrial exports encouraged. The plan produces some improvement, cutting inflation fifty percent by the following year. However, it also produces a recession, weakening support for Alfonsín's government.

1986   Argentina wins second World Cup title

Argentina defeats West Germany 2–1 to win its second World Cup soccer championship at games held in Mexico. Hundreds of thousands of fans celebrate nationwide, thronging the streets of Buenos Aires, Córdoba and other cities, stopping traffic and draping everything in sight in blue and white, the Argentine national colors.

1986   Former president jailed in connection with Falklands defeat

Former president General Leopoldo Galtieri is one of three men imprisoned for negligence in the Falklands war of 1982.

1987: September   Radical Party loses control of Congress

Alfonsín's Radical Party loses its majority in the chamber of deputies, as well as most provincial elections. Alfonsín tries to win Peronist backing for his policies but fails, and the Peronists stage a general strike in November.

1988–89   Civil unrest grows as Alfonsín's authority weakens

Public dissatisfaction and political strife grow as Argentina's economy continues to worsen. Two unsuccessful military revolts take place in 1988, and there is a renewal of terrorist activity by both right- and left-wing groups.

1988   Pope John Paul II draws huge crowds on visit to Argentina

Pope John Paul II's visit to Argentina marks the first time since the Middle Ages that a pope has held Palm Sunday mass outside Rome. Over one million attend.

1988   World's oldest dinosaur remains found in Argentina

A team of archaeologists from the University of Chicago find the skeleton of the oldest dinosaur ever discovered. Six feet in length and weighing approximately three hundred pounds, it is estimated to have lived 230 million years ago. The research team names the dinosaur *Herrerasaurus* after the Argentine farmer who first found the skeleton and led them to it.

1989: January   Guerrillas attack army barracks near Buenos Aires

Sixty-nine leftist guerrillas launch a raid on an army barracks near the capital. The siege lasts for thirty-two hours before the terrorists are defeated.

1989: January   Emergencies caused by public service breakdowns

Argentine cities suffer power blackouts lasting up to six hours, endangering the health of shantytown residents during the height of Argentina's summer.

1989: January 28   Argentine vessel causes major oil spill in the Antarctic

The Bahia Taraiso runs aground in the Antarctic and leaks oil that contaminates a number of islands, endangering thousands of seabirds in the area. The Argentine navy cleans up the spill between January and March.

1989: May   Menem wins presidential election

In the 1989 presidential election, Peronist Carlos Saúl Menem (b. 1935) wins a decisive victory over the Radical party candidate, Eduardo Angeloz.

1989: May–July   Economic and political crisis deepens, Alfonsín resigns early

At the end of May, food riots and looting lead the government to declare a one-month state of siege. Inflation reaches record levels, even for Argentina—over 1,000 percent. The situation becomes so dire that sitting president Alfonsín resigns from office on July 8, 1989, five months before the end of his term, and Menem quickly takes over.

1989: July   Menem assumes the presidency

Although a Peronist (traditionally the party of labor), Menem seeks the cooperation of business in carrying out his economic policies, introducing free-market reforms that include privatization of state-owned enterprises. The economy is also opened to foreign investment. Inflation is reduced, but social programs are cut and unemployment rises.

1990: December   Unsuccessful coup attempt by the military

Tensions between Menem and the military peak with an unsuccessful uprising aimed at unseating the president. To placate his opponents, Menem makes a major concession, granting pardons to military officers imprisoned for human rights violations during the period of military rule between 1976 and 1983. This action angers many in the country.

1991   National Women's Council created

The National Women's Council is formed to coordinate government policies on women. In 1993, President Menem officially bans sexual harassment in government jobs.

1992   Menem signs accord creating MERCOSUR

Argentina's president signs an agreement establishing the Mercado Común del Sur (MERCOSUR), or Southern Common Market, with Brazil, Uruguay, and Paraguay. This region accounts for nearly two-thirds of South America's economic activity. By 1995, almost all tariffs between member nations have been eliminated.

1992: January   Argentina returns to the peso

Argentina returns to the peso as its currency, replacing the austral. The new peso is pegged to the U.S. dollar.

1993   Menem announces new economic plan

President Menem announces short-term economic goals that include reducing inflation to four percent, creating a million new jobs, and growth in manufacturing and agriculture.

1994   Convention revises Argentina's constitution

A special Constituent Assembly modifies the nation's constitution, which has been in place for 140 years. It expands the senate, reduces the length of senatorial terms from nine to six years, and reduces the presidential term from six to four years, giving the president the right to run for two consecutive terms. It also provides for direct election of the president and reverses the requirement for the president and vice president to both be Roman Catholics.

1994: December   Devaluation of Mexican currency leads to recession

Mexico's devaluation of its peso results in a recession in Argentina, whose economy had stabilized under the policies of President Menem in the early 1990s.

1995: May 14   Menem wins second term

President Carlos Menem is elected to a second term, signaling voter approval of his government's free-market economic reforms. Menem's Peronist party also wins control of both houses of Congress. Menem's victory, made possible by constitutional changes enacted in 1994, make him only the second Argentine president in the twentieth century to win reelection (the other is Juan Perón).

Shortly after the election, Menem announces plans to increase exports to the other MERCOSUR countries (Brazil, Uruguay, and Paraguay), reform the nation's education and public health systems, and implement measures to reduce unemployment, which stands at fourteen percent as of July 1995. Menem is sworn in to his second term in office on July 8.

1995: June 23   Riots erupt in Cordoba over nonpayment of state workers

Residents of Cordoba, Argentina's second-largest city, riot to protest delays in payment of wages and pension benefits to government employees. Cordoba's provincial governor, and former Radical party presidential candidate, Eduardo Angeloz, resigns his post after thirteen years in office.

1995: December 10   Perónists retain Senate

In Senate elections the Perónist *Partido Justicialista* retains its majority by holding forty of the body's seventy-two seats.

1996: March Menem granted emergency powers

The Congress allows President Carlos Menem emergency leeway to raise taxes and cut public spending in order to forestall an anticipated budget shortfall and allow the country to meet requirements set by the International Monetary Fund.

1996: September 26 General strike called to protest austerity measures

Argentina's largest labor federation calls a thirty-six-hour general strike, the most wide-ranging protest against President Menem's economic policies since Menem took office in 1989. With unemployment standing at seventeen percent, the Menem government is still moving ahead with its planned economic reforms, including tax increases, reductions in public sector jobs, and legislation rescinding measures favorable to labor, such as collective bargaining. Between eighty and ninety percent of workers stay home, virtually paralyzing normal activity in the nation. About 70,000 protesters gather in the Plaza de Mayo in Buenos Aires, outside the presidential palace. Menem says that the strike will not affect his economic policies.

1997: March 25 Spain calls for arrest of former Argentine leader Galtieri

Spain issues an arrest warrant for General Leopoldo Galtieri, head of Argentina's military government between 1981 and 1982, for the murder of three Spanish nationals during the "dirty war" against government opposition in the late 1970s and early 1980s. In an unpopular 1989 pardon, President Menem had exempted Galtieri and other military officials from prosecution by Argentina for abuses during that period.

1997: March 19 Government returns Indians' ancestral homeland

The government of Argentina formally returns 309,000 acres (125,000 hectares) of land in the country's northwest, seized by Spanish conquistadors in the sixteenth century, to the Colla Indians. The move is part of a 1994 constitutional reform that has already returned some four million acres of land to Argentina's indigenous peoples. President Menem presents Colla leaders with deeds to the land, which is located in the Salta province.

1997: October 26 Perónists retain plurality in the Chamber of Deputies

In elections for the Chamber of Deputies, the Perónists capture 36.2 percent of the vote and hold 119 out of 257 seats.

Bibliography

American University. *Argentina: A Country Study,* 3rd ed. Washington, DC: Government Printing Office, 1985.

Biggins, Alan. *Argentina.* Santa Barbara, Calif: Clio Press, 1991.

Brysk, Alison. *The Politics of Human Rights in Argentina: Protest, Change, and Democratization.* Stanford, Calif: Stanford Univ. Press, 1994.

Calvert, Susan. *Argentina: Political Culture and Instability.* Pittsburgh: University of Pittsburgh Press, 1989.

Crassweller, Robert D. *Perón and the Enigma of Argentina.* New York: Norton, 1987.

Hastings, Max, and Simon Jenkins. *The Battle for the Falklands.* New York: Norton, 1983.

Masiello, Francine. *Between Civilization and Barbarism: Women, Nation, and Literary Culture in Modern Argentina.* Lincoln: University of Nebraska Press, 1992.

Rock, David. *Authoritarian Argentina: The Nationalist Movement, Its History, and Its Impact.* Berkeley: University of California Press, 1993.

Sunday Time of London Insight Team. *War in the Falklands: The Full Story.* New York: Harper & Row, 1982.

Timerman, Jacobo. *Prisoner Without a Name, Cell Without a Number.* New York: Knopf, 1981.

Bahamas

Introduction

The Bahama Islands (or Bahamas) were formed some 200 million years ago from the ongoing layering of sediments in the shallow marine waters off the Atlantic coast of North America. During the last Ice Age some two million years ago, the drop in sea level exposed large areas of the shallow banks. The prevailing easterly winds intensified, and large dunes were created from fine, red African sands carried across the Atlantic. As these dunes turned into rock, they formed the rocky cores for most of the islands.

The Bahamas consist of nearly 700 islands that extend across 600 miles in a northwest to southeast line, from Grand Bahama Island (at 27°N latitude) to Grand Turk (at 21°N). The climate is pleasantly subtropical, with daily average temperatures ranging from 73°F (23°C) in the winter to 81°F (27°C) in the summer, making the islands conducive to year-round tourism from North America. Although the Bahamas are located slightly north of the usual path of hurricanes, the immense storms occasionally strike the islands. The islands are all low-lying and relatively infertile, but the waters around the islands are rich in marine life: fish, mollusks, seals, turtles, sea grass, algae, and plankton thrive in Bahamian waters.

Only about thirty of the islands are inhabited, with New Providence the most densely populated. The islands that have few people are known as the Family Islands or the Out Islands and include Bimini, the Exumas, Andros, the Abacos, and many others. The total population is about 259,000, with some 195,000 living in Nassau, the capital, on New Providence. The only other metropolitan area in the Bahamas is Freeport, on Grand Bahama. The proximity to the American mainland and shipping channels has made the Bahamas a natural site for piracy and smuggling throughout its history.

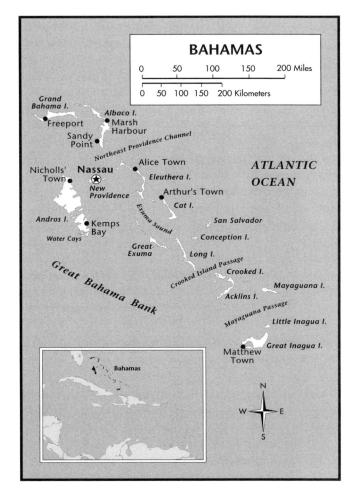

First Inhabitants and the Spaniards

The Lucayans were the first recorded inhabitants of the Bahamas, and had settlements on several islands where they lived for over 500 years. They were completely decimated only twenty-five years after Columbus arrived on the island now known as San Salvador. Passively accepting the Spanish presence, all twenty to forty thousand native Lucayans were abducted by Spanish traders and sold as slaves on other Spanish-controlled islands in the Caribbean. They worked the mines, where they later died. By the 1510s the islands were uninhabited and remained so for over a century.

Several attempts by the British to settle the Bahamas failed when they were driven out by the Spanish. The first permanent European settlement was finally founded in 1647 by a group from Bermuda known as the Eleutherian Adventurers. In 1670 the administration of the Bahama Islands was granted to the Lords Proprietors, six wealthy lords from the American mainland colony of Carolina. During their administration, law and order steadily deteriorated and the Bahamas gained a notorious reputation as a base for pirates, who cruised the seas looking for merchant vessels to plunder. By the first decade of the 1700s, piracy was at its peak in the Bahamas. The English attempted to control the rogue territory by making it a crown colony and installing a governor, Captain Woodes Rogers. By 1725 piracy was on the decline, and the slave trade was the predominant economic activity in the Bahamas.

Slavery and Loyalist Immigrants

The first large shipment of African slaves arrived in 1721, to become the chief source of labor for the developing plantations on New Providence. As the British increased the slave trade to the Americas during the eighteenth century, Bermuda and the Bahamas were often the first stop.

During the chaotic years of the American Revolution, settlers in the Bahamas were divided between loyalty to the Crown and support for the rebellious colonies. However, there was never any serious intention on the part of the Bahamians to join the American colonists in their rebellion. American forces occupied New Providence during the war, but Britain won back control of the Bahamas in the 1783 Treaty of Versailles which recognized the United States as an independent nation.

After the United States achieved independence, thousands of colonists on the mainland, known as Loyalists, emigrated either because they had supported the British during the war (and feared prosecution) or because they did not want to live under the new American government. Some 600 Loyalist families and their slaves moved to the Bahamas during the 1780s. This population boom prompted the British government to buy out the Lords Proprietors.

In 1834 Britain emancipated the slaves throughout its empire. The transition from slavery to freedom was less of a problem in the Bahamas than elsewhere in the British Caribbean because the plantation system had never been firmly established in the Bahamas. The plantations had already collapsed and a meager sponge industry was the main economic activity in the Bahamas.

Smuggling

During the U.S. Civil War, the Bahamas became a hub for smuggling goods through the Union blockade to the Confederacy, creating a short-lived economic boom. When the war ended, poverty returned. During the late nineteenth century, Bahamians experimented with various industries, including production of sisal (a plant fiber used for making rope, sacks, etc.), conch shells for making brooches, and the growing of citrus. During World War I (1914–18), trade with Britain was cut off, and poverty worsened.

During the 1920s, however, Prohibition in the United States brought another brief period of prosperity as the Bahamas once again returned to smuggling. During Prohibition (1920–33), U.S. law declared that manufacturing, selling, and transporting alcoholic beverages was illegal. The Bahamas' proximity to the United States made the islands an ideal base for transporting the illegal beverages into the United States. With the new prosperity in the Bahamas, Nassau became a glamorous hangout for criminals. When Prohibition ended, the Bahamian economy became dormant once again.

Independence

During World War II, Britain allowed the United States to build an airbase on New Providence. In 1942 one thousand black construction workers demonstrated against low wages and ransacked the center of one of its cities—Nassau. The riot marked the first time that blacks had confronted the power structure of the ruling white minority. During the 1950s and 1960s race relations in the Bahamas began to change. Until then economic opportunities were severely limited, and blacks were subjected to obvious forms of racism. The tourism sector was developing, and foreign companies were encouraged to invest in the Bahamas to avoid corporate taxes and take advantage of generous banking secrecy laws. The town of Freeport was established as a center for tourism and industry, and it became the islands' second metropolitan area.

In 1968 a constitutional conference gave the Bahamas the most autonomous form of colonial government possible and paved the way for independence when it came in 1973. Lynden O. Pindling, leader of the Progressive Liberal Party (PLP), became the nation's first prime minister.

During the 1980s, the Bahamas became an important smuggling center once again, but this time for cocaine shipments from Colombia destined for the United States. At the same time the Bahamian government cracked down on illegal Haitian immigrants, and during the 1990s this effort was redirected towards illegal Cuban immigrants.

In 1992 the country's first successful transfer of political power came when Hubert Ingraham's Free National Movement won the majority in the legislative elections. In the mid-1990s the government started selling off its state-owned hotel properties in order to revitalize the languishing tourism industry.

Timeline

A.D. 600–800   First human settlement in the Bahamas

Having been driven north by the Caribs, the Taino, an Arawak people originally from the Orinoco River basin of South America, venture into the Antilles by means of dugout canoes. The Orinoco River is a major river that flows through several countries of South America, and forms the border between Venezuela and Colombia. Their canoes are made from the hollowed-out trunks of trees, and are built in various sizes, some large enough to carry fifty people. The early settlers maintain inter-island communications and commerce.

The first settlers to arrive on one of the Bahama Islands probably migrate from Hispaniola to Great Inagua or the Caicos Islands looking for salt, or perhaps they migrate from Cuba. The Taino who settle permanently on the islands gradually develop their own sense of identity and become the Lucayans.

The Lucayans (whose name *Lukki-carri* means "island people") are the first recorded inhabitants of the Bahamas, and have settlements on Great Inagua, Mayaguana, Acklins Island, Crooked Island, Long Island, Great and Little Exuma, Rum Cay, Cat Island, Eleuthera, San Salvador, New Providence, Great Abaco, Grand Bahama, and the Caicos Islands. The largest community is located at Pigeon Creek on San Salvador and has a peak population of some 300. Their settlements are usually made along the leeward (away from the wind) beach coasts and near shallow waters with marine grass. Their communities are usually scattered and lack a distinct political structure. Each large village is headed by a *cacique* (chief), who is descended from the ruling class on his mother's side.

The Lucayans are a gentle people who go about naked or with only a waist covering, although they do wear cosmetic paint and earrings. They use personal fetishes (known as zemis) in rituals, and sometimes paint or tattoo these designs on their bodies.

Their diet primarily consists of seafood (fish and conch), but they also practice elementary farming and produce their own bread. The main crop they cultivate is manioc (plant with edible, starchy root), which grows well in the central islands. The Lucayans create Arawak-style pottery that is known as "Palmetto Ware." They inhabit many of the central islands of the Bahamas and are feeling the pressure of Taino migration and Carib invaders from the southeast.

The Lucayans live on the islands for over 500 years in peace, but only survive twenty-five years after their encounter with Columbus. Although the Spanish commit atrocities to native peoples throughout the Caribbean, few are wiped out as completely as the Lucayans.

European Exploration

1492: October 12   Columbus lands at San Salvador

Christopher Columbus (c.1451–1506), Italian explorer in the service of Spain, allegedly makes his first landfall on the island now called San Salvador (formerly Watling Island). The native Lucayans' name for the island is Guanahaní, which means "place of the iguana."

Columbus names the islands *baja mar* (literally, "low water" in Spanish) for the shallow sea that surrounds them.

The Spanish, however, make no permanent settlement here because the islands do not contain the mineral wealth found on other islands in the region.

The Lucayans, aware of the Spaniards' technology, try to placate them by passively accepting the Spanish and giving them gifts. Columbus demands that their leaders pay tribute. Only one of the Lucayans that Columbus abducts survives the voyage back to Spain.

Spanish traders soon capture the twenty to forty thousand native Lucayans living in the Bahamas and sell them as slaves on Hispaniola and other islands to replace the dwindling numbers of Tainos who are dying of exhaustion and European diseases. The Lucayans die out while working in the Spanish colonial mines. Within twenty-five years the islands are depopulated and remain so for over a century. The lack of fresh water, poor soil, and treacherous coral reefs keep European colonists away.

1499   Lucayans are taken as slaves by the Spanish

The explorers Alonso de Hojeda and Amerigo Vespucci (1454–1512) oversee the abduction of 232 Lucayans to sell as slaves. The crew has been looking for gold to no avail, and they decide to enrich themselves by conducting a slave raid. Columbus and his contemporaries believe that as conquerors, they have the God-given right to decide the fate of the peoples they conquer.

1502–09   Lucayan chiefs are subdued by the Spanish

Governor Ovando, an early Spanish colonial leader, systematically executes many of the leading caciques (chiefs) throughout the region. Cuba, Hispaniola, Jamaica, and Puerto Rico fall to the Spanish conquerors.

1509: May   Spain authorizes enslavement of Lucayan population

King Ferdinand (1452–1516) of Spain, the monarch who originally backed Columbus' first voyage to America in 1492, gives his official consent for the enslavement of any Lucayan living on any of the islands under Spanish exploration. Later that year he authorizes ships to conduct planned slave raids.

1513   Ponce de León sails through the Bahamas, Lucayan deportation is completed

In search of the "fountain of youth," Spanish explorer Juan Ponce de León (1460–1521) travels through the Bahamas.

The entire surviving Lucayan population lives on the island of Hispaniola, having been removed from their homes on the Bahamas. There they are treated as one of two types of slaves; a resisting Lucayan is regarded as a proprietary slave, while a passive Lucayan is a *naboría* (taken from the Arawak term that refers to someone who is a lifetime servant). They are arbitrarily divided into groups that break up families.

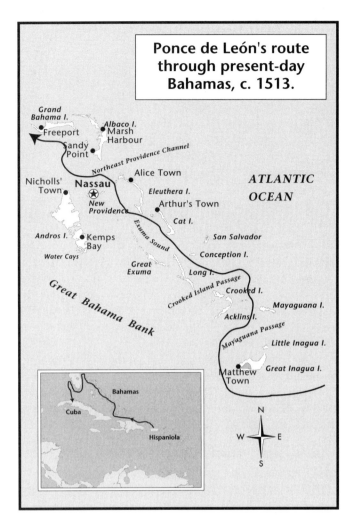

Ponce de León's route through present-day Bahamas, c. 1513.

Between 1512 and 1520, the remaining Lucayans are forced to dive for pearls.

English Colonization of the Bahamas

1629: October 30   England issues royal grant for the islands

Colonists in Bermuda and the Carolinas are interested in settling the Bahamas. The attorney general of England, Sir Robert Heath, is issued a royal grant for the settlement of the islands. The English attempt settlement at New Providence, but are driven out by the Spanish in 1641. The English return in 1666 but are again expelled in 1703.

1647: July 9   First permanent European settlement

A group of religious refugees from Bermuda known as the Company of Eleutherian Adventurers meets in London for the sole purpose of colonizing and developing the first European settlement in the Bahamas. William Sayle, former governor of Bermuda, leads the colonizing effort, which includes

twenty-five well-to-do businessmen and officials. The Eleutherian Adventurers propose to establish an aristocratic republic.

1649: August 31   Sayle and associates are given title to the islands

The British Parliament passes an act that names William Sayle and the other influential colonists as the proprietors of the islands. Sayle later returns to Bermuda in the mid-1650s.

During the civil turmoil in England, many republican sympathizers and Puritans are exiled to the Bahamas because they refuse to swear allegiance to the new king. One ship with about sixty refugees arrives, too ill-equipped to support its passengers. Puritan churches in the Boston area donate £800 for relief, and Sayle later sends a shipment of brazilwood to show his gratitude. The proceeds from the sale of this wood formed a large part of the original endowment of Harvard College.

1670: November 1   Charles II grants the islands to the Lords Proprietors

Six Lords Proprietors (the Duke of Albemarle, the Earl of Craven, Lord Ashley, Sir George Carteret, and Sir Peter Colleton) have the grant for the large colony of Carolina (now the states of North and South Carolina). King Charles II (1630–85) grants the islands to the Lords Proprietors as an addition. Prior to this royal grant, the colonists have already organized their own system of government, which includes an elected House of Assembly. Captain John Wentworth, already serving as the elected governor, is appointed governor by the Lords Proprietors in December 1671. The Bahamas continues its system of participatory government.

The Era of Pirates

1670s–90s   Bahama Islands used as bases by pirates

The Bahama Islands are largely unsupervised and officially under the control of the Lords Proprietors, who do very little for the islands' development. The lack of government restrictions and taxation makes the islands attractive as a lawless frontier, especially when farming, fishing, and other legitimate activities are difficult, and poverty exists. By the early 1700s, the judicial system in the Bahamas breaks down. Piracy continues to be a problem in the Caribbean even into the nineteenth century.

Pirates such as the notorious Blackbeard (Edward Teach or Thatch, c.1680–1718) prowl the waters of the Caribbean Sea from bases in New Providence, looking for opportunities to seize and plunder commercial vessels. Two famous female pirates, Anne Bonney and Mary Read, are also based on New Providence. Compared to merchant and naval crews of the time, pirate crews exhibit egalitarianism—captains are elected by the ship's council, crewmembers receive fairly equal shares of the plunder, and discrimination based on race or nationality is minimal.

1680–84   Spaniards try to take New Providence

The Spaniards attempt to seize the island of New Providence several times between 1680 and 1684. In 1684 almost all the British settlers are driven out. Many flee to Jamaica, Massachusetts, and Maine and return in 1688.

1688: July 12   Thomas Bridges is made governor

Reverend Thomas Bridges leads the return of the British colonists to the Bahamas. There are only a few hundred people inhabiting the scattered islands. Upon their return, the Lords Proprietors make him governor, and the settlement prospers until the French and Spanish attack in 1703, causing the British settlers to flee once again for about a year.

1690s   Captain Woodes Rogers fights against piracy

British navigator Captain Woodes Rogers (c.1679–1732), a revered sailor who has circumnavigated the world, arrives on New Providence to much fanfare. He initially convinces the people to stop the piracy, but this is only temporary. He later toughens his position and actively fights against piracy. Piracy is at its peak in the Caribbean during the first decade of the 1700s, when there are hundreds of privateer ships (ships sailed by civilians who seize and plunder ships that belong to the enemies of England) available and thousands of Royal Navy sailors out of work after military cutbacks.

1717: October 28   British establish a crown colony

England makes the Bahama Islands a crown colony, and decides to step in and reassert civil and military rule. At the time, there are more pirates than colonists on New Providence, and the pirates completely control Fort Nassau, the harbor, and the town.

1718   Rogers drives out pirates

Captain Woodes Rogers, a former pirate himself, becomes the first royal governor of the newly-created crown colony. He arrives in Nassau with four armed vessels escorting a ship carrying extra troops, settlers, and supplies. Initially Rogers offers clemency to pirates who renounce their lawless ways and later conducts public hangings of unrepentant pirates. The remaining privateers either leave or adapt to normal colonial society, leaving the slave trade as the main economic enterprise on the islands. By 1725 piracy is on the decline. The competition among predatory pirates, the inherent risk and violence, and the alliance of governments against its parasitic nature make it difficult for piracy to be profitable.

Rogers' efforts are immortalized in the colony's motto: *Expulsis piratis, restituta commercia* (the pirates having been expelled, commerce was restored). The motto remains on the official coat of arms until independence in 1973.

British Colonial Development in the Bahamas

1720   Bahamas repel Spanish attack

After Rogers expels the pirates, he sets out to reinforce New Providence's meager defenses. When the Spanish attack, the Bahamians successfully turn the Spanish away.

1721   First large cargo of African slaves arrives in the Bahamas

Under Governor George Phenney, nearly 300 Africans are brought as slaves to the Bahamas from Guinea in the *Bahama Galley*, which Phenney partially owns. The slaves form the core of laborers for the plantations on New Providence.

As the slave trade grows, British slave ships often stop at Bermuda or the Bahamas first, giving owners there the chance to buy the healthiest and strongest slaves. Both white and black slave owners in the Bahamas are less brutal than in many other islands of the Caribbean.

By the time the slave trade ends at the beginning of the nineteenth century, the ethnic diversity of slaves in the Bahamas includes Yoruba, Congo, Igbo, Mandingo, Fulani, and Hausa. The Africans bring their respective cultures and beliefs to the Bahamas, which eventually become interlinked. These beliefs include that of Obeah, a type of witchcraft or magic practiced in West Africa where the practitioner is a folk magician who commands powers over others.

1724   First known Bahamian literary work is written

Captain Charles Johnson writes a short play entitled *A General History of the Robberies and Murders of the most notorious Pyrates*. The play satirizes the English judicial system and business practices.

1729   Assembly convenes

The Bahamas' first legislative body is formed under the leadership of Woodes Rogers. Until independence in 1973, the Assembly mainly addresses only issues that affect New Providence.

1758–68   The development of Nassau

Governor William Shirley, who previously has served as governor of Massachusetts, plans the urbanization of Nassau. He authorizes surveyors to map the town, and he oversees the layout of streets and the drainage of nearby mosquito-breeding swamps.

1772   Britain outlaws slavery

A growing abolitionist movement causes Britain to ban slavery in its own domain. The ban, however, does not apply to its territories abroad.

1776   The Bahamas during the American Revolution

During the American Revolution, there are divided loyalties in the Bahamas; some sympathize with American rebels, while others remain loyal to the king. Commodore Hopkins of the Continental (later U.S.) Navy deprives New Providence of its artillery and supplies and takes the governor and others as prisoners. The Americans seize Ft. Montague on New Providence and capture Nassau. American forces occupy New Providence briefly. Two years later, John Peck Rathburne conducts a similar American raid on New Providence.

1780s   Loyalist immigration

During the 1780s the Bahamas receive an influx of Loyalists (colonists allied with the British government). When the British are defeated in the American Revolution (also called the War of American Independence), many colonists in the south leave the mainland colonies in America and flee (with their slaves) to British territories in the Caribbean. In 1783 there are only about four thousand people living in the Bahamas, but by 1789 the population reaches eleven thousand. Many Bahamians today trace their ancestry to the 600 Loyalist families that immigrate in the 1780s. In the years after the immigration, there is friction between the Loyalist newcomers and the "conchs" (Bahamian old-timers). The newcomers want to have a say in politics and want the British government to grant them land rights.

The Loyalists have a profound effect on Bahamian culture, drastically changing the racial distribution of the islands and encouraging institutional discrimination. Up until the arrival of the Loyalists, slaves in the Bahamas work mostly on small farms, as servants, salt panners, and laborers. Some also work on the crews of ships engaged in "wrecking" (see early 1800s) or as privateers (civilian sailors who seize and plunder ships that belong to the enemies of England).

The Loyalists bring a large population of slaves and set out to reproduce southern cotton plantations in the Bahamas. However, most of the plantations are poorly adapted to the new climate and fail within a decade. The main food crops of the time are guinea corn, peas, yams, sweet potatoes, snap beans, cabbage, and pumpkins. Domesticated animals include sheep, pigs, steers, hens, turkeys, mules, and horses. Fishing and forestry supplement plantation revenues. Unlike the sugar plantations on the islands to the south, each slave family is allowed to grow food on its own plot of land.

1782: May 8   Spain captures Nassau

During the scuffle between Britain and her American colonies, Spain takes advantage of the opportunity by attacking Nassau. Spanish sailors suddenly outnumber British colonists on New Providence, and the island comes under Spanish control for nearly a year. Although capture of the island is easy, the Spaniards have many administrative problems during their brief rule. This marks the only break in British sovereignty during the entire colonial period.

1783   Bahamas formally ceded to the English

Before the end of the American Revolution, Colonel Andrew Deveaux (b. 1758), a Loyalist from South Carolina, leads the British to recapture the islands. By shuttling the same 200 men back and forth by ferry, Deveaux tricks the Spanish into believing he has them outnumbered, and they surrender. The Treaty of Versailles declares the Bahamas as a British possession (in exchange for Florida becoming American territory). The British evict the Spanish forces still occupying the islands, and Spain formally gives up its long-standing unenforced claim to the Bahamas.

Lord Dunmore, the last Royal Governor of Virginia, becomes the new governor of the Bahamas. His administration is full of corruption. He appoints his mistresses' husbands to government positions and authorizes large building projects. His concern about the islands' security leads him to approve the construction of two large forts, extensive barracks, and gunnery units. His mismanagement of public finances causes the legislative branch to strip him of the power to levy taxes or spend public money.

1787: March 19   Lords Proprietors required to sell off their properties

With the large influx of Loyalist immigrants from the American mainland, the British government decides to buy out the Lords Proprietors.

Early 1800s   "Wrecking" reaches its peak

During the early years of the 1800s, Bahamians engage in a deceptive practice known as "wrecking." The shallow waters off many of the islands are rocky and dangerous, and there are few lighthouses to guide ships. By shining lights to simulate a haven, ships are lured to crash upon the rocks, thus enabling the deceivers to pilfer the wrecked ship's cargo. In the 1830s the British government authorizes the construction of lighthouses in the Bahamas to improve safety.

1800   Population reaches fifteen thousand

Due to the heavy immigration of Loyalists after American Independence, the population of the Bahamas reaches fifteen thousand. About two-thirds of the population consists of slaves.

1807   Slave Trade Abolition Act

There is a shorter history of slave trade in the Bahamas than in other parts of the West Indies, largely due to the historical lack of plantation agriculture on the islands. The British Parliament places a ban on importing slaves for all of its possessions. Abolitionists argue that the slave trade is still going on

In 1829 Governor Sir James Carmichael Smythe commissions a twelve-foot statue of Columbus for Nassau. (EPD Photos/CSU Archives)

after the ban and propose a system for all slave owners to register their slaves.

1820s   Origins of traditional Bahamian holiday

A foreign visitor to Nassau writes about a Christmas-time festivity that will later become the islands' best-known celebration. *Junkanoo* is originally held on Christmas as a day off given to slaves, and early *Junkanoo* costumes resemble those of a secret Yoruban religious sect. The African-influenced celebration later develops much along the lines of carnival festivities in Trinidad. In the modern *Junkanoo* celebration costumed groups, unified by specific themes, compete for prizes as they and other revelers parade through the streets to the accompaniment of whistles and goatskin drums called *goombays*.

1829   Statue of Columbus commissioned

Sir James Carmichael Smythe, Governor of the Bahamas, commissions a twelve-foot statue of Christopher Columbus and places it on display in Nassau where it still stands.

1830   Slave resistance on Exuma

Under the leadership of the slave Pompey (b. 1798), some forty-four slaves escape from the Rolle Estate into the woods of Exuma after they learn of a plan to transfer them to Cat Island as temporary laborers. For five weeks, the slaves evade capture. They seize Lord Rolle's boat and sail to Nassau, because they want to make their case known to the government. Upon their arrival in Nassau, the slaves are imprisoned, flogged, and not permitted to make their case. However, since the slaves' owner was involved in an illegal slave transfer, the slaves are returned to Exuma. The rebellion, which is minor compared to the bloody revolts on other islands in the Caribbean, rallies slaves throughout the Bahamas. Pompey becomes an heroic figure in Bahamian folklore.

1834: August 1   Emancipation of slaves

Although the British have prospered through the use of slave labor in the West Indies, the slavery is cruel and highly inefficient. Slaves are unwilling laborers and are forced to work through fear, suffering, and brutality. Slavery also desensitizes and demoralizes the British colonists, although most of the colonists oppose any change in the system. When the imperial government orders emancipation, many colonists throughout the British West Indies develop a distrust of the government back in England, because they believe that it is meddling in their way of life. Emancipation covers all of Britain's colonies, but each colony has its own history of slavery, with some more brutal than others. There are already some three thousand persons classified as Free Negroes in the Bahamas, some of whom own slaves. Slave owners are compensated for the loss of their slaves by the British government; nearly £129,000 is awarded to slave owners in the Bahamas. At the time of emancipation, the Bahama Islands has a slave population of over ten thousand.

Britain has control over most of the Caribbean and the high seas, and is now able to enforce an abolition of the slave trade. The transition from slavery to freedom is less of a problem in the Bahamas because the plantation system is not as firmly entrenched in society as elsewhere in the Caribbean. Moreover, the plantations that do exist in the Bahamas have already fallen apart by the time emancipation comes.

After the end of slavery, the Bahamas serve as a source of sponges and occasionally as a strategic location.

1837   First bank opens

Banking starts in the Bahamas when the first commercial bank opens in New Providence.

1838: August 1   Apprenticeship system abolished

After the slaves are emancipated, a system of apprenticeship is implemented as part of the same legislation that gave the slaves their freedom in 1834. Under the apprenticeship system, freed slaves work for their former masters as artisans and laborers. The apprentices work without pay, but receive food, clothing, and shelter as compensation.

1840s   Bay Street Boys control local politics

A ruling elite of white merchants and lawyers known as the "Bay Street Boys" gains control in Nassau (Bay Street is Nassau's main avenue) and remains the islands' chief local political force for nearly a century. The Bay Street Boys control the Assembly and are often hostile to the British monarch's appointed governor, who has virtually no legal authority in the Bahamas. These patricians exert considerable control over the islands' economy, and effectively keep the black population out of positions of power as well as colluding to keep out any competition to their businesses. The elite merchants provide the poor farmers and fishers of the Out Islands (Family Islands) with the loans and materials necessary to make a living, and acquire political loyalty in return. The merchants' involvement prevents the development of agriculture beyond the simple peasant level, and the Bahamians are unable to grow enough food to feed themselves. The influence of the Bay Street Boys continues until the 1940s.

1848   Turks and Caicos Islands removed from Bahamian administration

The Turks and Caicos Islands, which are part of the Bahama Islands chain, are claimed at various times by both Bermuda and the Bahamas. The islands had been placed under the Bahamian administration in 1797. The Turks and Caicos islanders are granted a charter of separation from the Bahamas after they complain about Bahamian taxes on their salt industry.

Economic Booms and Busts

1861   The tourism industry begins

The Bahamas premier industry, tourism, begins with the opening of its first hotel on New Providence. The Royal Victoria Hotel opens as an elegant accommodation for winter visitors. During the U.S. Civil War the hotel is a center for arms dealers and spies, and it remains a center for social gathering throughout the nineteenth century. The marvel of the hotel declines in the 1930s, and it eventually closes in 1971.

1861–65   The Bahamas during the U.S. Civil War

During the U.S. Civil War the Union navy carries out a blockade against the Confederacy in order to cut off trade to foreign countries and weaken the southern states economically. The Confederates utilize the strategic position of the Bahama Islands to get ships through the blockade. The islands prosper during the war years as a center for illegal commerce. A building boom occurs in New Providence, and new ware-

In the 1950s, tourism becomes a leading source of foreign revenue for the Bahamas. Ships such as this from Cunard Lines bring in millions of tourists each year. (EPD Photos/CSU Archives)

houses, docks, and houses are constructed to facilitate the smuggling industry.

1866: October 1   Hurricane strikes New Providence

After the end of the U.S. Civil War, the economy of the Bahamas slumps. The quick wealth brought by the opportunity to smuggle evaporates just as quickly, and Nassau becomes a virtual ghost town. The following year, a severe hurricane damages New Providence, further worsening economic and social conditions. During the late nineteenth century, there are several economic experiments with producing sisal, conch shells for making brooches, and pineapples. Many Bahamians, especially those living in the inhospitable Family Islands, go to Florida as migrant laborers in the citrus industry.

1901   Height of the sponge industry

At its peak, the Bahamian sponge industry employs about six thousand men, or over thirty percent of the labor force.

1914–18   The Bahamas during World War I

During World War I (1914–18), German submarines (known as U-boats) patrol the waters of the Atlantic and routinely sink British merchant vessels. The war effectively cuts off the Bahamas from its main source of trade. Nassau becomes impoverished, and some settlements in the Out Islands are abandoned as people continue to emigrate to southern Florida in search of work. Bahamians also volunteer for British military service during the war.

1920s   The Bahamas during Prohibition in the U.S.

Between 1920 and 1933 the manufacture, sale, and transportation of alcoholic beverages is constitutionally illegal in the United States. Many people engage in illegal activities to either produce liquor or smuggle it in from abroad, and the influence of organized crime grows. The Bahamas' proximity to the United States makes the islands ideal as a base for liquor smugglers (who are called "rum runners"). Smugglers operating out of Nassau typically are destined for the coast of New Jersey or the coves of Florida. The Bahamas, with its fancy hotels and the islands' first casino, becomes a hangout for gangsters. A one pound tax on every bottle of imported liquor causes the colonial government's revenues to quadruple during the early 1920s. When Prohibition ends, the smuggling industry crashes and sends the economy into the doldrums, just as it did nearly seventy years earlier when the U.S. Civil War ended.

1939   Collapse of the sponge industry

A harmful fungus infects sponges in Bahamian waters and wipes out the industry.

1940   Former King of England becomes governor of the Bahamas

Britain's ex-king, Edward VIII (1894–1972), abdicated in 1936 in order to marry an American divorcée, Wallis Simpson (1896–1986). The former king and his new wife are known as the Duke and Duchess of Windsor, and the situation is very awkward for the British monarchy. In a strategic maneuver by Prime Minister Winston Churchill (1874–1965), the duke is offered the governorship of the Bahamas, which has never had such a prominent royal governor. The British government believes that such a high-profile governor will give it more clout in the colony, and both the duke and the government can save face through this official appointment far away from England. The Duke of Windsor takes his job seriously and endeavors to improve the economy of the islands and undo the damaging policies of the Bay Street Boys. He remains governor until the end of World War II.

1939–45   The Bahamas during World War II

During World War II, the United States uses the islands for naval bases.

1942: June   Riots by black construction workers

During World War II Britain authorizes the United States to build an airbase on New Providence. In early June the one thousand black construction workers on the project believe that their wages are too low (the rate is based on a 1936 statute, but price increases during the war have eroded its value), and they start an unplanned protest demonstration. The demonstrators march on Bay Street, the historical center of the ruling white elite, and start ransacking and looting buildings. The police drive the protesters back to a primarily black residential area, and several men are killed during the skirmish. Ironically, the airbase project is a significant source of employment after the economic malaise and high unemployment of the 1930s. The riots, which start out as a labor dispute, mark the first time that blacks challenge the status quo of the ruling white elites. The Duke of Windsor, who is in the United States when the uprising starts, sets up a commission to investigate the riots and punish the instigators while investigating the workers' grievances. The government bans all public meetings (including the annual festive Junkanoo parade) until 1947.

1942: June 7   Wealthy Bahamian philanthropist is murdered

Sir Harry Oakes, a Canadian-born philanthropist residing in the Bahamas, is murdered in Nassau. Sir Harry is one of the richest men in the British Empire and known throughout the Bahamas as a prominent philanthropist and financier (he backed the construction of the islands' only airport before World War II). The murder, which is never solved, stuns Bahamians.

The Political Rise of Blacks and the Era of Tourism

1950s   Civil rights movement in the Bahamas

During the 1950s and 1960s race relations in the Bahamas begin to change. Historically, economic opportunities for blacks were severely limited, and they were subjected to obvious racism by being prohibited from theaters, shops, hotels, and other public places. Black leaders of the Progressive Liberal Party (PLP—see 1967: January) establish themselves as a growing political force. In a single decade, blacks move from the political periphery to center stage.

1950s   Development of tourism

After the war the Bahamas faces its ongoing problem of economic stagnation. Since the turn of the century, the islands have achieved more popularity as a tourist destination for Americans and Canadians. Sir Stafford Sands heads the newly found Development Board, which starts marketing the islands as a year-round tourist destination. The growing availability of air-conditioning makes the summer heat more tolerable to tourists. Foreign companies receive encouragement to set up offices in the Bahamas through the removal of corporate taxes and banking secrecy laws. By 1970 there are over one million tourists arriving in the Bahamas every year, and the number reaches three million by the late 1980s.

1955 City of Freeport created

Wallace Groves (1901–88), a Virginia financier, orchestrates the creation of Freeport, on the western part of Grand Bahama Island. The town's site starts as a clearing in the pine forest. Groves first wants to create a free port and develop industry. The harbor is dredged, and a shipping terminal and a cement plant are constructed. A 200-room luxury hotel opens in 1963, complete with a large casino.

1961 Women's suffrage granted

Bahamian women are given the right to vote for the first time. The relatively short history of women in politics means that men dominate the government, although there are some women in appointed positions.

1964 Bahamas achieves self-government

After revising its constitution the previous year, the Bahamas assumes authority for all local governmental matters. Foreign affairs and defense are still Britain's responsibility.

1967: January First election with universal adult suffrage

In the first election where all adults are allowed to vote, Lynden O. Pindling (b. 1930), leader of the Progressive Liberal Party (PLP, formed in 1953), forms a government with the Labor Party. The United Bahamian Party becomes the official opposition.

Pindling's leadership marks the first time that blacks wield significant political power in the Bahamas. The transfer of power to his administration is peaceful and without incident, and is later known as the Quiet Revolution. Blacks fully participate in the government, and educational and job opportunities start improving.

1968: September Constitutional conference is held

A constitutional conference establishes the Bahamas with the most autonomous form of colonial government possible without total independence. It has a bicameral legislature, and an executive branch with a prime minister and eight other ministers. The British monarch is still represented by a governor.

1970 The rise of "captive" insurance companies

The establishment of a large number of insurance firms in the Bahamas is encouraged by a law that permits companies to conduct part or all of their business out of the country, while benefiting from local tax advantages. The government encourages the formation of "captive" insurance companies, created to insure or reinsure the risks of offshore companies.

By the late 1990s there are about thirty captive insurers in the Bahamas.

1972 Poet Susan Wallace publishes first book of poetry

Susan Wallace, from the island of Grand Bahama, is the nation's best-known writer of poetry. Her first book of poetry, *Bahamian Scene,* is rich in its depiction of daily Bahamian life. She later edits *Back Home,* an anthology of Bahamian literature.

Independence

1973: July 10 Bahamas becomes an independent nation

The Bahamas achieves full independence and adopts a constitution. Under its constitution, the Bahamas adheres to a republican form of government, formally headed by the British monarch, who is represented by a governor-general. Executive authority is given to a prime minister and a cabinet. The constitution also calls for a bicameral legislature consisting of the Senate and House of Assembly. The Bahamas becomes the thirty-third member of the Commonwealth.

The country's first prime minister is Lynden O. Pindling, leader of the Progressive Liberal Party (PLP). The PLP emerges as the nation's majority party in the 1970s. Pindling rules for almost twenty years, during which the Bahamas benefits from tourism and foreign investment.

The Central Bank of the Bahamas is established as the country's central issuing and regulatory authority. Funds for local development are made available through the Bahamas Development Bank.

The Bahamas joins the United Nations on September 18th.

1974 College of the Bahamas founded

The College of the Bahamas provides a two-year/three-year program which leads to an associate degree. It also offers a Bachelor of Arts degree in education.

Mid-1970s Rastafarianism comes to the Bahamas

Rastafarianism began in Jamaica as an Afro-centric movement in response to growing social inequality. The beliefs are based on the teachings of the Jamaican Marcus Garvey (1887–1940), who in the 1920s advocated African racial pride and the eventual return of blacks to Africa. He believed that an African king would rise up and rescue Africans from oppression. In 1930 Prince Ras Tafari Makonnen (1892–1975) of Ethiopia was crowned Emperor Haile Selassie I (r. 1930–36, 1941–74). Garvey's adherents believe the Ethiopian emperor is the king that he had foreseen. Emperor Haile Selassie claims to be a direct descendent of King David. The Rastafarians reject the European culture imposed by colonists and seek a simpler life. Rastafarians are associated with the youth black power movement that is spreading across the Caribbean. By the late 1970s Rastafarians are recognized internationally for two of their characteristics: wearing their hair in dreadlocks (long cords of twisted and uncombed hair) and *ganja* (marijuana). The smoking of marijuana is a sacrament in Rastafarianism.

1980   Bahamian law requires decennial census

The independent Bahamian government passes a law requiring that a census be conducted once every ten years. The 1980 census figure for the islands' total population is 209,505.

Early 1980s   Bahamas role in drug trade

Just as the islands were once a center for smuggling liquor during Prohibition, the Bahamas are a major center for smuggling illegal drugs into the U.S. Some ninety percent of the cocaine entering the U.S. reportedly passes through the Bahamas. Colombian drug traffickers initially begin operating from Norman Cay in the late 1970s.

Mid-1980s   Haitian immigration grows

Because of its economic and political problems, Haiti has long been a source of immigration for the Bahamas, as well as for other islands of the Caribbean region and Florida. Between the mid-1950s and the mid-1980s, the number of Haitians living in the Bahamas rises from about one thousand to over forty thousand (about one-quarter of the islands' population). During the mid-1980s, Haitians are illegally smuggled into the Bahamas, often to one of the uninhabitable Family Islands. The government also begins a campaign to deport illegal Haitian immigrants.

1990   Second decennial census

The 1990 census counts 255,095 people living in the Bahamas, an increase of twenty-two percent from the 1980 figure. Some two-thirds of the population (171,542) lives on the island of New Providence. During the 1980s some 2.9 percent of all Bahamians emigrate to North America.

1992: August   Independent Bahamas' first transfer of political power

Hubert Ingraham becomes prime minister, after Lynden O. Pindling serves in the office for nearly twenty years. During Ingraham's first term, the country's reputation for political corruption and mismanagement improves.

In the 1992 elections the Free National Movement (FNM) wins thirty-two of forty-nine possible seats in the House of Assembly. The FNM is a union of the Free Progressive Liberal Party (founded in 1970) and the United Bahamian Party. It is the first time since independence that the Progressive Liberal Party is not the majority.

1995   Government divests itself of hotels

The government is seeking to attract more investment from abroad in order to rejuvenate its tourist industry. As an incentive, Prime Minister Ingraham authorizes the sale of state-owned hotels to foreign investors. Some 3.2 million tourists visit the Bahamas that year; the tourist industry is still recovering from the 1988–92 slump.

1997: March   Ingraham is reelected

Hubert Ingraham is reelected to another four-year term. The focus of his administration is economic development and job creation. During the campaign, Ingraham's campaign manager, Charles Virgill, is murdered. The murder brings crime and gang activity to the forefront in the Bahamas. The year before, capital punishment (by hanging) had been reinstated after a fourteen-year break.

Bibliography

Bloch, Michael. *The Duke of Windsor's War: From Europe to the Bahamas, 1939–1945.* New York: Coward-McCann, Inc., 1983.

Craton, Michael and Gail Saunders. *Islanders in the Stream: A History of the Bahamian People.* Athens, Ga.: University of Georgia Press, 1992.

Dahl, Anthony G. *Literature of the Bahamas, 1724–1992: The March Towards National Identity.* Lanham, Md.: University Press of America, 1995.

Hamshere, Cyril. *The British in the Caribbean.* Cambridge, Mass.: Harvard University Press, 1972.

Keegan, William F. *The People Who Discovered Columbus: the Prehistory of the Bahamas.* Gainesville, Fla.: University Press of Florida, 1992.

Kurlansky, Mark. *A Continent of Islands: Searching for the Caribbean Destiny.* Reading, Mass.: Addison-Wesley Publishing Co., 1992.

Lewis, James A. *The Final Campaign of the American Revolution: Rise and Fall of the Spanish Bahamas.* Columbia, S.C.: University of South Carolina Press, 1991.

Marx, Jenifer. *Pirates and Privateers of the Caribbean.* Malabar, Fla.: Krieger Publishing Company, 1992.Bahamas

Barbados

Introduction

Barbados is the most easterly of the islands in the Caribbean Sea. Barbados covers 166 square miles (430 square kilometers) and has a total coastline of 60 miles (97 kilometers). The island's coast is almost entirely encircled with coral reefs. The only natural harbor is at Carlisle Bay on the southwest coast. The topography is mostly flat or terraced, the highest point being Mount Hillaby at 1,102 feet (336 meters). A tropical climate is tempered by an almost constant sea breeze from the northeast in the winter and early spring, and from the southeast during the rest of the year. Most of the level ground on the island is devoted to agriculture, and the island has no large area of natural forest.

Some 257,000 people live on Barbados, giving the island one of the highest population densities in the world (five times the population density of India). Bridgetown, the capital, has an area population of about 123,000. About ninety percent of all Barbadians, who are also called Bajans (pronounced "BAY-juns," derived from "Barbajians," which is the common British pronunciation of "Barbadians") are descendants of former African slaves. Some five percent are mulattos (mixed race) and five percent are whites.

Approximately 2,000 years ago, the Saladoid/Barrancoid people came to Barbados from the Orinoco River Basin in present day Venezuela, South America. Arawak Indians eventually replaced the earlier settlers. By the thirteenth century, Carib Indians had attacked and wiped out the Arawaks. When the Spanish and Portuguese began exploring the Caribbean Sea in the 1500s, there was little interest in settling Barbados because the island was not in the main shipping channels. Sixteenth-century maps sometimes label Barbados as Saint Bernardo, Bernados, Barbudoso, Barnodos, and Barnodo.

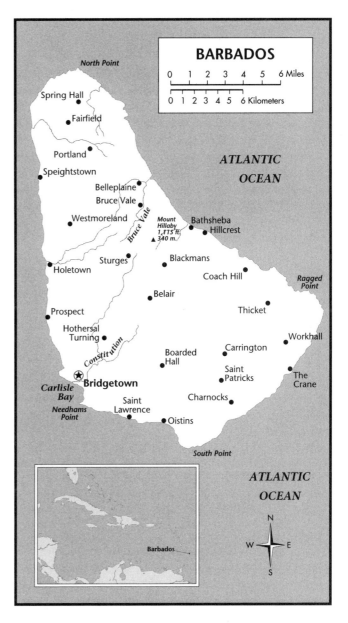

British Colonial Rule

An English vessel, the *Olive Blossom,* landed on Barbados by accident in 1605, and claimed it for King James I. The first settlers began to arrive around 1625. The early settlers grew tobacco, cotton, and indigo before concentrating on sugar cane as the island's primary export. Indentured laborers from Britain and slaves from Africa began arriving in the 1630s.

During the English Civil War (1642–52), Barbados supported the forces that were fighting for King James I and the maintenance of the English monarchy. By the 1660s, an elite class of sugar cane planters had emerged to dominate the island, and the island came under the official protection of the British government in 1663. Slaves eventually outnumbered the British settlers; slave rebellions occurred in 1675 and 1692. The colonial sugar cane industry reached its height in the early 1700s. In 1807, the slave trade was made illegal, but slavery still existed. In 1816, an organized slave rebellion challenged the ruling elite and was the first large-scale slave uprising among Britain's Caribbean territories.

Slavery Ends

In 1834, Britain abolished slavery in its territories, but the last slaves were not freed until 1838. A transitional system of apprenticeship went into effect until it was abandoned four years later. A cholera epidemic in 1854 killed 20,000. The island was in an economic depression in the 1880s and 1890s because of a slump in world sugar prices and inflation. A damaging hurricane in 1898 and epidemics of smallpox in 1902 and yellow fever in 1908 exacerbated the island's economic troubles. Between 1906 and 1914, thousands emigrated to Panama to work on the construction of the Panama Canal. A falling demand for labor during the Great Depression in the 1930s caused massive unemployment in many of the Caribbean island colonies and wages plummeted. Riots in Barbados and elsewhere in the British West Indies eventually led to the introduction of social and political reforms. Universal adult suffrage was granted in 1950, and during 1958–62 Barbados was a member of the short-lived Federation of the West Indies. Barbados became an independent nation in 1966 led by Errol Barrow, founder of the Democratic Labour Party. Barrow was defeated in 1976 by Tom Adams. In 1983, Barbados supported the U.S.-led invasion of Grenada, and the island developed its offshore business services during the 1980s and 1990s. In 1994, Prime Minister Erskine Sandiford dissolved the House of Assembly, and the Barbados Labour Party led by Owen S. Arthur, took power. They remained in control of the House of Assembly in the January 1999 elections, when BLP candidates won twenty-six of twenty-eight seats, and Owen Arthur continued as prime minister.

Timeline

1st century A.D.   Saladoid/Barrancoid settlers

Settlers from the Orinoco River Basin of South America arrive on Barbados some 2,000 years ago. These early settlers are known as the Saladoid/Barrancoid people, named after the place along the Orinoco River in present-day Venezuela where fragments of their pottery were first discovered. There

is little surviving archeological evidence of their community, but relics have been discovered at present-day Chancery Lane, Boscobelle, and Golden Green. The Saladoid/Barrancoid people primarily engage in farming and fishing, and their population is in decline by the mid-600s.

700–800   Arawak settlers

The Arawak Indians, who call themselves the Lokono, eventually replace the earlier settlers. The Arawaks do not have a developed system of agriculture and only cultivate small plots of cassava, corn, peanuts, squash, and papaya. The Arawaks live on Barbados for about 400 years, establishing their settlements near the island's natural springs.

1200s   Carib settlers

In the thirteenth century, Caribs from the Orinoco Basin of South America migrate northward and began to drive out the Arawaks. The Caribs may have been living on Barbados since the tenth century. Eventually they conquer the Arawaks. The Arawaks are skilled artisans, while the Caribs are experienced canoe builders. The Caribs are also more politically organized and have a greater knowledge of cultivation and fishing. (In the 1990s, however, archeologists discover that there were several waves of Amerindian immigrants to the Lesser Antilles and that the differences between the Arawaks and Caribs may not be so great.) The Caribs settle on Barbados (which they called Ichirougánaim) and other islands throughout the Caribbean Sea. The Caribs are named by the Spanish after the word *caribal* (cannibal), although it is unclear whether the Caribs actually eat human flesh. The Caribs have a reputation as fierce warriors, with war canoes able to hold a hundred men who can paddle fast enough to intercept a sailing vessel.

European Exploration

1518–Early 1600s   Portuguese and Spanish ships arrive

Although explorers from Iberia (Spain and Portugal) visit Barbados, none choose to settle there. The Spanish are more interested in settling larger Caribbean islands such as Cuba, Hispaniola, Jamaica, Puerto Rico, and Trinidad. Barbados is not in the shipping routes to those islands.

1529   Barbados recognized by European cartographers

Barbados is first shown on maps produced in Europe.

1536   Carib population disappears

The visiting Portuguese navigator Pedro a Campos records that the island is uninhabited.

There are two possible factors that explain this lack of inhabitants. During the early 1500s, Spanish and Portuguese ships had routinely abducted native peoples from their islands to work as slaves on mines and plantations. It is believed that this was the case on Barbados, and that the remaining population may have died from diseases brought by the Europeans.

1605: May 14   British land by accident

The English ship *Olive Blossom*, captained by John Powell, lands at Barbados by accident near present-day Holetown. Captain Powell claims the island for King James I of England, and the crew finds the island deserted. Upon their return, a colonizing expedition is organized.

Early British Settlement

1625   Sir William Courteen directs a settlement

British settlers arrive in Barbados under the direction of Sir William Courteen. King James I has granted the island to Lord Leigh, and he sponsors the first settlers.

1627: February 17   British settlers arrive

Sir William Courteen directs a party of settlers bound for Barbados. He and his financial backers possess the island, but do not have the official royal patent needed for colonization. It is believed that King James I has granted this patent to Lord Leigh, who may be among the sponsors of the settlement party. John Powell, who is employed by Courteen, sails the ship *William and John* to Barbados. They land with about eighty colonists and ten African slaves captured from a Spanish galleon. During the next two years, some 2,000 British settlers arrive. The first British settlers enlist the help of thirty-two Arawaks from Guyana to show them how to grow crops in the tropical climate.

1627–31   Tobacco cultivation

The colonists begin growing tobacco for export, but the market back in England for tobacco is saturated within a few years.

1629   The Great Barbados Robbery

Through clever strategy, James Hay (d. 1636), the Earl of Carlisle, manages to acquire possession of the island from Sir William Courteen in a deal which modern Barbadians call "the Great Barbados Robbery." There is a legal dispute involving the system under which Courteen had begun colonization, since he did not have the official royal patent needed for colonization. The early setters did not own their land, but worked as employees for Courteen's financiers. John Powell is the leader of Courteen's colonial government.

The Earl of Carlisle now receives a royal patent for all Caribbean islands. The two earls compete for political power while Powell's government defies both. The Earl of Carlisle receives a second patent, and begins granting land to his own group of merchant investors. The earl's agents keep anyone allied with Courteen out of public office, but Courteen loyalists manage to briefly take control in 1628. Feuding between the Earl of Carlisle and Courteen continues into the mid-1630s, and the instability from the dispute is a severe drain on the island's economy.

Plantation Agriculture

1630s–40s   Indentured laborers arrive

More than half the whites who arrive on Barbados during these years are indentured laborers. These individuals agree to work for five to seven years in exchange for the trip to Barbados and some initial provisions upon arrival. The use of indentured labor is profitable for planters who have already established themselves on the island. White servants have no recourse and are not allowed to leave plantations without written permission. There is very little difference between the treatment of white servants and black slaves.

1632–39   Cotton cultivation

After the market for tobacco is glutted, high prices in England entice many planters throughout the Caribbean to switch to cotton. By 1639, however, the market for cotton is saturated and the prices plummet. Many small farmers are ruined.

1636   Slavery officially sanctioned

Governor Henry Hawley issues a proclamation that all blacks and Amerindians arriving on Barbados will be considered slaves for life. Prior to this, blacks living on the island were not necessarily considered slaves, and some Amerindians lived among the Barbadians. Between 1640 and 1700, approximately 134,500 West Africans are brought to Barbados as slaves.

1639   First elected government convened

The House of Assembly (House of Burgesses) is organized under Governor Henry Hawley. The House of Assembly consists of eleven men elected to council and twenty-two appointed burgesses. The legislative body is largely an instrument for Hawley's own authoritarian rule.

1640–42   Indigo cultivation

Several planters start experimenting with indigo cultivation. Indigo, a dark blue dye that grows in tropical climates, is increasingly being used in the manufacture of textiles.

1640s   Sugar cane cultivation

After the markets for tobacco, cotton, and indigo dry up, the colonists search for another export commodity. Sugar cane is first brought from Brazil by the Dutchman Pieter Blower sometime between 1637 and 1640. At first, sugar cane is grown for fodder (food for livestock) and fuel. During the mid-1640s, a civil war in Brazil between Portuguese settlers and Dutch financiers causes sugar exports to drop and the supply in Europe weakens. The early settlers develop a sugar-based economy supported by slave labor, and the first sugar exports are sent to England in 1647. Within fifty years, the sugar-based economy transforms the island into a wealthy colony.

1642–52   English Civil War

Two opposing factions coexist in Barbados during the English Civil War. The Cavaliers are royalists who support King Charles I (Charles Stuart, 1600–49) and the Round-heads support the parliamentary system and the revolutionary Oliver Cromwell (1599–1658).

1649   Unrest among slaves

There is a small uprising on a few plantations due to food shortages.

1652: January 11   Charter of Barbados grants right to self-rule

During the English Civil War (1642–52), the colonial government supports the Cavaliers, and many Roundheads are deported and have their property seized. The Charter of Barbados, signed at Oistins, recognizes the rule of Parliament and the Commonwealth government of Oliver Cromwell back in England. In return, the residents of Barbados receive official recognition for the right to self-rule while remaining a British colony. Confiscated properties are returned and trade restrictions with the Dutch are eased.

1655   Slave population

The island's population of black slaves represents forty-six percent of the total, and is growing.

1660s   Social stratification and emigration

The total number of landowners in Barbados falls from over 11,000 in the 1640s to less than 800 in the 1660s. An elite class of planters emerges so that about 175 individuals control approximately sixty percent of all the property on the island. Some of the most prominent planters are made knights or barons by King Charles II of England. Smaller farmers are pushed out of business and forced to emigrate. In the 1660s, about 3,000 small farmers or white laborers emigrate to other Caribbean islands and to the American colonies on the mainland. Between 1670 and 1675, another 4,000–5,000 leave the island. Treatment of slaves is becoming harsher, with policing, repression, and executions increasing.

1660   Emigration to Carolina

Sir John Colleton, a wealthy planter in Barbados, makes a petition to King Charles II (1661–1700, r. 1665–1700) for a land grant within the area that later becomes the American colony of Carolina (now the states of North and South Carolina). An initial settlement has to be abandoned, but a second settlement is established near present-day Charleston, South Carolina. Plantation owners around Charleston attempt to establish indigo as a crop.

The Charleston area still exhibits influences of Barbadian heritage, and some of its residents are descendants of those Barbadian settlers. Many local place names reveal a Barbadian origin. The local Gullah dialect that is spoken along the coast of the Carolinas also resonates with a Barbadian legacy.

Crown Colony

1663   Barbados becomes crown colony

When the monarchy is restored to England in 1660 with King Charles II on the throne, there are attempts to take away some of the independence Barbados has received during the Commonwealth years. The system under which Barbados is privately owned is abolished and the island comes under direct control of the British monarch. A tax of four-and-one-half percent on all exports is introduced.

1663   Plague of locusts

A plague of locusts destroys much of the island's vegetation.

1665   Dutch attack

Michel DeRuyter (1607–76), a Dutch admiral, attacks Barbados. The island is the site of international friction during the Anglo-Dutch War (1672–74).

1667   Disasters in Bridgetown

Over 100 houses burn down in Bridgetown, the capital of Barbados, and a hurricane adds to the destruction.

1668   Drought

Crops are ruined by a severe drought, causing serious economic hardship.

1669   Disastrous floods

Disastrous floods sweep through the island.

1675: May   Aborted slave rebellion

A slave rebellion, large enough to impede all public affairs on the island, is begun. Most of the rebels are African-born and the planning for the event takes nearly three years and involves plantations all over the island. Eight days before its planned inception, authorities learn about the operation. Over 100 conspiring slaves are arrested and about 50 are executed.

1676   Slaves are more than half the population

Slaves account for sixty percent of the population. The white population is in decline and the number of slaves is increasing, making it more difficult to monitor and control the slave population.

1675   Hurricane

A devastating hurricane pounds the island.

1684   Census of population

Governor Richard Dutton orders a census of the population. There are 20,000 whites and 46,000 African slaves recorded.

1692: January   Aborted slave rebellion

Another conspiracy for a general rebellion is uncovered. It is a comprehensive and more intricate plot than the earlier rebellion (see 1675). The slaves had planned to strike the plantations first, when there were fewer British troops on the island, and then to descend upon Bridgetown. An unknown slave informant exposes the conspiracy, and about a hundred collaborators are executed.

1700   Slave population reaches three-fourths of total

About seventy-five percent of the population consists of blacks, a small number of whom are legally free.

1710   Codrington College founded

Colonel Christopher Codrington founds Codrington College for the purpose of "propagating the Gospel." It trains young men for religious orders.

1715–17   Height of the colonial sugar industry

The value of exports by Barbados to the United Kingdom is nearly the same as the exports of all the mainland American colonies combined.

1731   First newspaper published in Barbados

The *Barbados Gazette* is the first newspaper published in Barbados. Its editor is Samuel Keimer from Philadelphia who taught Benjamin Franklin to operate a press.

1750   Population of Barbados reaches 95,000

Barbados has approximately a population of 95,000 inhabitants, over eighty percent of which are blacks.

1760   Plague of sugar ants

Swarms of sugar ants consume the sugar cane crop.

1776   Bridgetown fires

Two fires destroy Bridgetown. The first is in 1776, followed by the second a year later.

1778   British relief sent to Barbados

Barbados suffers extreme shortages of provisions because of the Revolutionary War of Independence being fought between the American colonies and British forces. The people of Barbados are suffering such severe shortages that the British government sends relief supplies.

1780   Hurricane

About 4,000 people lose their lives in a devastating hurricane. Bridgetown is destroyed and only four churches remain standing in the city.

1807   Slave Trade Abolition Act

Barbados is one of the only British colonies in the West Indies that has stopped importing slaves before the ban on the practice comes in 1807. Planters give female slaves incentives to have babies. When the British parliament makes the slave trade illegal, few planters on Barbados are concerned because they already have a large and healthy slave population. Abolitionists argue that the slave trade is still going on after the ban, and propose a system for all slave owners to register their slaves.

1816: April 14   The Bussa Rebellion

On Easter Sunday, slaves set the cane fields and trash heaps on fire in a revolt. The insurrection begins in the southeast and spreads to cover half the island. An African-born slave, Bussa, and a mulatto, Washington Franklin, are the leaders of the rebellion. The revolt lasts only three days because it is squelched by a joint force of local militia, imperial troops, and black slave soldiers. The island is placed under martial law for three months. Approximately 1,000 slaves are either killed in battle or executed afterwards.

The uprising is ignited after Parliament rejects a bill that would have required slave owners to keep a registry of slaves. The Barbados slave rebellion is the first of its kind among Britain's Caribbean territories.

1831   Voting franchise expanded

The right to vote is granted to "free colored men" who own property. Jews and free non-whites receive full citizenship.

1831   The Great Hurricane

Over 1,500 people are killed by a hurricane, and property damage is extensive.

1834: August 1   Slavery abolished

Although the British prosper through the use of slave labor in the West Indies, the institution of slavery is cruel and highly inefficient. When it is abolished, slave owners are compensated for the loss of their slaves by the British government. Most of the freed slaves continue to work on the plantations as hired servants because the island has a high concentration of laborers and there are few other employment opportunities on the island.

Since Britain has control over most of the Caribbean and the high seas, it is able to enforce the abolition of the slave trade.

1838: June 2 Apprenticeship system abolished

After the slaves are emancipated, a system of apprenticeship is implemented as part of the same legislation that has given the slaves their freedom. Under the apprenticeship system, freed slaves work for their former masters as artisans and laborers. The apprentices work without pay, but receive food, clothing, and shelter as compensation. The apprenticeship system is designed as a period of transition between slavery and freedom, but work relations soon deteriorate between apprentices and owners. Once the apprenticeship system is abolished, most of the former slaves begin farming on small plots of land cleared from the forest.

1843 Samuel Jackson Prescod elected to House of Assembly

Samuel Jackson Prescod, born of a white father and a free black mother, is the first non-white man elected to the House of Assembly. Prescod is a prominent spokesman for the newly emancipated people of Barbados.

1845 Fire in Bridgetown

Some ten acres of Bridgetown are consumed in a blaze.

1847 Free library opens in Bridgetown

The government of Barbados opens a free library in Bridgetown. Over the next 150 years, it will open seven branches and operate bookmobile stops to serve all islanders.

1852 First lighthouse constructed

The island's first lighthouse is built at Sandy Point at the southern tip of the island.

1852 First postage stamps issued

The island's first postage stamps are issued. Prior to this event, postage reflected Barbados's status as a British colony.

1854 Cholera epidemic

A cholera epidemic strikes the island, killing 20,000. Bridgetown is one of the unhealthiest towns in the Caribbean, with open sewers and rotting trash littering its streets. The government had begun to address public sanitation and health issues in the early 1850s but it failed to prevent the tragedy of the cholera epidemic.

1870s First cricket clubs formed

Cricket—a game reminiscent of baseball, but which uses a flat wooden bat and heavy, round ball—is introduced to Barbados from England. The first cricket clubs form, but cricket is only allowed to be played by the elite class; the sportsmanlike principles of cricket are believed to be admirable guidelines for the social behavior of gentlemen. By the 1930s, however, cricket is the most popular game in Barbados, and all classes of people enjoy playing it.

1875 Bridgetown illuminated

Gas lights are introduced to Bridgetown.

1876: April 21–22 Riots opposing confederation

Abortive efforts to bring Barbados into confederation with the Windward Islands (including St. Lucia, St Vincent and the Grenadines, and Grenada) ends in riots. Planters oppose the confederation because they believe it would limit the autonomy of the island's existing government. The riots cause few fatalities, but hundreds are wounded.

1880s–90s Economic depression

The English sugar market is saturated as European countries begin cultivating sugar beets. Cuba and Puerto Rico are already supplying cheaper sugar to North America, and by the 1880s the plantation economy has eroded. Land prices in Barbados begin slipping, and the patterns of land ownership begin to change. A merchant class develops during this time as a new elite in Barbadian society.

1898: September 10 Hurricane strikes

A devastating hurricane sweeps through the eastern Caribbean. Some 18,000 houses on Barbados are damaged and eighty people are killed. Resulting outbreaks of typhoid and dysentery kill many, especially poor laborers. The British government sends relief.

1901 Population approaches 200,000

Barbados is reportedly one of the most densely populated of the Caribbean islands at the turn of the century. The published census reports the population at 195,558, or 1,178 inhabitants per square mile.

1902 Smallpox epidemic

A smallpox epidemic lasts for fourteen months and causes 118 deaths. During the epidemic, the whole island is placed under quarantine by nearby colonies.

1906–14 Emigration to Panama

The total population is reportedly 196,287. Thousands of people—as many as 45,000—leave Barbados. Most are men who go to work on the construction of the Panama Canal. The mass departure is the single largest wave of black migration in the colony's history. At first, sugar planters are relieved to be rid of surplus laborers, but soon there is a labor shortage and women take many of the physical jobs once held only by men. The remittances sent back to Barbados become known as "Panama Money." When the construction of the Panama Canal is completed in 1917, many people return to Barbados. Some of the migrants save enough money to have considerable social prestige upon return, and land ownership among blacks rises considerably.

1908   Yellow fever epidemic

Yellow fever, a deadly tropical disease, strikes the population.

1924   Political parties organized

The Democratic League is founded by Charles Duncan O'Neale (1879–1936). The party receives support from the black and mulatto middle class as well as the working poor. During the 1930s, the labor movement grows to challenge plantation dominance.

1927   Writer George William Lamming is born

Barbadian author George William Lamming is born. Like many Barbadian authors, Lamming writes African-inspired novels with stories of growing up in Barbados. Many are introspective works that deal with heritage and race and the impact of slavery. Lamming's most famous works are *In the Castle of My Skin*, *The Emigrants*, and *Season of Adventure*.

1930   Barbados Museum and Historical Society founded

A general museum, the Barbados Museum and Historical Society is founded. It features collections on the geology, history, and plantation and Arawak life of Barbados. It is located in St. Ann's Garrison in Bridgetown. The Barbados Museum is housed in a former British military prison built in 1817 and expanded in 1853.

The museum displays artifacts of the early inhabitants of the Caribbean region, furnishings from an eighteenth century plantation house, and historic maps and geneological records.

1934   Grantley Adams elected to House of Assembly

Grantley Adams (1898–1971) later becomes a prominent Barbadian leader through the Barbadian Labor Party.

1936   Birth of cricketeer Garfield Sobers

Sir Garfield Sobers, regarded by many as one of the world's greatest cricket players, is born in Barbados.

1937: July   Riots

A falling demand for labor during the Great Depression in the 1930s causes massive unemployment in many of the Caribbean island colonies and wages plummet. Between 1929 and 1935, the world price for sugar drops by half. Rising inflation since the 1870s steadily erodes the real income of agricultural workers, and the standard of living deteriorates. Riots in Barbados and elsewhere in the British West Indies result in the dispatch of the British Royal Commission to the West Indies and a gradual introduction of social and political reforms. The riots inspire the foundation of the Progressive League, the successor to the Democratic League (see 1924).

1944   Women's suffrage

Women are granted the right to vote in Barbados.

1950: April   Universal adult suffrage granted

Property and income requirements for voting and membership in the House of Assembly are lifted. As a result, the size of the electorate increases from about 30,000 to 100,000. The following year, the Barbados Labor Party wins the elections and Grantley Adams emerges as the island's leader.

1954   Adams becomes premier

Sir Grantley Adams is made the island's first premier under a ministerial government system.

1955: April 27   Democratic Labor Party formed

The Democratic Labor Party is formed by Errol Barrow (1920–87) as a rival to Adams' conservative Barbados Labor Party. The party is officially socialist and advocates the reduction of inequality and reforms in education, public health, and social security.

1956   Founding of Barbados Astronomical Society

The Barbados Astronomical Society is founded to foster the study of astronomy, and to offer fellowship for working astronomers.

1957   "Tie-head" religious movement begins

Bishop Granville Williams founds the "tie-head" movement in Barbados after having been in exile in Trinidad for sixteen years. Known as the Jerusalem Apostolic Spiritual Baptist Church, members wear colorful gowns. The colors represent various worthy characteristics: white (virtue), blue (holiness), cream (spirituality), brown (blissfulness), gold (regality), and pink (success). The tie-heads are so named after the cloths that cover the heads of male and female members.

1958–62   Federation of the West Indies

Sir Grantley Adams serves as premier of the short-lived Federation of the West Indies. The idea of creating a federation of British states in the Caribbean is first discussed in 1932 at a conference in Dominica, a nearby island. The federation's members include Grenada, Barbados, Trinidad, Jamaica, and

British traditions remain on Barbados after independence. Pictured here are the Barbados Mounted Police wearing British-style colonial uniforms. (EPD Photos/CSU Archives)

the islands of the Leeward and Windward Antilles. At first, the federation is popular and people see it as a way for the assortment of small islands to gain international recognition. However, after a short time, political differences between the islands intensify; Jamaica wants the federation to be a loose association, while Trinidad wants a strong federal government. Jamaica secedes in September 1961 and Trinidad follows in January 1962. The entire federation is dissolved on May 31, 1962.

1963   University of the West Indies opens branch in Bridgetown

The University of the West Indies opens the Cave Hill campus in Bridgetown. It offers study in medicine and the social sciences.

Independence

1966: November 30   Barbados proclaims independence

After the Federation of the West Indies falls apart, there is an attempt in 1963 and 1964 to reconstitute a smaller federation, with Barbados as the capital. Support for the revised federation also deteriorates. The government of the Democratic Labor Party immediately begins reforms that promote tourism and agriculture and prepares the island for independence. Barbados becomes an independent nation with its own constitution and joins the Commonwealth of Nations. Errol W. Barrow (1920–87) becomes the first prime minister (see also 1986: May 28).

1966: December 9   Barbados joins the United Nations

Barbados joins the United Nations (UN); other countries joining the UN in this year are Botswana, Guyana, and Lesotho.

1968   Barbados Community College founded

To provide training in science and technology, the Barbados Community College is founded.

1969: September 1   Local governments abolished

All district and municipal governments are abolished and their functions are subsumed by the national government.

1971   Democratic Labor Party wins elections

Errol Barrow (1920–87) is elected to retain the position of prime minister after the Democratic Labor Party wins a two-thirds majority in the election.

1972   Congor Bay Earth Station begins satellite communications

To provide satellite communications, Barbados opens the Congor Bay Earth Station.

1973: August 1   Barbados helps form CARICOM, issues own currency

The Caribbean Community and Common Market (CARICOM) is an intensification of the Caribbean Free Trade Association (CARIFTA) organized in 1968. The CARICOM countries seek regional integration through fiscal, tax, and tariff agreements. The Barbados dollar replaces the East Caribbean dollar with its own currency.

1975   Rastafarianism comes to Barbados

Rastafarianism begins in Jamaica as an Afro-centric movement in response to growing social inequality. The beliefs are based on the teachings of the Jamaican Marcus Garvey (1887–1940), who in the 1920s advocated African racial pride and the eventual return of blacks to Africa. Garvey believed that an African king would rise up and rescue Africans from oppression. In 1930, Ras Tafari (1892–1975) of Ethiopia is crowned Emperor Haile Selassie I (r. 1930–36, 1941–74). Garvey's adherents believe the Ethiopian emperor is the king that had been foreseen. Emperor Haile Selassie claims to be a direct descendent of King David. The Rastafarians reject the European culture that is imposed by colonists and seek a simpler life. By the late 1970s, Rastafarians are known internationally for two of their characteristics: wearing their hair in dreadlocks (long cords of twisted and uncombed hair) and *ganja* (marijuana). The smoking of marijuana is a sacrament in Rastafarianism.

1976   Barbados Labor Party wins elections

The Barbados Labor Party wins seventeen seats in the House of Assembly, and J.M.G.M. "Tom" Adams (Grantley Adams' son) (d. 1985) becomes the prime minister.

1978   Coup plot exposed

Robert Denard (b. France, 1929) and a group of mercenaries conspire to overthrow the government of Barbados. The coup is discovered in time and leads to the establishment of the Barbados Defense Force (see 1979). Denard later controls the nation of Comoros off the coast of eastern Africa between 1978 and 1989 through figurehead presidents.

1979   Barbados Defense Force established

The Barbados Defense Force includes the former Barbados Regiment and Barbados Coast Guard.

1979   Grantley Adams Airport opened

Grantley Adams Airport, 11 miles (18 kilometers) southeast of Bridgetown, opens for international airline service.

1980s   Offshore business center

Laws enacted in the 1980s lead to the development of Barbados as an offshore business center. This means that other countries, especially industrialized nations, locate businesses in Barbados to take advantage of favorable tax and tariff regulations.

1981   General elections

In general elections, the Barbados Labour Party retains its majority in parliament, and Tom Adams remains prime minister (see 1976).

1983: October   Barbados assists U.S.-led invasion of Grenada

Barbados assists the U.S.-led invasion of Grenada by acting as a staging area for the incursion. A Marxist military group allied with Cuba has usurped power from the Grenadian government. The U.S. government invasion successfully ousts the Marxist group and returns the Grenadian government to power.

1985: March 11   Prime Minister J.M.G.M. Adams dies while in office

After Prime Minister J.M.G.M. "Tom" Adams dies in office (see 1976), H. Bernard St. John assumes the post of prime minister. Adams' sudden death severely lowers morale within the ruling Barbados Labor Party.

1985   Regional Security System

After the government assists the U.S. in its invasion of Grenada, Barbados is designated as the center for the

Regional Security System by an alliance of nations in the region. The alliance includes Antigua and Barbuda, Dominica, Grenada, St. Kitts and Nevis, St. Lucia, and St. Vincent and the Grenadines. The alliance is U.S.-funded and in charge of conducting military exercises in the region.

1986: May 28   Democratic Labor Party wins elections

The Democratic Labor Party wins twenty-four of twenty-seven seats. It is the first time a landslide political victory occurs in Barbadian politics. Errol W. Barrow, head of the DLP, becomes prime minister, a position he also held from 1966–76.

1987: June 2   Prime Minister Errol W. Barrow dies while in office

Errol W. Barrow (1920–87), prime minister, dies while in office (see 1986: May 28 and 1966). Deputy Prime Minister Lloyd Erskine Sandiford (b. 1937), minister of education and leader of the House of Assembly, assumes the office of prime minister.

1994: May   Barbados hosts UN conference

In May 1994, Bridgetown hosts the United Nations Global Conference on Sustainable Development of Small Island Developing States. Some 120 nations are represented, and the conference helps to strengthen the solidarity of small nations.

1994: June 7   House of Assembly dissolved

After a no-confidence vote, Prime Minister Erskine Sandiford dissolves the House of Assembly, the first time such an action had taken place. Sandiford's administration is troubled by the growth of crime and taxes, while tourism is in decline.

1994   Barbados sends soldiers to Haiti

As part of a Caribbean military operation, Barbados sends soldiers to Haiti as peacekeepers.

1994   Drought

Barbados experiences a prolonged drought, with the lowest rainfall in over ten years. As a result, there are water shortages in parts of the island.

1995   Turner's Hall Woods is nature preserve

Turner's Hall Woods, a forty-six acre tract of forest, is the last remaining native forest on Barbados. Its trees include evergreens, cabbage palm, the jack-in-the-box tree, and Spanish oak.

1999: January 20   General election

A general election is called, and the Barbados Labour Party wins twenty-six of the twenty-eight seats in parliament. The other two seats are won by candidates of the Democratic

Labour Party. As leader of the Barbados Labour Party, Arthur Owen, prime minister since 1994, remains in office.

Bibliography

"Barbados: General Election Called." *New York Times,* December 28, 1998, vol. 148, p. A16.

Beckles, Hilary. *Barbadian Historic Struggle for Citizenship and Nationhood.* St. Michael, Barbados: Barbados Government Information Service, 1998.

———. *A History of Barbados: From Amerindian Settlement to Nation-State.* New York: Cambridge University Press, 1990.

Hamshere, Cyril. *The British in the Caribbean.* Cambridge, Mass.: Harvard University Press, 1972.

Mayers, Harry. *Against the Odds: The Story of the Nation Organisation.* Bridgetown, Barbados: The Nation Corporation, 1998.

Pariser, Harry S. *Adventure Guide to Barbados.* Edison, N.J.: Hunter Publishing, 1995.

Wilder, Rachel, ed. *Barbados.* Boston: Houghton Mifflin Co., 1993.

Belize

Introduction

Belize has often been called a backwater, a swampy land of mosquitoes and impenetrable jungles. It has even been referred to as a land with little value. British author Aldous Huxley, who traveled through Central America and the Caribbean in the 1930s, could not understand why Britain would want to hold on to "this strange little fragment of the Empire."

Yet, when Belize became a nation in 1981, its people celebrated their sovereignty in part by remembering the past, when Belize was home to the Maya. Today, these impenetrable forests and jungles and its once feared coastline have become a source of pride for this small nation, exploited for centuries by British loggers who built their wealth with slave labor.

For more than three centuries, Belize seemed to exist solely to create wealth for England. Wrested away from the Spaniards, who cared little about the area, Britain slowly took control of Belize. British buccaneers came in the 1600s and hid in the protected waters of Belize. From there, they attacked Spanish ships filled with riches. In time, these former buccaneers turned to cutting logwood, a tropical American tree whose heartwood was used to produce a dye before the invention of synthetic dyes.

Long before the Spaniards and British arrived in the area, Belize, Southern Mexico, Guatemala, and Honduras were home to one of the great civilizations of the ancient world, the Maya, who built great cities there. Mayan cities were built of elaborately decorated temples and homes surrounding a plaza. The Maya developed a system of writing, mathematics, and a calendar based on their study of the sun, moon, and stars. In its classical period (A.D. 250–900), as many as 400,000 Maya are believed to have lived in what is now Belize.

In 1981, when Belize became the world's newest nation, the remaining Maya were part of an ethnically diverse nation that included Protestants largely of African descent, Roman Catholic *mestizos* (of mixed Spanish and Amerindian descent), and Garifuna (of mixed African and Amerindian descent) and many other ethnic groups. Together, they continued to lift a poor nation that has struggled to break from its colonial past.

Today, about thirty percent of the population is Creole, a mix of African and European ethnicity. Nearly forty-four percent are mestizo, nearly fifteen percent are Maya and about seven percent are Garifuna. Many immigrants have come from China, India, Syria and Lebanon and European countries.

Belize, formerly known as British Honduras, is on the Caribbean coast of Central America. With an area of 8,865 square miles (22,960 square kilometers), Belize is slightly larger than the state of Massachusetts. It is bounded on the north by Mexico, on the east by the Caribbean Sea and on the south and west by Guatemala. Its capital city, Belmopan, is located in the center of the country.

North of Belmopan the country is mostly flat. To the south, the coast is flat and swampy, with many lagoons. The coastal waters are sheltered by a line of reefs and are dotted by islands and cays: small low islets composed largely of coral or sand. The interior is mountainous, with the Maya and the Cockscomb Mountains reaching heights of more than 3,500 feet. Belize is a warm, humid country, with seasons marked more by differences in humidity than temperature. Annual rainfall averages 50 inches in the north and 150 inches in the south.

Because of low population density (about twenty-five people per square mile), Belize has suffered fewer environmental problems than its neighbors. Yet, as much as eighteen percent of the population does not have access to pure drinking water. Pollutants threaten Belize's magnificent reef, the longest in the Western Hemisphere.

Endangered species include the tundra peregrine falcon, hawksbill, green sea turtle and leatherback turtles, American crocodile and Morelet's crocodile. The iguana, Larpy eagle, spoonbill, wood stork and several types of parrots and hawks also are on the endangered list. In the 1980s and 1990s, the country's natural beauty fueled a rapid growth in tourism. Belize has many forests, which consist of mixed hardwoods, including mahogany, cedar and sapodilla, the source of chicle used in the manufacture of bubble gum.

English is the official language of Belize, but Spanish is widely spoken. About eighty percent of the population speak English and/or a Creole patois. About sixty percent of the population speak Spanish. Other languages include Garifuna and Maya and Low German in the Mennonite colonies.

History

The great Maya civilization had collapsed well before Christopher Columbus arrived in the Americas in 1492. Why it collapsed remains a mystery to archeologists, but by the time of the Spanish conquest, only ruins and a peasant population of unknown numbers, remained. One of the oldest set of ruins was uncovered in Belize with archeological remains dating as far back as 2500 B.C.

Although Spanish conquistadors journeyed through Belize, they did not colonize the area. On the other hand, Spain did not want any other European nation establishing colonies and encroaching on its monopoly in the Western Hemisphere. But the English, French, and Dutch soon began to challenge Spanish supremacy. In the sixteenth and seventeenth centuries, these countries resorted to smuggling, piracy, and war to break Spain's stranglehold on the Americas. By the early 1600s, the British maintained a strong presence in the Caribbean and by 1638, the British had established the first settlement in Belize.

When the European powers agreed to suppress piracy in 1670, the British encouraged the buccaneers to cut logwood, a tropical American tree from which dye was made. They established British settlements, which the Spanish attempted to dislodge several times. Finally the question of possession was settled when the British soundly defeated the Spanish in a battle at St. George's Cay in 1798.

With the Spanish effectively out of the way, the British continued to log wood, including cutting down the more profitable mahogany trees. When they could not subjugate the Maya and force them into labor, the British began importing slave labor, bringing thousands of African slaves into the area. As they moved deeper into the country looking for mahogany, the British pushed the Maya off their land, burning their homes and destroying their crops.

When Belize, then known as British Honduras, officially became a colony in 1862, it was dominated by a very small group of Europeans who owned most of the private land and nearly all of the slaves. By the mid-eighteenth century, slaves made up the majority of the population. By the turn of the century, more than three-quarters of the population were slaves.

Slaves, who suffered tremendously under an inhumane system, expected an improvement in their situation when Britain abolished slavery instituting a five-year plan that lasted from 1833 to 1838. In Belize, the plan called for new "apprenticeship" laws that kept former slaves from buying land, forcing them to work for their former masters.

Over the next 100 years, Belize was little more than a logging camp, its stagnant economy heavily dependent on mahogany exports. But the makeup of the population was

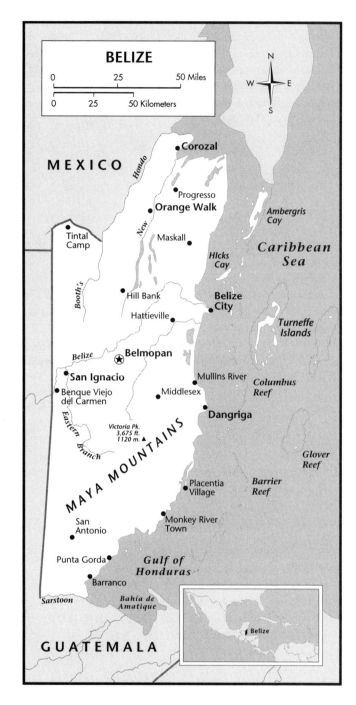

shifting and Belizeans were growing restless to bring change to a society stratified by race. A new, wealthier class of Creoles, people of mixed African and European descent, rose to prominence in the early 1920s. The country began to shift its allegiances and the United States became an influential trading partner.

The Great Depression of the 1930s gave rise to a militant labor movement whose goals included the legalization of trade unions, a minimum wage and health insurance. By the early 1950s, with much of the movement's goals and a demand for a semi-representative government unrealized, the drive for independence was unstoppable.

Britain granted self-rule in 1964, but complete autonomy for Belize was out of the question since Guatemala claimed rights over the area; in spite of an 1859 border treaty with Britain, Guatemala claimed Belize as part of its territory.

Achieving sovereignty was not easy. Initially, many nations favored Guatemala's claims, but Belize pressed its case in the international community. Finally in 1980, a United Nations resolution provided sovereignty for Belize.

The new nation faced many obstacles. For most of its history, Belize remained a largely undeveloped and often neglected colony heavily dependent on a single and volatile industry: logging. Logging interests controlled most of the land and discouraged farming as unprofitable. For decades, despite its rich land, Belize imported most of its food.

Most of the population had little education and those who could afford to pay for it had to leave the country to attend a university. The country's first university was not built until 1986 and a national library had only 150,000 volumes in 1997.

The Belize government attempted to diversify the economy, granting more land to peasants, investing in the sugar, citrus, and fishing industries and encouraging more foreign investment. Tourism also became an important sector of the economy starting in the 1980s.

Many challenges remain for Belize. Some of its troubles are easily seen in Belize City, a large, decaying town on the Caribbean coast. Here, unemployed people wander the streets and there has been an increase in urban crime. However, with a diverse culture that is working to overcome its colonial past, Belize continues to seek a democratic path that will bring a better future.

Timeline

2500 B.C.   Ruins indicate early settlements in Belize

Nomadic people, who may have crossed the Bering Strait as early as 35,000 years ago, begin to settle throughout the Americas and develop many distinct cultures. The Mayan culture develops in the lowland areas of the Yucatan Peninsula (in Mexico) and the highlands to the south in what are now southeastern Mexico, Belize, Guatemala, and western Honduras. This region is known as Mesoamerica.

The early people are hunters and foragers who settle in small villages and begin to cultivate corn, beans, squash, and chili peppers. Many languages and subcultures develop within the Mayan culture, which spreads throughout Mesoamerica.

One of the earliest Mayan settlements uncovered by archeologists is Cuello, in Belize, which dates to 2,500 B.C. Jars, bowls, and other pottery are among the oldest Mayan

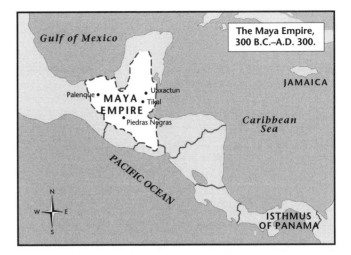

items unearthed in Mesoamerica. Cuello has building platforms arranged around a small plaza, which reflect a typical Mayan community.

300 B.C.   Ruins show a sophisticated community

The ruins of Cerros, where the tallest of its temples rises to seventy feet, show traces of a major trade and ceremonial center that develops between 300 B.C. and A.D. 100 along the Belize coast.

The architecture at Cerros shows distinguishing Mayan features, including temples and palatial residences organized in groups around plazas. The structures are built of cut stone and covered with stucco. They are elaborately decorated and painted. The Maya have a highly developed style of art and their buildings are decorated with carvings and paintings of gods, people, and animals. The Maya also carve geometric patterns on buildings. Large masks, representing a serpent god, decorate one of the temple platforms at Cerros.

200 B.C.   Maya build city near present-day Belize City

As many as 10,000 Maya may have lived in the Altun Ha area, near present-day Belize City. Excavations in the twentieth century lead to the discovery of one of the finest Mayan carved jade objects: the head of the sun god Kinich Ahau. The Maya are skilled at making pottery and carving jade.

Excavations at Altun Ha produce evidence that suggests a revolt, possibly by peasants against the powerful priestly class. The struggle at Altun Ha and other Mayan communities may have contributed to their downfall.

A.D. 250–900   The golden age of Mayan culture

The Maya flourish throughout Mesoamerica, embarking on great construction projects and developing a hierarchical society of farmers, skilled craftsmen, warriors, merchants, and priest-astronomers. The years between A.D. 250–900 are known as the classic period of Maya culture.

Mayan ruins, like these at Xunantunich, provide evidence of a thriving civilization. (Mary A. Dempsey)

Farmers develop large-scale agriculture using sophisticated irrigation techniques. Priests coordinate the work of farmers and other seasonal activities with a cycle of rituals at ceremonial centers. The Maya develop writing, mathematics, and a calendar. They observe the movement of the sun, moon, planets, and stars.

The majority of Maya are farmers living in simple thatched houses surrounded by gardens. They eat tortillas, beans, tomatoes, peppers, and other vegetables. The ruling classes have a more elaborate diet of turkey, fish, and game meat, and a chocolate drink made of cocoa.

A.D. 900   Inexplicably, the Maya abandon the cities

Sometime in the tenth century, the Mayan cities are abandoned. Factors contributing to the demise of the cities might be increased warfare among the city-states, revolts by the peasants against the ruling class, overexploitation of the environment, or a combination of these. In spite of their abandonment of the cities, the Maya continue to inhabit Mesoamerica. As many as 400,000 Maya live in Belize during this period.

Columbus and Conquest of Belize

1502   Columbus sails past Belize

In his fourth voyage to the Americas, Christopher Columbus sails past Belize on his way to the Gulf of Honduras. A few years later, two of his navigators, Martin Pinzon and Juan de Solis, sail northward along the coast of Belize to the Yucatan Peninsula.

Interest in the area grows after Hernan Cortez conquers Mexico in 1519 and Pedro Arias Davila founds Panama City in the same year. With a strong foothold, the Spanish send expeditions throughout Central America and begin the conquest of Yucatan in 1527.

The Spanish attempt to control Belize from Yucatan, encountering stiff resistance from the Mayan provinces of Chetumal and Dzuluinicob. The provinces become a refuge for escaping Maya who are battling the Spaniards in other parts of Mesoamerica. Some of these Maya bring with them diseases contracted from the Spaniards; smallpox, yellow fever, and malaria devastate the indigenous populations. Weakened by disease, they cannot defend themselves against the invaders.

Ironically, while the Spanish lay claim to it, Belize is never directly administered or settled by them.

1618 Franciscans build church in Tipu

Two Franciscan monks (members of a Roman Catholic religious order) build a church in Tipu, a Mayan stronghold.

Britain Begins to Challenge Spain's Supremacy in Belize

1638 British arrive in Belize

Little is known about early British settlements in Belize, although the year 1638 is generally accepted as the date of the first permanent settlement by shipwrecked English seamen. According to legend, an English buccaneer named Peter Wallace, called "Ballis" by the Spanish, settles near the Belize River and names it after himself. It is the mispronunciation of his name that gives way to the name Belize. Another possible origin of the name Belize is that it derives from the Mayan word, Belikin.

The Spaniards attempt to maintain a monopoly on trade and colonization in the Americas, but Britain, Holland, and France begin to challenge their supremacy, first resorting to smuggling, piracy, and war, especially in the Caribbean region.

England captures Jamaica in 1655 and uses the island as a base to support other settlements along the Caribbean Coast, from Yucatan to Honduras. The Spanish don't have firm control of Belize and gradually the English presence grows in the region.

The first English buccaneers begin using Belize's treacherous coastline to attack Spanish ships. In time, exporting logwood, a tropical American tree whose heartwood is used to produce a dye, becomes a profitable venture.

1667 Treaty leads to end of piracy

The European powers agree in a treaty to suppress piracy, bringing more stability to the Caribbean region. The British encourage settlement in Belize and promote the logwood cutting industry.

1670 Treaty grants land to English

Spain and England sign a treaty in 1670, giving England possession of lands where it has established colonies in the New World. But the treaty is ambiguous about where these colonies are located. This leads to more conflict between Britain and Spain, especially over logwood cutting.

1717 British expelled from Belize

The Spanish attempt to enforce their sovereignty over Belize and attack British settlements repeatedly, forcing them to leave. The first major attack occurs in 1717, followed by three others in 1730, 1754, and 1779.

1724 British bring first slaves to Belize

A Spanish missionary writes that the English are importing slave labor from Jamaica and Bermuda. The report is the earliest reference to black slaves in Belize.

1738 British Baymen begin to structure their society

The English settlers, known as Baymen, on their own initiative and without recognition by the British government, begin to elect magistrates and establish common law.

1763–86 Treaties give Britain right to cut logwood

During the eighteenth century, Spain continues to lose ground against the British in Belize. Despite attempts by the Spanish to dislodge them, the British increase their presence in the region and gain the right to cut more wood through a series of treaties.

The Treaty of Paris in 1763 gives England the right to continue logging in Belize while asserting Spain's sovereignty over the area. But the treaty does not specify where the English can cut logwood, creating a source of further friction between the two nations.

1779: September 15 Spanish attack British settlement

The Spanish capture St. George's Cay, where most of the settlers live. They capture 140 prisoners and 250 slaves and ship them to Havana, Cuba. The settlement is deserted until a new peace is declared in 1783 and the British return to Belize.

1784 Legal Code established for colony

By the mid-1760s, the British government begins to take a more active role in the affairs of the Baymen. They codify and expand the Baymen's loose regulations and in 1784, they name Colonel Edward Marcus Despard (1751–1803) superintendent of the Belize settlement.

1786 New treaty expands British logging

The British settlers petition for a new treaty and are allowed to cut mahogany and logwood. Yet that treaty prohibits the British from building forts, setting up any type of government, and developing agriculture.

1787 Treaty forces relocation of settlers

According to the new treaty (see 1786), the Spanish retain the right to inspect the British settlements twice a year. The treaty also forces England to give up its claims in eastern Nicaragua, home to more than two thousand settlers and their slaves. Those people are forced to move to the Belize settlement, strengthening the English presence.

1788   British suppress the Maya

With the increase in mahogany exports, the British push deeper into Belize beginning to encroach on Mayan communities. The Maya attempt to defend themselves and the British report they are attacked in 1788.

The British send troops and weapons to destroy Mayan villages, burning homes and fields, intending to drive them out of mahogany-rich areas by destroying their food supply.

How many Maya live in Belize at the turn of the century is not known, but clearly the British threaten their way of life and the Maya become refugees in their own land. The British are apparently unable to exploit the Maya for labor and must therefore rely on imported African slaves to do the heavy work in Belize.

1789   Settlers challenge British authority

By the late eighteenth century, a small group of wealthy settlers controls most of the British land. These settlers also own about half of all the slaves in the British settlement and begin to wield enormous power.

The wealthy loggers control imports, exports, and the wholesale and retail trade. They collect taxes, elect leaders from amongst themselves and resist any social or political changes. Superintendent Colonel Edward Marcus Despard, appointed in 1784, is suspended in 1789 when the wealthy loggers challenge his authority.

1796   The Garifuna immigrate to Belize

The Garifuna, descendents of native Caribs and African slaves, resist British and French colonialism for many years in the Lesser Antilles (the Leeward and Windward Islands). When the British put down a violent rebellion on Saint Vincent, they move as many as five thousand Garifuna across the Caribbean to the Bay Islands off the north coast of Honduras. The Garifuna begin to migrate to Central America, settling in present-day southern Belize, Guatemala, Honduras, and Nicaragua.

The language of the Garifuna is basically Carib and their music and dance are chiefly of African origin. Traditionally, Garifuna make a living from fishing and in Belize, most of them live in the coastal communities of Dangriga, Punta Gorda, Hopkins, Seine Bight, and Barranco. They attend Roman Catholic or Methodist churches, yet they retain many of their ancient beliefs and rituals.

1798: September 10   Final Spanish attack on Belize

Alarmed by the growing English presence in Belize, the Spanish attack from Yucatan with a flotilla of 30 vessels, 500 sailors, and 2,000 soldiers.

After a two-hour battle, the British drive off the Spanish at St. George's Cays. The battle is Spain's last attempt at controlling Belize.

1802   Garifuna settle in Stann Creek area

About 150 Garifuna settle in the Stann Creek (present-day Dangriga) area of Belize. Here, they farm and fish for a living. Other Garifuna migrate to other parts of Belize. Many of them find work alongside slaves cutting mahogany.

The British treat the Garifuna as squatters, forcing them to seek permits to sell their produce in Belize City. They are widely discriminated against and their lands are threatened.

1831: July 5   Civil rights extended to people of mixed ethnicity

The white government of Belize is forced to expand the rights of the growing Creole population, people of mixed European and African ancestry. Although some Creoles are legally free, they only have limited privileges. They are not allowed to hold commissions in the military or act as jurors and magistrates. They are allowed to vote only if they have more property and live in the area longer than whites.

1832: November 19   More Garifuna arrive in Belize

Many Garifuna leave Honduras, caught in the midst of a civil war. They land in Belize on November 19, 1832. Modern-day Belize commemorates their arrival with a national holiday: Garifuna Settlement Day.

1833   England takes first step to abolish slavery in its colonies

The English agree to abolish slavery over a five-year period. The act includes two measures that benefit the slave owners: it creates a system of "apprenticeship" that forces former slaves to continue working for their masters without pay while at the same time compensating the former owners for loss of property.

1838   Emancipation of slaves

Full emancipation is achieved in Belize, but former slaves have few opportunities to better their lives. The ruling white minority fosters a hierarchical and authoritarian society whose citizens are ranked by class and race. Limiting privileges for the underclasses, the Europeans manage to control the land and the country's economy for the next century.

Whites, who make up about one-tenth of the population, have built their wealth with slave labor and continue to control Belize after emancipation.

1838–1917   East Indian workers begin to replace slave labor

Between 1838 and 1917, Britain will bring 548,000 East Indians to the Caribbean colonies. As indentured workers, the

East Indians are contractually obligated to work for one employer for a given length of time with little or no compensation.

After serving their time, the workers are allowed to return to India or remain in the colonies as free men. However, many cannot return because they are indebted to their employers for living expenses incurred during their servitude. Several hundred are brought to Belize to work on sugar estates. In time, they are acclimated into the surrounding culture to such an extent that they are no longer distinguishable from the Creole population.

Today, some of their descendants live in the Corozal and Toledo districts in northern Belize and many of them are found in Calcutta, a small community named after the Indian city.

1846 Mahogany exports hit a peak

Belize exports 4 million linear meters of mahogany in 1846, but three years later, that figure drops to 1.6 million. By 1870, Belize exports only 8,000 linear meters of mahogany. The economy is completely dependent on the export of mahogany, and to some degree of logwood. When both their prices collapse, Belize's economy falls into a state of prolonged depression that lasts through the turn of the century.

Efforts to develop plantation agriculture like sugar, coffee, and bananas, fail. By the 1880s, demand for chicle, a gum taken from the sapodilla tree and used to make chewing gum, will help the economy.

1847–55 War in Yucatan affects Belize

In the mid-nineteenth century, the Maya rise against Mexican authorities in what will be known as the Caste War of Yucatan. More than half the population of Yucatan perishes in the struggle and thousands of refugees flee to Belize.

Most of the refugees are farmers who settle in what is now northern Belize and begin to grow large quantities of sugar, rice, corn, and vegetables. But their presence brings them into direct conflict with the British, whose laws do not allow Maya to own land.

1850 Americans and British agree to build canal in Central America

With the signing of the Clayton-Bulwer Treaty of 1850, Britain and the United States agree to promote the construction of a canal across Central America. They also agree not to colonize any part of Central America.

The British believe the colonization clause applies to future occupation only; they currently occupy Belize, the Bay Islands off Honduras, and the Mosquito Coast in eastern Nicaragua. But the U.S. government pressures Britain to give up its possessions. Britain agrees to move out of the Bay Islands and the Mosquito Coast, but presses for formal possession of Belize.

1854 Britain formally takes possession of the Settlement of Belize

By 1854, the British government has grown weary of trying to establish a formal government to rule Belize. At first, they had feared angering Spain and later the United States, with greater interest in Central America.

As early as 1738, without British government authority, Belize settlers developed a simple government and regulations to run their affairs. In time, these informal meetings gave way to the Public Meeting, with elected representatives. Participation in the Public Meeting depended on race, wealth and length of residency, ensuring that the rich loggers retained power.

The British government, wary of the settlers' autonomy, sought greater control of Belize. By 1765, Rear Admiral Sir William Burnaby, the commander-in-chief of Jamaica, had recommended that a superintendent be appointed to oversee Belize. The settlers resented the growing influence of the British government and prevented Britain from permanently naming a superintendent until 1796. In time, friction diminished between the settlers and the British government and by 1854, with changing social conditions throughout Central America and the Caribbean, including the end of slavery, Britain formalizes its possession of Belize by enacting a new constitution to govern the settlement.

The new constitution replaces the Public Meeting with a Legislative Assembly, composed of eighteen elected members. The superintendent, who is appointed by the British government, can appoint three colonial officials to serve in the assembly. Much like the Public Meeting, only the wealthiest of Belizeans can run for office and only residents who earn a certain income and possess land can vote.

Despite the appearance of local control, Britain holds the real power in Belize. The elected members of the Legislative Assembly have to be British citizens and own property. The British superintendent chairs the assembly and can dissolve it at any time, create legislation, and approve or reject bills.

1859: April 30 England and Guatemala attempt to define borders

Britain and Guatemala sign a treaty defining a border and acknowledging the intent of both nations to build a road from Guatemala City to the Caribbean Sea. However, Guatemala, which claims its independence from Spain in 1821, insists on its territorial rights to Belize.

1859 Partnership leads to powerful company

The British Honduras Company is formed and quickly emerges as the major landowner of the British colony. In 1875, the company is renamed Belize Estate and Produce Company and has its base in London. It owns half of all the private land in Belize.

The company, along with Scottish and German merchants, comes to dominate the economy by the 1890s, forcing

Belize to rely on imported goods. For the next century, it remains the dominant force in Belize's economy.

1862   Belize becomes a British colony

The Settlement of Belize in the Bay of Honduras is officially declared a British colony and is given a new name: British Honduras. The British government names a lieutenant governor, subordinate to the governor of Jamaica, to replace the superintendent and rule the colony. But British Honduras retains its elected Legislative Assembly.

1865   Hundreds of Chinese laborers come to British Honduras

The British bring 474 Chinese immigrants to work in Belize. More than 100 die and 100 others escape to Mexico because of cruel treatment, overwork, and bad food. Three years later, slightly more than 200 Chinese laborers remain in Belize.

1867   Law prohibits Maya from owning land

Conflict escalates between the Maya and the British. The governor of Belize rules that Indians (Maya) will not be permitted to live on or cultivate any land unless they pay rent to the government or landowner.

1871   British assume direct control of Belize

Between 1862 and 1871, elected members of the Legislative Assembly become deeply divided over the economy and security issues. Some want more taxes on land, others want increases in import duties and others fear that not enough is being done to keep the Maya in check.

Unable to move forward and facing a deepening political crisis, the Legislative Assembly allows Britain to take direct control of Belize in exchange for greater security. In 1871, the elected Legislative Assembly is dissolved and Belize becomes a crown colony under direct supervision of the Colonial Office in London.

In a crown colony, Britain retains control over defense, foreign affairs and internal security. The British also control administration and the budget. A British-appointed governor and a locally elected assembly govern a crown colony.

The new constitution of 1871 designates a lieutenant governor and a Legislative Council with five ex officio or "official" members and four appointed or "unofficial" members to govern Belize. The change in government has profound social consequences. The old logger and settler elite no longer holds power. From now, power rests in the Colonial Office and British companies doing business in Belize.

1872   British establish reservations for native peoples

Threatened by the Maya and Garifuna, the British establish reservations for them and force them to live there.

1872: September 1   Mayan leader falls in battle

Mayan leader Marcos Canul had risen to prominence in 1866, when he and a group of Mayas attacked a mahogany logging camp and demanded payment for the release of prisoners. They also demanded the British pay rent for land.

In 1872, Canul and his men occupy the town of Corozal. Canul is mortally wounded when his band attacks the British barracks in Orange Walk, fighting for several hours before the Maya retreat.

With this last defeat, the Maya are forced to give up their independence and live on reservations.

1880–1890   Maya from Guatemala settle in Belize

In the 1880s and 1890s, Mopán and Kekchí Maya escape from forced labor in Guatemala and settle in several villages in southern Belize. It is a remote area of the country and the Maya are able to maintain a more traditional way of life.

By the end of the nineteenth century, Belize has become a fascinating polyglot society, although separated by class, race and, to a great extent, geography. The ethnic patterns remain largely unchanged during the twentieth century. The Maya and Mestizos, who are overwhelmingly Roman Catholic, live in the north and west of Belize. They speak Spanish or Creole. Protestants, mostly of African descent, live in Belize Town and speak English or Creole. The Garifuna, also predominantly Roman Catholic, speak English, Spanish, or Garifuna and settle in the southern part of Belize.

1900   United States strengthens economic ties with Belize

Greater demand for mahogany in the United States briefly resurrects the industry in Belize. But Belize quickly depletes its resources. In the 1920s, Belize uses heavy machinery to reach mahogany forests deep in the country. During this time, mahogany, chicle and other forest products account for eighty-two percent of the colony's exports.

Trade with the United States is fostered by the Creoles, who have begun gaining more political power in Belize. In time, this emerging class of Creole businessmen and professionals begins to challenge the political and economic ties with Britain.

1927   Creoles gain political voice

By the turn of the twentieth century, representatives of wealthy landowners dominate the Legislative Council, created to rule Belize when it became a crown colony in 1871. Companies like the British-owned and powerful Belize Estate and Produce Company maintain a powerful hold on the country.

But by the mid-1920s, Creole merchants and other professionals begin to gain a foothold in the Legislative Council, changing once again the power structure of the colony.

The Struggle for Independence

1931: September 10   Hurricane worsens Belize's economic troubles

The Great Depression virtually ends mahogany and chicle exports, devastating the Belizean economy. The situation grows worse when a hurricane devastates Belize Town, killing more than 1,000 people and nearly destroying the entire town.

1934   Protests mark beginnings of independence movement

In the 1930s, workers begin to unite in large numbers to protest working conditions and demand their rights. In the mid 1930s, investigators report terrible working conditions at the Belize Estate and Produce Company. Workers are not paid cash wages. Instead, they are given coupons that can be redeemed for poor quality food in the company's commissaries. Prices are high and workers are constantly in debt and unable to leave. The poor quality food leads to malnutrition among workers and their families.

The colonial government repeatedly rejects proposals to legalize trade unions and make labor reforms, including the introduction of a minimum wage and sickness insurance.

1934: February 14   Workers demand more rights

A group calling itself the Unemployed Brigade marches through Belize Town to present demands to the governor. The march starts a broad movement, with the poor turning to the governor for help.

The governor creates a few civil service jobs to avoid civil disturbances. Some workers are hired for ten cents per day to build roads.

1934: October   Early labor leader lands in prison

Antonio Soberanis Gómez (1897–1975) leads several hundred picketers in a protest at a sawmill. Many are arrested after clashes with authorities and Soberanis is jailed for five weeks.

Soberanis, a barber who feels unsuited for the role of leadership, nevertheless begins to lead the workers' movement. Soberanis and other members of the Labourers and Unemployed Association attack the colonial government, repressive labor laws, and the Belize Estate and Produce Company, which owns half of the private land in the country.

1941–43   Labor movement makes progress

Trade unions are legalized in 1941, but employers are not required to recognize them. However, repressive labor laws are removed from the criminal code in 1943, allowing the new trade unions to continue fighting for improved labor conditions.

The General Workers' Union, created in 1943, quickly grows into a nationwide organization and provides crucial support for the nationalist movement that takes off with the formation of the People's United Party seven years later.

1945   Few Belizeans are allowed to vote

In 1945, only 822 people are registered to vote in a population of 63,000. Between 1939 and 1954, less than two percent of the population are allowed to vote in Legislative Council elections, and only six of thirteen members are directly elected.

The proportion of voters increases in 1945 simply because the minimum age requirement for women is reduced from thirty to twenty-one years. More people are eligible to vote by 1949, when devaluation of the British Honduras dollar reduces property and income voter-eligibility requirements.

1947   Leader of independence movement enters politics

George Price (b. 1919) begins his political career when he is elected to the Belize City Council in 1947. Described as an eclectic and pragmatic politician, Price quickly emerges as one of the leading nationalist leaders and becomes one of Belize's most powerful political figures.

Price comes from a middle-class family and had considered a career in the Roman Catholic Church. He studied with Jesuits in the United States in the 1930s and returned to Belize in 1942. Instead of entering the priesthood, Price went to work for Robert Turton, a Creole chicle millionaire who was an elected member of the Legislative Council between 1936–48. Turton introduced Price to the political process and Price quickly became involved in Belize City's affairs. By 1947, Price is well-known and is elected to the Belize City Council. By 1958, Price becomes Belize's mayor and in 1964, becomes Belize's first premier.

During his early political career in the Belize City Council, Price and his colleagues begin to press for constitutional changes. They want universal voting rights and an all-elected Legislative Council. Their demands for representative government gain them a large following.

1950   People's United Party founded

George Price co-founds the People's United Party (PUP), which challenges the monopoly of colonial administration.

1954   Voting rights are extended to more people

Pressure for a more representative government forces the British colony to approve a new constitution in 1954, which extends voting rights to all literate British subjects over the age of twenty-one. The constitution also replaces the Legislative Council with a Legislative Assembly (not to be confused with the Legislative Assembly abolished in 1871). The new Legislative Assembly has nine elected members, three officials and three appointed members. The constitution, which also creates an executive council headed by the governor to run Belize, is seen as a key component in the process of decolonization.

On April 28, more than seventy percent of the electorate vote in the elections and Price's party, the PUP, gains more than sixty-six percent of the vote. The PUP captures eight of the nine seats in the Legislative Assembly, and becomes one of the most influential parties in Belize.

1958   Mennonites begin arriving in Belize

About 5,000 German-speaking Mennonites begin to migrate to Belize from Mexico. They buy large tracts of land and establish a quiet agrarian lifestyle. The Belizean government agrees to allow them freedom to practice their religion, use their language, control their schools and organize their own financial institutions. The Mennonites also gain an exemption from military service.

The Mennonites practice complete separation of church and state and do not vote. Yet they contribute a great deal to Belizean life with their dairy industry, which supplies the domestic market with eggs, milk, cheese, poultry, and vegetables.

1964: January 1   New constitution leads to self-government

The country's newest constitution calls for self-government, but Britain keeps its troops in Belize, partly because of a border dispute with Guatemala that dates back to 1859.

1964   First premier

George Price becomes Belize's first premier. (See also 1947.)

1973   A new name for emerging nation

To prepare for independence, British Honduras formally becomes Belize, a name favored by nationalists who want to connect the nation to its Mayan past.

1970s   Oceanographer Jacques Cousteau visits Belize

Jacques Cousteau (b. 1910) sails his research ship, *Calypso,* to Belize, where he studies Belize's atolls, coral islands and reefs that nearly enclose a lagoon. His accounts of the region help spawn interest in tourism in Belize.

1980   U.N. resolution seeks Belize's independence

A United Nations resolution demands the secured independence of Belize, with all of its territory intact. The resolution, approved by 139 countries including the United States, calls for Britain to continue to protect Belize and for all countries to come to its assistance. Seven nations abstain from voting and Guatemala refuses to vote. For Belize, the resolution is the end to a long struggle for recognition as an independent nation.

1980s   Tourism grows

Tourism becomes a major industry for the new nation of Belize in the 1980s and 1990s.

1981: September 21   Belize is world's newest nation

Belize formally declares its independence, but Guatemala refuses to recognize the new nation, severs diplomatic ties with Britain, and declares the date a national day of mourning.

The new government consists of a governor-general appointed by the Queen of England, an elected prime minister who heads a cabinet, and a National Assembly with a Senate and House of Representatives. The governor-general appoints the senators and legislators are elected to office.

1982   Journalist publishes first novel

Novelist Zee Edgell is born Zelma Inez Tucker. She works as a journalist from 1959–69, and publishes her first novel, *Beka Lamb,* in 1982. The novel is set in 1951, when residents of British Honduras were launching a movement for independence. *Beka Lamb* was a joint winner of the British Fawcett Society Book Prize, awarded to works of fiction with strong perspectives on women's roles. *Beka Lamb* is required reading for all high school students by the Caribbean Exam Council (CXC). Edgell's second novel, *In Times Like These*, is published in 1991, and deals with the transformation of British Honduras into independent Belize.

1983   Pope visits Belize

Pope John Paul II visits Belize on his tour of South and Central America.

1984   Opposition breaks PUP's hold on power

Manuel Esquivel (b. 1940), former Belize City mayor and one of the founders of the United Democratic Party (UDP), wins in a landslide victory to become the country's second Prime Minister.

His victory ends a cycle of one-party dominance by the PUP. More importantly, UDP's victory signals Belize's commitment to democracy during a time of great turmoil in Central America.

Esquivel is a more conservative leader who favors a free market economy and is considered to be pro-United States.

1991   Belize becomes member of international organization

Belize is admitted to the Organization of American States (OAS), a move which Guatemala tries to block several times. In another development, Guatemala's president Jorge Serrano and Belize prime minister George Price reach an agreement that leads to full Guatemalan recognition of Belize's independence. A year later, the two countries sign a non-aggression pact.

1992 National Lands Act and reservation life

The National Land Acts sets forth new regulations for the reservation lands of the Maya and Garifuna. The Garifuna practice Roman Catholicism but incorporate into their worship rituals that include spirit helpers and *"gubida"*—dead ancestors. Garifuna have suffered persecution by Christians and others because of these traditional rituals. One of the most powerful of these ceremonies is the *"dugu,"* a feast that is staged to make peace with dead ancestors. Garifuna believe that the *gubida* may become angry with a living relative because of an insult or selfish lifestyle. Illness is believed to be a sign that the person has offended the *gubida*. To treat the illness, the Garifuna believe a *dugu* must be staged. The *dugu* lasts several days, and relatives and friends of the sick are invited to attend. Garifuna travel long distances to attend a *dugu*. The *"buyei,"* or spiritual leader, leads the guests in singing and dancing to drumbeats that call the *gubida*. When the *gubida* arrive, announced by the *buyei*, the guests make offers of food and drink. The ceremony ends with the burial of food or drink (or the dumping of food and drink in the ocean).

The Stann Creek district, one of the earliest Garifuna settlements near present-day Dangriga (see 1802), is the area in Belize where the *dugu* is most likely to be held. There is a Garifuna temple, known as a *"dabuyaba,"* there.

1994 Britain withdraws its troops from Belize

With full recognition of Belize's independence by Guatemala and a non-aggression pact signed by both nations, Britain no longer sees a need to maintain troops in Belize.

1995 Tourism becomes Belize's chief source of income

The number of tourists visiting the country grows from 64,000 in 1980 to 247,000 in 1992. In 1994, 357,385 tourists spend $71 million in the country. Mayan ruins and culture, the longest barrier reef in the Western Hemisphere, wildlife, and beaches attract visitors.

1998 Dominant party wins elections

The People's United Party wins twenty-six of twenty-nine seats in the House of Representatives. The country's new prime minister, Said Musa, is a PUP member.

Bibliography

Ball, Joseph W. *Cahal Pech, the Ancient Maya, and Modern Belize: The Story of an Archeological Park.* San Diego: San Diego State University Press, 1993.

Bolland, O. Nigel. *Belize: A New Nation in Central America.* Boulder, Colo.: Westview, 1986.

Clegern, Wayne M. *British Honduras, Colonial Dead End: 1859–1900.* Baton Rouge: Louisiana State University Press, 1967.

Edgell, Zee. *Beka Lamb.* London: Heinemann, 1982.

Fernandez, Julio A. *Belize: Case Study for Democracy in Central America.* Brookfield, Vt.: Avebury, 1989.

Grant, C. H. *The Making of Modern Belize.* Cambridge: Cambridge University Press, 1976.

Kerns, Virginia. *Women and the Ancestors: Black Carib Kinship and Ritual.* Urbana: University of Illinois Press, 1983.

Mallan, Chicki, *Belize Handbook.* Chico, Calif.: Moon Publications, 1991.

McClaurin, Irma. *Women of Belize: Gender and Change in Central America: Belize, Guatemala, Honduras, and El Salvador.* Norman, Okla.: University of Oklahoma Press, 1996.

Bolivia

Introduction

Bolivia is a landlocked country in the heart of South America. In the two-and-a-half centuries of its history as a nation, Bolivia has lost more than half its original territory, including coastal lands that today belong to Chile. The great majority of Bolivia's inhabitants are descended from the native Indian population that peopled the land at the time of the Spanish conquest in the sixteenth century. The snowcapped peaks of the Andes Mountains are Bolivia's most dramatic physical feature. However, the nation's terrain also includes tropical rain forests, swamps, and savanna. The majority of Bolivians live on the *altiplano,* a high plain that stretches between the country's mountain ranges. Bolivia's population is currently estimated at over nine million people. Its official capital is Sucre, but the actual seat of government is La Paz.

Advanced Indian civilizations, including the Tiahuanaco and Aymara, were living in present-day Bolivia in the first millennium A.D. The skilled stonework of the Tiahuanaco can still be seen in giant monoliths they left behind, and in the majestic Sun Gate, a portal with a carved picture of their sun god over the doorway. By the fifteenth century, the Incas, a Quechua-speaking group that originated in Cuzco (in present-day Peru), had extended their growing empire southward to what today is Bolivia. With an economy based on collective agriculture and an extensive road network to hold it together, the Inca empire flourished until the early sixteenth century. At its height, it stretched for over two thousand miles (three thousand kilometers).

The sixteenth century brought both internal strife and external conquest to the Incas. Already torn by a five-year civil war, they fell prey to the Spanish conquistador Francisco Pizarro (c.1475–1541) and his men, who landed on the shores of Peru in 1532. They killed the Incas' leader, Atahualpa (c.1502–1533), and within a year had vanquished and subjugated the native population, whose labor was to be exploited by Europeans for centuries. Silver mining, the major focus of this exploitation until the middle of the seventeenth century, began in 1545 with the discovery of a mountain of silver at Cerro Rico and the founding of the town of Potosí. La Paz was soon established as a center for the silver trade. Under Spanish colonial rule, present-day Bolivia, then called Upper Peru, became part of the Viceroyalty of Peru, with its capital at Lima. The Spanish later placed it under the jurisdiction of the Viceroyalty of the Río de la Plata, with its center at Buenos Aires.

The Fight for Independence

Bolivians were the first in South America to rebel openly against Spanish rule, but the last to win their freedom. The struggle for independence lasted from the 1809 armed uprisings at Chuquisaca and other cities to the final ouster of Spanish troops by Simón Bolívar in 1824. The new republic that was formed adopted the name Bolivia in honor of its principal liberator. Its first president was Bolívar's lieutenant, Antonio José Sucre (1795–1830) in whose honor the new nation's capital was renamed.

Bolivia suffered from political instability during its first forty years, with frequent changes of government. During this period, the nation's Indian population was disenfranchised yet further when their collective land holdings were broken up and divided into parcels, which were then sold to wealthy landowners. A particularly disastrous episode in the nation's history was the War of the Pacific (1879–1884), fought over nitrate-rich land on the Pacific Coast over which Chile eventually won control, wresting from Bolivia all of its coastal territory and removing its direct access to the sea. Bolivia also lost some of its rubber-rich eastern land to Brazil in 1903, and the Gran Chaco desert region in the southeast to Paraguay in the Chaco War of 1932–35.

In the 1880s rival political parties, representing mine owners and rural landowners, emerged, with the conservative interests of the mine owners predominating. The single greatest economic influence on Bolivia at the start of the twentieth century was the boom in tin, whose industrial uses had suddenly made it a valuable commodity. A group of tin barons known as the *rosca* controlled Bolivian politics during the first two decades of the new century, ignoring the welfare of the nation's poor, including its Indian population. Oil also became a sought-after resource, with the Standard Oil company setting up oil fields in the 1920s. Large oil concerns also played a role in the Chaco War: the Gran Chaco became dis-

puted territory only after it was rumored to contain oil deposits. (And such deposits were never found.)

In 1934, Bolivia's elected president, Daniel Salamanca (1869–1935), was deposed in a military coup. Over the next fifty years, the nation's government would alternate between civilian and military control. A major political force for much of this period was the Movimiento Nationalista Revolucionario (National Revolutionary Movement, or MNR), a pro-labor party that was founded in 1941 by Victor Paz Estenssoro. After gaining power briefly through a coup in the 1940s, the

MNR won control of the government in free elections in 1951, inaugurating a period of social and economic reform known as "the revolution of 1952." Paz Estenssoro, who served as president from 1952 to 1956, initiated numerous reforms, including universal suffrage, government social programs for the rural Indian population, and nationalization of the tin industry.

The MNR was ousted in a coup in 1964, and another period of dictatorships began. The longest-lived was that of Colonel Hugo Banzer Suárez, who suspended normal government activities and repressed dissent, even while implementing measures that significantly improved the nation's economy. After a succession of military governments in the late 1970s, the military finally agreed to the return of civilian rule by allowing the Congress elected in 1980 to convene. In 1985 former president Victor Paz Estenssoro was returned to office, twenty years after being forced from that position in a coup. Under pressure from the International Monetary Fund to carry out economic austerity measures, Paz Estenssoro instituted policies very different from those of his previous terms in office. Although these measures were unpopular with the nation's workers, they brought the nation's spiraling inflation under control within three years.

A major concern for Bolivia in the 1980s and 1990s has been cocaine production. Second only to Peru in the cultivation of coca leaves, Bolivia was producing $2 billion worth of cocaine annually by the mid-1980s. Government officials have been implicated in drug-related activities and a former minister has been extradited to the United States and convicted on drug charges.

In 1989 Bolivia had its first orderly transfer of power between civilian governments in twenty years, as Jaime Paz Zamora of the Revolutionary Left Movement (MIR) succeeded Victor Paz Estenssoro as president. In 1993 Paz Zamora's government was followed by yet another elected civilian administration, that of Gonzalo Sanchez de Lozada. Bolivia received approval for admission to the MERCOSUR in December 1997, joining Argentina, Brazil, Uruguay, and Paraguay. (MERCOSUR is an acronym for Mercado Común del Sur—Southern Common Market.)

Timeline

A.D. 500–1000   Tiahuanaco culture flourishes along Lake Titicaca

An early advanced civilization in the Andes, the Tihuanaco live on the altiplano (high plains in the west at about 4,000 meters or 13,000 feet above sea level) south of Lake Titicaca. They plant and harvest crops, perform stone and metal work, make pottery, and build a central city (today known as Tiahuanaco), whose ruins can still be seen at a site twelve miles (nineteen kilometers) south of Lake Titicaca. They represent only the central portion of a city that once covered at least 2,600 acres (1,000 hectares). Among the relics found there are a sunken temple, a fifty-foot (fifteen-meter) high pyramid, and an enormous stone platform.

The most prominent feature of these stoneworks is a stone gateway called the Sun Gate topped with a carved image of the Tiahuanaco sun god. A short figure with a headdress that fans out around him, he holds a staff on either side of him (thus, he is also known as the staff god). His image is found on other Tiahuanaco stonework and on their pottery.

The ruins at the site are made from enormous blocks—the largest weighs more than 100 tons (90 metric tons)—that the Tihuanaco managed to transport across Lake Titicaca.

A.D. 1000   Aymara culture flourishes on the Andean highlands

During the period when the Tihuanaco culture declines (around A.D. 1000), the Aymara flourish on the *altiplano* (highlands) of the Andes Mountains. They live in a number of loosely allied nations, each with its own leaders. Like the Incas, who eventually absorb the Aymara into their empire, they farm the land collectively, living in settlements called *ayllus*. They also raise llamas and alpacas. There is no money exchanged by the Aymara, but they still evolve a sophisticated system of trade. The dominant Aymara nations are the Colla and the Lupaca.

1200s   The Incas begin to expand their territory

The Incas are a Quechua-speaking tribe from Cuzco, in the southern Andean highlands of present-day Peru. In the thirteenth century, they begin expanding their territory. Expansion continues at a slow pace until the fifteenth century. Eventually the Incas build an empire that stretches for over two thousand miles (three thousand kilometers) along the Pacific and rules over five million or more people.

Using advanced agricultural techniques including irrigation and crop rotation, the Incas work the land, living in communities called *ayllus*. An advanced road network connects all the parts of their vast empire. They have neither a written alphabet nor the wheel, and keep records using sets of knotted strings called *quipus*. Aside from the ruins of their buildings, their greatest legacy is the Quechua language, which they bring with them to much of South America and which is still spoken by native groups today.

1460–1490   The Incas conquer Aymara lands

The Incas' expansion increases under the rule of their ninth emperor, Pachacuti Inca Yupanqui (1438–1471), and grows even more rapidly during the reign of his son, Topa Inca Yupanqui (1471–1493). In the latter part of the fifteenth century, the Incas conquer the Aymara living in the region of Lake Titicaca and rule over them until the Inca empire itself is overrun by the Spanish in the next century. The Aymara mount several rebellions against the Incas, but none succeed.

The incorporation of the Aymara nations into the Inca empire around 1460 actually brings relatively little change to the culture of the Aymara, who retain their language and customs. However, the Incas do introduce a number of their farming innovations, such as terracing, and other advancements, such as the *quipu* system of keeping records with knotted strings. The Incas take areas of land for themselves and require a certain amount of labor from the conquered

Aymara (whose unhappy history of forced labor continues for centuries under the Spanish).

1528   The Inca empire is torn by civil war

In the first part of the sixteenth century, the Incan empire is weakened by civil war between Atahualpa and Huáscar, the sons of emperor Huayna Capac (1493–1524). After their father's sudden death, war breaks out over the division of the empire. Atahualpa wins the five-year conflict, but it weakens the Incas just before Spanish conquerors, under the command of Francisco Pizarro (c.1470–1521), arrive in the New World.

The Spanish Conquest and Colonial Rule

1532: November 16   Pizarro's forces ambush Inca leader in Peru

Spanish explorer Francisco Pizarro lands at Tumbes, on the northern coast of present-day Peru with fewer than 200 men. He and his men attack the Incas and capture their leader, Atahualpa. Although the Incas gather a lavish ransom—enough silver and gold to fill two entire rooms—Atahualpa is still executed by the Spanish in 1533. After their emperor is slain, the Incas are easily conquered by the Spanish, who have, among other advantages, horses and guns.

1535   Almagro journeys to Lake Titicaca region

The first significant visit by a Spaniard to the Aymara region is that of Diego de Almagro (c. 1475–1528), who has joined with Pizarro in conquering the Incas. Juan de Saavedra, a member of Almagro's expedition, founds the first Spanish settlement at Paria.

1538   Hernando Pizarro defeats the Incas in Bolivia and founds settlements

Hernando Pizarro (c. 1475–1578), the brother of conquistador Francisco Pizarro, completes the conquest of the Incas by overcoming them in the region that today is Bolivia (then called the Audiencia of Charcas, or Upper Peru). The Spanish found the settlements of Chuquisaca (present-day Sucre), Potosí, and La Paz.

1538   First church is built in Bolivia

The church of San Lázaro is built in Chuquisaca (later La Plata/Sucre).

1539   Dominican missionary arrives in Aymara territory

The first missionaries arrive with the Spanish conquistadors and set about converting the native population to Christianity. In 1539 the first Dominican missionary arrives in the lands of the Aymara, on the shores of Lake Titicaca. Within ten years,

there are monasteries in all the major Aymara settlements. The Jesuits, who come after the Dominicans, build missions for the Indians of the lowlands, show them how to cultivate crops, and teach them how to read and write.

1542   Spain issues new laws governing policy in Peru

Charles I of Spain (1500–58, r. 1516–56) issues laws curbing the land grant system known as the *encomienda,* by which Spanish settlers are given large tracts of land previously occupied by the Incas, as well as the right to tribute (involuntary payments, similar to taxes) and free labor from the native population. This system makes the local leaders, or *encomenderos,* more powerful than the Spanish government wants them to be. Instead, Spain divides the land into administrative districts called *corregiemientos* governed by Spanish-appointed officials. However, the new laws are widely ignored and many encomiendas remain in place. Eventually they are replaced by yet another system of large estates called *haciendas.*

1545: April   Spanish discover silver at Cerro Rico

The discovery of Cerro Rico ("Rich Mountain"), a mountain that is nearly solid silver, provides the Spanish with the treasure for which they have come to the New World. Potosí is founded as a mining town at the base of the mountain, and thousands of Spaniards flock to the site. The mines are worked by the local Indian population. Under the forced labor system known as the *mita,* they work in the mines six days a week, virtually living underground and receiving little or no compensation for this difficult and dangerous work.

The mines in the Potosí area provide roughly half the total silver mined in Spanish colonial South America for the next hundred years. It is estimated that several hundred thousand dollars' worth of silver is extracted from these mines—enough to build a bridge a yard (one meter) wide between Spain and Bolivia.

1548   La Paz is founded

The city of La Paz is founded both as a base for the silver trade between Potosí and Peru and to protect the roads leading from Chuquisaca, Porco, and Potosí to the coastal city of Lima. Silver, silk, and other commodities are transported between the two regions by caravans of llamas. The new settlement grows quickly and becomes an important trading center and, by the end of the nineteenth century, the de facto capital of Bolivia.

1552   Bolivia's first bishopric created

Pope Julius II creates the bishopric of La Plata. This is Bolivia's first religious district presided over by a bishop. More follow in La Paz and Santa Cruz at the beginning of the seventeenth century.

1559   Local government is established under Spanish jurisdiction

The Audiencia of Charcas, with both judicial and executive powers, is created to administer Upper Peru under the control of the Viceroyalty of Peru. The seat of government is at Chuquisaca. The Audiencia governs an area that includes parts of present-day Peru, Chile, Paraguay, and Argentina. A series of regional governorships are established under its authority. The major one located within present-day Bolivia is that of Santa Cruz de la Sierra, established in the 1590s.

1570s   New process facilitates mining of low-grade silver

After thirty years the top-grade ores at the surface of Cerro Rico are depleted. Viceroy Francisco de Toledo (c. 1515–84) introduces the mercury amalgamation process for refining silver, developed in Mexico in 1555. Mercury amalgamation makes it possible to mine the remaining lower-grade ores. To acquire the additional labor that is needed, Toledo reintroduces the pre-Columbian (mita) system of forced labor. Adult males from sixteen highland districts are pressed into service and required to work in the mines one year in every six. The water required for the refining process is provided by a system of artificial lakes capable of holding several million tons of water. The refining process is completed by 1621. Cerro Rico remains a silver mining site for a total of nearly 200 years. Bolivia's seventeenth-century silver output peaks between 1610 and 1645 and then drops until the end of the century.

1572–75   Spanish viceroy inspects Upper Peru

As part of a five-year tour of his entire viceroyalty, Francisco de Toledo (c. 1515–84), the fifth Spanish viceroy of Peru, journeys to Upper Peru to survey conditions there. This action, and the policies that result from it, expand government influence in the region. Among other measures, Toledo orders that the Indian population be relocated to new, centralized communities where they can be more effectively controlled by the Spanish.

1590s   Major governorship is established at Santa Cruz de la Sierra

The Audiencia of Charcas, under the control of the Viceroyalty of Peru, governs an area that includes parts of modern Bolivia, Peru, Chile, Paraguay, and Argentina. A series of regional governorships are established under its authority. The major one located within present-day Bolivia is that of Santa Cruz de la Sierra.

1592   Famous sculpture crafted by Indian artist

Indian sculptor Tito Yupanqui produces a statue of the Virgin Mary that is still used for devotional purposes as of the late 1990s.

1607   Oruro founded by discovery of new silver deposits

With the discovery of silver deposits on the *altiplano,* or highlands, the settlement of Oruro is founded. Although never equaling the output at Potosí, Oruro is the second great silver discovery of the region, and the area remains an important mining site.

1610   Aymara-Spanish dictionary produced

The first Aymara-Spanish dictionary is written by Ludovico Bertonio, a Jesuit missionary. Bertonio praises the native Aymara language for its logic and simplicity.

1624   First university is founded

Bolivia's oldest university, and one of the oldest in all of South America, the University of San Francisco Xavier, is established at Chiquisaca (later Sucre). Many of South America's Jesuit priests are trained here. (See also 1681.)

1646   Birth of composer Juan de Araújo

Araújo (1646–1712) is Bolivia's foremost seventeenth-century composer, also serving as chapelmaster of the cathedral at Chuquisaca (later Sucre).

1650   Potosí is major population center

Potosí has a population of 160,000 and is one of the major population centers in the New World. Although silver output drops off from its highest levels after 1650, it continues to be mined there until the 1890s.

1650   Silver deposits start to run out

By the middle of the seventeenth century, the silver reserves of Upper Peru are becoming depleted, plunging the region into an economic depression that lasts a hundred years.

1660   Artist Melchor Pérez de Holguín is born

Pérez de Holguín is the preeminent Bolivian artist of the Spanish colonial period. He is known for his religious paintings.

1681   University of San Francisco Xavier introduces legal training

Bolivia's oldest university, the University of San Francisco Xavier, begins to offer training for lawyers. Jesuit priests are also trained there. (See also 1624.)

1719–20   Epidemic reduces native population

An epidemic of what is thought to have been pneumonic plague or influenza, originating in Buenos Aires, spreads over the central Andean region. From this point, the Indian population of Upper Peru (present-day Bolivia) stops growing for several decades.

1730   Mestizos stage uprising in Cochabamba

In response to the more centralized colonial administration of the eighteenth century, a number of revolts take place in Upper Peru. One of the largest is carried out in Cochabamba by *mestizos* (people of mixed Spanish and Indian ancestry) protesting an attempt to make them subject to the tribute payments imposed on full-blooded Indians. They are aided by *creoles* (native-born descendants of white settlers).

1739   Mestizo revolt in Oruro

Mestizos, joined by creoles and led by Juan Vélez, rise up in protest against higher taxes imposed by the Spanish government. They draft a document calling for equality among Upper Peru's different racial groups and abolition of repressive measures including tributes, forced labor, and the forced purchase of goods by Indians.

1760s   Jesuits expelled from South America

As part of their policy of secularization, Spain's Bourbon rulers expel the Jesuit missionaries.

1750   Population decline in silver region

Many inhabitants leave the silver-mining regions as the silver supply is depleted. The populations of Potosí and Oruro are only half of what they were in 1650 (see entry). One group, the local Indians, benefits from the slowing of economic activity, and their numbers actually increase during the mid-1700s.

1770s   Native and mestizo schools of painting emerge

Well-defined schools of painting develop among Indian and mestizo artists in the cities of Upper Peru, including Potosí and La Paz. The Colla school is active in the La Paz area. Most of the paintings have religious themes.

1776   Bolivia becomes part of the Viceroyalty of the Río Plata

Administrative control of Upper Peru (present-day Bolivia) is transferred to the newly formed Viceroyalty of the Río Plata, which has its center in the future Argentine capital of Buenos Aires.

1780–82   La Paz is besieged in Indian uprising

In response to their exploitation by the Spanish, the native population of Peru and Upper Peru engages in sporadic uprisings throughout the seventeenth and eighteenth centuries. The years between 1770 and 1780 see a new wave of rebellions. The most important one is led by José Gabriel, who takes the name of Túpac Amaru (the last Inca ruler), and gathers an army of thousands of Indians who stage a rebellion in Peru.

An Indian in Upper Peru named Julián Apasa leads a revolt in Upper Peru, under the name of Túpac Catari. Apasa's forces besiege La Paz for over 100 days in 1781.

Apasa is captured by Spanish troops later that year. Túpac Amaru is also captured in 1781, but the general uprising in the region continues until the following year.

1784   New centralized colonial administration is established

A new system of local government is put in place and Upper Peru is divided into four intendencies: La Paz, Potosí, Cochabamba, and Chuquisaca.

1803–25   Drop in silver output sparks economic crisis

Silver production declines by eighty percent between 1803 and 1825, adding economic troubles to the political turbulence of this period. (See also 1846.)

Rebellion and Independence

1809   Uprisings begin struggle for independence from Spain

The fight for Bolivian independence begins with armed uprisings by white and mestizo colonists in Chuquisaca, Cochabamba, Oruro, and Potosí. In La Paz, rebels capture the Spanish governor and name Pedro Domingo Murillo their president. The rebellions are quickly suppressed by Spanish troops dispatched by the viceroy in Lima. Rebel troops take to the hills and continue waging guerrilla warfare against the Spanish.

1813: January 26   Murillo is executed

Rebel leader Pedro Domingo Murillo is taken prisoner by the Spanish and executed. It will take another fifteen years for Bolivian independence to be won. Thus, Upper Peru, where the first colonial rebellions in South America have occurred, is ultimately the last region to win its freedom.

1824   Bolívar liberates the Peruvian highlands

Although Lima has been liberated from Spanish rule in 1821, royalist troops still maintain control of the altiplano, or highlands. In 1824, Simón Bolívar (1783–1830) and his lieutenant Antonio José de Sucre (1795–1830) lead a campaign to oust the last Spanish forces from the region. Sucre wins a decisive victory for Upper Peru when he defeats the Spanish troops at Junín on August 6, 1824. The last royalist troops are routed at Ayacucho on December 9.

1825: August 6   Bolivia declares its independence

An assembly declares Upper Peru an independent republic and adopts the name Bolivia in honor of liberator Simón Bolívar. The name of the capital city is changed from Chuquisaca to Sucre in honor of Bolivia's other war hero, Antonio José de Sucre. Bolívar drafts the constitution for the new republic, and Sucre is chosen as its first president.

Bolivia takes its name from Simon Bolivar (1783–1830), the leader of Latin American independence from Spain. In 1825, Bolivar helped draft Bolivia's first constitution. (EPD Photos/CSU Archives)

1825–28   Sucre serves as nation's first president

Sucre presides over a country devastated by a long struggle for independence and torn by divisions between whites, Indians, and mestizos. He implements administrative reforms, issues an order for primary, secondary, and vocational schools to be established throughout the country, and reduces the influence of the church, but the army is unhappy with his actions. Sucre resigns from office in 1828.

1829–39   Santa Cruz serves as president

Andrés Santa Cruz (c. 1792–1865), who fought in the war for independence, becomes Bolivia's second president. His major goal is the unification of Bolivia and Peru. In 1836, his troops overthrow the government of Peru and declare a confederation between the two nations (the Confederación Perú-Boliviana), becoming its leader. However, both Chile and Argentina, fearing the growth of Bolivian power, send forces to dismantle the confederation. The Bolivian forces are crushed at the Battle of Yungay in January 1839 and Santa Cruz is exiled.

1836   Birth of author Gabriel René Moreno

Bolivian historian and man of letters, Gabriel René Moreno (1836–1909), spends much of his life in exile in Chile. He authors *The Last Days of the Colony of Upper Peru* and other historical works, as well as literary criticism and sociological works. A university in Santa Cruz is named for René Moreno.

1841–80   Period of political strife and military government

After President Santa Cruz is deposed, the nation goes through a period of political instability, in which political control is seized by a series of local bosses (*caudillos*) who rule as dictators. The first of these strong leaders is General José Ballivián y Segurola (1804–52), who has led the Bolivian army in vanquishing Peruvian invaders at the Battle of Ingaví. Ballivián retains power from 1841 to 1847, working to expand Bolivia's frontiers and contain military threats by Indian tribes. The major political leaders after Ballivián are Manuel Isidoro Belzú (1848–55), José María Linares (1857–61), and Mariano Melgarejo (1864–71).

1846   Archaeology museum founded

Bolivia's National Museum of Archaeology, located in La Paz, is the nation's most renowned museum.

1864–71   Melgarejo destroys Indian communities

Mariano Melgarejo (1818–71) is one of a succession of *caudillos* (bosses) who gain control of the country in military coups between 1841 and 1880. During the period when he is in power, from 1864–71, he breaks up the Indians' system of communal agriculture, consisting of settlements called *ayllus*. Authorized by newly passed laws, his government disbands the ayllus, requiring their inhabitants to repurchase them within ninety days or lose them to the highest bidder. Since few Indians can afford to buy back their own land, most of it is seized by wealthy Spanish landowners. All resistance by the Indians is crushed by the military.

1867   Treaty with Brazil guarantees access to Atlantic Ocean

Bolivia signs an agreement with Brazil to gain access to the Atlantic by transporting goods on Brazilian rivers. In exchange for these rights, however, Brazil demands territorial concessions, and Bolivia cedes 40,000 square miles (104,000 square kilometers) of land to its neighbor.

1872   Author and diplomat Ricardo Freyre is born

Ricardo Jaime Freyre (1872–1933) is considered one of Bolivia's foremost writers. His works, which include short stories, show compassion for the country's Indians. Freyre also serves his country as a diplomat.

Bolivia prior to the War of the Pacific, c. 1878.

1874 Birth of anthropologist Arturo Posnansky

This Austrian-born anthropologist and archaeologist (1874–1946) produces groundbreaking studies of the Indian civilizations that flourished near Lake Titicaca.

1879–84 War of the Pacific

Bolivia's original borders stretch to the Pacific Ocean and include rich deposits of nitrates in the Atacama Desert, which

it exports to Europe to be used in fertilizers. Chile, which has an agreement to develop this land for Bolivia, disputes taxes imposed by Bolivia and, in 1879, seizes this land and claims it as its own. Peru is obligated to join Bolivia in fighting Chile, because of a mutual defense agreement, and the two nations jointly declare war on Chile. The Bolivians are defeated at the Battle of Tacna, after which, most of the fighting is between Chile and Peru.

Chile emerges victorious, winning all of Bolivia's land on the Pacific coast and turning Bolivia into a landlocked nation. However, in a 1904 agreement, provisions are made to ensure Bolivia access to the sea. The city of Arica is declared a free port, and a railway is built from it to La Paz. The War of the Pacific is only one of a series of nineteenth- and twentieth-century conflicts in which Bolivia loses land to its neighbors.

1879: July 15   Birth of author Alcides Argüedas

Argüedas (1879–1946) becomes one of his country's best-known authors, both at home and abroad. Author, diplomat, and politician, he is born in La Paz, where he studies law. Later, he studies sociology in Paris. For twenty-five years he pursues a diplomatic career, representing his government in London, Paris, Colombia, and Venezuela. During this time, he also writes articles for Bolivian newspapers. In 1916 he returns to Bolivia, where he becomes a leader of the Liberal party and wins a seat in the Chamber of Deputies.

Argüedas's literary career encompasses journalism, history, sociology, and fiction. His most famous literary work is the sociological study *Pueblo Enfermo* (A Sick People; 1909), which wins acclaim throughout South America, although his critical view of his nation's people arouses controversy. It is among the first literary works to draw attention to the situation of the South American Indians, detailing their customs and way of life and describing the poverty in which they live. Argüedas also wins acclaim for the realistic depiction of Indian life in his novels, especially *Raza de Bronce* (Bronze Race; 1919), which narrates the travels of a group of Indians. As well, Argüedas is the author of the 1922 historical study *Historia general de Bolivia* (General History of Bolivia) and a historical novel about the War of the Pacific, *Pisagua* (1903).

Democracy and Stability

1880   New constitution enacted and military government ends

Due to the disastrous War of the Pacific, Bolivia's military government is discredited and a new constitution is drafted. However, political turmoil continues as rival political parties—the Liberals and Conservatives—gain strength, representing the interests of landowners and mine owners. During the last two decades of the nineteenth century, power is in the hands of the Conservatives, who represent the silver mining interests. They encourage free trade and oversee construction of a railway network to transport minerals across the country for export.

1880s   Bolivia's first railroad is completed

A rail line covering 700 miles (1,127 kilometers) is built from Oruro to the Chilean port of Antofagasta. The construction is financed by Great Britain. It is later expanded from Oruro to other Bolivian cities including La Paz, Potosí, and Cochabamba. (See also 1913 and 1950.)

1892   Chiriguano uprising is defeated

The Chiriguano Indians of the Andean foothills in southeast Bolivia, numbering about 100,000, remain generally free from government interference until the late nineteenth century. After an uprising—put down in 1892—many Chiriguanos are forced into slavery, and their land is appropriated and distributed to those who have defeated them.

1898–99   Federalist War makes La Paz Bolivia's de facto capital

Threatened by the growing dominance and prosperity of La Paz, legislators from Sucre try to push through Congress laws that would ensure Sucre's continued position as the nation's capital. In response, government officials from La Paz obtain the support of the Liberals to unseat the current Conservative regime. An armed revolt also takes place involving the Liberal and Federalist parties as well as an Indian contingent led by Zárate Willka. Although Sucre remains the nation's official capital, the actual seat of government is moved to La Paz, where it remains to this day.

1899   Brazil seizes land in eastern Bolivia

An uprising by Indians and Brazilian settlers in the Acre region in the jungles of northeastern Bolivia, paves the way for Brazil to seize control of this land, which is in demand because of its rubber resources. A rubber boom (1871–1912) is under way in the Amazon region, and much of the native population has been forced into virtual slave labor by wealthy entrepreneurs. The 1903 Treaty of Petropolis awards Brazil 70,000 square miles (181,300 square kilometers) of this valuable land.

1900–80   Tin becomes mainstay of Bolivian economy

Industrialization in the West has created a demand for tin—found in abundance in Bolivia—just as the nation's silver industry is declining. The network of railroads created to transport silver now becomes a lifeline for the tin boom. Eventually, Bolivia will become the world's second largest tin producer, after Malaysia. With the shift from silver to tin, economic and political power moves from Sucre and Potosí in the south to La Paz, which becomes Bolivia's financial center. Simón Patiño (b. 1867), later known as the "tin baron," develops his first tin mines in the 1890s. Tin requires a much greater capital investment than silver, so Patiño and the other tin producers are forced to enlist the aid of foreign investors.

The tin industry grows rapidly in the first decades of the twentieth century, with its interests concentrated in the hands of a few wealthy entrepreneurs, including Patiño, his partner,

Mauricio Hochschild, and the Aramayo family, former wealthy silver miners.

1900–20   Politics dominated by *la rosca*

In the first decades of the twentieth century, Bolivian politics are controlled by a new aristocracy of tin and silver mining barons, known as *la rosca,* that dominates the political parties and their candidates. Through its political connections, the rosca backs railroad construction and other projects that further economic expansion. However, the interests of Bolivia's poor, especially of its large Indian population, are neglected as the government fails to undertake needed social reforms. During this period, the Conservative party, which represents the older silver mining interests and has been in power for the previous twenty years, is superseded by the Liberals, who remain in power until 1920.

1900–10   La Paz becomes musical capital of Bolivia

In the first decade of the twentieth century, important musical institutions are founded in Bolivia's new de facto capital. These include the Military School of Music (1904), the Conservatorio Nacional (1908), and the Círculo de Bellas Artes (1910).

1904   Newspaper *El Diario is founded*

Bolivia's oldest newspaper, *El Diario,* begins publication as the nation's tin boom starts.

1908   Birth of author Fernando Díez de Medina

Fernando Díez de Medina (b. 1908), like many other modern writers in Bolivia and elsewhere in Latin America, has also been involved in politics. He serves as minister of education in the MNR government (see 1941) of the 1950s. Other twentieth-century Bolivian writer-politicians include Ricardo Anaya (b. 1937).

1910   Birth of painter Jorge Carrasco Núñez del Prado

Twentieth-century Bolivian artist Jorge Carrasco Núñez del Prado is born. He achieves international fame as a painter. Perhaps his best-known painting is *Painting at 4,000 Meters* (the altitude of La Paz).

1910   Franz Tamayo publishes important work

Franz Tamayo (1879–1956) is a poet, philosopher, and statesman. Although born into Bolivia's landowning elite, Tamayo becomes a prominent advocate for Indian rights. He is best known for his poetry, which forms part of the *modernismo* movement.

Poet Franz Tamayo is also a respected social and political thinker and author. He publishes *Creación de la pedagogia nacional* (1910), which expresses his ideas about education and the need to recognize the value of Indian culture. (See also 1934.)

1913   Railway from La Paz to Arica, Chile is completed

The La Paz to Arica, Chile railway, like the other major railway in Bolivia, is financed by Great Britain. (See also 1880s and 1950.)

1919   Birth of Maria Núñez del Prado

An internationally acclaimed sculptor, Maria Núñez del Prado is born. She often uses native materials found in Bolivia, such as basalt, onyx, and guayacán wood. Her sculptures include *White Venus, Black Venus,* and *Spirit of the Cloud.* In her sculptures she tries to express the spirit of Bolivia, especially that of its ancient Indian civilizations.

1920   Republicans gain control of government

The Republican party, formed in 1914, comes to power in a bloodless coup, ending the twenty-year dominance of the Liberals. Although the Republicans are more sympathetic to the demands of the working class, their government, like that of the Liberals before them, still controlls the tin interests. However, tin prices are on the decline, and the world market will nearly collapse in the aftermath of the 1930s. Two presidents are in power during the 1920s: Juan Bautista Saavedra (1921–25) and Hernando Siles Reyes (1926–30).

1921–25   Saavedra institutes first social legislation

President Juan Bautista Saavedra institutes modest social welfare and labor programs, the first in Bolivia's history.

1926   Amelia Villa becomes Bolivia's first woman doctor

Amelia Villa earns her medical degree, becoming the first female physician in Bolivia. She pursues a distinguished career in pediatrics. The children's ward of a hospital in Oruro is named in her honor.

1926   Academy of Fine Arts is founded

Bolivia's National Academy of Fine Arts is created and fosters a nationalist aesthetic among the country's painters and sculptors.

1927   Standard Oil Company starts first Bolivian oil fields

The Standard Oil Company of New Jersey locates oil fields in southeast Bolivia. In the intervening years, control of oil production alternates between private companies and the government.

1928   Central bank is established

The Central Bank of Bolivia is created as the sole bank of issue. Reorganized in 1945, it is set up as a commercial bank.

1930   University of San Andrés is founded

Located in La Paz, the University of San Andrés is Bolivia's largest university.

1930   Guzmán de Rojas becomes head of the National Academy of Fine Arts

Painter Guzmán de Rojas (b. 1900) is named director of the National Academy of Fine Arts. Rojas is a leader in the use of native themes in Bolivian art. He paints many pictures portraying the lost Inca city of Machu Picchu. After studying in Madrid he returns to his homeland in 1929. Rojas is also known for restoring the work of painter Pérez de Holguín and for introducing mural painting to Bolivia.

1932–35   Chaco War

In the 1930s a border dispute arises between Bolivia and Paraguay over the Gran Chaco desert region in Bolivia's southeast region that separates the two countries. Both nations are interested in the land because it is rumored to contain oil reserves. An undeclared border war breaks out in the region even as the two sides are attempting to negotiate an agreement. The fighting lasts for three years, with each side receiving help from oil companies that will be involved in the future drilling if their side wins. As many as 100,000 Bolivians are wounded or killed, and Bolivia loses the disputed territory, which officially becomes part of Paraguay in 1938. Bolivia's territorial losses total 100,000 square miles (260,000 square kilometers), and, at the war's end, the country is $230 million in debt.

The Chaco War is socially significant because it breaks down many of the nation's racial barriers, as men from different social and ethnic backgrounds serve together in the armed forces. Those who have served in the war later participate in implementing social changes.

Military Dictatorship and Rise of the MNR

1934   Poet Franz Tamayo runs for president

Poet Franz Tamayo runs for the presidency with the backing of the Salamanca government and wins the election. However, a military takeover prevents him from taking office. (See also 1910.)

1934   Military seizes power

With Bolivia's government discredited by defeat in the Chaco War, the military ousts President Daniel Salamanca (1869–1935, president 1931–34) and takes control of the country, ushering in a period of alternating civilian and military rule that is to last until the 1980s. Colonel David Toro, who heads the military government during this period, nationalizes Standard Oil Company holdings in Bolivia and seizes their profits.

1937–39   Busch takes control

Colonel Germán Busch, a hero of the Chaco War, assumes control of the government and establishes a dictatorship. He implements reforms to better the nation's miners, including the encouragement of labor unions. Busch's regime ends when he commits suicide under mysterious circumstances in 1939.

1938   New constitution is adopted

Bolivia adopts a new constitution and provides for the process of presidential elections.

1940   Peñaranda elected president under new constitution

General Enrique Peñaranda (b. 1892) is elected president under the new constitution (see 1938). Under President Peñaranda, Bolivia allies itself with the United States and the other nations fighting the Nazis in World War II (1939–45). The country's main contribution is the mining of strategic raw materials used in the war effort.

1940   National orchestra is established

The Orquesta Nacional (National Orchestra) is founded by composer José Maria Velasco-Maidana.

1941   MNR political party is founded

The Movimiento Nacionalista Revolucionario (Nationalist Revolutionary Movement—MNR) is formed. Bolivia's first grass-roots political party, it has the support of both working class and white-collar middle class employees, as well as leading intellectuals. Its leaders include Víctor Paz Estenssoro (b. 1907), an economics professor, and Hernán Siles Zuazo, son of a former Bolivian president. Its platform calls for extensive social reforms and nationalization of the country's tin industry. The MNR becomes the leader of the congressional opposition.

1942   Striking workers are massacred at Catavi

In one of Bolivia's most violent labor confrontations, striking workers at a Patiño mine in Catavi, as well as their families, are attacked by government troops.

1943   MNR seizes power

The National Revolutionary Movement, or MNR, is one of a number of new political parties formed in the 1940s by people who have become critical of the country's political power structure during the Chaco War. The MNR, a nationalist party that advocates the rights of miners and other workers, ends up allying itself with pro-Nazi factions during World War II (1939–45) as a way to oppose the government's ties to the United States and its allies. In 1943 it stages a military coup and, in 1944, Major Gualberto Villarroel is named president

Villarroel establishes a dictatorial regime but implements social and labor reforms. He encourages miners to unionize and convenes a congress of native leaders to improve the lot of Bolivia's Indian population. Nevertheless, Villarroel's authoritarian style of governing arouses opposition that leads to his downfall. In July 1946 he is hanged from a lamppost outside the presidential palace.

1945   First indigenous congress is held

Over a thousand Indian leaders from the Andean highlands attend a national congress called under President Villarroel. During his tenure as president, Villarroel issues decrees to end the practice of unpaid Indian labor on large agricultural estates (*haciendas*), and establish schools in Indian communities. Unfortunately, they are not carried out.

1946   Conservatives seize power after death of Villarroel

Bolivia's Conservatives, controlled by the mine owners, came back into power after General Villarroel is ousted and killed, and try to reverse his policies. However, they meet with only limited success, and the opposition MNR party gains support during the postwar years, even though its activities are suppressed and many of its leaders exiled.

1946–52   Labor militancy grows during the *sexenio*

During the six years immediately following World War II (1939–45), referred to as the *sexenio*, the labor movement becomes more militant. The Thesis of Pulacayo, issued by workers in 1946, calls for armed revolution. Confrontations with an increasingly oppressive government escalate. A second uprising at Catavi in 1949 is brutally crushed by the military.

1947   Major discovery is made at Bolivian physics lab

An international team of physicists at Bolivia's Chacaltaya astrophysics laboratory discovers the pi-mesan, or pion. A pion is a fundamental particle that makes up the quark.

1950   Railway from Santa Cruz to São Paulo, Brazil, is complete

The railway is Bolivia's last major rail system. (See also 1880s and 1913.)

The "Revolution of 1952"

1951: May   MNR wins national elections

The MNR wins a majority of votes in the 1951 presidential election, but the defeated regime hands power over to the military instead. Popular opposition to military rule grows, fueled by serious social and economic problems including high inflation, low tin prices, and poor conditions for the nation's workers.

1952   Violinist Jaime Laredo makes his debut

Concert violinist Jaime Laredo (b. 1941) makes his debut at the age of eleven in a concert with the San Francisco Symphony Orchestra.

Laredo becomes a world-renowned soloist and chamber music performer. Born in Cochabamba, he moves to the United States with his family at the age of seven to pursue musical studies at the Curtis Institute in Philadephia, Pennsylvania, with renowned teacher Ivan Galamian. In 1959 he becomes the youngest person to win the Queen Elisabeth of Belgium Competition. The following year he launches his professional career with a well-received Carnegie Hall concert in New York.

Laredo performs throughout the United States and Europe. Particularly acclaimed as a chamber music player, he is a regular performer at the famous Marlboro Chamber Music Festival in Vermont. Bolivia issues a series of stamps in the musician's honor.

1952   New government implements sweeping reforms

Under Paz Estenssoro, the MNR implements numerous social, political, and economic reforms, ushering in the most politically stable period in Bolivia's history. The government introduces universal suffrage with neither property nor literacy requirements. By 1956 the Bolivian electorate has grown five-fold, from 200,000 to almost one million voters. Estenssoro's government also takes over the holdings of the tin mining companies, nationalizing them under a new state enterprise called COMIBOL (Corporaci'on Minera de Bolivia—Mining Corporation of Bolivia).

Civil rights are extended to the Indian population, and a Ministry of Peasant Affairs is established to improve health and educational facilities in Indian communities throughout the nation. In addition, land belonging to wealthy hacienda owners is divided and redistributed to the Indians, from whom it was originally taken away (see 1953). The MNR government also sponsors public works and construction projects including manufacturing plants, road building, and hydroelectric plants. Paz Estenssoro reduces the size of the military and creates a civilian militia.

In spite of the social programs introduced by the MNR, Bolivia continues to suffer from the economic effects of a decline in the tin industry and runaway inflation. As the 1950s progress, the military gradually gains back much of its former power. Dr. Hernan Siles Zuazo serves as president from 1956 to 1960, and Paz Estenssoro becomes president again in 1960.

1952   Female suffrage is enacted

Bolivian women win the right to vote.

1952: April 9   The "Revolution of 1952"

With the support of miners, peasants, and other groups, the MNR stages an armed revolt against the nation's military rulers that comes to be known as the Revolution of 1952. After several previous unsuccessful attempts to seize power by force, the MNR takes control of government arsenals in La Paz and passes out weapons to civilian supporters. The capital is besieged by armed miners and MNR militants lay siege to other cities as well. The military government conceded defeat within three days. There is widespread involvement in the fighting by ordinary citizens, and approximately 600 people are killed. Victor Paz Estenssoro, leader and founder of the MNR, becomes president.

1952: October 31   Major tin companies are nationalized

The Bolivian government takes over the country's three largest tin companies, turning over their assets to the state-run COMIBOL. As a result of this act, two-thirds of Bolivia's mining industry comes under government control.

1953   Agrarian reform is implemented

The government establishes the Agrarian Reform Commission in January 1953 and promulgates the Agrarian Reform Law in August. Property is taken from large landowners and distributed to the rural Indian population under a program of government reforms (see 1952). The owners are compensated with government bonds. Lesser amounts of property are expropriated from small- and medium-sized estates. Forced labor is abolished.

1960   Academy of Sciences is founded

The Bolivian National Academy of Sciences is established.

1961   Church and state are formally separated by the constitution

Bolivia's 1961 constitution ends all state support for the Roman Catholic Church, making church and state completely separate.

Return to Military Rule

1964   Estenssoro is ousted in military coup

A military coup led by the country's vice president, General René Barrientos Ortuño, removes Paz Estenssoro and the MNR government from power. From 1964 to 1966 Bolivia is ruled by a military junta with Barrientos as president. Barrientos, who is part Indian, works to better conditions for Bolivia's Indian population, providing agricultural aid and new technology and working to improve education. However, he alienates workers with anti-union measures and suppression of strikes.

1966   Barrientos is elected president

Barrientos forms the Movimiento Popular Cristiano (Popular Christian Movement—MPC) and competes in a presidential election, which he wins. As president, he is supported by conservatives, businessmen, and peasants. His primary opposition comes from labor. Other critics oppose his encouragement of U.S. private investment in Bolivia and his sale of natural resources to the United States. Criticism mounts when one of Barrientos's top aides, Minister of the Interior Colonel Antonio Arguedas, is revealed as an agent for the U.S. Central Intelligence Agency (CIA).

1966–67   Che Guevara leads communist guerrilla movement

An Argentinean who has been a leader in Cuba's communist revolution, Ernesto "Che" Guevara (1928–67) establishes a base of operations in southern Bolivia for a planned continent-wide revolution in South America. With the aid of troops from the United States, the Bolivian military crushes the guerrillas, capturing and killing Guevara in October 1967. (See also 1997.)

1969: April   Barrientos is killed in helicopter accident

After President Barrientos's death, his vice president, Dr. Luis Adolfo Siles Salinas, takes over the presidency but is ousted in September 1969 in the first of numerous right- and left-wing coups that shake the country over the next two years. General Alfredo Ovando Candía, chief of staff of the armed forces, serves as president in 1969–1970, canceling scheduled elections in 1970 and dissolving Congress.

1969–1971   Ovando and Torres regimes

Following the death of Barrientos, Bolivia is ruled by military regimes led by Alfredo Ovando (1969–70) and Juan José Torres González (1970–71). Ovando tries to make the military government an agent of social and economic reform, in a program he calls "revolutionary nationalism, " which includes nationalizing the Bolivian assets of Gulf Oil. However, he fails to win the backing of labor and other important groups and is ousted on October 7, 1970 and replaced by Torres. Torres is no more successful than Ovando at gaining the support needed to unify the country and is deposed in a coup in August 1971.

1970   Gulf Oil interests are nationalized

The Bolivian government takes over holdings of the Gulf Oil Company, but compensates the company for its interests.

1971–78   Banzer Presidency

A right-wing military coup in August 1971 brings Colonel Hugo Banzer Suárez (b. 1926) to power. In the early years of the Banzer government, the economy grows, as Banzer encourages foreign investment, and benefits from the development projects put in place by the previous MNR government. However, Banzer forcibly represses opposition to his government. A 1972 general strike against devaluation of the peso is violently crushed and martial law is imposed on the country. In 1974, peasants blocking roads to protest price increases are massacred. In the same year, in response to a coup attempt, Banzer replaces all his cabinet ministers and suspends the activities of political parties, trade unions, and student political groups.

After 1974 Banzer faces economic problems, as petroleum and cotton production decline. The government also faces setbacks abroad as negotiations with Chile over access to the Pacific fail and U.S. President Jimmy Carter pressures Bolivia to hold elections.

Banzer is unseated in July 1978 by his own protégé, General Juan Pereda Asbún. Military government continues, with further coups, until 1982.

1975 Indian languages are officially recognized

The Bolivian government makes the Quechua and Aymara languages official languages, together with Spanish. Both languages have been spoken by Bolivia's native population for centuries.

1979–80 First woman serves as Bolivia's president

Lydia Gueiler Tejada serves as interim president from November 1979 to July 1980, becoming the first woman to hold Bolivia's highest executive office.

1980: July 17 Meza seizes power

In the political chaos following the 1979 presidential elections, General Luis García Meza comes to power in Bolivia's 189th military coup. It is known as the "cocaine coup" because of its leaders' involvement in international drug trafficking. There are also alleged links to Nazi war criminal Klaus Barbie (b. 1914), who is living in Bolivia. Meza remains in power for fourteen months, suppressing all opposition through imprisonment, torture, and killing. During the Meza regime, Bolivia is said to have exported US$850 million worth of cocaine. Meza resigns in 1982, and is succeeded by General Celso Torrelio Villa and General Guido Vildoso Calderón.

Civilian Rule Restored

1982 Nature preserve is established in Beni

A 334,000-acre (135,000-hectare) expanse of tropical land in the department of Beni is set aside by the government as a reserve to protect its native population and endangered animal species from cattle ranching and other forms of encroachment.

1982: October 10 Zuazo becomes president as country returns to civilian rule

Congress appoints Hernán Siles Zuazo, winner of the 1980 elections, as president. However, his leftist government is unable to steer Bolivia out of its economic crisis. The strict austerity measures demanded by the International Monetary Fund are highly unpopular, resulting in worker strikes and protests. In 1984 the government stops paying its foreign debt.

1983 Nazi war criminal is extradited from Bolivia

Former Gestapo commander Klaus Barbie, alleged to have links to the ousted government of Luis García Meza, is extradited to Lyons, France, to stand trial for war crimes.

1983–84 Drought and flooding cause sharp drop in farm production

A severe drought, followed by flooding, reduces Bolivia's agricultural output by sixty percent.

1984: June President Zuazo is kidnapped

Bolivia's president is briefly kidnapped but emerges unharmed. The incident, which is believed to involve drug lords, leads to the arrests of 100 army officers and two former top government officials.

Mid-1980s Mega-inflation hits Bolivia

Bolivia's currency becomes virtually worthless due to runaway inflation, which eventually rises over 10,000 percent. In early 1986 Bolivians commonly carry as many as 200 million pesos with them to make purchases. A person can easily spend twelve million pesos on an ordinary restaurant meal.

1985 Paz Estenssoro is returned to office as president

The 1985 presidential elections are split between two former presidents, Banzer and Paz Estenssoro, and Congress chooses to return Paz Estenssoro (b. 1907) to office. Under pressure from the International Monetary Fund, the former president imposes strict economic reforms that, while unpopular, help stabilize both the government and the economy of his nation, which has the world's highest inflation rate when he takes office. A new currency, the boliviano, replaces the peso, with one boliviano worth one million pesos. Paz Estenssoro also increases taxes on wealthy Bolivians, ends government subsidies, and improves the administrative efficiency of the government.

Although the government economic measures raise unemployment, inflation drops, and an economic recovery is under way by 1988.

1985   United States renews military aid to Bolivia

After an eight year suspension, the United States renews military aid.

1985: September   General strike is called to protest government policies

Labor unions call a general strike to protest economic austerity measures by President Paz Estenssoro that have resulted in higher unemployment and lower wages. Government arrests and other measures end the strike by October. However, labor unrest in response to the austerity program continues through 1988.

1986   Paz Estenssoro suspends rights, orders crackdown on labor leaders

To deal with extensive labor unrest in response to his government's economic policies, President Paz Estenssoro has labor leaders arrested.

1986: July   U.S. military aids in drug crackdown

By 1985 illegal cocaine trade by Bolivians has reached $2 billion annually, with Bolivia second only to Peru as the world leader in the production of coca leaves. By the 1980s and 1990s it is estimated that a quarter of the land cultivated in Bolivia is devoted to growing coca, much of it illegally.

In 1986 soldiers from the United States help Bolivian authorities close down many of the secret labs where the coca leaves are turned into paste to produce cocaine. As part of Operation Blast Furnace, U.S. officers train 1,000 Bolivian military and police personnel in antinarcotics and counterinsurgency tactics. U.S. soldiers also provide helicopter transport to drug processing sites in the Chapare, Beni, and Santa Cruz regions. However, many of the sites have been deserted due to publicity about the operation. The United States and Bolivia extend the operation until November 1986. However, the drug trade continues to flourish.

1987   Acclaimed Bolivian novel is published

This highly regarded novel, *Jonah and the Pink Whale* by José Wolfango Montes Vannuchi is a humorous look at life in the city of Santa Cruz. It receives an award for the best Latin American novel of the year, and is translated into English.

1987: July   Government preserves natural areas in exchange for debt payment

The Bolivian government agrees to protect 3.7 million acres (1.5 million hectares) of tropical lowlands in a deal with Conservation International, which agrees to assume $650,000 of the nation's foreign debt. This makes Bolivia the first nation in the world to protect part of its environment in exchange for having its foreign debt reduced.

1988   Pope John Paul II visits Bolivia

Huge crowds turn out to greet Pope John Paul II when he visits Bolivia as part of a South American tour. In a speech in Cochabamba, the pope warns against involvement in drug trafficking.

1989   Former government official is extradited to United States on drug charges

Former Justice and Interior Minister Colonel Luis Arce Gómez is extradited to the United States on charges of involvement in the illegal drug trade. The Bolivian Supreme Court challenges the extradition and claimes it infringes on their authority. This action touches off a constitutional crisis that is only resolved in 1991 when representatives of the nation's five major political parties sign an agreement affirming the autonomy of the court. Gómez is eventually sentenced to thirty years in prison.

1989   Researchers discover ancient grave site

Archaeologists from the United States discover a mass grave near Lake Titicaca, where the Tihuanaco Indians flourished between A.D. 500 and 1000.

1989: May   New president is chosen in free elections

Three candidates tie for the presidency in the May 1989 elections. One of the candidates, former president Hugo Banzer Suárez, throws his support to Jaime Paz Zamora, who becomes the new president. Paz Zamora's party is the moderate Revolutionary Left Movement (MIR). The election of Paz Zamora represents Bolivia's first peaceful transfer of power between civilian governments in over twenty years.

1990   Government rulings favor indigenous population

A series of government policies affirms the rights of the native Indian population. A portion of Bolivia's northern rainforest is reserved for native peoples.

1990   Bolivia's lowland tribes demonstrate to protect their land

A major protest is staged by 500 members of the Guaraní, Guarayo, Chiquitano, and other tribes living in the lowland plains and tropical forests. They march on La Paz to seek government acknowledgment of their right to 4 million acres (1.6 million hectares) of forest land, and guarantee to protect the land from logging and other threats to the environment.

1991   Cholera is reported in La Paz region

Villagers living near the capital city of La Paz contract cholera. The government attempts to prevent the spread of the disease by banning certain food imports from Peru.

1991: July   Amnesty program for drug traffickers

Bolivia's government institutes an amnesty program by which the country's drug lords would be guaranteed protection against extradition to the United States in exchange for turning themselves in.

1992: January   Agreement with Peru grants sea access

Bolivia and Peru sign a pact that gives Bolivia access to the Pacific Ocean over an area stretching from the port of Ilo (in Peru) to the town of Desaguadero on the Peru-Bolivia border until the year 2091.

1993   Sanchez de Lozada is elected president

Gonzalo Sanchez de Lozada, a former planning minister, is elected president. He declares he will continue the free-market policies of his predecessor and work for the rights of the Indian population.

1994   Former government officials testify about drug deals

Officials of the Paz Zamora government are called to testify about their involvement in drug trafficking.

1994   Powerful underground earthquake hits Bolivia

Bolivia has one of the strongest earthquakes ever recorded measuring 8.3 on the Richter scale, but it occurs 400 miles underground, so injuries and property damage are minimal. However, the quake can be felt as far away as Canada.

1994: April   Former dictator Meza tried by Supreme Court

General Luis García Meza, whose dictatorial regime committed human rights abuses in 1980 and 1981, is tried by Bolivia's Supreme Court and sentenced to thirty years in prison. However, Meza is in exile in Brazil at the time and not expected to return to Bolivia. Other officials of his government have also received sentences, which they are serving.

1996: December 17   MERCOSUR nations vote to admit Bolivia

The member nations of the Mercado Común del Sur (MERCOSUR), or a Southern Common Market formed by Uruguay, Paraguay, and Brazil, vote to accept Bolivia as an associate member.

1997: June 1   Banzer comes in first in presidential election

Former dictator Hugo Banzer Suárez receives the highest number of votes in the Bolivian presidential election. However, the 22 percent showing is not sufficient to win him the election, and the contest is thrown into Congress, which in August gives the presidency to the seventy-one-year-old leader, who has put together a diverse coalition of political supporters.

1997: July 12   Government confirms that remains are Guevara's

Bolivian officials confirm the discovery of the remains of communist revolutionary Ernesto "Che" Guevara (1928–67), (see 1966–1967). His skeleton is discovered in a mass grave close to an airstrip in Vallegrande, together with the remains of six other people. The discovery of the skeleton attracts attention because the hands are missing, and Guevara's hands were severed when he was killed to prove that he was dead. The remains are returned to Cuba for burial and the funeral is attended by Fidel Castro, who fought together with Guevara in the Cuban revolution.

1997: December 9   Workers strike to protest rising fuel prices

Bolivia's unions declare a one-day general strike and hold demonstrations in opposition to large fuel price hikes announced by the government of President Banzer. Initial results of the price hikes can be seen in the bus companies' decision to raise fares 50 to 100 percent. The Banzer government claims the price hikes are needed to offset budget deficits passed on by the administration of former President Sanchez de Lozada.

1997–98   El Niño causes damages, loss of life in Bolivia

The Andean region of northwestern South America, including parts of Peru, Bolivia, and Ecuador, is hit harder by El Niño than any other area in the Western Hemisphere. El Niño's effects result in 130 deaths in Bolivia, leaving 70,000 people homeless. El Niño-related disasters in Bolivia also include a mudslide that causes the deaths of eighty miners in February 1998.

1998: May 22   Bolivia suffers its strongest surface earthquake of the century

An earthquake measuring 6.8 on the Richter scale strikes central Bolivia, killing at least thirty-three people in the colonial towns of Aiquile and Torora. One hundred people are reported missing in the aftermath of the quake. In the first twelve hours after the tremor, there are some one hundred fifty aftershocks.

Bibliography

Alexander, Robert Jackson. *The Bolivian Presidents: Conversations and Correspondence with Presidents of Bolivia, Peru, Ecuador, Colombia, and Venezuela.* Westport, Conn.: Praeger, 1994.

Blair, David Nelson. *The Land and People of Bolivia.* New York: J.B. Lippincott, 1990.

Gallo, Carmenza. *Taxes and State Power: Political Instability in Bolivia, 1900–1950.* Philadelphia: Temple University Press, 1991.

Hudson, Rex A. and Dennis M. Hanratty. *Bolivia, a Country Study.* 3rd ed. Washington, D.C.: Government Printing Office, 1991.

James, Daniel (ed.). *The Complete Bolivian Diaries of Che Guevara and Other Captured Documents.* New York: Stein & Day, 1968.

Klein, Herbert S. Bolivia: *The Evolution of a Multi-Ethnic Society.* 2nd ed. New York: Oxford University Press, 1992.

Malloy, James M. and Eduardo Gamarra. *Revolution and Reaction: Bolivia, 1964–1985.* New Brunswick, NJ: Transaction Books, 1988.

Morales, Waltrand Q. *Bolivia: Land of Struggle.* Boulder, Colo: Westview Press, 1992.

Lindert, P. van. *Bolivia: A Guide to the People, Politics and Culture.* New York: Monthly Review Press, 1994.

Parker, Edward. *Ecuador, Peru, Bolivia. Country fact files.* Austin, Texas: Raintree Steck-Vaughn, 1998.

Sanabria, Harry. *The Coca Boom and Rural Social Change in Bolivia.* Ann Arbor: University of Michigan Press, 1993.

Spitzer, Leo. *Hotel Bolivia: Culture and Memory in a Refuge from Nazism.* New York: Hill and Wang, 1998.

Yeager, Gertrude Matyoka. *Bolivia.* Santa Barbara, Calif.: Clio Press, 1988.

Brazil

Introduction

Brazil is the world's fifth largest country (after Russia, China, Canada, and the United States). Its population, currently estimated at approximately 165 million, is the largest of any country in South America—in fact, it is home to half the inhabitants of that continent. It also occupies nearly half the continent and shares borders with every other South American nation except Chile and Ecuador. Brazil is also the only Portuguese-speaking nation in Latin America. Brazil's vast territory encompasses great geographical diversity, ranging from mountains to tropical rain forests, and from grassy plains to ocean beaches. The country is commonly divided into the following five regions: the Amazon region in the north (home to only seven percent of the population); the northeast, a poor region with high unemployment and little industry; the west central region, a mostly low-lying area that is home to the modern capital city of Brasília; the southeast, where the major cities of Rio de Janeiro and São Paulo are located; and the cattle country of the south.

Portuguese Explorer Arrives in Brazil

In 1500 Pedro Álvares Cabral (1468–1520) arrived on the coast of Brazil and claimed the land for Portugal. At the time, Brazil was home to a native population of approximately four million, some living along the coast and some in the interior. With the advent of European settlement, the Indian population was greatly reduced through warfare, disease, and enslavement. The native population of Brazil today is estimated at about 200,000.

Portugal initially divided Brazil among private citizens, called *donatorios,* each charged with developing the land allotted to him. However, this arrangement did not last long. By 1549 Portugal assumed direct control of the territory and appointed a captain-general to rule over it. Brazil was to remain under Portuguese rule for three centuries, except for a period of joint Spanish-Portuguese rule (1580–1640), when the two European nations were united under a single monarchy. At various times both French and Dutch settlers attempted to carve out their own colonies, but both groups were eventually driven out.

The first important economic milestone in Brazil's history was the discovery of gold in 1693, which triggered a gold rush and brought new settlers to the region. By the 1700s, coffee was introduced and became a central factor in the nation's economy. By the middle of the nineteenth century, Brazil had become the world's largest coffee exporter.

Like the other nations of South America, Brazil won its independence at the beginning of the nineteenth century. However, it followed a path very different from that of its Spanish-speaking neighbors. When Napoleon Bonaparte invaded Portugal in 1807, the Portuguese prince regent, John VI, fled to Brazil with 15,000 of his subjects and stayed there for fourteen years. During this time, the royal court greatly enriched its New World territory, both culturally and economically. By 1815 it had recognized Brazil as an equal partner in a royal union. The Brazilians, however, eventually demanded complete independence from Portugal. When John VI returned to Europe in 1821, his son, Pedro, whom he had left behind to govern Brazil, declared the former colony independent and named himself emperor. Portugal eventually accorded the new country formal recognition.

Brazil remained an empire, governed by the same royal family, for nearly seventy years. In 1831 the crown passed to Pedro's son, Pedro II. The new prince was still a child, so the government took the form of a regency for the next nine years, until the sovereign was ready to ascend the throne. Under the empire, Brazil prospered, becoming the center of a profitable rubber boom as well as the world's foremost coffee producer. However, in 1889 Emperor Pedro II was overthrown in a military coup and a republic was established. For several years, it was threatened by revolts in various parts of the country. A strong military president, Floriano Peixoto (1842–95), kept these revolts under control and consolidated the new republic. By the mid-1890s, the nation's first civilian president, José de Morais (1841–1902), had been elected.

Vargas Comes to Power

In the early twentieth century, Brazil's cities grew, the nation solved boundary disputes with its neighbors, and malaria and yellow fever were eliminated from the capital city of Rio de Janeiro. During World War I, Brazil declared war on Germany and sent troops and ships to help the United States win the war. In the 1920s opposition to the political power of the rural landowners grew among left-wing movements and parts of the military. In 1930 backers of unsuccessful presidential candidate Getúlio Vargas (1883–1954), a provincial governor, overthrew the elected government and placed Vargas in power, beginning a nearly unbroken twenty-four-year period of dictatorship by one man. Vargas ruled, either as self-appointed president or through a "puppet," from 1930 to 1945. He strengthened Brazil's industries and expanded the role of government in the nation's economy. After 1937, Vargas's rule became even more absolute; he canceled scheduled elections and established the *Estado Nôvo* (The New State). In 1942, Brazil entered World War II on the Allied side and sent troops to fight in Italy.

With expanded influence and power after its participation in the war, the military succeeded in ousting Vargas from power in 1945. Democratic government was restored, a new constitution was approved, and free presidential elections were held. However, the economic policies of the new president, Eurico Dutra (1885–1974), were not successful, and Vargas was returned to office in democratic elections in 1950. During this period, Brazil was torn by political unrest and economic problems. When the military demanded Vargas's resignation in 1954, the longtime leader committed suicide.

Paving the Way to Democracy

Between 1956 and 1964, the nation had three democratically elected presidents. In 1960 the capital was moved from Rio de Janeiro to Brasília, a brand new city that had been built 600 miles northwest of Rio under the leadership of President Juscelino Kubitschek (1902–76) to shift more of the nation's population to the country's interior. Following increasing unrest between right- and left-wing political factions in the early 1960s, the government of President João Goulart (1918–76) was overthrown in a coup in March 1964, and the nation entered a twenty-year period of military rule. During the first part of this period, government policies that kept wages low and encouraged foreign investment led to dramatic economic growth averaging ten percent annually, in what became known as the "Brazilian miracle." However, Brazil's economy began to falter after 1974, partly due to rising oil prices, and the public became increasingly less willing to give up the political and civil freedoms that had been suspended by the government. By the late 1970s, the government of João Baptista de Figueiredo (b. 1918) had instituted a liberalization process known as the *abertura,* or opening, which paved the way for democratic elections in 1985. Tancredo Neves (1910–85) was elected, although he died shortly afterward and his vice president, José Sarney (b.1930), became president.

In 1990 Fernando Collor de Mello (b.1950) became the first Brazilian in twenty-one years to be elected president by direct popular vote. However, he was impeached two years later on charges of corruption. In 1994 Fernando Henrique Cardoso, a former finance minister, was elected to the presidency. At the time, the Brazilian constitution barred him from running for a second consecutive term, but in 1997 the congress approved a constitutional amendment that made it possible for Cardoso to seek a second term in 1998.

In May 1997 a controlling stake in Brazil's state-owned mining company, Companhia Vale do Rio Doce, was sold to a group led by steel manufacturer Companhia Siderurgica Nacional in the largest privatization in Latin American history.

Timeline

9000–8000 B.C.   Foraging people in Amazon River forest region

Archaeologists believe Paleoindians (early native people) live in the southern Amazon River region and forage for food. Evidence of these people is discovered in the 1990s (see 1996).

15th century A.D.   Native population inhabits present-day Brazil

At the time of the European conquest of what today is Brazil, the land is inhabited by a number of different native groups living along the coast and in the interior. Some are hunter-gatherers, while others farm. A small number are cannibals. Together they belong to four different language groups. Two of these, the Carib and the Arual, are found throughout the Caribbean region and Central America. It is thought that these native inhabitants number about four million in 1500 (their population has been halved by 1808).

European Exploration and Colonization

1500: April 22   Cabral explores Brazilian coast

The Portuguese explorer Pedro Alvares Cabral (c. 1460–1526) lands on the coast of Brazil and claims the land for Portugal. This claim is in accord with the Treaty of Tordesillas, signed in 1494, that has divided the Western Hemisphere between Spain and Portugal along a line demarcated 370 leagues west of the Cape Verde Islands off Africa. All land east of this line (including present-day Brazil) is to go to Portugal; everything to the west is reserved for Spain. Cabral and his men explore Brazil's coastal lands for ten days.

1501–02   Vespucci explores Brazilian coast

The Italian Amerigo Vespucci (1451–1512) sails along the Brazilian coastline on the way to the Río de la Plata. On January 1, 1502 his expedition sails into Guanabara Bay, which is mistaken for a river and named Rio de Janeiro (River of January). This becomes the name of the city that is eventually built in the area. Vespucci gives a number of geographic features the names of Catholic saints (such as the São Francisco

River, named for Saint Francis). "America" is also named after the explorer.

1530s   Brazil divides into captaincies

The Portuguese king, John III (1502–57, r. 1521–57), initially tries to settle Brazil by dividing it into captaincies (districts presided over by captains), which are granted to twelve private citizens and called *donatarios*. Each donatario is allotted a portion of land and given authority to rule over it, including civil and criminal jurisdiction over the inhabitants, the right to make land grants to individual settlers and to impose taxes on them, and the right to pass on their territory by inheritance. Most, however, eventually abandon the scheme, due in large part to resistance by the Indian population and to a shortage of labor on the sugar plantations. Only two lasting settlements come out of the donatario system: Pernambuco and São Paulo. The captaincies are officially dissolved in 1579.

1530   Portuguese found settlement

The Portuguese explorer Martim Afonso de Sousa (c. 1500–64), accompanied by about 400 men, founds the settlement of São Vicente, near the present site of Santos. The first immigrants are *degregados*, criminals set free on the shore and left to fend for themselves. Many survive and intermarry with the local Indians, giving Brazil the beginnings of its mestizo population.

1549   Portuguese king appoints first governor-general

The crown, dissatisfied with the success of individual captaincies, appoints Tomé da Souza governor-general of Brazil and charges him with developing settlements in the region. Souza sets up a capital at Salvador, on the Bay of All Saints. A local government is established, and military defenses put in place. His group is accompanied by an advance party of six Jesuit priests, who will be followed by more later. With the inauguration of direct colonial administration, immigration to Brazil increases.

Brazil remains under the authority of a governor-general or viceroy and the seat of government remains at Salvador until 1763 when it is moved to Rio de Janeiro. The governor-general has extensive authority: he makes appointments, organizes security against Indian attacks, presides over the court, distributes land, and carries out other administrative, military, and commercial duties.

1554   Jesuits form a settlement at São Vicente

Jesuit priests, sent to Brazil to spread Christianity among the Indians, move their center of operations from Bahia to the São Vicente region, where they found an *aldeia* (a missionary village) at Piratininga (later the site of São Paulo). The Jesuits, whose primary aim was the welfare of the Indians, clash frequently with Portuguese settlers, who wanted the Indians for slave labor, and other segments of the clergy, who defend the practice of slavery. Under the Jesuits, the Indians are taught handicrafts and agricultural skills in addition to religion. The Jesuits also establish their own plantations and found the first colleges in Brazil.

1555   French attempt to establish a colony

A settlement, called France Antarctique, is founded by some 600 French Protestants (Huguenots) led by Nicolas Durand de Villegagnon at Guanabara Bay, the future site of Rio de Janeiro, as a first step toward establishing a colony. However, they are unable to attract settlers in sufficient numbers and are driven out by the Portuguese in the mid-1560s.

1558–72   Mem de Sá serves as governor-general

Mem de Sá (c. 1500–72) plays an important role in strengthening Portuguese rule in Brazil. He drives the French Huguenots from their settlement near the Bay of Guanabara, founds Rio de Janeiro, and brings the Indians of the coastal area under control.

1562   Epidemic strikes Indian Population

The Indian population, having no immunity against European diseases such as measles, tuberculosis, and smallpox, is decimated by an outbreak of disease in 1562, followed by another one the following year. Together, these epidemics kill between one-third and one-half of the Indians in the parts of Brazil under Portuguese control. With their native economies shattered, many surviving Indians sell themselves to Portuguese settlers as slaves.

1570s   Transition begins from Indian to African slave labor

With a growing scarcity of Indians, as well as the official ban on enslaving them, the Portuguese in Brazil begin importing large numbers of African slaves, laying the groundwork for Brazil's multiracial society. African labor has been used in Pernambuco and São Vicente as early as 1550. The majority of slaves come from West and West-central Africa, including Ghana, Nigeria, Dahomey, Ivory Coast, the Congo, and Angola. At the same time that the availability of Indian labor is decreasing, the growth of the sugar industry is increasing the need for workers on plantations.

By 1600 roughly seventy percent of all plantation labor is done by black slaves, who number an estimated 13,000 to 15,000. The import of African slaves to Brazil increases until about 1680, reaching about 8,000 per year. By one estimate, over forty percent of all black slaves shipped to the Americas in the seventeenth century go to Brazil. After 1580 the importation of slaves drops, only to rise again with the gold rushes of the eighteenth century. During this period, between 5,000 and 8,000 slaves per year are brought to Bahia alone. Blacks account for about half the population in the northeast, and two-thirds in the sugar-growing regions.

1570 Portugal bans Indian slavery

The Portuguese monarchy issues a decree banning the enslavement of Indians in Brazil. However, many settlers ignore it, and the practice continues. The settlers capture Indians in raids (called *bandeiras*) in which they also hunt for gold and other valuable resources. The volume of Indian slavery decreases, however, as the Indian population is reduced by disease and migration, with many Indians fleeing to the interior to escape the European settlers.

1576 Publication of first book about Brazil

Portuguese historian Pero de Magalhães Gandavo publishes his *História da Província de Santa Cruz, a que vulgarmente chamomos Brasil* (History of the Province of Santa Cruz, Popularly Known as Brazil). With the goal of luring new settlers to the province, he describes its discovery, as well as its fauna and flora, government, and system of land distribution under the Portuguese. He also assures prospective settlers of finding gold and gems in the new land. A previous work, the *Tratado da terra do Brasil*, is not published until 1824, by the Royal Academy of Sciences in Lisbon.

Other literature of the colonial period comprises letters, journals, and logs by Portuguese travelers. Gabriel Soares do Sousa, who migrated to Brazil in 1567 and operated a sugar mill, writes *Tratado descritivo do Brazil em 1587* (Descriptive Treatise on Brazil in 1587).

1580s Sugar becomes Brazil's major export

Sugar surpasses brazilwood as Brazil's dominant export commodity. The first sugar plantations (*engenhos*) are established in the 1530s and 1540s at São Vicente and Pernambuco. By 1600 Brazil has a total of 120 sugar mills. Sugar production and export peak between 1600 and 1640, after which the islands of the Caribbean become significant sources of competition, and world sugar prices begin to decline. In the 1680s, Brazil's sugar industry is nearly wiped out by disease, drought, and economic recession. Although there is a recovery, by the 1690s sugar cedes its economic dominance to gold, newly discovered in Minas Gerais.

1580–1640 Brazil under joint Spanish-Portuguese rule

When Spain and Portugal are temporarily under a single monarchy, Brazil is technically governed by Spain.

1609 High court established at Salvador

A *relação,* or high court, is created at the colonial capital of Salvador. It is dissolved in 1626, and then re-established in 1652. In 1751, a second court is established at Rio de Janeiro. Decisions by these courts can be contested in an appeals court (*casa da suplicação)* in Lisbon, Portugal.

1624–54 Dutch occupy parts of Brazil

The Dutch West India company makes inroads into the Portuguese control of Brazil. They seize Brazilian ships, invade Bahia (1624–25), and gain control of the Pernambuco sugar-growing region in the northeast (1630–54). A Dutch general, John Maurice of Nassau, serves as governor-general during part of this period (1637–44). Portuguese resistance to the Dutch begins with a revolt in 1645, and residents of Rio de Janeiro finally drive the Dutch out of the area in 1654.

1677 First convent for women founded

Salvador is the site of Brazil's first convent. Convents house young unmarried women as well as those who are divorced, separated, or abandoned by their husbands. After a period of initiation, the women take their vows, joining one of two categories of nuns ("white nuns" and "black nuns," who differ in social status). Brazil's convents are allowed to engage in commercial activities, including owning property and loaning money.

1693 Gold discovered in Minas Gerais

The discovery of gold in this area north of Rio de Janeiro triggers the first great gold rush in the Western hemisphere and makes Brazil the world's largest producer of gold. Prospectors and slaves throng the region, and new towns spring up. Art and architecture flourish with the new wealth created by the gold boom, which also helps fund the Industrial Revolution in Europe–by one estimate providing as much as four-fifths of the gold circulating in Europe in the eighteenth century. Exploitation of the mines is so rapid that they are depleted within 100 years, reaching a peak between 1735 and 1745.

1694 Settlement of escaped slaves overrun

Palmares is one of a number of settlements established by escaped African slaves. Home to some 20,000 persons, the enclave withstands numerous raids. but is finally overrun and its inhabitants returned to slavery.

1708–09 War of the Emboabas

Hostilities break out between established settlers in the Minas Gerais region and newcomers arriving from other areas of Brazil and from Europe to prospect for gold. The *emboabas* (outsiders, or "greenhorns") prevail, and settlement remains open to all. Some of the established settlers, or *paulistas*, respond by moving farther into the interior and discovering gold at new sites, as well as diamonds in the Serro do Frio region.

1720s Diamonds discovered in Minas Gerais

Diamonds are found north of the Minas Gerais goldfields, and the town of Diamantina is founded. It achieves worldwide fame as a diamond center. Eventually competition from South Africa slows the demand for Brazilian diamonds, but the diamond industry survives through the twentieth century.

1720s–30s   Gold discovered in Mato Grosso and Goiás

Major gold deposits are found in the Mato Grosso and Goiás. However, the largest amount of gold is still mined in Minas Gerais.

1739–1814   Life of artist Aleijadinho

Aleijadinho (Antônio Francisco Lisboa), a native of Minas Gerais, is a sculptor and architect known for his church designs and soapstone carvings of religious figures. The name *Aleijadinho* (little cripple) refers to the disabling and disfiguring disease–variously described as leprosy, syphilis, or a virus–from which Lisboa suffers starting in 1777. One of Aleijadinho's most renowned designs is for the Church of São Francisco at Ouro Prêto, built between 1774 and 1794. His most famous carvings are in the church of Bom Jesus de Matozinhos at Congonhos do Campo, a pilgrimage site. There, between 1796 and 1799, he sculpts sixty-six life-sized figures as part of a depiction of the stations of the cross, as well as twelve states of the Old Testament prophets carved in soapstone (1800–1805). Aleijadinho executes these last statues with his carving tools strapped to his wrists.

1750s   Brazilian coffee industry is started

Coffee has first been introduced to Brazil at the beginning of the eighteenth century. In 1750, twelve tons of coffee beans are exported from Pará Brazil to Portugal. Coffee becomes a profitable commodity when world prices rise in the latter part of the century.

1759   Jesuits expelled from Brazil

Members of the Jesuit order are compelled to leave Brazil, and the influence of other religious sects is reduced as well. Secular administrators take over the *aldeias* (religious communities).

1763   Capital moves to Rio de Janeiro

Brazil's capital moves from Salvador in the sugar country of the northeast to Rio de Janeiro, closer to the gold boom in the southeast.

1767   Birth of composer Nunes García

José Mauricio Nunes García (1767–1830) becomes the most acclaimed composer of eighteenth-century religious music in Brazil, writing masses and a requiem. Nunes García is appointed chapelmaster of the Rio de Janeiro cathedral in 1798.

Late 18th century   Religious music flourishes in Minas Gerais

The province of Minas Gerais becomes a center of musical activity in the late 1700s. Prominent composers include Ignacio Parreiras Neves (c.1730–c.1793), José Lôbo de Mesquita (c.1740–1805), and Damião Barbosa de Araújo (1778–1856).

1789   Early revolt foreshadows independence

Led by a folk healer named Joaquim José da Silva Xavier (called "Tiradentes," or "tooth puller"), a group of Brazilians takes part in a revolt against Portugal. Their leader is arrested and hanged, and the other members of the group are exiled. Tiradentes becomes a national hero, however, and his rebellion is considered a preliminary stage in winning independence from Portugal.

1790   Vellozo completes major botanical study

Botanist José Mariano de Conceição Vellozo completes *Flora fluminensis*, a study of the plant life in the Rio de Janeiro area. It is not published until 1825.

Events Leading up to Independence

1807–08   Napoleon invades Portugal, king flees to Brazil

When Napoleon Bonaparte invades Portugal, the prince regent, John VI (c. 1767–1826; r. 1816–26), together with some 15,000 of his subjects, flees to Brazil for safety, establishing the seat of the Portuguese government in Rio de Janeiro in March 1808. He opens Brazil's ports to foreign shipping and establishes hospitals, educational facilities, a royal library, printing presses, and a botanical garden. His court remains in Brazil for fourteen years.

1810   National Library founded

Brazil's National Library is established in Rio de Janeiro. Housing some 60,000 volumes brought to Brazil by John VI in 1808, its total collection today numbers over eight million items, including a number of rare manuscripts.

1815   John VI declares Brazil on equal terms with Portugal

Even after the downfall of Napoleon, John VI remains in Brazil. In 1815 he gives the former colony equal status with Portugal as part of the United Kingdom of Portugal, Brazil, and the Algarves, and declares Rio de Janeiro the capital of the new union.

1816   Artists arrive in Brazil to further art education

Spurred by the presence of the Portuguese court of John VI, a group of European artists and architects, led by Joachim Lebreton, arrives in Brazil with the mission of educating young artists, and they soon founded an academy.

1818   National Museum founded

The National Museum becomes an institution renowned throughout South America. It is especially famous for its collection of Brazilian ethnography (Branch of anthropology dealing with mostly nonliterate cultures).

1821   John VI returns to Portugal

When liberals in Portugal threaten a revolution to unseat the monarchy, John VI is obliged to return home to Portugal, leaving his son Pedro in charge of the government of Brazil.

The New Empire

1822: September 7   Brazil declares independence

Pedro (1798–1834; r. 1822–31), the Portuguese prince regent, declares the independence of Brazil as a new empire on September 7, 1822, naming himself Emperor Pedro I. After three years of negotiations, Brazil is recognized by Portugal. Pedro drafts a constitution in 1824, giving the emperor extensive political powers, including naming a council of state and the power to dismiss the legislature and cabinet.

1823   Birth of poet Antônio Gonçalves Dias

Dias, regarded as the national poet of Brazil, is born in Caxias, Maranhão. He studies at the University of Coimbra in Portugal and begins a successful literary career in Europe. In 1846, Dias returns to Brazil, where he publishes *Primeiros cantos* (First Songs; 1847), *Leonor de Mendonça* (1847), and other works that combine the ideals of Romanticism with Brazilian nationalism, especially a love of the Brazilian landscape and appreciation of the Indians and their culture. His poem "Canção do Exílio" (Song of Exile) is one of the best-loved poems in Brazilian literature. Besides poetry, Dias also writes history, ethnology, and dramas, teaches, and holds government posts in Brazil and Europe. Dias dies on November 3, 1864 at the age of 41.

1825   Newspaper established in Pernambuco

The *Diario de Pernambuco* claims to be the oldest Latin American newspaper still in existence today.

1825–28   War for control of Uruguay

Pedro I embroils Brazil in an unsuccessful three-year campaign against Argentina for control of the Banda Oriental (which becomes the independent state of Uruguay in 1828). The Brazilian forces suffer their major defeat at the Battle of Ituzaingo on February 20, 1827.

1830s   Regional revolts threaten stability of Brazilian government

Government control remains weak in regions remote from the capital city of Rio de Janeiro. Two major regional revolts begin in the 1830s. Tariff policies and border disputes lead to the Farroupilha revolt (1835–45) in the southern province of Rio Grande do Sul. In the northern Pará province, enslaved blacks and Indians stage the bloody Cabanagem revolt (1834–40).

1831   Law schools founded in Recife and São Paulo

The two law schools established in 1831 have a major effect on Brazil's government by providing a literate class to run its administrative bureaucracy.

1831: April   Pedro I abdicates and is succeeded by Pedro II

As the 1820s progress, Pedro I becomes increasingly unpopular among the Brazilian people. Facing a mutinous military and rioting subjects, Pedro resigns, leaves for Portugal and turns the monarchy over to his son, Pedro Alcântara Brasileiro (Pedro II, 1825–91), Pedro II, who is still a child, reigns under a regency for ten years and is crowned in 1841.

The regency liberalizes the Brazilian government, in reaction to the authoritarian reign of Pedro I. The council of state is abolished, much of the regular military is replaced with a national guard, and political parties are allowed.

1834   First woman graduates from Brazilian medical school

Marie Durocher (1809–93) becomes the first woman in Brazil to receive a medical degree. A graduate of the Medical School of Rio de Janeiro, Durocher is one of the first women doctors in Latin America. She pursues a sixty-year medical career in Brazil.

1838   Historical and Geographical Institute of Brazil is founded

The founding of the Historical and Geographical Institute is a milestone in establishing a sense of Brazilian national identity distinct from that of Portugal.

1839   Novelist J. Machado de Assis born

Novelist and poet Joaquim Maria Machado de Assis (1839–1908) is widely regarded as Brazil's greatest writer. In addition, he is the founder and first president of the Brazilian Academy of Letters. Machado de Assis is born into poverty, to a black father and a Portuguese mother in the slums of Rio de Janeiro. He is afflicted with a stammer and epilepsy and receives little formal education. However, he embarks on a literary career at a young age, publishing his first book of poems by the age of twenty-five. Over the years, he works as a typesetter, journalist, editor, and civil servant.

The Marchal Floriano Park in Sao Paulo as it appeared in the early 1900s. (EPD Photos/CSU Archives)

In 1878, already the author of a substantial body of poetry and fiction, Machado de Assis takes a leave from his job and spends it at a resort near Rio de Janeiro. This hiatus inaugurates a new period in his literary career, and he writes his greatest works from this point on. These include the novels *Memórias Póstumas de Braz Cubas* (The Posthumous Memoirs of Braz Cuba; 1881); *Dom Casmurro* (1890); and *Quincas Borba* (1891) as well as a number of short stories. Joaquim Machado de Assis is celebrated for his insight into the human condition and his portrayal of nineteenth-century Brazilian society in Rio de Janeiro. His work is internationally recognized and has been translated into several foreign languages.

1840s–1910 Rubber boom in the Amazon

Charles Goodyear's (1800–60) 1839 invention of the vulcanization process makes it possible to transform the latex from rubber trees into resilient, heat- and cold-resistant rubber products. The growing world demand for rubber generated by this development creates an economic boom in Brazil's Amazon jungle region. Some 1,500 tons (1,364 metric tons) of rubber are exported in 1850, and the volume of exports grows to 30,000 tons (27,273 metric tons) by the beginning of the twentieth century. At the height of the boom, Brazil accounts for half of all the rubber produced in the world.

The population in Brazil's Amazon region is multiplied six times during this period. The wealth from the rubber boom, however, is largely enjoyed by a small group of entrepreneurs who enjoy a lavish lifestyle, while the Indians, whose labor they rely on to tap the rubber trees, live in conditions that are little better than slavery. The rubber boom ends when competition from Asia drives world prices down.

1850s Coffee is leading export

By the middle of the nineteenth century, coffee is the leading export commodity, surpassing both cotton and sugar. Coffee represents over forty percent of all exports and remains the leading export until the latter half of the twentieth century (see 1980).

1852 Feminist periodical founded in Rio de Janeiro

The feminist journal *O Jornal das Senhoras* is founded by Joana Paula Manso de Noronha (1819–75). It is published until 1855. Other nineteenth-century women's periodicals in Brazil include *O Domingo,* founded in 1874, and *A Familia,* founded in 1888.

1860–90   Immigration to Brazil grows

To meet its need for agriculture labor, Brazil has tried throughout the nineteenth century to attract immigrants from abroad. However, immigration does not increase dramatically until the 1860s and 1870s. Several thousand immigrants from the Confederate South in the United States arrive in Brazil during this period. Immigration from Portugal, Spain, Italy, Germany, and other European nations increases as well. Most immigrants are drawn to one of the country's two major coffee-growing centers, Rio de Janeiro and São Paulo.

1865–70   War of the Triple Alliance

In return for Brazilian military aid, Uruguayan president Venancio Flores (1809–68) is obligated to join Argentina and Brazil in their war against Paraguay's dictator Francisco Solano López (1790–1862). The Triple Alliance is victorious, and Paraguay is defeated.

1871   Government enacts free birth law

During the nineteenth century Brazil takes gradual steps to abolish slavery, moving slowly because of the economic importance of slaves in the growing of coffee. The Free Birth, or Rio Branco, law guarantees freedom for children born to slave mothers. In some cases, however, these children do not gain full freedom until they reach twenty-one years of age. The law gives their mothers' masters the option of using their labor until that point. The Rio Branco Law also gives slaves the right to save money and purchase their own freedom after a period of not longer than seven years. The Sexagenarian Law, which takes effect in 1885, frees slaves over the age of sixty.

1887   Leading composer Heitor Villa-Lobos born

Heitor Villa-Lobos (1887–1959), the most famous Latin American composer of the twentieth century, is born in Rio de Janeiro. He receives his early musical training from his father, a library administrator and amateur musician. In his late teens and early twenties, he travels extensively in Brazil, becoming familiar with folk music and customs throughout the country. Villa-Lobos returns to Rio in 1911 and begins writing compositions, which he premieres in 1915. At first he is criticized for his lack of formal training and his unorthodox compositional methods.

With the support of musician friends, including composer Darius Milhaud (1892–1974) and pianist Artur Rubinstein (1888–1982), Villa-Lobos makes two trips to Europe in the 1920s. After his works are performed in Paris in October and December of 1927, his career is launched. Conducting engagements and performances of his works follow in Vienna, London, Berlin, and other cities in Europe. In the 1930s Villa-Lobos returns to Brazil, where he directs a nationwide music education program for the Brazilian government. The composer makes his first visit to the United States in 1944.

Villa-Lobos is regarded as a key figure in the development of a national style in Brazilian classical music. His compositions are best known for their incorporation of native Brazilian musical traditions—Portuguese, African, and Indian. Perhaps his best-known works are the *Bachianas Brasileiras,* pieces that merge the essence of Johann Sebastian Bach's music with the Brazilian spirit.

1888   Pedro II's daughter abolishes slavery

Acting in her father's stead while he is abroad seeking medical treatment, Pedro II's daughter, Isabel (1846–1921), decrees the abolition of slavery, making Brazil the last nation in the West to abolish slavery altogether within its borders. Over half a million slaves are freed, with no compensation to their owners. The abolition of slavery in Brazil is a major factor in the overthrow of the monarchy and the establishment of the Brazilian republic.

The Republic of Brazil

1889: November 15   Brazil establishes republic

A military coup with support from wealthy planters and merchants, factory owners, opposition politicians, and other civilians, ousts Brazil's governing dynasty, the House of Braganza, and establishes a republic. Political stability is achieved over the following decades through an informal power-sharing agreement between the coffee-growing elite of São Paulo and the cattle barons of Minas Gerais. This partnership becomes known as *café com leite* (coffee with milk).

1889–91   Fonseca serves as president

To appease the military, the Brazilian congress appoints Manuel Deodoro da Fonseca (1827–92) the nation's first president. However, Fonseca dissolves congress and seizes dictatorial power. The military then stages a coup against him, and he is replaced by his vice-president, Floriano Peixoto.

1891   Constitution drafted

A new constitution is written by a constituent assembly. It provides for separation of the executive, judicial, and legislative branches of government and direct elections. The vote is restricted to literate men over twenty-one, excluding clergy and some military personnel. Separation of church and state is mandated, with a ban on religious instruction in the public schools and on the formation of new monastic orders. Basic freedoms of thought, assembly, and property are guaranteed. This constitution remains in effect until 1930.

1891–94 Peixoto assumes presidency

Brazil's second president, Floriano Peixoto (1842–95), also rules as a dictator, ignoring constitutional provisions for new elections. However, he strengthens the central government against regional revolts and threatened restoration of the monarchy.

1893–94 Anti-government revolts unsuccessful

The first five years of the republic are marked by civil unrest. In 1892 there is an uprising to protest President Peixoto's refusal to hold elections mandated by the constitution. The following year, there are revolts by the navy, and by forces in the southern area of Grande do Sul, who want more local autonomy. There are also a number of incidents in various cities, involving run-ins between civilians and the military or the police. By 1894 the central government has proven unified and capable of resisting all challenges to its authority.

1894–98 Morais presidency

Prudente José de Morais (1841–1902) presides over a relatively democratic, civilian government that has the support of Brazil's wealthy coffee growers. During Morais's presidency, the civil unrest in Rio Grande do Sul is finally quelled, although a new problem arises with the founding of the religious community of Canudos in Bahia.

1895 Soccer comes to Brazil

Charles Miller, a young man born in Brazil to British parents, learns how to play soccer during his education in England and brings it back to Brazil, teaching the rules to his friends in São Paulo. A soccer league, the Liga Paulista de Futbol, is formed in the city by 1901. Soon European-educated young men in Rio de Janeiro are setting up clubs of their own, which include Fluminense (1902), America and Botafogo (1904), and Flamengo (1911).

1896–97 Confrontation with Canudos

The most serious challenge to president José de Morais is the confrontation between the government and the religious community of Canudos, which has been founded in Bahia in 1893 under the leadership of a mystic named Antônio Vicente Mendes Maciel, popularly known as "The Counselor" (Conselheiro) among his followers. The 9,000 inhabitants of Canudos refuse to obey government laws or pay taxes. After repeated attacks, the Brazilian military overruns the community following a six-month siege, killing most of its members.

1896–1900 Brazil faces economic crisis

A decline in world coffee prices, beginning in 1896, weakens Brazil's economy as the new century approaches. Investment in industry falls, and in 1900 there is a bank panic. President Manuel Ferraz de Campos Salles (1846–1913, president 1898–1902) helps rescue the country from economic collapse, negotiating a loan that helps stabilize the nation's currency.

1898 Brazilian builds and flies early airplane

Pioneering aviator Alberto Santos-Dumont (1873–1932) constructs a gasoline-powered airplane and flies it to France. He builds several more prize-winning airships, including an aeronautically advanced monoplane in 1905.

1899–1976 Life of noted Brazilian feminist Bertha Lutz

Lutz, an activist for the rights of women and children, receives a biology degree from the Sorbonne (in Paris, France) and works for the National Museum in Rio de Janeiro in the 1920s, becoming the first woman to work for the Brazilian government. She attends the 1923 Pan-American Association for the Advancement of Women as a delegate and is also active in the campaign for female suffrage in Brazil. Lutz is elected to the Brazilian parliament in 1936.

1900 Brazilian border dispute with French Guiana settled

France, which claims the northern Brazilian territory of Amapá as part of French Guiana, agrees to submit the matter to international arbitration. In 1900 the president of the Swiss Confederation decides the issue in Brazil's favor, setting the Rio Oiapoque as the boundary line between the two sides.

1900 Birth of cultural historian Gilberto Freyre

Freyre's 1933 essay *Casa-grande & senzala* (The Masters and the Slaves), which applies theories of anthropologist Franz Boas to Brazil's multiracial society, is a seminal influence on thinking about culture and race in Brazil and abroad. Its emphasis on racial harmony is reflected in Brazil's political policy and popular culture. Other central themes in Freyre's work include the development of the patriarchal family, the history of slavery and plantation life, and the significance of folk traditions. Freyre's later books include *Sobrados e mucambos* (The Mansions and the Shanties; 1936) and *Ordem e progresso* (Order and Progress; 1959). Freyre, who lectured at universities throughout the world, was awarded an honorary knighthood by Great Britain in 1971. He died in 1987.

1902 Publication of *Os Sertões*

Author Euclides da Cunha (1866–1909) publishes his classic work *Os Sertões* (Rebellion in the Backlands), which

Territorial Growth
after 1777

☐ Brazil in 1777

⬚ Land gained,
c. 1777–1907.

describes the 1896–97 revolt of the Canudos religious community in Bahia and its violent suppression by the government of José de Morais. Cunha, a journalist, has been sent by his newspaper, *A Provincia de São Paulo,* to cover the warfare between the Canudos movement and the government, spending a month on the battlefield. The book that results is an acclaimed portrait of the frontier lands of the Brazilian interior, combining sociology, anthropology, philosophy, and

other disciplines into a literary work of art. *Os Sertões* is a bestseller upon publication. Internationally, it is probably the best-known Brazilian literary work.

1903   Treaty of Petrópolis settles border dispute with Bolivia

Under the leadership of the Baron of Rio Branco, a treaty is signed that settles the border dispute between Brazil and

Bolivia over Acre, the westernmost Brazilian state. Rio Branco, who serves as Brazil's foreign minister from 1902 to 1912 is regarded as one of the nation's great diplomats and statesmen. He also concludes agreements with Venezuela, Dutch Guiana (present-day Suriname), Venezuela, and Peru.

1903   Artist Cândido Portinari born

Cândido Torquato Portinari (1903–62), considered by many to be Brazil's greatest artist, is born in São Paulo, one of twelve children in an Italian immigrant family. He studies at the National School of Fine Arts in Rio de Janeiro and later in Europe, with help from the National Salon. Among his influences are Picasso, Chagall, and Miró. Portinari's social realist style evolves in the 1930s, when he depicts the lives of miners and coffee plantation workers, both in murals and on canvas.

His first international recognition comes when he wins a prize for a painting exhibited at the Carnegie International Exhibition in Pittsburgh. In 1941 Portinari is commissioned by the U.S. Library of Congress to paint a group of four murals on the theme *The Discovery of the New World*. The individual murals, which are permanently hung in the library, are *Discovery of the Land*, *The Entry into the Forest*, *The Teaching of the Indians,* and *The Mining of Gold*. In 1953 Portinari paints the *War and Peace* murals that hang in the UN General Assembly. Portinari receives the Guggenheim National Award in 1957. The artist dies in 1962 of paint poisoning.

Early 20th century   Rio de Janeiro conquers malaria and yellow fever

In a major medical advance that serves as a model for U.S. programs in Panama, malaria and yellow fever are wiped out in the Rio de Janeiro area through a public health program headed by Dr. Osvaldo Cruz (1872–1917). Yellow fever, which has killed 10,000 people a year in the latter half of the nineteenth century, is eliminated by 1906.

1906   Taubaté Convention sets "valorization" policy for coffee growers

Brazil's coffee-producing states of Rio de Janeiro, São Paulo, and Minas Gerais agree to stockpile coffee so that the market price won't drop below a certain point. This policy is carried out by the federal government, which buys the coffee from the growers and stores it, releasing it in subsequent years when production is lower and it can be sold at a higher price. This practice becomes known as the "valorization" of coffee and remains government policy until the late 1920s. Between 1900 and 1920 Brazil supplies three-quarters of the world's coffee.

1909   Birth of actress Carmen Miranda

Probably the best-known native Brazilian to become part of the North American entertainment industry, Miranda (1909–55) becomes a singer and dancer in Hollywood musicals including *Copacabana* (1947) and *Nancy Goes to Rio* (1950). She is known for her elaborate headgear, often decorated with tropical fruits. Many of her costumes are on display at the Carmen Miranda Museum in Rio de Janeiro.

1910   Indian Protection Service formed

Explorer and humanist Candido Rondon creates the Indian Protection Service in an attempt to preserve the native cultures of Brazil's Indian groups.

1911   *Contestado revolt begins*

A rural uprising similar to the Canudos movement of the 1890s takes place in the states of Paraná and Santa Catarina in the south. Like the earlier revolt, it is led by a religious idealist, José Maria, who is revered as a saint after his death. Government troops suppress the movement by 1915.

1912   Birth of modern author Jorge Amado

Amado (b. 1912) is Brazil's most famous living writer. Born in Bahia and raised on a cacao (type of tree from whose seeds cocoa and chocolate are made) plantation, he is educated at a Jesuit college and settles in the city of Salvador, publishing his first novel at the age of twety. Most of his novels are set in his native Bahia. The earliest ones, including *Cacau* (1933), portray the harsh lives of workers on cacao plantations. Until the 1950s, his novels continue to focus on social issues, paralleling his political activism, for which he is imprisoned and intermittently exiled from Brazil. These works include *Suor* (Slums; 1934), *Jubiabá* (1934), *Capitães de areia* (1937), *Terras do sem fim* (The Violent Land; 1942), and *São Jorge dos Ilhéus* (1944). Many are banned in both Brazil and Portugal.

Amado's works since the 1950s are less political and more celebratory of the color, humor, and sensuality of life in Bahia. These later works include *Gabriela, cravo e canela* (Gabriela, Clove and Cinnamon; 1962), *Dona Flor e seus dois maridos* (Dona Flor and Her Two Husbands; 1966), *Teresa Batista, cansada de guerra* (Tereza Batista, Home from the Wars*; 1972), and *Tieto do Agreste, pastora de cabras* (Tieta the Goat Girl; 1977). In 1946 Amado is elected to Brazil's Constituent Assembly. Amado is admitted to the Brazilian Academy of Letters in 1961. His works, which include over twenty novels, have been translated into more than two dozen foreign languages and several have been made into popular movies.

1914: September 20   First international soccer match

The same year that Brazil's national soccer organization is set up, the country participates in its first international match against Argentina.

1917: October   Brazil declares war on Germany

Brazil enters World War I after having several of its ships sunk by Germany. Some Brazilian doctors and pilots join the

Allied forces in the fighting. Brazil is also a participant in the peace conference at Versailles.

1920s–30s Rural bandit Lampião active in the interior

The legendary desperado Lampião, regarded as a sort of Robin Hood by the poor, carries out armed exploits in the *sertão* (backlands) of the northeast.

1920 Birth of artist Lygia Clark

Lygia Clark (1920–88), a founder of the Neo-Concrete group in Rio de Janeiro in the 1960s and 1970s, is born. Clark and fellow artist Helio Oiticica (1937–80) permanently altered the panorama of art in Brazil. They were founding members of the Neo-Concrete group in Rio, and during the 1960s and 1970s they broke away from traditional art forms and styles. Making the viewer an integral part of the work and criticizing what they saw as an overly rational conception of abstract structure, they created works which paradoxically transformed casual materials into agents of sensual involvement. Even after the Neo-Concrete movement dissolved, their works continued to develop in complementary ways and generated a synergy that has been extremely influential to subsequent generations of Brazilian artists.

1921 Birth of author and educator Paulo Freire

After studies in philosophy and law, Freire teaches adult literacy in Pernambuco between 1947 and 1959. In 1959 he is appointed to the faculty of the University of Recife to teach the history and philosophy of education. In 1963 President Goulart (1918–76, president 1961–64) appoints Freire to head a national literacy campaign. After the April 1965 military coup, Freire is arrested and exiled from Brazil. He settles in Chile, working as a consultant to UNESCO for agricultural training and leading a literacy campaign that wins a UNESCO award.

In the early 1970s Freire publishes his first two books, the widely read *Pedagogy of the Oppressed* (1970) and *Education for Critical Consciousness* (1973). After taking part in literacy campaigns in a number of developing countries, Freire is permitted to return to Brazil in June 1980 as part of President Figueiredo's general amnesty for political exiles. Freire's educational philosophy has been influential throughout the world, especially through the publication and translation of *Pedagogy of the Oppressed.*

1922: February "Modern Art Week" held in São Paulo

A group of São Paulo's artists, writers, academics, and other leading intellectuals organizes the *Semana de Arte Moderna* (Modern Art Week) to showcase Brazilian culture and (on the centennial of Brazilian independence) declare cultural independence from traditional European values and standards. The event, which takes place on February 13, 15, and 17 at the Teatro Municipal in São Paulo, includes lectures, paint-ing, sculpture, and architecture exhibits, readings, concerts, and dances. The focus is on modernism and art forms with roots in the popular culture of Brazil. Much of the general public reacts negatively to the avant-garde cultural event.

1922 Brazilian Association for the Advancement of Women is Founded

The Brazilian Association for the Advancement of Women is established to promote child welfare, education for women, and female suffrage. The first president and leading organizer of the group is Bertha Lutz.

1922 *Tenente* movement spurs military revolts

A democratic reform movement by junior army officers results in revolts, beginning with one on July 5, 1922 at Igrejinah Fort in Copacabana Beach. It is quickly put down, and sixteen officers are killed. There are further disturbances in 1924–25, beginning on the second anniversary of the initial incident, and again in 1926. The *tenente* movement, which is to play an important role in Brazil's political future, is allied with the emerging urban middle classes against the traditional power of the wealthy landowners.

1922–26 Economic boom ends

By the end of the nineteenth century, Brazil's economy had grown due to the rapid cultivation of coffee and an increased labor market from many immigrants (most notably Italian, German, Levantine, Polish and Portuguese immigrants). Yet by 1930 Brazil's boom economy collapses, creating crisis conditions. Coffee prices fall following World War I, and the government's "valorization" policy encourages overproduction on the part of coffee growers, who get paid to produce coffee that is then stockpiled. Government borrowing to pay for the valorization program drives up the nation's foreign debt. By 1930, servicing this debt will require one-third of the national budget. In an effort to deal with the looming economic crisis, President Arthur da Silva Bernardes (1875–1955) introduces an income tax and imposes economic austerity measures.

1925 Birth of author Clarice Lispector

Lispector (1925–77), a novelist and author of short stories, is one of the foremost literary figures of post-World War II Brazil. Her works are known for their lyrical, poetic style and use of interior monologue. Although highly personal, they also reflect an emphatic social conscience. Her first novel, *Perto do coraçao selvagem* (Near to the Savage Heart; 1944) has been praised for its sensitive portrait of adolescence. Other works include: *A maça no escuro* (The Apple in the Dark; 1957), *A paixão segundo G. H.* (The Passion According to G. H.; 1964), and *Agua viva* (Living Water; 1973). Although Lispector is credited with moving Brazilian fiction away from the regionalism that had been dominant since the 1930s, her last novel, *A hora da estrela* (The Hour of the Star; 1977) is concerned with life in the northeast of Brazil.

1929   Coffee prices plunge during the Depression

The world coffee market collapses with the Depression. The government is forced to end its twenty-year-old stockpiling policy when the loans needed to sustain it become unavailable.

1930: March 11   Prestes wins brief presidency

In the 1930 election, opposition to the political establishment centers around Rio Grande do Sul governor Getúlio Vargas. He is supported by the Aliança Liberal, a political coalition with strong support in the states of Minas Gerais, Rio Grande do Sul, and Paraíba. Julio Prestes, the candidate backed by current president Washington Luíz (president 1926–30), wins the election, allegedly tainted by fraud. The opposition responds with an armed revolt before he is inaugurated.

1930   Brazil participates in first World Cup game

The Brazilian national team takes part in the first World Cup soccer competition, held in Uruguay. Brazil will become the only country to qualify for every World Cup championship between 1930 and 1998 and will win four times within those years: 1958, 1959, 1970, and 1994. In addition, the Brazilian team is the runner-up in 1950 and 1998.

The 1930 Revolution

1930: October   Vargas seizes power

The event that triggers this historic revolt is the assassination of Getúlio Vargas's vice-presidential candidate, João Pessoa. Supported by the *tenente* movement of the 1920s, Vargas's backers rise up in his political strongholds and seize control of the country within two weeks. They oust elected president Julio Prestes and install Vargas as provisional president. Vargas (1883–1954) is to dominate Brazilian politics for twenty-four years.

1930–45   Vargas heads government

During these years, Vargas is head of the government, although he is only the elected president for a portion of this time. From 1930 to 1934 he rules as leader of a provisional government, dissolving the legislature. He schedules elections for a constituent assembly on May 3, 1933. A new democratically elected congress meets in November of that year, and a new constitution is drafted the following year. Vargas is elected president from 1934 to 1937, resuming his dictatorship from 1937 to 1945.

Vargas strengthens the federal government and expands its role in the nation's economy. He makes Brazil less dependent on coffee by strengthening its industries and other types of agriculture. His government also forms and taxes labor unions and oversees their activities.

1931   Vargas forms government agency to regulate coffee industry

The Conselho Nacional do Café (later renamed the Departamento Nacional do Café) is established by Vargas to regulate the nation's coffee industry. It begins by destroying the coffee that has been stockpiled previously. From 1931 to 1944, it remains in charge of the government's coffee policy. By forcing a reduction in the amount of coffee grown, the council reduces coffee growers' debt and insures a stable income for growers. Its policies also lower unemployment and help Brazil overcome the effects of the global economic depression. Recovery is evident as early as 1932.

1934: July 16   Vargas drafts new constitution

With the 1934 constitution, Vargas returns Brazil to democratic rule. The new constitution, the nation's third, gives women the vote and lowers the voting age to eighteen. It also reflects Vargas's labor policies, including the creation of regional minimum wages. The constitution establishes free, mandatory, universal education as well.

1935: November   Leftists try to overthrow Vargas

The opposition Aliança Nacional Libertadora, a group uniting communists and liberals, stages armed uprisings in Rio de Janeiro, Recife, and Natal but lacks the popular support to prevail against Vargas and the military.

1937: November   Vargas establishes Estado Nôvo

Vargas cancels the presidential election scheduled for November 1937 and, with the aid of the military, seizes dictatorial control once again, inaugurating an eight-year authoritarian period known as *Estado Nôvo* (The New State). Vargas's support for industrialization and his support of urban over rural interests has a profound effect on his country's history and economic development.

1940: October 23   Birth of soccer star Pelé

Pelé (Edson Arantes do Nascimento) is born on October 23, 1940 in Minas Gerais but grows up primarily in São Paulo, where he learns to play street soccer (*peladas*, the word from which his nickname is derived). He makes his debut as a professional player at the age of fifteen with the Santos Football Club in 1956. He joins the Brazilian national team the following year and helps Brazil win its first World Cup in 1958, scoring six goals. Injuries hamper his participation in the 1962 World Cup games, although Brazil keeps its title. Pelé plays in his last World Cup match in 1970, and is named the most valuable player in the games. After retiring from Brazilian soccer in 1974, he plays with the New York Cosmos in the North American Soccer League, sparking U.S. interest in the game. Becoming the only player in soccer history to score over 1,000 goals (1,300 career total), Pelé becomes a soccer legend and is revered as a national hero. After his soccer career, he goes on to successful activities in the fields of business and entertainment.

1942: August   Brazil enters World War II against Germany

Initially some of Brazil's political factions support the Allied nations, while others support Germany and the Axis powers. By 1942, however, the country enters the war on the U.S. side, partly in response to German attacks on its vessels and partly to gain financial support from the United States, which helps fund the construction of the Volta Redonda mill, the first steel mill in Latin America. A Brazilian force of 25,000 soldiers takes part in the Allied campaign in Italy—the only troops from Latin America to take an active part in the fighting. About 450 men lose their lives. Returning soldiers play an important role in removing the dictator Vargas from power.

1943–58   Geneticist Dobzhansky does research in Brazil

Over a period of fifteen years, Russian-born Theodosius Dobzhansky (1900–75) intermittently pursues studies of the tropical fruit fly, training future Brazilian geneticists in seminars at the University of São Paulo.

1945: October   Military ousts Vargas from power

Wielding newfound power and political influence after its participation on the Allied side in World War II, the military succeeds in removing Vargas from power and restoring democratic government to Brazil. Vargas's former minister of war, General Eurico Dutra, is elected president. A new constitution is approved in September 1946.

1946: March 20   Worst rail disaster in Brazil's history

An overcrowded commuter train derails near Aracaju in the worst train wreck to occur in Brazil. Carrying 1,000 passengers, the train is unable to ascend a steep incline and the cars telescope into each other as they roll to the bottom, injuring hundreds of passengers and killing 185. Enraged relatives turn on the train engineer and attempt to lynch him. Suburban commuter trains in Brazil continue to be chronically overcrowded in the 1950s, resulting in a string of derailments and accidents.

Late 1940s   First art exhibits by the Group of 19

A loosely united group of artists, known as Group 19, jointly exhibit their works in the Galeria Prestes Maia in São Paulo. The artists, known for their "magical surrealist" style, include Mario Gruber and Otavio Araujo.

1950–80   Brazil undergoes major urbanization

Between 1950 and 1980 Brazil turns from a mostly rural society to a primarily urban one, as Brazilians migrate to the cities in search of jobs. At the beginning of this period, rural dwellers account for sixty-four percent of the population, compared to thirty-six percent for urbanites. By 1980, these figures are almost exactly reversed (sixty-seven percent urban, thirty-three percent rural).

1950   Chateaubriand begins collecting art

Gilberto Chateaubriand begins collecting art. A large part of his extensive collection eventually becomes part of the Museu de Arte Moderna (Museum of Modern Art) in Rio de Janeiro. Chateaubriand retains about 800 works in his private collection.

1950   Brazil hosts World Cup soccer tournament

Brazil hosts its only World Cup soccer tournament in the newly built Maracanã stadium (the largest in the world). The tournament results in Brazil placing second to Uruguay.

1951   Scientific and technological council formed

The National Council of Scientific and Technological Development, created in 1951, is in charge of setting Brazil's policies on science and technology issues.

1951   First Bienal art exhibit held in São Paulo

The Bienal, a biennial art exhibition that ranks as one of Latin America's most important art events, is organized in São Paulo by arts patron Cicillo Matarazzo, art critic Sérgio Milliet, and Laurival Gomes Machado, director of the São Paulo Museum of Modern Art. Modeled on the Venice Biennale, its goal is to encourage art production in Brazil and provide exposure for its artists. The first Bienal includes 1,800 works by artists from nineteen countries. Today the three-month-long event includes painting, sculpture, video shows, films, plays, and lectures. Works by Picasso and other major artists have been exhibited in the Bienal alongside art by lesser-known Brazilian artists.

1951–54   Vargas returns to power

The economic changes attempted by Dutra are not successful, and Vargas is returned to office in democratic elections held in 1950. He continues his former policies of government intervention in economic and social development. His administration creates government enterprises including Petrobrás (the Brazilian petroleum corporation), Electrobrás (the electric corporation), and the national bank of social and economic development (BNDES).

Vargas's last years in office are marked by economic turmoil and political unrest. His years in public life end abruptly in August 1954, when the military demands his resignation from office, and he commits suicide.

After the Vargas Era

1956–61 Presidency of Juscelino Kubitschek

After Vargas's death, there is a brief period of political instability, and then Juscelino Kubitschek (1902–76) is elected president in free democratic elections. Encouraging foreign investment, Kubitschek initiates numerous economic programs. The government funds construction of highways, hydroelectric plants, and other major projects. Brazilian industrial production expands rapidly, including the manufacture of chemicals, machinery, and automobiles. Kubitschek also supports the development of the country's interior to lessen the population concentration in coastal areas. The most dramatic step taken in this area is the construction of the new capital city of Brasília in 1960.

The costs of Kubitschek's ambitious projects create new economic problems, including a large public debt and rising inflation.

1956 Publication of epic novel by Rosa

João Guimarães Rosa (1908–67), the foremost Brazilian novelist of the post-World War II period, publishes his masterpiece, *Grande Sertão: Veredas* (The Devil to Pay in the Backlands), an ironic variation on Euclides da Cunha's 1902 classic *Os Sertões* (Rebellion in the Backlands). Rosa employs Joycean narrative techniques to his account of the Brazilian outback and its inhabitants.

1958 Pelé leads Brazil to first World Cup soccer championship

Soccer phenomenon Pelé (Edson Arantes do Nascimento, see 1940: October 23) leads the Brazilian team to its first World Cup win. The team wins its second consecutive Cup four years later. (See 1962.)

1960s Emergence of Cinema Novo film movement

Young Brazilian film directors take part in the *Cinema Novo* (New Cinema), which appears during the populist Goulart presidency (1918–76, president 1961–64) and abruptly ends with the 1964 military takeover. The Cinema Novo directors produce movies noted for their exploration of the country's social issues as well as their artistic quality. Anselmo Duarte's 1962 film *O Pagador de Promessas* (The Given Word) wins top honors at the Cannes Film Festival. Another prominent director of this period, Nelson Pereira dos Santos, makes films that depict both rural poverty (*Vidas Sêcas*/Barren Lives, 1963) and urban slum life (*Rio 40 Graus*/Rio 40 Degrees, 1955).

1960 Nation's capital moves to Brasília

To encourage development of the nation's interior, the Brazilian government under President Kubitschek has Brazil's capital moved from Rio de Janeiro to the newly built city of Brasília, about 600 miles (965 kilometers) northwest of Rio. As early as the colonial period, the idea of moving the capital inland to protect it from attack by sea has been advanced. In 1821 statesman Andrada e Silva writes a resolution designating the name "Brasilia" for a new capital in the heart of the country. However, it takes until the 1950s for the idea to become a reality.

Soon after he is elected president, Kubitschek recruits urban planner Lúcio Costa and architect Oscar Niemeyer, with whom he has worked on other projects, to plan the city and design the major public buildings. (Officially twenty-six architects compete for the contract, but the hiring of Costa is widely regarded as a foregone conclusion.) The core of the city, which is known for its distinctive modern architecture, is constructed between 1957 and 1960 by a huge labor force called the *candangos*. Oscar Niemeyer's distinctive structures include the *Palácio da Alvorada* (Palace of Dawn), *Praça dos Três Poderes* (Square of the Three Powers), and the Ministry of Foreign Affairs. The nighttime view of the city, illuminating by modern lighting installations, is particularly spectacular. The new capital is officially dedicated on April 21, 1960.

1960 Publication of bestselling book about the urban poor

Cuarto de Despejo (Child of the Dark), by Carolina Maria de Jesus, is published in Rio de Janeiro. Describing the life of urban slum dwellers, it becomes a non-fiction bestseller, with 60,000 copies purchased within six months.

1960 Birth of artist Nina Moraes

Nina Moraes is born in Sao Paulo, where she lives and works as an artist. Her art is characterized by assemblages (a kind of sculptured collage), which she builds by filling oval or rectangular containers with items such as glass shards, powders, gels, and costume jewelry.

1961 Jânio Quadros takes office as president

Reformist candidate Jânio Quadros is elected to succeed Kubitschek as president. Quadros attempts to eliminate government corruption, but he arouses both military and political opposition for attempting to foster trade with the Soviet bloc, a policy that is perceived as pro-Communist. Right-wing opposition is intensified when Quadros awards left-wing revolutionary Ché Guevara (1928–67, Cuban revolutionary born in Argentina) Brazil's highest honor, the Order of the Southern Cross. Civil unrest is sparked, and Quadros is obliged to resign after only seven months in office.

1961 Birth of artist Leda Catanda

Leda Catanda is born in Sao Paulo. She works there as an artist, using fabric to substitute for the more traditional artist's canvas. She creates works on such items as tablecloths, shower curtains, blankets, and towels.

1961–64 Goulart succeeds Quadros

Quadros's vice-president, João Goulart (1918–76), is not able to directly succeed him in office due to opposition from the military and governs by an arrangement in which power is held by Congress. However, a special 1963 election gives him full presidential powers, enabling Goulart to move ahead with social and economic reforms. But his programs stall in the legislature, and Brazil is increasingly torn between civil unrest in right- and left-wingers. When Goulart attempts to preempt congressional power and act on his own, the military stages a coup and deposes him in March 1964.

1962 Brazil wins second World Cup

Reigning World Cup champion Brazil wins its second World Cup soccer championship in Chile. The runner-up is Czechoslovakia. (See 1958.)

1963 Rural Labor Statute passed

The Rural Labor Statute governs working conditions for farm workers and sharecroppers. It attempts to provide agricultural laborers with benefits comparable to those accorded urban workers. However, plantation owners try to circumvent the law by hiring increased numbers of temporary workers.

Military Rule

1964: March/April The military seizes power

The Brazilian military ousts President João Goulart in a virtually bloodless coup that has the support of some state governors and major economic groups. The coup leaders move swiftly to suppress any opposition, arresting over 7,000 persons within one week. A decree is issued suspending constitutional freedoms, asserting the regime's right to remove elected or appointed officials, and depriving individuals of voting rights. On April 15, the Congress, from which all opposition members have been ousted, elects Castelo Branco president.

1964–85 Military government continues under different presidents

Following the 1964 coup, Brazil remains under military government for twenty years under five different presidents: Humberto Castelo Branco (1964–67), Artur da Costa e Silva (1967–69), Emílio Garrastazú Médici (1969–74), Ernesto Geisel (1974–80), and João Baptista Figueiredo (1980–85). The political stability and low wages imposed by these politi-cally repressive regimes, together with large-scale foreign investment and loans, and export growth, produce economic results that become known as the "Brazilian miracle." Brazil experiences an average annual economic growth rate of ten percent.

To maintain its power, the military regime imposes a number of authoritarian measures. Congress is suspended in 1966, and the president is given the power to rule by decree. Military courts are given jurisdiction to rule in matters involving civilians, and the secret police becomes increasingly powerful. University professors are dismissed in 1969, and mandatory courses incorporating the government's moral and political doctrines are instituted. All opposition elements are removed from labor unions, and strict censorship is imposed.

1965: April Brazil's first television network founded

Roberto Marinho forms the Rede Globo (Globo Network) with help from financial backers in the United States. Avoiding programming that might attract the disapproval of the military regime, the conservative businessman builds a successful communications empire. By the mid-1970s the network attracts 60–70 percent of the total viewing audience for all programs (and nearly 100 percent for some of its prime time soap operas). By the mid-1990s Globo has over ninety affiliate stations throughout Brazil and is the world's fourth largest commercial television network. Marinho, who has been criticized for cooperating with the military government, provides extensive coverage of the pro-democracy demonstrations of the mid-1980s. The support of his powerful network is instrumental in the 1989 presidential victory of Fernando Collor de Mello.

1966–67 Mudslides kill 259 in Rio de Janeiro

Erosion of the hills above Rio de Janeiro leads to mass deaths in two separate mudslide episodes following heavy rains in 1966 and 1967. In the first disaster, which occurs between January 11 and 13 of 1966, shanties of poor urban dwellers, perched precariously on the hillsides leading down to the city, are crushed or swept away by gigantic mudslides. The middle-class residents of Rio itself are unaffected by the disaster, and city officials take no steps to shore up the crumbling terrain, reconstructing the shanties instead.

In 1967, eastern Brazil is inundated by tropical rainfall. When the second episode occurs in February 1967 after a eleven-inch (twenty-eight-centimeter) rainfall, not only are the shanties crushed once again, Rio's power plants are disabled and more than half the city's power supply is cut off, resulting in additional deaths throughout the city, as people with heart conditions are deprived of air-conditioning and elevators. This time the city shores up the hillsides with concrete, bars the construction of more shacks, and takes steps to prevent further defoliation and the resultant erosion of the hills.

1969: September 9   Terrorists kidnap U.S. ambassador

Terrorists from two militant left-wing groups, the National Liberating Action (ALN) and the October Eight Revolutionary Movement (MR-8), kidnap U.S. ambassador Charles Elbrick outside his residence in Rio de Janeiro. In return for his release, they demand publication of a 1,000-word manifesto denouncing Brazil's military government and the release of fifteen political prisoners within forty-eight hours. The Brazilian government agrees to the conditions, releasing the political prisoners, who are then flown to Mexico City. In return, the terrorists release Elbrick unharmed.

A 1997 docudrama about the incident, *Four Days in September*, is based on a memoir by Fernando Gabeira, one of the terrorists. It wins an Academy Award for best foreign film (see 1998: March).

1970   Brazil wins World Cup championship

The Brazilian national soccer team wins its third World Cup in Mexico. This is the last championship in which the legendary Pelé competes.

1971   Program reforms educational sustem

Brazil's educational system is restructured to improve access to education, decentralize educational administration, keep students in school longer, make educational opportunities more uniform throughout the country, and integrate vocational education universally into the curriculum. The availability of a university education is increased. Following the reforms, the nation's educational system is divided into four levels, or grades (*graus*): eight years of primary school; three or four years of middle school; undergraduate college education; and graduate school.

1973   First transcontinental railway links Brazil and Chile

South America's first transcontinental railroad, crossing the Andes Mountains between Brazil and Chile, provides rail links between the port of Santos and the Pacific Ocean.

1974   Economic decline weakens the military regime

While Brazil's military regimes have been legitimized by the economic boom from 1967 to 1974, the economic decline that follows, due partly to rising oil prices, erodes support for the government and forces it to become more democratic. In addition to discontent at home, there is international pressure over human rights violations. Under President Ernesto Geisel (b. 1907, president 1974–79), the government gradually begins returning to democratic rule.

1974: February 1   Skyscraper fire kills 220 in São Paulo

In one of the world's worst office building disasters, the twenty-five-story Joelmo building, housing the offices of the Crefisul Bank, is destroyed when plastic construction materials stored near an overheated air-conditioning vent catch fire. The fire rapidly spreads to other flammable construction materials used in the building, trapping hundreds of office workers on the upper floors of the building above the six-story parking garage. In spite of heroic efforts by firefighters, many occupants jump to their deaths. The final survivors are rescued by an army helicopter, which makes a succession of dangerous landings on the crumbling roof of the building, carrying eighty-five people to safety. Thousands of spectators stop to gape at the spectacle, clogging the streets with abandoned cars and impeding rescue efforts.

1974: March 24   Flood kills 200 in Tubarao

After days and nights of heavy rain, the Tubarao River overflows its banks, flooding this city of 65,000. Houses and other buildings are flooded to the tops of their first stories, and there is extensive damage to crops and livestock in the surrounding area. Two hundred people drown, and thousands flee to higher ground. The final property damage tally totals $250 million.

1975–82   Itaipu dam constructed

The Itaipu dam, the world's largest hydroelectric project, is built jointly by Brazil and Paraguay on the Paraná River.

Late 1970s   *Abertura* paves the way for return to democracy

By the late 1970s, a liberalization process known as the *abertura,* or political "opening," is underway. In 1978 a constitutional amendment revokes the most authoritarian measures set in place at the beginning of military rule, and freedom of expression increases. The government of President Figueiredo (b. 1918, president 1979–85) allows the reappearance of opposition parties.

1979: August 22   Congress passes amnesty act

The Brazilian congress passes a bill granting amnesty to political prisoners and exiles not convicted of violent acts, including murder, kidnapping, and other terrorist activities. The bill allows some 4,500 people to return to Brazil.

1978–80   Labor strikes spread

Labor leader Luís Inácio da Silva (popularly called "Lula") leads some 300,000 metal workers in São Paulo in a strike protesting the loss of real wages by government falsification of inflation figures. They demand that reparations be paid for wage losses going back to 1973. Eventually the strike spreas to Minas Gerais, Paraná, Acre, and other states, as metal workers are joined by employees from other sectors, including construction and sanitation workers, medical personnel, and teachers. After President Figueiredo is inaugurated in 1979, the São Paulo strike is declared illegal and crushed by armed troops.

1980   Film industry flourishes with the end of military rule

Brazil's movie industry, which has been strictly monitored by the military government since 1964, experiences a revival in the 1980s with the international success of movies such as *Dona Flor and Her Two Husbands* (1980) and *Pixote* (1982). *Dona Flor*, based on a work by acclaimed Brazilian novelist Jorge Amado (see 1912), becomes the most commercially successful film ever made in Brazil and launches a Hollywood career for its lead actress, Sonia Braga (b. 1951). *Pixote*, a shocking and heartbreaking portrayal of a young boy's life in the slums of São Paulo, wins a best film award at Cannes, and its director, Hector Babenco, goes on to a successful international career, directing such movies as *Kiss of the Spider Woman* (1986), *Ironweed* (1987), and *At Play in the Fields of the Lord* (1992).

1980   Soybeans replace coffee as leading export

Coffee, the leading export since the last century (see 1830), is surpassed by soybeans.

1982: December   Brazil seeks relief from foreign debt

The Figueiredo government, plagued by economic problems, seeks a moratorium on its growing foreign debt to avoid defaulting on loans totaling around $83 million. The following year, efforts to limit domestic interest rates result in the closing of twenty-five state banks. In July 1983 the government imposes strict economic austerity measures.

1984: April   Brazilians rally in support of direct popular elections

Roughly a million Brazilians demonstrate in support of a proposed constitutional amendment to allow the direct election of the president by popular vote. However, the legislature rejects the amendment, leaving the existing electoral college system in place.

The Return of Civilian Government

1985: January 15   Neves and Sarney presidencies

Under increasing political pressure, Brazil's electoral college, previously under the control of the military, elects opposition candidate Tancredo Neves president. But he becomes seriously ill just before his inauguration and dies in April 1985 without ever taking office. His vice-president, José Sarney (b. 1930), takes over the presidency. During Sarney's tenure in office, Brazil has continuing problems paying its foreign debt, and the nation's inflation rate tops 200 percent.

1986: February 28   Sarney implements economic plan

President Sarney imposes the "Cruzado Plan" to rescue Brazil's ailing economy. Brazil's currency is devalued—one cruzado is equal to 1,000 old cruseiros—and wage and price controls are instituted. Inflation is halted, but only temporarily. Similar economic plans in 1987 and 1989 also fail to control Brazil's runaway inflation.

1988   New constitution is adopted

A constituent assembly adopts a new constitution that gives the states more power, and enhances civil rights, outlawing censorship and recognizing Indian land rights. It also legalizes strikes, protects the environment, shortens the work week from forty-eight to forty-four hours, and provides for stronger worker benefits, including four-month maternity leaves.

1989   Cholera outbreak occurs

Cholera, which was thought to have been conquered in Brazil, reemerges, with 1.1 million people affected.

1989   Sarney unveils program to protect rain forest

President Sarney announces a program to protect Brazil's rain forest from excessive development by loggers, farmers, and others. Business interests have already damaged large sections of forest for development purposes. By 1992 the rate of destruction is reported to be slowing.

1989: March–April   Inflation spurs nationwide wave of strikes

Some 300 strikes are called in March and April 1989 to demand higher wages in the face of spiraling inflation. In the same year, Brazil is obliged to stop paying interest on its foreign debt until 1991.

1990   Brazil agrees to join MERCOSUR

Brazil, Argentina, Uruguay, and Paraguay agree to form the Southern Common Market (MERCOSUR). The region represented by this organization accounts for nearly two-thirds of South America's economic activity. By 1995 almost all tariffs between member nations have been eliminated.

1990   Collor de Mello elected president

Fernando Collor de Mello is the first Brazilian president in twenty-one years to be elected by direct popular vote. Two years later, however, he becomes the first to be impeached, for tax evasion and corruption in office. At the conclusion of the impeachment process, Collor de Mello resigns in December 1992. The remainder of his term is completed by his vice-president, Itmar Franco. In 1994, Collor de Mello is acquitted of corruption charges by the Supreme Court but forbidden to hold public office again before the year 2000.

1990s   Rapid spread of AIDS epidemic in Brazil

HIV/AIDS spreads rapidly in the 1990s, especially among heterosexuals. By 1992 women account for one-third of all new AIDS cases. By the mid-1990s, with 40,000 cases officially diagnosed, Brazil has the third highest incidence of AIDS in the world. Almost fifty percent of Brazil's AIDS cases are in the Rio de Janeiro and São Paulo metropolitan areas. A major factor contributing to the spread of the disease is Brazil's under-regulated blood supply—almost twenty percent of AIDS cases are the result of blood transfusions, compared with one to two percent in Europe and the United States.

1994   Brazil wins World Cup championship

Brazil wins its fourth World Cup soccer title at games held in the United States, triumphing over Italy.

1994   Romario wins Golden Ball trophy

Romario, Brazilian soccer player, is voted the best player of the 1994 World Cup Soccer tournament, and receives the Golden Ball trophy.

1994: October   Cardoso elected president

Former finance minister Fernando Henrique Cardoso, architect of a successful plan to stabilize Brazil's economy by reducing inflation, is elected to the presidency.

1996: January 9   Cardoso reverses 1991 decree on Indian land decisions

President Cardoso signs a decree nullifying a 1991 decree barring business interests from challenging allocations of Indian land made by Funai (the government Indian protection agency). Since 1991 some 200 reservations have been created on land set aside and protected from encroachment by mining, logging, and ranching concerns. However, 300 more reservations are still in the process of being created and could be threatened by challenges from business groups.

1996: April   Archaeologists publish description of discoveries

The journal, *Science,* publishes an article by archaeologist Anna Roosevelt of Chicago's Field Museum and the University of Illinois and her colleagues. They describe findings from an excavation in a group of caves on a high plain in the hills of Monte Alegre (Happy Mountain in Portuguese).

Roosevelt and her colleagues theorize that early Paleoindians were living and foraging for food 11,000 years ago in the region. Their site, known as Caverna da Pedra Pintada, features sandstone walls covered with red and yellow handprints and paintings. During their excavations, they find evidence of cooked fruits, seeds, and small fish and animals, as well as stone tools.

1996: April 17   Police attack land reform protesters, killing 19

A national furor is aroused when police in the northern state of Pará open fire on a crowd of demonstrators, killing nineteen persons and injuring at least fifty others. A highway near the town of Eldorado is blocked by 2,500 landless peasant farmers, and 100 police are sent in to disperse the protesters. Police claim that the demonstrators opened fire on them first, but a videotape of the incident does not back up their claims. Members of the landless peasant movement claim that local police are often controlled by the wealthy landowners whose estates would be broken up by the reforms that the protesters are demanding.

1997: May 6   State-owned mining company privatized

In the largest privatization in Latin American history, the Brazilian government sells its controlling stake in Companhia Vale do Rio Doce (CVRD) to a consortium headed by steel manufacturer Companhia Siderurgica Nacional (CSN). The sale arouses widespread controversy, with many arguing against giving control of Brazil's natural resources to private interests. The privatization also raises fears that foreign interests could eventually come to control CVRD, which is the third largest mining company in the world.

Sale proceedings are delayed by the filing of over 120 legal challenges, to which the government is obliged to respond. On April 29, more than 1,000 protesters demonstrate in Rio de Janeiro against the privatization, and CVRD employees file a final legal injunction to try to halt the sale.

1997: May 21   Scandal mars approval of constitutional amendment

The Brazilian senate approves a constitutional amendment allowing presidents, governors, and mayors to run for two consecutive terms in office. The amendment would allow current president Fernando Henrique Cardoso to run for reelection in 1998. The victory in the upper house is muted by newspaper accounts that two legislators have taken bribes in exchange for approving the measure in the Chamber of Deputies (the lower house). The deputies have been caught on tape admitting that they and some of their colleagues have accepted thousands of dollars from government officials in exchange for approving the amendment. The two legislators whose conversation has been taped resign their posts. The amendment is signed into law by President Cardoso on June 5.

1997: December 31   Riot troops storm prison, free 650 hostages

About 250 government troops end a three-day siege at the maximum-security Sorocaba prison, where twenty inmates have held some 650 guards and visitors hostage since a failed escape attempt. The inmates release about 35 of the hostages, and one inmate and a visitor are killed in the initial hostage-taking.

1998 Brazil finishes second in World Cup soccer tournament

The Brazilian Soccer Confederation, disappointed when their team finishes second to France in the World Cup soccer tournament, fires the team's head coach Mario Zagallo and his entire staff. Zagallo was the first person in World Cup history to win the championship as both a player (he was a member of the 1958 and 1962 World Cup champion teams) and as a coach (he coached Brazil's 1970 World Cup champion team). Zagallo's career record is 107 wins, 29 ties, and 12 losses.

Zagallo's replacement is Vanderley Luxemburgo, who was coaching the club, Corinthians, before assuming the national team coaching duties.

1998 Ronaldo wins Golden Ball trophy

Ronaldo (b. September 22, 1976), a soccer forward from Brazil is voted the best player of the 1998 World Cup Soccer Tournament and receives the Golden Ball trophy. As of the late 1990s, Ronaldo is paid $4.2 million per year by Inter (an international soccer consortium) and earns another $1.7 million in endorsements from Nike and Brahma.

1998: March Brazilian docudrama wins Academy Award

Four Days in September, based on the 1969 terrorist kidnapping of American ambassador Charles Elbrick, wins the 1997 Academy Award for Best Foreign Film. Directed by Bruno Barreto, it is based on material in *What's Up, Comrade*, a memoir by Fernando Gabeira, one of the terrorists who participated in the kidnapping (see 1969: September 9).

Bibliography

Baaklini, Abdo I. *The Brazilian Legislature and Political System.* Westport, CT: Greenwood Press, 1992.

Bacha, Edmar L. and Herbert S., eds. *Social Change in Brazil, 1945–1985: The Incomplete Transition.* Albuquerque: University of New Mexico Press, 1989.

Becker, Bertha K. *Brazil: A New Regional Power in the World Economy.* New York: Cambridge University Press, 1992.

Burns, E. Bradford. *A History of Brazil.* 3rd ed. New York: Columbia University Press, 1993.

Carpenter, Mark L. *Brazil, an Awakening Giant.* Minneapolis, MN: Dillon Press, 1987.

Chacel, Julian, et al. *Brazil's Economic and Political Future.* Boulder, CO: Westview, 1985.

Gibbons, Ann. "First Americans: Not Mammoth Hunters, But Forest Dwellers?" *Science* 272, no. 5260 (April 19, 1996): 346+.

Hinchberger, Bill. "Gilberto Chateaubriand: The Price of Oranges; Rio de Janeiro, Brazil." *ARTnews* 93, no. 6, (Summer 1994): 146+.

Kinzo, Maria D'Alva G., ed. *Brazil, the Challenges of the 1990s.* New York: St. Martin's Press, 1993.

Mahar, Dennis J. *Government Policies and Deforestation in Brazil's Amazon Region.* Washington, D.C.: World Bank, 1989.

Payne, Leigh A. *Brazilian Industrialists and Democratic Change.* Baltimore: Johns Hopkins University Press, 1993.

Poppino, Rollie E. *Brazil: The Land and People.* New York: Oxford University Press, 1973.

Roop, Peter, and Connie Roop. *Brazil.* Des Plaines, IL: Heinemann Interactive Library, 1998.

Roosevelt, A.C., et al. "Paleoindian Cave Dwellers in the Amazon: the Peopling of the Americas." *Science* 272, no. 5260 (April 19, 1996): 373.

Schneider, Ronald M. *Brazil: Culture and Politics in a New Industrial Powerhouse.* Boulder, CO: Westview Press, 1996.

Canada

Introduction

The ancestors of the modern Inuit are believed to be the first inhabitants of Canada. They probably crossed from eastern Siberia to Alaska along a now-submerged land bridge, making their way across Canada and Greenland. People settling in the south learned to hunt game animals with stone-tipped spears. The Great Lakes region gradually changed from scrub brush to forest. As the climate warmed and the glaciers retreated, hunters gradually moved north. The Blackfoot, Blood, Piegan, Cree, Gros Ventre, Sarcee, Kootenay, Beaver, and Slavey Indians, speaking a variety of Athapaskan and Algonkian languages, inhabited the vast wilderness territory. The Algonkian-speakers of the land that later became known as Ontario lived mostly in the north, and were the ancestors of the Algonkin, Cree, and Ojibwa (Chippewa) peoples. Iroquoian-speakers lived farther to the south, along the St. Lawrence River and in present-day southwestern Ontario, and were the ancestors of the Five (later Six) Iroquois Nations, the Cayugas, Mohawks, Oneidas, Onondagas, Seneca (and later Tuscaroras).

Early Explorers

The first European explorers to visit Canada were the Icelandic Norsemen, who sailed to the eastern Arctic from their settlements in Greenland and reached a land they called Vinland (Newfoundland). The Norsemen established a seasonal settlement at L'Anse aux Meadows in Newfoundland, but only remained there for a few seasons before abandoning the settlement.

In 1497 John Cabot's ship landed on Newfoundland. Europeans began fishing expeditions off the coast of Newfoundland in the 1500s. War between England and France for colonial control shaped the history of Newfoundland during the 1600s and 1700s. Other famous early explorers of the sixteenth century and early seventeenth century included Jacques Cartier in what is now Québec, Martin Frobisher in the Arctic northeast, Samuel de Champlain in what is now New Brunswick and southern Ontario, and Henry Hudson in the large bay that now bears his name. Québec developed as

the center for a growing fur trade. In 1663 the colony of New France (much of present-day southeastern Canada) was placed under the administration of the French monarch. France issued large land grants to estate owners by royal decree. In 1670 England granted a huge tract of the Canadian interior to the Hudson's Bay Company, which began establishing fur-trading posts in the wilderness.

By the early eighteenth century, the English and French began to challenge land claims made by each other, and animosity between the two cultures grew. In 1754 the English deported the Acadians, French-speaking settlers who had been living in a British-controlled area of what is now New Brunswick. In the late 1750s the British seized control of several key French forts and captured Québec City. In 1760 France surrendered its North American colony to Britain, and the British permitted the French to retain their language and religion rather than assimilating them into British culture. The conquest of France's colonies marked the beginning of the distinct dual society (English and French) in Canada. In 1774 Britain officially granted recognition to French civil laws, guaranteed religious freedom, and authorized the use of the French language in its recently acquired territory.

Influence of the American Revolution

During the American Revolution, Continental armies (armed forces of the American colonies) attacked Québec. After the war, about 40,000 colonists wishing to remain loyal to Britain fled the American colonies and settled in what is now Nova Scotia, New Brunswick, Prince Edward Island, Ontario, and western Québec. The influx of the Loyalists (loyal to the British monarchy) altered the demographic and political character of Canada. Although the American Revolution freed the United States from British rule, the matter of independence and boundaries was not entirely settled until after the War of 1812.

Development of Western Canada

In the first half of the nineteenth century, development of the westernmost regions began, as explorers traveled into the Northwest Territories and the Yukon. In the east, railroad and

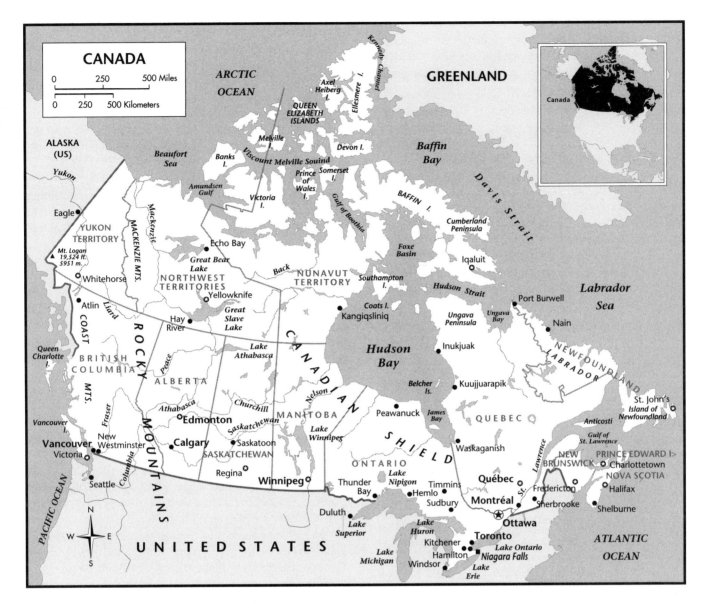

canal construction opened up trade routes and commerce. In 1856 the Grand Trunk Railroad opened, eventually connecting Québec City and Montréal to Toronto. The 1857 discovery of gold in the Pacific northwest prompted thousands to seek their own fortunes. Although the Grand Trunk Railroad was a money-losing venture at the time, it stimulated Canada's industrial growth. As the railroads moved west during the later part of the nineteenth century, small towns arose by the tracks to provide goods and services to nearby farmers. Immigrants arrived in large numbers to farm, work as contract laborers, and work in the towns.

On July 1, 1867, the Dominion of Canada was created as a confederation of four provinces: Ontario, Québec, Nova Scotia, and New Brunswick. Confederation helped finance railroads, which opened up the interior of the continent. Manitoba joined in 1870, British Columbia in 1871, and Prince Edward Island in 1873.

Industrialization and Economic Expansion

From the 1810s until the 1880s, most items were produced in small workshops by skilled artisans. By the 1890s, however, industrial corporations had become quite large, and farming was no longer Canada's largest job sector. Many young people left farms to work in factories, forests, and mines. Large industrial and manufacturing enterprises became prominent, and the banking system developed. In 1896 gold was discovered in the Klondike region of the Yukon. Prospectors came from all across North America and even from abroad with the hope of striking it rich. The swell in population during the gold rush prompted the Canadian government to create the Yukon Territory to ensure Canadian jurisdiction.

Canada's greatest period of economic expansion occurred between 1900 and 1913, and was a time of intensive increases in domestic agriculture, manufacturing, and immigration. Labor unions began to expand across the country, and a confrontational pattern developed between labor unions and owners in the west. Saskatchewan and Alberta became provinces in 1905.

World War I

Over 600,000 Canadians served in Europe during World War I. Canada lost over 68,000 soldiers in the war (with some 60,000 killed in action). Canada's international prestige was considerably elevated after the war. During the war, the cost of living significantly increased from runaway inflation. The women's suffrage movement gained popularity, because many women started working in factories to replace the men who were fighting in the war. Returning veterans faced a bleak future of scarce, low-paying jobs, while tariffs on imports kept prices for consumer goods high. Farmers in the prairie provinces prospered from high wheat prices during the war, but global grain markets collapsed after the end of the war.

During the 1920s grain prices recovered, and Canada experienced a period of rapid industrialization. Railways and roads enabled commercial opportunities to expand, and automobiles, telephones, electrical appliances, and other consumer goods became widely available. Provincial and municipal governments went on a spending spree, and suburban growth was aided by the construction of streetcar lines and paved streets for the growing number of automobiles. As in the United States, consumer confidence led to the rapid expansion of credit, which created a volatile stock market.

Every province but Québec had banned the drinking of alcoholic beverages by the end of World War I. However, by the early 1920s there was massive evasion of the law, with widespread forgery of medical prescriptions, bootlegging, and home-brewing. Furthermore, Canada served as an excellent base for liquor smugglers to the lucrative U.S. market, which also had prohibition. As enforcement of prohibition in Canada became impossible, provincial governments started replacing prohibition with government liquor control boards.

The Great Depression

Unemployment, farm bankruptcies, social distress, and hunger became widespread during the Great Depression, which began in 1929 and lasted until the start of World War II (1939–45). Canada's economy relied heavily on foreign trade, and markets for exports collapsed. The prairie provinces (Alberta, Manitoba, and Saskatchewan) became the most impoverished part of Canada during the Great Depression and suffered from droughts and frequent crop failures as well. Social welfare programs rapidly expanded during the 1930s.

Canada during and after WWII

More than one million Canadians took part in the Allied war effort during World War II, and over 32,000 were killed. The military was quickly reduced after the war, and improvements for reconstruction and the expansion of services resumed. In the following years, the monitoring of northern Canadian airspace served a vital role in the defense of North America against a possible nuclear attack from Soviet bombers. Newfoundland became a Canadian province in 1949.

During the 1950s Canada experienced high rates of employment, birth, and immigration. The political climate became more liberal and welfare programs expanded. The federal and provincial governments launched a national hospital insurance program in 1957.

The "Quiet Revolution"

In 1960 Québec began a period of change and secularization known as the "Quiet Revolution," an era of rapid economic development, expansion of welfare, cultural pride, and restructuring of political institutions. The significance of the French language increased as a mark of cultural identity. The Quiet Revolution also marked the beginning of political tension and federal-provincial bickering as the province sought to assume greater control over its economy and society. In 1970 Québec separatist terrorists kidnapped the British Trade Commissioner and killed a Québec cabinet minister.

For Canada's centennial anniversary in 1967, the world's fair was held in Montréal. Expo '67 became one of the most exciting national events in Canadian history. Montréal later hosted the Summer Olympics in 1976.

In 1976 the citizens of Québec elected the Parti Québécois (PQ) as a majority, a party wanting independence for Québec. French was made the sole, official language of Québec.

In 1981 Canada finally got control over its own constitution. Beforehand, the British North America Act had given the United Kingdom ultimate authority over Canada's constitution. The popular defeats of the Meech Lake Accord in 1987 and the Charlottetown Accord in 1992 failed to resolve the situation of Québec's role in Canada. Both proposed recognizing Québec as a "distinct society." In 1995 Québec held its own provincial-wide referendum regarding the possibility of secession. The measure was defeated by a slim majority of less than one percent.

Timeline

15,000–10,000 B.C.   **First inhabitants of Canada**

The ancestors of the modern Inuit are believed to be the first inhabitants of Canada. They probably cross from eastern

Siberia to Alaska along a now-submerged land bridge, making their way across Canada and Greenland. However, there is no account of such a migration in the mythology of the native peoples now living in Canada. A sub-Arctic climate extends all the way to southern Ontario. People in the south learn to hunt game animals with stone-tipped spears. The Great Lakes region gradually changes from scrub brush to forest. As the climate warms and the glaciers retreat, the hunters gradually move north. Around 2,000 B.C., native peoples begin using seashells for money and as jewelry. The seashells, called wampum, come from the eastern seaboard of what is now the United States, and are traded and carried into the woodland and prairie areas of Canada.

Approximately 10,000 years ago the ancestors of the Dene Indian people live along the Mackenzie Valley in what is now the Northwest Territories. Their ancestors are thought to have crossed the Bering Sea from Siberia thousands of years ago. The Blackfoot, Blood, Piegan, Cree, Gros Ventre, Sarcee, Kootenay, Beaver, and Slavey Indians, speaking a variety of Athapaskan and Algonkian languages, are the sole inhabitants of what is a vast wilderness territory. The woodland tribes of the central and northern regions later become valuable partners of the European fur traders who arrive in the eighteenth century.

700 B.C.–A.D. 1300   Dorset culture

The descendants of a second wave of Inuit immigrants are known as the Dorset people. Their Inuit ancestors came across the Bering Strait either in boats or on the winter ice around 3,000 BC. They inhabit the central Canadian arctic and are primarily hunters of walrus and seal.

A.D. 900   Early inhabitants of Ontario

The Algonkian speakers of the land that later becomes known as Ontario live mostly in the north, and are the ancestors of the Algonkin, Cree, and Ojibwa (Chippewa) peoples. Iroquoian speakers live farther to the south, along the St. Lawrence River and in present-day southwestern Ontario, and are the ancestors of the Five (later Six) Iroquois Nations, the Cayuga, Mohawk, Oneida, Onondaga, Seneca (and later Tuscaroras). The Iroquois speakers live in large villages and grow corn.

A.D. 1000   Norse explore Newfoundland

The first European explorers to visit Canada are the Icelandic Norsemen, who sail to the eastern Arctic from their settlements in Greenland and discover Vinland (Newfoundland). The Norsemen know about the existence of lands to the west because the Icelandic mariner Bjarni Herjolfsson had sighted the coast of Labrador by accident in A.D. 896. They establish a seasonal settlement they call Leifsbudir at L'Anse aux Meadows in Newfoundland. (The Norse settlement at L'Anse aux Meadows is the world's first cultural discovery location

to receive recognition as a UNESCO [United Nations Educational, Scientific, and Cultural Organization] World Heritage Site.) The Norsemen coexist and trade with the Inuit of Greenland, Baffin Island, and possibly Labrador, but only remain at L'Anse aux Meadows for a few seasons before abandoning the settlement. Icelandic sagas, however, dramatically recall armed confrontations between the Inuit and the Norsemen.

1200–1500   Thule culture

The Thule culture possibly assimilates the Dorset and only lasts until the time when Europeans start arriving. Although most Inuit live near the coast, some follow caribou herds to the interior and develop a culture based on hunting and inland fishing.

European Exploration

1497: June 24   Cabot lands on Newfoundland

Italian seafarer John Cabot (Giovanni Caboto, c. 1450–c.1498) sails to investigate the northwestern Atlantic Ocean. Cabot lands on a large island on the feast day of St. John the Baptist, and thus calls the new land "St. John's Isle" (part of present-day Newfoundland) in honor of the saint. Cabot claims the island for Henry VII of England, his patron and employer. Cabot also lands on the northern Cape Breton shore of Nova Scotia in June 1497. When Cabot arrives in Nova Scotia, there are approximately 25,000 native Micmac Indians living there.

War between England and France for colonial control shapes the history of Newfoundland during the 1600s and 1700s, until the Treaty of Utrecht in 1713 (see 1713).

1500s   Europeans begin fishing off the coast of Newfoundland

Basques, Portuguese, Spanish, British, and French begin fishing expeditions off the coast of Newfoundland. The Grand Banks, a shoal in the North Atlantic off Newfoundland's southeastern coast, becomes famed for its bounty of fish.

1534   Cartier explores Québec and Prince Edward Island

Québec is originally inhabited by Algonquin and Iroquois peoples. The northern part of the province is inhabited by the Inuit (previously known as Eskimos).

The European history of Québec begins with the arrival of the French explorer Jacques Cartier (1491–1557). Cartier's expedition lands at the tip of the Gaspé Peninsula. The following year the expedition sails up the St. Lawrence River, reaching as far as the future site of Montréal. Afterwards a thriving fur trade is established, there are relatively friendly relations with the aboriginal people, and continuous rivalry between French and English colonists.

When Cartier lands on Prince Edward Island, he describes it as the most beautiful stretch of land imaginable. However, no permanent colony exists on the island until the French establish one in 1719. By 1750 only 700 people are living on Prince Edward Island.

1570s   Frobisher explores the north

Martin Frobisher's expeditions in the 1570s are the first recorded visits to the Northwest Territories by an explorer.

1604   Champlain lands in New Brunswick

Europeans have known about the existence of the area that later becomes New Brunswick since the 1400s, when intrepid Basque fishermen trawled off Miscou in its northeastern part. At that time the region is inhabited by the Malecite and Micmac Indians. The Micmacs are the first to meet Samuel de Champlain and the French when they land, and the French name the area Acadia (which includes parts of present-day Québec, New Brunswick, and Maine). The Indians establish good relations with the French right away, helping the French settlers, known as Acadians, to adapt to their new country and assisting them in their attacks on New England.

The British and French feud over the territory for a century. Control passes back and forth until the Treaty of Utrecht in 1713.

1605   French establish first European settlement north of the Gulf of Mexico

Pierre de Monts (Pierre du Guast, Sieur de Monts) establishes the first successful agricultural settlement in Canada at Port Royal (now Annapolis Royal, Nova Scotia). It is the first permanent settlement of Europeans north of the Gulf of Mexico. Samuel de Champlain (c.1567–1635) accompanies him and organizes the famous Order of Good Cheer, the first social club in North America.

Development of Frontier Trade and Commerce

1608   Québec City founded

Québec City is the capital of New France, which includes much of present-day eastern Canada, the Great Lakes region, and the Mississippi River valley. The fortified city is an important center of trade and development and is regarded as the cradle of French civilization in America. Between 1608 and 1756, some 10,000 French settlers arrive in Canada. (In 1985 Québec City is named a World Heritage City by UNESCO.)

1610   Hudson explores Ontario and Northwest Territories

Sailing into the large bay that bears his name, Englishman Henry Hudson (d. 1611) becomes the first European to land on the shores of present-day Ontario and the Northwest Territories. Hudson is looking for the Northwest Passage, a route through the icy waters of northern Canada that will ultimately link the Atlantic and Pacific oceans. Many explorers believe that such a passage will improve trade with Asia. His discovery opens the door for further exploration of the interior of the continent.

1612   Exploration of northern Manitoba

English navigator, Captain Thomas Button (d. 1634) stays through the winter with two ships at the mouth of the Nelson River on Hudson Bay. In their search for the Northwest Passage, Europeans reach Manitoba through Hudson Bay. Unlike most of the rest of Canada, the northern parts of Manitoba are settled before the south.

The Assiniboine Indians are the first inhabitants of Manitoba. Other tribes included the nomadic Cree, who follow the herds of bison and caribou on their seasonal migrations. The name Manitoba likely comes from the Cree words *Manitou bou*, which mean "the narrows of the Great Spirit." The words refer to Lake Manitoba, which narrows to less than a kilometer (about five-eighths of a mile) at its center. The waves hitting the loose surface rocks of its north shore produce curious bell-like and wailing sounds, which the first aboriginal peoples believe came from a huge drum beaten by the great spirit.

1632   Company of One Hundred Associates is created

During the frequent skirmishes between New France and New England, the English capture Québec in 1629. Three years later the territory is restored to France. Québec, together with the rest of New France, is placed under the absolute control of the Company of One Hundred Associates. The company's purpose is to exploit the fur trade and establish settlements.

1635   Jesuit college opens

The *Collège des Jésuites*, the first postsecondary educational institution in America north of Mexico opens in Québec City.

1640s   Calamities strike the Huron

In the early 1600s there are some 18,000–32,000 Hurons living in present-day Ontario south of the Georgian Bay. An epidemic in 1638 wipes out about half the population. The Iroquois and the Huron are in competition to assist the Europeans in the development of the fur industry. In the late 1640s the Iroquois confederation (Mohawk, Cayuga, Oneida, Onondaga, and Seneca) starts invading the Hurons' territory, destroying its villages, and massacring the population. The

Huron disperse as a people, fleeing to the west and south. The western Huron refugees later become known as the Wyandotte. The Iroquois also eliminate the Erie, Neutral, and Petun tribes in order to gain control of commerce in southern Ontario.

1643   City of Montréal founded

The city of Montréal, originally named Ville Marie, is established as an outpost for the growing trade in beaver furs. The Sulpicians, a Roman Catholic order of priests, are partly responsible for the founding of the city and later become the *seigneurs* (lords) *of the Ile de Montréal*.

1650s–90s   Era of the *coureurs de bois*

Since the Hurons have been eradicated, the French are looking for other tribes farther west to supply them with furs. The *coureurs de bois*, French independent fur traders, start out initially as agents to ensure a steady supply of pelts. Native peoples trap and skin the animals, which consist primarily of beaver (other pelts include deer, marten, bear, moose, seal, and lynx). The French government starts sending agents to find new trade routes and expand commerce. However, officials in Montréal outlaw the *coureurs de bois,* because they are believed to be depleting the colony by not farming. The *coureurs de bois* establish contacts with many other tribes. By the 1680s the monopoly on the fur trade and licensing requirements push out the independent *coureurs de bois* in favor of government employees. Later, this grueling lifestyle is romanticized as the embodiment of hardy individualism, much like the cowboy of centuries later.

1663   New France becomes a royal province

New France is placed under the administration of the French monarch. Royal administration involves three important officials: the royal governor, the intendant, and the bishop. Each competes for control of the government.

France issues large land grants to seigneurs (estate owners who are granted land by royal decree). Under the seigneurial system, the seigneurs oversee large estates, and then make other grants to settlers. The farmers pay feudal dues (the *cens et rentes*) to the seigneurs and can only sell the property by paying a large duty (the *lods et ventes*).

1670   The Hudson's Bay Company is created

Early European interest in Manitoba centered on the fur trade. King Charles II of England (1630–85) grants to the Hudson's Bay Company a large tract of land named Rupert's Land. The company establishes fur-trading posts.

1670s   Origins of the Métis

In the early years of French settlement, men significantly outnumber women. The royal French government even resettles 770 young women to the colony during the 1660s, but the imbalance between males and females remains and prevents many men from finding a wife within the community. Since farming requires the work of both men and women, many men cannot farm but instead work as fur traders in the west. These men take native women to be their wives, and their decendants are absorbed into their mothers' cultures. The children of the Europeans and the natives become known as the Métis (French, meaning mixed), who eventually develop into a distinct cultural group.

1690   European exploration of Saskatchewan begins

The earliest European explorer of what is now Saskatchewan is Henry Kelsey (d. 1729), a Hudson's Bay Company agent who follows the Saskatchewan River to the southern plains of Saskatchewan. Soon after, trappers come to the area, with fur-trading companies and trading posts to follow.

The first European explorers and trappers to visit Saskatchewan find established settlements of indigenous peoples. The Ojibwa (Chippewa) live in the north; the nomadic Blackfoot roam the eastern plains, while the Assiniboine inhabit the west. The territory of the Cree, who are long-time residents of the north, also extends southward to the plains.

The Rise of British Power

1713   Treaty of Utrecht awards land to England

The English seriously contest the French claims to land in Acadia, which includes parts of modern-day Québec, New Brunswick, Nova Scotia, Newfoundland, and Maine. Control passes back and forth between the British and French until 1713, when the treaty awards all of Acadia, including the fishing banks off Newfoundlad, to the British.

Mid-1700s   Preindustrial development

Merchants and traders prosper from domestic and foreign trade. Merchants often hold key positions in local governments. Import-export company representatives dominate Canadian trade. Out of this prosperity rise famous affluent merchants such as Alexis Lemoine Moniére (1680–1754), Simon McTavish (1750–1804), and François-Augustin Bailly de Messein (1709–1771). Early merchants often invest their profits in real estate. Most domestic manufacturing is performed by individual artisans (such as blacksmiths, carpenters, and millers) operating small family businesses.

1750s–early 1800s   Nova Scotia's population booms

In 1753, 2,000 Protestants from Germany establish the town and county of Lunenburg. In 1760, twenty-two shiploads of New England planters arrive to occupy lands left vacant by the deported Acadians. In the 1770s, eleven shiploads of people from Yorkshire settle in Cumberland County, where many of their descendants still farm. After the American Revolution, 25,000 Loyalists arrive from the newly independent New England states of the United States of America. The influx of Loyalists doubles Nova Scotia's population; and in 1784, the territory is divided into the two colonies of New Brunswick and Cape Breton Island. After the War of 1812, several thousand blacks, including the Chesapeake Blacks, settle in the Halifax-Dartmouth area. Early in the 1800s, Highland Scots start arriving and within 30 years 50,000 settle on Cape Breton Island and in Pictou and Antigonish counties.

1754   First European exploration of Alberta

Anthony Henday becomes the first European explorer to reach what is now Alberta.

1755   Deportation of Acadians

Britain has established its dominance as a colonial power. In 1754 the Seven Years War begins, pitting the British against the French for colonial control. For a century the Acadians, French-speaking settlers in the Minas Basin area, prosper in their trade with the New England colonies, while England and France continue their battle for the territory. Fearing that the Acadians are a security threat, the British deport all Acadians who will not swear allegiance to the British Crown. The separation of families, loss of property, and deportations are immortalized in Longfellow's poem "Evangeline" (Henry Wadsworth Longfellow, American poet, 1807–1882). They are sent south to Louisiana and Virginia (the Cajuns of Louisiana are the descendants of Acadian immigrants). Their exile lasts eight years, and after British control is firmly established, many are allowed to return to their homeland.

1758   The fall of Louisbourg

Louisbourg is a French fortress at Île Royale (present-day Annapolis Royal, Nova Scotia) which is the largest and most impressive of its imperial colonial fortresses in New France. Governor William Shirley of Massachusetts authorizes English troops to seize Louisbourg in 1740 in order to get the French out of the Gulf of St. Lawrence area. Undermanned, the fortress falls, and the joint effort of New England militiamen and Royal Navy vessels prevents the French from recapturing the fortress. However, the Peace of Aix-la-Chapelle in 1748 returns Louisbourg to the French. As tensions escalate between the British and the French, the British impose a blockade, and France is unable to reach the fortress. Louisbourg falls again to the British in 1758, signaling the upcoming end of France's control over its North American colonies.

1759   British seize Québec City

French-English rivalry in North America culminates with the Seven Years' War (1756–63), which sees the fall of Québec City to British forces in 1759. The area around the city is devastated by the war.

1760s   Britain tries to control colonial expansion

In 1763 western tribes under the leadership of the Ottawa chief Pontiac (1720–69) attack British traders around the Great Lakes and seize all British forts except Detroit. The British government is concerned about future confrontations between westward-moving colonists and the Indians already living there. Britain begins regulating westward expansion, rather than allowing the colonial governments to determine policy. The colonial governments, however, resent the policy and long for greater autonomy from Britain.

1760s   Development of Prince Edward Island

Ironically, the population of Prince Edward Island increases after the British deport Acadians from Nova Scotia in 1755. In 1766 Captain Samuel Holland prepares a topographic map of the island, then known as the Island of Saint John, dividing it into 67 parcels of land and distributing it by lot to a group of British landowners. The influx of Britons more than compensates for the loss of the Acadians. The absentee landlords, many of whom never set foot on the island, create numerous problems. Some refuse to sell their lands to their tenants, while others demand exorbitant purchase or rental prices. In 1769 the Island of Saint John becomes a separate colony (in 1799 it is renamed in honor of Prince Edward of England).

1760   Britain conquers France's North American colonies

The French Army surrenders at Montréal in 1760, ending France's dominion over its colonies in North America. The British permit the French to retain their language and religion rather than assimilating them into British culture. The clergy's power increases, as they often serve as intermediaries between the French and British.

1763   The Treaty of Paris; forfeiture of New France to Britain

Under the Treaty of Paris, New France becomes a colony of Britain. The new British province of Québec primarily has French inhabitants who are unfamiliar with the English language and culture. The conquest of France's colonies marks the beginning of the distinct dual society (English and French) in Canada. For some French Canadians, however, the conquest is seen as the beginning of the friction that still exists between the two cultures.

1774   Spaniards visit British Columbia

In contrast with eastern Canada, Spain and Russia are the first countries to claim ownership of certain parts of what is now British Columbia. The Spanish claim the west coast from Mexico to Vancouver Island, while the Russians have an overlapping coastal claim from Alaska to San Francisco.

At the time, the aboriginal peoples of British Columbia have one of the most developed cultures north of Mexico. Because of the diversity of the Pacific coast—mild to cold climate, seashore to mountains—the tribes that settled in this area developed completely different cultures and languages. The coastal inhabitants are experts at wood sculpture and whaling and their social system is marked by occasions such as the *potlatch* (a ceremony in which gifts are given to guests) and theatrical performances.

Canada and the American Revolution

1774   The Québec Act

Under the Québec Act, Britain grants official recognition to French civil laws, guarantees religious freedom, and authorizes the use of the French language. (After the British conquest of 1760, Catholics were not allowed to hold public office.) The concession reflects a sympathy by the British ruling class for the French upper classes, and institutes the separateness of French-speaking society as a distinctive feature of Canada. The compromise also secures the loyalty of the French clergy and aristocracy to the British crown during the American Revolution (1775–1783). The act sets new boundaries that cut off Albany from the fur trade and prohibit Pennsylvanians and Virginians from settling in the Ohio River valley to the west. Poorer French settlers sympathize with the Revolutionists, but armed efforts to join them fail.

1775: November   American armies invade Québec

Two Continental armies attack Québec, seeking to drive out the British. American rebels under General Richard Montgomery (1736–1775) capture Montréal without a struggle, and Benedict Arnold's forces hold Québec City under siege. The following spring the Americans retreat from Québec when British reinforcements arrive.

1778   Cook charts British Columbia

Captain James Cook (1728–1779) of Great Britain becomes the first person to map the coast of British Columbia.

1778   First trading post in Alberta established

Peter Pond, of the Montréal-based North West Company, establishes the first fur-trading post in what later becomes the province of Alberta. The Hudson's Bay Company gradually extends its control throughout a huge expanse of northern North America known as Rupert's Land and the North West Territory. Development of the region is fought over by the Hudson's Bay Company and the North West Company.

1783   Loyalists take refuge in eastern Canada

Some 40,000 Loyalists flee from the American colonies after the American Revolution. These American colonists, wishing to remain faithful to the British Crown, found communities in what is now Nova Scotia, New Brunswick, Prince Edward Island, and Ontario (some 6,000 also settle in western Québec). The Loyalists consist of former soldiers and civilians that had sought protection from the king's army. Many of the Loyalists settling in Canada are recent immigrants from Britain, farmers, Indians from the Iroquois confederacy, or members of religious or ethnic minorities. The influx of Loyalists also alters the political character of Canada. The surge in population creates a rift between Nova Scotia and New Brunswick, and New Brunswick becomes a separate province in June, 1784.

1791   Constitutional Act splits territory

After 1774 the British were in control of southern Ontario, then part of the British colony of Québec. As an amendment to the Québec Act, the Constitutional Act of 1791 divides Québec in two. Upper Canada comprises present-day southern Ontario, and what is now southern Québec becomes known as Lower Canada. The eastern territory adheres to French law and landownership, while the western half follows English law and land tenure. Voting rights are based on property and not religion, allowing French Canadian Catholic men to elect their own people into the assembly.

1796   Birth of writer Thomas Haliburton

Thomas Chandler Haliburton (1796–1865) is born in Windsor, Nova Scotia. He becomes a member of the bar (legal profession) in 1820 and serves as a judge in both the court of common pleas (1828) and the supreme court (1842). He begins a writing career while serving as a judge. His works are known for their humor, and include *The Clockmaker* (1837), *The Attaché* (1843), *Wise Saws and Modern*, and *Nature and Human Nature* (1844), *The Old Judge* (1849), and *Instances* (1853).

Industrialization and Urbanization

1809   First steamboat service established

Canada's first steamer, the *Accomodation*, operating between Montréal and Québec City, becomes the first mechanized transportation in Québec.

1812–15   Canada during the War of 1812

During the War of 1812, the United States fights against Britain. Although the American Revolution had freed the United States from Great Britain, the matter of independence and boundaries is not entirely settled. During 1812–14, the United States is on the offensive against Great Britain, but by 1814 the United States is put on the defensive.

1812   Settlers establish farms in Manitoba

Manitoba's first European agricultural settlement, Assiniboia, is established in the area around the junction of the Red and Assiniboine Rivers by Lord Selkirk, a Scottish nobleman.

Scottish Highlanders begin to settle on the land Selkirk had secured form the Hudson's Bay Company. The colony suffers through floods, inexperience with the environment, and rivalries within the fur trade and in 1836, the Selkirk family transfers Assiniboia back to the Hudson's Bay Company.

1812: August 8   Invading U.S. forces defeated

The U.S. army under General Hull invades Canada but is defeated by Canadian and British soldiers at Brownstown, near present-day Toronto. They formally surrender eight days later.

1813: April 27   York captured by U.S. forces

York (present-day Toronto) is captured by the United States.

1813: May-June   Fighting continues at Fort George, Stony Creek, and Beaver Dams

Fighting continues between Canadian and British soldiers and the U.S. army forces in Canada.

1813: October 5   Battle of the Thames

The U.S. Army, under the command of General (and future President) William Henry Harrison (1773–1841), defeats a British force under the command of Colonel Henry Proctor. During the battle, one of Proctor's chief allies, Shawnee Indian chief Tecumseh (1768–1813) dies.

1813: November 11   Battle of Chrysler's Farm

Battle of Chrysler's Field takes place on the north shore of the St. Lawrence River, ninety miles north of Montréal.

1814: July 5   Americans defeated at Battle of Chippewa Plains

U.S. forces are defeated at the Battle of Chippewa Plains, Upper Canada.

1814: July 25   Battle of Lundy's Lane (or Bridgewater), Upper Canada

Although this battle is indecisive, U.S. forces are forced to retreat when the British receive reinforcements.

1814: December 24   Treaty of Ghent

The Treaty of Ghent formally ends the War of 1812. The war does not alter the boundaries between the United States and Canada; both sides give back captured territory (the United States has western Upper Canada, and the British have Fort Niagara and part of Maine).

1818   Dalhousie University founded

Dalhousie University in Halifax, Nova Scotia, is founded.

1821   North West Company merges with Hudson's Bay Company

The rivalry for developing fur-trading posts ends when the Hudson's Bay Company and the North West Company merge. The Hudson's Bay Company now is the undisputed authority over most of the north and west.

1821   McGill University is founded

McGill University is founded at Montreal, Québec. It is noted for its strong curricula in practical and applied science, medicine, and theology.

1825   Exploration of the Yukon

The first modern European visitors were Russian explorers who traveled along the coast in the eighteenth century and traded with the area's indigenous peoples. In 1825 Sir John Franklin anchors off the Yukon's arctic coastline. The Hudson's Bay Company moves into the interior in the 1840s.

1832   Cholera epidemic strikes Québec

Many infectious diseases (such as typhus, measles, and cholera) are transported to Canada via immigrant ships. Health conditions on such ships are often deplorable, and many passengers die at sea or at quarantine stations such as the Grosse Île station near Québec City. Cholera epidemics strike Lower Canada from 1832 to 1867. The 1832 outbreak results in over 5,200 deaths in Québec City and Montréal, and many more in rural areas. Riots occur during the epidemics because people are fearful of the authorities and immigrants.

1837   The Rebellion of 1837

William Lyon Mackenzie, a vocal opponent of Upper Canada's government, owns a newspaper and gains a reputation for seeking to uncover conspiracies and scandals. In the early 1830s Mackenzie becomes a legislative representative for York County, but is repeatedly expelled from the assembly. In 1837 he organizes a group of supporters in central and western Upper Canada who demonstrate on the outskirts of York (Toronto). They plan an attack on the city, but are driven off. Mackenzie flees to the American border at the Niagara River. For the next several years, Mackenzie's supporters launch raids from the U.S. border.

1838   *Literary Garland* begins publication

An early and respected magazine, the *Literary Garland*, is published from 1838–47. Contributors include Susanna Moodie (1803–52) and Katharaine Parr Traill (1802–99), both of whom contribute stories and serialized novels.

1840s The Roman Catholic Church gains influence in Québec

Until the 1840s, church services in Québec sometimes have interruptions from drinkers and hecklers. Churchwardens are then empowered to arrest people causing disruptions. There is a dwindling number of priests because during the early years of development, members of the clergy are supported only by the seigneur landowners. After 1840, the Catholic church starts to grow across Québec, and many religious orders from France establish themselves. Parishioners establish their own lay, burial, and philanthropic societies.

1840: February 10 Upper and Lower Canada reunite

After rebellions in both Upper and Lower Canada in 1837, the two are reunited by the Act of Union and become the Province of Canada. Britain believes that reunification will help Anglicize the French Canadians. Although the two regions unite, it takes nine years to put a functioning government in place. When the Dominion of Canada is created in 1867, the two regions, Canada West and Canada East, become the provinces of Ontario and Québec.

1843 Canada's canal system flourishes

Canada's first canals are built around Montréal during 1779–83. The Rideau Canal opens in 1832 (at a cost of £1 million), providing a new route between Montréal and Lake Ontario. In 1843 the Cornwall Canal opens, providing a more direct route to Lake Ontario. The Erie Canal in the United States opens in 1825 and provides another route for Canadian exports.

1843 First colony established in British Columbia

Britian establishes the first permanent colony in present-day Victoria.

1845–48 *Histoire du Canada*

François-Xavier Garneau (1809–66) writes the four-volume *Histoire du Canada* (History of Canada) that particularly focuses on the distinct history of the French Canadians. The work becomes one of the most important books to come from Québec during the nineteenth century and helps shape how the people of Québec see themselves even until the mid-twentieth century. Among Garneau's other publications, he is best known for his poetry.

1846 Western border set

In the west U.S. expansionists want the northern border of the United States fixed at 54°40'N. As a compromise, the border is set at 49°N, where it remains.

1847 Establishment of College of Physicians and Surgeons

Until the mid-nineteenth century, doctors and other professionals typically learn their skills as apprentices. By the mid-nineteenth century, however, doctors in Lower Canada are publishing medical journals and controlling hospitals. The regulation of medicine reduces the power of midwives. The College of Physicians and Surgeons is established to provide professional credentials to practitioners. Laws are passed requiring future doctors to attend accredited medical schools.

1849 Women's right to vote prohibited in Québec

Before a law passes taking away women's right to vote, there are a few women who are eligible to vote. (During the 1830s, some thirteen percent of the constituency of Montréal West's electorate consists of women.) The law is passed under the premise that it protects women from political violence. Women in Québec are not permitted to vote again until 1918 for federal elections and 1940 for provincial elections.

1849: April 25 Riots in Montréal

The governor general, Lord Elgin, permits exiled rebels of the 1837–38 rebellion to return home and makes French an official language of Québec. This angers the English-speaking traders and business owners of Montréal, who believe they are threatened as a minority in Lower Cananda. A mob descends upon the parliament house and torches the building.

1850s The role of women in urbanization

As Canadian industry develops, women play an important role in urbanization. Women typically work alongside their husbands in family businesses, learning commercial skills. Women who are deserted, widowed, or single are able to earn a livelihood in the shops and markets of the cities. In Montréal almost ten percent of the city's taverns are operated by women.

1856: November 12 Major railroad opens

The 850-mile Grand Trunk Railroad opens, eventually connecting Québec City and Montréal to Toronto. When finally completed in 1860, the railroad extends from Sarnia, Ontario all the way to the ice-free port of Portland, Maine. The railway's headquarters are in Montréal, and many maintenance and support shops open in the suburb of Pointe Saint-Charles. By 1859 the railway reaches its terminus at Detroit, Michigan. On its own the Grand Trunk, however, is a money-losing venture, but even before it is completed, it is already stimulating Canada's industrial growth.

1857 Gold discovered in the Pacific Northwest

When gold is discovered in the lower Fraser Valley of what later becomes British Columbia, thousands of people come looking for instant wealth. To help maintain law and order, the British government establishes the colony of British Columbia in 1858.

1858: August Ottawa made capital

By a royal appointment, the city of Ottawa, formerly known as Bytown, becomes the capital. The capital is moved partly to avoid further political chaos, such as the riots that occurred in Montréal earlier (see 1849: April 25, p. 29).

1860s Territorial expansion of Canada

In the 1860s the provinces of Canada, anxious to expand into the great northwest, offer to buy up land from the Hudson's Bay Company. Negotiations for the transfer of sovereignty over the Hudson's Bay Company lands to Canada follows, with little regard to the wishes of the inhabitants.

1861 British send troops to Canada

With the start of civil war in the United States, Britain sends 3,000 troops to Canada to protect its territory and avoid involvement in the war. The movement for confederation is spurred by the need for common defense.

1865 Collection of French-Canandian folksongs published

Ernest Gagnon launches a published collection of folksongs popular among French-speaking Canadians entitled *Les Chansons populaires du Canada* (Popular songs of Canada).

1866 Gold rush ends in British Columbia

When the frenzy of the gold rush ends, the colony of Vancouver Island joins British Columbia. The new colony is isolated from the rest of British North America by thousands of miles and a mountain range. The promise of a rail link between the Pacific coast and the rest of Canada convinces British Columbia to join the Confederation in 1871.

The Beginning of Canada as a Nation

1867: July 1 Canadian confederation-the Dominion of Canada is created

The provinces of Canada (Ontario and Québec), Nova Scotia, and New Brunswick join together to form the Dominion of Canada. The union creates a parliament consisting of the 72-member Senate and the 181-member House of Commons. The new government recognizes the British monarch as its head of state. Confederation helps finance railroads, which open up the interior of the continent.

1870s North American sports gain popularity

Baseball, hockey, and football gain in popularity and replace cricket as the preferred sport of Canadians. Middle class workers set up curling, bowling, lacrosse, and hunt clubs.

Lacrosse and hockey are two sports native to Canada. Lacrosse is based on the game *baggawatay* that the French saw Indians playing in the sixteenth century. Hockey's first notable team, the McGill Hockey Club, is established in 1880.

1870 Canadian government buys Northwest Territories

For 200 years the Hudson's Bay Company has owned and administered the vast Northwest Territories (which at the time includes Saskatchewan). Realizing their agricultural potential (especially the fertile belt north of the Palliser Triangle) and the opportunities for colonization, the Government of Canada purchases the territories for £300,000. The government then encourages homesteaders to settle in the area through the Dominion Lands Act of 1872. Immigration and the new railway begin bringing settlers to farm the rich lands.

The Europeans reshape the Arctic north, bringing with them a new economy and way of life. Communities grow around trading posts, mission schools, and Royal Canadian Mounted Police stations.

1870: July 15 Manitoba becomes a province

The Hudson's Bay Company negotiates the transfer of what is now Manitoba to the Canadian government. The movement of American and Canadian settlers into the territory leads the Métis (people of mixed aboriginal and European blood) to fear for the preservation of their land rights and culture. The Métis, under the leadership of Louis Riel, oppose the Canadian proposals in an insurgency known as the Red River Rebellion. Riel succeeds in establishing a locally-elected, provisional government in December 1869. Delegates of this provisional government negotiate terms with the new federal government of Canada, and Manitoba becomes a province of the Dominion of Canada on July 15, 1870.

The new "postage stamp" province (so named because of its square shape and small size) consists of 14,000 square miles (36,000 square kilometers) surrounding the Red River Valley. However, the province's boundaries are stretched in 1881 and again in 1912. With the construction of the railroad, thousands of settlers from eastern Canada and from all over the world settle in Manitoba during the next fifty years.

1871 British Columbia joins the Dominion of Canada

Largely on the promise of acquiring a transcontinental railroad, British Columbia joins the Canadian confederation. The construction of the Canadian Pacific Railway is a major achievement during the 1860s. The railway is granted large tracts of land in return for its promise to help settle these areas.

Volume 2: Americas   **WORLDMARK CHRONOLOGY OF THE NATIONS**

1871: June 29   British North America Act

The new parliament receives the authority from Britain to expand the Dominion of Canada through the establishment of new provinces. The British North America Act (BNA) serves as Canada's constitution and designates which powers are to be held by the federal (dominion) government and which are to be held by the provincial governments.

The British realize that it is not possible to defend Canada from a future attack by the United States. Troops are first withdrawn from central Canada in 1871. By the turn of the century, British troops are withdrawn from the last of the imperial fortresses at Nova Scotia.

1872   John Richardson publishes *Wacousta*

Canadian writer John Richardson publishes *Wacousta,* an exciting story of the War of 1812. *Wacousta* is regarded as the first uniquely Canadian novel.

1873   Prince Edward Island joins the Dominion of Canada

Prince Edward Island joins the Dominion of Canada as its smallest province. Prince Edward Island is known as the Cradle of Confederation, since Charlottetown, its capital, was the site of the planning conference that made the confederation possible in 1867.

1874   Montreal Stock Exchange opens

The Montreal Stock Exchange charter creates the institution for raising capital in Québec.

1876   Supreme court created

Under Alexander Mackenzie (1822–92), who served as premier from 1873–78, a supreme court is established. It is comprised of five judges presided over by a chief justice.

1877: June 20   The Great Fire of St. John

A devastating fire destroys over 1,600 homes and leaves 15,000 people homeless in St. John, New Brunswick. The fire exacerbates the city's faltering economy and dwindling shipbuilding industry.

Urbanization in other Canadian cities between the 1850s and 1870s also leads to several other major fires in Québec City and Montréal. Houses in poor urban areas are typically built close to each other and are made of wood. There is often no public lighting, water, or sewers.

1879   Fréchette publishes prize-winning poem

Poet Louis Fréchette (1839–1908) publishes "Les Fleurs boréales," which wins the Prix Monthyon from the French Academy. He is one of the few poets writing during the nineteenth century in Canada to earn an international reputation. Another well-known work is *La Légende d'un peuple* (1887), an epic about the Canadian people.

1880s   Québec's population exceeds 1.3 million

Québec's population is four times higher than it was in 1815, at nearly 1,360,000. Montréal and several other cities have grown to prominence while rural Québec has lost political power. By the 1880s about 28 percent of Québec's population lives in cities. Despite immigration by English speakers, the province is still primarily inhabited by French speakers.

1880   Population of Manitoba and Northwest Territories reaches 120,000

By the 1880s, about one-third of the 120,000 people living in Manitoba and the Northwest Territories are recent immigrants. During the 1870s, immigrants begin settling in the prairie regions of Canada. In 1874, 7,500 German-speaking Mennonites from Russia arrive in Manitoba, followed by 1,250 Icelanders in 1875. During the 1880s, Swedish farmers immigrate from the Dakota territories in the south, English planters establish farms, and Jewish refugees from Russia settle in Winnipeg.

The Canadian government shifts its immigration policy. Until the late nineteenth century, the government prefers Anglo-Saxon immigrants, followed by Germans and Scandinavians. In order to attract people to the rugged climate of the prairie steppes, the government allows subjects of the Russian and Austrian empires to immigrate, thus significantly changing Canada's demography. Many of the immigrants are Slavs who are familiar with steppe farming and harsh climatic conditions. By 1911, some 60,000 immigrants arrive, including Ukrainians, Czechs, Slovaks, Poles, Hungarians, Serbs, and Croats.

In order to gain title to an area given for homesteading, farmers are required to construct a house and maintain a large section of land under crops within three years of occupation.

1881   Royal Society of Canada founded

The Royal Society of Canada, an academic society, is founded. Its members are drawn from the leaders in academia, the arts, and sciences.

1883   Population of Alberta rising

With the construction of the railroad in 1883, Alberta's population grows quickly. New strains of wheat particularly suited to the climate of the Canadian prairies are discovered, making farmland more productive.

1887   Experimental farms established

Experimental farms are established by the Department of Agriculture to provide assistance and advice to farmers across Canada. The main experimental farm is located near Ottawa while others are located in Nova Scotia, Manitoba, Saskatchewan, and British Columbia.

1890s Establishment of the *Ecole Littéraire de Montréal* (Montréal Literary School)

As industrialization spreads across Québec, literature changes to reflect its sense of identity. The most well-known *Ecole* poet of the day is Emile Nelligan (1879–1941). The *Ecole* writers emphasize a more open and philosophical outlook, while earlier works stress family and country.

1890s The emergence of large corporations

From the 1810s until the 1880s, most industrial goods in Canada are produced in small workshops by skilled artisans. Industrial capitalism, however, changes the methods of production, management, and labor across Canada. Industrial corporations become larger, and farming is no longer Canada's largest sector of employment. Many young people leave farms to work in factories, forests, and mines.

Canada also begins large-scale development of its hydroelectric potential, with generators and transmission lines constructed at Niagara Falls, Ontario. In 1898 a second large hydroelectric plant is built on the Saint-Maurice River at Shawinigan, Québec.

1896 Gold discovered in the Yukon

With the discovery of gold near Dawson City, the Klondike becomes one of the most populous regions in northwestern Canada.

1898 Yukon made a Canadian territory

The sudden increase in population during the Klondike gold rush prompts the Canadian government to give the Yukon more control over its affairs. The Yukon Territory is officially established to ensure Canadian jurisdiction; the Yukon Act provides for a commissioner and an elected legislative assembly.

1899 Canadian participation in the Boer War

When the Boer War (1899–1902) begins in South Africa, there is debate over whether Canada should send troops to help Britain fight what is inherently a colonial war. The Boer War poses no threat to Canada or Britain, and so many Canadians (especially French-speaking Canadians) oppose sending troops. Canada's premier, Sir Wilfred Laurier, arranges initially to send 1,000 volunteers to South Africa. Some 7,300 Canadians fight in the Boer War.

1900–13 Canada's Great Boom

Canada's greatest period of economic expansion is a time of intensive increases in domestic agriculture and manufacturing. Ontario is the leader among the provinces in cultivation of wheat, with about 1.5 million acres producing about 55.5 million.

The value of exports increases by 100 percent, and construction rises 400 percent as new technology, increased demand, and capital improvements (such as railroads) fuel the economy. Large industrial and manufacturing enterprises become prominent, and the banking system develops. It is also a time of large-scale population growth. Between 1901 and 1911, Canada's population increases by 34 percent (from 5.4 million to 7.2 million), the largest growth spurt in population in Canadian history. Some 39 percent of the population growth is attributable to immigration.

1901 Census reports literacy rate

The government reports that 86 percent of the population over age five can read and write.

1903 The expansion of labor unions

At the turn of the century, union activity rapidly expands throughout Canada. Within five years, the number of union locals tripled in Ontario and quintupled in British Columbia.

1905: July 1 Saskatchewan joins Canadian confederation

Both Alberta and Saskatchewan separate from the Northwest Territories. Saskatchewan (from the Cree word *kisiskatchewanisipi,* which means "swift-flowing river") enters the Confederation first, with Regina as the provincial capital. The years following are prosperous until the 1929 economic crash and years of poor harvests in the 1930s.

1905: September 1 Alberta joins Canadian confederation

Alberta (named for Princess Louise Caroline Alberta, fourth daughter of Britain's Queen Victoria) becomes a province of Canada with Edmonton as its capital

1907 New strain of wheat permits greater crop production

Charles Saunders develops the Marquis strain of wheat, a fast-growing type that will thrive in the short but intense growing season of the northern prairie. The new wheat strain expands farming and settlement in northern Saskatchewan.

1911–19 Immigration increases

The numbers of immigrants coming to Canada steadily rises, reaching 400,870 in 1913. Some 27 percent of immigrants that year are from southern and eastern Europe. Before the turn of the century, immigration was offset by emigration of Canadians abroad. In the early years of the twentieth century, emigration decreases as the economy booms and opportunities expand in Canada. As immigrants arrive, ethnic communities in large cities such as Montréal and Toronto develop.

Due to the large influx of immigrants, Canada's population distribution changes. By the early 1920s, over twenty percent of all Canadians live in the prairie provinces (Manitoba, Saskatchewan, and Alberta), up from just eight percent

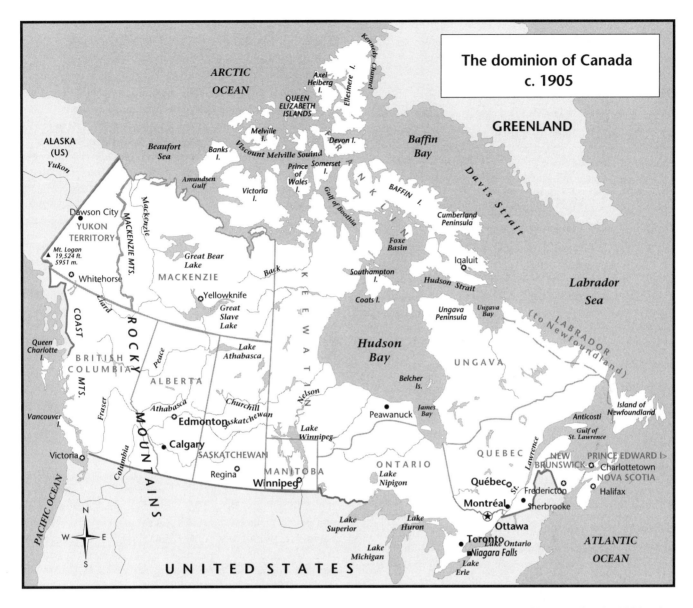

The dominion of Canada
c. 1905

in 1911. Nearly one-third of all Canadians now live west of Ontario.

1912   Boundaries of several provinces are set

The provinces of Manitoba, Ontario, and Québec are enlarged and the Northwest Territories assumes its current boundaries. All territory west of Hudson Bay, south of 60°N, and east of Ungava Bay is split up among the three provinces.

The Great War and the Great Depression

1914   Riots in British Columbian coal town

At the beginning of the twentieth century, a confrontational pattern develops between labor unions and owners in British Columbia, especially in the logging and mining sectors (the province's main industries). The coal town of Nanaimo, on

Vancouver Island, is the scene of intense riots in 1914, when strikers seize the town to protest the hiring of Chinese and Japanese miners to replace them. The British Columbian militia occupies Nanaimo for a year to maintain order.

Unlike other areas of Canada, British Columbia has a significant population of Chinese immigrants. Many had come during the gold rush of the late 1850s, and many others had come from Canton in the 1870s as contract laborers. In the 1880s, 10,000 Chinese men are brought from Hong Kong for hazardous work on the Canadian Pacific Railway. As railroad construction winds down, the Chinese begin working as servants and as inexpensive laborers. Racial prejudice against them builds, and organized labor sees them as a threat.

1914: July 23   Sikhs denied entry to Vancouver

A Japanese steamer carrying nearly 400 Sikhs (Hindu religious sect) as prospective immigrants, anchors in Vancouver harbor, but is denied permission to land. Although the Sikhs are British subjects, the policy of racial discrimination is enforced, as the Royal Canadian Navy escorts the vessel out to the open sea.

1914–18   Canada during World War I

At the outset of World War I (known then as the Great War, 1913–18), the Canadian government passes the War Measures Act, which permits the government to act independently in order to help win the war. The act broadly increases the government's power to issue decrees. In 1917 the government passes a controversial law that allows it to draft men into military service. The war is the last time that Britain maintains a common imperial foreign policy for its territories.

Over 600,000 Canadians serve in Europe during World War I. Canada experiences losses of over 68,000 soldiers in the war (some 60,000 are killed in action, 12,000 more than the United States), and her international prestige is considerably higher than it had been before the war. Canada becomes a founding member of the League of Nations in 1919.

1916   Women's suffrage granted in Manitoba

At the turn of the century, most provinces permitted only male British subjects age twenty-one and older to vote. (Native peoples living on reservations are not allowed to vote in federal elections until 1960.) The women's suffrage movement gains popularity during World War I, when many women begin working in factories to replace the men who are fighting. The Women's Christian Temperance Union, a popular prohibitionist movement, also supports allowing women to vote. Activist and author Nellie McClung (b. Chatsworth, Ont. 1873–1951) from Manitou, Manitoba, is instrumental in obtaining women's suffrage in Manitoba in 1916. Saskatchewan and Alberta also permit women to vote that year. British Columbia and Ontario follow in 1917. In 1918 the right to vote in federal elections is extended to women. The Atlantic provinces grant women the vote during 1918–22, Newfoundland (still a dominion of Britain) in 1925, and Québec in 1940.

In 1929 women win the legal right to serve as provincial senators in Alberta, and the first female senator is appointed in 1930.

1919: May 15   General strike called

Workers in the metals and building trades in Winnipeg stage a strike to protest problems of collective bargaining and union recognition by management. The event, known as the Winnipeg Strike, involves 22,000 sympathetic picketing workers and is the only general strike in Canadian history.

At first the strike spreads through the Winnipeg area, but workers in other cities soon begin sympathy strikes. After six weeks of strikes, the federal government orders the arrests of strike leaders and promises to investigate the strikers' grievances.

1920s   Rapid industrialization

During the 1920s, grain prices recover, and Canada experiences a period of rapid industrialization. Infrastructural improvements (railways and roads) enable commercial opportunities to flourish, and automobiles, telephones, electrical appliances, and other consumer goods become widely available. Provincial and municipal governments have money to spend, and suburban growth is aided by the construction of streetcar lines and paved streets for the quickly growing number of automobiles. As in the U.S., consumer confidence leads to the rapid expansion of credit, which creates a volatile stock market. American investments, tourists, and culture (radio programs, magazines, and movies) begin permeating Canadian culture.

1920s   Prohibition in Canada

Every province except Québec has banned the drinking of alcoholic beverages by the end of World War I. However, by the early 1920s there is massive evasion of the law, with widespread forgery of medical prescriptions, bootlegging, and home-brewing. Furthermore, Canada serves as an excellent base for liquor smugglers to the lucrative market of the United States, which also is under prohibition. As enforcement of prohibition in Canada becomes impossible, provincial governments start replacing prohibition with government liquor control boards that have a monopoly on the sale of wine and spirits.

1923   E. J. Pratt has first book published

Edwin John Pratt (1882–1964), one of Canada's most prominent poets of the twentieth century, has his first book published, *Newfoundland Verse*. Pratt's style is influenced by Browning and Tennyson, and he is best known for his long narrative poems. Some of Pratt's other works include *Titans* (1926), *The Iron Door* (1927), *The Titanic* (1935), and *Brébeuf and His Brethren* (1940).

1925   United Church of Canada forms

Presbyterian and Methodist congregations unite to form the United Church of Canada, representing the majority of Canada's Protestants. The United Church becomes a political and economic force, especially in rural Canada. To counter urban secularism, the United Church supports enforcing laws that restrict business on Sunday prohibit alcohol.

1930s   Canada during the Great Depression

Unemployment, farm bankruptcies, social distress, and hunger become widespread during the Great Depression, which begins in 1929 with the crash of the stock market in New York. Canada's economy relies heavily on foreign trade, and markets for exports collapse as demand for finished goods and raw materials withers.

The prairie provinces become the most impoverished areas of Canada during the Great Depression. In addition to the problems with prices during the early 1920s, droughts and frequent crop failures devastate the economy. Social welfare programs rapidly expand during the 1930s. In 1933, a half million people in Ontario alone are dependent on local governments for relief. Montréal, where most residents are tenants, is especially hit hard by the depression—many laborers are unskilled and social services are spread thin. In the cities many people are forced to live in the streets.

1931   Rapid urbanization

By 1931 some sixty-three percent of Québec's population lives in cities, up from thirty-six percent in 1901. During those 30 years, the population of Montréal doubles to over 818,000, accounting for over 28 percent of the province's growth.

Canada's two largest cities, Toronto and Montréal, emerge as rival metropolises. After World War II, however, Montréal becomes the economic and cultural center of French-speaking Canada, and Toronto takes on the same role for English-speaking Canadians. Large metropolises, such as Montréal and Toronto, become decentralized clusters of smaller cities and towns rather than single urbanized areas.

1931: February   Norway recognizes Canadian title of Arctic islands

Once Norway formally recognizes Canadian sovereignty over the Sverdrup group of Arctic islands (now called Queen Elizabeth Islands), Canada controls the whole Arctic sector from the Canadian mainland all the way to the North Pole.

1933   Rise of the Cooperative Commonwealth Federation

The Cooperative Commonwealth Federation (CCF) is established in Regina, Saskatchewan, as a representative body uniting farm, labor, and socialist political groups. The CCF advocates socialized planning and the eradication of capitalism. The CCF is strongest in Saskatchewan and British Columbia, and by the next year there are hundreds of local chapters.

1936   Canadian Broadcasting Corporation founded

The Canadian Broadcasting Corporation (CBC) is founded as a Crown (public) corporation. The CBC is largely modeled after the British Broadcasting Corporation (BBC) in England. In its early years, the CBC advances the literary arts by commissioning Canadian writers for drama scripts during a time when live theater is languishing as a result of the Depression.

World War II and the Post-war Years

1939–45   Canada during World War II

Before war breaks out in Europe, the governments of Canada and the United States promise to defend each other in case of an attack. More than one million Canadians take part in the Allied war effort, and over 32,000 are killed. The War Measures Act, left over from World War I, permits the federal government to govern the country with centralized power until 1947. After the Germans seize control of Norway and France in 1940, Canada's role in supplying war machines and supplies greatly increases. By 1943, 1.2 million Canadians are working for the war effort, typically in factories that were built only after the war started.

During the war, racial prejudice leads the Canadian government to remove 19,000 Japanese Canadian civilians to internment camps.

Canada emerges from the war with enhanced national prestige, actively involved in world affairs, and committed to the Atlantic alliance. Canada becomes a charter member of the United Nations. Unemployment insurance and other social welfare programs are also created following the war, when Canada rapidly demobilizes. Urbanization spreads quickly by means of the National Housing Act, which makes home ownership more accessible. Housing construction in 1947 proceeds faster than people are getting married, which helps make up for the lack of housing construction during the war. By 1948 university enrollment is double that of 1945. Under the leadership of Prime Minister Louis St. Laurent, old age pensions are increased in 1951.

1942   Poet Earle Birney publishes his most famous poem

Earle Birney (b. 1904) becomes western Canada's first notable poet with the publication of "David" in *David and Other Poems*. Some of Birney's other popular poems are inspired by his trips abroad to places such as India, Japan, and Colombia. Birney later publishes novels such as *Tuvey* (1949) and *Down the Long Table* (1955).

1944   Poet Al Purdy publishes first book of poems

Al Purdy (b. 1918) self-publishes his first book of poetry *The Enchanted Echo*. In the 1950s Purdy's poems are widely read by Canadians in popular magazines such as *Maclean's, Weekend,* and *Saturday Night.*

1944: June   First socialist body of government established in North America

The CCF, under the leadership of Tommy Jones, wins control over the legislature of Saskatchewan, becoming the first socialist government in North America.

1945   Novelist Gabrielle Roy publishes her most famous work

Gabrielle Roy (1909–83) becomes a prominent francophone (French speaking) author with the publication of her novel *Bonheur d'occasion* (English translation, *The Tin Flute*), which portrays the plight of a large, poor Montréal family during the Great Depression and at the beginning of World War II. Other prominent works by Roy include *La petite*

The Canadian parliament meets in special session on September 7, 1939 to support Great Britain in the war against Germany. Lord Weedsmuir, governor-general reads from the throne, while Prime Minister Mackenzie King stands to his left. (EPD Photos/CSU Archives)

poule d'eau (*Where Nests the Water Hen*), 1950; *Alexandre Chenevert, caissier* (*The Cashier*), 1963; *Rue Deschambault* (*Street of Riches*), 1955; and *La route d'Altamont* (*The Road past Altamont*), 1966.

1947 Robertson Davies publishes book of sketches

Robertson Davies (b. 1913), a writer for the *Petersborough Examiner*, compiles the brief sketches of his fictitious character, Samuel Marchbanks, in *The Diary of Samuel Marchbanks*. Marchbanks is a grumpy and cynical old man from the imaginary town of Skunk's Misery, Ontario. Davies uses Marchbanks to make social commentary and expose the idiosyncrasies of Canadians. Davies writes two sequels to the popular "diary" in 1949 and 1967 and becomes noted as a playwright, critic, and novelist.

1947 Discovery of Leduc oil field in Alberta

Before the war, modest amounts of oil and gas are produced in Alberta's Turner Valley. In 1947 an enormous oil field is discovered at Leduc, near Edmonton, Alberta. Oil transforms the province's modest agricultural economy into one of the most prosperous in the country. By 1956 Alberta's oil production meets 75 percent of Canada's demand. Much of the oil is exported to the United States.

1947: January 1 Universal health care begins in Saskatchewan

Saskatchewan's provincial government begins its own universal health care program under the Hospital Insurance Act. The province's plan later serves as the model for Canada's national insurance program. Saskatchewan's plan is successful but proves to cost much more than policymakers expected.

1949   Union membership reaches one million

Membership in trade unions expands rapidly in the post-war years, reaching one million by 1949. (There had only been 166,000 union members in 1914.) The popularity of unions, however, also leads to an increasing number of strikes. Strikes typically involve matters of wages and working conditions.

1949: March 31   Newfoundland becomes a province

Newfoundland rejects joining the Dominion of Canada several times from the 1830s to the 1910s, but in 1948, with the campaign to join Canada led by Joseph R. Smallwood, the Newfoundlanders are enticed by a prosperous Canada and the promises of social programs. After a public referendum Newfoundlanders vote in favor of joining the Canadian Confederation. Newfoundland, which had been Britain's oldest colony (and one of its poorest), is Canada's newest province.

1949: April 4   Canada joins NATO

Canada joins the United States, the United Kingdom, France, Italy, Portugal, Norway, Denmark, the Netherlands, Belgium, and Luxembourg in a treaty that assures the defense of Western Europe in the event of a possible Soviet attack.

1950s–60s   Canada's role in the Cold War

When North Korea invades South Korea in June 1950, Canada becomes involved in the United Nations operations to protect South Korea, and its armed forces double in size. Canada's Lestor Pearson (1897–1972), president of the United Nations General Assembly during 1952–53, has an important role in the discussions that bring about an end to the war.

In 1952 large quantities of uranium are discovered in central Ontario. At the time, uranium is a strategic mineral to possess because it is mainly used to make nuclear weapons.

The monitoring of northern Canadian airspace serves a vital role in the defense of North America against a possible nuclear attack from the Soviet Union. However, when ballistic missiles replace bombers as the means of delivering nuclear warheads, this strategy becomes obsolete.

1950s   Post-war urbanization

During the 1950s Canada experiences high rates of employment, birth, and immigration. Between 1945 and 1956, one million immigrants come to Canada. About one-third are Italians, and one-third are British. Many others are "displaced" persons from Eastern Europe. Canada begins accepting large numbers of refugees, and the government does away with the racial discrimination policy of the past.

Canadians begin relocating mainly to two provinces: Ontario and British Columbia. The number of Canadian farms decreases from 623,000 in the early 1950s to 570,000 by the early 1960s.

1951: June 1   Massey Report recommends state support for the arts

The Massey Commission, organized by the government to report on the status of Canadian culture, issues its report. According to the report, Canadian culture is not a prominent part of Canadian society. The report recommends the establishment of a federal council for the promotion of the arts, literature, humanities, and social sciences. The federal government establishes the National Library in 1953 and the Canada Council in 1957. The Canada Council later serves as the federal government's main promoter of Canadian culture and becomes the precursor to many other programs that sponsor Canadian artists. Some of the cultural insititutions that arise after the late 1950s include the Royal Winnipeg Ballet; the National Ballet of Canada and Les Grands Ballets Canadiens; the Shakespeare Festival at Stratford; the Shaw Festival at Niagara-on-the-Lake; the Citadel Theatre in Edmonton; the Neptune Theatre in Halifax; the Playhouse in Vancouver; and Tarragon and Passe Muraille in Toronto.

1957   National hospital insurance plan introduced

After the end of the war, the political mood in Canada becomes more liberal, and welfare programs expand. The federal government and the provincial governments agree on the details of launching a national hospital insurance program. The Hospital Insurance and Diagnostic Services Act is approved by the federal legislature.

The Rise of Québec Separatism

1960s   The "Quiet Revolution"

Beginning in 1960, Québec enters a period of transition and secularization known as the "Quiet Revolution" (la Révolution silencieuse). It is an era marked by rapid economic development, expansion of welfare, cultural pride, and restructuring of political institutions. The significance of the French language increases as a mark of cultural identity because Québec continues to resemble the rest of North America socially and economically. Until 1960, Canadians who spoke French, or were of French origin, were often disadvantaged compared with those who spoke English or were of British origin (by the 1980s, however, French Canadians and British Canadians generally had about the same incomes). The Quiet Revolution is also the beginning of a period of political tension and federal-provincial bickering as the province seeks to assume greater control over its economy and society. Regrettably, acts of terrorism bring into sharp relief the issue of Québec's status in Canada.

Early 1960s   Canada's foreign policy becomes more independent

During the rule of the Conservative party government under Prime Minister John Diefenbaker, Canada seeks to assert its own foreign policy rather than depend wholly on the United States for direction. Canada refuses to join the Organization of American States, fails to recognize Communist China, and approves of neither the Soviet initiative nor the American response during the Cuban missile crisis. Canada also cancels production of some American misssiles and stops buying nuclear warheads for missiles.

1964: December 11   Auto Pact trade agreement

The federal governments of Canada and the United States negotiate the Canada-United States Automotive Agreement (also known as the Auto Pact), an agreement that permits the free trade of automotive parts and vehicles between the two countries. As a result, Ontario's automotive industry expands through the mid-1970s.

1965: November   Trans-Canada Highway completed

Extending through every province, the Trans-Canada Highway, begun in 1948, is completed as the country's main interprovincial highway system. The project costs $924 million.

Mid-1960s   American draft dodgers flee to Canada

As the United States escalates its military campaign in Vietnam, a substantial number of young American men opposed to the war cross the border into Canada to avoid being drafted into military service. There is a steady flow of Americans into Canada after the end of World War II; by the 1980s about 500,000 Americans have arrived in Canada, and many have remained to become a part of Canadian society.

Mid-1960s   Expansion of "near-bank" institutions

Although no bank had collapsed in Canada since 1923, the federal government is concerned about the solvency of "near-banks" (trust companies, mortgage companies, and credit unions), which have grown rapidly during the 1950s and 1960s. The near-banks are less closely regulated. In 1965 the near-bank institution, Atlantic Acceptance Corporation collapses, and the Bank of Canada provides emergency help to avoid a loss of public confidence and the possibility of setting off a rush of withdrawals.

Banks provide increased access to consumer credit through issuing general purpose credit cards.

1966   University student population reaches nearly 300,000

The number of students attending universities in Canada reaches almost 300,000, up from just 56,000 in 1956. The federal government increases funding for university education throughout the 1960s. New universities are created, such as Carleton (1957), York (1959), Waterloo (1959), and Trent (1963). Two-year colleges are converted into four-year universities, such as the University of Victoria (1963) and the University of Calgary (1966).

1967   World's Fair held in Montréal

The world's fair in Montréal is known as Expo '67, which coincides with Canada's centennial celebration. Expo '67 becomes one of the most exciting national events in Canadian history. It is also one of the most successful of the World's Fairs.

Expo '67 also reinforces the rise of *Québecois* nationalism. French President Charles DeGaulle (1890–1970) visits the fair and irks Canadian federalists with his statement, "*Vive le Québec libre!*" ("Long live free Quebec!"). President DeGaulle's comments sour Canadian-French relations for several years.

1968   Parti Québécois established

A new middle-class political party forms, known as Parti Québécois (PQ). The PQ forms as a liberal nationalist party that advocates required public usage of the French language in the province and political sovereignty for Québec while maintaining an economic union with Canada. Under the charismatic leadership of René Lévesque (1922–87), the PQ gains prominence in the 1970s. Attempts to mandate the usage of French, however, provoke bitter opposition by Québec's English-speaking minority.

1968   Nationwide medicare begins

Saskatchewan had already begun its own provincial medicare plan in 1962, and Alberta had introduced voluntary medicare in 1963. In 1968 the provinces agree to accept a federal program of Medicare. Spending on health and social welfare programs (including pensions) increases from $1 billion in 1950 to $9 billion by 1971.

1968   Writer Alice Munro publishes first collection of short stories

Alice Munro (Alice Laidlaw, b. 1931) publishes her first collection of short stories, *Dance of the Happy Shades*. During the 1970s Munro becomes renown for her carefully crafted stories and as a contributor for magazines such as the *New Yorker, Ms.,* and *Redbook*. Other works by Munro include: *Lives of Girls and Women* (1971), *Something I've Been Meaning to Tell You* (1974), and *Who Do You Think You Are?* (1978).

1969   Royal Commission on Bilingualism and Biculturalism concludes report

Before the 1960s, federal civil servants are usually English-speakers, even in French-speaking areas of Canada. The Royal Commission on Bilingualism and Biculturalism, established in 1963, documents the ways that the French language and culture are given a secondary position in Canadian society. After the commission issues its findings, there is a tide of legislation requiring the use of both English and French in public activities, which stirs up opposition.

1970: October 17   Terrorism by Québec separatists

Terrorists belonging to the Front de Libération du Québec (FLQ) kidnap James Cross, the British Trade Commissioner, and kill Québec cabinet minister, Pierre Laporte, leaving his corpse in the trunk of an abandoned car. As a result, Prime Minister Trudeau invokes the War Measures Act to declare a state of emergency and imposes martial law, which lasts until April 30, 1971.

1971   Expansion of unemployment insurance

The federal government removes the income ceiling for eligibility to receive unemployment payments. The government already subsidizes a complex social security system, which includes pensions, family allowances, and health insurance. Self-employed persons and seasonal workers become eligible to receive payments.

1971   Unionization reaches 2.2 million

Canada's legal system shelters labor unions and fosters their growth: membership increases from 360,000 in 1939 to 711,000 in 1945 and to 1,459,200 in 1960 before reaching 2.2 million in 1971. Union membership among government and office workers expands significantly during the 1960s and early 1970s, bringing many women into labor unions.

1972   Margaret Atwood's most famous novel is published

Margaret Atwood (b.1939), one of Canada's most prominent contemporary literary figures, publishes *Surfacing,* her most famous novel. By this time Atwood has already had one novel and five books of verse published. She also gains literary prestige with the publication of *Survival: A Thematic Guide to Canadian Literature* that same year. Other novels by Atwood include *Lady Oracle* (1976), *Bodily Harm* (1981), and *The Handmaid's Tale* (1985).

1976: November   Pro-independence Party gains power in Québec

Québeckers elect the Parti Québécois (PQ), a party wanting independence for Québec. The PQ makes French the sole, official language of Québec, and in 1980 leader René Lévesque (1922–87) proposes that Québec becomes politically independent from Canada.

1980   Damage from acid rain extensive across northeastern Canada

For decades, tons of sulfur dioxide and other chemicals are put into the air by factories in the American Midwest. Prevailing winds carry the toxins northward across Ontario, Québec, and the northeastern United States. As a result, poisonous rainclouds release acidic rain on the lands and lakes to the north. Hundreds of lakes in Ontario are severely damaged; forests and farms are affected as well.

Early 1980s   Provincial objections to federal control over natural resources

Newfoundland and the federal government disagree on the development and revenue-sharing aspects of the vast Hibernia offshore oil and natural gas field during the early 1980s. Alberta later objects to federal control over oil pricing. Disagreements also arise between native peoples in the Northwest Territories and the federal government concerning the uses of natural resources.

1981: December   Constitution Act passes

The British North America Act is replaced by the Constitution Act, thus giving Canada control over its own constitution. Before the act passes, the British North America Act, which gives the United Kingdom authority over Canada's constitution, serves as Canada's primary political document (see 1871: June 29).

1987   Meech Lake Accord defeated

The country-wide popular defeat of the Meech Lake Accord fails to determine the situation of Québec's role in Canada. Québec was to sign the new 1981 constitution, after winning the inclusion of a clause acknowledging that Québec is a "distinct society."

1990: July   Oka land dispute

Native communities in more densely populated southern Canada pose different problems than the sparsely populated areas of the north. In July 1990, the proposed expansion of a golf course onto lands considered sacred by the Mohawk sparks violence between Mohawks and the government of Quebec. The Mohawks, who have had a reserve at Kanesatake (Oka) since 1721, claim that their territorial rights to the land were never transferred by treaty to the Canadian government. They then block a bridge in Kahnawake in protest. When the provincial police come to remove the barricade, an officer is shot and killed. The army is called in and a deadlock persists through the summer.

1992   Northwest Territories to be divided

The Inuits and other residents agree to divide the immense territory into two smaller administrative units. The division is scheduled for April 1, 1999, when the eastern part will become the semi-autonomous Nunavut Territory, which will serve as an Inuit homeland. The name of the western half is undetermined. Other native groups begin advancing land claims.

1992: October 26   Charlottetown Accord defeated

The popular defeat of the Charlottetown Accord again fails to clarify the situation of Québec's role in Canada. The Charlottetown Accord additionally proposed acknowledging aboriginals' inherent right to self-rule.

1992: December 17   Canada joins NAFTA

Canada joins the U.S. and Mexico in signing the North American Free Trade Agreement (NAFTA), which is based on the earlier U.S.–Canada Free Trade Agreement. NAFTA, which is implemented in 1994, seeks to create a single market of 370 million people.

1995: October 30   Secession referendum narrowly defeated

Québec holds a provincial-wide referendum regarding the possibility of secession. The measure is defeated by a majority of less than one percent.

1998: May   Red Cross and government at fault for tainted blood supply

An independent commission concludes that the government did not adequately screen the nation's blood supply during the mid-1980s, and responsibility for the blood system is taken away from the Red Cross. Some 20,000 people had received tainted blood during transfusions between 1986 and 1990, and many later contract AIDS and hepatitis C as a result.

Bibliography

Bothwell, Robert, Ian Drummond, and John English. *Canada Since 1945: Power, Politics, and Provincialism.* Toronto: University of Toronto Press, 1989.

Bothwell, Robert, Ian Drummond, and John English. *Canada 1900–1945.* Toronto: University of Toronto Press, 1987.

Bothwell, Robert. *A Short History of Ontario.* Edmonton: Hurtig Publishers Ltd., 1986.

Coleman, William D. *The Independence Movement in Quebec 1945–1980.* Toronto: University of Toronto Press, 1984.

Dickinson, John A. and Brian Young. *A Short History of Quebec.* Toronto: Copp Clark Pitman Ltd., 1993.

Forbes, E. R. and D. A. Muise, ed. *The Atlantic Provinces in Confederation.* Toronto: University of Toronto Press, 1993.

McNaught, Kenneth. *The Penguin History of Canada.* London: Penguin, 1988.

Morton, Desmond. *A Short History of Canada.* 2nd ed.. Toronto: McClelland & Stewart Inc., 1994.

Rudin, Ronald. *Making History in Twentieth-Century Quebec.* Toronto: University of Toronto Press, 1997.

Silver, A. I., ed. *An Introduction to Canadian History.* Toronto: Canadian Scholars' Press, 1991.

Stouck, David. *Major Canadian Authors.* 2nd ed. Lincoln, Nebr.: University of Nebraska Press, 1988.

Woodcock, George. *A Social History of Canada.* Markham, Ont.: Penguin Books Canada, 1989.

Chile

Introduction

Chile is a long, narrow country that reaches to the southernmost tip of South America. One of the world's narrowest nations, it is only 56 miles (90 kilometers) wide at its thinnest point. Chile has three major geographic features that run from north to south down the length of the country: the Andes Mountains that border Bolivia and Argentina to the east; a central valley; and, to the west, a low coastal mountain range that runs along the Pacific Ocean. In 1997 Chile had an estimated population of 14.6 million, of which close to 5 million lived in the capital city of Santiago. Although it has a longer history of stable, democratic government than most other Latin American nations, Chile has recently emerged from a period of strict military rule. The first free presidential elections in nearly twenty years were held in 1989, and there was another peaceful change of government in 1994.

Chile's native Indian groups did not develop an advanced civilization such as that of the Incas or Aztecs. In the century preceding the arrival of the first Europeans, there were groups dispersed throughout present-day Chile, living either as hunter-gatherers or in settling farming communities. In the 1400s, the groups living in the northern part of the region were conquered by the Incas, who expanded their empire southward from Peru. Although they paid tribute to the Incas, each Indian group, for the most part, maintained its own culture and way of life.

In 1535 the Spanish conquistador Diego de Almagro led an expedition across the Andes from Peru to Chile but did not attempt to settle this new territory. The first settlements were established by a party of explorers led by Pedro de Valdivia five years later. Santiago, the capital of the future nation, was founded in 1541. The same year Spain created the Captaincy General of Chile as a colony to be administered by the Viceroyalty of Peru. A major obstacle to Spanish settlement was the fierce resistance of the Araucanian Indians, who were not fully subdued for over three hundred years. For a long time, the Araucanians successfully kept all European settlers out of the land south of the Bío-Bío River.

Chile Declares Independence

Like the other nations of South America, Chile won its independence at the beginning of the nineteenth century, spurred by Napoleon's conquest of Spain. In the years between the initial rejection of Spanish rule (1810) and the final declaration of independence (1818), Chilean rebels were driven into exile in Argentina, where they found an ally in liberator José de San Martín. In 1817 San Martín led an army over the Andes from Argentina to aid the Chileans in their fight against the Spanish. He won an important victory at Chacabuco, and Spain was decisively ousted from Chile by the following year.

The new republic of Chile was formally proclaimed in February 1818. Its first president was revolutionary leader Bernardo O'Higgins, who remained in office for five years before being forced to resign by the military. The new nation's political instability continued for several more years. By 1830, however, a conservative government was securely in power, inaugurating one hundred years of stable democracy in Chile. Thirty years of conservative rule were followed by thirty years of liberal government. During this time, Chile enlarged its territory by more than one-third in the War of the Pacific (1879–1884), winning nitrate-rich coastal land that had belonged to Bolivia and territory belonging to Peru as well. From 1891 to 1925, Chile was under an unusual type of democratic rule (called the "Parliamentary Republic") in which congress took over most of the political power of the executive branch.

In 1925 a new constitution, approved during the presidency of reformist president Arturo Alessandri Palma, restored the power of the executive branch and instituted other reforms, including separation of church and state. Chile experienced a period of authoritarian rule by Carlos Ibáñez del Campo in the late 1920s and early 1930s, followed by a year of political stability. Alessandri was returned to office in 1932, and his policies helped the nation's economy recover from the Great Depression. Chile remained neutral for part of World War II but threw in its lot with the Allies by 1943. After the war, the nation's economy was buffeted by world prices for its copper and nitrate exports. Increased industrialization created labor problems, and inflation rose rapidly.

However, Chile's tradition of peaceful transitions between democratically elected governments was maintained.

In the 1960s, increasing political support went to parties with reform agendas that included redistribution of land and nationalization of Chile's mining industries. Eduardo Frei of the Christian Democratic party won a decisive victory in the 1964 presidential election and began a program of extensive social and economic reforms. In 1970 Chilean voters went even further and elected a socialist president, Salvador Allende Gossens. However, Allende won the presidency with a plurality rather than a clear majority, and his policies were opposed by large segments of Chilean society. During his presidency, Chile was increasingly torn along class lines. Middle class opposition to Allende increased as Chile's economy weakened due to falling copper prices and a suspension of credit by banks in the United States.

Military Rule Fosters Political Repression

On September 11, 1973, Allende was removed from office in a military coup. He died during the coup, allegedly committing suicide in the presidential palace. The coup leaders formed a four-member ruling junta led by General Augusto Pinochet and proceeded to take over the government, suspending congress indefinitely. In the early and mid-1970s, the junta embarked on a campaign of political repression, exiling, arresting, torturing, or imprisoning thousands of Allende supporters and others whom it regarded as enemies. Censorship was imposed on the press, political parties and unions were disbanded, and all opposition was silenced in the universities. In 1980 a new constitution was adopted, extending Pinochet's rule for another eight years but providing for a return to civilian government at the end of that period.

The Chilean economy worsened in the 1980s, strengthening opposition to the military government. Falling copper prices and heavy military spending contributed to a severe economic recession at home, and a large foreign debt creating problems abroad. Between 1983 and 1985 anti-government protests spread, both among the middle and lower classes. The Roman Catholic Church added its voice to those demanding a return to democratic government. By 1988, in response to mounting pressure, the government allowed activity by opposition political parties. In the presidential plebiscite held in October 1988, voters were asked to either retain or oust Pinochet, and they voted against him overwhelmingly. This action paved the way for democratic presidential elections—the first in nearly twenty years—in 1989.

On December 14, 1989, Patricio Aylwin Azócar, a moderate Christian Democrat, was elected president of Chile. Aylwin reconvened congress first the first time in nearly twenty years and issued a general amnesty for most political prisoners. During Aylwin's presidency, human rights abuses by the Pinochet regime were investigated. Mass graves were

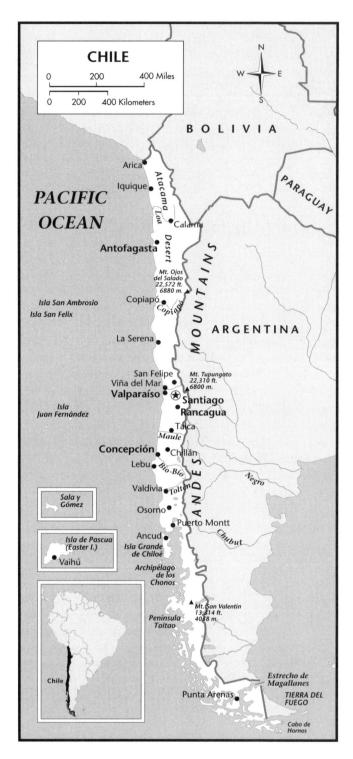

discovered containing the bodies of political prisoners, and it was estimated that a total of as many as 3,000 people had been killed as part of the government's campaign of political repression. In December 1993 the second free presidential election was held since the downfall of Pinochet. Eduardo Frei Ruiz-Tagle, the son of former Chilean president Eduardo Frei Montalva, won a majority of the popular vote. Frei's center-left coalition controlled the lower house of congress, or Chamber of Deputies, but the Senate (the upper house) remained under the control of conservative interests with the power to block electoral and constitutional reforms and other legislation.

In 1995 the United States announced its support for Chile's admission to the North American Free Trade Agreement (NAFTA). Chile also sought membership in the Southern Common Market Customs Union (MERCOSUR), a trade alliance between Argentina, Brazil, Uruguay, and Paraguay.

Timeline

1400s   Inca conquest of Chilean Indian groups

The Incas, a Quechua-speaking people from Cuzco, begin expanding their territory in the thirteenth century. In the fifteenth century, the Incas advance southward from present-day Peru to conquer the native groups living in the northern area of present-day Chile. They develop an advanced civilization, although they have neither written records, money, or the wheel. They practice agriculture, use crop rotation, terracing, and irrigation systems. People work the land collectively, living in small communities called *ayllus*.

Although they are politically subject to the more sophisticated Incas and have to pay tribute to them, the variety of native groups dispersed throughout Chile maintain their diverse cultures and ways of life. The Diaguita, living in the interior, are herders, raising flocks of llamas and alpacas. Coastal tribes include the Chonos, Yamanes, and Alacalufe. Tierra del Fuego and Chile's other island possessions south of the mainland are home to hunting and fishing groups, including the Ono, Chono, Yamaná, Tehuelche, and others.

The most important group are the Araucanians, who comprise the Mapuche and the Picunche. The Mapuche are the most numerous, and they offer fierce resistance to the Inca conquerors. There are still Mapuche living in Chile today. When the first Europeans arrive at the beginning of the sixteenth century, it is estimated that, all together, there are between 500,000 and 1,000,000 Indians living in the region that will become Chile.

Spanish Exploration and Conquest

1520: October 21   Magellan sails up the Chilean coast

The first recorded European sighting of Chile is that by the Portuguese explorer Ferdinand Magellan (1480?–1521) in 1520.

1535   Almagro leads Spanish expedition to Chile

Diego de Almagro (1475?–1538), who had joined with Francisco Pizarro in the conquest of Peru, is the first to attempt the European settlement of Chile. He leads a party of explorers across the Andes to the Chilean interior. However, finding no mineral riches there and encountering resistance from the native Indians, he and his men return to Peru in 1537.

1540   Valdivia founds first Spanish settlements

In 1540, Pizarro sends another of his compatriots, Pedro de Valdivia (1500?–1553), to explore and colonize Chile. Like Almagro before him, Valdivia finds the Araucanians fiercely resistant to appropriation of their land. However, he manages to establish several settlements, including Santiago, which is founded on February 12, 1541, on the banks of the Mapocho River.

1541   Captaincy General of Chile created

Spain creates the Captaincy General of Chile, subordinate to the Viceroyalty of Peru, whose center was Lima. Its first governor is Pedro de Valdivia.

1541: September 11   Female pioneer Inés de Suárez foils Indian attackers

Inés de Suárez, a thirty-three-year-old widow, is a member of Valdivia's Chilean expedition. She rescues the entire group from disaster by finding a source of freshwater in the Atacama Desert. When a chief named Michimalongo and his warriors (said to number in the thousands) attack the newly founded settlement of Santiago while Valdivia was away, Inés masterminds the defense of the town. Wearing chain-mail armor, she fights alongside the men in a full day of battle, and it is she who has the idea of decapitating seven captured Indian chieftains and throwing their heads at the attacking warriors. Then she decapitates the first one herself. The Indians are driven away.

Suárez also plays a crucial role in the settlement's survival by saving remnants of its grain and livestock. By 1543 the settlers are living on the pigs and chickens she raises and baking bread from the grain she saves.

1550s   Araucanian uprising

In 1553 Pedro de Valdivia, founder of Chile's first settlements and its first colonial governor, is killed in the aftermath of an Araucanian attack on the fort of Tucapel led by Lautaro, a Mapuche chief. The Spanish employ Lautaro and he uses his knowledge of them to gain the upper hand in battle. The Araucanian attack is part of a legendary uprising that lasts until 1558. By 1599 the Araucanians have driven the Spanish from all settlements south of the Bío-Bío River, which becomes the boundary line between the two groups for the rest of the colonial period.

1550 Concepción founded

The city of Concepción is founded by Valdivia at a site that later becomes Penco, while Concepción itself moves farther inland in 1754 to its present site. Located in an agricultural area in the southern part of the Central Valley, it serves as both a port and a military base from which the settlers defend themselves against the Araucanians. In 1551, Valdivia founds the settlements of Imperial, Villarrica, Angol, and one named for himself—Valdivia.

1557–61 Mendoza appointed new governor of Chile

Following the death of Pedro de Valdivia, the Viceroy of Peru names a new governor of Chile, García Hurtado de Mendoza (1535–1609). Mendoza puts down Indian revolts in the environs of Concepción, making the city safe for Spanish settlers. The new settlements of Osorno and Cañete are founded under Mendoza's rule.

1568–1618 Church of San Francisco built

The Church of San Francisco, in Santiago, is constructed.

1569–89 Composition of Chilean national epic *La Araucana*

Although the Araucanians are their enemies, the Spanish have great respect for their military prowess. The Indian chief Lautaro, who killed Valdivia (and was himself murdered in 1557), is treated as a hero in the epic poem *La Araucana*. The Spanish soldier and poet Alonso de Ercilla y Zúñiga (1533–94) writes Chile's national epic on scraps of bark and leather. The three-part poem, composed of thirty-seven cantos, is called the Chilean *Aeneid*. The poem is a classic within Chile. It also wins the acclaim of Europeans, including Voltaire and Cervantes, who mentions it favorably in *Don Quixote*. The poem is so popular that eighteen editions appear by 1632. Portions of it are still recited today in Chilean schools.

1578 Sir Francis Drake leads raid on Valparaíso

The Englishman Sir Francis Drake (1543–96) raids Chile's major port, Valparaíso. Drake is only one of a number of pirates and other adventurers who prey on the early Chilean settlers.

1606 Spanish decree enslaves captured Mapuche Indians

Spain issues a decree ordering that all Mapuche captured in battle are to be made slaves. This decree is withdrawn in 1674.

1646 Treaty mandates Spanish-Araucanian truce

The Spanish and Araucanians continue fighting in the seventeenth century. The Spanish formally establish a small standing army, or militia, in 1603. It is based in the south to protect the frontier. The war against the Indians even receives the blessing of Pope Paul V in 1609. After heavy fighting in the 1620s and 1630s, with numerous Indian victories, Chile's governor, Francisco Lopez de Zúñiga, calls a peace conference in 1641. Two years later, the Araucanians actually help the Spanish expel other European explorers from southern Chile.

The end result of the Spanish-Araucanian talks is a truce formalized in the 1646 Pact of Quillon, the only treaty signed by the Spanish in South America that recognizes Indian rights to territory. However, the terms of the truce are broken, and warfare resumes. In 1655 the Indians destroy all Spanish settlements in the Bío-Bío Valley. Nevertheless, the Spanish continue to push southward.

1647 Earthquake destroys Santiago

The first recorded earthquake to level Santiago is only one in a series of tremors that decimates the city in the course of its history.

1730 Earthquake rocks Chile

A major earthquake causes widespread destruction in the rebuilt capital city of Santiago and throughout central Chile. Concepción, one of the cities that is hardest hit, will suffer serious earthquake damage again in 1835, 1837, and 1960.

1740 Trade boosted by removal of restrictions

Chile is allowed to trade with Spain and the provinces of the Viceroyalty of Río de la Plata directly, instead of having to go through the Viceroyalty of Peru. Other measures giving Chile greater independence from the Viceroyalty include the authorization to mint its own coins (1750) and the establishment of its own consultate (1796).

1758 University of Chile founded

The University of Chile (then called the Universidad Real de San Felipe) is established. Today it is a national institution with branches throughout the country.

1759–96 Bourbon monarchs reform Spanish colonial policy

In the eighteenth century, the Habsburgs lose the Spanish throne, and it comes under the control of the Bourbon dynasty, which implements a number of reforms in Spain's administration of its colonies. Most of these occur between the coronation of King Charles III in 1759 and the end of the eighteenth century. The Chileans are given greater independence from the Viceroyalty of Peru. They are allowed to trade directly with European nations instead of going through Peru. Spain also banishes the Jesuits, a move that finds favor with Chile's Creole (native-born descendants of Spanish settlers) population.

1767   Jesuits banished from South America

In keeping with the secular philosophy of Europe's eighteenth-century Enlightenment, the Bourbon monarchy ousts the Jesuit religious order from its South American colonies. The Creoles favor this move because the Jesuit missionaries help the Indians in ways that protect them from enslavement by the Europeans. In addition, once the Jesuits are gone, the Creoles are able to take their lands. However, the quality of education in the colony suffers once the priests leave.

1780   Birth of Chile's first composer

Manuel Robles (1780–1837) is Chile's first native-born composer.

Independence From Spain

1810: September 18   Creole leaders declare independence

The cursor Napoleonic conquest of Spain in 1808 serves as the impetus for independence movements throughout Spain's colonies in the New World. In 1810, Santiago's *cabildo* (town council) votes to declare independence from Spain and forms its own ruling junta. However, the leaders of the breakaway government can't agree on whether to revolt only against the king that Napoleon has installed, or against Spanish rule altogether.

1810–14   Self-government under the Patria Vieja

After it first declares its independence, Chile's government is called the *Patria Vieja* "Old Fatherland", and its principal leader is José Miguel Carrera Verdugo (1785–1821), who seizes power in November 1811. Carrera, a dictatorial ruler with a military background, arouses much opposition among the Chileans. His principal rival is Bernardo O'Higgins (1778–1842), son of a former governor of Chile under the Spanish. The disagreement between the two leaders escalates into civil war, making it easier for the Spanish to attack and defeat their forces.

1814: October 2   The Reconquest

Between 1813 and 1814 the Viceroy of Peru, José Fernando de Abascal y Sousa, sends three detachments of troops to Chile to challenge the rebels. The first invasion takes place on March 26, 1813, when 2,000 troops from Peru are joined by royalists from the cities of Valdivia and Chiloé.

The royalist troops regain control of Chile at the Battle of Rancagua (October 1–2), inaugurating a two-year period known as the *La Reconquista* "the Spanish reconquest". A number of the Chilean rebels flee to Argentina. The severe treatment of the remaining Chileans by the victorious Spanish turns Chilean thoughts increasingly toward total independence from Spain. There is widespread sympathy for rebels such as Manuel Rodríguez, who leads guerrilla raids against the Spanish.

1817   Birth of prominent writer and political figure José Lastarria

José Victorino Lastarria (1817–88) is a leading liberal thinker, writer, and politician in mid-nineteenth-century Chile. Lastarria is the driving force behind the "Generation of 1842" literary and cultural movement and also serves his country as a legislator and diplomat. During the Conservative era, Lastarria's political views get him arrested and exiled to Peru in 1850, and dismissed from congress in 1851. All together, he is elected to six terms in the Chamber of Deputies and serves as a senator from 1876 to 1882. He is also minister of finance and of the interior and carries out diplomatic missions to other South American nations in the 1860s, '70s, and '80s.

Lastarria publishes numerous newspaper and magazine articles, as well as fiction, political works, and an autobiography, *Recuerdos literarios* (1878). Lastarria urges other Chilean writers to develop a distinctive national literature.

1817: February 12   Argentine forces defeat the Spanish at Chacabuco

Rebel leader Bernardo O'Higgins, in exile in Argentina, allies himself with Argentine liberator José de San Martín, who believes that the liberation of Chile is a necessary step on the way to liberating Peru, the stronghold of the Spanish forces. Early in 1817, San Martín leads an army of about four thousand men over the Andes into Chile and defeats the Spanish royalists in an important victory at Chacabuco on February 12. However, Spain rallies and fighting continues over the following year.

1818: January 1   New government of Chile declares independence

The rebel forces under Bernardo O'Higgins form a new government and declare Chilean independence. O'Higgins is chosen as "supreme director" of the new nation.

1818: April 5   Decisive defeat for the Spanish at Maipú

The decisive victory over the Spanish occurs at the Battle of Maipú, although Chilean independence has already been formally proclaimed earlier in 1818. San Martín and his forces then go on to liberate Peru. Spanish forces in the south of Chile are not entirely defeated until 1926.

The Chilean Republic

1818–23   O'Higgins serves as head of new republic

Revolutionary hero Bernardo O'Higgins serves five years as special director (executive head) of the new Chilean republic.

In spite of his initial popularity, O'Higgins soon faces opposition from many sides. His authoritarian rule alienates liberals. Conservatives are unhappy with his attempts to reduce the power of the church, eliminate aristocratic titles, and change the rules governing inheritance of wealthy estates. He is also criticized for using Chilean resources to help San Martín in his campaign to liberate Peru, although this is done to help secure Chile itself from any future threat of Spanish reconquest.

During his time in office, O'Higgins works to promote education and foreign trade, and build Chile's infrastructure. However, after five years, the military forces O'Higgins to resign. The former leader, disillusioned, retires to Peru, where he remains until his death in 1842.

1820–70   Copper boom

Industrialization in European countries produces a rising demand for copper. Chile's copper exports rise from 60 tons in 1826 to 2,000 tons by 1831, and to 12,700 by 1835. Copper accounts for more than half of all Chilean exports by the 1860s, and Chile is producing 44 percent of all the world's copper. However, reliance on copper production also makes Chile vulnerable to the instability of foreign markets. When industrial growth abroad declines in the 1850s and 1870s, so does Chile's copper market. By the 1880s, production drops due to mine depletion and foreign competition.

1822: November 19   Earthquake kills 10,000

An earthquake centered in Valparaíso hits the Chilean coast, destroying entire villages and towns. The worst damage is in Valparaíso, where thousands are killed by fires and collapsing buildings. The elevation of the city is permanently raised by four feet, and the ocean bed in the area rises so much that large numbers of fish perish on dry land.

1823   Chile abolishes slavery

Chile becomes one of the first countries in the Western Hemisphere to abolish slavery.

1823–30   Instability continues after O'Higgins's departure

For seven years after O'Higgins's departure from office, factionalism continues to create political instability in Chile. The primary issues of contention are how much authority the central government should have over the states (centralism versus federalism) and how closely the church and state should be linked. Chile has numerous constitutions and a total of ten presidents in the 1820s (compared with one president for each of the following decades). By 1830 the Conservatives gain control and they proceed to stabilize the Chilean government.

1827   El Mercurio founded

El Mercurio, which claims the distinction of being the world's oldest Spanish-language newspaper, is founded in Valparaíso. Now published in Santiago, it is Chile's most influential newspaper, and the paper of record, in which all major government news appears.

1830–1920   Life of author and diplomat Alberto Blest Gana

Often compared to the French novelist Balzac, Blest Gana writes sweeping, detailed novels that realistically portray Chilean society of his time. Like many Latin American authors, he is also active in public affairs.

1830–61   Conservatives in power

Under the Conservatives, Chile becomes the first nation in South America to have a stable civilian government. The Conservatives favor a strong central government with power concentrated in the executive branch, and a strong church. A merchant named Diego Portales Palazuelos is the dominant political figure at the beginning of this era, even though he never serves as president. He spearheads the drafting of the constitution of 1833, which reflects many of his political views.

In the thirty years of Conservative rule, Chile has only three presidents: Joaquín Prieto (1831–41), Manuel Bulnes (1842–52), and Manuel Montt (1851–61).

1833   New constitution approved

The constitution of 1833 lays the groundwork for the autocratic rule of the Conservative Party over the next thirty years, and it remains in force until 1925. It provides for a strong central government with a powerful president who appoints his own cabinet and has absolute veto power. Voters still have to meet literacy and property qualifications. Roman Catholicism is declared the official state religion, and the practice of all other religions is outlawed.

1835   Major earthquakes rock Chile

The nation is shaken by extensive earthquakes accompanied by tsunamis (giant tidal waves). The city of Concepción sustains severe damage and lives are lost.

1836–39   Chile breaks up confederation between Peru and Bolivia

Andrés Santa Cruz (1792–1865), Bolivia's second president, tries to unify Bolivia and Peru. In 1836 his troops overthrow the government of Peru and form a confederation between the two nations (the Confederación Perú-Boliviana). However, both Chile and Argentina, feeling threatened by the growth of Bolivian power, send forces to oppose the confederation. In January 1839, Chile wins decisive victories over the Peruvians at Casma and over the Bolivians at the Battle of Yungay.

1839   Chile prints first bank notes

Chile prints its first currency in 1839, and its first postage stamps in 1853.

1840s   Chilean exports flourish with California gold rush

The California gold rush of the 1840s proves to be a "gold mine" for Chile as well, expanding the market for its grain and flour. However, this export boom is short-lived, lasting only until the mid-1850s.

1840s   "Generation of 1842" influences culture and politics

The "Generation of 1842" is Chile's first native literary movement. It is founded by a group of leading intellectuals, from both Chile and other South American countries. Through articles in the prominent newspapers and periodicals of the 1840s, they take an active role in encouraging social and educational reform. (The movement takes its name from the year that the National University of Chile was founded.)

The Generation of 1842 is composed of poets and other writers, including prominent scholars and philosophers. José Victorino Lastarria (see 1817) is prominently associated with the group and generally considered its founder. Other leading thinkers in the group include Francisco Bilbao, Santiago Arcos, and Eusebio Lillo. The movement helps spur the resurgence of the Liberal party, which gains growing political support in the 1850s and controls the government from 1861 to 1891.

1842   University of Chile founded

The University of Chile (Universidad de Chile) is established, taking the place of the University of San Felipe, established in 1758. The first directors are renowned Venezuelan author and educator Andrés Bello and minister of education (and later Chilean president) Manuel Montt.

1845–52   Chile's first railway is built

An American, William Wheelwright (1798–1873), raises the financing for construction of Chile's first railroad, between Copiapó and the port of Caldera. The first 51 kilometers are completed by 1851, and construction is completed by the following year. A second railway line, running from Santiago to Valparaíso, is finished in 1863.

1852   Santiago and Valparaíso linked by telegraph

Chile's first telegraph line goes into operation. Within fourteen years there are forty-eight lines linking points within Chile, as well as lines to Peru and Argentina.

1861–91   Period of Liberal rule

During the 1850s the power of the Liberals grows and, after two armed rebellions, they take control of the government.

Presidents during this period are José Joaquín Pérez (1861–71), Federico Errázuriz (1871–76), Aníbal Pinto (1876–81), Domingo Santa María (1881–86), and José Manuel Balmaceda (1886–1891).

The political strength of landowners and other wealthy elites declines, the power of the church is limited, immigration increases, and education expands. Financial qualifications for voting are removed for literate males, and to limit the power of the executive branch, presidents are barred from serving two consecutive terms.

1864–66   Chile joins Peru in war with Spain

Spain seizes Peru's uninhabited Chincha Islands, the source of the nation's wealth from exporting *guano* (bird droppings) for fertilizer. The Spanish claim they are entitled to the islands as payment of outstanding debts owed by Peru. Chile, together with Ecuador and Bolivia, join forces with Peru and defeat the Spanish by 1866.

1879–84   War of the Pacific

Part of Chile's present-day territory includes land that once belonged to Bolivia, whose original borders stretched to the Pacific Ocean. This land includes rich deposits of nitrates, which Bolivia exports to Europe to be used in fertilizers. Chile has an agreement to develop the land for Bolivia. However, after a dispute over taxes imposed by Bolivia, Chile sends in military forces and lays claim to it in 1879. In response, Peru and Bolivia jointly declare war on Chile. By the end of 1879, Chile captures the nitrate-rich Tarapacá province. In 1880 Chile wins Tacna and Arica provinces, decisively defeating the Bolivians at the Battle of Tacna. Afterward most of the fighting is between Chile and Peru. In 1881 Chilean forces seize Lima, the Peruvian capital, and occupy it for three years in the face of scattered Peruvian resistance.

Finally, Peru agrees to a peace treaty in exchange for ceding land to Chile. It gives up the province of Tarapacá and also cedes the rights to Tacna and Arica for the next ten years. In addition to valuable Peruvian territory, Chile also wins all of Bolivia's coastal land, including the Atacama desert. (Bolivia, in turn, becomes a landlocked nation.) Chile's land gains in the War of the Pacific enlarge its area by more than one-third and usher in an era of mining wealth and newfound prosperity.

1880–1930   Nitrate industry generates prosperity for Chile

The territory won in the War of the Pacific provides Chile with a nitrate-exporting boom whose effects are felt in many areas of its economy. In the early 1880s, Chile owns less than one-third of the nitrate industry; by 1920 Chilean ownership has increased to fifty-six percent. Traders and the shipping industry prosper, touching off increased demand for coal and agricultural products. The government itself is enriched by

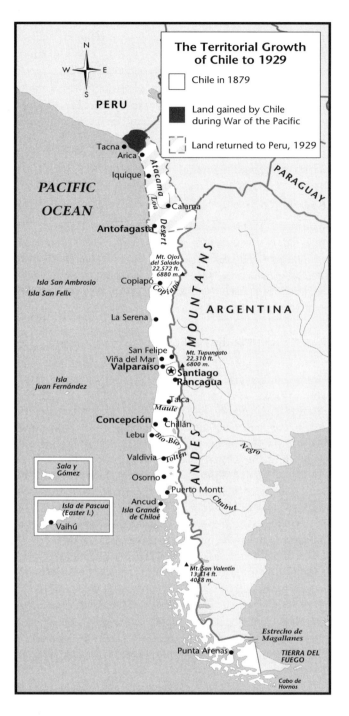

The Territorial Growth of Chile to 1929

Chile in 1879

Land gained by Chile during War of the Pacific

Land returned to Peru, 1929

1889–1957 Life of Nobel Prize-winning poet Gabriela Mistral

Gabriela Mistral (1889–1957) is one of Chile's foremost literary figures. In 1945 she becomes the first South American (and the first Latin-American) to win the Nobel Prize for Literature. Mistral (born Lucila Godoy Alcayaga) grows up in a rural village and becomes a schoolteacher at the age of 15. The life of the poor people in the countryside is the subject of many of her poems. Throughout her life, she remains an educator, and also serves her country as a diplomat and cultural ambassador. Mistral first wins professional recognition with her 1914 "Sonetos de la muerte" (Sonnets of Death). Her poetry collections include *Desolación* (1922), *Lagar* (1954), and *Lagar II,* published posthumously in 1991.

1891 Political crisis leads to civil war

The ouster of President José Manuel Balmaceda (1840–1891) is the one time between 1830 and 1924 that political power in Chile is not transferred peacefully from one government to the next. Since the 1830s, Chile's congress has gradually exercised more and more power over the cabinet and over government spending. In 1890 Balmaceda challenges the congress in both of these areas. He appoints a cabinet without congressional approval, and the congress subsequently refuses to approve his budget. He then defies them by deciding to extend the existing budget without their approval, and a constitutional crisis ensues. Chile's Liberal and Democratic parties back the president; the Conservative party sides with the rebels. Balmaceda also has the political (but not military) support of the United States.

Congress, backed by the navy, establishes a military headquarters at Iquique and authorizes an armed revolt against the president, defeating his forces in battles at Placilla and Concón. The rebel forces then seize the capital city of Santiago. Ultimately, Balmaceda is defeated and the legislature gains the upper hand over the presidency (and retains it until 1925). Balmaceda is granted political asylum by the Argentine embassy. When his term as president is officially over, he commits suicide. He later becomes a hero of the Chilean left.

the taxes on nitrate exports. Between 1890 and 1930 over half its income comes from this source, and these taxes are instrumental in paying for the construction of Chile's railroads and for improvements in its roads and educational system.

1882 Final defeat of the Araucanian Indians

The Chilean army crushes any remaining threat from the Araucanian Indians, permanently ending more than three hundred years of intermittent warfare with them and opening the southern frontier to further settlement. The Indian population is restricted to reservations.

1891: October 16 *Baltimore* incident raises threat of war with U.S.

Angered by United States support of President Balmaceda, a Chilean mob in Valparaíso attacks U.S. sailors from the vessel *Baltimore,* resulting in two deaths. The U.S. demands an apology. Chile refuses but eventually agrees to pay reparations.

1898 Public employee pension system established

Chile becomes one of the first nations in Latin America to provide retirement benefits to workers.

Congressional Rule

1891–1925   The Parliamentary Republic

The downfall of President Balmaceda ushers in a period, known as the Parliamentary Republic, when the majority of political power rests with congress, which, in turn, represents the interests of the wealthy landowners. Chile is the only Latin American country in which this type of situation occurs. Although the government of this period is widely faulted for its corruption and ineffectiveness, it allows the growth of a multiparty political system. The country's economy flourishes, due to the incomes from copper and nitrate exports, but only the traditional elites enjoy the prosperity. The standard of living for farm workers and laborers remains poor.

Widespread urbanization is a major social and political development of this period, and the middle class gains power as a political force, one that the working classes begin to recognize as an ally.

1893–1948   Life of poet Vicente Huidobro

Huidobro starts an avant-garde poetry movement called *Creacionismo* (Creationism). Born to an upper-class family, Huidobro moves to Paris in 1920. There he meets influential avant-garde poets including Guillaume Apollinaire, and works on the ground-breaking magazine *Nord-Sud* with Apollinaire and Pierre Reverdy. Huidobro also spends time in Madrid, where he helps found a poetic movement similar to *Creacionismo*. Traveling between Europe and Chile, Huidobro brings the post-World War I French avant-garde spirit of experimentation to his homeland as well.

Huidobro's poetic style is characterized by its unusual, striking, and even seemingly irrational combination of images. Rather than imitate reality, Huidobro believes that he should share whatever is in his mind.

Early 20th century   Chile's second copper boom

Technology developed in the United States, including open-pit mining methods and a system of concentrating the ores (froth flotation), makes possible the mining of lower-grade copper ore, and a new copper boom begins. Deposits of these lower-grade ores are located at El Teniente, Chuquicamata, Potrerillos, El Salvador, and Rio Blanco. Production begins at Chuquicamata, the site of the largest deposit, in 1912. Another major mining site is established at Potrerillos in 1926. Foreign interests own all the mines at first. On July 16, 1971, Chile's copper mines are nationalized.

Early 20th century   Beginning of labor movement

By the late 1800s, Chile's workers begin to organize. In the first decade of the new century, significant strikes or other types of protests take place in Santiago, Valparaíso, and Anto-fagasta. In Iquique, government troops open fire on thousands of miners and their families gathered in a school building with their strike leaders. An estimated two hundred people are killed, and many hundreds more are injured.

By 1920 tens of thousands of laborers belong to unions of all kinds. The traditional neglect of Chile's poorer classes is referred to in political circles as the "social question," and congressional committees begin to probe labor conditions and formulate plans for improving them.

1902   Border dispute with Argentina resolved

Chile and Argentina sign an agreement putting an end to a border dispute. This agreement is commemorated with a statue "Christ of the Andes," built near the Uspallata Pass two years later.

1903   Birth of pianist Claudio Arrau

Arrau (1903–91) is an internationally renowned Chilean concert pianist and one of the best-known Latin American musicians of the twentieth century. Born in Chillán, Arrau, who is a child prodigy, studies in Berlin from 1912 to 1918. He performs in North and South America and in Europe in the 1920s and 1930s. After World War II begins, he emigrates to the United States.

Arrau is known for the honesty and lack of pretention or showiness in his playing, which emphasizes the inherent beauty of the music rather than the prowess of the performer. He is especially well known for his interpretation of Beethoven's music. Arrau wins the UNESCO International Music Prize and is honored as a commander in the French Legion of Honor. In addition, a street in his home town of Chillàn (as well as one in Santiago) are named for him.

1904–73   Life of internationally renowned poet Pablo Neruda

Pablo Neruda, widely regarded as Chile's greatest poet, wins the Nobel Prize for Literature in 1971 and the Soviet Union's Lenin Prize for Peace in 1953. Neruda, whose real name is Neftalí Ricardo Reyes Basoalto, is born in southern Chile to the family of a laborer. He moves to Santiago in 1921 and begins publishing his poems in a literary magazine. His first book appears two years later. In 1924 he publishes his best-known book, *Veinte poemas de amor y una canción desesperada* (Twenty Love Poems and a Song of Despair).

Neruda's diplomatic career begins in 1927, when he is named honorary consul and dispatched to a series of Asian countries. He later serves as consul to Mexico, Argentina, and Spain. In 1945 Neruda is elected to the Chilean Senate, where he serves for three years. However in 1948 a right-wing crackdown ends the political career of Neruda, who is a communist. Neruda goes into exile and writes the epic poem *Canto General* (General Song). He returns to Chile in 1952.

1906 Government passes low-cost housing bill

Chile's government begins subsidizing low-cost housing in 1906 with legislation that grants tax exemptions to builders.

1906: August 16 Earthquake levels Valparaíso

The city of Valparaíso is destroyed by an earthquake for the fifth time in two hundred years. The quake, felt throughout Chile's entire length, occurs four months after the infamous San Francisco earthquake, and inflicts far more destruction. Fifteen hundred people die and more than 100,000 become homeless. Property damage totals over $200 million.

1910 Chileans play in Argentine soccer tournament

Chile participates in a tournament organized by the Argentine Football Association to celebrate the one hundredth anniversary of Argentine independence. The event inspires the formation of the world's first regional soccer organization: the Confederación Sudamericana de fútbul (CONMEBOL) in 1916.

1912 Birth of leading painter Roberto Matta Echaurren

Roberto Matta Echaurren (b. 1912) (known simply as "Matta") is generally considered the foremost Chilean painter. Born in Santiago, he receives a Jesuit education and goes on to earn an architectural degree. In 1935 he moves to Paris, where he associates with the famed architect Le Corbusier and also with the surrealist painters. Matta then moves to New York from 1939 to 1948. He continues to live abroad, spending part of each year in Paris. His surrealistic paintings and prints win international acclaim for their originality, and they influence many younger painters.

1913 U.S. investment in El Tofo iron ore mine

The Bethlehem Steel Company takes out a lease on high-grade iron ore deposits at El Tofo in Coquimbo province.

1914–18 Chile exports nitrates for explosives during World War I

Officially, Chile remains neutral in World War I. However, it helps the Allies by providing nitrates for explosives. During the war, the Germans learn how to manufacture synthetic nitrates, a development that ultimately has a disastrous effect on the Chilean economy. In 1919, the nation supplies 90 percent of the world's nitrates; within ten years, this figure drops to 24 percent. Declines in the prices of both copper and nitrate further contribute to Chile's problems following the war.

1917 Birth of sculptor Marta Colvin

The work of Marta Colvin (b. 1917), Chile's best-known sculptor, reflects the cultural influences of her country's native Indian traditions going back to the period before the Spanish conquest.

1917–67 Life of singer and folklorist Violeta Parra

Parra, a popular singer-songwriter and folk song performer, begins her career singing in cafes and bars in Santiago. In the 1950s, she begins to collect and perform Chilean folk songs and, later, to write her own music in the style of the folk songs she performs. She lives in Paris in 1954–56 and 1961–65. In the 1960s, she starts a folklore center in a Santiago suburb. Violeta Parra commits suicide in 1967. Parra's music is one of the driving forces behind Chile's New Song movement of the late 1960s and early 1970s.

1920 Alessandri elected president

Reformist candidate Arturo Alessandri Palma (1868–1950) is elected president, and his strong popular support forces the conservative congress to approve his election. However, they block his political programs, creating a stalemate in government. Military right-wingers force him out of office in September 1924, but his own supporters in the military reinstate him in March 1925.

1924 Birth of novelist José Donoso

José Donoso (b. 1924) is Chile's best-known contemporary novelist, and one of Latin America's most renowned writers. His narrative innovations are compared to those of his Latin American contemporaries Gabriel García Marquez and Carlos Fuentes. Donoso is born and educated in Santiago and also attends Princeton University in the United States, where he studies English literature. His first novel, *Coronación* (Coronation) is published in 1957 and later wins the William Faulkner Foundation Prize for the Latin American Novel. In 1965–67, Donoso is invited to the prestigious Iowa Writers' Workshop in tne U.S. as a visiting faculty member.

Donoso has published both novels and short stories. His fiction is known for its exploration of the unconscious and the fantastic and its inclusion of themes from Chilean folklore and myth. His most acclaimed novel is *El obsceno pajaro de la noche* (The Obscene Bird of Night) (1970). Other works include *Este domingo* (This Sunday) (1966); *Casa de campo* (A House in the Country) (1978); *El lugar sin limites* (Hell Has No Limits) (1966); and *Tartuta: naturaleza muerta con cachimba* (Tartuta: Still Life with Pipe) (1990).

End of the Parliamentary Republic

1925 New constitution ends parliamentary rule

A new constitution ends parliamentary power and embodies many of Alessandri's reforms. It returns power to the executive branch, forbidding the removal of cabinet ministers by the legislature. The presidential term is extended from four

Most Chileans practice Roman Catholicism. Here over 100,000 children take communion at the eighth National Eucharistic Congress in 1941. (EPD Photos/CSU Archives)

years to six, but immediate reelection is forbidden. The new constitution also reflects the ongoing erosion of the Catholic Church's power by calling for the separation of church and state. It also provides for compulsory public education at the primary level, provides a legal basis for social welfare programs, expands voting rights, and recognizes the right of workers to unionize. The constitution of 1925 remains in effect until it is rewritten by the Pinochet regime following the military coup of 1973.

1927–1931 Ibáñez presidency

Once his constitution is approved, Alessandri resigns to make way for a new elected president. His successor, Emiliano Figueroa, is unseated after two years by Carlos Ibáñez del Campo (1877–1960), who presides over a four-year period of authoritarian rule. Chile's economy is relatively stable during the first part of this period, and Ibáñez implements educational and labor reforms.

However, conditions deteriorate once the Great Depression of the 1930s begins. Declining copper and nitrate prices destabilize the economy, and unemployment rises to 25 percent. The nation stops paying its foreign debt in 1931 and takes its currency off the gold standard the following year. Widespread strikes force the resignation and exile of Ibáñez in July 1931, and a year of political unrest and turmoil follow, in which seven different presidents hold office.

1929 Chile and Peru reach agreement on Tacna town

Chile agrees to return the town of Tacna, won during the War of the Pacific (1879–84) to Peru. The 1893 Treaty of Ancon initially stated that it was to be returned within ten years.

1932–38 Alessandri returned to the presidency

Chilean voters return Alessandri to office in the October 1932 elections. This time he is supported by a coalition of Liberals and Conservatives. His government backs social reforms and guides the nation out of the Depression. During his six years in office both communist and fascist parties rise in Chile. An attempted fascist revolt in 1938 is defeated.

1938–41 Cerda presidency

Backed by the Popular Front, a coalition of radical left-wing groups, Pedro Aguirre Cerda is elected president. During his three years in office (he dies before the end of his term), he institutes part of an ambitious social reform program, much of it meant to benefit the working class. His government also launches the Chilean Development Corporation (CORFO).

1939: January 24 Earthquake causes record number of deaths

Five historic towns are decimated in a cataclysmic earthquake that kills more people than any other earthquake in South America's history. Chillan, which is nearest to the epicenter, suffers the worst damage. Only a third of its buildings remain standing; hundreds are destroyed. Among those killed in the disaster are 300 theatergoers who are crushed during a performance when the theater building collapses.

Other damaged towns include Concepción (which loses 70 percent of its buildings and has hundreds of coal miners buried in collapsing mine shafts), Coiheuco, Coronal, and Angol. All together, 50,000 people are killed (70 percent of them children), 60,000 are injured, and over 700,000 lose their homes.

1939–45   World War II

Like its neighbor, Argentina, Chile remains neutral for part of World War II. Although there is substantial Chilean support for the Nazis, ties with the U.S. and the Allied alignment of most other Latin American countries influence Chile to declare war on the Axis powers by 1943.

1942   Birth of novelist Isabel Allende

Allende is one of the first Latin American women to gain widespread recognition as a novelist. Born in Lima, Peru, Allende is the niece of former Chilean president Salvador Allende. After the military coup that deposes her uncle, Allende, who has worked as a journalist in Chile, is forced into exile in Venezuela. A letter she starts writing to her grandfather eventually turns into her first novel, *La casa de los espíritus* (The House of the Spirits; 1982). Works that follow include the novels *De amor y de sombra* (Of Love and Shadows;1984) and *Eva Luna* (Eva Luna; 1987), and a short story collection, *Cuentos de Eva Luna* (Stories of Eva Luna;1990). In 1995 Allende publishes *Paula,* a memoir in the form of a letter to her dying daughter.

Stylistically, Allende's novels follow the "magical realist" tradition of such Latin American authors as Gabriel García Márquez and Julio Cortazár. In most cases, their content reflects the political realities of Allende's native land and of the role of women there.

1945   Poet Gabriela Mistral wins Nobel Prize for Literature

Gabriela Mistral (see 1889–1957) becomes the first South American (and the first Latin-American) to win the Nobel Prize for Literature.

1946–52   González presidency

Gabriel González Videla, backed by a leftist coalition, is elected president in September 1946. Although initially supported by the Communists, he breaks with them because of their disruptive tactics, including organizing strikes and demonstrations. In 1948 González breaks off diplomatic relations with the Soviet Union and outlaws the Communist Party in Chile. (It continues clandestine activities and is legalized again ten years later.)

Chile's economy remains troubled following World War II, and depends heavily on exports of copper and nitrate to the U.S. The nation's inflation rate remains high, and growing industrialization leads to labor unrest.

1949   Women are granted the right to vote

After a lengthy suffrage campaign, Chilean women win the vote. However, they are required to vote separately from men.

1949   University of Santiago de Chile founded

The second of Chile's two state-run universities, the Universidad Técnica del Estado (now called the University of Santiago de Chile) is established.

1952–58   Carlos Ibáñez returned to the presidency

Voters return 75-year-old former dictator Carlos Ibáñez to office in hopes that he will take measures to improve the nation's ailing economy. Although he spends the first part of his term fighting off political opposition, he imposes a strong anti-inflation policy that has dramatic effects between 1957 and 1958.

1958–64   Alessandri presidency

Jorge Alessandri Rodríguez, son of former president Arturo Alessandri Palma, is elected president. His government funds major public works programs to reduce unemployment, adopts anti-inflation measures, and implements education and housing programs. However, Alessandri's efforts are marred by the economic effects of natural disasters—volcano eruptions, earthquakes, and tidal waves—in 1960.

1960   Chilean woman holds top international labor post

Ana Figueroa (1908–70) is appointed Assistant Director-General of the International Labor Organization (ILO), becoming the first woman in the history of the organization to hold this post.

1960: May 21–30   Ten-day series of natural disasters kills thousands

A series of earthquakes in rapid succession sets off an unprecedented chain reaction of natural disasters that victimize thousands of people, destroys 20 percent of the nation's industrial facilities, and brings emergency relief aid from thirty-five foreign nations. Starting just after 6:00 A.M. on May 21, five earthquakes occur within a forty-eight-hour period. Concepción, which has suffered major earthquake damage in the past, is leveled once again, as are Osormo, Valdivia, and Puerto Montt.

The earthquakes are followed by tsunamis (giant tidal waves), the largest attaining a height of twenty-four feet and reaching speeds of up to 520 miles per hour. It flattens fishing villages and floods the entire coast. After receding from the Chilean coast, the tsunami reverses direction, eventually rolling into Japan, where it kills 150 people. The series of earthquakes also sparks nine volcanic eruptions. Overflowing lava causes Lake Ranco to burst over its banks, resulting in mudslides that are lethal for both people and livestock.

In the immediate aftermath of the disaster, 100,000 Chileans desperate for food and shelter engage in violent rioting that has to be subdued by the military and the police. The damaged areas are eventually rebuilt with millions of dollars in international aid, but 5,700 people lose their lives in the ten-day cataclysm.

1962   Chile hosts World Cup soccer tournament

Chile is the site of the World Cup competition, in which it places third. Chilean teams qualify for the World Cup games in Munich in 1974 and in Spain in 1982.

1964   Scientific academies founded

Chile's two major scientific organizations, the Academy of Medicine and the Academy of Sciences, are both founded in 1964. In addition to these groups, there are dozens of learned societies in various scientific fields.

Chile's Government Moves to the Left

1964–70   Eduardo Frei serves as president

In September 1964 voters elect moderate Eduardo Frei (1911–82) to the presidency by a wide margin, giving his Christian Democrat party control of the congress in the March 1965 legislative elections. Frei begins the process of nationalizing U.S.-owned copper mines and carries out educational, labor, tax, and land reforms. In 1966 the government passes a law giving Chile fifty-one percent ownership of all copper mines. The U.S. government strongly supports Frei's policies as a model of social change that offers an alternative to communist revolutions like the one that has taken place in Cuba. Frei's government receives large amounts of U.S. aid and is a major participant in the Alliance for Progress program.

Late 1960s–early 1970s   New Song cultural movement flourishes

The New Song movement is a powerful cultural force that grows up around the work of poet/singer/songwriter Violeta Parra, who dies in 1967. Steeped in the traditions of native Chilean folk music, her songs express the spirit of the Chilean people and serve as a rallying point for many during the Pinochet era.

1969   Voting age lowered to eighteen

The Chilean voting age is lowered to eighteen as one of the reforms carried out by the administration of Eduardo Frei.

Salvador Allende (1908–73), socialist president, waves a Chilean flag. His attempts to make Chile a socialist country are resisted and eventually leads to his ouster and death in a military coup. (Corbis Corporation)

1970–73   Salvador Allende presidency

In 1970, dissatisfied with the moderate pace of change under the Frei administration, voters give socialist Salvador Allende Gossens (1908–73) a plurality in the presidential elections. Allende proceeds to move Chile toward socialism by a series of peaceful and gradual measures that include nationalizing its copper and nitrate mines, banks, and other industries. He also establishes diplomatic relations with communist countries including Cuba and the People's Republic of China.

1970–71   Growing opposition to Allende's policies

Allende's administration is plagued by economic problems, including sharply declining copper prices, inflation that reaches 300 percent by September 1973, and suspension of credit by banks in the U.S. and other countries. Chilean society is increasingly divided along class lines, with the working classes favoring Allende's policies and the middle class opposing them. Middle-class groups stage a series of mass protests in Chile's major cities in 1971 and 1972, and small businessmen and professionals hold a week-long general strike. A truckers' strike in October 1972 leads Allende to impose martial law in thirteen Chilean provinces.

1971   Poet Pablo Neruda wins Nobel Prize for Literature

Pablo Neruda wins the Nobel Prize for Literature. Neruda, whose real name is Neftalí Ricardo Reyes Basoalto, was born in southern Chile (see 1904–1973).

The Pinochet Regime

1973: September 11   Allende ousted in military coup

After an abortive attempt in June, the military succeeds in ousting Allende from power in a violent coup that is supposedly aided by the destabilization efforts of the United States Central Intelligence Agency (CIA). President Allende dies in the coup. He allegedly commits suicide in the ruins of the bombed presidential palace. The coup leaders form a four-member ruling junta led by General Augusto Pinochet Ugarte (b. 1915) and disband congress indefinitely.

1973–90   Pinochet regime

General Augusto Pinochet, leader of the military coup that ousts Salvador Allende, remains in power through the 1970s and 1980s as both president and commander-in-chief of the military. His government consolidates its power by a campaign of severe political repression, especially in the early and mid-1970s. Thousands of Allende supporters are imprisoned, exiled, or "disappeared." Many are tortured and killed. The press is strictly censored, all opposition parties and labor unions are outlawed, and strikes are banned. Pinochet discontinues Allende's social reform programs and sells nationalized industries back to private investors. Although condemning the worst excesses of the Pinochet government, the U.S. supports its reversal of Allende's socialist agenda.

Economically, the Pinochet regime implements "neoliberal" economic policies that dramatically reduce the government's role in the economy in terms of both regulation and ownership. Price controls are lifted, tariffs slashed, and government spending for social programs cut. State-owned industries are returned to private ownership, and rural land is returned to large landowners. By 1980 state ownership of the nation's industries has fallen from 77 percent to less than 30 percent. Although these economic policies produce rapid economic growth between 1977 and 1981, unemployment increases dramatically and real wages fall.

1976   Chilean exile assassinated in U.S.

Relations between Chile and the United States are temporarily strained when Orlando Letelier, a Chilean who has been part of the Allende government, is assassinated in Washington, D.C. by agents of Pinochet's secret police. U.S. support for Chile also wanes during this period because of the human rights policies of President Jimmy Carter. In response to U.S. pressure, the Chilean ruling junta lifts its state of siege in 1978 and holds a tightly monitored plebiscite (vote by qualified voters to decide an important issue) to test popular endorsement of Pinochet's rule.

1980   New constitution approved

After a campaign that doesn't allow any open opposition, 68 percent of Chilean voters approve a new constitution legitimizing Pinochet's rule and extend his presidency for eight more years. Political parties remain outlawed under the "authoritarian democracy" outlined in the new constitution, which also provides for a return to partial civilian rule after Pinochet's term ends.

1981   New social security program inaugurated

Chile's new social security program is managed by private organizations. Organized in the form of an individual savings plan, it does not depend on large contributions by workers.

1981   Hydroelectric project opens at Colbun-Machicura-Chiburgo

A major hydroelectric project—one of South America's largest—is completed.

1982   Chile is once again world's top copper producer

Chile eclipses the United States by once again becoming the world's largest producer of copper. Chile lost the number one spot to the U.S. exactly one hundred years before, in 1882.

1982   Economic crisis fuels opposition to Pinochet regime

Falling copper prices, heavy military spending, and large international debt payments contribute to an economic recession and general crisis that weakens Chile's banks and causes many businesses to fail. By 1982, crowds show their anger toward Pinochet at the funeral of former president Eduardo Frei. The following year, widespread demonstrations take place in response to a ninety-day government moratorium on payment of its foreign debt.

By 1983 unemployment in Chile reaches 30 percent. Between 1983 and 1985 anti-government protests spread to diverse segments of Chilean society, from the middle class to impoverished shantytown dwellers, and win the support of the Roman Catholic church. In May 1985 the government imposes a ninety-day state of siege to restore order. By the time the economy begins to improve in the mid-1980s, support for the Pinochet regime is seriously eroded.

With the promised 1988 plebiscite approaching, Pinochet faces increased opposition, both at home and abroad, including an assassination attempt in September of 1986. The murder of a recently returned nineteen-year-old political exile in 1986 spurs renewed international condemnation, and Pope John Paul II, visiting Chile in April 1987, openly criticizes the Pinochet government. By 1987, members of Pinochet's own ruling junta are demanding a return to civilian rule. In 1988 opposition political activity is once again allowed, and the suspension of constitutional freedoms ends.

1983   Female-run radio station established

Radio Tierra is an all-female radio station that addresses itself to the concerns of women. It claims to be the first such station established in the Americas.

1985: March   Major earthquake strikes central Chile

An earthquake measuring 7.4 on the Richter scale causes hundreds of deaths and injures over two thousand people. Many thousands are left homeless, as houses and public buildings collapse. Santiago, Viña del Mar, and Valparaíso are among the cities sustaining the worst damage.

1985: September 5   Women protest repressive tactics of Pinochet regime

Chilean women organize a demonstration in Santiago to protest the repressive rule of the country's military government. The protest is broken up by troops with water cannons.

1987   Pope John Paul II visits Chile

Pope John Paul II makes a four-day visit to Chile. At the National Sanctuary of Maipú he crowns the statue of the Virgin del Carmen. In speeches, he refers to Chile's military government as "dictatorial" and "transitory."

1987   Opposition newpapers appear

Toward the end of the Pinochet regime, opposition voices find expression in the media. *La Época* is founded by the Christian Democrats. *Fortín Mapocho,* an existing leftist publication, becomes a daily newspaper.

1988   Labor confederation formed

The Confederación Unica de Trabajadores (Unitary Confederation of Labor, or CUT) is established as an umbrella organization for mining, industrial, and professional unions. It replaces the older Coordinadora Nacional de Sindicatos (CNS).

1988: October 5   Pinochet defeated in plebiscite

The long-awaited presidential plebiscite is held in 1988, with voters offered the choice of approving or rejecting a single candidate—Pinochet. Without the resources to win the election by force or fraud, Pinochet is decisively defeated and democratic elections are scheduled for 1989.

Return to Democratic Rule

1989: December   Independent central bank established

A central bank free of government control—the first such institution in Latin America—is founded.

1989: December 14   Aylwin elected president in democratic elections

Patricio Aylwin Azócar, a moderate Christian Democrat, is elected president with backing from both centrist and leftist parties. Aylwin, who takes office in March 1990, appoints a cabinet with a broad political base and announces a general amnesty for nonviolent political prisoners. Congress, dissolved since 1973, reconvenes in its new Valparaíso headquarters.

1990   Aylwin government probes human rights violations

The administration of President Aylwin begins to investigate human rights abuses by the Pinochet regime. Mass graves with bodies of political prisoners are found in the Atacama Desert. It is estimated that the Pinochet regime had as many as 3,000 people killed.

1991   Mudslides cause death and destruction in Antofagasta

Mudslides in Antofagasta kill 116 people and leave about 20,000 persons homeless after entire shantytowns are washed away.

1991   Chile and Mexico sign free-trade pact

The leaders of Chile and Mexico sign an agreement to allow free trade between their nations, the first such agreement in Latin America.

1993   Former head of secret police sentenced for political assassination

Manuel Contreras Sepúlveda, a former head of DINA, the Chilean secret police, is sentenced for the 1976 assassination of former Allende government official Orlando Letelier, while the latter was in exile in the United States.

1993   First Chilean saint canonized

Teresa de Jesus (Juana Esrequeta Josefina Fernandez Solar), a Carmelite novice who died in 1920, becomes the first Chilean to officially be made a saint when she is canonized by Pope John Paul II.

1993: December   Frei elected president

In the second free presidential election since the end of the Pinochet regime, Eduardo Frei Ruiz-Tagle, the son of former Chilean president Eduardo Frei Montalva, receives 58 percent of the popular vote. Frei runs with the backing of a center-left coalition. However, the upper house of congress remains under the control of conservative interests with the power to block electoral and constitutional reforms and other legislation.

1995   U.S. backs Chilean admission to NAFTA

The United States announces that it will seek to have Chile admitted to the North American Free Trade Agreement (NAFTA). Chile also seeks admission to the Southern Common Market Customs Union (MERCOSUR), a trade alliance between Argentine, Brazil, Uruguay, and Paraguay

1996: November 10–11   Summit meeting held in Chile

The sixth annual Ibero-American Summit is held in Viña del Mar. It is attended by the heads of state of most Latin American countries as well as the leaders of Spain and Portugal.

1997: June   El Niño blamed for casualties and damage in Chile

El Niño is linked to three weeks of flooding and mudslides that result in 17 deaths and $200 million worth of damage.

1997: September 8   Chamber of Deputies approves divorce bill

By a vote of 55 to 26, Chile's lower house approves a bill to legalize divorce, in spite of strong opposition by the Roman Catholic Church. The divorce legislation is not due for a Senate vote until 1998.

1997: October 14   Earthquake strikes north-central Chile

Eight people die and about 100 more are injured when an earthquake measuring 6.8 on the Richter scale strikes north-central Chile. The town of Punitaqui, about 250 miles (400 km) north of the capital city of Santiago, receives the brunt of the damage.

1997: December 11   Ruling party retains majority in midterm elections

Chile's ruling Concertacíon coalition is victorious in midterm congressional elections, winning just over 50 percent of the votes cast, compared to 36 percent for its nearest competitor. Concertacíon thus maintains the same number of seats in the Chamber of Deputies (70 out of 120). Even though voting is mandatory in Chile, close to 20 percent of registered voters stay home, facing possible fines. Others leave their ballots blank or deface them. In addition, nearly one million eligible young people do not register to vote.

Concertacíon fails to gain a majority in the Senate (the upper house of congress), and thus remains unable to amend provisions of the 1980 constitution approved during the Pinochet regime. This constitution bars the president from removing military officers and requires that four Senate seats be held by former military commanders.

Bibliography

Arriagada Herrera, Genaro. Trans. Nancy Morris. *Pinochet : The Politics of Power.* Boston: Allen & Unwin, 1988.

Blakemore, Harold. *Chile.* Santa Barbara, Calif.: Clio Press, 1988.

Caviedes, César N. *Elections in Chile: The Road Toward Re-democratization.* Boulder, Colo.: Rienner, 1991.

Collier, Simon. *A History of Chile.* Cambridge: Cambridge University Press, 1996.

Falcoff, Mark. *Modern Chile, 1970–89: A Critical History.* New Jersey: Transaction Books, 1989.

Fleet, Michael. *The Rise and Fall of Chilean Christian Democracy.* Princeton: Princeton University Press, 1985.

Hudson, Rex A., ed. *Chile, a Country Study.* 3rd ed. Washington, D.C.: Federal Research Division, Library of Congress, 1994.

Kaufman, Edy. *Crisis in Allende's Chile: New Perspectives.* New York: Praeger, 1988.

Lowden, Pamela. *Moral Opposition to Authoritarian Rule in Chile 1973–90.* New York: St. Martin's Press, 1996.

Nunn, Frederick M. *The Military in Chilean History: Essays on Civil-Military Relations, 1810–1973.* Albuquerque: University of New Mexico Press, 1976.

Petras, James F. *Democracy and Poverty in Chile.* Boulder, Colo.: Westview Press, 1994.

Politzer, Patricia. Trans. Diane Wachtell. *Fear in Chile: Lives under Pinochet.* New York: Pantheon Books, 1989.

Silva, Eduardo. *The State and Capital in Chile : Business Elites, Technocrats, and Market Economics.* Boulder, Colo.: Westview Press, 1996.

Smith, Brian H. *The Church and Politics in Chile: Challenges to Modern Catholicism.* Princeton: Princeton University Press, 1982.

Colombia

Introduction

In spite of its history of democratic government and rich cultural tradition, many people today primarily associate Colombia with the political and drug-related violence that has gripped the country for over two decades. Columbia was once the centerpiece of Simon Bolívar's dreams of a peaceful, united South America but has become the scene of its longest and most intractable guerrilla rebellion, and the producer of four-fifths of the world's cocaine.

Colombia is the fourth-largest country in South America and is second only to Brazil in population (estimated at 36.8 million in 1996). Anyone traveling from North to South America by land must pass through Colombia, and it is also the only South American country with coastlines on both the Atlantic Ocean (by way of the Caribbean Sea) and the Pacific. The Andes mountains dominate the western part of Colombia, breaking into three chains, or *cordilleras,* with the country's two rivers, the Cauca and the Magdalena, running between them. To the east of the mountains lie grasslands (*llanos*) and rain forest (*selva*), and there are lowlands along both the Caribbean and Pacific coasts.

The Colonial Era

Little is known of Colombian history before the Spanish Conquest. Landing on the Caribbean coast in 1499, Alonso de Ojeda encountered an advanced culture of Chibcha-speaking natives. The Muisca cultivated corn and potatoes utilizing irrigation and terracing techniques and the Tairona produced fish and salt and lived in towns connected by a system of paved roads.

The promise of riches, rumored to exist in abundance, encouraged the Spanish to settle there. Although the settlement founded by Ojeda in 1510, Santa María la Antigua de Darién, lasted only through the years of exploration, Santa Marta, founded in 1525 by Rodrigo de Bastidas is still Colombia's oldest settlement.

During the colonial era (1500–1810), Colombia was called New Granada. In 1717 the colony, which had been part of the Viceroyalty of Peru since the sixteenth century, became part of the Viceroyalty of New Granada. This new viceroyalty also included the colonies of Venezuela, Ecuador, and Panama, an administrative link that was to persist beyond the colonial era. The factional strife that was to plague Colombian politics was evident from the beginning as the newly formed colony divided along racial lines. At the top were the *peninsulares*, the Spanish-born who held political power and the highest social prestige. Below them were *criollos,* of Spanish descent but born in the colonies, and the *mestizos* of mixed Spanish and Indian descent. These divisions transformed into political factions, prevented the creation of a unified central government, allowing Spain to retake areas liberated between 1810 and 1815. Final victory came only through the efforts of the South American liberator Simón Bolívar (1783–1830) in 1819. To help fulfill Bolívar's vision of a political union that reached throughout the continent, Venezuela, Panama, and Ecuador became part of the new Republic of Gran Colombia. By 1830, however, both Venezuela and Ecuador had seceded to form nations of their own. Panama, remained a province of Colombia until the twentieth century.

By 1850 the two political groups that would dominate Colombia through the nineteenth and twentieth centuries were well-defined. The Conservatives were identified with the nation's privileged elite and sought to maintain existing class structures.

They were staunch defenders of the Catholic church and advocated a strong central government. The Liberals sought equality for the rising middle class and the poor and were in favor of giving less power to the central government and more to the individual provinces (later to become states). They also opposed the special ties between the government and the Catholic church and fought to eliminate the privileged position traditionally held by the church.

The nineteenth century saw a long period of Liberal rule (1863–80), plagued by civil unrest and political instability. In 1880, President Rafael Nuñez (1825–94), a moderate politician with a Liberal background, adopted many aspects of the Conservative agenda and brought unity and stability to a country formerly torn by civil strife and rebellion. Under Nuñez, Colombia adopted the Constitution of 1886 which reflected conservative principles, including making Roman

COLOMBIA

0   50   100   150   200   250 Miles

0   50 100 150 200 250 Kilometers

NETHERLANDS
ANTILLES

Caribbean Sea

Uribia

Santa Marta

Barranquilla

Cartagena

Cristóbal Colón Pk.
18,947 ft.
▲ 5775 m.

*Gulfo
de Venezuela*

Valledupar

*Lago de
Maracaibo*

PANAMA

Sincelejo

Montería

Turbo

*Gulfo
de Panamá*

Guapá

Atrato

Cauca

Cúcuta

La Fria

Bucaramanga

Barrancabermeja

Arauca

Elorza

V E N E Z U E L A

Juradó

Baudó

San Juan

Medellín

C O R D I L L E R A   O C C I D E N T A L

C O R D I L L E R A   C E N T R A L

Magdalena

Tunja

Yopal

Puerto
Carreño

L L A N O S

PACIFIC

OCEAN

Ambalema

Bogotá ✪

Ibagué

Villavicencio

Meta

Buenaventura

Palmira

Cali

Neiva

C O R D I L L E R A   O R I E N T A L

Guayabero

San José del
Guaviare

Guaviare

Puerto
Inírida

Popayán

Patío

Vaupés

Mitú

Apaporis

Uaupés

Tumaco

Pasto

Tres Esquinas

Caas

Monclar

Caquetá

E C U A D O R

B R A Z I L

Rocafuerte

Napo

El Encanto

Putumayo

PACIFIC
OCEAN

NICARAGUA

Providencia

San Andrés

Caribbean Sea

COSTA
RICA

PANAMA

COLOMBIA

Malpelo

P E R U

N
W   E
S

Amazon

Leticia

Colombia

Catholicism the state religion. The country finally became known as the Republic of Colombia.

After the period of peace overseen by Nuñez, tensions between Liberals and Conservatives again erupted into the War of a Thousand Days (1899–1902), in which 100,000 people died. Following this devastating war, Colombia was unable to prevent the secession of Panama (1903), aided by the United States, seeking to build a canal across the isthmus.

Under a succession of Conservative presidents, Colombia entered a period of national reconciliation and prosperity that also saw the rapid growth of the coffee industry. However, prosperity was still felt largely by the country's privileged classes. In the twentieth century, the demands of the nation's disenfranchised poor—its peasants and laborers—to have a voice in the political process and a share in the nation's wealth fueled an active labor movement that led to a period of Liberal reform (1930–45).

After World War II these long-neglected demands became an element in a tide of violence that swept the nation, beginning with a period of widespread civil unrest between 1947 and 1958 known as *La violencia*. Following the failure of either the Conservatives or Liberals to contain the violence, an unprecedented agreement was forged between the two groups. They agreed to a sixteen-year power-sharing plan known as the National Front, by which the presidency alternated every four years between Liberals and Conservatives, and members of both parties played an active role in every government. The National Front plan was followed successfully from 1958 to 1974, but just as it was bringing a measure of stability to the political process, a new threat emerged. Left-wing guerrilla groups such as M-19 and the Colombian Revolutionary Armed Forces (FARC) began terrorist campaigns of kidnappings, bombings, and assassinations which have persisted into the 1990s. A recent hopeful sign, however, has been the transformation of the M-19 organization from a guerrilla group to a legitimate political party.

Another source of violence in Colombia has been the growth of the international cocaine trade since the 1980s. In the late 1970s, drug dealers based in Medellín airlifted large amounts of cocaine to the United States and other countries. In the following years, Colombia became the supplier of eighty percent of the world's cocaine, and the drug lords of the Medellín and Cali cartels amassed enormous wealth and power. Government efforts to stop their activities have resulted in a bloody war of retribution involving bombings, assassinations, and other forms of violence. In 1989 the Liberal presidential candidate, Luis Carlos Galán, was assassinated by drug traffickers as part of this ongoing war. In the late 1990s, the drug trade continued to have strong repercussions for Colombian politics, as President Ernesto Samper was accused of accepting money from the Cali cartel to fund his 1994 campaign. The United States, in response, cut off economic aid designed to help Colombia fight the cartels.

Timeline

20,000–10,000 B.C.   First inhabitants settle in Colombia

Colombia is first inhabited by Indians traveling south from North and Central America.

1200 B.C.   New wave of migration from Central America

Indians migrating from Central America introduce agriculture, primarily the growing of corn, to Colombia.

400–300 B.C.   The Muisca settle in Colombia

Migrating from parts of Central America that today are Nicaragua and Honduras, the Muisca (also called the Chibcha, which is the name of their language) settle in the highlands of the Cordillera Oriental and develop an advanced civilization. The Muisca live in villages and towns and farm the region's fertile soil using terracing and irrigation techniques. Their staple crops include maize (corn), potatoes, and a grain known as *quinoa*. Their leadership is organized by matrilineal succession (descent through the mother). They are sun worshippers and skilled artisans, producing pottery, goldwork, and textiles as well as rock paintings and carvings.

5th century A.D.   The Tairona populate Northern Colombia

The Tairona Indians settle in the northern portion of the Sierra Nevada de Santa Marta and the Caribbean lowlands, forming two distinct groups. Among the more advanced early Indian civilizations, they build stone roads, and live in towns. The Tairona who live on the coast fish and produce salt.

Spanish Exploration

1499   Alonso de Ojeda sails along the Colombian coast

European exploration of Colombia begins with Spanish navigator Alonso de Ojeda (1465–1515), a veteran of Christopher Colombus's second voyage, who sails along South America's Caribbean coast, accompanied by Juan de la Cosa, and comes ashore on the Guajira peninsula. He explores the Sierra Nevada de Santa Marta, discovers the local Indian population, and hears tales of fabulous wealth which give rise to the legend of El Dorada, the fabled land of emeralds and gold that is to haunt many future Spanish explorers of the New World.

1500   Rodrigo de Bastidas leads Colombian expedition

Rodrigo de Bastidas(1460–1526) leads the second Spanish voyage to Colombia, initially called Tierra Firme. He sails the entire length of Colombia's Caribbean coast, from Cabo de la Vela to the Gulf of Urabá.

1505 Juan de la Cosa sails to Colombia

Explorer Juan de la Cosa(1460?–1510), who has accompanied both Ojeda and Bastidas on previous expeditions, returns to Colombia. He seizes Indian towns and transports some of their inhabitants to the island of Hispaniola for enslavement.

1508 First European attempts at settlement are made

An expedition that includes Alonso de Ojeda and Juan de la Cosa, who have previously sailed to Colombia, is organized to found the first settlements. These include San Sebastián de Urabá and Santa María la Antigua del Darién on the shores of the Gulf of Urabá. Neither settlement survives past the initial period of exploration.

1520 Church begins conversion of Indians

The Catholic Church begins evangelizing among the native population of Colombia. Dominicans and Franciscans are the first orders to arrive in Colombia., but they are followed by others.

1525 Santa Marta is founded

Santa Marta, Colombia's oldest surviving settlement, is founded by Rodrigo de Bastidas.

1533 Cartagena is founded

The town of Cartagena is established on the Caribbean coast by Pedro de Heredia and becomes the main center for trade and travel in the region and provides security from Indian attacks.

1534 First dioceses are founded by the Catholic Church

The first Catholic dioceses in the New World are established in Santa Marta and Cartagena.

1536 Trio of explorers set out to probe the interior

Motivated by Francisco Pizarro's conquest of Peru in 1533, three different explorers—Gonzalo Jiminéz de Quesada (1495–1579), Sebastián de Belalcázar (c. 1495–1551), and Nikolaus Federmann—independently launch expeditions into the Colombian interior hoping that they will yield similar riches. Coming from different starting points and traveling away from the coast, all three expeditions eventually meet near the newly founded city of Bogotá.

1536 Belalcázar founds Cali and Popayán

Sebastián de Belalcázar, traveling north from Quito (the capital of present-day Ecuador) through the Cauca Valley, founds the city of Cali. A year later, he founds the city of Popayán.

1538 Quesada founds the city of Bogotá

After conquering the Muisca Indians living in the region, Gonzalo Jiminéz Quesada founds the city of Santa Fe de Bogotá (present-day Bogotá) and claims the surrounding territory for Spain, calling it the Nuevo Reino de Granada (Kingdom of New Granada), To Quesada's disappointment, his men find little gold in the area. Shortly after the founding of Bogotá, the paths of Belalcázar's and Federmann's expeditions converge there as well, and the three men negotiate to divide the rights to the newly conquered territory among themselves. These expeditions lay the groundwork for exploration and settlement of the Colombian interior.

Colonization

1550 Spain forms Audiencia of New Granada

The Real Audiencia del Nuevo Reino de Granada, based in Bogotá, is created by King Charles V of Spain to administer the territory of New Granada . Through a series of governors, it controls the cities, which also elect *cabildos*, or local councils. In addition, New Granada as a whole is placed under the jurisdiction of the Viceroyalty of Peru, whose headquarters are in Lima.

1550–1640 Colombia's first gold boom

The first great period of gold mining in Colombia begins with the discovery of a deposit in the northwest. Additional deposits, mostly in Cáceres and Zaragoza, are mined after 1580, using the labor of black slaves. After peaking in the 1590s, output from this gold boom declines in the first half of the seventeenth century.

1560 Cartagena is attacked by French pirates

Beginning close to two centuries of aggression and plunder by British, French, and Dutch buccaneers in search of gold and silver, the coastal city of Cartagena is attacked by Jean and Martin Côte, French pirates.

1560 Colombia's population stands at 6,000–8,000

An estimated 6,000–8,000 colonists are widely scattered throughout Colombia, including the coastal cities of Santa Marta and Cartagena, the city of Bogotá in the interior, and the province of Popayán to the south and west.

1561 Death of musician Juan Pérez Materano

Conductor Juan Pérez Materano (birthdate unknown), the first to perform sacred music in Cartagena, dies.

1572 National University founded

The National University in Bogotá is founded. It is one of the first universities to be established in the Americas.

1580   University of Santo Tomás is founded

The University of Santo Tomás is established in Bogotá.

1584   Gutierre Fernández is appointed chapelmaster of Bogotá cathedral

Well-known composer Gutierre Fernández Hidalgo (1553–1620) becomes the chapelmaster of the cathedral at Bogotá, the musical center of colonial Colombia. He conducts a chorus of seminarians from the Seminario Conciliar de San Luis. Born in Andalusia, Fernández becomes the foremost sixteenth century composer to live and work in South America. The polyphonic style of his compositions reflects the influences of Spanish composers including Tomás Luis Victoria and Cristóbal de Morales. He later lives in Quito (in present-day Ecuador), Cuzco (Peru), and Sucre (present-day Bolivia).

1586   Sir Francis Drake attacks Cartagena

The English pirate Sir Francis Drake (c. 1540–96) attacks Cartagena, but its residents have already fled the city, taking their treasure with them.

17th century   Colonial economy shifts to agriculture and textiles

As gold mining declines, agricultural output expands and textile production is introduced. In one region, Indian laborers are put to work making woolen cloth, blankets, and clothing, which is then traded. Elsewhere, rudimentary cotton textiles are produced for domestic use and trade. These advances make Colombia less reliant on commodity imports from Spain as well as less dependent on the ups and downs of gold mining.

1608   Fugitive slaves build *palenques*

Fugitive slaves, known as maroons, join together and build fortified communities known as *palenques*. One of the oldest is founded near Cartagena under the leadership of the fugitive slave and former African king, King Benkos.

1611   Spanish Inquisition established in Colombia

A tribunal of the Spanish Inquisition is set up in Colombia, primarily to guard against the influence of the Animist religions practiced by the Africans being imported into the region as slaves.

1615–1738   Figueroa artists' workshop is in operation

The foremost training ground for Colombian artists in the seventeenth and early eighteenth centuries is the workshop run by Gaspar de Figueroa (d. 1658) and his family, where young artists learn their craft through apprenticeships. The foremost artists of this period, including Gregorio Vázquez de Arce y Ceballos (see 1638), receive their training here.

1622   Jesuit university is founded

The Univeridad Javeriana is founded by the Jesuit order (a Catholic sect).

1634   Cartagena builds a wall to keep out invaders

Residents of Cartagena, besieged by pirates for over eighty years, begin constructing a thick wall that will encircle the city and protect it. In most places, it is at least thirty-nine feet (twelve meters) high and fifty-six feet (seventeen meters) thick. It is nearly a century before the wall is completed.

1638   Birth of artist Gregorio Vásquez de Arce y Ceballos

Ceballos (1638–1711) is the foremost artist of the colonial era. He lives and works in Bogotá, leaving a total of over 500 paintings, which are displayed in the city's museums and churches. A collection of his sketches is on display at the Museo de Arte Colonial in Bogotá.

Late 17th century   Second gold boom begins

New gold discoveries, mostly in the Pacific lowlands and the Central Cordillera, prompt a second major period of gold production that lasts until the end of the colonial era.

1697   Baron de Pointis invades Cartagena

Breaking through the city's partially completed defenses, French Admiral Baron de Pointis attacks Cartagena with a force of ten thousand men and plunders the city.

1698   Scots found a colony

A Scottish colony is established at Darién, challenging Spanish rule.

1711   Death of painter Gregorio Vásquez de Arce y Ceballos

Colombia's foremost colonial painter dies. (See 1638.)

1717   Viceroyalty of New Granada is formed

Splitting Colombia off from the Viceroyalty of Peru, Spain founds the Viceroyalty of New Granada, which includes not only Colombia but also present-day Panama, Venezuela, and Ecuador. Bogotá becomes the capital of the new viceroyalty. Except for a period between 1723 and 1739 when the viceroyalty disbanded, it administers all affairs in Colombia until independence from Spain is won. Under its rule, trade restrictions are relaxed. The region's economy grows, and cultural and intellectual activity flourish.

1732: April 6   Birth of renowned botanist and physician José Celestino Mutis

Mutis is born in Cádiz, Spain, and pursues studies in botany, medicine, and other sciences. In 1760 he is sent to New Granada as personal physician to the viceroy, Pedro de Messía de

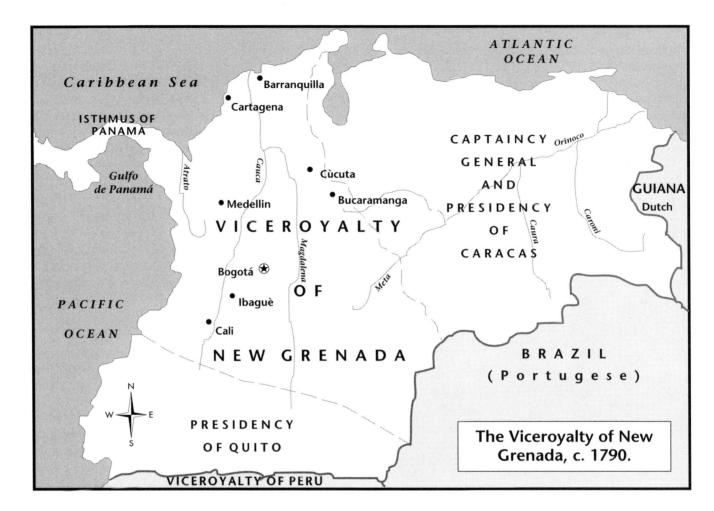

The Viceroyalty of New Grenada, c. 1790.

la Cerda. He is one of only two doctors in the colony. He undertakes an extensive study of the medicinal value of plants found in the region. He also implements a number of public health projects, including measures to control epidemics. In 1783 Mutis is appointed First Botanist and Astronomer of the King and leads the historic Expedición Botánica documenting the plant life of New Granada. Mutis plans to publish thirteen illustrated volumes of findings but his medical duties prevent their completion. However, after his death in 1808, his manuscripts and drawings are archived in Madrid.

1770 Birth of botanist José de Caldas

Botanist José de Caldas (1770–1816) is born. He develops a way to calculate altitude using the boiling point of water. He is generally credited with producing the first scientific publications in Colombia.

1781 Early colonial rebellion is crushed

The *comunero* revolt originates as a protest against measures implemented by Spanish envoy Juan Gutiérrez de Piñeres to reform taxes for higher yields. At its height, the revolt involves 20,000 people. Although its slogan is "Long live the king and down with bad government," many still see it as a precursor of the struggle for independence from Spain that will take place a few decades later. The revolt is brutally suppressed and its leaders are executed.

1783–1810 Expedición Botánica records native species

The Expedición Botánica is the first of three great botanical surveys of South America. It is led by distinguished botanist and physician José Celestino Mutis (see 1732), together with eighteen scientists sent from Spain. In addition to amassing voluminous notes and illustrations for 130 families of plants and 20,000 herbs, the Expedición also studies the fauna and minerals of the region and establishes an astronomical observatory.

In addition to its scientific import, the expedition is a milestone in the development of both art and literature in Colombia. A special class is set up to teach the illustrators, who are officially employed by the Spanish monarchy. Over 5,000 illustrations are still preserved in the Real Jardín Botánico in Madrid, Spain, together with 4,000 manuscript pages from the Expedición.

1783: July 24   Birth of South American liberator Simón Bolívar

Simón Bolívar Palacios (1783–1830) is born in Venezuela. Known as Simón Bolívar, he plays a central role in winning Colombian independence and in the nation's early political history—including serving as its first president. When his newly created government in Venezuela is overthrown by royalists in 1814, Bolívar takes refuge in New Granada (Colombia) and allies himself with one of the rebel factions there, helping them take Bogotá. After further military campaigns in Venezuela, Bolívar undertakes a march across the Andes mountains to liberate New Granada and wins a decisive victory at the Battle of Boyacá on August 7, 1819. He is instrumental in uniting New Granada, Venezuela, and Quito (Ecuador) into the Republic of Gran Colombia and takes part in the constitutional congress of 1821, which names him as the first president of the new republic. However, he leaves soon afterwards to direct further military campaigns, and plays a part in the liberation of Peru and Bolivia (which was named for him), and the full liberation of Ecuador.

Returning to New Granada in 1926, he finds the government facing a rebellion and finds himself increasingly at odds with his vice president, Francisco de Paula Santander. Faced with heavy opposition to his political agenda, Bolívar assumes dictatorial control of the country. In September 1828 he survives an assassination attempt, and in 1830 he resigns. Disillusioned by the failure of his grand schemes for South American political unification, he dies as he is setting off for a self-imposed exile. In succeeding years, however, his fame grows and he becomes renowned as the foremost figure in the struggle to liberate South America from colonial rule. Bolívar dies in 1830.

1794   Colombian translates the Rights of Man into Spanish

Antonio Nariño's translation and publication of Thomas Paine's *The Rights of Man* in Spanish is a major catalyst of the movement for independence from Spain. Nariño is one of several Creoles (Colombian-born persons of Spanish descent) who are arrested for sedition. Nariño is imprisoned in an African jail, but he escapes and secretly returns to Latin America, becoming a prominent participant in the campaign for liberation of the Spanish colonies.

1808   Napoleon Bonaparte names his brother king of Spain

When French leader Napoleon Bonaparte (1769–1821) invades Spain, overthrows the Spanish monarch Ferdinand VII (1788–1833), and places his brother Joseph (1768–1844) on the throne, many colonists throughout the Spanish empire begin to agitate for independence.

1808: September 11   Death of botanist and physician José Celestino Mutis

Pioneering botanist and physician, and leader of the Expedición Botánica, José Celestino Mutis (1732–1808) dies. (See 1732.)

1809   Quito revolts against Spanish rule

Quito, Ecuador, becomes the first entity of the Viceroyalty of New Granada to rebel against the Spanish. The rebellion is violently suppressed.

1810: June 14   Cartagena declares independence

Cartagena is the first city in Colombia to declare independence from Spain, replacing its royal governor with a ruling *junta* (committee).

1810: July   Independence movement spreads

The cities of Cali, Pamplona, and Socorro follow Cartagena in declaring independence and setting up their own governments.

1810: July 20   Bogotá declares independence

Bogotá throws off the rule of the Spanish viceroyalty. However, the city fails to establish a viable central government for New Granada because of disagreements among the representatives elected for this purpose. The region breaks down into competing governments that eventually go to war with each other, facilitating the Spanish reconquest later in the decade.

1812   Bolívar begins his struggle for Latin American independence

Venezuelan-born Simón Bolívar (1783–1830; see 1783: July 24) launches the military campaign that will eventually free Colombia, as well as Venezuela, Panama, Ecuador, Peru, and Bolivia from Spanish rule. He wins a series of battles against the Spanish military, some of them in Colombia.

1815–16   The Spanish reconquest

Once Napoleon is defeated and the Spanish monarchy is restored, Spain undertakes a vigorous fight to recapture its colonies. An expedition headed by Pablo Morillo is sent to South America to reconquer New Granada. With the help of royalist forces among the colonists themselves, Cartagena is retaken, and the Spanish soldiers move on to the interior of the region, where they reestablish control. However, the repressive government measures taken in the wake of the rebellion create widespread discontent. Pro-revolutionary sentiments spread even more widely than before the reconquest, preparing the way for victory by Bolívar's forces when they return.

1817   Birth of poet José Eusebio Caro

Poet José Eusebio Caro (1817–53) is born. He is among the important Romantic poets of Colombia, and is significantly influenced by the poets of England.

1819: August 7   Bolívar wins a decisive victory at Boyacá

After crossing the Andes from Venezuela with a force of 2,500 men, Bolívar defeats Spanish troops at the Battle of Boyacá, with assistance from forces led by Francisco de Paula Santander (1792–1840). After the battle, Bolívar marches into Bogotá, and independence for Colombia and the other former provinces of the Viceroyalty of New Granada is assured.

1819: December   New republic is officially formed

The Congress of Angostura officially founds the Republic of Gran Colombia, composed of the colonies that had made up the Viceroyalty of New Granada: Panama, Venezuela, New Granada, and (after 1822) Ecuador.

1821   Congress of Cúcuta draws up constitution

A constitution for the new republic is drawn up at the Congress of Cúcuta. A strong central government is created, with a president and a bicameral legislature. Freedom of the press and other civil rights are guaranteed, but only educated persons with property can vote.

At the Cúcuta Congress, divisions emerge between the centralists, who want a strong central government and the federalists, who want a looser confederation. The former are led by Simón Bolívar (1783–1830) and the latter by Francisco de Paula Santander (1792–1840). Bolívar prevails, and he is named president of the new republic, with Santander as vice president. However, Santander takes charge of the government when Bolívar leaves to fight for Peruvian independence. The unity of the Gran Colombia soon unravels as its member provinces reject central control by Bogotá.

1823   Museo Nacional is founded

Colombia's first museum is established in Bogotá. Its immediate purpose is to house the written and artistic works resulting from the Expedición Botánica (see 1784). By the late twentieth century, it houses all types of art work produced throughout Colombia's history, as well as significant archaeological and ethnological collections.

1825: September 28   Birth of writer and statesman Rafael Nuñez

Rafael Nuñez (1825–94), who, as president, will inaugurate over forty years of Conservative rule, is born in Cartagena and earns a law degree in the 1840s. After pursuing careers in law and literature, and being active in Liberal politics, Nuñez is dispatched abroad as a diplomat when the Liberals come to power in 1863. He spends twelve years living in Europe and the United States, where he serves as a consul and also writes for Spanish-language newspapers. Nuñez returns to Colombia in 1875 and launches a political career as a member of the Liberal party, running unsuccessfully for president.

In 1880 Nuñez is elected president and dismays his constituency with his Regeneration program, which abandons the Liberal program of the past twenty years. He calls for a new constitution. Nuñez presides over the creation of the Constitution of 1886, under which the Republic of Colombia is formed. This constitution remains in effect until 1991.

Except for two years between 1882 and 1884, Nuñez serves as president from 1880 to 1888 and remains the dominant figure in Colombian politics until his death in 1894. In addition to his achievement in unifying and stabilizing his country's government, he is also renowned for his poetry and for prose works including *Ensayos de critica social* and *La reforma politica en Colombia*.

1826   Venezuela withdraws from Gran Colombia

Under the leadership of separatist José Antonio Páez (1790–1873), Venezuela breaks away from Gran Colombia to form its own independent nation.

1828   Bolívar takes control of the government

Having returned to Colombia, Simón Bolívar (1783–1830; see 1783: July 24) assumes control but is unable to unify the nation's feuding political factions. His proposed new constitution for Gran Colombia is unpopular, and he resigns the presidency within two years.

1830   Ecuador secedes from Gran Colombia

Ecuador withdraws from Gran Colombia, leaving the territory that will become present-day Colombia, plus Panama, which does not achieve independence until 1903.

1830: December   Death of Simón Bolívar

The man revered as the liberator of Latin America dies in Santa Marta. (See 1783: July 24.)

1832   The Republic of New Granada is formed

The remaining territory of Gran Colombia, after the secession of Venezuela and Ecuador, becomes the Republic of New Granada, the forerunner of the present-day Republic of Colombia.

1832   Santander becomes president

War hero and vice president under Bolívar, Francisco de Paula Santander (1792–1840) is elected president. He institutes policies to establish economic stability, reform education, and reconcile the different sides in the debate over the separation of church and state.

1833   Birth of poet Rafael Pombo

Rafael Pombo (1834–1912) is among the foremost Colombian Romantic poets. He is also famed as the author of children's books including *El Renacuajo Paseador.*

1834   First Colombian coffee is exported

Colombia exports its first coffee. Coffee does not become a leading export until later in the century.

1837   Birth of writer Jorge Isaacs

Writer Jorge Isaacs (1837–95) is born. His most famous work is the Romantic novel, *María* (1867).

1839–42   War of the Supremes

Tensions over separation of church and state come to a head when the government orders the closing of convents in the city of Pasto. Opponents lead a three-year revolt against the government. This revolt, in conjunction with other similar revolts, leads to the Constitution of Rionegro of 1863. This Constitution allows for a high degree of autonomy among the differing Columbian states, and helps to further perpetuate the priviledges of the regional oligarchies.

1840s   Conservative and liberal positions are defined

The centralist/federalist split that has divided Colombians since the early days of independence evolves into an opposition between two organized political parties: the Conservatives and the Liberals. The Conservatives, who have inherited the centralist position, advocate a strong centralized government and authoritarian rule, support the traditional Spanish alliance between the state and the Catholic church, favor the continuation of slavery, and defend traditional distinctions between social classes. The Conservative constituency is made up of large landholders, slaveowners, and persons affiliated with the church.

The Liberals, many of whom are members of the rising merchant, manufacturing, and artisan classes, favor decentralization of government and, in particular, less power for the chief executive; separation of church and state; freedom of the press and other civil liberties; and equal voting rights for persons of all classes. Liberal governments dominate Colombian politics for most of the nineteenth century until 1880.

1847   Sociedad Filarmónica founded

A society, Sociedad Filarmónica (Philharmonic Society), is formed to promote symphonic music.

1853   New constitution is adopted

The 1853 Constitution reflects the liberal government that has adopted it. It abolishes slavery, mandates universal suffrage, provides for the popular election of governors and other officials, and expands the authority of local officials. It is also the first Latin American constitution to proclaim the separation of church and state. Civil marriage and divorce are legalized.

1855   Constitutional amendment authorizes creation of states

A constitutional amendment makes it possible for the country's provinces to declare themselves separate states. The first province to do so is Panama, and others soon follow.

1858   New constitution creates confederation

Reflecting the greater power now accorded to the individual states, the country's official name is changed from the Republic of New Granada to the Granadine Confederation.

1860–63   Confederation is destroyed by civil war

Civil war erupts between the Conservatives and Liberals over the degree of power accorded to states, and the confederation is dissolved.

1863: February   New constitution is adopted

With the Granadine Confederation disbanded, a constitutional convention at Rionegro establishes the United States of Colombia. The nine individual states have a great degree of autonomy, including the right to maintain their own armies and ratification rights over any constitutional amendments. In line with the Liberal opposition to a strong executive branch, the president serves only a two-year term.

1867–80   Intermittent civil war

The Liberal governments that rule during this period are plagued by political instability and repeated civil unrest. In addition to battling opposition from the Conservatives, the Liberals fight among themselves, reflecting the conflicting interests of the coastal region and the interior. Altogether, there are more than fifty rebellions between the 1860s and the 1880s.

1873: October 29   Birth of poet Guillermo Valencia

With the publication of his first volume of poems, *Ritos* (Rites;1898), Valencia is hailed as a leader of the experimental Modernismo movement in poetry. He is also active in politics and diplomacy and runs for president of Colombia in 1918 and 1930. In his later career Valencia turns mostly to translations, including a translation of Oscar Wilde's "The Ballad of Reading Gaol," and essays. He dies in 1943.

1875: May 15   Earthquake kills 16,000

A 45-second earthquake strikes Colombia, killing 16,000 people, most of whom live near the Venezuelan border. Near Cucuta, a volcano erupts as the earthquake occurs, launching fireballs onto the city.

1878 Feminist journal *La Mujer* is founded

The feminist journal *La Mujer* is founded by Soledad Acosta de Samper, who serves as its editor through 1881. Samper is also the author of historical and biographical works and fiction.

1880 Nuñez is elected president

Lawyer and poet Rafael Nuñez, a politically moderate Liberal, is elected to the presidency and forms a Liberal-Conservative coalition that reverses many of the Liberal policies of previous governments. Restoring a strongly centralized government to Colombia, Nuñez's *Regeneración* program brings stability to the country. Except for two years between 1882 and 1884, Nuñez serves as president until his death in 1894.

1882 Academia Nacional de Música is founded

The Academia Nacional de Música (National Academy of Music) is established by Jorge Price.

1884–85 Liberals revolt against Nuñez government

Liberals opposed to the policies of President Rafael Nuñez carry out an unsuccessful rebellion.

1886 Conservative constitution adopted

A new constitution institutionalizes much of the Conservative political agenda, establishing the Republic of Colombia, a centralized state with provinces called departments. Roman Catholicism becomes the official state religion. The presidential term is extended to six years, and the authority of the president is increased.

1886 Escuela Nacional de Bellas Artes is founded

Colombia's first school of art, Escuela Nacional de Bellas Artes (National School of Fine Arts), is established in Bogotá by Alberto Urdaneta (1845–87).

1886 Gran Exhibición showcases Colombian art

The Gran Exhibición brings together 1,200 artworks by Colombian artists of all periods. It includes the first display of pre-Columbian and colonial art produced in Colombia.

1887 Concordat is signed with the Vatican

Nuñez signs a *concordat* (agreement) with the Vatican (leaders of the Roman Catholic church) that insures the privileged position of the church in Colombia. Only Catholic marriages are recognized as legal, and Catholicism is included in the educational curriculum for all students.

1894 Rafael Nuñez dies

President Rafael Nuñez dies (See 1825.)

Coffee becomes a leading Columbian export in the late nineteenth century. (EPD Photos/CSU Archives)

1899–1902 War of a Thousand Days

Colombia is once again torn by civil war, this time between its Conservative government and Liberal guerrilla forces. (The government is also assisted by guerrillas.) The Conservatives win back their power, but the demoralizing war is a national disaster. It leaves a legacy of roughly 100,000 dead, widespread destruction, and enduring bitterness and hostility. The chaos in Colombia also paves the way for Panama to break away and gain its independence.

20th century Coffee production grows rapidly

Coffee production undergoes a major expansion. Exports grow from one million bags, each weighing 132 pounds (60 kilograms) in 1913 to three million bags by the early 1930s and over five million by 1943.

1900–10 Industry gains new importance

Significant industrial growth occurs. New facilities are opened for textile and pottery production and brewing.

1903: November 3   Panama declares independence

Following Colombian opposition to a treaty granting right to the canal in Panama (which is still a province of Colombia) to the United States, a group of Panamanians rebel against Colombia, The United States frustrates Colombian efforts to put down the revolt and signs a treaty with Panama which, in effect, becomes a U.S. protectorate for roughly thirty years.

1904   Reyes is elected president

General Rafael Reyes (1850–1921), a Conservative, becomes president and oversees a period of national reconciliation after the devastating War of a Thousand Days (see 1899). He gives the Liberals a voice in the government, including cabinet appointments. However, Reyes also strengthens the role of the president by becoming thoroughly involved in not only politics, but in the economy and foreign relations as well. Under Reyes, Colombia's economy and foreign relations improve. He attracts foreign investment, expands roads and railways, stabilizes the nation's currency, puts Colombia back on the gold standard, and encourages increased coffee production. He also tries to normalize relations with the United States, strained because of the U.S. role in helping Panama break away from Colombia (see 1903). However, this is an unpopular move that arouses strong political opposition.

1909   Reyes resigns

Reyes loses his political support after backing the Thompson-Urrutia Treaty calling for renewed relations with the United States, and he resigns. However, the cooperation between Conservatives and Liberals that begins with his government continues for the next two decades under five more Conservative presidents, giving the nation a degree of political stability unknown in the nineteenth century.

1911   Newspaper El Tiempo is established

The Liberal daily newspaper El Tiempo is founded. It will become Colombia's most widely read newspaper and one of the most influential in the Spanish-speaking countries of the Americas.

1912   Death of poet Rafael Pombo

Acclaimed Romantic poet Rafael Pombo dies (see 1833).

1918   First major labor strikes

Transportation and dock workers hold Colombia's first major labor strikes. Industrial expansion leads to the growth of the labor movement in the 1920s and the creation of a socialist political party. Ultimately the development of labor as a political constituency is a factor in the return to Liberal government in the 1930s.

1919   National airline is founded

Avianca, one of the oldest privately operated national airlines, is founded.

1920: June 4   Birth of artist Alejandro Obregón

Alejandro Obregón (1920–92) is born in Barcelona, Spain, and studies art in France, England, and the United States before settling in Barranquilla, Colombia, where he gains renown for his depiction of the people and landscape using idiosyncratic symbols and bold brush work. In 1948–49 he serves as director of the School of Fine Arts at the National University in Bogotá. Obregón's work is shown in prestigious exhibitions in Spain and the Americas and he receives the grand prize at the São Paulo Bienal in 1967. In 1985 a substantial retrospective of his work is shown in Bogotá before traveling to Paris and Madrid. Obregón dies in 1992.

1922   Treaty reconciles Colombia and the United States

The Urrutia-Thomson Treaty provides for $25 million in reparations from the United States to Colombia for the U.S. role in separating Panama from Colombia.

1924   First soccer league is formed

Colombia's first soccer league, the Liga de Football del Atlantico (Atlantic Football League, or LFA) is founded in Barranquilla. Its name is later changed to the Asociación Colombiana de Fútbol (Colombia Football Association).

1924   Publication of groundbreaking novel La vorágine

Poet and novelist José Eustasio Rivera (1880–1929) publishes La vorágine (The Vortex; 1924), a novel about the exploitation of laborers in the caucherías (rubber-gathering camps) of the upper Amazon jungle. In addition to acclaim within Colombia, Rivera's novel wins international praise and recognition of its author as a prominent figure in the Latin American literary world.

1925: July 14   Birth of composer and diplomat Luis Antonio Escobar

Escobar pursues studies in composition in Bogotá before going abroad to study in the United States, Germany, and Austria. Returning to Colombia in the early 1950s, he teaches composition and harmony at the National conservatory and is later appointed director musical programming for Colombian television. He wins two Guggenheim fellowships and is commissioned to write Preludios para percusión for the New York City Ballet. In addition to his musical work, Escobar also pursues a diplomatic career, serving as a consul in Bonn, Germany and as chargé d'affaires in Miami. He dies in 1993.

1927   Coffee growers' federation is formed

The National Federation of Coffeegrowers (Fedecafe) is created to take charge of the nation's coffee marketing.

1927: March 6   Birth of author Gabriel García Marquez

García Marquez (b.1927), the Nobel Prize-winning novelist and short story writer, is the most renowned Colombian literary figure of the twentieth century and a dominant figure in modern Latin American literature. García Marquez is born in Aracataca, which becomes the model for the author's fictional town of Macondo. He studies law and journalism at the National University of Colombia and during the following decade works as a journalist in Colombia, Venezuela, Europe, and New York. His first published work of fiction is the novella *La hojarasca* (Leaf Storm), which appears in 1955. In 1961 García Marquez moves to Mexico and publishes *El coronel no tiene quien le excriba* (No One Writes to the Colonel).

García Marquez's most famous work, *Cien años de soledad* (One Hundred Years of Solitude) is published in 1967. It tells the story of seven generations of the Buendia family, who live in the fictional town on Macondo. More than any other single work, this novel popularizes the style known as "magical realism," which comes to be associated not only with García Marquez but with an entire generation of Latin American authors. It is characterized by its mingling of myth and fantasy with the realistic concrete details of everyday life and political realities familiar to the author. In 1982 García Marquez wins the Nobel Prize for Literature. Since 1975 the author has divided his time between Colombia and Mexico City. His other works include *El otoño del patriarca* (The Autumn of the Patriarch; 1975), *Crónica de una meurte anunciada* (Chronicle of a Death Foretold; 1982), *El amor en los tiempos del cólera* (Love in the Time of Cholera; 1985), *El general en su laberinto* (The General in His Labyrinth; 1989), and *Del amo y otros demonio* (Of Love and Other Demons; 1995).

1928: November–December   The Great Banana Strike

A strike by plantation workers in the Santa Marta area is brutally suppressed. About twenty thousand workers go on strike against the United Fruit company and Colombian growers, demanding that they be given the status of regular employees to make them eligible for benefits. They also demand higher wages and collective contracts. The government declares martial law and sends troops to disband the strike. They fire into a crowd of 1,500 strikers, killing thirteen. After reprisals by the strikers, more are attacked and killed. The total number dead has been estimated at between sixty and seventy-five.

The repressive methods used to break up the strike are a factor in the defeat of the Conservative government in the 1930 elections, leading to a period of Liberal reform (see 1930). The strike is portrayed in a section of Gabriel García Marquez's novel *Cien años de soledad* (One Hundred Years of Solitude; 1967).

1929   Coffee prices plummet as global depression begins

The beginning of the worldwide Great Depression of the 1930s causes a steep fall in coffee prices and launches an economic crisis. The price of coffee drops about two-thirds between and 1928 and 1930.

1930–45   Liberals return to power

Widespread public discontent in the face of a depression linked to the worldwide economic collapse of the 1930s brings the Liberals back to power, where they implement a wide-ranging reform agenda.

1930s   Marijuana cultivation introduced

The first marijuana is brought into Colombia from Panama. It is grown in limited amounts, mostly by farmers in coastal areas.

1932   Birth of painter Fernando Botero

Botero, one of the best-known twentieth-century Colombian artists, is born in Medellín and begins his career as a journalist and illustrator. Moving to Bogotá in 1951, he has the first one-man show of his paintings. Botero then goes abroad to study art in Madrid and Florence, returning to Bogotá in 1955. Between 1958 and 1960, he is on the faculty of the Escuela de Bellas Artes at the National University.

Botero becomes known for his satirical paintings, with their puffed up bodies. In 1961 the Museum of Modern Art in New York purchases Botero's most famous painting, *Mona Lisa Aged Twelve*, a humorous depiction of Leonardo Da Vinci's famous *Mona Lisa*. In 1973 Botero moves to Paris but his paintings continue to include Colombian themes and subjects. In 1976 Botero wins Venezuela's prestigious Order Andrés Bello, given to outstanding figures in the arts in Latin America. Botero's work has appeared in museums worldwide.

1934–38   "Revolution on the March"

The most important president of the Liberal period between 1930 and 1945 is Alfonso López Pumarejo, whose reform program becomes known as the "Revolution on the March." Amendments to the 1886 constitution guarantee the right of laborers to strike, authorizes the government to intervene in the economy, and mandates the provision of public assistance to those in need. Educational, agricultural, labor, and tax reforms are also implemented, and the separation of church and state is reinforced.

1938 Colombia joins Caribbean soccer association

The Colombian soccer league (the Asociación Colombiana de Fútbol) becomes a charter member of the Central American federation, the Confederación Centroamericano y del Caribe de Fútbol. Its first international games are played in the Central American and Caribbean Games.

1939–45 World War II

In spite of internal dissension, Colombia allies itself with the United States.

1942 Museo de Arte Colonial is established

This state-run museum is founded in Bogatá for the preservation and display of artifacts from the colonial era, including paintings, sculpture, furniture, and silverware.

1943: July 8 Death of poet Guillermo Valencia

Valencia, an important figure in the experimental poetry movement known as Modernismo, dies in Popayán, Colombia. (See 1873.)

1944 Catholic Social Action is established

Catholic Social Action is formed to implement social and educational programs.

1948 Professional soccer is introduced

Colombia becomes one of the last South American countries to introduce professional soccer.

1948: April 9 Assassination of Liberal leader sparks decade of violence

The assassination of leftist Liberal mayor of Bogotá and prominent national political leader Jorge Eliécer Gaitán sets off the worst rioting in Colombia's history (it later becomes known as the "Bogotázo"). Gaitán's populist politics have drawn the support of both urban workers and rural dwellers, and many believe he would have become president in 1950. A large portion of downtown Bogotá is destroyed as angry crowds break windows, set fires, and engage in looting. An estimated 2,000 people are killed.

1948–57 *La Violencia*

The riot begins a long period of civil unrest known as *La Violencia*, which spreads throughout the country and eventually encompasses not just political violence between Liberals and Conservatives but also random rural banditry. Altogether, as many as 200,000 people are killed. Neither the Conservative nor the military regimes that are in power during this period are able to bring the violence under control.

1949 Liberals withdraw from the national elections

Following the assassination of Jorge Eliecer Gaitán (see 1948), the Liberal party does not field a candidate in the 1950 presidential elections.

1949 Conservative Laureano Gómez is elected president

Laureano Gómez wins the presidency in an uncontested election. His government brutally attempts to quell the political violence erupting throughout the country. By 1953 an estimated 160,000 people have died in *La Violencia* (see 1948).

1949 First exhibition of abstract art takes place

An exhibition of paintings by Marco Ospina (1912–83) introduces abstract art to Colombia.

1953 Military coup installs Rojas Pinilla

General Gustavo Rojas Pinilla becomes president in a military coup but is unable to achieve order in Colombia in spite of his authoritarian rule. Thousands of Colombians continue to die in nationwide political violence.

1954 Television is introduced

The first television station opens, run by the government-controlled National Institute.

1956: August 7 Truck explosion kills 1,200 in Cali

Seven trucks full of dynamite parked near downtown Cali explode simultaneously in the middle of the night, killing 1,200 people. No cause for the blast is ever found, but many assume the incident is part of the political violence that has shattered Colombia for nearly a decade.

1957 The National Front is formed

In an all-out effort to end the violence Liberals and Conservatives agree to join in an unprecedented coalition called the Frente Nacional (National Front). They create a plan that calls for free elections in 1958, after which the presidency will alternate between the Conservatives and Liberals every fours years until 1974. In addition, both the legislature and the cabinet will be equally divided between members of the two parties. Colombians overwhelmingly approve this plan in a special nationwide election.

1957 Women get the vote

Women win the right to vote on the same basis as men.

1957: May Coup ousts Rojas

President Rojas Pinilla is overthrown in a coup that promises a return to democracy.

1958 Camargo wins first election under the National Front

Liberal Alberto Camargo is chosen as president in democratic elections as specified in the National Front plan (see 1957). The presidency is then transferred between parties every four years as specified in the agreement.

1960s Women begin to achieve prominence in Colombian art

For the first time, women play an important role in the country's art scene. The textiles of Olga de Amaral (b.1936) win recognition, as do the experimental sculptures of Felisa Bursztyn (1933–82), whose materials include the symbolic use of rusted objects and other debris. However, widespread acceptance does not arrive for feminist painter Débora Arango until the 1980s, when Arango (b.1910) is in her seventies. For most of her life, her controversial expressionist paintings have been called immoral and obscene for their frank treatment of social and political issues, including the status of women.

1960s Champeta, music with African roots, emerges

Champeta, music with roots in Africa emerges in Basilio, Colombia, home to people who trace their ancestry to fugitive slaves. The music, also known as *terapia criolla,* is heard blaring from huge sound systems, sometimes mounted on pickup trucks that are colorfully painted. A deep, throbbing bass accompanies guitar and percussion. Popular initially among gangs of teenagers, champeta music enjoys a wider audience now.

1960 First woman is appointed to a Cabinet-level position

Esmeralda Arboleda de Cuevas Cancino, appointed minister of transport, becomes the first woman in Colombian history to hold a Cabinet-level position. She later serves as ambassador to Austria and Yugoslavia, and is also the first woman to be elected to the Senate.

1962 Colombia competes in final World Cup rounds

Colombia competes in the final rounds of the World Cup. It achieves a 4–4 draw with the Soviet Union and loses to Uruguay (2–1) and Yugoslavia (5–0).

1965–80 Cultivation and export of marijuana grow

With growing demand in the United States, Colombia's marijuana industry expands rapidly. Crackdowns on marijuana trade across the Mexican border in the early 1970s shift the bulk of the trade to Colombia, which supplies seventy percent of the marijuana imported into the United States by 1980. Eventually between 30,000 and 50,000 small farmers come to depend on the growing of marijuana for their survival, and an estimated 50,000 more Colombians in other walks of life also depend on the marijuana business for their livelihood. Marijuana trade brings unprecedented wealth to Colombia's Caribbean coast, but at the price of drug-related violence and police and government corruption.

Late 1960s Cocaine trade begins

Cocaine smuggling to the United States begins with the assistance of Cuban exiles based in Miami. The drug is mostly transported in small amounts by smugglers traveling on commercial airline flights.

1968 Singer Toto la Momposina launches professional career

Singer Toto la Momposina (born Sonia Bazanta) is inspired by the bolero and rumba of Latin America to forge her own unique Colombian style of singing. She performs around the world, with her own band accompanying her. She performs on two recordings, *Toto La Momposina* (recorded and released by Auvidisc) and *Colombia, Music from the Atlantic Coast* (recorded and released by Aspic). Before returning to Colombia to concentrate on her singing career, she studies history of dance at the Sorbonne Institute in Paris, France, and sings on street corners there.

1970s Guerrilla violence grows

Left-wing guerrilla groups challenge the failure of the National Front governments to bring the common people into the political process, claiming that it still protects the interests of the wealthy. These groups include the Colombian Revolutionary Armed Forces (FARC), the National Liberation Army (ELN), the Popular Liberation Army (SPL), and M-19. They launch campaigns of terror, including political kidnappings, assassinations, and bombings, that continue into the 1980s and beyond. In spite of opposition by the military, the guerrilla groups continue to attract followers among the poor and disadvantaged.

1971 Cali hosts Pan-American Games

Cali is the site of the Pan-American Games.

1973 Concordat with Vatican reduces influence of the Catholic church

A *concordat* (agreement) between Colombia and the Vatican limits the power of the church in a variety of areas. The constitutional description of Catholicism as the state religion is changed to read "Roman Catholicism is the religion of the great majority of Colombians." Catholic missionaries cede their jurisdiction over mission territories (Indian homelands). The church also loses the right to censor textbooks used at public universities, and the use of the Catholic catechism is no longer required in public schools. In addition, Colombians are allowed to enter into civil marriages without being expelled from the church.

1974   Liberals return to power

Once the National Front period ends, the two-party system resumes peacefully, and Liberal Alfonso López Michelsen is elected president.

1975   State of siege is imposed

The government declares a state of siege in response to guerrilla violence, and strikes by students and labor unions. The state of siege continues for one year.

1978–79   Medellín drug cartel begins drug sales to U.S.

Drug traffickers in Medellín make use of large-scale drug distribution via airlifts. Having established dominance in the South American market, they turn to the United States, setting off the "Cocaine Wars" in southern Florida. By the 1980s the Medellín distributors have organized themselves into a group that becomes known as the Medellín Cartel. Eventually becoming the source of eighty percent of the world's cocaine, they begin amassing tremendous wealth and influence. The drug lords undertake a campaign of violence against the Colombian government to intimidate officials who may have the power to extradite them to the United States to stand trial on drug charges.

1982   Belisario Betancur is elected president

Conservative Belisario Betancur Cuartas brings an end to the Liberal rule that has followed the National Front period.

1982   Gabriel García Marquez wins the Nobel Prize for Literature

Novelist and short story writer Gabriel García Marquez, acclaimed author of *Cien años de soledad* (One Hundred Years of Solitude), is awarded the Nobel Prize for Literature.

1983   Mammoth cocaine laboratory begins operations

The largest cocaine-processing laboratory ever built, called Tranquilandia, launches operations in Los Llanos. With its own roads, housing, and power supply, it is rumored to produce 3,500 kilograms of cocaine monthly.

1984: March   Police raid Tranquilandia cocaine lab

Police raid the country's largest cocaine laboratory, confiscating not only cocaine but also a variety of transport vehicles and weapons. Following this raid, the cocaine bosses, who publicly flaunt their wealth and influence, go underground.

1984: May   Minister of Justice is assassinated

Minister of Justice Rodrigo Lara Bonilla, the main adversary of the Medellín cocaine cartel, is murdered. In response, the government activates a long-neglected extradition law and sends four cocaine traffickers to the United States to stand trial on drug charges. The cartel declares war on the government, eventually murdering the publisher of the nation's top evening newspaper (for his endorsement of the extraditions) and the country's attorney general.

1984: August 24   Cease-fire negotiated with guerrilla groups

A special commission appointed by President Betancur reaches a cease-fire agreement with representatives of the M-19 and Popular Liberation Army (ELP) guerrilla groups.

1985   Most guerrilla groups accept government amnesty

All of the country's five major terrorist groups except M-19 accept the amnesty offered by President Betancur.

1985: June   M-19 violates cease-fire

M-19 breaks its cease-fire agreement with the government.

1985: November   M-19 terrorists seize Palace of Justice

Around sixty M-19 guerrillas take over the Palace of Justice in Bogotá, taking the people inside hostage. Government troops stage an attack, in which more than ninety people die, including eleven supreme court justices, and the building is completely leveled by fire.

1985: November 14   Erupting volcano causes 25,000 deaths

The volcano Nevado del Ruiz erupts, creating mudslides that kill an estimated 25,000 people. The explosion melts glaciers atop the mountain and creats massive mudflows.

1986   Virgilio Barco is elected president

Virgilio Barco Vargas assumes the presidency, returning the Liberals to power. The Conservatives refuse to participate in his government as cabinet members or in other posts, the first time since the inception of the National Front in 1957 that power is not shared by both parties. Barco's government is not able to bring the country's drug traffickers under control, but he does take steps to allow guerrilla groups participation in the political system.

1986   Historian Germán Arciniegas published America in Europe

Historian Germán Arciniegas (b. 1900) publishes *America in Europe: A History of the New World in Reverse*. In it he argues that the exploration of the Americas by Europeans caused European scholars to revise their views of geography and astronomy. He puts forth the view that the Americas have influenced Europe more than Europeans have influenced the Americas.

1987   Top cartel leader captured and extradited to the U.S.

With help from the U.S. Drug Enforcement Agency (DEA), Medellín cartel leader Carlos Lehder is captured by narcotics forces and extradited to the United States.

1987: September 27   Mudslide kills over 500 in Medellín

Over five hundred people in the impoverished Villa Tina section of Medellín are killed by a mudslide brought on by heavy rains. The bodies of 183 dead are found, but 500 are permanently missing and presumed dead. Seven of the 43 children killed are in church receiving their first Communion.

1987–88   Warfare erupts between rival drug cartels

Hostility between the competing Cali and Medellín drug cartels turns into overt violence.

1989: August 18   Presidential candidate is assassinated

Liberal presidential candidate Luis Carlos Galán is assassinated by members of the Medellín drug cartel. The government confiscates $250 million worth of property owned by the cartel, in addition to drugs and weapons. Almost 500 arrests are made by the Colombian police in 1989, and nine Colombians are extradited to the U.S. Meanwhile the cartel pursues a campaign of bombings in Bogotá, Medellín, Cali, and other locations. Bombings in 1989 total 265, and kill 187 people.

1989: September   Newspaper building is destroyed in bombing

The headquarters of the newspaper *El Espectador*, whose publisher has been murdered by drug lords, is bombed and destroyed as part of the cartel's campaign of terror.

1989: November   Airplane bombing kills 107

A mid-air bombing of an airline flight from Bogotá to Cali kills all 107 aboard. Drug lords Pablo Escobar and Rodriguez Gacha are widely believed to have ordered the bombing.

1989: December 15   Drug lord Gacha killed by police

Cartel leader José Gonzalo Rodriguez Gacha is gunned down by police, together with his seventeen-year-old son and fifteen bodyguards, near Cartagena.

1990   Gaviria Trujillo is elected president

César Gaviria Trujillo, another Liberal and former finance minister under Virgilio Barco, is elected to the presidency, winning close to fifty percent of the vote. He continues former president Barco's efforts to bring the guerrilla groups into the political system as legal parties.

1990   Colombia competes in final World Cup rounds

For the first time since 1962, Colombia competes in the final rounds of the World Cup soccer games. When Colombia defeats Argentina on the way to the World Cup finals, about twenty people die during the all-night celebration that results.

1990: March 9   M-19 guerrillas sign peace treaty with government

The notorious M-19 guerrilla group signs a peace agreement with the government, including provisions for the group to take part in national elections. After the group disarms, its leader, Carlos Pizarro, is assassinated onboard a commercial airline flight. Pizarro, a presidential candidate, is mourned by thousands of people. He is the second major candidate of the 1990 campaign to be assassinated (see 1989). The newly renamed Alianza Democrática M-19,selects another candidate, Antonio Navarro Wolff, who wins twelve percent of the vote in the presidential election.

1990: September 6   International soccer play banned

The South American Soccer Federation enacts a one-year ban that prohibits international soccer matches from being played in Colombia. This action results from an incident on August 26 when six armed men threaten the referees with submachine guns unless Colombia is guaranteed a win over Brazil at a match in Medellin.

1990: December 9   National plebiscite on constitutional change

Colombia's voters overwhelmingly approve the election of a constituent assembly to reform the nation's government by drawing up a new constitution.

1991   M-19 guerrilla group becomes a political party

In response to efforts by presidents Barco and Gaviria, the M-19 guerrilla group becomes a legitimate political party that can participate in electoral politics.

1991: June 19   Medellín cartel leader surrenders to government

Medellín drug cartel head Pablo Escobar Gaviria surrenders to authorities and is imprisoned in a "luxury jail" near his home town of Envigado.

1991: July 5   New constitution is adopted

A new constitution drafted by a specially elected constituent assembly includes measures to further democratize the Colombian political process and end control by the country's traditional elites. The size of the legislature is reduced, and civil divorce is declared legal for Catholic couples. Also included is a ban on extradition of Colombians to face charges in other countries.

1992: April 11   Death of artist Alejandro Obregón

Obregón, one of Colombia's foremost twentieth-century painters, dies. (See 1920.)

1992: July 22   Pablo Escobar escapes from prison

Drug lord Pablo Escobar escapes from jail and vows to renew his group's war on the government.

1992: November   Government announces new anti-terrorist policies

President Gaviria responds to the refusal of most guerrilla groups to disarm and join the political process by issuing new directives aimed at suppressing terrorist activities, as well as drug trafficking.

1993   Colombia signs trade pact with Mexico and Venezuela

Colombia, Mexico, and Venezuela sign a trade agreement designed to eliminate all tariffs among these three nations within twenty years.

1993: September 11   Death of composer and diplomat Luis Antonio Escobar

Composer Luis Antonio Escobar dies while serving in a diplomatic post in Miami. (See 1925.)

1993: December 2   Pablo Escobar is shot by police

A year after escaping from prison, Medellín drug cartel leader Pablo Escobar is shot dead by police in Medellín.

1994   Ernesto Samper is elected president

Liberal Ernesto Samper Pizano becomes the latest in a succession of Liberal presidents elected since 1982. Samper's effectiveness is hampered by allegations that his presidential campaign accepted contributions from drug traffickers, although Samper is later exonerated.

1994: June 6   Earthquake hits in the southwest

Southwestern Colombia is rocked by an earthquake measuring 6.8 on the Richter scale. The damage is aggravated by mudslides, avalanches, and floods triggered by the quake. Some 13,000 people are left homeless, and it is estimated that between 300 and 1,000 are dead.

1994: July 2   Soccer star is killed in Medellín

After mistakenly making a goal for the United States in a World Cup match, soccer star Andres Escobar is shot and killed by disgruntled fans while leaving a nightclub.

1995: June   Government begins crackdown on Cali cartel

The Cali drug cartel becomes the focus of a government crackdown that includes the arrest of four drug lords between June 19 and July 8. The most senior is José Santacruz Londono, said to be one of the founders of the cartel.

1995: August 6   Alleged cartel leader captured

Miguel Rodriguez Orejuela, the alleged co-leader of the Cali drug cartel, is captured by police at his apartment in the city of Cali. Rodriguez, becomes the sixth highly placed cartel leader to be arrested since June.

1995: December 20   American Airlines crash kills 167

An American Airlines jet crashes near Buga, killing 167 people. Investigation reveals the cause to be navigational errors by the crew and miscommunication between the crew and air traffic controllers, due in part to language barriers.

1996   Tuberculosis outbreak

The town of Buenaventura is the site of an outbreak of tuberculosis. The nationwide incidence of tuberculosis is 26.5 per 100,000 people, whereas it is 90.5 per 100,000 people in Buenaventura. Researchers believe the outbreak is due to lack of prevention and inadequate treatment.

1996: February 14   Charges are filed against President Samper

Federal prosecutors formally charge President Ernesto Samper with accepting money from the Cali drug cartel during the 1994 presidential campaign. However, Samper is cleared by Congress late in the year.

1996: March 1   U.S. decertifies Colombia as ally in war on drugs

Citing growing evidence that President Samper accepted campaign donations from the Cali drug cartel, the United States removes Colombia from the list of cooperative countries in the war on drugs, a move that will cost Colombia most of its foreign economic aid from the United States.

1996: July 11   U.S. revokes Samper's visa

The United States revokes the entry visa of President Samper, making Samper the first foreign leader since Austria's Kurt Waldheim to be barred from entering the country. The action is taken in response to allegations of $6 million in campaign contributions made to Samper by drug traffickers.

1996: August 30   Guerrillas launch major attack

Members of the Revolutionary Armed Forces of Colombia (FARC) carry out some of the worst guerrilla attacks in recent history, including a raid on a military base by 300 FARC rebels. At least ninety-six soldiers and civilians are killed in the hostilities.

1997: July 21   Residents of rural village are evacuated after massacre

Residents of the village of Mapiripan in eastern Columbia are evacuated following a massacre attributed to right-wing paramilitary forces in which more than thirty people are murdered.

1997: August   U.S. agrees to aid Colombian antidrug efforts

After removing Colombia from the roster of countries qualified for antidrug assistance in 1996, the United States agrees to provide $70 million worth of military equipment for antidrug operations on condition that the Colombian military improves its record with regard to human rights abuses.

1997: October 26   State and municipal elections held despite violence

Following three months of guerrilla violence, elections are held for state and municipal elections. The murder of at least 40 candidates and the kidnapping of over 200 others, mainly in rural areas, has caused over 2,000 candidates nationwide to withdraw from municipal elections. Campaign headquarters have also been destroyed in several towns. President Samper proposes peace negotiations with the FARC rebel group but is rebuffed. Terrorism also keeps many Colombians from voting. An estimated forty-nine percent of registered voters turn out at the polls.

1997: December 4   Colombia qualifies for 1998 World Cup games

Colombia qualifies for the World Cup finals to be held in France.

1998: March 8   Liberal Party triumphs in legislative elections

The Liberal party keeps its majority in both houses of the Colombian Congress in nationwide elections in which all seats are contested. The Liberals win forty-five percent of the vote, the Conservatives twelve percent.

Bibliography

Ambrus, Steven. "Spinning Threads of Progress. (Silk-production in Colombia). *Americas* (English edition), November–December 1996, vol. 48, no. 6, pp. 14+.

Bushnell, David. *The Making of Modern Colombia: A Nation in Spite of Itself.* Berkeley: University of California Press, 1993.

Cohen Alvin, and Frank R. Gunter. *The Colombian Economy: Issues of Trade and Development.* Boulder, Colo.: Westview Press, 1992.

Davis, Robert H. *Colombia.* Santa Barbara, Calif.: Clio Press, 1990.

Drexler, Robert W. *Colombia and the United States: Narcotics Traffic and a Failed Foreign Policy.* Jefferson, N.C.: McFarland, 1997.

Giraldo, Javier. *Colombia: The Genocidal Democracy.* Monroe, Maine: Common Courage Press, 1996.

Hanratty, Dennis M., and Sandra W. Meditz. *Colombia: A Country Study.* 4th ed. Washington, DC: Federal Research Division, Library of Congress, 1990.

Hartlyn, Jonathan. *The Politics of Coalition Rule in Colombia.* New York: Cambridge University Press, 1988.

Kline, Harvey F. *Colombia: Democracy Under Assault.* Boulder, Colo.: Westview Press, 1995.

McFarlane, Anthony. *Colombia Before Independence: Economy, Society, and Politics under Bourbon Rule.* New York: Cambridge University Press, 1993.

Pearce, Jenny. *Colombia: The Drugs War.* New York: Gloucester Press, 1990.

Costa Rica

Introduction

When Costa Rican President Oscar Arias Sanchez (b. 1908) received the prestigious Nobel Peace Prize for engineering an end to war in Central America in 1987, it was as much a tribute to him as it was to his small nation.

In many ways, the prize reflected the nation's democratic ideals and its struggles to build a more equitable and just society. In helping construct a peace plan to end conflicts in Guatemala, El Salvador, Honduras, and Nicaragua, Costa Rica, the only nation in the Americas without an army, showed it had the moral authority to lead these countries out of decades of bloodshed.

Costa Rica is by no means a perfect nation, and Ticos, as Costa Ricans call themselves, will be the first to say so. It is, after all, a poor nation, sustained by three industries: coffee, bananas, and tourism. But it is not simply a "banana republic," a negative term often used to designate poor and backward Latin American countries economically controlled by the United States.

Geography

Costa Rica is the third smallest country in the Americas with an area of 19,730 square miles (51,100 square kilometers). Slightly smaller than the state of Virginia, it is bordered on the north by Nicaragua, on the east by the Caribbean Sea, on the southeast by Panama, and on the southwest by the Pacific Ocean. San Jose, its capital city, is located in the central highlands of the country.

Costa Rica has three main topographic regions. The central highlands, which extend through the middle of the country, reach elevations of more than 12,000 feet (3,660 meters) above sea level. Costa Rica has active volcanoes, such as Irazu, which had a period of destructive eruptions in 1963–65. Nestled in the highlands is the Meseta Central, a 770-square-mile flat area that rises to 4,000 feet. It is rich in volcanic soil and home to four of the six main cities and half of the country's 3.2 million people. Most of the country's coffee beans are harvested here. The Atlantic Coastal Plain, on the Carib-bean side of the highlands, is low, swampy, and quite hot, as is the Pacific side of the highlands.

Population and Economy

The country's population is highly homogenous, primarily of Spanish descent with a small *mestizo* (mixed white and Amerindian) minority. About three percent are black and two percent from East Asia. Spanish is the national language and English is widely spoken by the middle class. Descendants of the Jamaican blacks speak an English dialect. More than ninety percent of the population is Roman Catholic.

While there are potential mineral deposits yet to be exploited, much of the country's economy relies on volatile commodities such as coffee, bananas, and sugar. World price fluctuations can severely affect the local economy, as they did in 1991, when coffee prices around the world dropped fifty percent. Costa Rica's coffee earnings dropped to $180 million, from about $300 million the previous year. One of the country's goals is to diversify the economy. So far, that means tourism. Costa Rica's national park system is among the most extensive and well-developed in Latin America and a strong magnet for this form of income. In 1996 tourism revenues reached $626 million, easily surpassing the combined revenues brought by bananas and coffee in previous years.

But tourism also is a volatile market, which can adversely affect a nation. In the 1990s Costa Rica became one of the fastest growing destinations for ecotourism. Ecotourism is considered a more responsible way to travel. It is supposed to contribute to conservation of natural environments and sustain the well-being of local people by promoting rural economic development. Costa Rica has heavily marketed its natural splendors and wildlife, but it has not done as well in regulating the travel industry. There are fears that efforts to boost tourism dollars could lead to rampant development of large resorts.

Democracy in Costa Rica

Yet, it is not simply the nation's natural beauty that attracts visitors. Long the most stable nation in Central America, Costa Rica has earned a reputation as a peaceful and friendly

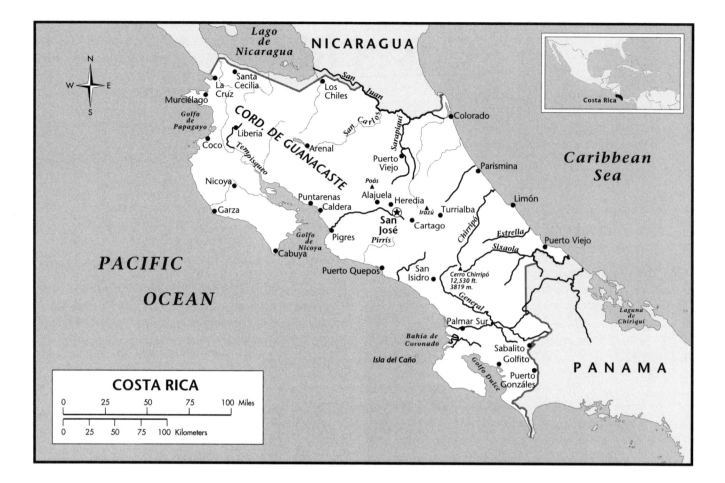

nation with a long tradition of democracy. While other Central American nations were mired in bloody conflicts, Costa Rica's people lived in a relatively prosperous and democratic nation. They didn't get to this point without a fight. Since the arrival of the Spanish Conquistadors in 1502, Costa Ricans have fought many times to ensure their freedom, both against foreign invaders and home-bred dictators.

Spanish Colonization and Independence

The Spaniards mostly ignored Costa Rica after the arrival of Christopher Columbus (1451–1506) in what is now Limon. They tried to settle the region without success for many decades but were kept at bay by mosquito-infected swamps on the coasts, brutal heat and humidity, and native people unwilling to be subjugated. It wasn't until 1563 that the Spaniards managed to settle permanently on the cooler and fertile central highlands. From here, Costa Rican society developed slowly and mostly in isolation. When Central America declared its independence on September 15, 1821, it took more than one month for Costa Rica to get the news.

The roots of democracy developed in this isolation perhaps, where self-reliance and close relations with neighbors meant survival. The conquistadors had few native peoples to work the land, so they had to do it themselves. The native peoples, who numbered 25,000 in 1502, were quickly decimated by war and disease brought by the conquering Europeans.

Large colonial-style haciendas like those in other Spanish enclaves didn't develop in Costa Rica. Land distribution was more equitable, unlike other Central American nations where a very small percentage of the population owned most of the land.

Civil War and Eventual Peace for Region

A rich class of coffee growers developed in the 1800s and began to dominate politics. Enlightened leaders stood in their way and curbed their power. By the late 1800s, a new class of intellectuals pushed for social progress and a more just society.

Those struggles continued through the twentieth century, leading to a civil war in 1948 that shaped modern Costa Rica. The country faced a different challenge entering the 1980s. The economy was in shambles, and Ticos feared they would be drawn into the conflicts of neighboring nations. President

Arias and many of his compatriots understood that only regional peace would ensure stability for Costa Rica.

From 1990 to 1996, the Costa Rican economy was in decline because of several years of bad weather and increasing international competition. By the late 1990s, the Costa Rican government had some success in tackling inflation, and the economic picture was improving slightly. In a changing world, the primary exports—coffee and bananas—face increasing competition, making it difficult to support the Costa Rican economy without economic diversification.

Timeline

A.D. 1000   The Pre-Columbian era

Before the arrival of Spanish explorers in 1502, several autonomous and poorly organized tribes live in what is now known as Costa Rica. As many as ten thousand people may have lived in Guayabo, about forty miles east of the capital city of San Jose. Swampy lowlands and high mountains provide a natural border against incursions by more powerful native groups, including the Mayas to the north.

Little is known about these small tribal groups and their social structures, although signs of human habitation date back more than ten thousand years. Archeologists have found few monuments and no signs of a written language. Excavations in the 1990s at an archeological site in the highlands of Costa Rica reveal a city with streets, aqueducts, and causeways.

Among the native groups are the Corobicis, who live in the highland valleys, and the Nahuatl, who arrive from Mexico shortly before Columbus. Highland groups produce intricate pottery and metalworking, and their gold ornamentation is believed to be more advanced than other regions of Central America.

Among some of the other native groups are the Caribs on the Caribbean seaboard and the Borucas and Brunkas in the southwestern part of the country. They are semi-nomadic hunters and fishermen. They raise yucca and squash and live in communal village huts surrounded by fortified palisades. The Brunkas, a matriarchal society (family units headed by women), are a warlike people and turn their prisoners into slaves. They are accomplished goldsmiths, live in round palm-leaf huts in a circular compound, and cultivate corn. They are believed responsible for nearly perfect round stone balls found at several burial sites.

The Borucas produce a brown textile woven in an intricate fashion that allows the design to appear only on the right side of the material.

A.D. 1300   Native group from Mexico settles in the Nicoya Peninsula

The Chorotegas (fleeing people), a group of native people from Mexico, arrive in the Nicoya Peninsula, a northwestern region of Costa Rica. The Chorotegas develop towns and bring with them a sophisticated agricultural system based on beans, corns, squash, and gourds. They also raise cotton, tobacco, and cacao (tropical American tree from which cocoa and chocolate are derived). They have a calendar, write on deerskin parchment, and produce jade figures and ceramics. The Chorotegas have a rigid class hierarchy with a ruling class of high priests and nobles. They are militaristic and have slaves. They believe in one supreme force and the immortality of the soul, but they also worship the sun and the moon. (See 1977.)

From Columbus to Conquest

1502: September 18   Columbus arrives in present-day Costa Rica

Christopher Columbus (1451–1506), the great explorer at the service of Spain, arrives near present-day Puerto Limon. He is searching, as he has since his arrival in the Americas in 1492, for gold. This is his fourth and last voyage to the New World. He lands in Costa Rica when a sudden tempest drives his ship into Cariari Bay.

Gold ornaments worn by some of the native people impress Columbus. At this time, the region's population stands at no more than 25,000 indigenous people.

After Columbus, who stays for seventeen days, the prospect of gold attracts many Spanish explorers. They begin calling the area Costa Rica (for "rich coast"). But there is little gold to be found, and the country's swamps, tropical diseases, and floods—and unfriendly natives—make life difficult for the European explorers.

1506   Spanish try to colonize Costa Rica

Ferdinand of Spain (1452–1516) sends Diego de Nicuesa to colonize the Atlantic Coast, but he runs aground off the coast of Panama. Nicuesa marches north by land, fighting the natives along the way. The natives prefer to burn their crops rather than see them fall to the Spaniards.

1522   Another expedition moves into Costa Rica

Gil Gonzales Davila sets off from Panama on another expedition to settle the region. With him are Catholic priests who appear to convert many natives to Christianity. But the expedition fails, with more than one thousand of Davila's men dying from sickness and starvation. Other attempts at settling the region fail. Coastal settlements vanish as quickly as they are established, from internal problems and outside pressures, including resistance by native people and pirate raids.

1524 Spanish try to settle Pacific coast

Francisco Fernandez de Cordova founds the first Spanish settlement on the Pacific, at Bruselas, near present-day Puntarenas. It lasts less than two years.

Unable to control it, the Spaniards virtually abandon Costa Rica. The conquerors are busy bringing other more wealthy lands under their control. In South America Francisco Pizarro (c.1478–1541) leads the conquest of the Inca Empire in Peru in 1532. And they consolidate their power in Mexico, where they discover great silver deposits.

1540 Costa Rica becomes a province

The Spaniards briefly turn their attention back to Central America and turn Costa Rica into a province.

1543 Spanish colony in Guatemala becomes administrator of Costa Rica

Costa Rica falls under the jurisdiction of present-day Guatemala, which becomes an important administrative center for the Spaniards.

1563 Cartago is founded

By the early 1560s several Spanish cities consolidate their power north of Costa Rica. Phillip II of Spain (1527–98) decides it is time to settle Costa Rica and Christianize the natives. But most indigenous people in the area have died from smallpox and tuberculosis. Those who remain have gone into hiding deep in the jungles and valleys of the country. The Chorotegas, the only large remaining block of natives, are unable to put up much resistance to the Spaniards.

Juan Vasquez de Coronado (1523–1565) is considered the true conquistador of Costa Rica. He arrives as governor and quickly sets about bringing order, moving existing Spanish settlers into the Cartago Valley and far from the disease-ridden coasts. Here the climate is temperate, and rich volcanic soil is perfect for agriculture. Coronado reportedly treats the surviving native people more humanely.

Spaniards rely heavily on their soldiers for the economic and social development of their provinces. The soldiers are granted *encomiendas*, land holdings that allow them to use a number of indigenous serfs to work the land. In exchange, the conquistadors teach them the Christian faith. In essence, the encomienda is a slave system. And many of those who rely on it abuse the indigenous population.

In Costa Rica, the conquering Spaniards face a serious labor shortage and the Spaniards have to work the land themselves. Even Coronado, the governor, works the land to survive. They have little to offer to neighboring Spanish cities, and trade virtually dies. Costa Rica practically vanishes from the map for many decades.

1635: August The patron saint of Costa Rica appears

According to tradition, a woman gathering wood finds a small black statue ("La Negrita") of the Virgin Mary in the black section of the town of Cartago. The woman takes the statue back to her home, where it disappears and then reappears at the same spot where she originally finds it. She takes it to the parish priest and the same thing happens. Miracles begin to be attributed to the statue, which becomes known as the Virgin of the Angels, and a church is soon built. Churches are built and rebuilt on the same spot; six churches on the site are destroyed by earthquakes. A church built in 1921 survives to the late 1990s. The virgin is declared the patron saint of Costa Rica. Every year on August 1 people make pilgrimages (religious journeys) on foot from the San Jose area to Cartago to fulfill religious promises.

While more than ninety percent of Costa Ricans are Catholic and most are baptized and married by the Catholic Church, Costa Ricans don't observe or expect rigid conformity to Church rules. They often are described as having a "lukewarm" Catholicism.

1650 First cacao exports

First exports of cacao are sent to neighboring Nicaragua. The seeds, called beans, of cacao are used to make chocolate and cocoa. The cacao plant is an evergreen tree that grows to over twenty feet (six meters). The fruit resembles a melon, and the seeds are found inside.

1660–95 Cacao production flourishes, then stagnates

Cacao becomes the first major source of wealth for Costa Rica. Governors, who also own cacao trees, actively support cacao production. There are about 130,000 trees by 1678.

Before the turn of the century, cacao production stagnates for many reasons. Spanish colonial laws and inheritance practices break up large tracts of land into smaller lots. New laws prohibit the use of indigenous people to harvest cacao, reducing the workforce. Illegal trade grows substantially by 1690. Piracy and attacks by the Zambo-Mosquitos (former slaves who intermarry with Mosquito tribal members in Nicaragua) also bring the industry to a standstill.

1709 Cacao beans used as currency

The cultivation of cacao trees remains important through the next two centuries. The scarcity of currency forces Costa Ricans to use cacao beans as legal tender for small purchases from 1709 to the mid-1800s.

1717 Heredia founded

Heredia is founded in the central highlands. Heredia becomes one of the four major cities located in the highlands. About one-half of the country's population lives in this region.

1723 Irazu volcano erupts

Meager economic and social conditions never allow for the development of large colonial-style haciendas of other Spanish enclaves, and Costa Rica doesn't develop a feudal system. Economic progress also stagnates because of Spain's mercantile policy, which prevents the colonies from trading between themselves or trading with any nation other than Spain.

Cartago in the center of the country is a barely functioning town when it is destroyed by an eruption of the Irazu volcano.

1737 San Jose founded

San Jose, which will become the modern-day capital, is founded between Cartago and Heredia.

1782 Alajuela founded

Alajuela is founded west of Heredia and northwest of San Jose. In time, the towns of Costa Rica strengthen their economies with wheat and tobacco exports. These cities also develop their own social and political structures, leading to competition and a battle for control of Costa Rica.

Building a Nation

1821: September 15   Spanish colonies declare independence

Shortly after Mexico declares independence from Spain, the Spanish colonies of Central America also proclaim independence. News of the independence movement takes more than a month to reach Costa Rica, where its leaders decide to join Mexico. The Spanish governor resigns, and a provisional government is established.

1821: December 7   Costa Rica's first constitution

Costa Rica proclaims its first constitution, calling it *Pacto de la Concordia* (Pact of Harmony). Ironically, there's little harmony in the newly independent province.

By this time, the four major cities in Costa Rica—Cartago, Heredia, San Jose, and Alajuela—operate much like city-states. They maintain separate governments and their economies and social structures are relatively independent of each other. The wealthy and conservative leaders of Cartago and Heredia cannot agree on political and social decisions with the more progressive republican leaders of San Jose and Alajuela.

1823 Former colonies formalize their relations

Disagreements among the four leading cities quickly lead to civil unrest. After a short battle in the Ochomogo Hills, the republican forces of San Jose emerge as leaders of the province.

One of their first decisions is to dissolve their alliance with Mexico and join the United Provinces of Central America, along with Guatemala, El Salvador, Honduras, and Nicaragua. The provinces set up their capital city in Guatemala. San Jose becomes the provincial capital of Costa Rica. The following year, the Guanacaste region in northwest modern-day Costa Rica secedes from Nicaragua to join Costa Rica.

1824   Costa Rica elects first chief of state

Juan Mora Fernandez (1784–1854) becomes the nation's first chief of state, ruling for nine years. He establishes a judicial system, founds the nation's first newspaper, and expands public education. He also encourages coffee cultivation and gives free land to anyone willing to grow coffee.

1824: May 25   Costa Rica abolishes slavery

Few slaves are imported into the country and there are fewer than 200 blacks when the country abolishes slavery.

Blacks in Costa Rica suffer discrimination and do not receive the right to vote until 1949, the same year women gain the right to vote. They are forbidden to migrate from the Atlantic coastal area until 1948, when a government decree grants them citizenship.

1829   Coffee is leading export

Coffee becomes the nation's chief export, despite a modest start sometime in the early 1800s. At least three different people have been credited with being the first to grow coffee—Father Felix Valverde, Cuban-born Tomas de Acosta, and Father Juan Francisco Carazo.

1835: September   San Jose under siege

Forces from the cities of Cartago, Heredia, and Alajuela attack San Jose in what becomes known as the War of the League. Despite improvements and attempts by Fernandez to hold the country together, rivalry among the cities remains deeply rooted. San Jose prevails and Braulio Carrillo (1800–45), who has taken power as a benevolent dictator, establishes an orderly public administration and new laws to replace colonial Spanish law.

1838   Federation falls apart

Carrillo withdraws Costa Rica from the federation, United Provinces of Central America, and proclaims complete independence.

The federation, thanks to continued opposition to its leader, the conservative Honduran General Francisco Morazan, falls apart. Morazan seeks refuge in South America.

1842   Carrillo overthrown

Braulio Carrillo has been declared president of Costa Rica for life, but his decisions are growing unpopular. He dissolves

Congress and suspends personal freedoms during his administration. Honduran General Francisco Morazan returns to Central America. At the request of some Costa Ricans, he topples Carrillo. Carillo is publicly shot in San Jose. Morazan imposes a military draft and direct taxes, and attempts to revive the federation. He is murdered five months after he overthrows Carrillo. For the next five years, a series of weak leaders head Costa Rica's government.

1847   Jose Maria Castro becomes president

Jose Maria Castro becomes president, a post he holds until 1849.

c.1849   Coffee revenues create wealthy *cafetaleros*

Coffee revenues have created a new social class of wealthy *cafetaleros* (coffee growers), who in time come to dominate national politics. The export boom transforms the country, with the government actively encouraging coffee cultivation. Most of the public works, including roads and railroad, are constructed to benefit the crop.

1849   Coffee growers emerge as the ruling class

Coffee growers have become successful and wealthy. Political competition grows among the wealthiest of the families. Several families of powerful coffee growers conspire to overthrow the nation's president, Jose Maria Castro. Castro's administration founds a school for girls and supports freedom of the press. He is considered an enlightened leader, but the coffee growers want someone in the presidency that they can control.

Juan Rafael Mora Porras (1814–60) comes to power on the shoulders of the coffee aristocracy. Under his ten-year leadership, Costa Rica grows economically. Mora establishes the first national bank, a system of street lights, and many public schools. His establishment of the national bank, however, alienates coffee growers because they fear the bank will undermine their control of credit to coffee producers. During his second term, Mora becomes a national hero when he organizes an army of several hundred civilians to defend the nation against the notorious American adventurer William Walker, who wants to turn Central Americans into slaves (see 1855).

1855   American adventurer invades Nicaragua, threatens Costa Rica

The long history of overt U.S. intervention in Central America gets under way more or less when William Walker (1824–1860), declares himself leader of Nicaragua and then threatens the sovereignty of Costa Rica.

Walker is a soldier of fortune, a racist who believes, like many Americans of his time, that it is his nation's "manifest destiny" to control other peoples. Born in Nashville, Tennessee, Walker first tries to provoke a secessionist movement in

Sonora, Mexico and fails. In June 1855—at the invitation of liberals engaged in a fight for power with conservatives—he arrives in Nicaragua with a private army.

Walker declares himself president of Nicaragua after defeating the conservatives, threatening the interests of another American: financier Cornelius Vanderbilt.

Vanderbilt secures concessions from Nicaragua to set up a profitable transit route for goods and people through the country long before the opening of the Panama Canal. The route consists of water travel up the San Juan River, a boat trip across Lake Nicaragua, and a portage across the isthmus of Rivas to the port of San Juan del Sur. The route becomes profitable after gold discoveries in California.

Vanderbilt is concerned Walker will disrupt his many business ventures in Nicaragua and threatens to oust him from the country. But Walker's power grows when Southern slaveholders offer him men, weapons, and money. In return, Walker promises to institute slavery in Nicaragua. His dream is to conquer the weak Central American nations and turn them over to his supporters as part of a Confederacy of Southern States.

1856–57   Costa Ricans fight Walker's army

Vanderbilt encourages Costa Rican President Juan Rafael Mora Porras to declare war on Walker, who invades the Guanacaste region in March 1856. Mora's army pursues Walker into Nicaragua, where half his men lose their lives. Other Central American forces join the two-year battle, and Walker's forces are finally defeated in April 1857.

Walker is escorted out of Nicaragua on an American ship bound for Panama. But Walker is not done. He tries to return twice—the first time he is arrested by an American captain. In 1860 he attempts a landing in Honduras but is forced to surrender to a British sea captain who turns him over to Honduran authorities. He is promptly executed.

Inadvertently, Walker manages to do what Costa Rican independence had not: unite its people. Costa Rican nationalism is strengthened by the war, giving the country many of its national heroes and myths.

1859   Mora ousted

Costa Rica's gratitude for Juan Rafael Mora Porras for his role in resisting American William Walker (see 1855) is short-lived. Mora is ousted from power when citizens blame him for a cholera epidemic that claims one in ten people. Mora attempts to recapture the government, but fails and is executed on September 30, 1860.

1870: April   Military man brings order to chaotic Costa Rica

After Mora's execution, the powerful coffee families, with support from their respective military cronies, struggle throughout the 1860s to gain control of the nation.

General Tomas Guardia (1831–82), who distinguishes himself in the war against Walker, ends the bickering by toppling the government and ruling for the next twelve years. Guardia builds a powerful centralized government that sets the stage for shaping the modern liberal-democratic state. Unlike other military strongmen, Guardia is a progressive thinker who sets to improve the lives of those he rules.

He abolishes capital punishment, revises the constitution to provide free and obligatory primary education for boys and girls, and uses taxes and coffee revenues to build roads and public buildings.

Just as importantly, he curbs the power of the coffee barons and brings the nation's army under control, lowering its influence in national politics. Guardia also seeks to connect the highlands of Costa Rica to the Caribbean by rail to bring coffee to the world market. The trip to the coast at this time is a painful one through challenging terrain.

1871 Atlantic Coast Railroad constructed

American Minor C. Keith (1838–1929) travels to Costa Rica to supervise construction of the Atlantic Coast Railroad. In return, Keith receives considerable land along the railroad right-of-way and decides to plant bananas to make money.

West Indian blacks, many from Jamaica, are brought to Costa Rica to build the Atlantic Coast Railroad and work on the banana fields.

1872 Banana exports to the United States begin

Minor C. Keith (see 1871) exports 230 stems of bananas to New Orleans, Louisiana. This is his first shipment of banana, and his banana operations help to make him one of the most powerful men in Costa Rica. He establishes two ventures—Tropical Trading Company and Boston Fruit Company—to market his agricultural products.

1880s Liberals, conservatives, and the church

In the 1880s a new class of politicians emerges. In Costa Rica and Central America, liberals believe in the separation of church and state, with the state in a dominant position. Liberals believe formal education can improve society's problems, and they favor public education, freedom of the press, and religious tolerance.

However, liberals don't necessarily favor extreme social legislation to benefit the underprivileged. Those who do are called progressives. Conservatives, on the other hand, tend to align themselves more closely with the Roman Catholic Church, often belong to the upper class, and do not generally favor social legislation.

Liberal governments in Costa Rica in the 1880s expel the Jesuits (a Roman Catholic religious order), secularize the schools and cemeteries, and close a university because it is deemed too close to the church.

1888 Birth of novelist and educator Carmen Lira

Maria Isabel Carvajal (1888–1949), a novelist who writes under the pen name Carmen Lira is born. Lira, considered one of Costa Rica's most important women, is a teacher and women's political leader in addition to being a novelist. She is one of the co-founders of *Centro Germinal*, an influential political group that rises after World War I and becomes increasingly Marxist-oriented.

Lira opens the first kindergarten in the country and pushes for the development of the National Library (opened in 1887). She publishes a magazine titled *El Maestro* (The Teacher), which deals with social and education issues. As novelist, she writes *Las Fantasias de Juan Silvestre* (1918), *En Una Silla de Ruedas* (1918), and *Cuentos de mi tia Panchita* (1922). She is most famous for a collection of children's stories.

She flees the country after the 1948 civil war and dies in exile in 1949. The government later honors Lira for her contributions to national culture.

1889 Democratic ideals are tested

President Bernardo Soto (1854–1931) calls for elections. When he attempts to install his own candidate, Costa Ricans rebel. Farmers and townspeople march to the presidential palace and demand that Soto respect the electoral results. Soto agrees and steps down, naming Dr. Carlos Duran Cartin (1852–1924) to finish his term.

Duran ensures free elections, with Jose Joaquin Rodriguez (1838–1917) winning the presidency. While women and blacks are not allowed to vote, Rodriguez's victory is considered the country's first "honest" election.

For the next three decades, leading to the military coup of 1917, electoral honesty and democracy withstand many blows. Rodriguez suspends constitutional guarantees and postpones Congress in his first year in power. His handpicked successor, Rafael Yglesias Castro (1861–1924) also administers a semi-dictatorial government. He is reelected in 1898 when the opposition refuses to field a candidate, fearing the government is not going to conduct free elections.

1889 Banana exporter consolidates operations

American banana grower Minor C. Keith merges his Tropical Trading Company and Boston Fruit Company (see 1872) to create the United Fruit Company. Under his leadership, the company has a virtual monopoly in the Atlantic zone of the country. In the 1930s, various diseases attack the bananas and the company is hit by strikes. Later, it transfers its operations to the Pacific coast, before ceasing banana production in Costa Rica in 1985.

1904   Indentured black population tops 1,000

Indentured (under contract to work for a specific amount of time) blacks brought to Costa Rica to provide labor for the construction of a railroad (see 1871) number about 1,200.

1905   Historian Ricardo Fernandez Guardia

Ricardo Fernandez Guardia (1867–1950), a historian and novelist, is one of the country's most important historians. He studies in Paris and brings a new approach to researching Costa Rican history, which relies on biographies and the retelling of myths. Among his historical books are *Discovery and Conquest* (1905), *Cartilla Historica de Costa Rica* (1909), *Colonial Chronicles* (1921), and *Independence: The History of Costa Rica* (1941).

Guardia is instrumental in introducing the *costumbrismo* literary style, which describes the ordinary day-to-day lives of Costa Ricans. Its literature is punctuated by the vernacular speech of its characters, who live mostly in rural settings. His *costumbrista* novels include *Cuentos Ticos* and *Hojarasca* (1894).

1913–18   World War I ends era of coffee exports to England

Prior to World War I (1913–18), England imports nearly eighty percent of its coffee from Costa Rica, which grows a high-grade arabica bean. Coffee in England is the drink of choice for the wealthier classes willing to pay higher prices for quality coffee. After the collapse of the British market, Costa Rica begins to sell its beans to the United States and Germany.

1913   Painter Margarita Berthea is born

Margarita Bertheau (1913–76) is born. She becomes one of the country's most important exponents of modern painting. She is a member of the so-called Nationalist Generation, a group of artists who seek to depict Costa Ricans from all walks of life.

Bertheau paints in oils and watercolors and specializes in landscapes. She is also one of the prime forces behind the development of classical ballet in Costa Rica.

1914   New president faces many obstacles

Power struggles continue through 1914, when three candidates seek the presidency. Two of the three candidates resign, and Congress declares the third ineligible, choosing instead Alfredo Gonzalez Flores (1877–1962) to lead the country. Gonzalez, a lawyer and politician, had not been a candidate originally and Costa Ricans have a difficult time accepting him.

Economic strains of World War I (1913–18) also make it difficult for Gonzalez, with his country facing declining exports and rising debts. To offset decreasing revenues, Gonzalez proposes a new tax law.

Most tax revenues in Costa Rica come from the lower working classes. The upper classes are affected minimally by taxation and Gonzalez's proposal is immensely unpopular with them.

American and British oil interests also put pressure on Gonzalez, who refuses to sign any oil concession contracts.

1915   Feminist journal *Figaro* founded

Angela Acuna de Chacon, Costa Rica's first woman lawyer, founds the national feminist journal *Figaro*. In 1923, she forms a national feminist group. (See also 1949.)

1917   Gonzalez Flores comes under attack

Gonzalez prepares to defend his presidency in the upcoming elections, and his opponents gather strength to oppose him. But before elections are held, Gonzalez's minister of war, Federico Tinoco (1870–1931), ousts him from power.

The dictatorial Tinoco, along with his brother Joaquin, rule the country harshly, filling the jails with political prisoners and clamping down on the press.

1919: May 5   Tinoco opponents attack from Nicaragua

Costa Rican exiles living in Nicaragua launch an invasion to overthrow the Tinoco regime. But the Sapoa Revolution—known by that name because the rebels cross the Sapoa River to attack military posts along the border—fails.

The attempted revolution galvanizes opinion against Tinoco and sets the stage for his eventual ouster.

1919: August 12   Tinoco forced to resign

High school students and their teachers, mostly women, organize protests against Tinoco. The demonstrations turn violent when they set fire to a pro-Tinoco newspaper plant, and he responds by sending government troops against them. Some protesters take refuge at the United States consulate, which is hit by government fire.

Diplomats convince Tinoco that a coup against him is imminent and encourage him to give up power. He does so only after his brother Joaquin is assassinated. Tinoco dies in exile in Paris.

1919: September 2   United States exerts its influence

When he resigns, Tinoco turns over the government to his vice-president, Juan Bautista Quiros Segura (1853–1934), which angers the United States. American consul Valentine Chase tells the Costa Rican government his country will not accept the nomination. To make his point, he says the cruiser *Denver* is standing by in Puntarenas harbor on the Pacific coast. Quiros is forced to resign on September 2.

At the insistence of the United States government, Francisco Aguilar Barquero (1857–1924), a lawyer and professor of law, is selected president of the country.

Barquero helps the country return to normalcy and supervises an honest election in 1920. The winner of that election is Julio Acosta Garcia (1872–1954), leader of the revolution against Tinoco.

1930s   Banana plantations attacked by disease

American banana grower Minor C. Keith (see 1872) and his United Fruit Company suffer losses when the banana crops are damaged by diseases.

1931   Artist Isidro Con Wong is born

Isidro Con Wong (b. 1931) becomes one of the most important Costa Rican contemporary painters. His work flourishes during the 1960s, during a period of *Arte Nuevo*, a school of painting dominated by abstract and expressionist work.

Wong, of Chinese ancestry, is a poor farmer before he starts painting with his fingers and *achiote*, a red paste made from seed. His work is found in several U.S. and French museums and sells for thousands of dollars.

1934   Banana strikes paralyze country

More than ten thousand workers strike against United Fruit Company (see 1872 and 1889) in the Atlantic region of the country. Communist Party leaders Carlos Luis Fallas and Jaime Cerdas lead the strike. While banana revenues have made many men rich, banana workers remain extremely poor and live in horrible conditions. After several weeks of violence, the government sends troops and the company, owned by Minor C. Keith (see 1871), ultimately agrees to better working conditions and higher wages. It is the first successful strike against a foreign company.

Civil War Shapes Modern Costa Rica

1940s   Social conditions inspire Generation of 1940 writers

A narrative style in literature develops when the nation's intellectuals worry about social and political conditions. The authors of this era are known collectively as Generation of 1940, and one of its most important exponents is Carlos Luis Fallas, novelist and labor leader.

Fallas, one of the founders of the Costa Rican Communist Party, works in the banana fields and later organizes and helps direct a major banana strike (see 1934). He is a military leader of the government forces in the 1948 civil war. The strike inspires him to write *Mamita Yunai* (1940), one of the best-known works in Costa Rican literature. The book exposes the shocking and terrible working conditions of the banana plantation.

1940–48   Political reforms, ideological differences lead to civil war

Exploitation and the growing divide between wealth and poverty set the stage for violent conflict in the 1940s. Although the 1920s and 1930s are relatively peaceful, the country's underprivileged make few social gains, and their numbers grow due to economic depression. By 1948 the country is on the verge of a civil war.

1940   New president initiates social reforms

Rafael Angel Calderon Guardia (1900–70) becomes Costa Rica's president in 1940, ushering in a new era of social reform for the country.

Calderon is a deeply religious man, a physician who tries to improve the life of poor Costa Ricans. His actions take place at a time when many Central American and Caribbean nations are ruled by dictators who have no concern for the poor.

During his term Calderon attempts land reform to give land to poor farmers, establishes a minimum wage, paid vacations, unemployment compensation, and many other rights designed to protect workers. He even creates a program to distribute shoes to poor children.

His social reforms scare the wealthier classes, but are hailed by the political left. Yet Calderon compounds serious economic problems by funding new social services during the lean years of World War II (1939–45). With high inflation, he loses the support of the middle and working classes. The communists, who at first criticize Calderon for being too slow to implement changes, come to his support under an alliance with the Roman Catholic Church to support social gains for the underprivileged. But Calderon begins to behave undemocratically, and appears poised to steal the upcoming elections of 1944 and 1948.

1942: July   Exile turns common man into opposition hero

Jose Figueres Ferrer (1906–90), a coffee grower, enters public life when he criticizes the policies of President Calderon during a radio broadcast. Figueres dreams of social revolution and a second republic. In his broadcast he accuses Calderon of being a demagogue who is leading the nation to economic and social disaster and criticizes his close relations to the communists. Calderon turns autocratic and sends Figueres into exile in Mexico.

From the moment he is exiled, Figueres begins to plan an armed revolution against the government. He portrays himself as a successful capitalist with socialist tendencies, a man who wants to liberate Costa Rica from the tyranny of Calderon. Somehow, he has come to believe that Calderon is part of the network of dictators who rule Central America and the

The harbor of Puerto Limon is the sight of a German submarine attack in 1942. (EPD Photos/CSU Archives)

Caribbean, and he must bring him down. Others feel Figueres is simply trying to settle a personal vendetta.

During exile Figueres meets other Central American liberals who share his ideas. They create a loose federation called the Caribbean Legion in Guatemala, where they come under the protection of President Juan Jose Arevalo.

Arevalo shares their dreams. He wants to get rid of the totalitarian governments, but he also wants to bring the small nations of Central America under a single democratic government. He has the weapons and men. Figueres tries to convince him that Costa Rica is the right place to begin the revolution. Yet Arevalo remains cautious and uncertain about the young and fiery Figueres.

With help from his party, Accion Democrata, Figueres begins an endless political assault against the government after returning from exile in 1944. He waits for the right

opportunity to begin an armed insurrection while publicly calling for the government's overthrow.

1942: July 2   German submarine enters Puerto Limon and sinks a U.S. ship

A German submarine enters the harbor of Puerto Limon and sinks an American freighter, the *San Pablo*. The sinking brings the war home to Costa Ricans.

1944   Calderon's hand-picked successor becomes president

In an election widely believed to have been fraudulent, Teodoro Picado Michalski assumes the presidency. Picado, who is in Calderon's camp, fails to quiet growing discontent during his tenure and faces stiff opposition from intellectuals, businessmen, and even laborers. Picado's term in office is considered simply a presidential campaign for Calderon, who wants to regain power in the 1948 election.

1946: June 24   Insurrection against Picado fails

A poorly organized group, unconnected to Figueres, attempts an armed insurrection against the government. Curiously, Picado personally travels to the police station to free the frustrated rebels. His gesture is seen as a sign that Picado believes Costa Rica can work itself out of a crisis without resorting to violence. But Picado begins to lose control, with all sides resorting to violence.

1947: December 16   Figueres pledges to fight dictatorships

Figueres and other exiled leaders sign the Pact of the Caribbean, pledging to wage war to free Central America from tyrannical governments. Jose Figueres convinces Arevalo, the Guatemalan president, that the battle should begin in Costa Rica, where he has 500 trained men but no weapons.

1948: February 8   Elections are held

Early election returns are inconclusive, but give Otilio Ulate, a newspaper publisher, a slight lead over Calderon, who is running for a second term.

The final vote tally shows Ulate as the winner by a slight margin, but Calderon and his supporters refuse to accept the results. Picado claims fraud and refuses to step down. The Calderon-dominated legislature annuls the election results. Calderon plays a waiting game in hopes of retaining the presidency in an upcoming congressional vote.

The communists remain firmly aligned with Calderon's forces, believing that the opposition will take away all social gains made over the past decade. But they have no weapons to defend the government.

1948: March 10   Civil war

Jose Figueres, who has been waiting for an excuse to begin armed insurrection, launches the "War of National Liberation" when election results are not respected. His forces capture Cartago and Limon and prepare to attack San Jose. Already, more than 2,000 people have lost their lives, and Picado fears massive losses if Figueres attacks San Jose. Picado is also aware that the United States is ready to invade Costa Rica to suppress communist forces holding on to San Jose.

1948: April 19   Conflict ends

Picado and opposition leaders sign a pact ending the conflict. After winning the war, Figueres does not immediately allow Otilio Ulate to assume the presidency. Instead, Figueres becomes the president of the Junta Fundadora de la Segunda Republica (a founding council of the second republic).

With the *junta* at his service, Figueres bans the press and communist party, gives women and blacks the right to vote as well as full citizenship for blacks, revises the constitution to ban a standing army (including his army), and nationalizes the banks. He surprises some by continuing Calderon's early social reforms. But his most enduring legacy is ensuring free and honest elections. He even establishes presidential term limits.

Figueres is not quick to forget his enemies and exiles Calderon and many of his followers to Mexico. Many important leftist leaders are abducted and murdered. Others lose their jobs and property. Calderon tries to return to power by force twice, invading Costa Rica from Nicaragua. A volunteer army of six thousand, including high school students, repels Calderon for a second time in 1955. Calderon is allowed to return in 1962, when he unsuccessfully runs for the presidency.

Many of those who support the uprising against Calderon simply want to re-establish democracy and don't agree with Figueres' philosophy. To his credit, he allows Ulate to assume the presidency after eighteen months.

Yet the 1948 civil war gives Figueres immense power, and he dominates politics for the next two decades. He founds the *Partido de Liberacion Nacional* (National Liberation Party), which becomes one of the most important forces in contemporary Costa Rican politics. He is elected to the presidency twice (1953–58 and 1970–74). He is considered a national hero.

From War to Stability

1948: December 1   Costa Rican Army officially disbanded

Costa Rica becomes the first country in the Americas without an army. Policing duties are divided between several agencies governed by political appointees. By 1995 the country has a three thousand-member Civil Guard and a two thousand-member Rural Assistance Guard performing security and police functions. An anti-terrorist battalion serves as the presidential guard. The country spends only $26 million in defense in 1996.

1949   Pioneer feminist helps Costa Rican women gain right to vote

Angela Acuna de Chacon dedicates her life to fight for women's rights. Her political writings and speeches help women achieve the right to vote. (See also 1915.)

1949: November 8   Otilio Ulate becomes president

As promised, Jose Figueres hands back the presidency to Otilio Ulate (1892–1973), who manages to serve a fairly quiet term. Ulate is a conservative newspaper publisher and politician. He serves several times in Congress and uses his newspaper, *El Diario de Costa Rica*, to attack Calderon and to support his own candidacy in the presidential elections of 1948.

In later years Ulate turns against Figueres, who builds a powerful political machine. In the most unlikely alliance, Ulate joins forces with his nemesis, Calderon, to create the National Unification Party. But he withdraws from the party during the 1970 elections, won by Figueres.

1952 Figueres elected president

Figueres strengthens Costa Rica's democratic ideals, and the country prospers during his tenure as president. The country breaks diplomatic relations with several Latin American countries with dictatorial regimes. Figueres also launches a number of social programs to help the underprivileged. Later presidents attempt to reduce Costa Rica's social welfare system, mostly without success.

1970 Writer earns National Theater prize

Jose Basilio Acuna (b. 1897), one of the most important Costa Rican writers of the twentieth century, wins a national theater prize for his work. Acuna studies psychology and philosophy and during World War I, he fights with the French Foreign Legion. His most important stage works are a trilogy dealing with Inca themes called *Intiada: The City of the Golden Doors, Fantastic Intermission, and the Inca Empire.* In 1983 he receives the highest national award for literature.

1977 Government creates a reserve for Chorotegas

Only 100 or so Chorotegas families, descendants of early inhabitants of the region, remain. The government provides a reservation for them at Matambu, in the Guanacaste region. (See 1300.)

1980 Artist's work put on a stamp

Lola Fernandez (b. 1926), a Colombian-born painter who comes to Costa Rica at age three, wins many awards for her paintings. One of her works is selected to appear on a postage stamp. Fernandez is one of the leading members of the *Vanguardista* school of painting, a movement of the late 1950s that stresses abstract painting.

Fernandez becomes a member of Grupo 8, a group of artists who call for a new emphasis on art starting in 1961. They advocate leaving behind classic Old World styles to concentrate on Costa Rican themes. They hold several shows for the next three years. Another artistic movement evolves from Grupo 8. From 1961 to 1970 these artists encourage artistic dialogue and promote the plastic arts (art that is three-dimensional, such as sculpture and ceramics) with art shows and public lectures.

1982 New president declares Costa Rica's perpetual neutrality

President Luis Alberto Monge Alvarez (b. 1926) declares that Costa Rica will remain neutral forever in all warlike conflicts. His act reaffirms national sovereignty and the continued abolishment of the armed forces. His declaration is driven by regional conflicts that threaten to involve Costa Rica. Monge faces serious domestic problems as well.

Severe inflation and currency devaluation, coupled with growing debt and constant disruptions to trade caused by the ongoing Nicaraguan war, spin the country into an economic crisis. To make things worse, banana and coffee prices plummet almost at the same time that large international debts come due.

With economic aid from the United States and the International Monetary Fund, Monge directs Costa Rica to a slow recovery.

1985 United Fruit Company ceases operations

At one time employing more than ten thousand workers, the United Fruit Company ceases banana production in Costa Rica. Prior to shutting his operations down, Keith transfers banana production from the Atlantic coast where the labor force is hostile, to the Pacific Coast in an attempt to save the business.

1986 Coffee brings more than $354 million in revenues for Costa Rica

Coffee remains one of Costa Rica's strongest exports, with the harvest of November 1987–February 1988 employing more than 200,000 people (see 1829.) After being controlled in its early years by foreign businessmen, by the late 1980s and 1990s, eight-five percent of coffee properties are owned by Costa Ricans.

1987 Costa Rican President wins Nobel Peace Prize

President Oscar Arias Sanchez (b. 1940) is awarded the Nobel Peace Prize for structuring the Guatemala Peace Plan. On August 7, 1987, five Central American nations agree to end armed conflict gradually.

Arias is a politician, writer and university professor, and president of the country from 1986 to 1990. He holds many other governmental posts and presides as Secretary General of the National Liberation party. The Nobel Peace Prize is considered one of the most distinguished worldwide awards.

During the 1980s Costa Ricans deal with severe economic problems and increasing political violence. The government resists pressures from the United States to become involved in the Nicaraguan war, where the U.S.-supported "contras" are fighting the leftist Sandinista government. Costa Rica refuses to support the contras or build up its military strength.

Arias becomes heavily involved in foreign relations, understanding that national security is connected to peace in Central America. His peace plan, signed by Guatemala, Nicaragua, Costa Rica, El Salvador, and Honduras provides for free elections in all countries, a cease-fire by contra and government forces in Nicaragua, an end to outside aid to the contras, and amnesty for the contras. It also calls for repatriation or resettlement of all war refugees and an eventual reduction in the armed forces.

1988   Home of renowned author becomes museum

The home of author Joaquin Garcia Monge (1881–1958) becomes a museum. Monge is considered one of Costa Rica's chief exponents of *costumbrismo*, a literary movement that began in the 1890s (see 1905).

Monge is an author and journalist who is influenced by Marxist philosophy. He serves as minister of education and publishes a renowned international review titled *Repertorio Americano*. In 1958 he receives some support for the presidency, but his party is ruled illegal because of its communist ties. His most important works are *La Mala Sombra, El Moto, Hijas del Campo,* and *Abnegacion*.

Among other important *costumbristas* are Manuel Arguello Mora, Juan Garita, Manuel Gonzalez Zeledon, Carlos Gagini, and Ricardo Fernandez, who is instrumental in introducing this literary style.

1990s   Blacks are three percent of population

Blacks first came to Costa Rica from Caribbean islands such as Jamaica to provide labor for railroad construction (see 1880s). By the 1990s their population increases to about thirty thousand, or about three percent of the total population.

1990   Literacy rises

Costa Rica enjoys one of the highest literacy rates in Latin America. By 1990 adult illiteracy among males drops to 5.2 percent and among women to 5.3 percent, according to government statistics. The government spends about 21 percent of its budget on education. Primary and secondary education is free.

1990: February   Another Calderon becomes president

Rafael Angel Calderon Fournier takes office, promising to restore the country's economy. He is the son of Rafael Angel Calderon Guardia (1900–70), one of Costa Rica's most important presidents. (See 1940.)

Unlike his father, a social reformer who sought to improve the lives of working-class people, Calderon Fournier is a conservative lawyer and member of the Social Christian Unity Party. He seeks to liberalize the economy and reduce the country's extensive social welfare system.

1994   Coffee and bananas lead exports

Coffee beans and bananas account for forty percent of the country's exports, bringing revenues of $310 million and $528 million, respectively. Since the 1970s the banana industry produces more than one million tons of bananas annually.

1994: February 6   The Figueres political legacy continues

Jose Maria Figueres Olsen, son of 1948 revolution leader and former president Jose Figueres Ferrer, takes office and espouses political measures that represent a reversal from his father's party and legacy.

His reform package, introduced in 1995, focuses on cutting government expenses and privatizing state-owned enterprises. The government is the country's largest employer, and labor leaders fear cutbacks in government jobs. His proposed reforms touch off a month-long strike by the nation's teachers. Figueres alters his economic plan, but he has lost public support.

1996   Economy falls into a recession

The economy declines for the fourth year in a row, and annual inflation stands at nearly fourteen percent in 1996. By the end of the year, Figueres announces a tax hike, privatization of some state-owned banking and telecommunication systems, and an end to the state insurance monopoly.

1998   President vows to improve economy

Miguel Angel Rodriguez Echeverria, who loses to Figueres in the 1994 elections, becomes the country's new president. He belongs to the conservative Social Christian Unity Party (PUSC). The party is formed in 1983 from several parties to offer a unified opposition against Figueres' National Liberation Party.

Bibliography

Bell, John Patrick. *Crisis in Costa Rica: The 1948 Revolution.* Austin: University of Texas Press, 1971.

Benson, Elizabeth P., ed. *Between Continents, Between Seas: Pre-Columbian Art in Costa Rica.* New York: Henry N. Abrams, 1981.

Biesanz, Richard, Karen Zubris Biesanz and Mavis Hiltunen Biezanz. *The Costa Ricans.* New Jersey: Prentice Hall, 1982.

Honey, Martha. *Hostile Acts: U.S. Policy in Costa Rica in the 1980s.* Gainesville, Fla.: University Press of Florida, 1994.

Stone, Doris. *Pre-Columbian Man in Costa Rica.* Cambridge, Mass.: Peabody Museum Press, 1977.

Todorov, Tzvetan. *The Conquest of America.* New York: Harper & Row, 1984.

Weinberg, Bill. *War on the Land: Ecology and Politics in Central America.* London: Zed Books Ltd., 1991.

Williams, Philip J. *The Catholic Church and Politics in Nicaragua and Costa Rica.* Pittsburgh, PA: University of Pittsburgh Press, 1989.

Winson, Anthony. *Coffee and Democracy in Modern Costa Rica.* New York: St. Martin's Press, 1989.

Cuba

Introduction

Cuba is a long, narrow island in the Caribbean Sea. It is the largest of the Caribbean islands, and the seventh largest island in the world. Strategically located at the passageway between the Atlantic Ocean and the Gulf of Mexico, it is also only ninety miles from the Florida Keys. Its 2,500-mile (4,022-kilometer) coastline has a number of excellent harbors, including Havana and Guantánamo. Much of Cuba's terrain is flat or rolling, with fertile valleys. Some areas are mountainous or hilly, and there are lowlands and swamps along the southern coast. Cuba's population is mostly descended from the Spaniards who colonized the island and the African slaves they brought there. Many Cubans are of mixed Spanish and African ancestry. There are few if any descendants of the island's original Indian population, which was mostly wiped out within the first century of colonization. Cuba's estimated population in 1997 was roughly 11 million.

Early History of Cuba

Spain colonized and began to settle Cuba early in the sixteenth century, developing both the island's mineral and agricultural resources. The Taínos and other Indian groups who were living on the island at the time of the Spanish conquest were forced to labor in gold mines, and the deprivations they endured, together with the effects of warfare and disease, decimated the native population so rapidly that African slaves were brought in as early as the 1520s to bolster the declining slave population. The sugar industry, which began in the late 1500s, came to dominate the island's economy. Through most of the colonial period, Cuba's trade in commodities such as sugar and tobacco was tightly regulated by Spain. However, the brief British occupation of Havana, the principal port, in 1762–63 brought with it freer trade policies that ultimately led to long-term changes by the Spanish, who liberalized their own trade policies in 1778. When the Haitian revolution of 1796 devastated that country's sugar trade, Cuba became the chief exporter of sugar among Spain's New World possessions.

Unlike the Spanish colonies in Central and South America, Cuba did not attain its freedom in the period of revolutionary upheaval following Napoleon's conquest of Spain in 1808. However, Cubans dreamed of independence and tried to achieve it in various ways for much of the nineteenth century, including a long but unsuccessful military rebellion from 1868 to 1878. Finally, as the century was ending, a campaign inspired and organized by exiled poet and patriot José Martí—and eventually aided by the United States—succeeded in ousting the Spanish from Cuba. However, even though Cuba was no longer a colony, it was confronted with the prospect of domination by another foreign power—the United States, which set up a military occupation for four years (1898–1902) and legislated for itself the right to intervene in Cuba's internal affairs, a prerogative it would hold until 1934. After nearly four centuries of colonial rule, the new republic rapidly fell into political turmoil once the U.S. withdrew, and, at the request of the Cuban government, there was a second period of U.S. military government (1906–1909) and further military interventions in the first two decades of the century. In addition, U.S. investment in Cuba grew rapidly during this period, a development resented by many Cubans.

Batista and Castro

Cuban nationalism grew in the 1920s, but the young nation soon found itself under yet a new type of domination—authoritarian rule by a strong single political leader. Except for a few brief periods, this was the form that Cuban political power was to take for the remainder of the twentieth century, and its history during that time was to be dominated by two men: military dictator Fulgencio Batista, who controlled Cuba for much of the period between 1934 and 1959, and revolutionary leader Fidel Castro, who has led the nation since 1959. Under Castro's communist government, life in Cuba has changed dramatically in a number of ways. Traditionally Catholic, Cuba was officially declared an atheistic nation and church property was confiscated. Private landholdings were broken up and reorganized to form collective farms run by the state. Free education was made widely available at all levels, but its content was strictly determined by the government.

Contemporary Cuba

With the breakup of the Soviet Union in 1991, Cuba entered a difficult period of uncertainty and change, having lost its major trade partner, political ally, and source of foreign aid as well as privileged trade status with other countries of the former Soviet Bloc. The nation plunged into an ongoing economic crisis from which it is still struggling to recover. It has faced severe shortages of crucial commodities; in addition, sugar production has dropped dramatically and unemployment has risen. There have been some modifications in economic policy designed to ease the situation for the economically hard-pressed population. In contrast to the previous government stance, tourism—a mainstay of the Cuban economy before the communist revolution—is once again being encouraged and foreign investment is being sought. In addition, some modest market reforms were implemented, such as permission for the creation of farmer's markets operated independently by the growers themselves.

However, Cuba could not turn to its neighbor to the north for aid—or for relief from an economic embargo of over thirty years—unless its government made fundamental concessions in its political philosophy and methods of maintaining power. There was a brief thaw in Cuba-U.S. relations in the early years of the Clinton administration when there was discussion of lifting the U.S. travel ban to Cuba and other conciliatory measures. The Cubans held out the promise of possible free-market reforms and the United States hoped these might lead to democratic reforms as well. This period of conciliation ended, however, when two planes carrying Cuban-American civilians were shot down by Cuban military aircraft over international waters in February, 1996, and the U.S. Congress responded by passing the Helms-Burton Act, which restricted travel to and investment in Cuba and further tightened the trade embargo. Thus, the two nations retained their uneasy stalemate as the twentieth century neared an end, with Cuba still firmly hewing to its revolution and remaining among a dwindling number of nations still committed to the political ideology that inspired it.

Timeline

1000 B.C.   Ciboney Indians arrive in Cuba

The Ciboney Indians begin migrating to Cuba. In successive waves of migration over the next two thousand years, they become the largest group of Indians inhabiting the island. The Ciboney are simple hunter-gatherers living in temporary camps and fashioning tools from shells. They settle mostly in coastal areas, where they gather wild fruits and nuts and live on fish and crustaceans.

A.D. 1200   Taíno Indians settle in present-day Cuba

The Taíno Indians migrate to Cuba. Unlike the Ciboney and other groups of hunter-gatherers on the island, the Taíno, an Arawak-speaking people, live in settled villages in huts with thatched roofs called *bohíos*. They raise potatoes, manioc (called cassava or yuca in Central America), maize, and other crops. The Taíno are also known for their fishing skills and sturdy wooden canoes. They give the island of Cuba its name, which means "the central place" in Arawak.

1492: October 27   Christopher Columbus lands on Cuba

Christopher Columbus (1451–1506) arrives in Cuba on his first voyage to the New World and claims the island for Spain.

1508–09   Ocampo surveys the island

Sebastián Ocampo is charged by the Spanish Crown with measuring the perimeter of the island of Cuba.

The Colonial Era

1511   Velásquez conquers Cuba for Spain

Diego Columbus, Christopher Columbus's son and the governor of the island of Hispaniola, commissions Diego Velásquez to conquer Cuba for Spain. Velásquez leads a party of 300 men in subduing the Indians on Cuba and taking control of the island. Led by a chief named Hatuey, the local Taíno Indians put up resistance but are ultimately subdued and their leader is burned at the stake. Velásquez establishes the first settlement at Baracoa.

1512–17   Settlements are founded

Additional Spanish settlements are established at Bayano, Trinidad, Sancti Spíritus, La Habana (later Havana), Puerto Principe, and Santiago de Cuba.

1513   Economic development begins

The *encomienda* system is established on Cuba, allotting land and Indian slaves to settlers. Gold mining is begun, and the Indian population is forced to labor in the mines. Agriculture is another important source of wealth for the early colonists.

1520s   Settlers leave Cuba for Mexico

The rapid early development of Cuba is stalled when greater riches are discovered in Mexico and many settlers leave to exploit its mineral wealth.

1520s   First black slaves are brought to Cuba

The first African slaves are brought to Cuba to replace the rapidly dwindling Indian population. The slaves are from West Africa—most are Bantu, Congolese, Dahoman, Mand-

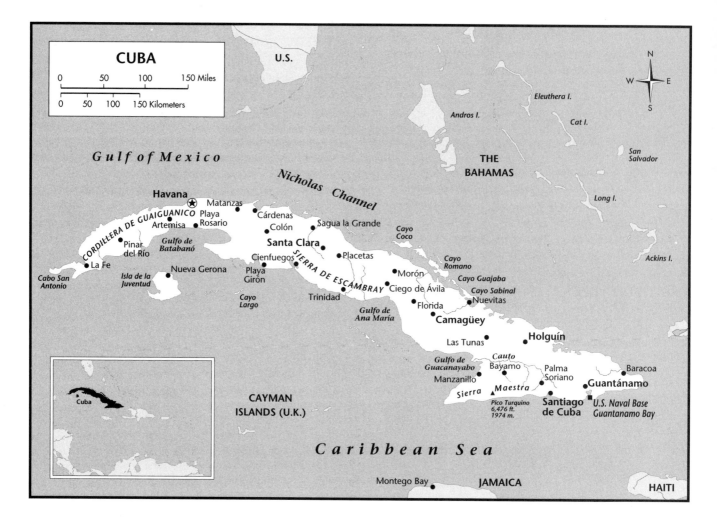

CUBA

ingo, or Yoruba. By the end of the sixteenth century, the native Indian population has been virtually destroyed through a combination of warfare, slavery, and disease, dwindling from an estimated 100,000 to 4,000 in the course of the century.

1524–1730   Mines are among world's greatest producers of copper

The Cobre copper mines near Santiago are among the greatest producers of copper in the world. In the 1730s the mines are abandoned and are idle for about a century.

1553   Governor's residence is moved to Havana

Initially utilized mainly as a maritime center and supply station, Havana grows in importance when the governor's residence is relocated there, The city is fortified against attack. Nevertheless, it remains vulnerable, and much of Havana is burned down by French pirates a few years later.

1561   Fleet system is established

To protect its ships from attack at sea, the Spanish organize a system whereby all ships sailing between Europe and the

Americas sail together at the same time every year. Havana, which serves as a meeting point for different ships and their crews, benefits greatly from the increased business brought to the island by these visitors, some of whom stay for months waiting until the entire fleet is re-assembled.

1570s   First sugar mills are built

The first sugar mills are built in Bahía Honda, not far from Havana. Toward the end of the sixteenth century, sugar production expands to the Santiago region in the east and grows rapidly.

1607   Havana becomes capital

Havana is named the colonial capital of Cuba.

17th century   Cuba suffers pirate raids

Like other parts of the Americas that have coastlines on the Caribbean or the Atlantic, Cuba becomes an object of pirate attacks, especially along its eastern coastline. French pirates attack the cities of Santiago and Bayamo in 1603, 1628, and 1633, and occupy Baracoa and Remedios in 1652. British adventurers sack Santiago in 1662, Puerto Príncipe is

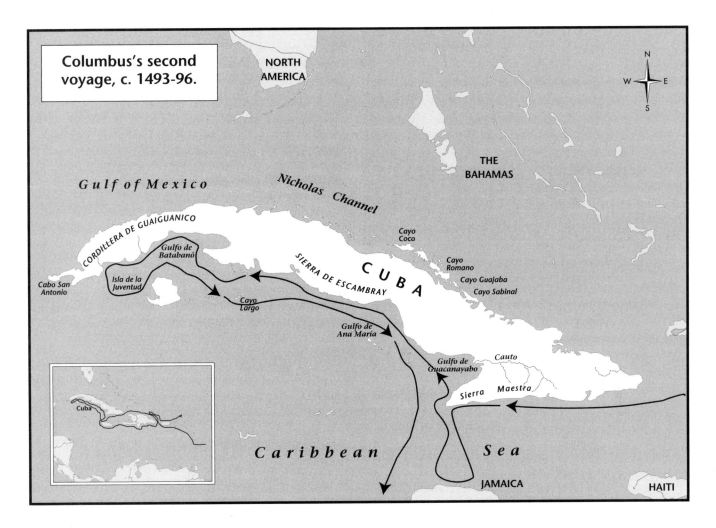

Columbus's second voyage, c. 1493–96.

destroyed by British pirate Henry Morgan in 1668, and John Springer seizes the city of Trinidad in 1675.

1717   Spain creates government monopoly on tobacco

Tobacco production becomes a government monopoly with the creation of a purchasing agency, the *Casa de Contratación,* to control all aspects of the tobacco trade. Tobacco growers can only sell their produce to Spain and only through this agency. The monopoly is temporarily suspended following an uprising that occurs the same year.

1720   Tobacco growers revolt

Tobacco growers carry out a second revolt against Spain's imposition of a government monopoly on their business. In response, Spain modifies the monopoly system. A third revolt three years later meets with severe repression by the military, but the monopoly ends.

1728   University of Havana

The University of Havana is established.

1762–1763   British occupy Havana

British forces numbering over 10,000 invade Havana under the command of Lord Albemarle and Admiral Pocock. The British seize the city, which they occupy for several months, and capture vast amounts of weapons, merchandise prepared for shipping, and other booty. During their brief occupation, they give the island a taste of free trade in contrast to trade restricted only to Spain. They also introduce the concept of increased religious tolerance, and further the growth of the sugar industry.

1778   Free trade is introduced

A regulation introduces free trade between Spain and the Indies. Tariffs are reduced, duties on Cuban sugar are removed, trade between colonies is permitted, and access to Spanish ports is expanded.

1790   Luis de las Casas is appointed governor

With Luis de las Casas, Cuba experiences its most renowned governor. Las Casas implements widespread improvements, including the establishment of schools, libraries, newspapers, and an orphanage. He also orders a census to be undertaken and founds an economic development organization, the *Real Sociedad Económica del País*. He is governor until 1796.

1792–1815 Napoleonic Wars benefit Cuba

The Napoleonic Wars in Europe speed Cuban economic development by increasing demand for sugar, coffee, and tobacco.

1796 Haitian revolution spurs Cuba's sugar industry

The Haitian revolution benefits Cuba, eliminating its principal rival in the sugar trade, as Haiti's sugar industry declines due to violence and destruction of property. An estimated 300,000 French refugees, including many sugar growers, flee to Cuba to escape the carnage, bringing with them advanced processing methods and skilled laborers. Cuba becomes the major sugar exporter in Spanish America.

1797 Birth of José Antonio Saco

Writer and historian José Antonio Saco is born. His major work is a six-volume history of slavery in Cuba.

1803: December 31 Birth of Romantic poet José Heredia

José M. Heredia y Heredia, the leading figure in the Romantic movement in Latin America, is educated largely by his politician father. Heredia moves to Mexico with his father, who is posted there in 1819. While there, he publishes his first book of poems, *Ensayos poéticos*. Heredia returns to Cuba, receives a degree from the University of Havana Law School in 1823, and publishes three more volumes of poetry: *La inconstancia, Misantropía,* and *El desamor.* On a trip to North America in 1924, Heredia visits Boston, where he meets with prominent Cuban liberals, and also travels to Niagara Falls, where he writes his famous poem "Niagara." Back in Cuba, Heredia takes part in pro-independence activities, and in December, 1824, the government banishes him from the country. Heredia dies in Mexico in 1839.

1814: March 23 Birth of writer Gertrudis Gómez de Avellaneda

One of Cuba's most highly regarded poets, as well as a novelist and playwright, Gómez de Avellaneda spends part of her life in Cuba and part in Spain, and is eventually claimed as an important literary figure by both countries. The daughter of a Cuban mother and a Spanish father, she grows up in Cuba, but her family moves to Spain when she is twenty-two. There, she establishes a literary career and forms friendships with a number of the country's most distinguished writers and political figures. In 1845, she wins two major Spanish literary prizes, one for a work that she has submitted under her brother's name. She also has a number of her plays produced. In 1852, Gómez de Avellaneda is denied admission to the Royal Spanish Academy because she is a woman.

In 1854, Gómez de Avellaneda, already married once and widowed, marries a prominent politician who is later assigned a post in Cuba. She returns there with him in 1859 and is welcomed as a revered literary figure. Her husband dies four years later, and Gómez de Avellaneda leaves Cuba once again in 1865. After traveling in the United States, London, and Paris, she returns to Spain, where she remains until her death in 1873.

1818 First school of fine arts is established

The Academia de San Alejandro, Cuba's first fine arts school, is founded in Havana.

1821 Secret society formed to work for independence

The secret organization for the creation of a Cuban republic (*Rayos y Soles de Bolívar*) is founded and attracts six hundred members, including poet José María Heredia. When it is discovered, the members are arrested and sent into exile.

1823 Cuban newspaper is founded in New York City

Father Félix Varela establishes a newspaper in New York to spread his reformist ideas about Cuban politics, most notably his views on the abolition of slavery.

1823 Monroe Doctrine supports Spain's claim to Cuba

The Monroe Doctrine is issued by the president of the United States, James Monroe (1758-1831). To the disappointment of Cubans favoring independence, the Doctrine supports Spain's right to its remaining possessions in the New World, Cuba and Puerto Rico, even as it recognizes the rights of the newly liberated Spanish American republics.

1830s–1840s Slave revolts

The growth of antislavery feelings in Cuba is signaled by a series of slave rebellions. A conspiracy to carry out an additional revolt in Matanzas—which comes to be known as *la escalera*—is also discovered, and the would-be rebels are brutally tortured by royalist troops.

1834–39 Miguel Tacon is governor

Miguel Tacon becomes governor. Formerly a well-known, forceful soldier, he is a tyrannical ruler. Tacon makes many improvements in Havana, particularly, and is noted for strengthening the police.

1836 Spanish constitution is proclaimed

Although Miguel Tacon, governor of Cuba, does not give his consent, the Spanish constitution is proclaimed to be the law of Cuba. Tacon resists this, and the next year his representatives are not admitted to the Cortes of Spain (a governing council).

1838–40 Copper mining renewed

The Cobre copper mines near Santiago are reopened, and about $2 million in ore is shipped each year to the United States.

1842 First public art collection is organized

Cuba's first art collection open to the public is established for teaching purposes at the Academic San Alejandro in Havana. It consists of European paintings.

1847: July 31 Birth of composer Ignacio Cervantes

Ignacio Cervantes Kawanagh is a leading figure in the introduction of nationalism to Cuban music. A protégé of American composer Louis Moreau Gottschalk, Cervantes studies in Paris. After returning to Cuba, he becomes involved in revolutionary activities and is exiled from Cuba by the Spanish government during the Ten Years' War (1868–1878). Cervantes later becomes a friend and follower of revolutionary leader José Martí (1853–95). The use of nationalistic elements in Cervantes's music is most notable in his Cuban dances titled *contradanzas*. Cervantes dies in Havana in 1905.

1850 Revolutionary forces launch an attack

An expeditionary force under the command of General Narciso López arrives in Cuba and unsuccessfully attempts to start a revolution.

1851 López leads second revolt

A second revolt led by López falters after failing to gain enough support. The participants, including López, are tortured and killed. The flag they carry with them later becomes the flag of Cuba.

1853: January 28 Birth of José Martí

Renowned both as a writer and a revolutionary leader, José Martí (1853–95) is Cuba's national hero. He is born in Havana to Spanish immigrant parents. By the age of sixteen, Martí is in trouble with the authorities for publishing a pro-independence newspaper. He is first sentenced to prison and then sent into exile in Spain, where he completes a university education. After his banishment in 1871, Martí spends most of the next two decades outside Cuba, working in Mexico, Venezuela, and the United States as a journalist for Spanish-language newspapers. During this period, Martí becomes known for his brilliant essays and articles as well as for his poetry. His first poetry collection, *Ismaelillo* (1882), is considered a forerunner of the Modernist movement. *Versos sencillos* (1891) is another acclaimed volume. Martí also publishes a Spanish-language newspaper for children and is an accomplished orator.

During his long, self-imposed exile from his homeland, Cuban independence continues to occupy Martí. In 1892, while living in New York, he founds the Cuban Revolutionary Party and masterminds the invasion that finally liberates Cuba from Spanish rule in 1895. Although not a military man, Martí insists on taking part in the campaign and is killed in an early skirmish a month after the initial landing. He does not become famous immediately after his death. However, with the wave of nationalism that sweeps the country in the 1920s and 1930s, a personality cult evolves around the slain patriot, and he becomes the country's most revered historical figure.

1866 Baseball is introduced to Cuba

Nemesio Guillot, a young Cuban going to school in the United States, introduces baseball to Cuba when he returns home.

1868–78 Ten Years' War

Rebels seeking independence for Cuba battle the Spanish government for ten years. Early in the war they control half the island, but the Spanish mount an intensive military campaign to recapture it and the revolt ultimately proves unsuccessful. Generals who distinguish themselves in this conflict include Máximo Gómez, Calixto García, and Antonio Maceo (1848–96). Cuba's national anthem is composed and first performed during the Ten Years' War. Having killed an estimated 50,000 Cubans and 208,000 Spaniards, the war exhausts both sides, resulting in a military stalemate and a peace treaty. Cuba remains under Spanish control.

1868: October 10 Ten Years' War is launched by planters

The Ten Years' War is launched by a group of landowners. Led by Carlos Manuel de Céspedes (1819–74) in the province of Oriente. The landowners set their slaves free and issue a declaration of independence; the slaves then join them to fight for Cuba's freedom. A republic is proclaimed and Céspedes is declared its president.

1869 Rebel government adopts a constitution

The leaders of the rebel government sign and adopt a constitution. Many Latin American countries formally recognize the rebel republic, but the United States does not.

1871 Arrest of medical students stirs anti-Spanish sentiment

A key event that helps extend the Ten Years' War is the mistaken arrest of forty-eight medical students from the University of Havana for tomb desecration. Eight receive death sentences and the rest are sent to prison. This instance of unjust arrest and punishment further mobilizes the Cuban population against Spain and prolongs the war.

1872   First professional baseball team is founded

Cuba's first professional baseball team, the Habana Baseball Club, is founded. The Matanzas club is formed a year later. Most players on both teams have been educated in the United States and learned the game there.

1873: February 1   Death of writer Gertrudis Gómez de Avellaneda

Poet, playwright, and novelist Gertrudis Gómez de Avellaneda (1814–73) dies in Spain (see 1814).

1874   First professional baseball game is played

Cuba's first professional baseball game is played between the Habana and Matanzas clubs. It is organized by Esteban Bellán, a Cuban who has attended Fordham University and played professionally in the United States on the Troy Haymakers team.

1878   Pact of Zanjón ends Ten Years War

Under the leadership of Spanish governor General Arsenio Martínez Campos, the Spanish governor, the government and the rebels reach an agreement, cemented with the pact of Zanjóhn, and the war ends. The rebels remain under Spanish rule but win many of the reforms they have sought.

1878   García forms a revolutionary committee in New York

General Calixto García, who has refused to sign the Pact of Zanjón, establishes the Cuban Revolutionary Committee in New York City and draws the support of other veterans of the Ten Years' War. They publish a manifesto denouncing the Spanish and plan a new revolt.

1878   First professional baseball league is organized

The first professional league is organized by the Habana and Matanzas teams and joined by the newly formed Almandares Club.

1879: August 26   La Guerra Chiquita begins

General Calixto García and his rebel forces attack, launching what comes to be known as *La Guerra Chiquita* (the Small War).

1880: February 13   Spain abolishes slavery

Slavery is abolished in Spain and in its possessions. However, the abolition law institutes the *patronato* system that requires freed slaves to spend eight more years working for their masters without pay.

1880: September   End of La Guerra Chiquita

The new uprising, *La Guerra Chiquita* (see 1879: August 26), is suppressed by Spanish brigades after just one year of skirmishes.

1880: October 7   Cuban slaves are freed

A special Spanish decree abolishes the *patronato* system in Cuba (see 1880).

1881   Cuban doctor discovers the cause of yellow fever

Cuban physician Carlos Juan Finlay discovers that yellow fever is caused by mosquitoes, making possible the eventual elimination of the disease from Cuba.

1886   Cubans play baseball against U.S. pro team

The Philadelphia Athletics play a series of exhibition games against Cuban teams.

1892: March   Cuban rebels begin publishing newspaper in New York

Cuban revolutionaries organized by José Martí (see 1853: January 28) begin a newspaper outlining their philosophy and goals.

1895   Spain bans baseball in Cuba

Spain bans baseball in Cuba during the Spanish-American War, as it already has once in the past. The Spanish feel that baseball expresses Cuban nationalism and rejection of colonialism and European culture.

The War of Independence

1895: February 24   War of Independence begins

Troops led by General Juan Gualberto Gómez land in Cuba and begin the final Cuban war for independence.

1895: April   Rebels meet at Playitas

Revolutionary forces led by Antonio Maceo (1848–96) rendezvous with troops under the command of Juan Gómez and José Martí (1853–95; see 1853: January 28).

1895: May 19   Death of José Martí

José Martí (1853–95; see 1853: January 28), who had insisted on taking part in the fighting, dies at Dos Rios in his first battle. (See 1853.)

1895: August 6   Birth of composer Ernesto Lecuona

Composer and pianist Ernesto Lecuona y Casado (1895–1963) begins playing the piano at a young age and demonstrates the gifts of a child prodigy. Early in his life, the legendary Polish pianist Jan Paderewski (1860–1941) comments on Lecuona's remarkable natural ability, which is also evident in his composing. The compositions of Lecuona's first creative period are influenced by European traditions, although their style is already distinctively Cuban. Around 1920, he begins composing music with a strong African influence. Lecuona's final group of works reflect a strong Spanish influence and include the famous piece "Malagueña" and the song "Siboney." Lecuona dies in 1963 in Spain. In 1985, a recording of one of Lecuona's compositions sung by the tenor Plácido Domingo wins a Grammy Award.

1895: September   Rebels proclaim a Cuban republic

Cuban revolutionaries proclaim a republic. Salvador Cisneros Betancourt is appointed president and Bartolomé Masó is named vice president.

1896: January 5   Birth of painter Amelia Peláez

One of the foremost Cuban artists of the twentieth century, Peláez receives her training in Cuba and Paris, graduating from impressionism to a variety of other styles and also working in ceramics and stained glass. She is a lifelong resident of Cuba. Her most famous works include *Gundinga* (1931), *Las dos hermanas* (1943), *Las muchachas* (1943), and illustrations for the poem "The Agony of Petronius" by her uncle, symbolist poet Julián del Casal.

1898: January 1   Spain offers Cuba autonomy

Under pressure from the United States, Spain modifies its repressive policies and grants Cuba self-rule, but the Cubans press on in their campaign for full independence.

1898: February 15   Sinking of the *Maine*

Because of widespread violence in Cuba, the United States sends the battleship U.S.S. *Maine* to Cuba for the protection of American citizens there. The ship explodes and sinks, and 266 people are killed. The United States blames the explosion on a Spanish mine.

1898: April   U.S. declares war on Spain

The United States Congress demands full independence for Cuba and officially declares war on Spain. U.S. military personnel contact the Cuban rebels. The U.S. and the rebels arrange to cooperate in a joint war effort. In addition to the help they receive from the Cubans, the rapid American victory that follows can be attributed at least partly to the preceding years of warfare, which have worn down Spain and drained its resources. As a result of the Spanish American War, the United States also gains control of Puerto Rico, the Philippines, and Guam.

1898: June   U.S. troops land in Cuba

Following the establishment of a naval blockade, roughly 17,000 U.S. troops arrive in eastern Cuba. Within a month, the Spanish forces are defeated at the battles of El Caney and San Juan Hill, and the Spanish fleet in Santiago Harbor is destroyed. Spain surrenders and sues for peace.

U.S. Military Rule

1898: December 10   Paris Treaty is signed

Representatives of the United States and Spain meet in Paris and sign a treaty drawn up by a special peace commission. Cuba ceases to be a Spanish colony and, in effect, becomes an American protectorate. The treaty includes a provision for four years of direct American rule; Cubans are not asked to take part in the negotiations.

Cuba is governed by two leaders during the four years of military occupation: General John Brooke and General Leonard Wood. Following the long and difficult war, the Cuban economy is in shambles, and many people have died of diseases and famine brought on by the forced relocation of much of the population by the former Spanish governor. The U.S. military government begins rebuilding Cuba's infrastructure and setting up an educational system. It also begins a sanitation program to eradicate yellow fever.

Early 20th century   U.S. investment in Cuba grows rapidly

In the years of military control and the early years of Cuban independence, private American investment in Cuba increases. By 1905, U.S. investors control twenty percent of Cuban sugar production and own twenty-nine sugar mills. Areas of other U.S. holdings include tobacco, mining, and railroads.

1901   Platt Amendment is adopted

Before the U.S. relinquishes military rule of Cuba, it insists on the adoption of the Platt Amendment, an addition to the newly adopted Cuban constitution. The amendment gives the United States the power to intervene in Cuba's internal affairs and requires Cuba to lease land to the United States for the establishment of naval bases.

1901   José Martí National Library is founded

The José Martí National Library is established in Havana. Cuba's national library, it also offers reference, lending, and children's services. In the mid-1990s, its collection totals over 2.4 million volumes.

The Republic of Cuba

1901: December 31   Estrada Palma is elected Cuba's first president

Tomás Estrada Palma (1835–1908) is elected the first president of Cuba. He is a leader of integrity who charts an ambitious course of reform and rebuilding for the nation, with plans for health care, education, and economic development. However, the Cuban government as a whole is plagued by widespread corruption and torn by factional strife.

1902: May 20   The United States withdraws from Cuba

The U.S. occupation of Cuba ends, and the Cuban flag is flown in place of the American flag.

1902: July 10   Birth of poet Nicolás Guillén

Cuba's most famous twentieth-century poet, Guillén is born in Camagüey. He becomes interested in poetry as a youth. Guillén begins publishing his poetry in literary magazines while earning a living, first as a printer and then as a typist at a government ministry. In the 1920s, he begins writing poems about Cubans of African descent which are published in 1930 in the volume *Motivos de son*. In the 1930s, Guillén continues writing and publishing poetry, edits two newspapers, and becomes increasingly involved in leftist politics, for which he is dismissed from his government job in 1935. In 1937, he joins the Cuban Communist Party.

In the 1940s, Guillén travels extensively, visiting the Soviet Union and Eastern Europe as well as many Latin American countries. He also runs unsuccessfully for two political positions: mayor of Camagüey (1940) and senator, on the Communist Party platform (1948). During this period, he publishes a volume of his collected poems, *El son entero: Suma poética, 1929-1946*. Guillén lives outside Cuba during the Batista dictatorship of the 1950s and returns shortly after the Cuban Revolution in 1959. He receives important government appointments in arts- and education-related posts and becomes a member of the Central Committee of the Communist Party. His later volumes of poetry include *El diario que diario* (1972) and *Sol de domingo* (1982). Guillén dies in 1989.

1902: December 8   Birth of artist Wifredo Lam

Cuba's foremost modern painter, Wifredo Lam is born to a Chinese father and a mulatto mother. He studies painting at the Escuela de Bellas Artes in Havana and later in Spain, where he lives from 1923 to 1938. After fighting on the Republican side in the Spanish Civil War, Lam leaves for Paris, where the world-renowned painter Pablo Picasso (1881–1973) becomes a friend and mentor.

At the outbreak of World War II, Lam is one of some 300 artists and intellectuals who choose to leave France rather than remain under the Vichy regime. After fleeing to the island of Martínique, Lam returns to Cuba. In 1942–43, he produces his most famous painting, *The Jungle*. After his return, Lam's paintings increasingly begin to reflect the influence of African sculpture and religions, including voodoo, which he researches on a special trip to Haiti. Lam returns to Paris in 1952 and settles there but continues to travel extensively. His painting *The Third World* (1966) is displayed in Cuba's Presidential Palace. African influences are evident throughout Lam's work, integrated into paintings whose styles range from postimpressionism to surrealism.

1903: May   Treaty reduces tariffs between the U.S. and Cuba

The Treaty of Relations reduces the tariff on Cuban sugar exports to the United States by twenty percent in return for favored tariff treatment for American exports to Cuba.

1904: December 26   Birth of novelist Alejo Carpentier

Leading Cuban novelist and short-story writer Alejo Carpentier is born in Havana. He develops an early love of music that will be evident in his writing and other pursuits throughout his life. In the 1920s, Carpentier helps introduce the works of Stravinsky, Poulenc, and other contemporary composers to Cuba in a series of new music concerts. In 1928, he flees Cuba for Paris to escape imprisonment for his opposition to the Machado regime. His first novel, *!Ecue-Yamba-Ó* (1933), depicts the lives and culture of Afro-Cubans. Carpentier returned to Cuba from 1939 to 1945. During this time he taught music history at the National Conservatory. He then left for Venezuela and did not return to Cuba until the Revolution of 1959. While in South America, he traveled widely in the Amazon region, recapturing this experience in his masterpiece, *Los pasos perdidos* (*Lost Steps*), which was published in 1953.

Carpentier returned to Cuba in 1959. From 1963 to 1968, he directed Cuba's national publishing house. From 1968 until his death in 1980, Carpentier served as Ministerial Counsel for Cultural Affairs at the Cuban embassy in Paris. Carpentier's other works include the novels *El reino de este mundo* (The Kingdom of This World, 1949), *Guerra del tiempo* (War of Time, 1958), and *El siglo de las luces* (Explosion in a Cathedral, 1962); the essay collection *Tientos y diferencias* (Acts of Feeling and Differences, 1964); and *El recurso del método* (Reasons of State, 1974), a volume about the Cuban dictator Gerardo Machado. Carpentier dies in Paris in 1980.

1905: April 29   Death of composer Ignacio Cervantes

Nationalist composer Ignacio Cervantes dies in Havana (see 1847.)

1906–1909   U.S. military intervention

The new republic is beset by political turmoil caused by clashes between liberals and conservatives and the shady practices of corrupt government officials. President Estrada Palma, reelected in 1906, faces a Liberal uprising soon afterward and, citing the Platt Amendment (see 1901), calls on the United States to intervene. Marines and other troops are sent to Cuba and set up a provisional government. Secretary of War William Howard Taft becomes acting governor until the appointment of Charles E. Magoon, who remains in charge until a new president is elected and the occupation ends.

1907   Manufacturing centers on sugar and tobacco

During the early 1900s, about sixteen percent of the Cuban workforce is engaged in manufacturing. Leading products are sugar, tobacco products (cigars and cigarettes), rum, and whiskey. Government estimates put the cigar production at 500 million per year, but international sources regard this estimate as too high. During 1904–06, about 230 million cigars and 14 million packages of cigarettes are exported to the United States each year.

1909: January 28   End of U.S. administration

Newly elected president José Miguel Gómez (b. 1856), leader of a faction of the Liberal party, assumes office. His vice president is Alfredo Zayas.

1909: April 1   Last U.S. troops are withdrawn

The remaining U.S. troops are withdrawn from Cuba.

1912   U.S. intervenes to quell revolts

Following uprisings by the Independent Colored Union (an organization formed to oppose racism), U.S. marines once again intervene in Cuba.

1913–1921   Presidency of Mario García Menocal

Menocal, a conservative, is elected president by a large majority. He introduces a national currency and welcomes increased U.S. investment in the Cuban sugar industry. He is reelected in 1916 but the legitimacy of the election is questioned. The United States orders new votes in districts whose votes are under dispute, but the Liberals still revolt. The United States is set to intervene once again, but the Cuban government itself contains the rebellion.

1913   Museo Nacional is founded

The Museo Nacional is established in Havana. It houses a part of the Academia San Alejandro collection permanently and also acquires modern Cuban art.

1914–1918   Sugar prices rise rapidly during World War I

World War I brings great wealth to Cuba as the demand for sugar grows and prices rise. This period becomes known as the "Dance of the Millions" (millions of dollars).

1917: April 7   Cuba in World War I

Cuba sides with the Allies and declares war on Germany.

1920   U.S. envoy intervenes in disputed election

The United States sends envoy Enoch Crowder to Cuba to keep peace in an election dispute between the Liberals and Conservatives. However, the Liberals boycott the election and conservative Alfredo Zayas becomes president. Crowder remains in the country throughout the Zayas administration (1920–25) to assure political stability and advise on economic matters. He wields considerable power, arousing nationalistic feeling and resentment against the United States among many Cubans. By 1924, American investors control half of Cuba's sugar industry.

1920–1921   Sugar prices plunge

Cuba enters a grave economic crisis as inflated World War I sugar prices fall. The price of sugar drops seventy percent between 1920 and 1921, throwing the country into financial chaos. The banks fail, many other businesses go under, and thousands of people are left unemployed. This period awakens many Cubans to the dangers of reliance on a single export and, especially, on a single market for it (the U.S.).

1921: March   Capablanca wins world chess championship

The Cuban chess master José Raúl Capablanca (1888–1942) becomes the world chess champion, defeating Emanuel Lasker. He becomes one of the few non-Europeans (and the only Latin American) ever to hold the title, which he retains until 1927, when he is defeated by Alexander Alekhine. Capablanca is universally recognized as one of chess's all-time greatest players.

1924   First Cuban soccer club is formed

Cuba's first soccer club, the Asociación de Football de la Republica de Cuba, is established.

1925   Communist Party is formed

Cuba's first Communist Party is formed by a small group of activists, including Julio Antonio Mella and Carlos Baliño.

The Machado Era

1925–1933   Machado regime

General Gerardo Machado y Morales is elected president. He implements major education and public works programs, including the Central Highway, which runs the entire length of the country, and the capitol building in Havana, closely modeled on the U.S. Capitol in Washington, D.C. Significant modernization is carried out in Havana and other cities. Although Cuban law prohibits Machado from running for reelection, he is not willing to cede power when his term is over. He decrees an order, the *Prórroga de Poderes*, that indefinitely extends the power of his government and delays election of a new one. His regime eventually becomes one of the most authoritarian in Cuba's history. Opposition parties are banned, members of civil rights movements are persecuted, and political enemies are assassinated.

1926: August 13   Birth of Fidel Castro

Fidel Castro Ruz, revolutionary and longtime political leader, is born to a well-to-do family in Oriente Province. He attends Catholic school and a Catholic college before studying law at the University of Havana. His political involvement begins in college. In 1952, he runs for Congress. Backed by the Orthodox Party, he runs as a nationalist rather than as a Communist or socialist. However, the elections are scrapped when Fulgencio Batista overthrows the government and institutes his dictatorship. Convinced that the only road to change is through revolution, Castro heads an ill-fated attack on the Moncada barracks in Santiago on July 26, 1953. Although he and his followers are apprehended, arrested, and ultimately jailed, the effort wins publicity for Castro and inspires what becomes known as the 26th of July Movement (see 1953: July 26). After his release from prison, Castro goes to Mexico with his political supporters, including his brother Raúl and Argentine revolutionary Che Guevara, and they plan their next attack.

On December 2, 1956, Castro leads a band of about eighty men in a landing on Cuba's southeastern coast. Most are killed, but the small group of survivors eventually ally themselves with other rebel groups. The Batista dictatorship is overthrown on January 1, 1959. Castro has remained the head of Cuba's government ever since, serving as premier (1959–76) and president of the Council of State (1976–present). He has also been head of the Cuban Communist Party since 1965.

1926: October 20   Hurricane kills 650

Touching down off the coast of Havana, a hurricane with winds reaching 130 miles per hour causes 650 deaths, leaves 10,000 people homeless, and results in $100 million in property damage. Waves reaching twenty-five feet sink dozens of ships in Havana Harbor. Extensive damage is done in the provinces of Matanzas and Pinar del Rio, as people are crushed in their own homes and entire villages virtually disappear. In Havana, people, cars, and even trolleys are carried off by giant waves, and an American monument to the men killed on the battleship *Maine* is destroyed.

1929   National soccer championship established

A national soccer championship is introduced. The first winner is the Real Iberia Football club.

1930–1933   "Revolution of 1930"

Beginning with chronic student unrest, widely scattered strikes, demonstrations, and other protests aimed at overthrowing Machado's dictatorship, a movement forms that becomes known as the "Revolution of 1930." It lasts for three years.

1930   Cuba plays its first international soccer games

Cuba enters international soccer competition at the Central American and Caribbean Games, which it also hosts in Havana. The Cubans win the championship after defeating Jamaica, Honduras, Costa Rica, and El Salvador.

1932: November 9   Hurricane kills over half the population of Cuban town

A hurricane with recorded winds of 210 miles per hour creates a giant tidal wave that flattens the city of Santa Cruz del Sur, killing some 2,500 people, or over half the city's inhabitants.

1933   U.S. ambassador negotiates with Machado

Newly appointed U.S. ambassador Sumner Wells attempts to facilitate negotiations between Machado and the political opposition.

1933: August   Machado is overthrown

Following a general strike, dictator Gerardo Machado is overthrown and goes into exile in the Bahamas. Carlos Manual de Céspedes, the son of a nineteenth-century leader in Cuba's struggle for independence, is appointed interim president.

Batista Comes to Power

1933: September   Coup is led by Fulgencio Batista

Céspedes is ousted in Cuba's first military coup, led by Sergeant Fulgencio Batista y Zaldívar. A five-member governing panel is appointed. Quelling a rebellion by his fellow officers, Batista places himself in control of the armed forces. From this point on, it is Batista who holds the real power in Cuba, although the country has a succession of different presidents in the 1930s.

1934   United States nullifies the Platt Amendment

The United States agrees to repeal the Platt Amendment, giving the country the right to intervene in Cuba's internal affairs. However, the United States retains the lease on its naval base at Guantánamo.

1934   Cuban women get the vote

Women win the right to vote on the same basis as men.

1937–38   Artists sponsor Free Studio for Painters and Sculptors

Cuba's avant-garde artists run the Free Studio as an experiment in unconventional, non-academic artistic training. Inspired by a similar school in Mexico, the Escuelas de Arte Libre, it promotes nationalism in Cuban art.

1938   Cuba qualifies for final round of World Cup games

Cuba makes it to the final round of the World Cup championship, where the Cuban team plays Romania and Sweden. This is the only time that Cuba, for whom soccer is secondary to baseball, competes in the World Cup games.

1939: March 1   Birth of composer Leo Brouwer

Leo Brouwer (born in Havanna in 1939) studies at the Juilliard School of Music with Vincent Persichetti and Stefan Wolpe and teaches theory and composition at the National Conservatory in Havana. An internationally esteemed guitarist as well as a composer, he records a wide range of classical and contemporary music. Brouwer's early compositions use elements from Cuban folk music in traditional settings. His later works are more experimental. Besides music for orchestra and chamber ensembles, Brouwer has also written theatrical and film scores and collaborated with popular artists. Among his compositions are *Danzas concertantes* for guitar and string orchestra (1958), *Homage to Mingus* for jazz combo and orchestra (1965), *Es el amor quién ve* for voice and chamber ensemble (1972), and *El decamerón negro* (1981).

1940   New constitution is drafted

A constitutional convention writes and adopts a new and progressive constitution that includes significant labor and social policy reforms.

1940   Batista is elected president

After controlling Cuban politics from behind the scenes for seven years, Batista finally runs for president, with the backing of a coalition that includes the Communist party, and is successful. For four years, Batista heads a democratic government.

1944   San Martín is elected president

In accordance with the Cuban constitution, which states that a president cannot succeed himself in office, Batista does not attempt to remain in office when his term is over. Dr. Ramón Grau San Martín, who was ousted in the coup led by Batista in 1934, is elected president. With his main adversary in office, Batista leaves the country with his family, spending several years in Miami. Grau's government implements major health, housing, and education programs.

1944: October 13   Hurricane pummels Cuba and Florida

A hurricane with wind speeds as high as 167 miles per hour wreaks havoc in Cuba for twelve hours before moving to Florida and finally up the Atlantic coast.

1946   Havana baseball stadium

The Cerro Stadium in Havana is completed.

1948   Carlos Prío Socarrás is elected president

Carlos Prío Socarrás, minister of labor in Grau's administration and a member of Grau's Auténtico political party, becomes Cuba's next president. Like Grau, Prío governs democratically, allowing full freedom of speech, but his government is marred by corruption. During Prío's term, a National Bank is founded, and Cuba, which has been using U.S. currency, begins circulating its own money, which is equivalent to the U.S. dollar.

1948   Ballet company gives its first performances

The National Ballet of Cuba, founded by internationally acclaimed dancer Alicia Alonso (b. 1921), debuts in Havana. The troupe disbands temporarily when its funding is withdrawn in 1956 by the Batista regime but is reestablished after the Cuban revolution of 1959.

1952: March 10   Batista seizes power

After being defeated in the presidential election, Fulgencio Batista, who has returned to the country, leads a military coup and installs himself as president. Opposition to his dictatorship is mobilized almost immediately in several different arenas, including among students, the government, the military, and members of the Ortodoxo Party, (a Cuban socialist party, headed by Fidel Castro after it's founder, Eduardo Chiba, commits suicide in 1951) who try to organize a civil war at the grass roots level.

1953: July 26   Castro seizes army barracks

Fidel Castro Ruz (see 1926), one of a group of young rebels from the Ortodoxo Party, leads an attack on the Moncada military fortress in Santiago, beginning the resistance activities that come to be known as the 26th of July Movement (see

1926: August 13). Castro and his followers are sentenced to three years in prison, although the sentence is later shortened and they are released in less than two years. Upon their release they leave for Mexico, where they organize another revolt against the government.

1953: December 5   Women battle police in Havana

Members of the Women's Martí Centennial Civic Front, which is allied with Fidel Castro's revolutionary 26th of July Movement, are beaten by police and arrested as they attempt to march to a political rally.

1956: December 2   Castro rebels begin guerrilla war

Having returned from Mexico, Fidel Castro and a group of about eighty followers begin guerrilla activities in the countryside surrounding the Sierra Maestra mountains. Among the guerillas are Castro's brother, Raúl, and the Argentine-born revolutionary Ernesto "Che" Guevara. Their campaign gains momentum and support as the Batista regime becomes increasingly brutal and repressive. Two additional military fronts are opened in different areas of the country by members of the 26th of July Movement, while other groups organize an effective underground organization in the cities.

1958   U.S. withdraws support for Batista

As the Batista government becomes more violent in dealing with its opposition, the United States withdraws military aid.

The Castro Government

1959: January 1   Batista flees Cuba

Following a decisive battle at Santa Clara, the Batista government is overthrown and the dictator flees the country, together with his family, friends, and top aides. Fidel Castro's 26th of July Movement takes control of the country, and Castro forms a revolutionary government.

1959: May   Cuban sugar industry is nationalized; tensions with U.S. grow

A land reform law nationalizes Cuba's sugar industry, causing losses to American investors. In response, the United States places an embargo on sugar imports from Cuba. The Castro government then seizes all U.S. property on the island. (The U.S. government later values these seizures at $5.8 billion.)

1960   Federation of Cuban Women is formed

The Federation of Cuban Women is established in Havana to raise the political involvement of the country's women and combat illiteracy. Its first head is Vilma Espin (b. 1930), who

Fidel Castro (b. 1926) overthrows the government of Fulgencio Batista (1901–73) in 1959 and establishes a communist dictatorship. Although opposed by much of the world, including the United States, Castro's regime continues in power a decade after the end of the Cold War. (EPD Photos/CSU Archives)

later works for the Castro government and is a member of the Central Committee of the Cuban Communist Party.

1960: March 4   Grenades explode as cable breaks on Belgian ship

A hoist cable breaks aboard the Belgian ammunition ship *La Coubre,* docked at Havana Harbor. A net full of grenades falls to the deck, where it explodes on contact, blowing up the ship and killing 100 crew members and longshoremen. Fidel Castro, who is in a helicopter hovering over the ship at the time of the explosion, blames the incident on sabotage by the U.S. Central Intelligence Agency (CIA).

1961: January   U.S. breaks diplomatic ties with Cuba

The United States breaks off diplomatic relations with Cuba.

1961: April 17   Failed Bay of Pigs invasion

Concerned about Castro's leftist policies and his government's growing ties with the Soviet Union, the United States government trains and equips Cuban exiles living in the United States for a military invasion of Cuba to overthrow the Castro regime. The invasion is a failure and a major embarrassment to the new Kennedy administration. About 120 men die in the attack and almost 1,200 are captured. They are later returned to the United States in exchange for medicine, food, and cash payments.

1961: May   Castro embraces Communism

Fidel Castro proclaims Cuba a Marxist-Leninist state.

1961: July 26   Single-party government is officially formed

All of Cuba's revolutionary organizations are unified into a single official government party, the Integrated Revolutionary Organizations, later renamed the Communist Party of Cuba.

1962   U.S. decrees trade embargo with Cuba

The United States implements a trade embargo on almost all imports from Cuba. In response to pressure from the United States, the Organization of American States (OAS) expels Cuba and later imposes diplomatic and trade sanctions. Cuba establishes closer ties with the Soviet Union.

1962   Castro takes over Communist Party leadership

Fidel Castro takes over the position of secretary general of Cuba's Communist Party from Anibal Escalante, who had organized the party but then fell into disfavor.

1962   Escuelas Nacionales de Arte are founded

The Escuelas Nacionales de Arte are established in Havana as the beginning of a nationwide free system of art education.

1962: October 24-26   The Cuban Missile Crisis

After U.S. surveillance aircraft find evidence of intermediate-range Soviet missiles in Cuba, President John F. Kennedy (1917–63) demands that the Soviets dismantle the missile sites and withdraw the missiles. The United States begins a naval blockade of Cuba and threatens nuclear retaliation if the U.S.S.R. does not remove the missiles. Soviet premier Nikita S. Krushchev (1894–1971) agrees to remove the missiles in return for guarantees that the United States will not invade Cuba.

1963: September 30   Hurricane Flora strikes Cuba

Passing over Cuba repeatedly over a ten-day period, Hurricane Flora wreaks massive destruction, killing more than 1,000 people and making 175,000 homeless. It also destroys one-fourth of the island's sugar crop and nine-tenths of its coffee crop. In spite of his denunciation of the United States for not providing warnings about the storm, Cuban leader Fidel Castro applies for disaster aid from the United States.

1963: October   New law creates collective farms

After an early emphasis on rapid industrialization, the Castro government begins concentrating on sugar production. It issues an agrarian reform law that nationalizes property from small farms and creates large state-owned agricultural collectives.

1963: November 29   Composer Ernesto Lecuona dies

Composer and pianist Ernesto Lecuona dies in Spain. (See 1895.)

1966: January   Cuba endorses Latin American revolution

Cuba hosts an international congress whose representatives declare their support for revolution by guerrilla warfare in other Latin American nations. The Latin American Solidarity Organization is formed.

1966: September 24-29   Hurricane Inez leaves 150,000 homeless

The Caribbean hurricane Inez causes mass destruction along the coast. Flooding submerges homes and farms, forcing 150,000 Cubans to evacuate their homes and killing hundreds more. The storm's total death toll from Cuba, Haiti and the Dominican Republic tops 2,500.

1970   Disruptions in sugar market increase Cuban dependence on Soviets

When the Cuban economy is disrupted by the government's failure to meet a projected sugar production goal, Cuba becomes increasingly dependent on aid from the Soviet Union, which comes to control more and more of the Cuban economy.

1970   Chilean president recognizes Cuba

Chile's Marxist president, Salvador Allende, affords diplomatic recognition to Cuba, paving the way for other Latin American nations to do so as well.

1975   OAS lifts embargo on Cuba

The Organization of American States (OAS) lifts its trade embargo on Cuba. The U.S. embargo remains in place.

1975   Cuban troops sent to Angola

Cuba sends about 15,000 troops to Angola to support the Marxist Popular Movement for the Liberation of Angola (MPLA), which, despite Soviet aid, fails to defeat rival groups aided by the United States and South Africa.

1975   U.S. eases Cuban trade embargo

President Gerald Ford (b. 1913) orders three provisions to modify the U.S. embargo on trade with Cuba. Subsidiaries of U.S. companies may be licensed to do business in Cuba, countries that give aid to Cuba will no longer forfeit their own aid from the United States, and ships engaged in Cuban trade may refuel in U.S. ports.

1976   Instituto Superior de Arte opens in Havana

This advanced school for the arts offers a five-year degree program and also has facilities for postgraduate work.

1977 Cuba sends troops to Ethiopia

Cuban troops are dispatched to help Ethiopia's government battle rebellions in the Ogaden region and Eritrea. The Ogaden revolt is suppressed, but the Eritreans eventually triumph and form their own country.

1977 Relations with the U.S. improve

Cuba's relations with the United States improve with the beginning of the presidency of Jimmy Carter (b. 1924). Partial diplomatic contacts are resumed and maritime agreements are signed.

1978 Political prisoners are allowed to emigrate to the U.S.

In the climate of improved U.S.-Cuban relations in the late 1970s, Castro agrees to let political prisoners, and 50,000 other Cubans with relatives outside the country, leave Cuba.

1980: April-September   Mariel boat lift strains U.S.-Cuba relations

The Cuban government allows 125,000 Cubans to emigrate to the United States through the port of Mariel in "freedom flotillas" organized by Cuban exiles. Angered and embarrassed by the large number of Cubans seeking to leave the country, Castro includes common criminals in the emigration transports.

1980: April 24   Novelist Alejo Carpentier dies

Cuba's leading novelist dies in Paris. (See 1904.)

1981–1982   Tensions rise between Cuba and Reagan administration

With the election of President Ronald Reagan (b. 1911), the openness that had characterized U.S.-Cuban relations during the Carter administration ends. Talks break off, and the Reagan administration tightens economic sanctions against Cuba as well as restrictions on travel and investment there.

1983   U.S. ousts Cuban forces from Grenada

The United States invades the island of Grenada and ousts nearly 1,000 Cuban troops, worsening the already strained relations between the U.S. and Cuba.

1985   U.S. sends back over 2,500 undesirable boat lift refugees

Cuba agrees to let the United States send back 2,746 criminals and other undesirable refugees who had entered the country during the boat lifts of the early 1980s in exchange for U.S. agreement to accept 3,000 political prisoners.

1985–1987   Drought affects sugar and tobacco production

A continued drought sharply reduces Cuba's sugar and tobacco output.

1989–1991   Cuban troops are withdrawn from Angola

In the wake of a peace agreement, Cuban troops are reduced from their peak level of 50,000.

1989   Cuban officers found guilty of drug trafficking

Cuban military personnel are found guilty of taking bribes from Colombian drug traffickers in order to let them use Cuban air strips to smuggle cocaine and other drugs into the United States. Fourteen officials receive jail sentences and four senior officers are executed.

1989: April 2–5   Soviet leader Gorbachev visits Cuba

Soviet leader Mikhail Gorbachev (b. 1931) visits Cuba. Castro criticizes Soviet moves toward democracy and a free market economy.

1989: July 16   Death of poet Nicolás Guillén

Guillén, considered Cuba's national poet, dies at the age of eighty-seven. (See 1902).

1990   Cuba wins a place on the U.N. Security Council

Cuba begins its first two-year term as a member of the United Nations Security Council.

1991   Collapse of the Soviet Union leads to recession in Cuba

Economically, the demise of the Soviet Union is a disaster for Cuba. The U.S.S.R. has been its main trade partner and customer for Cuban sugar as well as a supplier of consumer goods. Castro claims that Cuba has lost eighty-five percent of its trade due to the collapse of Communism in the U.S.S.R. and Eastern Europe.

1992   Cuban Democracy Act expands U.S. trade embargo

The Cuban Democracy Act, signed by President George Bush (b. 1924) during the 1992 U.S. presidential campaign, extends the U.S. government embargo on trade with Cuba and expands its reach. The embargo is to remain in place until Cuba establishes a multiparty democratic government.

1993   The last Soviet troops leave Cuba

The last troops from the former Soviet Union leave Cuba.

1994   New wave of Cuban emigrés leaves for U.S.

In response to the economic crisis brought on by the collapse of the Soviet Union, a new wave of emigration is launched with tacit permission from the Cuban government. Many of the refugees die on flimsy rafts. Others are captured by the U.S. Coast Guard and brought back to the Guantánamo naval base. Most are eventually allowed to enter the United States.

1994   Cuban film receives Oscar nomination in U.S.

Fresa y chocalate (translated "Strawberry and Chocolate," but the English language version of the film is *Like Water for Chocolate*) by Cuban filmmaker Tomás Gutiérrez Alea (1982–1996) receives an Academy Award nomination in the United States for Best Foreign Film.

1995   U.S. and Cuba reach emigration agreement

In response to a new wave of Cuban emigration, the United States and Cuba agree that 20,000 Cuban immigrants will be admitted to the U.S. every year and that Cuba will take steps to prevent immigrants leaving the country in unsafe boating conditions.

1995: September 5   Investment law reform is passed

An investment law reform is passed by the Cuban National Assembly to attract foreign investment to the country in the wake of the Soviet Union's demise. Marking a move toward free-market economics, the new law allows foreigners (including Cuban exiles) to own a majority share (up to 100 percent) of Cuban assets. It also opens all areas of the Cuban economy except defense, health, and education to foreign investment.

1996   Helms-Burton Act reinforces Cuban embargo

The Helms-Burton Act passed by the U.S. Congress and signed into law by President William Clinton (b.1946) expands and strengthens the U.S. trade embargo by posing new restrictions on foreign investment in and travel to Cuba. It also allows lawsuits in connection with properties nationalized by the Cuban government after the Cuban Revolution. The law is strongly criticized by other nations, and Canada expands its own trade with Cuba in response. The World Trade Organization is also critical of the U.S. measure.

1996: February 24   Cuban pilots shoot down civilian aircraft

The Cuban air force downs two civilian airplanes on a mission for the Cuban-American organization Brothers to the Rescue. The men aboard the planes had been searching for Cuban refugees at sea. All four men on the two planes are killed. The United States claims that the planes have been shot down over international waters, while Cuba insists that they were flying over Cuban territorial waters.

1996: November 19   Castro meets with Pope John Paul II

Pope John Paul II (b. 1920) receives Cuban leader Fidel Castro at the Vatican in a private meeting. For Castro, whose Communist government has pursued policies hostile to the Catholic church and officially declared Cuba an atheist nation, the conference is a significant move toward reestablishing better relations with the Church. The pontiff accepts Castro's invitation to visit Cuba.

1997: December 17   U.S. court finds Cuba liable in plane downing

A U.S. District Court in Miami orders Cuba to pay $187.6 million in damages to the families of Cuban-Americans killed when Cuban fighter jets shot down two planes over international waters in February 1996.

1997: December 30   Cuban baseball star defects

Star pitcher Orlando Hernández and seven other Cuban defectors are given protection by the U.S. Coast Guard, which retrieves them from Anguila Cay in the Bahamas. Hernández's half-brother, Livan Hernández, defects to the United States in 1995; after this, Orlando Hernández is cut from the Cuban national team and works as a physical therapist before his defection. (Livan Hernández, also a pitcher, joined the Florida Marlins, after coming to the United States. The Marlins won the 1997 World Series, in which Hernández was named most valuable player.)

Bibliography

Balfour, Sebastian. *Castro.* New York: Longman, 1990.

Brundenius, Claes. *Revolutionary Cuba: The Challenge of Economic Growth with Equity.* Boulder, Colo.: Westview Press, 1984.

Brune, Lester H. *The Cuba-Caribbean Missile Crisis of October 1962.* Claremont, Calif.: Regina Books, 1996.

Cardoso, Eliana A. *Cuba after Communism.* Cambridge, Mass.: MIT Press, 1992.

Gonzalez, Edward. *Cuba: Perilous Waters.* Santa Monica, Calif.: The Rand Corporation, 1996.

Jordan, David C. *Revolutionary Cuba and the End of the Cold War.* New York: University Press of America, 1993.

Perez, Louis A. *Cuba Between Reform and Revolution,* New York, Oxford University Press, 1995.

Rudolph, James D., ed. *Cuba: A Country Study,* 3rd ed. Washington, D.C.: U.S. Government Printing Office, 1985.

Simons, Geoffrey Leslie. *Cuba: From Conquistador to Castro.* New York: St. Martin's Press, 1996.

Suchlicki, Jaime. *Cuba: From Columbus to Castro and Beyond.* Washington, D.C.: Brassey's, 1997.

Wyden, Peter. *Bay of Pigs: The Untold Story.* New York: Simon & Schuster, 1979.

Dominica

Introduction

Lying between Guadeloupe to the north and Martinique to the south, Dominica is part of the Windward Islands in the middle of the Lesser Antilles chain of islands. Dominica has an area of 290 square miles (750 square kilometers) and is 29 miles (47 kilometers) long (north to south) and 16 miles (26 kilometers) wide (east to west). Its terrain is the most rugged of all the islands in the Lesser Antilles, with many peaks, ridges, and ravines. There are several mountains with peaks that are over 4,000 feet (1,200 meters) above sea level. Dominica's climate is fairly tropical with temperatures averaging 77°F (25°C) in the winter and 82°F (28°C) in the summer. Annual rainfall ranges from 80 inches (200 centimeters) along the coast to 250 inches (635 centimeters) in mountainous inland areas. Almost one fourth of the land is planted with crops.

With a population estimated at 83,000 in 1995, its population density is one of the lowest in the West Indies and over ninety percent of that population is descended from African slaves brought to the island in the seventeenth and eighteenth centuries. About six percent of the population is of mixed origins. Rosseau, with a population of 21,000 is the capital of Dominica. Due to the historic influence of the French, about eighty percent of all Dominicans are Roman Catholic. Smaller groups include Anglicans, Methodists, Pentecostals, Baptists, Seventh-Day Adventists, Baha'is, and Rastafarians. The Caribs' religious beliefs combine features of Christianity and nature worship. English is the official language of Dominica, but most of the population also speaks a French-based dialect called *kwéyòl*. Dominicans are increasingly using kwéyòl, which is unique but has elements in common with the dialects of St. Lucia and other islands with cultures influenced by France. Language in Dominica exhibits characteristics of Carib dialect and African phrases.

Dominica was the first island sighted by Christopher Columbus on his second voyage on Sunday, November 3, 1493. The island was named for the Latin phrase *dies dominica* "God's day"—Sunday. At that time, the island was inhab-

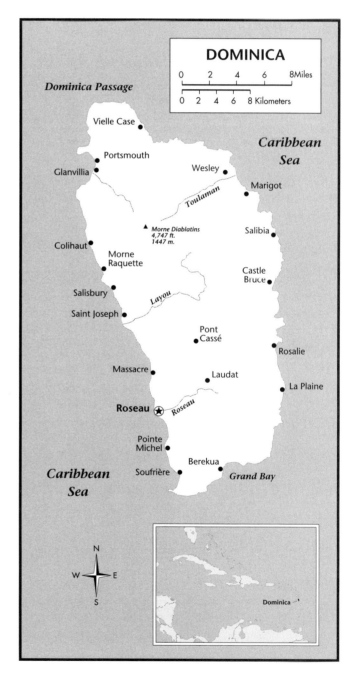

ited by Carib Indians, whose ancestors had originally come from the Orinoco Basin of South America. The Caribs had seized the island from the indigenous Arawaks in the fourteenth century. The Caribs fought against conquest, and the Spanish lost interest in the island because it apparently had no mineral wealth. Carib resistance also prevented the French and English from settling on the island in the early 1600s. In 1660, England and France agreed to let the native Caribs control the island without interference, but within thirty years Europeans began settling there. France took possession of Dominica in 1727 but forfeited it to Great Britain in 1763 after the Seven Years' war. Agricultural plantations became the foundation of the economy. Coffee was the main crop during the French colonial era, and sugar production was later introduced by the British. Dominica was governed by Great Britain as part of the Leeward Islands from 1871 until 1939. Between 1940 and 1958, it was administered as part of the Windward Islands. From 1958 until 1962, it belonged to the short-lived Federation of the West Indies. After that federation broke apart, Dominica became an associated state of the Commonwealth of Nations in 1967 and an independent republic on November 3, 1978.

Independent Dominica's first ruler, Prime Minister Patrick John was ousted following massive popular protests against his corrupt and authoritarian regime. Following a one-year interim administration, Mary Eugenia Charles became prime minister in July 1980. With her rise to the leadership, Prime Minister Charles became the first female head of government in the Caribbean and the only black head of government in the world. Charles took over at a perilous time for her country. Not only was Dominica coping with the aftermath of Prime Minister John's corrupt rule, it was also recovering from the devastating effects of Hurricane David in August 1979 which left the majority of the island's population homeless.

Over the course of the next two decades, Dominica embarked upon a program of development that included the continued development of tourism, the creation of a science and technology council, a hydroelectric power station, and the installation of the world's first country-wide digitally-operated telecommunications system.

Political stability also returned to the island after the mid-1980s with the arrest, conviction, and imprisonment of former Prime Minister John for treason in 1985. In 1995, Prime Minister Charles's Dominica Freedom Party lost is parliamentary majority and was replaced by a government led by Edison James of the United Worker's Party.

Timeline

3000 B.C.   First known inhabitants of Dominica

The first people to inhabit Dominica arrive around 3000 B.C. The earliest settlers migrate up the Antilles island chain from South America and establish small independent communities that gradually become distinct.

First century A.D.   Saladoid/Barrancoid settlers

Settlers from the Orinoco River Basin of South America arrive on Dominica. These early settlers are known as the Saladoid/Barrancoid people, named after the place along the Orinoco River in present-day Venezuela where fragments of their pottery are first discovered. There is little surviving archeological evidence of their community. The Saladoid/Barrancoid people primarily engage in farming and fishing, and their population is in decline by the mid-600s.

1200s   Carib settlers

The Caribs are the last Amerindian group to settle on the islands of the Caribbean before the advent of European exploration. The Caribs are a fierce people who maintain control by subduing the Arawaks who are already present on the island. The Caribs are able to displace and replace their enemies by killing male opponents and assimilating captured women and children into their culture. The Caribs expand their control over the Caribbean until the arrival of the Europeans.

European Exploration

1493: November 3   Christopher Columbus sights Dominica

On his second voyage to the Americas, Dominica is the first island sighted by the explorer Christopher Columbus (c. 1446–1506). Its name is taken from the Latin *dies dominica* (God's day) because it is discovered on a Sunday.

The Caribs fight against conquest, and the Spanish lose interest in the island because it apparently has no mineral wealth.

1500s   Spanish visit Dominica

The Spanish do not settle on Dominica, although they visit the island occasionally (1514, 1526, 1565). At times they do not encounter the Caribs, but at other times the Spanish are met with opposition. In 1567, a Spanish ship is wrecked off the coast of the island and every Spaniard that goes ashore is killed. During the 1500s, the Caribs of Dominica conduct raids on Spanish settlements in Puerto Rico, sometimes taking Spaniards or Africans as slaves.

1511   Spanish crown authorizes enslavement of Caribs

The Spanish government issues a royal decree that gives permission for Spanish ships in the Caribbean Sea to take Caribs as prisoners for enslavement. The only requirement is that the slaves must remain in the West Indies. During the next decade, thousands of Caribs and other Amerindians are killed by the Spanish.

Early 1600s   Caribs resist European colonization

The Spanish regularly raid islands across the Lesser Antilles and abduct Caribs for use as laborers on the early sugar plantations. These raids decrease the native population on many of the islands, and many of the remaining Caribs go into hiding in the mountainous Windward Islands (St. Lucia, St. Vincent, Grenada, and Dominica). These islands, which are sparsely settled by Amerindians because of their rugged terrain, are ideal as retreats. From these islands, the Caribs assemble a military force to resist the Spanish. Their opposition stalls the development of colonization by European powers. Carib resistance also prevents the French and English from settling on Dominica.

1627: July 2   Earl of Carlisle given title to islands in the Lesser Antilles

Several English merchants arrange to make their patron, James Hay (d. 1636), Earl of Carlisle in Scotland, owner of St. Vincent and most of the Lesser Antilles. The earl is well-liked by King Charles I (1600–49) of England and owes the merchants a great deal of money. The king issues a royal patent (title to a public land) to the earl, making him "Lord Proprietor of the English Caribbee Islands." The patent covers twenty-one territories, incuding Dominica.

1632   First European colonists arrive

France is the first European nation to send colonists to Dominica. The French begin settlements on Dominica, Martinique, and Guadaloupe at about the same time. The French settlers bring African slaves with them, and live peacefully with the Caribs.

1635   France claims Dominica

France claims the island. The same year that France makes its claim on the island, 1,500 Caribs from Dominica and St. Vincent attack the French settlement on Martinique.

1642   French missionaries visit Dominica

Following the French claim to the island, French missionaries begin visiting Dominica. They establish missions and plan to introduce Christianity to the Caribs.

1653   Caribs raid Antigua

Caribs from Dominica launch a surprise attack against an English settlement on Antigua. Most of the houses are burned and the English settlers are massacred.

France and Britain Compete for Control

1660   Europeans yield island to Caribs

England, France, and the Caribs reach an agreement that allows the native Caribs to control Dominica and neighboring St. Vincent without interference. Soon after the agreement is reached, Europeans, notably the British, continue efforts to settle on Dominica, but they are unsuccesful.

1686: February 7   England and France declare Dominica neutral

In their struggle for dominance over the Caribbean, England and France formally make Dominica a neutral island. However, the agreement is not honored; French settlers ignore the agreement and the British still attempt to gain control of the island.

1727   France takes possession

The French take possession of Dominica and claim it as their colonial territory.

1748   Treaty of Aix-la-Chapelle

French control over the island is tenuous. Under the Treaty of Aix-la-Chapelle, the island is again officially declared neutral by Britain and France, and settlers are required to leave. St. Lucia and St. Vincent are also declared neutral. The negotiated peace, however, is fleeting and the two colonial powers continue to fight over possession.

1750   Demographic changes

There are an estimated 3,300 Europeans and slaves living on Dominica. Slaves outnumber Europeans by more than two to one.

By the middle of the eighteenth century, the Carib population has decreased to only a few hundred. Wars, diseases, mass suicides, and emigration have severely depleted their numbers, which stood at approximately 5,000 a century earlier. Environmental damage such as deforestation, erosion, and pollution are also increasing during this time.

1759   British take control of the island

During the Seven Years' War (1756–63), Britain captures Dominica, which French settlers have developed into a plantation culture. At this time, the French colonial system is complete with an administrator, court, customs office, prison, and missions.

1763: April   France forfeits Dominica to Great Britain

At the end of the Seven Year's War, the Treaty of Paris restores British rule to Dominica, and the French officially surrender the island. The French plantations are expropriated, or taken possession of by the British government and sold to English planters. Most of the French remaining on the island own only small plots of land, and many are free blacks or mixed-race persons. The British deny these French-speaking settlers their civil and voting rights. Britain then develops fortifications for the island's defense and designs the layout of Roseau, which is to become the capital of the island. The British strongly encourage sugarcane agriculture on the island, to replace the coffee plantations of the French.

1778: September 7   French seize control

During the American War of Independence, trade between Britain and its Caribbean colonies is blocked. As a result, supplies and arms are often in short supply in Britain's Caribbean possessions. The French support the Americans in their fight against the British.

Taking advantage of the opportunity that short supplies have created, the French seize Dominica with the help of French troops from Martinique. The French General de Bouillé lands his force near Fort Cachacrou and proceeds to take the whole island.

After France reasserts its control over the island, many escaped slaves begin to arrive on Dominica. The escaped slaves, known as Maroons, settle in the rugged inland jungle where they are able to avoid detection.

1780s   Dominica is first stop for slave trade in the Caribbean

During the 1780s, ships bringing slaves from Africa routinely stop first in Dominica before sailing on to Guadaloupe and Martinique. Most of the slaves are re-exported, but many remain on Dominica, causing the slave population to steadily increase.

1781: April 16   Fire in Roseau

A devastating fire consumes six hundred houses and destroys the town's main buildings and markets.

1782: April 11–12   The Battle of Dominica

The Battle of Dominica (also called the Battle of the Saints, named after Isles des Saintes), marks the end of French naval power in the Caribbean. The battle occurs in the channel between Isles des Saintes and Dominica. The French fleet, commanded by Admiral de Grasse, tries to avoid confrontation with the British so that it can unite with an approaching Spanish fleet. The French want to drive the English out of the Caribbean. Britain's Admiral Rodney (1719–92) has a fleet of thirty-six ships that encounters the French fleet of thirty ships and defeats them.

1783   Treaty of Versailles restores British rule

With the Treaty of Versailles, British rule is restored in 1783. The French dominion over Dominica is over. This treaty is the settlement of a general British-French war that includes the American War of Independence.

When the British return, the Maroons (escaped slaves) are coaxing plantation slaves to run away and raiding the coastal communities. The British try to organize a militia to defend their settlements, but the Maroons are able to remain hidden in the forest and continue their raids.

Colonial Development

1783–1814   Maroon Wars

The Maroons, taught by the Caribs how to survive in the dense forest, engage in guerilla warfare against the British for nearly three decades. The Maroons remain hidden in the inland forest and use an intricate system of trails to avoid capture. Some of these trails are carved out of solid rock, and many are still used by hikers today.

The Maroons probably number between 300 and 600 and are organized into small camps, each led by its own chief. There are thirteen principal Maroon chiefs: Balla, Congo Ray, Gicero, Hall, Juba, Jupiter, and Zombie have camps in the south; Mabouya is in the east; Pharcell has a camp near Colihaut in the northwest; and Greg, Gorée, Jacko, and Sandy have camps near the source of the Layou River in the west. The Maroons establish refugee camps to assist escaped slaves. The last campaign to annihilate the Maroons takes place from 1812 until 1814, under the leadership of Governor Ainslie. The Maroon chief Jacko (d.1814) is killed after eluding apprehension for more than forty years.

1793   Yellow fever epidemic

An epidemic of yellow fever (also called the black vomit) strikes Dominica. The disease is brought to the island by French monarchist refugees from Martinique. These French refugees are fleeing the forces who support the French Revolution, which, by now, is in its most radical phase. Some 800 refugees and 200 British die from the disease. Dysentery, malaria, and tuberculosis (consumption) are common diseases in the early 1800s. Blacks, who have greater immunity from previous exposure, are much less likely to contract malaria and yellow fever than whites.

1795   French force attacks from Guadeloupe

French forces from Guadeloupe descend upon Dominica in an unsuccessful attempt to win back control of the island for France.

1805   French General La Grange takes Roseau

French General La Grange leads 4,000 French troops into Roseau. They pillage the island, but are unable to gain control over the British settlers there.

1807   Slave Trade Abolition Act

The British parliament makes the slave trade illegal. Dominica's economy is stagnating, and the ban on the slave trade makes it even worse.

Since Britain has control over most of the Caribbean and the high seas, it is able to enforce the abolition of the slave trade.

1816: September 21   Hurricane

The third hurricane within three years strikes Dominica. The abolition of the slave trade and this series of natural disasters seriously affects the plantations' productivity. Although sugar production had been on the rise in the years following the ban on the slave trade, many estates are on the verge of ruin. Some estates are abandoned, yet whenever this occurs the estate owner is still required to provide for the slaves.

1831   "Free coloreds" are granted social and political rights

Many "free coloreds" (persons of mixed European and African ancestry—or mulattos—who are not slaves) had emigrated from Martinique before the British took control in the late eighteenth century. The mulattos, who are often educated and wealthy property owners with slaves, develop into a political force. The Brown Privilege Bill of 1831 grants them the same rights the whites have. Within seven years, mulattos make up the majority in the House of Assembly.

1834: August 1   Slavery abolished

Although the British have prospered through the use of slave labor in the West Indies, they come to believe that the institution is cruel and highly inefficient. Slaves are unwilling laborers and have to be forced to work through fear, suffering, and brutality. When the British abolished slavery, slave owners are compensated for the loss of their slaves by the British government. At the time of emancipation, Dominica has over 14,000 slaves, and the planters receive some £275,000 in compensation. Many freed slaves move inland to farm small plots of land.

After emancipation, Dominica becomes a magnet for slaves escaping from French controlled islands such as Martinique and Guadaloupe. These escaping slaves come on make-shift rafts, and many die en route. The ones that survive settle on the island and generally farm small inland plots.

1838: June 29   Apprenticeship system abolished

After the slaves are emancipated, a system of apprenticeship is implemented as part of the same legislation that had given the slaves their freedom. Under the apprenticeship system, freed slaves work for their former masters as artisans and laborers. The apprentices work for five years without pay, but receive food, clothing, and shelter as compensation. The apprenticeship system is designed as a period of transition between slavery and freedom. Work relations soon deteriorate between apprentices and owners, who deeply mistrust each other. The British government initially sends ombudsmen (public officials who are appointed to investigate citizens' complaints) to monitor the program, but it is soon clear that the freed slaves are still exploited, powerless and gain nothing from apprenticeship. Once the apprenticeship system is abolished, most of the former slaves begin farming on small plots of land cleared from the forest. Estates begin to shrink in size and sharecropping becomes more common. Some estates are carved up into small parcels and sold off.

1840s   Peasant agriculture develops

As the plantations become smaller, peasant farmers begin to adapt their own agricultural techniques. These methods reflect the earlier influence of the Arawaks and Maroons, who practiced swidden (slash-and-burn) agriculture. The freed slaves combine the agricultural techniques learned from the Amerindians with some of their own from Africa. The crops they usually grow include several types of manioc (or cassava, a grain), yams, dasheen (a starchy root which is a staple of the Dominican diet), sweet potatoes, plantains (tropical bananas, with coarser fruit), bananas, cocoa, arrowroot, papayas, pineapples, calabashes (a gourdlike fruit), sugar cane, ginger, peppers, cotton, citrus, coffee, guavas, and mangos. Plantains and manioc are the staples of their diet.

The peasants also develop into a new social class. During the sugar plantation years, racism had increased, and skin color is used as a factor in establishing social status. At the time of emancipation, there is already a small group of wealthy mulattos (also called "coloreds") established on the island. The mulattos develop into an elite class and a political force on the island, but not all mulattos are a part of this power structure.

1840   Government starts relief program

After emancipation, the former slaves adapt to freedom. Some former slaves become prosperous, but others sink into poverty. Roseau's vagrant population is on the rise, and a shelter is opened to assist the town's beggars. The government passes the Act for the Relief of the Poor of the Island in 1840.

1844: June   Blacks rebel in protest to census

The government conducts a census, and many of the black peasants believe that the census is part of a white conspiracy to put them back into slavery. Census takers are threatened and attacked, and some villages are temporarily abandoned just before census officials arrive. The peasants' obstruction leads to general rebellion, with planters, who are identified with the white conspiracy, as their targets. The government tries to reassure the population that the census is for statistical purposes only, but eventually troops are brought in from Barbados to put down any incidents of civil unrest. Hundreds are arrested, and a few of the leaders are hanged.

1854: February–March   Black squatters removed from coastal lands

At the time of the British take over in 1763, three areas of land around the island are set aside as government land for the purposes of defense. The military, however, never uses the land because the existing forts are sufficient. The land sits idle until after emancipation, when former slaves squat there (settle the land without right or title) and start farming. A group of planters wants them evicted in order to gain control of the land for themselves. The government gives the squatters a choice to either pay rent or leave. When the peasants refuse to leave, the military is called upon to enforce the government's decision.

1858   Poor Law Act passes

The legislature passes the Poor Law Act to offer relief to the growing segment of the population living in poverty. This legislation, however, does little to relieve the situation.

1864: April 12   Dominica becomes a Crown colony

In the 1860s, several political groups compete for power in the House of Assembly. The colonial government realizes that this situation reduces its power to make local decisions and it declares Dominica a Crown colony. As a Crown colony, the government maintains a single House of Assembly, rather than a legislative body with directly-elected representatives. This system is unpopular, and some Dominicans riot after the bill is passed.

1871   Dominica joins Leeward Islands federation

The Dominican government joins the Federation of the Leeward Islands, thus the federal colony is under the administration of a local president but led by the governor in Antigua. At first, the federation is an attractive idea because Dominica's economy is in decline and people believe that a federal government will be more effective. However, very few Dominicans hold office in the federal government and disillusionment soon increases. (See also 1893 and 1939.)

1890s   Sugar production plummets

Sugarcane loses its monopoly status in the 1840s, as European farmers begin cultivating sugar beets. Prices for sugarcane steadily fall throughout the second half of the nineteenth century. In the 1890s, sugar production is less than half that of the 1880s, and only two sugar factories are still in operation.

1890   Botanical Gardens founded

The Botanical Gardens of Dominica lie just outside Roseau below the Morne Bruce hill. The gardens cover forty acres of a former sugarcane plantation. Since the gardens receive over eighty-five inches of rain per year, a wide variety of tropical ornamental plants can be grown. In the 1960s and 1970s, the gardens are a popular site for cricket matches.

1893   Proposal to leave the Leeward Islands Federation

A proposal is made to remove Dominica from the Federation of the Leeward Islands, just over twenty years after joining the group. The proposal is unsuccessful, and Dominica remains part of the loosely organized group. (See also 1871 and 1939.)

1900s–1920s   Lime industry

Limes have been growing on Dominica for nearly a century, but are not commercially exploited, although the island's climate is more conducive to growing limes than sugarcane. To prevent sailors from getting scurvy (a disease caused by Vitamin C deficiency), the British stock their ships with limes ("limey" becomes slang for a British person). A new market is created for limes on ships and in soft drinks.

American millionaire Andrew Green—whose engineering firm is working on the construction of the Panama Canal—visits Dominica. Green buys a sugarcane estate, converts it to limes, and starts one of the most intensive citrus processing operations in the world. By 1914, Dominica is the world's leading lime producer. The ensuing economic boom makes the island prosperous. The country's total reliance on the lime industry as the basis for its economy replaces its total reliance on the sugarcane industry. In 1922, however, two tree diseases devastate the island's lime orchards, and production falls by more than fifty percent.

1903   The Carib Reserve

The Carib Reserve (now known as the Carib Territory), some 3,700 acres on the northeast coast of the island, is set aside for the descendants of the original inhabitants of the Caribbean islands. The territory has approximately 3,400 Carib residents.

1924: September   New constitution grants elective government

A new constitution permits a return to elective representation.

1929: October   New York Stock Exchange crashes

Dominica's economy stagnates after the lime industry collapses in the 1920s. The New York Stock Exchange crashes, driving the world's economy into a depression as commerce withers. Many plantations on Dominica declare bankruptcy. People barter, fabricate their own materials, and resort to subsistence agriculture (agriculture not for profit but for food only) throughout the 1930s.

1930: September 19   The "Carib War"

Five police officers enter the Carib Reserve early in the morning, looking for smuggled goods. Much of the island had been devastated in a hurricane a few weeks earlier. The police confiscate some property they believe to have been smuggled, and a fight erupts. Police officers fire upon several Caribs, and two are killed. The policemen are then stripped of their weapons, beaten, and expelled from the reserve.

The government requests troops from other nearby islands, and marines are sent into the reserve in the days following to ensure there is no further rebellion. Chief Jolly John and five other Caribs are charged with wounding the officers and taking confiscated goods. The trial lasts for most of January 1931 and the jury acquits all six Caribs when the prosecution's evidence is deemed inadequate.

1938   Roseau Conference

Leaders from around the Caribbean meet in Roseau to discuss the rise of nationalism in the region. Socialism and the labor movement are growing in popularity.

1939   Dominica leaves the Leeward Islands

Dominica leaves the federation of the Leeward Islands, a group to which it had belonged for nearly seventy years. (See also 1871 and 1893.)

1940–58   Dominica joins Windward Islands

After it leaves the federation of the Leeward Islands, administration of Dominica transfers to the Windward Islands.

1948   Banana industry develops

A new type of banana is introduced to the island. The strain is hardier and easier for peasant farmers to grow. Banana plots typically cover only a few acres and the plants require a great deal of care. Over the next two decades, the number of small holdings increases and only a few large estates remain.

Despite their appearance, bananas are not true trees. They must be cut back to the ground each time a crop is harvested—a cycle which takes about nine months. The plants require constant pruning and fertilization, and maturing fruit often needs to be tied in bags to keep birds and insects away. A single banana plant produces a stem (or bunch) of about twenty hands (bananas are usually sold in stores by the hand

or partial hand); each hand may yield about twenty individual fruits.

The Decline of Colonial Power

1951   Adult suffrage

Universal adult suffrage is granted, marking the beginning of Britain's forfeiture of colonial control. During most of the 1950s, Dominican politicians campaign on their personal accomplishments and reputations rather than as members of a particular party.

1955   Dominica Labor Party founded

The Dominica Labor Party is founded by Phyllis Shand Allfrey and Robert Allfrey, a well-to-do planter family that has been living in Britain since the 1930s. The party's agenda is to advance socialism, which attracts working class Dominicans. In opposition to the Dominica Labor Party, the Dominica United People's Party organizes from a core of middle class independents.

1958–62   Federation of the West Indies

Dominica becomes part of the short-lived Federation of the West Indies, with Sir Grantley Adams (1878–1971) of Barbados as its premier. (The idea of creating a federation of British states in the Caribbean had first been discussed in 1932 at a conference in Dominica.) The federation's members include Grenada, Barbados, Trinidad, Jamaica, and the islands of the Leeward and Windward Antilles. At first, the federation is popular and people see it as a way for the assortment of small islands to gain international recognition. However, political differences between the islands intensify; Jamaica wants the federation to be a loose association, while Trinidad wants a strong federal government. Jamaica secedes in September 1961 and Trinidad follows in January 1962. The entire federation dissolves on May 31, 1962.

1967   Dominica joins Commonwealth of Nations

After the Federation of the West Indies is dissolved, Dominica becomes an associated state of the Commonwealth of Nations. Britain controls only Dominica's defense and foreign policy.

1970   Dominica's youth population increase

The number of Dominicans under fifteen years of age is twice the number it was in the 1940s. Dominicans under fifteen account for half of the island's population.

Mid-1970s   Rastafarianism comes to Dominica

Rastafarianism had begun in Jamaica as an Afro-centric movement in response to growing social inequality. Its beliefs are based on the teachings of the Jamaican Marcus Garvey (1887–1940), who, in the 1920s, advanced African racial pride and advocated the eventual return of blacks to Africa. His beliefs that an African king would rise up and rescue Africans from oppression seemed to have been confirmed when in 1930, Ras Tafari (1892–1975) of Ethiopia was crowned Emperor Haile Selassie I (r. 1930–36, 1941–74). Emperor Haile Selassie claimed to be a direct descendent of King David. The Rastafarians reject the European culture imposed earlier by colonists and seek a simpler life, embracing the youthful Black Power movement that is spreading across the Caribbean. By the late 1970s, Rastafarians become known internationally for two of their characteristics: wearing their hair in dreadlocks (long cords of twisted and uncombed hair) and *ganja* (marijuana). The smoking of marijuana is a sacrament in Rastafarianism.

In Dominica, the government perceives Rastafarians (locally called "Rasta" or "Dreads") as disaffected troublemakers. There are clashes between the Dreads and the government during the late 1970s, but violence escalates in the early 1980s.

1975: July   Dominica Defense Force organizes

The Dominica Defense Force forms to counter increasing the violence. Dreads (Rastafarians) in the island's interior commit kidnappings and murders to protest against the government.

Independence

1978: November 3   Dominica declares independence

Dominica becomes an independent republic on the 485th anniversary of Columbus's first sighting of the island.

1979: May   Dominica's "little revolution"

Patrick John (b. 1937) has been prime minister since 1975, but his popularity is low. He is involved in an abortive secret land deal and has a poor relationship with the Dominica Civil Service Association, an important labor union. His government attempts to restrict freedom of the press by legally requiring journalists to cite sources whenever people oppose the administration. As a result of this legislation, about 10,000 protestors demonstrate outside the government's headquarters. The Dominica Defense Force first uses tear gas and then shoots into the crowd, killing one person. The event becomes known as Dominica's "little revolution."

1979: August 29–September 1   Hurricane David

Hurricane David causes 56 deaths on Dominica and leaves 60,000 people homeless. Production of bananas (the main crop of Dominica) plummets. Nearly the entire fishing fleet is destroyed, as are nearly all schools. The hurricane also strikes Puerto Rico and the Dominican Republic, where there are 2,000 deaths. The next year, Hurricanes Allen and Frederick set back recovery on the island.

1980   Council for Science and Technology created

The government of Dominica creates a Council for Science and Technology under the Ministry of Education. However, no clear mission or budget is established for the Council, so it has little impact on technology development in Dominica.

1980: July   Mary Eugenia Charles made prime minister

The corrupt and tyrannical rule of Premier Patrick John ends with his ouster in 1979. After a year of interim rule, Mary Eugenia Charles (b. 1919) of the Dominica Freedom Party becomes prime minister, the first female prime minister in the Caribbean, and the only black woman to hold the office of head of state anywhere in the world. She remains in office for fifteen years.

During her first year in power, Charles uncovers a plot by Patrick John to stage a military coup on Dominica. The conspiracy involves a collaboration between some Dominica Defense Force members, some Dreads, and mercenaries from the United States and Canada.

1983: October   U.S.-led invasion of Grenada

Dominica supports the U.S.-led invasion of Grenada to defend against a Marxist military government, allied with Cuba, that had usurped power from the government on Grenada.

1985   Regional Security System

After Barbados assists the United States in its invasion of Grenada, it is designated as the center for the Regional Security System. This alliance includes Antigua and Barbuda, Dominica, Grenada, St. Kitts and Nevis, St. Lucia, and St. Vincent and the Grenadines. The alliance is U.S.-funded and provides for American or British defense from a foreign attack.

1985   Former prime minister imprisoned

Patrick John has been convicted in lower courts of planning to overthrow the constitutional government. His appeal goes to the Judicial Committee of the UK Privy Council in 1982, and the charges are dismissed. However, in 1985 he is retried, convicted of treason, and sentenced to prison for twelve years.

1987   New telecommunications system installed

Dominica becomes the first country in the world to operate a telecommunications system that is entirely digital.

Early 1990s "Economic citizenship"

Controversy flares over the practice of granting "economic citizenship" to Asian nationals who invest $35,000 or more in the country. In response, the government implements stiffer requirements for this citizenship, including licenses, waiting periods, and additional financial outlays.

1991 Hydroelectric power facility begins to operate

The Trafalgar Hydroelectric Power Station begins operation, harnessing power from the many streams on the island. With the opening of the facility, the island's power generation capacity nearly meets its total energy—in terms of electric power generation, the island is nearly self-sufficient.

1993: August 9 Bay Front opens

A new promenade and waterfront area opens in Roseau. The new site is built on reclaimed land that is made possible by the construction of a new seawall. The seawall reduces the impact of storm surge from hurricanes.

1995 Dominica Freedom Party loses majority

Prime Minister Charles's Dominica Freedom Party loses its majority in the 1995 elections. A new government is formed by the United Workers' Party under the leadership of Edison James.

1996 Islamic fundamentalist deported from United Kingdom to Dominica

The British government deports Saudi Mohammed al-Masari. Since 1994, he has lived in England. He is secretary-general of the Committee for the Defense of Legitimate Rights in Saudi Arabia. He says that his organization is dedicated to ending corruption and dictatorship in Saudi Arabia. International observers believe that the group is really a front for radical international Islam, and is funded by Iran and Sudan. His organization moves with him to Dominica.

1998 Expansion of Melville Hall Airport to promote tourism

Improvements at Melville Hall Airport are scheduled to be completed in time for the 1999 winter tourist season. These include construction of a 7,500-foot (2,250-meter) runway to accommodate larger airplanes. The government also expects to develop a model Carib Indian village.

Among tourist attractions in Dominica are Soufriere Sulphur Springs in the village of Soufriere, Freshwater Lake, Middleham Falls, and the Emerald Pool, all located within the 17,000-acre Morne Trois Pitons National Park. Also to be developed are Trafalgar Falls and the Morne Diablotin National Park.

1998 Springfield Centre for Environmental Protection operates former planation

Springfield Center for Environmental Protection, Research, and Education (SCEPTRE) operates a large former plantation as part wildlife sanctuary and part training center for ecological land management, organic farming, and agro-forestry. Students from the United Kingdom and North American universities can conduct tropical research studies at the site.

Bibliography

Baker, Patrick L. *Centering the Periphery: Chaos, Order, and the Ethnohistory of Dominica*. Montreal and Kingston: McGill-Queen's University Press, 1994.

Caribbean and Central American Databook, 1987. Washington, D.C.: Caribbean/Central American Action, 1986.

Honeychurch, Lennox. *The Dominica Story: A History of the Island*. London: Macmillan, 1995.

James-Bryan, Meryl. "Eugenia Charles." *Essence*, September 1994, vol. 25, no. 5, p. 73+.

Myers, Robert A. *Dominica*. Santa Barbara, Calif.: Clio Press, 1987.

Philpott, Don. *Caribbean Sunseekers: Dominica*. Lincolnwood, Ill.: Passport Books, 1996.

Trouillot, Michel-Rolph. *Peasants and Capital: Dominica in the World Economy*. Baltimore: Johns Hopkins University Press, 1988.

Whitford, Gwenith. "Mining on 'Nature Island': the Dominican Government's Resource Extraction Plans Anger Conservationists." *Alternatives Journal*, Winter 1998, vol. 24, no. 1, p. 9+.

Dominican Republic

Introduction

In 1998, two events brought the Dominican Republic into focus for North Americans who had given little thought to this small and impoverished nation in the middle of the Caribbean. While a Dominican chased one of the United States' most revered records in sports, a massive hurricane was crushing the small island nation.

Sammy Sosa, a charismatic Chicago Cubs slugger, was caught in a race with Mark McGwire to break the all-time single-season record of sixty-one home runs set by Roger Maris in 1961. With the national media focused on Sosa, Americans learned he had made a living shining shoes as a child to help support his family. His experiences gave his fans a glimpse of life in his poor island nation. Yet few news accounts bothered to explore beyond Sosa's hard childhood or explain why more than 200 Dominicans have come to the United States over several decades to play professional baseball. Or why there were as many as one million Dominicans living in the United States by the 1990s.

There's an intimate, if unequal, relationship that bonds the United States and the Dominican Republic, a nation that boasts the oldest city in the Western Hemisphere, Santo Domingo. The United States came close to annexing the tiny nation in 1870, and President Theodore Roosevelt tested his new foreign policy first on the Dominican Republic at the beginning of the twentieth century. The United States invaded in 1916 and ruled the country for eight years. U.S. troops landed on the island again in 1965.

Geography

The tiny republic takes up the eastern two-thirds of the island of Hispaniola, the name it received from explorer Christopher Columbus in 1492, during his first voyage to the New World. The country includes the islands of Beata, Catalina, Saona, Alto Velo and Catalinita in the Caribbean Sea and several islets in the Atlantic Ocean. With an area of 18,815 square miles (48,730 square kilometers), the Dominican Republic is slightly more than twice the size of the state of New Hampshire. Haiti lies on its western border and the Mona Passage

separates it from Puerto Rico. The country's capital city, Santo Domingo, is located on the southern coast.

Population

With about eight million citizens in 1996, the Dominican Republic has experienced significant emigration starting in the 1960s and this has helped alleviate overpopulation. During the early 1980s, 104,000 Dominicans entered the United States legally. Most Dominicans living in the United States settled on the eastern seaboard. Estimates for the Dominican population in the U.S. range as high as one million.

The Dominican Republic also has faced immigration problems of its own. As many as one million Haitians lived in the country illegally in the 1990s, and the Dominican government has had little success in repatriating them. Many Haitians have come across the border to work in the sugarcane fields under harsh and exploitative conditions or to escape brutal dictatorships.

Spanish is the official language of the country, and Roman Catholicism is the state religion. About sixteen percent of Dominicans are white, eleven percent are black, and seventy-three percent are mulatto. Most Dominicans are descendants of early Spanish settlers and black slaves from West Africa.

The indigenous Amerindians of Hispaniola, the Tainos, numbering perhaps as many as half a million, were virtually eliminated by the Spaniards. Despite their absence, race played a significant role in shaping the Dominican Republic, although never to the degree it shaped society in Haiti. In Hispaniola the color of someone's skin generally revealed that person's place in society, with lighter skin associated with higher social and economic status. While that distinction was easier to make in Haiti, where language also played a prominent role in determining social status, race distinctions were more difficult to make in the Dominican Republic, because there was more mixing between the races.

Colony of Spain

Spanish settlers had high hopes for Hispaniola when they arrived in 1492. The Tainos who welcomed them wore gold

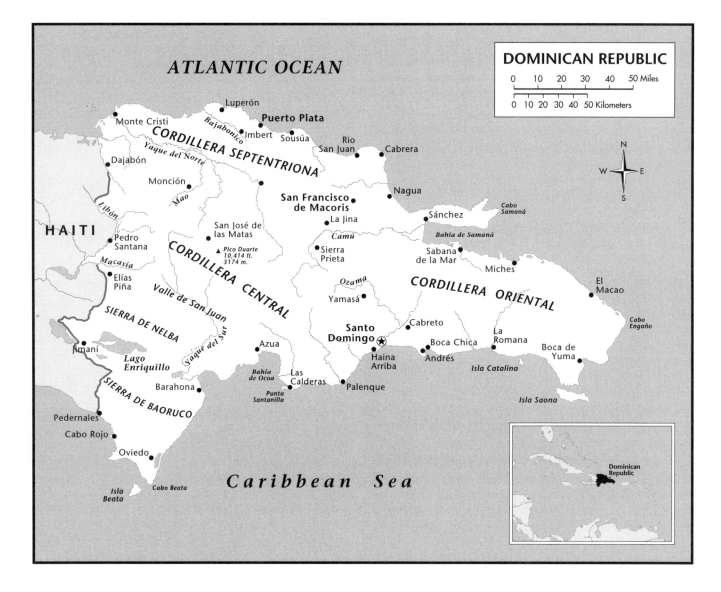

jewelry. From the start, the avarice of the conquerors set them on a collision course with the island's inhabitants. The Spanish quickly subdued the Tainos with extreme force and took control of Hispaniola.

Most Tainos were killed, died from disease, or committed suicide to avoid falling into the hands of the Spaniards during the first twenty years of occupation. By the mid-1550s, there were about 500 Tainos left on the island. Facing a severe shortage of labor, the Spaniards brought West African slaves to work the land.

Ultimately the Spaniards allowed Hispaniola to languish economically as they sought wealth in other parts of the Western Hemisphere. Santo Domingo simply became a stepping stone for conquering Spaniards traveling to the New World. In time, their carelessness cost them one-third of the island, lost to the French who also wanted a piece of the New World.

By the mid-1700s, hundreds of thousands of slaves had turned the French side of Hispaniola, known as Saint-Domingue, into one of the richest colonies in the world. At the same time, the Spanish side of the island continued to lag behind, in isolation.

Independence from Haiti

The slave and mulatto revolution that started on the French side of the island in 1791 cemented the fate of Hispaniola and the two nations that would rise from the conflict. The histories of Haiti and the Dominican Republic have crossed paths many times, often in a complicated and painful series of events. From those encounters, both nations built mistrust and disdain for one another, a legacy that has not been shed completely as the island approaches the twenty-first century. The Dominican Republic, unlike all other Spanish-speaking nations in the Western Hemisphere, celebrates independence from Haiti and not Spain. Haiti invaded the Dominican side of the island in 1822 and was forced to give it up in 1844. The nation's heroes arose from this conflict.

Series of Dictators and U.S. Invasion

Yet, much like Haiti, the Dominican Republic's constant battles for power, the legacy of colonialism, and meddling by other nations prevented the development of social structures that could ensure the rise of society.

In the Dominican Republic there were firm expectations for representative government after Haiti's retreat. But any dreams for democracy were crushed when the revolution's heroes, who had called for free elections, were arrested and later exiled by the first in a series of dictators that would rule the nation well into the late twentieth century. With General Pedro Santana Familias, the era of the *caudillo*, the "strongman," got under way, reaching its zenith with Rafael Trujillo, a despot who has been compared to Hitler and Mussolini. In the 1930s, well before the horrors of World War II got underway, Trujillo ordered the assassination of more than twenty thousand Haitians living along the border, largely to preserve what he claimed as the Spanish purity of his country.

Bullied by one dictator after another, Dominicans also suffered through two American invasions, the first to force the country to pay its debts and the second to prevent a Dominican president from regaining office after being deposed in a coup. During the twentieth century, the United States played a pivotal role in the Dominican Republic's internal politics, supporting dictators and withdrawing that support when they fell out of favor.

As the nation shifted towards democracy starting in the 1960s, with major political reforms again in 1978, the *caudillos* would not go away. The same two men who opposed each other in the 1966 election again faced each other in the 1990 election.

As the nation enters the twenty-first century, debilitating national problems—including wide-scale poverty, chronic shortages of electricity and other necessities and a fluctuating economy—threaten its stability. So far the country has survived largely on tourism and the remittances of Dominicans living in the United States. A weakened social structure could bring back the *caudillos* to a country that continues to struggle with its identity.

"We have a problem," nationally renowned photographer Polibio Diaz told *Americas* magazine in February 1997. "We haven't seen ourselves the way we really are, that we are a mixture of races."

His country faces a fundamental national dilemma, he tells the magazine. "We were in Queens (New York), and my cousin was explaining that one avenue basically divided residential neighborhoods where blacks and whites live," he said in the interview. "From my perspective as a Dominican, it meant that my mother lived on one side of the street, my father lived on the other, and I lived in the middle of the street. That's when I realized that I was a mixture of races, and I've been searching ever since to arrive at a racial definition for our country."

Timeline

Premodern Regional Chronology

Pre-Columbian Era: The Tainos

Long before Columbus reaches the island that he names Hispaniola—now home to the Dominican Republic and Haiti—indigenous peoples known as the Tainos have been living there for several hundred years. The Tainos belong to the language group known as Arawak, which means "meal" or "cassava eater." The Arawaks come from South America and over a long period of time, they move up the chain of Caribbean islands from Trinidad to Cuba.

On the island of Hispaniola, as many as 400,000 Tainos live in five different provinces, each with its own *cacique* (leader). The native people call the island *Ayti*, or *Hayti* (mountainous). The eastern part of the island is known as *Quisqueya* (mother of all lands), a name sometimes used by contemporary Dominicans to refer to their country.

The Tainos diet consists of yucca, sweet potatoes, chilis, peanuts, and corn. They supplement their meals with fish, shellfish, turtles, and even manatee.

Their farm tools are made of sharpened sticks and ax heads of polished stone. Land is owned and worked communally. They live in simple huts covered with woven straw and palm leaves. They have many gods and practice polygamy (more than one spouse at the same time). They are skilled in sculpture, ceramics, and basket weaving. They wear little or no clothes at all, but they wear jewelry, often made of gold.

The Tainos share the island with a small group of Caribs. The Caribs, who are thought to be cannibalistic, prey on the Tainos.

Columbus and Conquest

1492 Tainos welcome Columbus

Christopher Columbus (1451–1506) sights the island of Hispaniola on his first voyage to the "Indies." On Christmas Day, 1492, Columbus' flagship, the *Santa Maria,* runs aground in the vicinity of what is now Cap Häitien, in Haiti, on the northern coast. Taino natives welcome Columbus, who names the place Navidad (Christmas) and establishes a makeshift settlement.

The island is fertile, but Columbus and his fellow explorers are more interested in gold. The Spaniards discover that

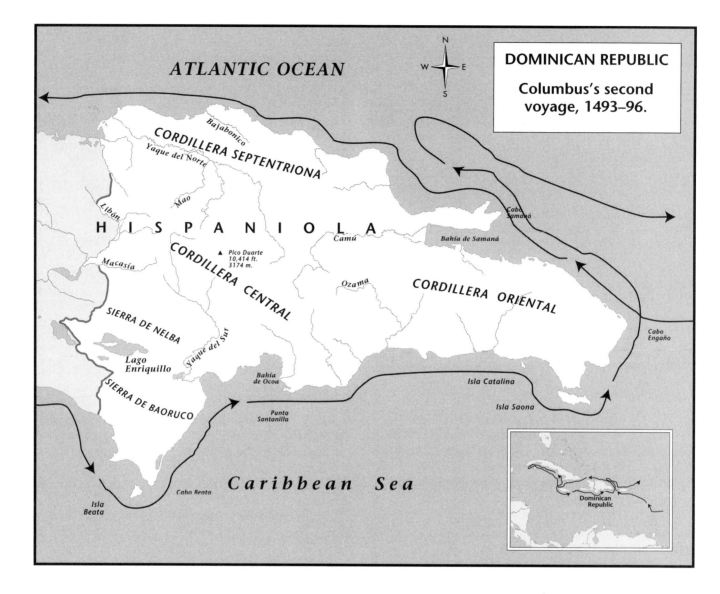

ATLANTIC OCEAN

DOMINICAN REPUBLIC
Columbus's second
voyage, 1493–96.

Bajabonico
CORDILLERA SEPTENTRIONA
Yaque del Norte
Mao
Libón
*Cabo
Samaná*
H I S P A N I O L A
Camú
Bahía de Samaná
▲ Pico Duarte
10,414 ft.
3174 m.
CORDILLERA CENTRAL
Macasía
Ozama
CORDILLERA ORIENTAL
SIERRA DE NELBA
Yaque del Sur
Lago
Enriquillo
Cabo
Engaño
SIERRA DE BAORUCO
Bahía
de Ocoa
Isla Catalina
Punta
Santanilla
Isla Saona
Cabo Beata
Caribbean   Sea
Isla
Beata

Dominican
Republic

they can barter for gold with the Tainos, who wear golden jewelry. The Spaniards also obtain gold from alluvial deposits on the island.

The friendly *cacique* allows thirty of Columbus's men to remain in the area. Columbus returns a year later to find all his men dead, killed by the Tainos for abusing their hospitality.

1494 Tainos rebel against invading Spaniards

Columbus establishes a second colony on Hispaniola, named Isabela. The settlers plant melons, wheat, and sugarcane, but spend most of their time searching for gold.

The presence of gold, the base for the new mercantilist system in Europe, attracts many Spaniards to the "New World." Instead of settling the land, most of them are initially interested in getting rich quickly.

As soon as they settle in Hispaniola, the Spaniards begin to abuse the Taino Indians. The conquerors take their food, abuse their women, and force them into labor. The formerly peaceful Indians rebel, but they are quickly crushed by the Spaniards.

1496 Columbus's brother founds oldest city in the New World

Bartholomew Columbus, the explorer's brother, succeeds in founding a permanent settlement on the island. He names it Santo Domingo, the oldest city in the New World. It is here that the New World's first cathedral, first university, first hospital, and first insane asylum are built.

Before the turn of the century, the surviving Tainos have been subdued. Thousands have died, killed by the Spaniards or diseases brought by the Europeans. Countless others commit suicide to prevent their capture. Many others attempt to flee. By the 1550s the island's Taino population has virtually disappeared.

Hispaniola becomes a logistical base for the conquest of most of the Western Hemisphere and remains an important way station for the next thirty years.

1499 Columbus makes changes before giving up governorship

Columbus rules the colony as royal governor until 1499, when he attempts to end the more serious abuses against the Taino Indians. He prohibits foraging expeditions against them and regulates the informal taxation imposed by the settlers. The Spanish settlers resent Columbus's new rules and force him to create the *repartimiento* system. Under the system, a settler is granted in perpetuity (forever) a large tract of land and the services of the Indians living on it. This system gives Spain's royalty little control over the settlers and the lands they occupy.

1500 Columbus accused of poor management

Columbus and his brother fall out of favor with the majority of the colony's settlers, mostly because of jealousy and greed. The crown accuses them of failing to maintain order. A royal investigator orders both to be imprisoned briefly in a Spanish prison.

1503 Spanish Crown takes control

Feeling that they have too little control over the settlers, the Spanish Crown replaces the *repartimiento* system with the *encomienda in 1503*. Under this new system, all land in the New World becomes property of the Crown, and the native people living on it are considered tenants on royal land.

The settlers are granted *encomiendas,* land holdings that allow them to use a number of indigenous serfs to work the land. In exchange, the conquistadors teach them the Christian faith. In essence, the encomienda is a slave system. And many of those who rely on it abuse the indigenous population.

In theory, settlers can lose their rights to the land if they abuse the native people under the *encomienda* system, but it simply serves to strengthen local authority and centralize power.

The new system does not curb the death of native peoples, and the Spaniards begin to import African slaves into Hispaniola and other holdings in the New World. By the 1520s Spain relies almost exclusively on slave labor. Only 500 Taino out of several hundred thousand who lived on the island before Columbus' arrival are alive by the late 1540s.

1509 Columbus's son becomes Hispaniola's new governor

Columbus's oldest son, Diego, becomes the new governor and serves until 1515. He also holds the important position of viceroy of the Indies from 1520 to 1524. Diego marries into Spanish aristocracy and builds a magnificent palace in Hispaniola, the Alcazar.

His ambitions arouse the suspicions of the crown, which establishes the *audiencia* in 1511 to diminish the governor's power. The *audiencia* is a tribunal composed of three judges whose jurisdiction extends over all of the West Indies. It is the highest court of appeals. The *audiencia* eventually spreads throughout the New World under Spanish rule.

1524 Audiencia's power grows

The tribunal's influence grows as the Spaniards begin to conquer lands in the Americas. Its name is changed to the Royal Audiencia of Santo Domingo, which has jurisdiction in the Caribbean, and Mexico and along the Atlantic coast of Central America and the northern coast of South America.

The audiencia represents the crown, with expanded powers in administration, legislation, and other functions. Its decisions in criminal cases are final, but civil suits are appealed to the Royal and Supreme Council of the Indies in Spain.

Charles V (1500–58) creates the Council of the Indies in 1524 to direct colonial affairs. During most of its existence, the council exercises almost absolute power in making laws, administering justice, controlling finance and trade, supervising the church, and directing armies.

Despite these changes, Santo Domingo's influence begins to decline, first with the conquest of Mexico by Hernán Cortés (1485–1547) in 1521 and the discovery of gold in large quantities both in Mexico and later in Peru. At this time, alluvial deposits of gold have been nearly exhausted in Hispaniola, and many settlers leave for Mexico and Peru. New settlers bypass Santo Domingo, and the island's population begins to fall.

1564 Earthquake destroys two cities

An earthquake destroys Santiago de los Caballeros and Concepcion de la Vega, the island's two main inland cities.

1586 Santo Domingo is attacked

The Western Hemisphere has been parceled by papal decree between Spain and Portugal, a decision that is not accepted by the other European powers.

The Spaniards attempt to maintain a monopoly on trade and colonization in the Americas. But Britain, Holland, and France begin to challenge Spanish supremacy, first resorting to smuggling, piracy, and war, especially in the Caribbean region.

The English Admiral Sir Francis Drake (1540–96) sacks, burns, and leaves Santo Domingo in ruins in 1586, but the Spaniards are more concerned about the French, who try to establish several colonies along Hispaniola's northern coast.

1655 English fleet bypasses Santo Domingo and captures Jamaica

An English fleet commanded by Sir William Penn (1621–70) tries to take Santo Domingo. After meeting heavy resistance, the English sail farther west and take Jamaica. The English use Jamaica as a base to support other settlements along the Caribbean. Gradually English, French, and Dutch presence grows in the region.

1659 French establish permanent settlement near Hispaniola

The first settlers of Tortuga Island, just off the northern coast of Hispaniola, are reportedly expelled by the Spanish from Saint Christopher (Saint Kitts). English and French begin to inhabit the island of Tortuga as early as 1625.

The French settlers sustain themselves by curing the meat and tanning the hides of wild game and attacking Spanish ships. Soon the French acquire the name buccaneers, derived from the Arawak word for the smoking of meat.

As Spain weakens, the buccaneers grow in strength and numbers, and Tortuga Island is officially declared a permanent French settlement in 1659 under the commission of King Louis XIV (1638–1715).

1664 French send clear indications they plan to take Hispaniola

The Spaniards withdraw from the northern coastal region of Hispaniola and concentrate their power in Santo Domingo. In the meantime, French settlers have been moving into the northwestern coast of Hispaniola.

The Spaniards destroy French settlements several times, but they continue to return. In 1664 the creation of the French West India Company to direct trade between the colony and France signals that France intends to colonize the western end of the island. For the next three decades, Spain and France fight for control of western Hispaniola.

1670 French establish first major settlement in Hispaniola

The French establish their first major community, Cap François (later Cap Français, now Cap Haïtien), in Hispaniola. During this period, the western part of the island is commonly called Saint-Domingue.

1697 Treaty gives western third of island to France

In 1697 under the Treaty of Ryswick, Spain cedes the western third of the island to France. The exact boundary of this territory (Saint-Domingue—now Haiti) is not established at the time of cession and remains in question until 1929.

1790 Population reaches 125,000 on Spanish side

In the early years of the eighteenth century, pirates force many of the Spanish landowners to abandon sugar plantations along the southern coast and foreign trade all but ceases on the Spanish side of the island.

But developments in Spain under the Bourbon dynasty begin to revive trade in Santo Domingo. The Crown relaxes rigid restrictions on commerce, allowing more trade between the colonies. By the end of the eighteenth century, Spain opens up trade to all neutral vessels.

These economic changes benefit Santo Domingo, where the population grows from about 6,000 people in 1737 to 125,000 by 1790. About 40,000 are white landowners, 25,000 are black or mulatto freedmen, and about 60,000 are slaves.

In contrast, the neighboring French colony of Saint-Domingue has a similar population of whites and freedmen but black slaves number 500,000. Relying on slave labor, the French turn their half of the island into one of the richest colonies in the world, producing sixty percent of the world's coffee and about forty percent of France's sugar imports.

1791–1804 Haitian revolution affects Santo Domingo

The Haitian revolution, which culminates in the independence of Haiti in 1804 (see Haiti) has profound effects on the Spanish side of Hispaniola.

At first the Spaniards, in an alliance of convenience with the British, attempt to take advantage of the slave uprising against the French. Spain wants a piece of Haiti, but its efforts fail, and the Spaniards are forced to cede all of Santo Domingo to the French in the Treaty of Basel in 1795.

By 1800 all of Hispaniola falls to the ex-slave liberator of Haiti, Toussaint L'Ouverture (1746–1803). In 1804 when Haiti declares its independence under Jean Jacques Dessalines (1758–1806), the French manage to hold on to the former Spanish colony. Dessalines attempts but fails to capture Santo Domingo.

The brief unity of all Hispaniola under L'Ouverture sets a precedent for the relationship between Haiti and the Dominican Republic. Some Haitian leaders come to believe that all of Hispaniola belongs to Haiti, leading to war, occupation, and several attempts to retake Santo Domingo during the nineteenth century.

1808 Spanish landowners return to Santo Domingo

A more stable political situation allows many émigré Spanish landowners to return to Santo Domingo. However, these royalists don't want to live under French rule, and they seek help to restore Spanish sovereignty.

The Haitians, who have reasons to fear and despise the French, provide weapons. So do the British, who occupy Samaná and blockade the port of Santo Domingo. By 1809 the French are gone.

1809 Restoration of Spanish rule backfires

The restoration of Spanish rule to the eastern half of the island coincides with an era referred to by some historians as

España Boba (Foolish Spain). Santo Domingo's economy deteriorates under the despotic rule of Ferdinand VII (1788–1833), and Dominicans begin to ponder independence.

1821   Dominican Republic declares independence from Spain

Several South American nations have been fighting for their independence from Spain. The Spanish lieutenant governor of Santo Domingo, José Núñez de Cáceres, announces the colony's independence as the state of Spanish Haiti on November 30, 1821.

The Dominicans seek to become part of Simon Bolivar's newly independent Republic of Gran Colombia (consisting of what later become Colombia, Ecuador, and Venezuela). But while the request is in transit, the president of Haiti, Jean-Pierre Boyer (1776–1850), invades Santo Domingo and reunites the island under the Haitian flag in 1822. For the next twenty-two years, the Haitians rule with an iron fist.

1822–44   Dominicans under Haitian rule

Under Haitian rule the economy declines, and resentment against the invading forces grows among Dominicans. The Haitian army abuses its power and takes what it needs without compensation. Dominicans see it as theft.

Many among the Dominican elite, including landowners, begin to flee the island rather than live under Haitian rule. Haitian administrators encourage such emigration because they confiscate holdings and redistribute them to Haitian officials. Dominican peasants, like their Haitian counterparts, begin to practice mainly subsistence cultivation with little or no production of export crops.

Racial animosities affect attitudes on both sides. Dominicans associate the Haitians' dark skin with the oppression and the abuses of occupation, while the black Haitian soldiers mistrust the light-skinned Dominicans (see Haiti).

The Haitians also associate the Roman Catholic Church with French colonialism. The church is powerful in the Dominican Republic, and the Haitians proceed to confiscate all church property in the east, deport all foreign clergy, and sever the ties of the remaining clergy to the Vatican. On its side of the border, Haiti does not formalize relations with the Vatican until 1860.

The attacks against the church insult the Dominicans, who have stronger ties to Catholicism. The Haitian elite also consider French culture superior to Spanish culture, which has shaped Dominican life. And Haitian soldiers disregard Spanish customs.

1844: February 27   Dominicans proclaim independence from Haiti

In 1838 Juan Pablo Duarte (1813–76) forms a secret revolutionary society of nine men known as *La Trinitaria* (The Trinity). Duarte, Francisco del Rosario Sanchez (1817–61), and Ramon Mella (1816–64), begin plans against Haiti.

Duarte, a member of a prominent Santo Domingo family, returns home in 1838 after spending seven years studying in Europe. He is shocked by the conditions of Santo Domingo under Haitian rule. He names his movement Trinity because the original members of his liberation group organize themselves into cells of three. The cells act as separate organizations to maintain secrecy.

Duarte, who designs the modern Dominican flag, inspires many young Dominicans who deeply resent Haitian occupation. Duarte leaves the country to seek support for a revolution from neighboring countries. In 1843, when a revolution in Haiti brings in a new president, supporters ask Duarte to return to lead an uprising. But Duarte falls sick during his return voyage, and his colleagues launch the uprising without him.

On Feburary 27 nationalist forces led by Sanchez and Mella seize control of Santo Domingo. The Haitian forces, unprofessional and undisciplined, quickly capitulate to the rebels. Duarte recovers and returns to a hero's welcome in March, when Haitians attempt to recapture the eastern half of the island. Haitian forces invade the Dominican Republic in 1849, 1850, 1855, and 1856. One of the underlying reasons for the invasions is the attempt by the Dominican Republic to seek annexation to the world's powers, including France, the United States and Spain. Haitians fear that if Dominicans are annexed, they will be next, leading once again to their slavery.

Historians refer to Duarte as an idealist, a man of principle, a romantic, and a genuine nationalist. Although he plays no significant role in ruling the nation, he is considered the father of the country.

1844: July   National heroes imprisoned by new dictator

Duarte and his followers have high hopes for the new republic. But two men more concerned about personal power than the welfare of their nation will derail their plans. Between 1844 and 1864, General Pedro Santana Familias and Buenaventura Báez Méndez take turns exiling each other, dominate politics, remove their enemies, divide the nation, and attempt to have the country annexed by foreign powers. Their dominance entrenches the tradition of *caudillo* (tough boss) rule in the Dominican Republic.

In July, a few months after the Dominican Republic declares independence, Duarte is asked by his colleagues to assume the presidency. But Duarte wants to hold free elections, an arrangement that is unacceptable to Santana. The general believes only the protection of a great power can save the Dominican Republic from falling to Haiti again.

On July 12 Santana's troops take Santo Domingo, and Santana is proclaimed the new ruler of the country. He imprisons Mella, Duarte, and Sanchez, all national heroes. After a period in prison, all are exiled.

1844–48   Santana rules with iron fist

Santana consolidates his power by executing his enemies and rewarding his friends with lucrative positions in government. He severely devalues the nation's currency by printing paper money to pay his large standing army. Throughout his term, Santana remains concerned about Haiti and continues to look for a protector. But the United States, France, and Spain are not interested in annexing the Dominican Republic.

With a stagnant economy, Dominicans begin to complain, and Santana responds by resigning and retiring to his ranch. In 1849 the Haitians invade again, and Santana leads his troops to victory. In May of that year Santana once again assumes the presidency by force.

1849: August   Santana steps aside for new president

Santana steps aside and allows an election to take place, but he remains in control of the army and thus, a powerful political player. Báez, who has been serving as president of the legislature, wins a second ballot to become the country's second president.

Like Santana, Báez attempts to interest the United States or France in annexation. But both nations are unwilling to take the entire country. Instead, they want to open a military or commercial port on the bay of Samaná.

While those two countries ponder offers by the Dominican Republic, Britain signs a commercial and maritime agreement with the Dominicans and helps mediate a peace agreement between the two nations of Hispaniola in 1850.

Domestically Báez attempts to curb Santana's power by purging his followers from government. He pardons many political prisoners and attempts to set up a militia to serve as a counterforce against Santana's army.

But Báez's efforts fail, and Santana assumes the presidency in 1853. He labels Báez a traitor and a Haitian collaborator and expels him from the country. But Báez is not done, nor is Santana. Unrest and criticism lead to Santana's resignation. Within months Báez returns, becomes president, and exiles Santana.

1850   *Merengue* becomes the national dance

By the mid-nineteenth century, a new style of dancing called the *merengue* appears in the Dominican Republic. In time, it becomes the national dance, replacing an older style of dance known as the *tumba*.

Some music historians agree that *merengue* originates from the "*perico ripiao*" ensembles from the Cibao region of the country. "*Perico ripiao*" originally refers to a method of cooking the "*perico*" or parrot. After catching a parrot, families would celebrate and dance, with musicians playing the "*tres*," a three-string guitar, "*guira*," "*maracas*," "*tambora*" (a two-headed drum), and "*marimba*" (similar to a vibraphone). The words to the *merengue* song deal with themes of love, women, social issues, and the country's colorful landscape.

As *merengue* transforms itself, the original *tres* is replaced by a guitar, which in turn is replaced by an accordion. By the late twentieth century, *merengue* begins to resemble big-band music similar to the salsa music of other Caribbean nations. Yet with its distinctive African and Spanish rhythms, *merengue* retains its unmistakable sound.

In the twentieth century, *merengue* becomes a symbol of nationalism thanks to Dictator Rafael Trujillo, who loves the music and promotes it as something distinctively Dominican. It is during his dictatorship that the *merengue* orchestra develops, with piano and brass instruments added to the mix. (See also 1990s.)

1857   Santana returns to power after bloody conflict

Báez doesn't learn from Santana's earlier blunders, and his government floods the country with worthless paper money. Farmers in the Cibao don't want to sell their cocoa, tobacco, and other crops for the devalued currency and rise against the government in what becomes known as the Revolution of 1857.

Santana, who has been pardoned by a provisional government established in the town of Santiago, returns to the country, quickly assembles an army and joins the revolution. After an expensive and bloody struggle, Santana takes power again, and Báez flees in exile. After a year of war the country's economy is severely depressed, and Dominicans are desperate.

1861–65   Spain again rules the Dominican Republic

In 1861 Santana finally convinces Spain to annex the Dominican Republic in an unusual return to colonialism.

In 1823 the U.S. adopts the Monroe Doctrine, which warns European countries not to intervene in the newly independent nations of the Western Hemisphere. But the Spaniards don't fear retaliation from the United States, which is occupied by its own Civil War. Spain advocates renewed imperial expansion while many Dominicans remain concerned about their Haitian neighbors and the state of the economy.

Yet sentiments for annexation are not universal. Santana, now representing the Spaniards against his own people, breaks up a rebellion against Spanish rule two months after annexation. A month later, Santana quells another revolt.

1863: September 14   Dominicans launch War of Restoration

Dominican nationalists set up a provisional government in the town of Santiago and proclaim independence, launching a war against the Spanish forces. Santana again is called to defend Spain, this time with a mercenary army. Popular at one point, now Santana is a hated man. The provisional government declares him a traitor and condemns him to death. Santana dies of unknown reasons on June 14, 1864.

1865: March 3   Queen of Spain repeals annexation

By 1865 Spanish military forces are unable to control the growing insurrection. In Spain renewed aspirations for an empire are crumbling, and the United States is poised to reinforce the Monroe Doctrine with the conclusion of the Civil War. Finally the Queen of Spain repeals the annexation.

On December 8 Báez returns from exile and assumes the presidency. But much has changed in the island nation. Unlike his previous presidencies, he is dealing with several new political parties and a diffused power structure. There's also no national army to control. By 1866 Báez is again forced to flee the country, only to return later as president.

1870   U.S. on island of Hispaniola

During his latest presidential tenure, Báez continues to attempt to sell portions of the country to foreign powers. In 1869 he negotiates an annexation agreement with President Ulysses S. Grant, but the U.S. Senate rejects the proposal.

By the late nineteenth century the United States its on its way to becoming a world power, especially after emerging victorious in the Spanish-American War of 1898. One of the spoils of war is Puerto Rico, but some policymakers in the United States favor a base in Haiti or the Dominican Republic to protect growing economic interests in the Caribbean and the Panama Canal, which is completed in 1914.

1876   Báez returns to power for last time

By now Báez has given new meaning to the word exile, leaving the country again in 1874 after a successful rebellion against his rule. But he is resilient and returns to power two years later. By now the Dominicans have used rebellions instead of polls to change governments, and Báez, for the last time, is forced from the presidency in 1878. He dies in exile in 1882.

1882–99   Heureaux's rule ends with his assassination

During a four-year period following Báez's exile, seven Dominicans claim regional, national, or interim leadership. Ulises Heureaux, the illegitimate son of a Haitian father and a mother who is from the island of St. Thomas, emerges from the chaos and assumes the presidency on September 1, 1882. He serves a quiet two-year term and reassumes the presidency in 1886, consolidating his power and ruling as dictator until his assassination on July 26, 1899.

Heureaux keeps his enemies at bay by promising clean elections while actually controlling the voting by keeping stuffed ballot boxes to ensure victory at the polls. He creates a network of secret police and eliminates the competition with assassinations, imprisonment, and exile. During his tenure the country's external debt grows at a rapid rate. He makes some

remarkable improvements with the money he borrows from other nations but also becomes rich by siphoning some of the money for himself.

It is under Heureaux that the remains of two national heroes, Duarte and Mella, are returned to the country and interred at the Santo Domingo Cathedral. He builds schools and sets aside scholarships, allowing students to study in Europe.

1899–1905   Foreign debt leads to U.S. intervention

Newly elected president Juan Isidro Jiménez Pereyra faces a fiscal crisis when French and Italian creditors begin to call in their loans incurred by the Heureaux government. Jiménez's negotiations with foreign powers are unpopular, and the political situation quickly degenerates into a fierce struggle for power.

In the meantime the influence of the United States increases considerably during the first few years of the twentieth century, and American forces are dispatched with regularity to protect U.S. economic interests in the Caribbean and Central America.

By 1904 the United States begins to take a greater interest in the stability of Caribbean nations, particularly the Dominican Republic and Haiti. That year, President Theodore Roosevelt introduces a new policy usually called the "Roosevelt Corollary" to the Monroe Doctrine. Roosevelt declares that if Latin American nations cannot pay their debts to European creditors, the United States would police them and collect payment to prevent the European countries from invading.

The Roosevelt Corollary first goes into effect in the Dominican Republic, preventing France and Italy from invading to collect $22 million in unpaid loans.

1916   United States troops invade Dominican Republic

Revolutions, uprisings, and political assassinations continue to plague the country between 1905 and 1916, despite U.S. intervention in the economy.

On May 16, 1916, the United States, which a year earlier takes full control of Haiti, sends troops to the Dominican Republic. Within two months, American troops control the Spanish side of the island and in November proclaim a military government. The United States justifies its occupation by citing the country's instability and the threat of German meddling. But some historians believe President Wilson wants to stabilize the nation to benefit U.S. sugar barons looking to expand sugar-cane plantations.

The Americans do not touch most Dominican laws or institutions, but they can't find Dominicans willing to serve under martial law and U.S. occupation. Military governor, Rear Admiral Harry S. Knapp, is forced to fill several key cabinet posts with United States naval officers who know little about the country and don't even speak Spanish.

1917–21 Dominicans fight the invading forces

While the United States makes many physical improvements in the country, including the construction of roads and many other public facilities, the majority of Dominicans greatly resent foreign rule.

Fierce opposition rises in the eastern provinces of El Seibo and San Pedro de Macorís, where the U.S. Marines engage a guerilla movement known as the *gavilleros,* from 1917 to 1921.

The *gavilleros* enjoy a great deal of support among Dominicans and manage, for a while, to survive the capture and execution of their leader, Vicente Evangelista. But the American troops eventually gain the upper hand with superior firepower, including warplanes, and often brutal counterinsurgent methods.

1920 Poet lands in jail for writing against U.S. occupation

Fabio Fiallo, a well-known poet, is accused of writing against the occupation, violating U.S.-imposed censorship, and is sentenced to a prison term of three years. Fiallo becomes a celebrity in his country and abroad. The Americans release him after a few months, and Fiallo continues to criticize the occupation.

1924: July 12 Dominican Republic recovers its sovereignty

Following World War I, opposition to imperialism grows in the United States, and policy makers begin to look for ways to get out of the Dominican Republic. President Warren G. Harding earns a victory at the polls in part by campaigning against the occupations of the two island nations.

Harding's secretary of state, Charles Evans Hughes, engineers the withdrawal of the United States from the Dominican Republic, and on July 12, 1924, an interim president, Horacio Vasquez, assumes the leadership of the small nation.

Allowing the Dominican Republic and not Haiti to control its destiny shows the biases and duplicity of U.S. foreign policy in the early twentieth century. Hughes is concerned that the occupation of the Dominican Republic is hurting U.S. relations with other Latin American countries. But policy makers don't think of Haiti as being part of Hispanic America. Nor do they believe that Latin American countries care about the fate of Haiti.

Once the Americans are gone, the Dominican Republic again fractures into many different political camps, giving rise to one of the most virulent dictators in the country's history. Dictators later will use the new Dominican army, created and trained by the Americans, to terrorize the country's citizens.

1930–61 Trujillo, the ultimate *caudillo,* shapes the nation

During the 1920s Rafael Leonidas Trujillo Molina (1891–1961), a Dominican of modest means, rises through the military ranks, building a network of allies and supporters. By 1930, Trujillo has set in motion a series of events to undermine the government and allow him to assume the presidency.

At his request Congress issues an official proclamation announcing the commencement of "the Era of Trujillo" when he becomes president in 1930. For the next thirty-one years, he will rule much like a feudal lord, brutally suppressing human rights, killing opponents, and exiling or jailing them. He only allows a single political party and completely controls the press.

He holds the office of president from 1930 to 1938 and from 1942 to 1952. Yet even when he is not presiding as president, Trujillo exercises absolute power. He leaves the ceremonial affairs of state to puppet presidents such as his brother, Héctor Bienvenido Trujillo Molina, president from 1952 to 1960, and Joaquín Balaguer, an intellectual and scholar, president from 1960 to 1961.

When Trujillo renames the city of Santo Domingo after himself, it boasts 1,870 monuments to the dictator. He gives himself the title of "The Benefactor of the Fatherland." But it is Trujillo who benefits, amassing a fortune estimated at $900 million to $1.5 billion in stolen government funds.

Trujillo is not an ideologue (advocate of a particular political system), historians point out. Unlike other dictators—and he is often compared to Hitler and Mussolini—he lacks a political vision.

1937: October Dictator orders death of thousands of Haitians

In 1937 Trujillo, who is seeking to expand his power over all of Hispaniola, sends his army to murder as many as twenty thousand Haitians, mostly unarmed men, women, and children. He orders their assassination because the Haitian government has discovered and executed Trujillo's most valued spies. But there's another reason as well: Trujillo seeks to maintain the Spanish purity of the Dominican Republic by murdering the black Haitians.

The massacre has great political implications on the Haitian side, where Trujillo may have supported an abortive coup attempt. Haitian President Stenio Vincent purges the officer corps of all members suspected of disloyalty, and he dismisses his top commander, Colonel Démosthènes Pétrus Calixte. The former colonel later accepts a commission in the Dominican military.

1960 Trujillo agents attempt to murder Venezuelan president

In his final years in power, Trujillo has built a large list of enemies, including leaders of other countries. Trujillo is concerned with Venezuela's President Rómulo Betancourt,

Demonstrators in Santo Domingo riot and burn the car of the United States ambassador. Many residents view American influence as the source of the Dominican Republic's political, social, and economic problems. (EPD Photos/CSU Archives)

whom he accuses of supporting dissident Dominicans who want to overthrow his government.

Trujillo supports several plots of Venezuelan exiles to overthrow Betancourt, who responds by filing a formal protest with the Organization of American States (OAS). Trujillo is furious and orders the assassination of the president. On June 24 Trujillo's foreign agents shoot and injure Betancourt. The shooting turns world opinion against Trujillo, and members of the OAS vote unanimously to sever diplomatic relations and to impose economic sanctions on the Dominican Republic.

1961: May 30   Trujillo loses U.S. support and is assassinated

Despite his horrible human rights record and dictatorial rule, Trujillo enjoys the support of the United States during most of his political career. He earns that support by declaring himself an anti-Communist, making generous campaign contributions to members of the U.S. Congress, and by cultivating a benign image fueled by U.S. public relations firms and lobbyists.

But by 1960 Trujillo becomes an embarrassment to his U.S. supporters. After his attempt on Betancourt's life, the United States Embassy in Santo Domingo is downgraded to consular level.

According to journalist Bernard Diederich (in his book *Trujillo: The Death of the Goat*), President Dwight Eisenhower asks the National Security Council's Special Group (the organization responsible for approving covert operations) to consider aiding Trujillo's enemies.

On May 30 Trujillo is on his way to visit his mistress when his car is ambushed, and he is assassinated. According to Diederich, the United States Central Intelligence Agency (CIA) supplies the weapons used by the assassins.

1961: November   Trujillo's son attempts to hold on to power

Trujillo's oldest son, Rafael Trujillo, Jr. (also known as Ramfis), returns from Paris and becomes the *de facto* (in existence, but not legally) leader of the nation. He avenges his father's death by capturing and killing most of the conspirators.

In November the United States sends warships just outside Dominican waters to prevent Trujillo, Jr. and two of his uncles from staging a coup. The Trujillos and others leave the country, taking with them $70 million.

1963: February   Free elections and a short-lived democracy

Juan Bosch, a scholar and poet, wins the presidential election with sixty-four percent of the vote and his party—the Dominican Revolutionary Party—captures two-thirds majorities in both houses of the legislature. Bosch assumes the presidency in February.

Under his tenure the 1963 constitution separates church and state, guarantees civil and individual rights, authorizes divorce, and endorses civilian control of the military. Bosch also advocates land reform and introduces a bill to confiscate all wealth unlawfully acquired by the Trujillo family.

Bosch's government is an anomaly for a nation accustomed to dictatorships. He expresses concern for the poor and seeks to make changes to improve their lives.

But his proposed changes anger military officers, conservative landholders, and the church. The elite begins to blame Bosch for opening the door to communist influence and claim it is only a matter of time before the country becomes another Cuba. Their opposition leads to a military coup that deposes Bosch on September 25, seven months after he assumes power.

1965: April   United States invades Dominican Republic

The country's economic situation deteriorates rapidly under a three-man *junta* (group of military men in power after a coup) that rules after Bosch is forced from office. U.S. President Kennedy doesn't recognize the new Dominican government and has privately supported Bosch. But two months after Kennedy's assassination, President Lyndon Johnson renews diplomatic relations and restores economic aid.

Bosch supporters don't give up, and within the army a group of young officers leads a revolt to bring him back to power. Virtual civil war breaks out in Santo Domingo. The United States privately fears that Bosch will return to office and bring communism with him. President Johnson dispatches 23,000 troops to the Dominican Republic.

Within weeks U.S. troops control the country and along with the OAS set up an election.

1966   A suspect election brings back Trujillo puppet

In elections held in June 1966, Joaquín Balaguer (b. 1907), a former president and Trujillo puppet, returns from exile in New York and campaigns vigorously throughout the country.

Bosch, who fears he will be killed if he campaigns, stays home.

Some historians point out that Bosch never has a chance to get reelected. President Johnson doesn't want him in office, and the American invading force controls the election process. The election result, which gives Balaguer 57.2 percent of the vote, is considered suspect. In 1970 and 1974 Balaguer runs for office essentially without opposition. Most parties withdraw from the campaign in response to rising political violence.

1978: May   Balaguer attempts to steal election

Early returns in the May elections favor Silvestre Antonio Guzmán Fernández over Balaguer.

On May 17 military units occupy the Central Electoral Board and impound the ballots. It is a clear signal that Balaguer is attempting to nullify the balloting or to falsify the results in his favor. Balaguer allows the resumption of the vote count when President Jimmy Carter, known for his tough stance against dictators, deploys navy ships in Dominican waters. Two weeks later Guzmán's victory is officially announced.

Guzmán releases political prisoners and practically abolishes press censorship. He embarks on reforms to reduce the influence of the military on government such as the introduction of a formal training course for officers and enlisted personnel stressing the nonpolitical role of the armed forces in a democratic society.

1979   Hurricanes leave widespread destruction

Two hurricanes kill at least 1,300 people and leave more than 100,000 without homes, aggravating a mounting economic crisis.

1982: July 4   President commits suicide

Guzmán commits suicide six weeks before his term expires. He is apparently depressed by accusations of nepotism (favoritism towards relatives regarding government positions) and corruption in his administration. Salvador Jorge Blanco, a left-wing senator who wins the elections in May, assumes the presidency in August.

Jorge promises to expand the democratic reforms begun by his predecessor. He promises "economic democracy" as well, but worldwide economic problems force him to adopt tough austerity measures.

1986–96   Balaguer returns to politics

Balaguer, now in his late seventies and legally blind, continues to enjoy widespread support. With the Dominican Republic deeply troubled by the economy, Balaguer returns to office with 41.6 percent of the vote. During his presidency, he embarks on an ambitious public works program that creates

100,000 jobs. But within two years inflation is again on the rise, and the country is hit by nationwide strikes in 1989.

Despite his problems, Balaguer wins another presidential term, yet with only 35.7 percent of the vote in the 1990 election. In 1994 Balaguer again wins the election, but charges of fraud spin the country into a deep electoral crisis. Under intense pressure from the United States, Balaguer agrees to shorten his term to two years.

1990s   Guerra becomes international music star

Among the many *merengue* stars (see 1850) is singer-writer Juan Luis Guerra, who attains international status. His 1993 concert in Rotterdam, The Netherlands, was sold out with 10,000 in attendance. Guerra's recordings are top sellers in The Netherlands. His success is attributed, in part, to the close ties between the Dominican Republic and the islands of the Dutch West Indies, namely Curucao, Aruba, and Bonair.

Born in the late 1960s, Guerra stands six feet, four inches tall and wears a beard. For concerts, he dresses in loose-fitting black clothing and stands without moving on stage while performing. His lyrics, accompanied by rhythmic island *merengue* tunes, deal with social issues of the Caribbean.

1990   U.S. census reveals large number of Dominican migrants

The 1990 U.S. census lists a half million Dominicans in the United States, but that number is clearly underinflated because it does not count undocumented immigrants.

Many Dominicans favor New York City as their new home, and the amount of money they send back to the island has been estimated at nearly $1 billion a year.

1991   Tourism increases

During the 1970s the Balaguer government begins to promote tourism, which grows dramatically during the 1980s, with more than one million visitors by 1987. Earnings increase as well, from $100 million in 1980 to $570 million by 1987. Tourism replaces sugar as the country's top foreign-exchange earner. By 1991 the Dominican Republic boasts 22,555 hotel rooms, more than any other location in the Caribbean. More than 1.71 million tourists arrive and spend $1.14 billion in 1994. In contrast, sugar, coffee, cocoa, tobacco, and cigars generate $345 million in earnings in 1995.

1994   Economic reforms create problems

Balaguer's continued economic reforms improve growth and lower inflation, but unemployment remains high, leading to substantial emigration to Puerto Rico and the United States. In March and June 1995 Dominicans riot against unauthorized increases in public transportation fare.

1996   One island, two countries, democracy on both sides

The election of Dominican President Leonel Fernandez marks the first time in the history of Hispaniola in which both Haiti and the Dominican Republic have democratically elected presidents at the same time.

Both Fernandez and Rene Preval, Haiti's president, have pledged to work together to ensure peace between their countries. It is a remarkable change for two nations that have historically despised one another.

In March 1996 outgoing Dominican President Balaguer receives Preval and awards him with the nation's highest honor, the Order of Juan Pablo Duarte (the Dominican hero who helps liberate his country from Haiti in 1844).

Balaguer, who historically has disdained the Haitians, makes another conciliatory gesture by arresting and deporting two Haitians who have threatened that country's democratic government.

1996   Western Hemisphere's oldest university struggles to stay open

The state-run Autonomous University, founded in 1538, suffers from a lack of resources and is on the verge of bankruptcy by 1996, when supporters (alumnae, businesses, and others) secure an emergency grant of $3.2 million to keep it open. By 1997, the university faces a deficit of $18.5 million and tenured faculty members earn only about $500 per month.

The Dominican Republic has managed to educate its population better than neighboring Haiti. By 1995 the adult illiteracy rate is 17.9 percent. In recent years the country's main educational objective is the enrollment in school of all Dominicans between the ages of six and fourteen.

1996: May 1   International observers arrive to monitor elections

A group of twenty-five OAS observers arrives in the country, where they will stay until May 20 to monitor elections.

1996: May 16   National elections

Jose Francisco Pena Gomez of the center-left Partido Revolucionario Dominicano wins over forty-five percent of the vote, but no candidate wins a majority. A second round of voting was scheduled (see 1996: June 30).

1996: June 30   Second round of election voting

When no candidate wins a majority in the first round of voting in presidential elections (see 1996: May 16), a second round of voting is scheduled. Leonel Fernandez of the Dominican Liberation Party (PLD) wins the presidency.

1997: March   Archeologists discover remnants of ancient city

Researchers exploring around a sinkhole in the country's East National Park find three large ceremonial plazas and the remains of a substantial settlement that appears to have been home to thousands of Tainos.

Archeologists believe there's a strong possibility the city is the same one whose brutal destruction in 1503 is described in an account by the missionary Bartolome de Las Casas.

Though the Taino are all but forgotten, remnants of their culture live on. The English word barbecue comes from the Taino term for the rock slabs they used to cook bread. The hammock is also a Taino invention.

1998: May 10   Death of Pena Gomez

Revolutionary Party leader Jose Francisco Pena Gomez dies.

1998: June   Elections

The opposition Revolutionary Party wins twenty-four of thirty seats in the senate, and a majority of the seats in the house. These results signal voter dissatisfaction with the current government's plans to privatize state enterprises, raise sales taxes, and reduce import tariffs.

1998: September 22   Hurricane Georges devastates country

Hurricane Georges, with winds of up to 125 miles per hour, wreaks havoc from the Caribbean to the Mississippi coastline, hitting the Dominican Republic with full force and killing more than 200 people. Hundreds are missing and 100,000 are homeless.

1998: December   Sammy Sosa brings honor to Dominican Republic

Sports Illustrated, one of the most influential U.S. sports magazines, names Sammy Sosa and Mark McGwire as Sportsmen of the Year for 1998.

A month earlier Sosa had won the National League's Most Valuable Player in a landslide over McGwire.

McGwire and Sosa (b. 1969) both shatter the all-time single-season record of sixty-one home runs set by Roger Maris in 1961. McGwire finished with a record seventy for the St. Louis Cardinals, while Sosa led the Chicago Cubs to the postseason with sixty-six.

Sosa, who was born in San Pedro de Marcoris, where many baseball players hail from, is one of more than 200 Dominican players who have played in the Major Leagues in the United States.

Unlike most Western Hemisphere countries, where soccer is the national sport, baseball is the national sport of the Dominican Republic. The game originally attained popularity and grew during the American occupation of 1916–24.

Many Dominicans have succeeded in the Major Leagues, including Juan Marichal (b. 1938), who attained fame as a pitcher from 1960 to 1975. Marichal was inducted into the Baseball Hall of Fame in 1983.

Sosa, who as a child shined shoes for a living, begins a foundation to raise money for underprivileged children in Chicago and the Dominican Republic. He is already considered a national hero.

Bibliography

Baud, Michiel. *Peasants and Tobacco in the Dominican Republic*. Boulder, Colo.: Westview, 1981.

Black, Jan Knippers. *The Dominican Republic: Politics and Development in an Unsovereign State*. Winchester, MA: Allen & Unwin, Inc., 1986.

Calder, Bruce J. *The Impact of Intervention: The Dominican Republic During the U.S. Occupation of 1916–1924*. Austin: University of Texas Press, 1984.

Diederich, Bernard. *Trujillo: The Death of the Goat*. Boston: Little, Brown, 1978.

Horowitz, Michael M. *Peoples and Cultures of the Caribbean: An Anthropological Reader*. New York: Natural History Press, 1971.

Logan, Rayford W. *Haiti and the Dominican Republic*, New York: Oxford University Press, 1968.

Lowenthal, Abraham F. *The Dominican Intervention*. Cambridge, MA: Harvard University Press, 1972.

Pons, Frank Moya. *The Dominican Republic: A National History*. New Rochelle, N.Y.: Hispaniola Books, 1995.

Safa, Helen Icken, *The Myth of the Male Breadwinner: Women and Industrialization in the Caribbean*. Boulder, Colo.: Westview Press, 1995.

Ecuador

Introduction

Ecuador, as its name suggests, straddles the equator in South America and covers nearly 110,000 square miles (284,000 square kilometers). Located northwest of present-day Peru, Ecuador once formed part of the Inca Empire. (The Inca were a South American native—or Amerindian—people. Their empire lay along the coast of the Pacific Ocean on the western edge of the South American continent. It included parts of a number of modern-day nations: Colombia, Ecuador, Peru, Bolivia, Chile, and Argentina.) At the center of the Inca Empire was the city of Cusco in modern-day Peru. (See Peru) The Ecuadorian city of Quito, located in the northernmost region of the Inca Empire, acted as its secondary capital.

Ecuador covers many geographical regions and, therefore, has a rich diversity of plants and animals. The three broad geographic areas are the coast, the *sierra* (mountains), and the jungle lowlands. Off the Pacific coast are the Galápagos Islands, renowned for their unique wildlife. Ecuador's population is approximately 11.5 million, with 1.8 million living in Guayaquil, the major port, and 1.3 million in Quito, the capital.

The first inhabitants of what is now Ecuador probably lived in the Andes Mountains region some 50,000 years ago and survived by hunting, fishing, and foraging. Over the millennia, their descendants permanently settled in the area. The archeological discovery of primitive tools suggests that by 9000 B.C., a number of primitive communities existed around what is now Quito. Archeological evidence also indicates that people were living in communities along the coastal regions of Ecuador as early as 4500 B.C. The Valdivian civilization flourished along the coastal plains from 3500 B.C. until about 1500 B.C. Archeologists have discovered hundreds of Valdivian houses along the Ecuadorian coast. Valdivian culture is known for its sculpture, especially the small earthen statues in the shape of females. The Valdivians may have engaged in trade with the people living in the Amazon River basin. Where the terrain was hilly or mountainous, terraces were

first constructed for agriculture in about 500 B.C. From about 500 to 300 B.C., several large irrigation systems were built by the tribal communities in the river valleys; these societies also fabricated copper tools and weapons. In the last few centuries B.C., the coastal region was inhabited once again by dozens of communities made up of small tribes, and remained so until the first few centuries A.D. By A.D. 1000, several highland groups had also formed a loose confederation.

By the 1400s, there were numerous independent groups living in the Andes region of present-day Ecuador. The most prominent peoples included the Cara and Quitu (Quito), the Puruhuas (Ambato), and the Cañari (Cuenca). In the mid-1400s, however, the Inca Empire began expanding into Ecuador, often encountering fierce opposition. The Incan occupation of Ecuador started in the South, with the subjugation of the Cañari. By the late 1400s, the Cara and Quitu peoples were conquered. Maintaining a large professional army, the Incas captured many diverse peoples, and assimilated them through colonization and strict central control. At the same time, the conquerors utilized the most successful social and technological contributions of these peoples. Notably, the Incas built an extensive footpath system that linked Cusco, the capital of the Inca empire in Peru, to Quito, over 1,000 miles away. Today, parts of the Pan-American Highway that run through Ecuador lie on top of the Incas' original paths. The influence of the Incas became imbedded in the Ecuadorian culture. A significant portion of the population today speaks Quechua, the language of the Incas. The Quechua language also spread into the lowland jungle areas at the time of Spanish conquest as a result of migration and the influence of Quechua-speaking missionaries.

Spanish Colonial Rule

The Inca Empire was severely weakened by events that were set in motion by the death of the emperor, Huayna Cápac in 1527. One of Cápac's two sons, Huáscar, was designated heir to the empire, but his brother, Atahualpa (1500?–33), controlled a large army in Ecuador. Civil war broke out, and the Inca Empire was severely weakened by the conflict between the two brothers. The war lasted until Atahualpa defeated Huáscar in 1532. Two years later, in 1534, the Spanish were able to defeat the weakened Incas. The Spanish took over the region and governed as the *audencia* (court of justice with political power) of Quito, part of the viceroyalty of Peru, 1563–1710 and 1722–39. Ecuador became a part of the viceroyalty of New Granada (later Colombia), during 1710–22 and 1739–1822. Spain's colonial system was based on the type of feudalism common to Europe during the late Middle Ages and involved the establishment of large estates. Amerindians (native people) of the highlands were forced to work for the Spanish lords. Coastal estates used slaves brought over from Africa. During this period of colonial exploitation, the Spanish were vastly outnumbered; some 2,000 Spaniards ruled over about 500,000 Amerindians.

Though there are many regional differences in Ecuador both in climate and society, musical style is similar across the country. It is the music of the Andes mountains that is regarded as "typically Ecuadorian." The music is descended from the native, highland rhythms and can be played by diverse ensembles, such as brass bands, guitar trios, or groups of wind instruments. However, it is the *rondador*, a small panpipe, that gives Ecuadorian music its distinctive sound.

For over three centuries, Spain was able to control the culture and economy of its colony, Ecuador. The first challenge to Spain's authority in South America occurred in Quito in 1809, but the rebellion was quickly put down. In 1810, however, when Napoleon Bonaparte of France dethroned the Spanish king, the resulting instability in Spain created the opportunity for its colonies to seek independence.

The colonies united against Spain. The locally born descendants of the ruling class (referred to as *criollos*) set up ruling councils in cities all across Spanish South America: in Caracas (Venezuela), Bogatá (Colombia), Buenos Aires (Argentina), Cartagena (Chile), and Santiago (Chile). Liberators Simon Bolívar (1783–1830) and José de San Martín (1778–1850) rapidly gained control over the northern and southern colonial areas of Spanish South America. By 1822, Spanish royalist forces had been driven out of Quito, and in 1830, the Republic of Ecuador was founded.

Independence

Throughout the 1800s, Ecuador's government was unstable, resulting in a rapid turnover of leadership. The Roman Catholic Church established many schools and many monasteries and churches were built in Quito, which became known as the "Cloister of America." From 1860 until 1875, the country was ruled by one of South America's most religious and conservative leaders, Gabriel García Moreno (1821–1875). His policies sought to bring about national unity through the authoritarian influence of the Roman Catholic Church. During the late 1800s, the pendulum shifted away from conservatism and the Church to liberalism and the state. The leading citizens of the coastal city of Guayaquil became associated with free-market economics and liberalism, while those of Quito became associated with state intervention and conservatism. By 1895, the Radical Liberals had assumed power under the leadership of General Eloy Alfaro (1864–1912). His policies greatly limited the power of the Roman Catholic Church and encouraged capitalist development but his regime was as authoritarian as his predecessor's and he was overthrown in 1911. When he tried to reassert himself in 1912, he and his lieutenants were executed by a mob.

The economy of the early 1900s rose and fell with fluctuations in world prices on cocoa and bananas, Ecuador's leading agricultural commodities. The end to a cocoa boom in the mid-1920s caused the economy to deteriorate and brought about renewed political instability causing a rapid change of leadership throughout the 1920s and 1930s, with nearly two dozen different presidents, dictators, or military councils. Out of the confusion emerged José Maria Velasco Ibarra (1893–1979), a populist who became Ecuador's leader five different times over the course of thirty years.

In 1941, war broke out with Peru over a long-standing border dispute that dated back to the colonial era. Peru wanted the region in the upper Amazon and Ecuador's military proved no match for Peru's superior forces. In 1942, a treaty was forged between the two countries granting half of Ecuador's territory to Peru. Since the treaty, called the Protocol of Rio de Janeiro, was supported by powerful entities—the United States, Chile, Argentina and Brazil—Ecuador was in no position to resist. Between 1948 and 1960, Ecuador experienced a period of relative peace and political stability. During that time, three presidents were freely elected and completed their terms. In 1960, however, social unrest grew as the banana industry declined. Velasco, who was in power at that time, lost control of the government and the army stepped in fearing that Cuban-style communism might spread to Ecuador.

The military government of the mid–1960s began a policy of economic and land reforms. Foreign companies developed oil resources in the Ecuadorian Amazon region. The government tried to encourage domestic manufacturing by restricting imports and establishing state-owned industries. The population grew rapidly during the 1960s, and the government focused on improving education and literacy in rural areas.

When elections were finally held in 1968, Velasco again emerged as president. Protests and economic problems plagued his administration, and he proclaimed himself dictator in 1970. Two years later, the military again stepped in and removed Velasco from power. The new ruling military council saw the value of exploiting the country's oil wealth, and the oil industry became state-controlled. Ecuador joined OPEC (the Organization of Petroleum Exporting Countries) and prospered from the high world oil prices of the 1970s. The development of the oil sector focused the economy away from its reliance on agriculture, however, Ecuador's environmental problems became more pronounced as the oil industry began discharging larger amounts of toxic chemicals into the air, soil, and water.

Democracy

In 1979, Ecuador became the first Latin American government to peacefully replace its military totalitarian government with a democracy. Armed hostilities flared up again with Peru in 1981 over the unresolved border dispute. Heavy rains and flooding in 1982 affected 4.5 million people and caused widespread crop and property damage. By the early 1980s with the oil boom over, the country was racking up foreign debt.

During the mid-1980s, the government faced many challenges. President León Febres Cordero instituted free market policies designed to advance export and foreign investment while reducing government interference. These policies did not win him favor among Ecuadorian citizens, and they retaliated with frequent labor strikes. Human rights violations increased as the government sought to silence protesters through the use of force. To complicate the economic difficul-

ties, oil prices fell in the mid-1980s, causing a shift away from free market economics. Then in 1987, the oil industry was further weakened when a serious earthquake caused hundreds of fatalities and crippled the industry for months.

A moderately liberal government came to power in 1988, but President Rodrigo Borja Cevallos was unable to reconcile the desires of the labor movement with those of industry. Inflation was running over fifty percent, and social unrest grew. In 1990, Ecuador's indigenous peoples revolted against the Borja government's policy that opened large tracts of land occupied by the Huaorani (an indigenous group) to foreign oil companies. In the early 1990s, the initial privatization (transfer of government services to privately owned companies) plans of President Sixto Durán-Ballén's government were met with stern opposition by the military, trade unions, and indigenous groups. As part of the privatization program, the government began eliminating its subsidies on consumer goods, and thousands of government jobs were cut. Ecuador signed free trade agreements with Colombia, Venezuela, and Bolivia.

By the mid–1990s, about eighty percent of the population was in poverty, roughly the same proportion as in the 1940s before the boom years of the banana and oil industries. Although Quito and Guayaquil are modern cities with contemporary buildings, small-scale farmers in the rural highland areas live in one-room houses with tiled or thatched roofs. There is an ongoing cultural division between Ecuador's different racial groups, especially between Amerindians and *mestizos* (persons of mixed Spanish and Amerindian blood). Violent clashes between the two groups have erupted during the 1990s, especially in the South.

In 1995, Ecuador's border dispute with Peru once again escalated into violence. Ecuadorian troops attacked a Peruvian post, which led to a five-week war. The war damaged Ecuador's economy, but it also increased support for President Durán-Ballén. After a cease-fire was reached in 1996, the two countries agreed to begin negotiations that would settle the border dispute once and for all. Durán-Ballén's administration lost its popularity when his vice president fled the country amid charges of political corruption. In 1995, the populist Abdalá Bucaram was elected to the presidency. Bucaram, nicknamed *el loco* ("the madman"), was from a prominent Lebanese family in Guayaquil and wanted to initiate tough economic reforms that were unpopular. By early 1997, Bucaram's popular support had all but vanished. By an act of Congress, he was removed from the presidency on the grounds of "mental incapacity." In a special session, Congress elected Fabián Alarcón as interim president.

Timeline

1000s–early 1400s   Diverse cultures arise in Ecuador

In the highlands of the Andes Mountains, several cultures emerge before the coming of the Inca Empire. (The Inca, a South American native or Amerindian people, lived along the coast of the Pacific Ocean on the western edge of the South American continent.) Moving from the south to the north, these groups include the Pasto, Otavalo-Caranqui, Cayambe, Panzaleo, Puruhua, and Palta. These mountain civilizations flourish in the centuries before the arrival of the Incas.

Before the Incan conquest, the most powerful groups in the highlands are the Quitu and Cara peoples (also known as the Quitucaras), living around what is now the modern-day capital city of Quito about ten miles south of the equator, the Puruhuas living around modern-day Ambato, and the Cañari of modern-day Cuenca.

The coastal groups typically consist of small, closely related tribes and include (from north to south) the Malaba, Cayapa, Nigua, Campaz, Caraque, Manta, Huancavilca, Chono, and Puna. Peoples living in the Ecuadorian Oriente (the eastern lowland tropical rain forests) include the Cofán, Coronado, Quijo, Macas, Jívaro, and several other groups.

1400s   Rise of the Inca Empire

The Inca Empire expands from the Andean city-state of Cuzco to cover most of what is now Peru and Ecuador, and it also covers areas of present-day Colombia, Bolivia, and Chile. From Cuzco, the empire first conquers the Aymara kingdoms of the Lupaqa and Colla peoples that live in the Lake Titicaca area of Peru. As the empire grows, the army has to deal with an increasing number of local rebellions. Although warfare is common in the region before the Incas, the threat of Inca domination often unites previously independent groups. The Incas impose strict rules and punishments on the societies they conquer, whom they rule through a strong centralized government.

The Inca Empire Conquest

1438   Beginning of expansion of the Inca Empire

There are many small kingdoms scattered throughout the Andes, with the Quitucaras, Kingdom of Quito being the largest in the north and the Cañaris controlling the largest in the south. Under the leadership of Yupanqui and his son and successor, Túpac Yupanqui (r. 1471–93), the Inca Empire expands northward into what is now Ecuador.

1492   Kingdom of Quito absorbed into the Inca Empire

The Inca Empire takes over much of present-day Ecuador in the late 1400s; the empire also encompasses what is now southern Colombia and northern Chile, as well as Bolivia and Peru. When the Incas finally conquer Quito, they make it into a base from which they spread into the north. Many different tribes are absorbed by the Incas, who often resettle their own colonists in newly conquered regions. The Quitucaras who have resisted the Incas for about twenty years, are finally overwhelmed. Over the course of fifty years, the people are assimilated into the Inca Empire and begin speaking Quechua, the Incas' language.

1493   Huayna Cápac leads consolidation of the Inca Empire

Huayna Cápac (r.1493–1525; d. 1527),who is the son of Túpac Yupanqui, and is part Cañari, first expands the Inca Empire into the south before heading north for Quito. In the north, Huayna Cápac's fighters are met by strong resistance. First Cápac kills Quiloga, the chieftainess of the Quitucaras, by impaling her in a pit of spears. Her death shatters the morale of the Quitucaras. Then Cápac orders the beheading of all the rebel leaders along the shore of a lake that he renames Yahuarcocha (Lake of Blood). The battle there results in 20,000–50,000 casualties.

1526   Spaniard sites Ecuadorian coast

Bartolomé Ruiz becomes the first European to see the Ecuadorian coast. Ruiz is piloting a ship under the command of Spanish conquistador Francisco Pizarrro (1474–1541). After an initial exploratory landing, Pizarro's expedition does not return to Ecuador until five years later. Pizarro is a cunning and seasoned conquistador who has recently gained experience subduing Panama. He later goes on to conquer what is now Peru.

1527   Inca emperor Huayna Cápac dies

Upon the death of Huayna Cápac, the Inca Empire deteriorates when his legitimate heir, Huáscar, is challenged by his brother, Atahualpa (d. 1533), who controls a large army in Ecuador. Civil war breaks out, and lasts until Atahualpa defeats Huáscar in 1532. Two years later, in 1534, the Spanish are able to defeat the weakened Incas. Many Inca nobles are dying from either malaria or smallpox, diseases brought by the Europeans

1529: July 26   Royal concession granted to Pizarro

Pizarro is officially granted permission by the Spanish monarch to leave Panama in order to seize the lands in the Inca Empire south of Santiago in present-day Peru. Ecuador's city of Quito is to the north, and is not part of the initial target for Pizarro's attack.

1531   Spanish forces gain control of coastal lands

Pizarro's ship's pilot, Ruiz, travels to the north and lands on the Ecuadorian coast. The Spaniards name their first landing spot Esmeraldas because they find a few emeralds there. However, the Spanish find the Ecuadorian shores valuable only as a stopover on the way to the riches of the Incas in Peru. Once the Incas are conquered, Sebastián de Benalcázar (c. 1495–1550), Pizarro's lieutenant, begins a campaign to extend Spain's control by conquering lands northward from Peru. A few of the Amerindian (native) groups actually ally with the Spanish, having only recently been conquered by the Incas. These smaller kingdoms view both the Spanish and the Incas as aggressive outsiders in their region.

1532   Atahualpa emerges as leader of the south

Atahualpa gains control over the southern region after a decisive battle near Riobamba where he defeats Huáscar's forces. Atahualpa executes Huáscar and his entire family and becomes the last Incan emperor with his administration based in what is now Ecuador.

1533   Pizarro executes Atahualpa

Atahualpa enjoys a great deal of popularity among the Incas, which only serves to irritate the Spanish. Atahualpa's refusal to convert to Christianity makes Pizarro furious. After ambushing and killing Atahualpa's forces, Pizarro has him imprisoned even though Atahualpa has handed over several tons of gold and silver to Pizarro. Atahualpa is charged with adultery, crimes against the king of Spain, worship of false gods, and murder (of his half-brother Huáscar). He is then baptized and garroted (strangled to death with an iron collar). Atahualpa had planned to double-cross and kill the Spanish, but they beat him to it.

1534   Quito School of Art founded

Franciscans (members of the religious order founded by St. Francis of Assisi in the thirteenth century) found the Quito School of Art to train native Ecuadorans in European arts and crafts. The Franciscans especially focus on teaching sculpture and painting in order to produce the altarpieces and works of art needed by the many newly founded churches and monasteries in the region.

The teachings of the Quito School result in an established population of craftsmen. By the end of the colonial era, art production flourishes, and Quito exports works to many other regions of Spanish South America.

1534: June 17   Rumiñahui orders the burning of Quito

General Rumiñahui is one of the most famous Inca leaders during the Incan resistance against Spanish domination. When he learns that the Spanish are going to attack Quito, he orders that the city be torched. Thus, he prevents the conquest of the city by the Spaniards.

1534: December 6   City of San Francisco de Quito founded

When Sebastián de Benalcázar (lieutenant to Spanish conquistador Pizarro) discovers the northern capital of the Inca Empire in ashes, he founds the Spanish city of San Francisco de Quito (later to become the capital of the republic) upon that site.

Located in the cool highlands, the city soon becomes a cultural center for the Spanish, with ornately decorated churches and monasteries. (Guayaquil, the major seaport, grows more slowly because of its unhealthy tropical climate.) The December 6 anniversary becomes a local holiday in the new Quito.

1534: late December   Last Inca attack led by Rumiñahui

Rumiñahui leads a band of guerrillas from the mountains on an attack of the new city of San Francisco de Quito, but he is soon captured and executed. With the death of Rumiñahui, organized resistance by the Incas ends.

1535   City of Guayaquil founded

Sebastián de Benalcázar establishes a Spanish city on the west coast of Ecuador at the site of a Puña settlement. He calls the city Guayaquil, named for the resident Puña chief Guaya and his wife Quill. Guaya and Quill both commit suicide when Benalcázar's forces converge on the site. Guayaquil becomes Ecuador's chief port and is the country's main connection to the rest of the world until air transport becomes available in the twentieth century.

1537   Spanish search for gold and cinnamon

Benalcázar goes north into Colombia looking for the legendary land of gold (referred to as El Dorado). Franciso Pizarro's brother, Gonzálo (c. 1506–48), sets off to the east of Quito in search of a fabled land of cinnamon, but gets lost and returns to Quito with nothing.

The Spanish Colonial Period

1541   Expeditions in search of gold

Spain's conquest of the Americas has long been motivated by the hope of finding gold. Spanish explorer Francisco de Orellana (c. 1500–49) leads an expedition in search of the mythical gold. He leaves Ecuador, heading east with over 300 troops and 4,000 Amerindians. By the time he reaches the Atlantic coast a year later, the treacherous jungle and tropical climate has killed all but 50 of the people making the trek.

1546   Spanish women arrive in Ecuador

Spain's method of taking over South America involves the initial placement of the conquistadors, who are all male. The

Spaniards typically start having children by local Amerindian women before any Spanish women ever arrive. From this situation arises an entire class called the *cholos* (persons of mixed blood).

When women from Spain arrive, it creates a class system where Spaniards with Spanish wives are given preferential treatment. The *cholos* often are skilled in the trades, and the Amerindians become peasants. Spanish and Amerindian culture officially remain separate from each other, paving the way for discrimination.

1550   Viceroyalty of Peru founded

Spain formally establishes its authority in Peru. Lima is the center of what is known as the Viceroyalty of Peru. Ecuador becomes a province of the viceroyalty.

1563   Quito becomes seat of royal audencia of Spain

Quito becomes the seat of Spanish colonial government in Ecuador. Known as the Audencia of Quito, under the viceroyalty at Lima, Quito will remain the seat of the government until 1710 (see 1710). The colonial government establishes a feudal system with Amerindians living on estates forced to work for Spanish lords.

1586   Library founded

The library of the Central University of Ecuador in Quito is the oldest and most important one in the country. The library is started eight years before the university is founded.

1594   Central University of Ecuador is established

Building on the scholarly resources of the library founded at Quito in 1586, the Central University of Ecuador is established.

1600s   Quito is base for Roman Catholic missionaries

The Roman Catholic Church sends missionaries to South America, and many Amerindians (native people) are converted to Christianity. Quito becomes the base from which Roman Catholic missionaries journey across the Amazon River.

1642   Colonial territory expands

The Roman Catholic Church is an important institution during the colonial period. As the Roman Catholic influence grows, the territory of the Audencia of Quito is enlarged to include the Marañón River Basin, located in what is now Peru.

1700s   Early stirrings of Ecuadorian independence

During the early 1700s, Ecuador experiences a recession. The Amerindian population dwindles due to imported diseases and the rigors of slave labor. Agricultural production and land ownership are in the hands of the Spanish. Many Amerindians are only given small plots of infertile land on which to grow their own crops.

Many of the Ecuadorian people are dissatisfied with their situations, and anti-colonial sentiments begin to develop.

1710   Center of Spanish colonial rule shifts between Quito and New Granada

Since 1563, the center of Spanish rule in the audencia has been at Quito, but now shifts to New Granada (modern-day Colombia). The power will remain there until 1722, when it will shift back to Quito. The Spanish colonialists rule from Quito until 1739, when the power again moves in 1740 to New Granada when it becomes a viceroyalty.

1740   New Granada becomes a viceroyalty of Spain

New Granada (modern-day Colombia) becomes a viceroyalty of Spain. The territories of modern-day Colombia, Panama, Venezuela, and Ecuador are all consolidated under the rule of New Granada. The region supplies Spain with gold and emeralds.

1745   French expedition visits Quito

A team of French scientists visits Quito during a trip across the world, gathering information for their experiments. The French expose the Ecuadorans to Europe's Age of Enlightenment (an eighteenth-century philosophical movement that stresses nationalism, individualism, and the questioning of authority). Later, these ideas will inspire Ecuador's fight for independence.

1747   Francisco Espejo, Ecuadorian national hero, is born

Ecucadoran writer and activist Francisco Javier Eugenio de Santa Cruz y Espejo (1747–95) inspires much of the independence movement through his political writings. He advocates complete emancipation from Spain, autonomous government for each colony, and nationalization of the clergy. Although he does not survive to take part in the war for independence (he is imprisoned by the Spanish for his political activities, and dies while in prison), he is an important figure in its philosophical development.

1767   Jesuits banished from Ecuador

The Roman Catholic Church has occasionally been assisting Amerindians in Ecuador and is, therefore, regarded as a nuisance to some of the estate holders. In addition, the Age of Enlightenment, is firmly established in Ecuador. As a result, an anticlerical movement gains strength, and the Jesuits, an order of Roman Catholic priests, are forced to leave.

Audencia of Quito, 1642

1778 Chuiza Baltazara leads rebellion against Spanish

Chuiza Baltazara is the first Ecuadorian woman to organize a revolt against the Spanish. Many women in Central and South America participate in the struggle for freedom from Spanish rule.

1780 Writer and statesman José Joaquín de Olmedo born

Joaquín de Olmedo is involved in the independence movement and also in the creation of the republic of Ecuador. He writes many famous poems—including *La Victoria de Junín, Canto a Bolívar*, a heroic poem which glorifies the liberator of the South American revolution Simón Bolívar (1783–1830)—and political works as well.

1795 Birth of artist Antonio Salas

Salas (1795–1860) becomes the unofficial portrait painter of the Ecuadorian independence movement. Salas paints heroes, military leaders and notable churchmen in response to popular demand for such works. The struggle for independence creates interest in art with subjects of local and national significance.

Antonio Salas's son, Rafael (1828–1906), will be among the first to make the Ecuadorian landscape a source of nationalism and pride.

1809: August 9 First major revolt against Spanish rule

When France invades Spain in 1808, Spain begins bearing down on its colonies as a means of support, but finds itself losing control. The Ecuadorian-born descendants of the ruling class (referred to as *criollos*) feel alienated as Spanish bureaucracy dominates daily life, and Spanish nationals are given preferential treatment. A group of criollo rebels storms Quito and demands that Spain make some changes. Spanish troops are sent in from Peru and Colombia to stop the rebellion, but the rebels are not to be stopped for long. The plan of independence is emerging.

1810: October 10 Second major revolt against Spanish rule

A second revolution attempts to throw off Spain's control, but the short-lived government, led by the Marquis of Selva Alegre (1758–1821), is overcome in December 1812.

1820: October 9 City of Guayaquil declares its independence

With the proclamation of an independent Guayaquil, Ecuador's decision to seek independence becomes more compelling. Many Ecuadorian people are inspired by the liberation movements in other South American countries, such as Bolivia, Peru, and Argentina.

1822: May 24 The Battle of Pinchincha

The forces of General Antonio José de Sucre (1795–1830) defeat the Spanish near Quito at the Battle of Pinchincha. The victory marks the end of Spain's control over Ecuador and unifies the liberation movements of South America. Ecuadorans honor Sucre for securing the country's independence.

1822: July 26–27 Bolívar and San Martín meet in Guayaquil

Two of South America's most famous liberators, Simon Bolívar and José de San Martín (1778–1850), meet in Guayaquil to consider the future of newly freed areas across South

Antonio José de Sucre

Antonio José de Sucre (1795–1830) is a companion of the great South American liberator Simón Bolívar (1783–1830). Born on the northeastern coast of Venezuela, Sucre grows up at a time when Spanish control in Venezuela is declining. He joins the revolutionary movement as a teenager, is captured in a skirmish with royalists, and is allowed to emigrate to the island of Trinidad (then a British colony) just off the coast of Venezuela. In 1813 he returns to Venezuela with nearly two dozen other revolutionary exiles. The group fights in eastern Venezuela, defeating Spanish royalist forces. Spanish reinforcements drive back the revolutionaries in 1814–15, forcing them into exile again. By 1820, Sucre is a successful colonel under Bolívar. After the victory at Pinchincha, Sucre goes on to beat the Spanish at the Battle of Ayacucho in 1824, which ends Spain's military strength in South America. He then becomes the first constitutional president of the Republic of Bolivia (1826–28). He resigns from the post to live with his Ecuadorian wife in Quito. Dissension and distrust divide the unity of the revolutionary movement. In 1830, Sucre is assassinated, but nobody ever identifies the killer.

America. Both want control over Ecuador, and, eventually, Bolívar prevails, and San Martín returns to his native Argentina.

Late 1822   Republic of Gran Colombia formed

Bolívar's hope of a liberated Ecuador finally comes true. Ecuador becomes part of the short-lived Republic of Gran Colombia, which consists of what is now Ecuador, Colombia, Venezuela, and Panama.

1823   Birth of journalist and essayist Juan Montalvo

Juan Montalvo (1823–89), whose works are influenced by French Romantic writers, is born. Montalvo is known for his attacks on those he considers Ecuador's unjust and unwise leaders. He is a liberal, and in opposition to the conservative leadership, he writes passionate essays condemning all injustice, including that against the native Ecuadorans. In the 1860s and early 1870s, Montalvo's essays oppose the government of the religious conservative Gabriel García Moreno (see 1860).

1830   Ecuador secedes from Gran Colombia

In 1829, Venezuela secedes from the union of Gran Colombia, and Ecuador follows in 1830. The federation is a disappointment to many Ecuadorans because of its military rule by Venezuela and Colombia and the taxes it pays to Bogotá. When the union falls apart, the traditional name Quito is dropped in favor of *La República del Ecuador* (The Republic of Ecuador).

The Republic of Ecuador

1830   Juan José Flores serves as Ecuador's first president

One of Bolívar's aides, Juan José Flores (1801–64), becomes the first president of the new republic. During his fifteen years in power, he is noted for his iron-handed conservative rule. At this time, two rival ideologies are developing in Ecuador, with the conservative one centered in Quito and the liberal one in Guayaquil.

1832   Flores orders "invasion" of Galápagos Islands

President Flores orders the military to seize the uninhabited Galápagos Islands, The islands lie about 700 miles off the coast of Ecuador in the Pacific Ocean and are famous for their unusual wildlife, which includes giant tortoises. Several of the islands occasionally serve as penal colonies until the practice is discontinued in 1959, when the islands become a national park.

1832   Literary figures Numa Pompilo Llona and Juan de León Mera are born

Two of Ecuador's leading literary figures, poet-philosopher Numa Pompilo Llona (1832–1907) and poet and novelist Juan de León Mera (1832–94), are born.

1835   Darwin's visit to the Galápagos Islands

During his voyage on the *HMS Beagle*, English naturalist Charles Darwin (1809–82) visits the Galápagos Islands. His observations of the Islands' wildlife make important contributions to the development of his theories of evolution and natural selection written about in his work *Origin of Species*, published in 1859.

1842   Birth of artist Joaquín Pinto

Artist Joaquín Pinto is born. He is known for his humorous paintings and sketches depicting people, landscape, and daily life. Pinto also documents the plight of the native Ecuadorian, especially in the cities. His work influences much of twentieth century Ecuadorian art, especially the "social realist" style of artists such as Oswaldo Guayasamin Calero (see 1919).

1845–60 Political instability after Flores

President Juan José Flores is ousted in 1845, and goes to Spain in exile (he returns in 1855 and is made commander of the army). During the fifteen years after Flores' presidency, Ecuador goes through eleven presidents and *juntas* (military takeovers). The nation is split between pro-clerical conservatives and the more secular liberals. Regional bosses also compete for power.

1860–75 Gabriel García Moreno rules Ecuador

From 1860 until 1875, Ecuador is ruled by the fervently religious conservative Gabriel García Moreno (1821–75), Ecuador's first great statesman. García Moreno seeks peace and consolidation for the divided country through a rigid, theocratic government. Beyond his religious zeal, García Moreno is also known for developing roads and public education, beginning the Guayaquil-Quito railway, and putting Ecuador on firm financial footing.

1860s Underground resistance to Garcia Moreno grows

Ecuadorans in exile begin to organize an underground resistance against García Moreno. The journalist Juan Montalvo (1823–89) bitterly and intelligently opposes García Moreno's conservatism in his essays and other works, such as *The Eternal Dictatorship*.

1865 Ecuador, Peru, Chile, and Bolivia declare war on Spain

The former Spanish colonies of Ecuador, Peru, Chile, and Bolivia join together to declare war on Spain.

1868: August 13–15 Earthquakes destroy towns in Ecuador and Peru

About 25,000 people are killed and 30,000 made homeless when earthquakes strike the cities of Arequipa, Iquique, Tacna, and Chincha.

1871 Armistice reached in war against Spain

After over half a decade of military conflict, peace is established between Spain and its former colonies in South America, Ecuador, Brazil, Peru, and Chile.

1873 García Moreno grants privileges to Roman Catholic Church

By an act of Congress, García Moreno dedicates the republic to the "Sacred Heart of Jesus." He feels that a Roman Catholic–based government will cause the country to be more moral. Moreover, he believes the Catholic clergy should control the country's school system.

1875 Ecuadorian Academy founded

Ecuador's outstanding contemporary learned society, the Ecuadorian Academy, is founded. It is the second academy founded in Spanish America. It is a correspondent of the Royal Spanish Academy.

1875 García Moreno assassinated

García Moreno's religious conservatism causes bitter strife, which ultimately leads to his assassination. He is stabbed and shot to death in front of Quito's cathedral. The conservatives are unable to continue García Moreno's program in the following years and the liberals gain more power over the next twenty years.

1895 Radical Liberals take over

General Eloy Alfaro (1841–1912) takes command after a brief civil war, and the Radical Liberals remain in power for nearly fifty years. Alfaro and his successors are able to counteract much of García Moreno's program. Church and state are separated, and liberty of thought, worship, and press are established.

1897 Poet Benjamín Carrión is born

Benjamín Carrión (1897–1979), a popular twentieth-century Ecuadorian poet, is born. His works are included in the 1928 collection, *Los creadores de la nueva America*. In addition, he publishes *El nuevo relato ecuatoriano: crbitica y antologbia, San Miguel de Unamuno; ensayos*, and *Santa Gabriela Mistral; ensayos*.

1903 Poets Gonzalo Escudero and Jorge Carrera Andrade are born

Gonzalo Escudero (1903–71) and Jorge Carrera Andrade (1903–78) become two of Ecuador's most famous poets of the twentieth century. Escudero's works include the 75-page volume of poems, *Autorretrato*, published in 1957.

1904 Ecuador cedes territory to Brazil

Ecuador allows Brazil to annex the eastern part of the Amazonian territory.

1906 Novelist Jorge Icaza is born

Jorge Icaza (1906–78) becomes a popular twentieth-century Ecuadorian novelist. He is one of a group of realistic novelists whose works, in harsh and crudely realistic style, describe the life of Ecuador's struggling poor people. Icaza's novel, *Huasipungo*, written in 1934, is considered one of the most controversial novels in Latin American history. Among his other works is *Cholos*, published in 1946.

1908 Guayaquil-Quito railway completed

With the help of U.S. investors, General Alfaro's administration has a railway completed between the country's two major cities, Guayaquil and Quito. Originally started during García's rule some forty years earlier, the task had seemed impossible with mountainous terrain and prohibitive costs. Once completed, the railway cuts travel time from twelve days to two hours.

1911 Alfaro ousted from power

Political division among the Liberals leads to Alfaro's downfall. Although he orders the confiscation of the Church's lands, conservative landlords still remain in control at the local level, and the Amerindians are no better off than before the liberals' takeover.

1912: January 28 Alfaro killed

Alfaro masterminds a coup in Guayaquil to return him to power, but it fails. He and his lieutenants are imprisoned in Quito, lynched by an angry mob, and their bodies are burned. Today, Alfaro is considered a national hero; he is believed to be the first leader to attempt the modernization of Ecuador by separating church and state, abolishing capital punishment, and creating civil liberties.

1916 Ecuador loses territory to Colombia

Ecuador gives up the part of the Amazonian jungle south of the Caquetá and Putumayo Rivers to Colombia. Colombia, however, then hands over some of the land to Peru in return for open passage to the Amazon River. This land is later strategically used by Peru in 1941 to launch an offensive against Ecuador.

1918 Poet César Dávila Andrade is born

César Dávila Andrade (1918–67) is a popular twentieth-century Ecuadorian poet.

1919 Painter Oswaldo Guayasamin Calero is born

Oswaldo Guayasamin Calero (b. 1919) becomes one of Ecuador's most prominent painters of the twentieth century.

1920s Bananas and cocoa become leading export crops

During the 1920s, Ecuador becomes the world's leading exporter of bananas and cocoa. However, this situation reverses when a fungus destroys the cocoa crop and causes a decline in the economy.

1922: November Hundreds killed in civil disturbance

Ecuador's economy is sinking. The currency is rapidly losing its value, and a fungus is ruining the cocoa crop. Workers begin unionizing in Guayaquil, and riots ensue. The military steps in and massacres hundreds of civilians.

1925: July Army takes over, political fragmentation results

The army seizes control of the country and attempts to make land reforms. However, local politicians prevent the reforms from being carried out. Ecuador has twenty-two different administrations over the next twenty-three years. Factionalism and party splits are common.

1927 Central Bank of Ecuador founded

Begun as a private bank, the Central Bank of Ecuador later becomes the country's chief financial institution. At this time, the *sucre* (Ecuadorian form of currency) is linked to the U.S. dollar instead of the pound sterling, the currency of the British empire.

1928 National Development Bank founded

The government-owned National Development Bank is founded to provide credit for agricultural and industrial development.

1929 Economy suffers during the Great Depression

When the world market for agricultural products collapses in 1929 during the Great Depression, Ecuador's economy plummets. Earlier in the decade, Ecuador had led the world in production of such products as bananas and cocoa.

1930s Immigrants arrive from Europe

In the years just before World War II (1939–45), Ecuador experiences an influx of European refugees as many people flee Europe.

1930s–1940s Development of school of realist literature

Several schools of writers—the most famous located in Guayaquil—begin to produce realist literature. Realist writers are heavily influenced by the styles of European and North American writers such as French writers Emile Zola (1840–1902) and Guy de Maupassant (1850–93) and Americans John Steinbeck (1902–68) and Ernest Hemingway (1899–1961). The Ecuadorans write highly political works about poor or underprivileged people, often using crude language (faithfully copying local dialects) and stark, harshly realistic prose. Many of these works will influence a future generation of writers of protest literature.

1932 Four thousand killed in "Four Day War"

The government has experienced instability since 1925, and the economy is stagnant. A battle erupts between soldiers in Quito and Liberal forces from Guayaquil that lasts for four days and results in over 4,000 deaths. This conflict is known as the "Four Day War."

During the 1920s, Ecuador becomes the world's largest banana exporter. Here three Ecuadorians harvest bananas using plastic bags in the 1940s. (EPD Photos/CSU Archives)

1934 Ecuador joins the League of Nations

1934–35 José María Velasco Ibarra serves as president

José María Velasco Ibarra (1893–1979) eventually serves as Ecuador's president on five separate occasions over the course of nearly thirty years. Early in his first term, he dissolves the Congress and makes himself dictator (a move which he later repeats). The military quickly overthrows him, and he goes into exile.

1939 First national beauty contest heightens local rivalries

The political and cultural rivalry between the coast and the *sierra* (highlands) permeates every aspect of Ecuadorian life. Guayaquil, the tropical port city, has become a cosmopolitan commercial center known for its liberal attitudes. Quito, the capital in the cool highlands, has become known for its conservative values. In 1939, when Ecuador holds its first national beauty contest in Guayaquil, contestants from Quito send photos out of modesty. They are shocked at the Guayaquil contestants' immodesty in presenting themselves in person.

1941 War with Peru

Ecuador's border dispute with Peru dates back to the colonial period. The border dispute finally escalates into a war when Peru invades territory claimed by Ecuador, based on its redefinition of the vague boundaries drawn during the colonial era. Ecuador's military, however, is quickly defeated.

1942 Rio de Janeiro Protocol awards Ecuadorian territory to Peru

The war reaches a climax when Peru invades Ecuador's southern and Oriente (Amazon Basin) provinces. The Rio de Janeiro Protocol awards Peru the greater part of the Amazon Basin territory claimed by Ecuador. Ecuador signs the treaty under pressure from the major countries of the Americas, which are busy with issues relating to World War II. Ecuadorans have been trying to revoke that treaty ever since, and the border dispute with Peru remains an ongoing source of tension.

1944 Velasco wins second term as president

José María Velasco Ibarra comes to power as a nationalist, denouncing the Rio de Janeiro agreement. He blames the Liberal Congress for the betrayal of the Ecuadorans through this treaty. Socialists, communists, and conservatives support him, but when these groups dissolve, Velasco makes himself dictator. He is deposed by the military in 1947.

1945: October 21 Ecuador joins United Nations

Ecuador joins the United Nations as a charter member.

1948 Central Bank of Ecuador nationalized

The Monetary Board of Ecuador declares that the Central Bank of Ecuador belongs to the state. The bank is owned by the national government and by private banks, which are required to invest at least five percent of their reserves in it.

1948 Galo Plazo Lasso elected president

After three ineffective presidents in less than one year, Galo Plaza Lasso (1906–87) serves as president for four years. He is later made chief of the Organization of American States (OAS). As the government stabilizes, the economy improves, agriculture increases, and exports flourish, particularly bananas, cocoa, and coffee.

1952 Ecuador signs Declaration of Santiago

Ecuador proclaims sovereignty (announces that it has jurisdiction) over its coastal waters to a limit of 125 miles (200 kilometers). The Declaration of Santiago acknowledges the enforcement of these rights. The declaration is initially signed by Ecuador, Peru, and Chile, and later by Colombia.

1952–56 Velasco's third term as president

Velasco is elected president once again. He gains support because of his strong progressive platform. In spite of his authoritarian practices, he actually completes his four-year term.

1956–60 Camilo Ponce Enríquez serves as president

Camilo Ponce Enríquez of the Christian Social Party becomes president. His is a splinter group of conservatives that does not retain power for long. By the end of his term, people are demonstrating against Ponce because of his strict policies.

League of Nations

Formed in the wake of World War I, the League of Nations—the forerunner of the United Nations—was the world's first international organization in which the nations of the world came together to maintain world peace. Headquartered in Geneva, Switzerland, the League was officially inaugurated on January 10, 1920. During the life span of the organization, over sixty nations became members, including all the major powers except the United States. Like the United Nations, the League of Nations pursued social and humanitarian as well as diplomatic activities. Unlike the United Nations, the League was primarily oriented toward the industrialized countries of the West, while many of the regions today referred to as the Third World were still the colonial possessions of those countries.

The League's structure and operations, which were established in an official document called the *League of Nations Covenant,* resembled those adopted later for the United Nations. There was an Assembly composed of representatives of all member nations, which met annually and in special sessions; a smaller Council with both permanent and nonpermanent members; and a Secretariat that carried out administrative functions. The League's procedures for preventing warfare included arms reduction and limitation agreements; arbitration of disputes; nonaggression pledges; and the application of economic and military sanctions.

Although the League of Nations solved a number of minor disputes between nations during the period of its existence, it lost credibility as an effective peacekeeper during the 1930s, when it failed to respond to Japan's takeover of Manchuria and Italy's occupation of Ethiopia. Ultimately, the organization failed in its most important goal: the prevention of another world war. However, it did make contributions in the areas of world health, international law, finance, communication, and humanitarian activity. It also aided the efforts of other international organizations.

By 1940 the League of Nations had ceased to perform any political functions. It was formally disbanded on April 18, 1946, by which time the United Nations had already been established to replace it.

1960s International recognition of the Latin American novel

In the 1960s, Latin American writers such as Gabriel García Márquez (Colombian, b.1928) and Carlos Fuentes (Mexican, b.1928) popularize the Latin American novel among international readers. Among Ecuadorian writers, one of the best-known is Jorge Enrique Adoum (b.1923). His works include *Entre Marx y una mujer desnuda* (1976), a novel about Marxism, politics, love, and Ecuador.

1960–61 Velasco's fourth term as president

José María Velasco Ibarra is elected president for a fourth time. He formally renounces the Rio de Janeiro Protocol (see 1942), and embarks on an economic program of "growth through inflation." The economy, however, begins to fail again, and the government has to reduce spending. Social unrest results.

1961 The military revolts against Velasco

By 1961, Ecuador's currency is in a slump, and consumers are heavily taxed. The air force revolts and sends Velasco into exile, thus ending the country's streak of elected governments.

1961: November 7 Carlos Julio Arosemena Monroy becomes president

Carlos Julio Arosemena Monroy, Velasco's vice-president, assumes the presidency after the military revolts against Velasco. Arosemena, too, has his problems. Some of his ministers are considered too liberal, while he is thought to be irresponsible.

1963: July Arosemena removed from office

Arosemena's presidency lasts less than two years. The military arrests him for "drunkenness" and sends him into exile. The military suspends the constitution and claims it is protecting the country from being overtaken by communists. This is a sensitive issue in Central and South America, since Soviet communists have cemented their influence over Fidel Castro and his government in Cuba since late 1960. (See Cuba.)

Military Governments

1963 Captain Ramón Castro Jijón begins military rule

A group of four military officers, led by Captain Ramón Castro Jijón, takes over the government and rules until March 1966. The army acts because they are afraid that the Cuban revolution will influence Ecuador. They launch a modernization program and land reforms. The land reforms, however, fail; without adequate funding, it is an uphill climb against the established landowners.

1965 Ecuador seeks closer economic ties with Colombia

Ecuador and Colombia establish a permanent Colombian-Ecuadorian economic integration commission. This allows the two countries to trade goods freely with one another.

1965: May Ecuador signs Andean Pact

Ecuador's agreement to the Andean Pact makes the country a member of the Andean Common Market. The Andean Common Market (or Ancom) is comprised of Bolivia, Chile, Colombia, Ecuador, and Peru. Venezuela joins Ancom in 1977.

1966 Interim civilian president appointed

Clemente Yerovi Indaburu is appointed interim civilian president by the military. He takes office after students protest military government. The *junta* is overthrown by high-ranking military officials; an assembly is elected to serve with the interim president.

1966: October Constitutional assembly elected

Elections are held to choose a constitutional assembly. Otto Arosemena Gómez, cousin of former president Carlos Julio Arosemena Monroy (see 1961: November 7), becomes the provisional president.

1967 Constitution ratified

Under the presidency of Arosemena Gómez, the constitution is written and ratified.

1967 Oil is discovered

Oil is discovered in the Andes mountains.

1968 Velasco elected president a fifth time

New elections are held and Velasco wins the presidency for the fifth time (see 1934–35).

1969 Stock exchanges open

Ecuador opens two stock exchanges, at Quito and Guayaquil. Purchase and sale of government and some private securities become the function of the National Financial Corporation, which, along with several hundred individuals, owns the two stock exchanges.

1970 Construction of highways

Ecuador begins building east-west routes linking the Oriente (eastern interior rain forests) with the Sierra (highlands), and Guayaquil with the hinterland. Also, construction of the five-nation Bolivarian Highway begins.

1970: June 22   Velasco begins rule as dictator

Velasco falls out of favor once again, this time with the wealthier people as he raises taxes. After a financial crisis, Velasco suspends the constitution, and assumes dictatorial powers. He dismisses the congress, reorganizes Ecuador's supreme court, and begins to rule by executive decree.

1971: June   Velasco promises new elections

Velasco promises new presidential and congressional elections, which are scheduled for June 1972.

1972: February 15   Velasco overthrown

Velasco is overthrown in a bloodless coup by senior army officers after he refuses demands to postpone the elections. Velasco is deported to Panama, and granted asylum by Venezuela. Velasco has ruled five different times since the 1930s as a populist and still has a tremendous cult following.

1972: February 16   General Guillermo Rodríguez Lara heads new military government

General Guillermo Rodríguez Lara ends Velasco's dictatorship and initiates his own. The new government, the Revolutionary National Government, promotes a number of changes, including the introduction of state control of oil and the renegotiation of contracts between the government and foreign companies. In addition, the government agrees to allow the U. S. company, Texaco, to assist the government in efforts to find new sources of oil.

1973   Ecuador joins OPEC

Ecuador becomes a full member of the OPEC (Organization of Petroleum Exporting Countries) cartel. The previous year, the country had begun exploiting some of its huge oil reserves. Oil is an important part of Ecuador's economy during the 1970s, and the country prospers as a result of the oil price increases on the world market in 1974 and 1979.

Return to Elected Government

1976: January 12   General Rodríguez ousted

General Rodríguez rules for four years before being removed from power. A three-member Supreme Council assumes power, promising to return to civilian government within two years. The military government opens the country's oil industry to foreign investors.

1978: July   Presidential election held

Ecuador holds its first presidential elections in nearly ten years. However, none of the candidates receive a majority, and a runoff election is necessary (see 1979).

1979   Environmental problems in the highlands

The Ecuadorian Institute of Water Resources estimates that since 1954, the amount of Ecuador's arid (dry) land has increased by over thirty percent. A program of reforestation and maintenance is initiated, but only 4,000 hectares are reforested annually during the early 1980s.

1979: April   Runoff presidential election held

Jaime Roldós Aguilera wins the presidential runoff election. This victory marks the first time a Latin American government has peacefully ended a military dictatorship. Roldós is a populist running under the banner of the Concentration of Populist Forces. Christian Democrat Osvaldo Hurtado is made vice-president.

1979: August 10   New constitution goes into effect

The new constitution (Ecuador's sixteenth), approved in 1978, goes into effect on the same day that Roldós and Hurtado are inaugurated.

1981: May 24   Roldós killed, Hurtado becomes president

President Roldós is killed in a plane crash. Hurtado becomes president and serves for three years. Hurtado's term is marked by modest gains in the economy. By 1984, however, the economy is faltering, caused in part by widespread flooding in 1982. Another factor influencing the unsteady economy is a decreased world need for Ecuadorian oil.

1982   Widespread flooding

Exceptionally heavy rains late in the year cause extensive flooding, affecting 4.5 million people and causing rampant crop and property damage.

1984   León Febres Cordero Rivadeneira elected president

With the economy faltering, the Ecuadorian people call for a change in government. Elections are held, and León Febres Cordero Rivadeneira becomes the new president. Febres is a conservative Social Christian who advocates a free enterprise economic policy. He forms a coalition government (characterized by cooperation among differing parties) and presses his platform of reducing state intervention in the economy, making it more responsive to market forces.

1986   World oil prices plunge

Just as Febres' fiscal policies are about to bring widespread benefits to Ecuador, a steep drop in world oil prices cuts revenues by thirty percent.

1986: June 1   Febres' coalition defeated

Leftist parties gain control of parliament after Febres's coalition partners are defeated in the elections.

1987: January   Government suspends foreign debt payments

As a result of the drop in revenue, the government stops repayments on its $11 billion foreign debt.

1987: March 5   Devastating earthquake strikes Ecuador

An earthquake destroys 24 miles (38 kilometers) of oil pipeline and a pumping station, kills hundreds of people, and leaves 90,000 destitute. The damage, estimated at more than $1 billion, cuts off oil exports for four months. The earthquake deals another blow to the economy and postpones debt repayment even longer. The quake measures six on the twelve-point Mercalli scale.

1988: January 31   Borja and Bucaram win presidential election

Rodrigo Borja Cevallos of the Democratic Left Party and Abdalá Bucaram Ortiz of the Roldista Party (honoring former president Roldós) win the most votes in a field of ten candidates. Borja wins the runoff election and takes office along with a strong contingent in Congress. The government makes improvements in Ecuador's human rights endeavors, and reaches an accord with a terrorist group—Alfaro Vive, Carajo (whose name means, literally, "Alfaro lives, damn it!").

1990   Democratic Left Party loses popular support

Economic troubles, particularly inflation, result in the Democratic Left Party losing half of its congressional seats in midterm elections. A number of grass-roots organizations protest government action and call for reforms in land and human rights.

1990: June   Uprising by indigenous people

A general uprising, called Levantamiento Indígena, by about two million indigenous people of Ecuador is staged. They are distressed about their living conditions and the loss of their land to developers. They organize to demand that the government grant them control of their lands, and to end exploitation of the rain forest.

1990: July and August   Government argues against autonomy for indigenous people

President Borja and his associates argue openly against the demands of indigenous people that their land be protected from development.

1992: April 11   Indigenous people stage 230-mile protest march

A march, led by the Organization of Indigenous People of Pastaza (Organización de los Pueblos Indígenas de Pastaza, OPIP), begins in Puyo and advances to Quito. Estimates of the number of participants range from 800 to 5,000 fluctuating as the marchers make their way from the upper Amazon River basin to Quito. The route passes through spectacular mountain scenery at altitudes of around 12,000 feet (3,600 meters). Marchers from the Amazon region carry lances made of palmwood, a symbol of independence, or headdresses made from colorful toucan feathers, another symbol of Amazonian freedom. The marchers from the Andes mountains wear ponchos that identify their home region.

1992: April 24   Protesters reach Quito

When the protest march of the indigenous people reaches Quito, the march leaders meet with President Borja in the National Palace. The president, who had announced an expansion of bilingual education just days before, offers to guarantee land rights for the indigenous people. Since elections are just one month away and the president is unlikely to remain in office, skeptics regard his offers as hollow.

1992: July   Sixto Durán-Ballén becomes president

Voters elect a conservative government, headed by President Sixto Durán-Ballén of the Republican Unity Party and Vice-President Alberto Dahik of the Conservative Party. Durán-Ballén imposes severe economic measures in an effort to improve Ecuador's situation. These measures prove economically successful but socially unpopular, and support for Durán-Ballén begins to decline.

1993   Court modernization begins

Because its judicial system has flaws and is susceptible to political pressures, Ecuador begins an initiative to modernize the court system. Despite attempts at reform, the court system remains notoriously slow and inconsistent.

1993   Indigenous people sue U.S. Texaco for $1 billion

Ecuadorean Indians bring a lawsuit against U.S. oil company, Texaco. The suit alleges that Texaco is dumping an estimated 3,000 gallons of oil a day into lagoons in Ecuador.

1993: January   Ecuador quits OPEC

Ecuador officially leaves OPEC, claiming that their participation in the group is costly and of little benefit. Although Ecuador is a small oil producer compared to some of the members from the Middle East, their departure from OPEC sets a precedent that is potentially damaging to the cartel's future. Ecuador immediately increases its output of crude oil since it is no longer bound by the production limits agreed to between OPEC members.

1994: March   Water pollution tied to oil industry

The Center for Economic and Social Rights, a New York–based health and human rights group, releases a report indicating serious pollution of Ecuador's water. The report blames the poor water quality on thirty years of oil exploration and production in the Amazon region. Parts of the jungle have become heavily contaminated with toxic chemicals that are byproducts of oil production. Skin problems and cancer occur more frequently in people that live near these oil production facilities.

1994: May 2   New agreement to pay Ecuador's foreign debt

Ecuador and its creditor banks come to an agreement that will restructure its foreign debt of $7.6 billion over thirty years.

1995   Amendment grants right of accused to attorney

An amendment to the constitution provides for the right of a detainee to have an attorney present at the taking of testimony if the testimony is to be used as evidence in court.

1995: January 21   Ecuador attacks Peru, war ensues

The long-standing border dispute with Peru springs to life when Ecuadorian troops attack a Peruvian post. The attack precipitates a war between the two countries.

1995: March 1   War with Peru ends

The five-week war with Peru leaves 80 dead and 200 wounded. The war creates further economic difficulties for Ecuador, but also stirs national pride and support for President Durán-Ballén.

1995: July 25   Peru and Ecuador agree to demilitarize disputed area

The two countries agree to remove troops from the disputed 200-square mile area of the Amazon rain forest. The agreement takes effect on August 1, but the border's exact location remains undetermined. Under the agreement, neither country can send troops or military aircraft into the area without approval from international observers.

1995: October   Support for Durán-Ballén erodes

Durán-Ballén once again falls out of favor with the public, due to charges of political corruption that involve him and Vice President Dahik. Dahik flees the country, and Durán-Ballén serves the rest of his term with little support.

1996   Indigenous political movement gains support

In the 1996 national elections, an indigenous political movement called Pachakutik (Quechua for "cataclysmic change") runs candidates for offices on the national, provincial, and local levels. Pachakutik candidates win eight seats in Congress as well as several mayoral positions throughout the country. Their successes, although small, help focus attention on the concerns of Ecuador's indigenous population.

1996: January   Electricity workers protest

Electricity workers opposed to the government's plans to privatize their industry commandeer seven hydroelectric plants and drastically cut production. A serious drought the year before has already affected hydroelectric generating capacity, and daily blackouts become common.

1996: July 7   Abdalá Bucaram elected president

Abdalá Bucaram, a showy and eccentric populist, is elected to the presidency, with Rosalia Arteaga as vice-president. Bucaram comes from one of the country's influential Lebanese families in Guayaquil. Bucaram quickly alienates most of the political establishment. He had campaigned as an irreverent populist but his economic reform plan, which involves privatization and currency reform, is unpopular.

1997: February 6   Bucaram removed from office

By an act of Congress, Bucaram is removed from the presidency on the grounds of "mental incapacity." Congress elects its own presidential successor, Fabián Alarcón, but Vice President Arteaga objects. Congress had been planning to revise the constitution to provide for just such an occurrence—the selection of a presidential successor when the elected president dies or is removed. They had been scheduled to meet the following week.

1997: February 9   Arteaga appointed interim president

At the request of the military, Rosalia Arteaga is made interim president, becoming Ecuador's first female president. Arteaga's term lasts only two days, while Congress creates rules for presidential succession in case a sitting president is removed from office. Arteaga resigns when the Congress bars her from making an address before its session begins.

1997: February 11   Fabián Alarcón elected interim president

In a special session, Congress elects its speaker Fabián Alarcón as interim president after altering the constitution to keep him in the position until the next presidential elections in 1998.

1998: May 31   Ecuador holds presidential and congressional elections

In the first round of presidential elections, no one candidate receives a majority. A runoff is scheduled for July 12 between the top-two finishers: left-wing Christian Democrat Jamil Mahoud Witt who has received 35.3 percent of the vote, and conservative Social Christian Alvaro Noboa Ponton who has garnered 26.9 percent

In elections to the 125 seat National Congress, Witt's Popular Democracy-Christian Democratic Union finishes with thirty-five seats, Panton's Social Christian Party wins twenty-six, the populist Ecaudorian Rodoldist Party garners twenty-five seats, and the Party of the Democratic Left picks up seventeen seats. The remaining seats go to four minor parties.

1998: July 12   Witt elected president

Jamil Mahoud Witt defeats Alvaro Noboa Ponton in the presidential runoff election. Witt wins with 51.3 percent of all votes cast while Ponton receives 48.7 percent.

Bibliography

Andrien, Kenneth J. *The Kingdom of Quito, 1690–1830: The State and Regional Development*. Cambridge: The Cambridge University Press, 1995.

Hemming, John. *The Conquest of the Incas*. San Diego: Harcourt Brace Jovanovich, 1970.

Mörner, Magnus. *The Andean Past: Land, Societies, and Conflicts*. New York: Columbia University Press, 1985.

Newson, Linda A. *Life and Death in Early Colonial Ecuador*. Norman, Okla.: University of Oklahoma Press, 1995.

Pineo, Ronn F. *Social and Economic Reform in Ecuador: Life and Work in Guayaquil*. Gainesville, Fla.: University of Florida Press, 1996.

Powers, Karen Vieira. *Andean Journeys: Migration, Ethnogenesis, and the State in Colonial Quito*. Albuquerque, N.M.: University of New Mexico Press, 1995.

Rathbone, John Paul. *Ecuador, the Galápagos, and Colombia*. London: Cadogan Books, 1991.

Roos, Wilma, and Omer van Renterghem. *Ecuador in Focus: A Guide to the People, Politics and Culture*. New York: Interlink Books, 1997.

El Salvador

Introduction

El Salvador has long been a nation facing serious problems of social inequality. During the 1980s, few countries in the world drew more international attention and for a more prolonged time than did El Salvador. A civil war that began in 1980 thrust El Salvador into the east-west conflict of the cold war. Now, after twelve years of war and billions of dollars of foreign money having been poured into the conflict, El Salvador finds itself alone again, dealing with its still unresolved social problems.

El Salvador is located on the western side of the Central American *isthmus* (land bridge) that connects the continents of North and South America. The nation is shaped roughly like a rectangle with dimensions of 170 miles by 65 miles. Because El Salvador is located on the angled portion of the isthmus, the Pacific Ocean makes up the nation's entire southern border. Honduras lies to the north, Guatemala is to the west, and the Gulf of Fonseca and Nicaragua are to the east. El Salvador is one of the smaller countries in the world at 8,124 square miles (21,040 square kilometers), approximately the size of the state of Massachusetts. However, with a population of 5.5 million people El Salvador is also one of the more densely populated nations at more than 650 people per square mile.

El Salvador is distinct from many other countries in the Middle American region because of its relative ethnic homogeneity. Historically, El Salvador was home to a mixture of Europeans, Indians and Africans. But today this ethnic and cultural diversity has eroded away such that most people consider themselves to be *mestizo,* a mixture of Indian and European heritage. The most recent statistics place the Indian population at barely over ten percent, and the number of people who still lay claim to European heritage at less than one percent. The remaining eighty-nine percent is counted as mestizo. This does not mean that racial stereotypes have disappeared, or that the population is equal in social standing. El Salvador's sharp divisions are not based on ethnicity to the extent of its neighbors in Guatemala and Mexico.

Spanish Conquest

El Salvadoran history can be divided into three periods: precolonial (the era prior to the arrival of the Spaniards in 1524); colonial (1524–1821); and independence (1821 to the present). The precolonial civilizations of El Salvador had high population densities. The largest Indian group at the time of the Conquest was the Pipiles, a Nahuatl-speaking group that migrated to the region between 900 and 1350.

The Spanish first arrived in El Salvador in 1524 as an invading army under the command of Pedro de Alvarado (c. 1485–1541), who had just completed the initial conquest of Guatemala. After a series of conflicts that took place over two years, the Spaniards finally subdued the Pipiles and other groups and assumed control of the region. As in all their newly conquered lands, the Spaniards exploited Indian labor to extract wealth from the land. Central America never became a major source of wealth for Spain, but El Salvador did prove to be an important economic region within Central America. It exported indigo (a blue dye), cacao (chocolate) and balsam (a type of wood).

Throughout the colonial era, El Salvador was under the administrative control of Guatemala. In fact, until 1786 when El Salvador became an Intendency—an administrative subunit—the region was called San Salvador and was part of Guatemala. This close historic relationship proved to be a source of conflict after Central America gained independence in 1821. El Salvador accused Guatemala of trying to maintain its traditional dominance. Verbal sparring rapidly deteriorated into military conflict, which in turn dashed hopes of political unity in the region. Throughout the remainder of the nineteenth century, the Central American nations constantly meddled in one another's affairs, exacerbating their already depressed economies. Not until the rise of the coffee market in the late nineteenth century did El Salvador achieve a noticeable degree of political and economic stability.

Coffee proved to be a highly valuable crop for El Salvador, but its economic rewards did not trickle down to the majority of the population. A small handful of plantation owners controlled the industry and grew fabulously wealthy. They ensured that successive governments did not infringe upon their economic interests. Little changed when the mili-

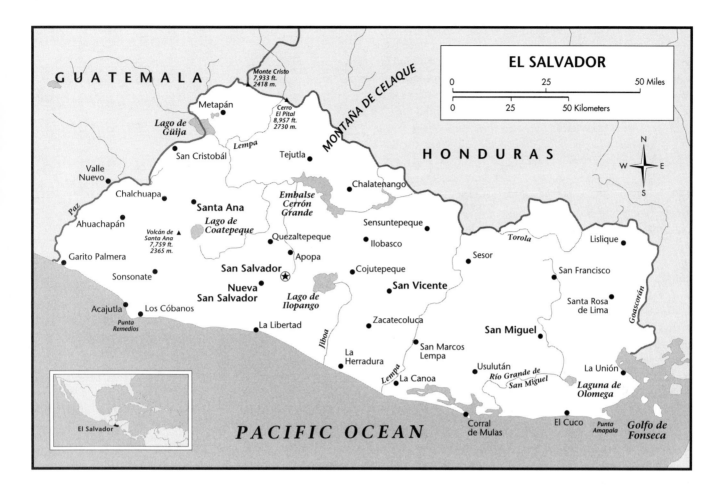

tary directly controlled the government between 1931 and 1979. Over the past century, governments of El Salvador have been notoriously authoritarian and repressive in economic policy making. They have clamped down on political dissent and refused to carry out reforms which might have lessened social disparity. As a result, public services for the vast majority of the population are sorely lacking. Not surprisingly, such conditions of inequality gave rise to sharp political divisions and eventually civil war.

Civil War

Contemporary El Salvador has been inexorably shaped by its civil war (1980–92) known as the "Twelve Years War." In the 1960s and 1970s there was no shortage of opponents to the military and the coffee growers, but the political system was so heavily controlled that the protestors were silenced. By the 1970s, more and more of the opposition figures concluded that violence was the only means left to effect change. They formed small guerrilla movements and began military attacks against the government.

What began as an internal civil conflict rapidly turned international in scope. In the United States, the Reagan administration argued that the conflict in El Salvador was being exploited by the Soviet Union, Cuba, and the new leftist Sandinista government in neighboring Nicaragua. The United States sent millions, and eventually billions, of dollars to the beleaguered El Salvadoran government and military. By the mid 1980s, the United States was contributing more to El Salvador's national budget than was the Salvadoran government. The Soviet Union, Cuba and Nicaragua, all allies of the guerillas, were outspoken opponents of U.S. policy. Even some of the United States's traditional allies in Europe questioned why the United States supported El Salvador's military, given the military's abysmal record in the arena of human rights.

The war in El Salvador ultimately ended as a stalemate after nearly 70,000 deaths and as many as 1,000,000 persons were displaced. Despite a brutal counter-insurgency campaign, the Salvadoran military was unable to eradicate the small guerrilla armies from the countryside. Neither were the guerrillas able to gain enough strength or popular support to defeat the military. The end of the cold war after 1989 helped bring the war to a close as external support evaporated. In January 1992 the government and the guerrillas finally signed a peace accord.

Since 1992 the international community has left El Salvador to its own devices. The war had devastating costs in both economic and human terms, but at least some things have changed. The size and power of the military has been vastly reduced and the police and security forces have been overhauled. Democratic elections were held in 1994 and are scheduled to be held again in 2000, but a fundamental dilemma still remains for the nation's political leaders: How to stimulate economic growth without leaving behind the vast majority of the population? One consequence of the war is that a consensus has been reached by all political parties that the economic issue cannot be ignored.

The Salvadoran economy is no longer as dependent on coffee as it once was. Business interests have been steadily diversifying their holdings to include investments in small industries such as food manufacturing. While such initiatives hold out promise for a more viable, diversified economy, but there is still reason to be pessimistic. The single largest source of revenue is money sent back to family members from Salvadorans working in the United States. Not only is this an unreliable source of capital, but studies also show that the recipients, who live at a subsistence level, use this money only to increase their buying power rather than invest in business ventures. The Salvadoran government has tried to promote tourism as a source of revenue, but the country does not offer much in the way of tourist attractions, particularly in comparison to neighboring Guatemala's pre-Colombian ruins.

Timeline

c. 3000 B.C.   Migration of native Americans

Native Americans migrate into the region of modern-day El Salvador and establish the first permanent settlements.

500 B.C.   Demographic increase

A significant demographic expansion of the Indian population in El Salvador begins, particularly in those regions below 3,000 feet in elevation. Many new settlements are established. These include Cara Sucia, Atiquizaya and Acajutla in the west; Los Flores, Río Grande, Cerro del Zapote and Loma del Tacuazín in the center; and Quelepa in the east.

400 B.C.   Chalchuapa increases in importance

The established settlement of Chalchuapa in the west increases its ties with lowland Maya settlements in Guatemala and emerges as one of the more important producers of a pottery type known as "Usulután," an important article of commerce in the Mayan world. Chalchuapa also gains control over the trade in obsidian from the volcanic source in Ixtepeque in Guatemala. Calendar and writing systems are in use in Chalchuapa.

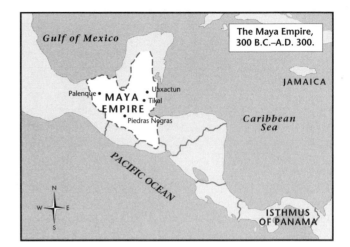

The Maya Empire, 300 B.C.–A.D. 300.

100 B.C.   Burial of war captives

In the modern era, archaeologists uncover thirty-three skeletons at the Chalchuapa site dating to approximately 100 B.C. The remains are those of people who were bound and decapitated, suggesting that the Indian inhabitants of Chalchuapa not only engaged in a form of sacrifice, but also were involved in some form of military conflict with neighboring groups.

c. A.D. 250   Ilopango volcano erupts

The Ilopango volcano near the modern city of San Salvador erupts dumping large amounts of ash on western and central El Salvador. Human habitation in the Zapotitán valley, the large plain to the west of modern-day San Salvador, is greatly changed. The valley is almost completely depopulated for the next 200 years. The eruption also breaks Chalchuapa's commercial ties with the rest of the country and eventually causes the settlement to decline. The inhabitants of Chalchuapa break into smaller groups and move to the highland region around Apaneca. Later they will return to Chalchuapa and establish settlements at Casa Blanca and Tazumal, but Chalchuapa never regains the prominence it had prior to the eruption.

c. 600   Eruption of the Loma Caldera volcano

The Loma Caldera volcano near modern-day San Salvador erupts, dumping between four and six meters of ash on the Zapotitán Valley. The eruption lasts up to two weeks, completely burying a settlement at the site known as Joya de Cerén. This site, which was not discovered until the 1980s, suggests that the eruption surprised the residents. Tools and household objects were last lying as if the inhabitants suddenly abandonded them.

c. 650   San Andrés emerges as political center

After the volcanic eruptions in the central region, settlements break up and reestablish themselves. One of the new settlements is San Andrés, which consists of one large pyramid and other smaller ones. San Andrés emerges as a political and economic center for the next 400 years. The population of the Zapotitán Valley during San Andrés's ascendance is around 40,000 people.

c. 900   Pipil Indians migrate to El Salvador

Nahuatl-speaking Indians from Mexico begin a slow but steady migration southward that lasts approximately 400 years. The region of El Salvador is the main area of settlement for these migrants within Central America. They come to be known as Pipiles, which is a derivation of the Nahuatl word *pipiltin* meaning "noble person." The regions within El Salvador that show heavy settlement by the Nahuatl-speakers include the Chalchuapa valley, the central portion of the country around modern-day San Salvador, the lower portion of the Lempa River, the valley of Sonsonate, the coastal area around Acajutla and balsam coast just to the east of Acajutla.

c. 1000   Quelepa abandoned

The site at Quelepa in eastern El Salvador is abandoned for reasons that remain unclear.

1100   Chalchuapa shows Nahuatl presence

Archaeological research from the civilization of Chalchuapa at this time show strong influence from the Nahuatl region of central Mexico. This includes the typical Nahuatl ball court game as well as green obsidian and ceramic styles from central Mexico. The new Nahuatl forms are intermingled with prior cultural designs indicating that the Pipiles did not displace prior settlers, but rather were incorporated into their society.

1200   Cihuatán is abandoned

The site at Cihuatán is destroyed and abandoned. Details of the conflict are unclear, but the signs of destruction are powerful evidence regarding the importance and degree of conflict between Indian groups.

1350   Ethnic composition of El Salvador reaches stability

Nahuatl migration to El Salvador comes to an end, bringing the ethnic composition of El Salvador to a relatively stable level until the arrival of the Spaniards. The Nahuatl-speaking Pipiles are the largest group. The second largest is the Lenca-speakers of the east. They likely were the original inhabitants until the Pipiles began arriving in 900. Other groups include the Chortí speakers in the north central region, the Cacoapera of the extreme northeast, the Mangue in the extreme east and the Ulúa in the southeast.

Colonial Era

1524   The first Spanish expedition of conquest

Pedro de Alvarado (1485–1541) arrives to conquer El Salvador. Alvarado was the lieutenant to Hernán Cortés during the Conquest of the Aztec Empire in Mexico (1519 to 1521). He arrives in El Salvador from Guatemala after having just completed the initial defeat of the Indian population in that region. The two main city states of the Pipiles, Cuscatlán and Tecpan Izalco, resist the Spanish.

1525   Founding of San Salvador

A second Spanish expedition arrives in El Salvador from Guatemala to further Spanish hegemony. The city of San Salvador is founded near the site of Cuscatlán. The Pipiles push the Spaniards out of the new city several times over the next three years.

1528   San Salvador settled

The Spaniards begin permanent occupation of San Salvador.

1530   San Miguel founded

The city of San Miguel is founded in the east.

1532   Inspection by Marroquín

Guatemalan priest Francisco Marroquín comes to El Salvador to conduct a survey of the region and its Indian population. The resulting compilation is one of the first and most comprehensive looks at the early colonial era in El Salvador.

1558   Establishment of Sonsonate

The western-most portion of El Salvador which the Pipil Indians called "Izalcos" and today consists of the departments of Sonsonate, Ahuachapán and Santa Ana is organized as an autonomous region under the name Sonsonate. It is not to be officially incorporated into El Salvador until 1823.

1567   Caluco Church inaugurated

Construction is completed on the church in Caluco, located approximately fifty kilometers to the west of San Salvador. This church serves as a main center of clerical activity for western El Salvador.

1582   Indian population decline

The Indian population has declined by ninety-seven percent. Most deaths are due to a disease brought by the Spanish, such as smallpox.

c. 1600   The rise of indigo

Faced with a declining cacao industry, Spanish merchants and entrepreneurs turn to cacao. Production of the blue dye increases steadily and soon it permanently surpasses indigo as the most important export crop for Central America. The high quality of El Salvador's indigo is recognized throughout European markets.

1713   Ana Guerra de Jesus Dies

Ana Guerra de Jesus (1639–1713) married young and was deserted by her husband. She entered a convent where she led a humble life and reportedly performed miracles. Today she is worshipped as a saint.

1736   Church at Metapán begun

Construction begins on the church in Metapán, one of many churches constructed in El Salvador during the 1700s with wealth from the indigo industry. The church at Metapán is considered to be one of the most impressive of the churches built in this period.

1760–92   Indigo production reaches high point

Stimulated by the Bourbon Reforms and a market boost in Europe, indigo production increases to its highest level.

c. 1760   Bourbon reforms

The Bourbon Reforms, named for the royal dynasties of Spain and France, begin to be implemented in Central America. The reforms, which include a variety of economic and political initiatives, are an attempt by the Spanish Crown to make its American dominions more profitable.

1782   Formation of Indigo Grower's Society

The Indigo Growers' Society (Sociedad de Cosecheros de Añil) is established. Its purpose is to give indigo growers an organization to stimulate the indigo trade. It was allowed to set prices at the annual indigo fair in San Vicente. The society was dissolved in 1826 after the indigo market collapsed.

1786   Region restructured

The region of "San Salvador," including the regions of San Miguel, San Vicente, and Santa Ana, are restructured as an Intendency. When the region of Sonsonate is annexed in 1823, the modern day borders of El Salvador are established.

1811: November 5   Revolt for independence

Father José Matías Delgado (1767–1832) declares independence from the Church of La Merced in San Salvador. A few days later he and Nicolás Aguilar launch a revolt against the local Spanish authority, Antonio Gutiérrez Ulloa. The Spanish Governor in Guatemala City sends a military force to crush the rebellion and restore the Crown officials.

1814: January 24   Second anti-Spanish revolt

Political leaders in San Salvador, among them Manuel José Arce (1787–1847), launch a second rebellion against the local Spanish officials, led this time by the Intendent José María Peinado. Once again troops are sent from Guatemala and the revolt fails.

1821: September 15   Independence

Political leaders in Guatemala declare Central America's independence from Spain.

Independent Era

1822: January 5   Annexation to Mexico

The Mexican Empire declares the annexation of Central America and sends a force of 600 men to Guatemala City to enforce its claim. Two municipal councils, San Salvador and San Vicente in El Salvador, refuse to recognize Mexico's authority.

1823: February   Mexico invades San Salvador

Mexican Troops arrive to San Salvador to crush an opposition movement to annexation. Shortly thereafter Agustín Iturbide (1783–1824) is deposed as Emperor of Mexico and the Empire collapses.

1823: July 1   Declaration of independence from Mexico

The United Provinces of Central America declares independence from Mexico.

1824: July   First newspaper in El Salvador

The first edition of a new periodical, *Semanario Político Mercantil* is published in San Salvador making it the first newspaper to be edited in El Salvador. The paper supports efforts to establish a bishopric in El Salvador and thereby lessen the religious and cultural influence of Guatemala.

1824: November 22   Government of Central America established

The federal government of Central America is officially declared with a constitutional mandate.

1825: April 29   National university founded

San Salvador's National University (Universidad Nacional) is founded. It will serve as the main institute of higher education in El Salvador for the next century and a half. Full operations will not begin until the 1840s.

1825: April 30   Arce as president of federation

The Salvadoran Manuel José Arce becomes President of the federal government of Central America after a hard-fought election against José del Valle of Guatemala.

1826–29   Civil War

Political competition within the federation, particularly between El Salvador and Guatemala, results in a civil war that lasts three years. The war ends when Guatemala City falls to an army under the command Francisco Morazán on April 13, 1829.

1833: February   Indian rebellion in central zone

A massive Indian rebellion erupts in the region of San Vicente under the direction of Anastasio Aquino (?–1833). The causes of the rebellion include grievances over military recruitment, taxation and land policies on the part of the government in San Salvador. The rebels occupy San Vicente City on February 16. Government troops force the insurgents out of the city on February 28. The rebellion is later crushed. Aquino is captured on April 23 and executed on July 24.

1837: April   Epidemic strikes

Cholera epidemic breaks out in San Salvador.

1839   Federation collapses

After years of warfare, bankruptcy and overall instability, the federal government of Central America collapses, giving rise to the five modern-day nations.

1841   Constitution adopted

El Salvador adopts its first Constitution. Its contents offer an impressive array of civil and political liberties such as democratic elections and the separation to powers, design to create a system of checks and balances. The reality of politics seldom matches the content of the Constitution.

1841–61   Period of political instability

Depressed economic conditions and ongoing conflict with neighboring countries help to create a highly unstable political environment. During this twenty year period, the presidency changes hands forty-two times.

1843   Archbishop Established in El Salvador

Pope Gregory XVI establishes a bishopric in El Salvador and appoints Monseñor Jorge Viteri y Ungo (1802–53) as the first Archbishop.

1854: April   Earthquake destroys capital

The capital city of San Salvador is destroyed by an earthquake. The city of Nueva San Salvador is established a few miles to the west as a replacement site. In the meantime, the capital is moved to the city of Cojutepeque until 1858. When the time nears to move the capital, political tensions increase. Those persons who own property in San Salvador stand to lose much money and prestige to their adversaries in Nueva San Salvador. After intense political wrangling, the capital is returned to San Salvador.

1857: June 7   Salvadoran Army returns from Nicaragua

Under the command of Gerardo Barrios (1813–65), the Salvadoran Army returns to El Salvador from its campaign in Nicaragua against the American freebooter William Walker. Barrios refuses to relinquish command of the army and begins marching to the capital to oust President Rafael Campo (1813–90). A negotiated settlement results in Barrios surrendering his army and Campo stepping down.

1859: March 9   Barrios as president

After two years of political intrigue, Gerardo Barrios becomes president. Barrios is seen as a one of the main liberals in El Salvador's history.

1863: March to October   Invading army from Guatemala

Barrios's main political opponent, Francisco Dueñas forms an alliance with President Rafael Carrera (1814–65) of Guatemala. Carrera sends an invading army into El Salvador to oust Barrios. After six months of intermittent fighting, Carrera's forces drive Barrios back to San Salvador. In October the invading troops take the city as Barrios flees eastward. Dueñas becomes President.

1865   Barrios executed

After escaping from the siege of San Salvador in 1863, Barrios is captured in Nicaragua trying to get to El Salvador to aid a revolt led by Trinidad Cabañas (1806–71). Barrios is extradited to El Salvador and executed on the orders of Dueñas on August 29, 1865.

1871   González and Liberals come to power

General Santiago González ousts Dueñas to become President. The rise of González marks the beginning of the liberals' dominance of politics. Over the next six decades they will institute free market policies, privatize land and oversee the establishment of the coffee industry.

1881   Land Law ends communal property holding

The National Assembly passes a comprehensive land law that stipulates that all communal property must be owned by individuals or it will be seized and sold. The primary target of the law is the Indian and peasant communities, particularly those located in the potential coffee growing regions of the west.

The motive is to promote the production of export crops like coffee. Over the next three decades Indian and peasant communities will lose much of their common lands, raising political tensions.

1886   Coffee economy

Coffee permanently surpasses indigo as El Salvador's most valuable export crop. Coffee is the engine of the Salvadoran economy until well after World War II (1939–45).

1898: January   The last of the violent coups

On the eve of the presidential election of 1898, General Tomás Regalado (d. 1906) overthrows President Rafael Gutiérrez. Although it seems like a typical political revolt, the Regalado coup is the last time that president will be overthrown by violence until the military coup of 1931. Each president now picks his successor. Although the system is non-democratic, it makes an era of political stability.

1913   President Araujo assassinated

President Manuel Enrique Araujo (?–1913) is assassinated by two machete-wielding men while listening to music in in San Salvador park. The motive for the killing is still unclear, but it appears not to have been politically motivated. A rich and well-established politician, Carlos Meléndez (1861–1919), succeeds Araujo as President. Over the next thirteen years only Carlos, his brother Jorge (1871–1953) and his brother in law Alfonso Quiñónez (1873–?), will serve as President. The era, known as the Meléndez-Quiñónez dynasty, offers a classic example of elites running the country through non-democratic methods.

1919   Formation of Liga Roja

As part of their efforts to retain political control in the 1919 election, Jorge Meléndez and Alfonso Quiñónez create the Red League (Liga Roja). It is a paramilitary organization in which poor peasants are recruited by cash payments or coercion. The recruits are placed under the authority of army officers loyal to Meléndez and Quiñónez and are then used to manipulate elections and repress any political opposition in the rural areas. The Red League is dismantled in 1923 after Quiñónez uses it to ensure his victory in the 1923 election.

1924   Labor union founded

The Regional Federation of Salvadoran Workers (Federación Regional de Trabajadores Salvadoreños, or FRTS) is founded in San Salvador. It is the central labor union designed to coordinate the nearly two dozen smaller unions spread throughout the country. Although the union deals largely with the concerns of urban workers, particularly those in San Salvador, its formation marks an important step in the political mobilization of the working class.

1926: November   Quiñónez selects Pío Romero as successor

President Quiñónez names his vice president, Pio Romero Bosque (d. 1935), as his successor instead of choosing a family member. Bosque is a loyal ally and a long-standing member of Quiñónez's political party.

1927–31   Romero supports democracy

Contrary to expectations, Romero emerges as a strong advocate of reform, primarily electoral reforms designed to establish genuinely democratic elections. It is not clear why Romero supports these reforms, but he emerges as a steadfast defender of them. Over the next four years he and a handful of other officials work tirelessly to support democracy by overturning those elections proven to be fraudulent. The task is overwhelming and success is often limited. But the highpoint is the January 1931 presidential election which most onlookers accept as being free and fair.

1930: March   Communist Party of El Salvador founded

The Communist Party of El Salvador (Partido Comunista Salvadoreña, PCS) is founded by approximately a dozen radical students and former members of the FRTS. The party announces its intention to organize rural workers in the west, but its capacity to carry out this program is limited. The party is small in size, urban in orientation, and bogged down by internal conflict. The party will be decimated by a military assault in January 1932 but it will survive to be one of the five main guerrilla movements in the 1980s; most of the other guerrilla organizations will be splinter groups from the party.

1930: November 23   Woman candidate for president

Prudencia Ayala goes before the Supreme Court to request that she be recognized as a full citizen of El Salvador with the corresponding rights to vote and run for office. Ayala's motive is to be allowed to run for president in the election of January 1931. The Court rejects her request, but her path-breaking case represents a major step in the cause of women's rights.

1931: January   Araujo elected president

Arturo Araujo (1878–1967), a rich but reform-oriented landowner from western El Salvador is elected President in elections which are uniformly accepted as free and fair. Araujo's election is a clear sign of success for Pío Romero's reform initiatives. However, Araujo's tenure in office is short-lived. Burdened by the economic pressures of the Great Depression, Araujo adopts policies which quickly alienate powerful sectors such as the military. He will be overthrown by a military revolt and the era of democratic reforms will come to a sudden halt.

1931: December 2   Military comes to power

A military revolt ousts President Araujo. The revolt is led by a group of military officers who hand power over to Vice President General Maximiliano Hernández-Martínez (1882–1966) less than one week later. The coup marks the beginning of nearly five decades of uninterrupted military control over government (1931–79). Hernández-Martínez remains in power until 1944.

1932: January 21–25   Peasant insurrection in the west

Peasant communities throughout the five western-most departments rise up in revolt. The rebels attack more than one dozen villages and gain control over half of those. The targets of the attacks are local elites and local military posts. The rebels kill approximately fifty people and destroy a dozen or more buildings and homes. Although the causes of the rebellion are still unclear, most of the municipalities at the center of the revolt are heavily populated by Indian. It is likely that the origin of the rebellion lay in long-simmering grievances by Indians over loss of land and abuses from local non-Indian elites.

1932: January 25–30   Military repression

In response to the peasant rebellion the military sends troops to the western region to carry out a rapid and intense counter-revolutionary campaign. They sweep through the insurgent regions killing indiscriminately. Resistance from the poorly armed rebels is negligible. The total number of persons killed by the military is still debated, but conservative estimates place it at around 10,000. Order is restored in less than one week. The military organizes a civilian militia known as the Guardia Cívica, which has the duty of monitoring for further outbreaks of revolt. In conjunction with its campaign in the west, the military attacks Communist Party cells in San Salvador and forces the party underground.

1932: February 1   Martí executed

One of the communists executed during the repression is Agustín Farabundo Martí (1893–1932). A former law student, Martí turned to radical politics and became perhaps the most recognized activist in El Salvador. He was exiled and jailed repeatedly in the 1920s and early 1930s. In the late 1920s he served alongside the famous Nicaraguan revolutionary Augusto César Sandino (1893–1934) during his war against the U.S. Marines. Martí became immortalized in 1980 when the coalition of guerrilla armies named their organization after him, the Farabundo Martí Front for National Liberation (Frente Farabundo Martí para la Liberación Nacional, or FMLN).

1932   Masferrer dies

Alberto Masferrer (1868–1932), one of El Salvador's most well-known writers and philosophers passes away due to nat-ural causes. Masferrer was a strong advocate of reform. One of his more well-known works is *The Vital Minimum* (El mínimum vital), in which he argued that the poor should be provided with at least a vital minimum of livelihood. His ideas were not well received by the elite.

1944: May   Hernández-Martínez resigns

A massive strike in San Salvador forces President Hernández-Martínez to surrender power on May 8. The immediate cause of his demise is the rumor that he intends to revise the Constitution to allow him to be re-elected again. His downfall is also part of a short-lived, but continent-wide rejection of dictatorship associated with the allied victories in World War II. Martínez turns over power to another General who promises to hold democratic elections. However, hard-line military officers seize power and put an end to the process.

1948   "Majors Revolt"

A group of army officers led by Major Oscar Osario (1910–69) leads a coup dubbed the "Majors Revolt" whick overthrows the government of General Salvador Castañeda Castro. Osario and his fellow conspirators announce their revolt as reformist in nature and declare their intent to carry out reforms. This pattern—revolt followed by unkept promises of reform—becomes standard for the military regimes over the next three decades.

1950   Women's suffrage adopted

A new constitution is written that established women's right to vote.

1955   Gavidia dies

Poet, dramatist and essayist, Francisco Gavidia (1863–1955) passes away. Gavidia is considered one of the founder's of modern Salvadoran writing. In 1932 he was named "Salvadoreño Meritísimo" (a title of disguishment) by the legislature.

1960   Christian Democratic Party founded

The Christian Democratic Party (Partido Demarcate Cristiano-PDC) is founded. It emerges as the main centrist political party. It is heavily associated with its leader and founder José Napoleón Duarte, who will twice be elected mayor of San Salvador. He will also lose the presidential election in 1970s on account of fraud by the military before finally winning it in 1984.

1960   Central American Common Market founded

The Central American Common Market is established with four countries. Costa Rica joins three years later. The market is intended to stimulate economic activity by creating a larger market for businesses to sell goods.

Salvadoran students protest a pending visit from United States President Lyndon B. Johnson (1908–73) in 1968.
(EPD Photos/CSU Archives)

1966   ORDEN founded

A right-wing paramilitary organization known as the Nationalist Democratic Organization (Organización Democrática Nacionalista, or ORDEN) is founded under the direction of the army to counter the growth of leftist views in the countryside. The military describes ORDEN as a benign aid organization for its members, but in reality it is used to monitor and attack other peasants who express opposition to the military. By the mid-1970s, ORDEN has more than 100,000 members. It vastly increases the scope and scale of political violence in the rural areas.

1968   Students protest LBJ visit

Salvadoran students protest a visit by United States President Lyndon B. Johnson (1908–73). The demonstrators oppose American foreign policy which they view as imperialistic.

1969: July 14–18   War with Honduras

El Salvador and Honduras go to war for four days. The conflict is improperly dubbed the "soccer war" by western journalists because of El Salvador's soccer victory over Honduras on a disputed referee's call. The deeper causes of the conflict were economic decline in Honduras and political strife in El Salvador. As many as 4,000 people are killed in the few days of fighting. El Salvador claims military victory and indeed extends its territorial limits at Honduras's expense. Long-term victory is less clear. The conflict shatters the Central American Common Market and with it any hopes of regional economic cooperation.

1972: March   Army wins fraudulent elections

The military's candidate for the March, 1972, elections, Colonel Arturo Armando Molina, wins fraudulently over José Napoleón Duarte of the Christian Democratic Party. In the midst of counting the votes, the military suddenly suspends tabulation and declares a recount which eventually results in Molina's victory. The government later arrests, beats and exiles Duarte.

1975: May 10   Poet Roque Dalton is assassinated

The young leftist poet Roque Dalton (1933–75) is assassinated by his former colleagues in the People's Revolutionary Army (Ejército Revolucionario Popular, or ERP). Dalton had recently decided to leave the ERP and although he continued to be a committed leftist, the leaders ERP would not tolerate his departure and falsely accused him of being an agent of the United States. During the Truth and Reconciliation proceedings after the end of the civil war in 1992, the guerrilla leader Joaquín Villalobos will confess to having ordered Dalton's execution. In his poetry, Dalton, who studied at the National University in San Salvador and in Mexico and Chile, evoked sympathy for the poverty-stricken masses of El Salvador as well as criticisms of the power structures that kept them down. Dalton's work was somewhat ignored immediately after his death. Conservatives villified him and even many leftists were obligated to ignore him as part of the justification for his murder. Currently, however, Dalton's work is undergoing a renaissance. His published works are being reprinted and many of his unpublished poems are being put into print for the first time. It is safe to say that he has become the *de facto* poet laureate of the nation.

1977   Romero becomes Archbishop

Archbishop Luis Chávez y González resigns after having held the office since 1939. He is replaced by Monsignor Oscar Arnulfo Romero y Galdames (1917–80).

1977: March   Army wins fraudulent elections

The army's candidate, General Carlos Humberto Romero (b.1924), wins another presidential election through fraudulent means.

1979: October 15   Reformist military coup

A group of military officers allied with a coalition of reform-oriented politicians overthrow President Romero and establish a *junta* composed of civilians and military officers. The junta announces its intent to initiate reforms, including land reform and the breaking up of right-wing paramilitary organizations. This junta will be succeeded by two others over the next two years. None of them succeed in stemming the surge in violence.

1980: March 24 Archbishop Romero is assassinated

Archbishop Romero is shot by a sniper while saying mass at a church in San Salvador. The perpetrators of the crime are never captured, but death squads under the direction of Roberto D'Aubuisson (1944–92) are strongly suspected. D'Aubuisson is a former army officer turned politician who is reputed to be the mastermind behind much activity. After the war, the Truth and Reconciliation Commission confirms D'Aubuisson's culpability. In the months prior to his death, Romero had become a powerful advocate of El Salvador's poverty-stricken masses and a vociferous opponent of the military's escalation of violence. As a man of the cloth and a person of unquestioned character, he became a powerful symbol of hope. His pleas for peace from the pulpit and over the airwaves threatened the military and its right-wing backers. Part of Romero's appeal extended from the fact that prior to his appointment as Archbishop he was considered moderate, even conservative in his political views. Landowners actually supported his appointment as Archbishop. But once confronted by the social realities of El Salvador, his views changed toward that of liberation theology and he began championing the cause of the poor. In death he is lifted instantly to martyrdom.

1980: March 30 Romero's funeral incites repression

Tens of thousands of people turn out for Romero's funeral celebration in San Salvador's central plaza. The peaceful protest turns violent. Who fired the first shots is still debated, but security forces at opposite ends of the plaza fired into the dense crowd leaving dozens dead and wounded. The events are played out on international television.

1980: October 10 FMLN formed

In an attempt to put an end to their factionalism and create a more united front against the army, the five main guerrilla organizations unite as the Farabundo Martí Front for National Liberation (FMLN). Despite this moment of unity, factionalism continues to a major problem for the guerrillas. Each movement has its own territorial base and the various leaders of each group hesitate to take orders from one another. Later testimony reveals that Cuba's leader Fidel Castro (b.1927)was personally involved in convincing guerrilla leaders to form the FMLN.

1980: December Murder of U.S. churchwomen

Four U.S. churchwomen—three nuns and a layworker—who had just arrived to El Salvador are detained on the highway leading into San Salvador. They are raped, murdered and buried in a shallow grave. The incident is another in a string of seemingly out of control right-wing violence and prompts the United States to suspend military aid. Four army soldiers are eventually arrested, convicted of the crime and sentenced to lengthy prison terms. High-ranking officers insist that the soldiers acted without orders. But eighteen years later, in 1998, still languishing in prison, the soldiers finally come forward with the names of the officers who gave the orders to kill the churchwomen. At least two of the identified officers are by then residing in the United States under asylum. Further release of information reveals that U.S. officials in the embassy and the State Department had been informed of the officers orders' shortly after they were given.

1980: December Third reformist junta founded

José Napoleón Duarte is installed as President of the of the third reformist junta, making him the first civilian head of state since 1931. The junta lasts until May 1982 and announces a range of reform initiatives which fail to stop the escalating war.

1981: January 10 FMLN final offensive

Anxious to take power before Ronald Reagan assumes office in the U.S. and emboldened by the success of the Sandinista victory in Nicaragua, the FMLN launches its "final offensive." It launches coordinated attacks throughout the nation in hopes that key military barracks will turn to its side and the general populace will rise up in support. The offensive fails. The insurgents are under-equipped, the military is able to hold its ground and the civilian population does not respond. The FMLN retreats to its strongholds in Morazán and Chalatenango departments and begins a protracted guerrilla war.

1981: September ARENA party founded

The National Republican Alliance Party (Alianza Republicana Nacionalista) is founded with D'Aubuisson as its leader. ARENA quickly becomes the leading party of the right and vies for power with the Christian Democrats of the center and FMLN-FDR alliance on the left. ARENA is a powerful political force but it is burdened by its constituency. It draws together "old-line" conservatives like D'Aubuisson with links to death squads and more "new-line" conservatives who are motivated more by free-market economics. The U.S. government does not support ARENA because of its hard-line elements and instead backs the Christian Democrats. ARENA controls the National Assembly until 1985, but is unable to win the presidency until 1989, when its more moderate elements ascend the party's ranks.

1989: March Cristiani wins presidential election

Alfredo Cristiani (b. 1947), a conservative yet charismatic businessman, wins the presidency as the candidate of ARENA. Despite his firm anti-communist beliefs, Cristiani is committed to bringing the war to a close and opens dialogue with the FMLN.

1989: November 11   FMLN launches second final offensive

Amid government boasting that the guerrillas are defeated, the FMLN launches its second final offensive. Not only does the attack come as a surprise, but it is launched in the heart of the army's stronghold, the capital city of San Salvador, where fighting had been effectively nil for the past decade. Guerrilla troops hold out for up to two weeks before retreating to their strongholds in the countryside. Although the offensive failed to topple the government, it delivered a clear message that the army had not won and that the war was at a stalemate.

1989: November 16   Army unit murders Jesuits

During the fighting of the second final offensive, a military unit enters the Jesuit University in San Salvador and murders the University's six Jesuit leaders. The killings are condemned throughout the world.

1991: December 31   Peace Accord

The Cristiani government and the FMLN sign a peace accord, the Chapultepec Accord, in Mexico. The fighting mostly ends, but it will be almost one year before combatants officially lay down arms. Major provisions of the peace treaty include vastly reducing the military, restructuring the security forces, reincorporating combatants into society and democratic elections.

1992: February 1   Cease fire

Official cease-fire takes effect as part of the peace process.

1992: December 15   United Nations teams detroy weapons surrendered

FMLN officially lays down its arms. Teams sponsored by the United Nations (UN) destroy the weapons.

1993: March   Findings of Truth Commission released

The United Nations-sponsored Truth Commission publishes the findings of its investigation into the war. The report states that the army and its paramilitary units were responsible for more than ninety percent of the deaths and human rights violations in the war.

1994: March–April   ARENA wins first post-war elections

The first elections after the cessation of civil war are held under U.N.-sponsored observation teams. The two leading candidates are Armando Calderón Sol (b. 1948) of ARENA and Rubén Zamora (b. 1942) of the FMLN. Calderón is the former mayor of San Salvador. Zamora is a long-standing political representative of the guerrilla alliance. Calderón Sol wins in a run-off election. Observers declare the elections to have been mostly fair although the FMLN contends that as many as 30,000 of its supporters in the North-central part of the country failed to receive registration cards in time to vote.

1995: April   New Archbishop chosen

Fernando Sáenz Lecalle becomes the new archbishop of El Salvador after Monseñor Arturo Rivera y Damas dies of a heart attack in November 1994. The Catholic Church in El Salvador is facing a crisis and Sáenz is a highly controversial choice. After the murder of Archbishop Romero in 1980, the Church steered away from politics, which caused it to lose face in the eyes of many of its poorer members. Church membership declined as Protestant and evangelical sects grew rapidly. Sáenz is an ultraconservative who believes that the mission of the church should be solely spiritual rather than material.

1997: March   FMLN gains in elections

The FMLN makes a strong showing in the municipal and legislative elections. It wins mayorships in a number of municipalities, including six of fourteen departmental capitals and San Salvador. It will have only one less deputy in the 1997–2000 National Assembly.

1997   Romero's canonization?

Archbishop Sáenz reports that discussions are ongoing in the Vatican about the possibility of Archbishop Romero being canonized as a saint.

1997   El Salvador has highest murder rate

Reports from the World Bank and other international organizations reveal that El Salvador has one of the highest murder rates in the world. Murders occur at the rate of 140 per 100,000 inhabitants, equal to South Africa as worst in the world. By comparison the rest of Central America has an average of just over 30 murders per 100,000 inhabitants.

Bibliography

Browning, David, *El Salvador: Landscape and Society*. London: Clarendon Press, 1971.

Byrne, Hugh, *El Salvador's Civil War: A Study of Revolution* Boulder: Lynne Reinner Publishers, 1996.

Ching, Erik, "From Clientelism to Militarism: The State, Politics and Authoritarianism in El Salvador, 1840–1940," Ph.D. Dissertation, University of California, Santa Barbara, 1997.

Flemion, Philip, *Historical Dictionary of El Salvador* Metuchen, N.J.: The Scarecrow Press, 1972.

Fowler, William, *Caluco: historia y arqueologia de un pueblo pipil en el siglo XVI* (San Salvador: Patronato Pro-Patrimonio Cultural, 1995.

Fowler, William, *The Cultural Evolution of Ancient Nahua Civilizations: The Pipil-Nicarao of Central America.* Norman: University of Oklahoma Press, 1989.

Gavin, Glenn, "El Salvador: Tiny Nation One of Most Homicidal in World," *Miami Herald*, August 3, 1997.

Haggerty, Richard, *El Salvador: A Country Study.* Washington, D.C.: Department of the Army, 1990.

Historia de El Salvador, 2 Vols. San Salvador: Ministerio de Educación, 1994.

Lauria, Aldo, "An Agrarian Republic: Production, Politics and the Peasantry in El Salvador, 1740–1920," Ph.D. Dissertation, University of Chicago, 1990.

Lindo-Fuentes, Héctor, *Weak Foundations: The Economy of El Salvador in the Nineteenth Century.* Berkeley: University of California Press, 1990.

MacLeod, Murdo, *Spanish Central America: A Socioeconomic History, 1520–1720.* Berkeley: University of California Press, 1973.

Prisk, Courtney (ed.) *The Comandante Speaks: Memoirs of an El Salvadoran Guerrilla Leader.* Boulder, Colo.: Westview, 1991.

Rohter, Larry, "File Focuses on Salvador Colonel in U.S. Women's Deaths," *New York Times*, June 25, 1998, p. A3.

Rohter, Larry, "The Guerrilla Poet, Lionized at Last but Still Elusive," *New York Times*, June 9, 1998, p. B2.

United Nations, *De la locura a la esperanza: la guerra de 12 años en El Salvador.* The Report of the Truth Commission, New York and San Salvador: United Nations, March 15, 1993.

Wortman, Miles, *Government and Society in Central America, 1680–1840.* New York: Columbia University Press, 1982).

Grenada

Introduction

Grenada assumed international importance quite out of proportion to its small size when the United States invaded the island in 1983. The invasion was an outgrowth of Cold-War politics. Since then Grenada has reassumed its status of relative obscurity, but the events of the early 1980s drew attention to the rather troubled history and rich cultural heritage of the island.

Grenada is located in the Lesser Antilles, the easternmost chain of Caribbean Islands running north-south between Venezuela and the Virgin Islands. The island of Grenada proper is oval in shape, approximately twenty miles long and twelve miles wide. Its small size of 133 square miles (344 square kilometers), approximately twice the size of Washington, D.C., is typical of the Lesser Antilles. The nation of Grenada actually consists of more than one island. The southern half of the Grenadines, a string of many hundreds of tiny islets and reefs running north-south between Grenada and St. Vincent, are dependents of Grenada. The largest of these is Carriacou, about twenty miles to the northwest and thirteen square miles in size. After Trinidad and Tobago, Grenada is the southern-most nation in the Caribbean. Its population is only 100,000 people, 30,000 of whom are to be found in the capital city of St. George's.

The island's origin is volcanic. With a ridge of mountains running north-south. The steep side is to the west, with the eastern side having a more gradual incline. Fortunately the island lies to the south of the most common hurricane path, but sizable storms struck in 1955, 1979, and 1980.

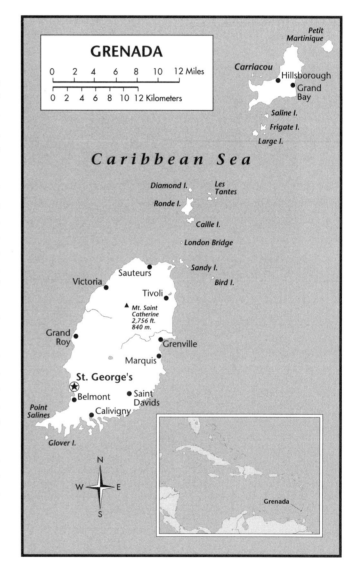

Population

Grenada's population is relatively homogenous. Roughly eighty-five percent is of African descent, a legacy of the island's slave history. Something less than half of these African-descended people are officially classified as "coloreds," being of mixed heritage. Like much of the rest of the Caribbean, subtle differences in skin color can have important social implications in Grenada. The remaining fifteen percent of the population consists of East Indians, Asians, and Europeans. English is the main language, but many people, mainly of African descent, still speak a patois or dialect that combines French, English, and African languages. Religious practices on the island are about as diverse as the population. A majority is Roman Catholic, reflecting the historic presence of the French. Protestants and a variety of African-based religions, such as Rastafarianism (see Timeline entry, "1998: June"), round out the religious faiths.

Pre-Columbian Peoples

Because Grenada's independent era has been so short, the history of the island is best understood in two parts, before and after the arrival of the Europeans in the late fifteenth century. By virtue of its proximity to Trinidad and Tobago, Grenada played an important role in the pre-European history of the Caribbean. With the exception of one migratory group that arrived at the Caribbean from the north around 4000 B.C., every major migration of Indian peoples into the Caribbean began in South America and passed through Grenada. The first of these occurred around 2000 B.C. The second, and arguably more important, began around 500 B.C. This latter migration was undertaken by peoples termed "Saladoid." They carried with them the technology for pottery and tool production and pushed aside the previous migrants. These Saladoids later became known as the "Tainos" and settled on the northern islands of Cuba, Hispaniola (modern-day Haiti and the Dominican Republic), and Puerto Rico. These were the peoples that Columbus encountered in his landing of 1492. Two later migrations from the south that passed through Grenada were those of the Arawaks and the Caribs, which occurred in A.D. 100 and 1200 respectively. Caribs came to dominate the island of Grenada by the time the Europeans arrived.

France and England Vie for Control

Columbus sighted Grenada during his third trip in 1498, but he did not land. The island remained unsettled by Europeans for another hundred years, partly because of stiff resistance by the Carib inhabitants. Finally in 1650 the French acquired the island from the British and established the first permanent settlement. The settlers conducted a campaign of extermination against the Caribs which resulted in a tragically famous last stand in 1654. The last remaining Caribs, having been pushed to the edge of a cliff by French colonists, leaped to their death rather than surrender.

Over the next 150 years, Grenada was a target of international rivalries between England and France, being traded back and forth between the two world powers on numerous occasions. In 1795 during one British occupation, a sizable rebellion broke out among the French inhabitants and the African slaves. The rebellion combined anti-British and pro-emancipation sentiments. Just one year earlier, the new revolutionary government in France had abolished slavery throughout its Caribbean possessions, and the slaves of Grenada were anxious to have the declaration apply to them. The rebellion lasted more than one year, and at one point the rebels had pushed the British back to a solitary stronghold at St. George's. But the British managed to cut the rebel supply line from neighboring islands and eventually put down the insurrection.

Grenada remained in the hands of the British for more than 150 years. British planters and colonial officials constantly searched for viable agricultural commodities for the island. Such tropical products as cacao and nutmeg did well, giving the island its nickname of the "spice isle" of the West Indies. However, the island has remained chronically poor throughout its history.

Leftist Governments

Partly because of its economic deprivation, the island has been the source of both authoritarian and radically leftist political ideas. Eric Gairy ran the island as a virtual dictatorship in the 1950s, 1960s, and 1970s. He gained even more control after independence from Great Britain in 1974. Finally in 1979 a group of opposition politicians overthrew the government while Gairy was out of the country. Maurice Bishop emerged as the leader of a socialist-oriented regime. The new government endeavored to improve the lives of the common people by installing social programs, but it also censored the press and refused to hold elections. The government also established close ties with Cuba which brought it into conflict with the United States.

U.S. Invasion

In 1983 a group of even hard-line leftists overthrew the government and had Bishop summarily executed. The Reagan administration took this as an opportunity to invade the island and set it back on a course more in line with its own interests. In October, 1983, over seven thousand U.S. troops occupied the island, arrested the coup leaders, and set up a new government. The legacy of this invasion has been a predominant feature of the island ever since. United States' interest in the island has waned significantly, and Grenada remains a poor and underdeveloped country, but the country has close ties to the United States and is considered a loyal and cooperative partner in the United States' Caribbean initiatives, such as the drug war. But political lines in Grenada can still be drawn around support for or against the U.S. action in 1983.

Importance of Economy

Whatever political party is in power in Grenada, the challenge before the government remains unchanged—lifting the island out of the depressed economic state that it has been in for most of its history. The island's per capita income is among the lowest in the Caribbean, and unemployment remains high at around twenty percent. Grenada's economic structure is heavily dependent on the fickle nature of international prices for a handful of tropical agricultural commodities, such as nutmeg and cacao, although it is the world's second largest exporter of nutmeg. An attempt at banana production seems to have failed on account of low quality merchandise and the possibility that the European Union will abandon its subsidies for Caribbean banana producers. The government has vigorously promoted a tourist industry, the vast majority of which is based on U.S. visitors. Nearly half of the work force is currently employed in tourist and service-related industries.

Timeline

c. 4000 B.C.   First major migration into Caribbean

The first inhabitants of the Caribbean begin migrating into the region from the north, probably from the Yucatán peninsula. They are Lithic peoples and are termed "Casimiroid" by archaeologists. They settle on the islands of Cuba, Hispaniola and Puerto Rico.

c. 3500 B.C.   Proto-Arawakan

The language known as Proto-Arawak emerges deep in the Amazon basin of modern-day Brazil. This is the parent language of Indian groups who will eventually inhabit the Caribbean, including Grenada, at the time of Columbus's arrival.

c. 2000 B.C.   Second major migration into Caribbean

People's known as "Ortoiroids" begin the second major migration into the Caribbean, arriving from South America. These people are "Archaic," meaning that they have advanced technological capacities beyond stone chipping to include the production of stone, bone, and shell artifacts.

c. 2000 B.C.   Emergence of pottery civilizations

In the upper reaches of the Orinoco River in modern-day Venezuela, a civilization known as "Saladoid" emerges. Its distinct characteristic is the development of pottery. Migrants carrying this new innovation will follow in the footsteps of their Ortoiroid predecessors and migrate northward into the Caribbean through the island of Grenada. These peoples are the ancestors of most of the Caribbean Indians that the Europeans will encounter.

c. 1000 B.C.   Frontier with Casimiroids

The northward migrating Ortoiroid peoples arrive on the island of Puerto Rico, come into contact with the Casimiroids, and establish a frontier region between Puerto Rico and Hispaniola.

c. 1000 B.C.   Ceramic peoples arrive at the coast

From their homeland in the interior of South America, the pottery-bearing Saladoid migrants arrive at the coast, near Trinidad, and establish a foothold that will serve as their stepping off point to the Caribbean. The language of these migrants, Proto-Northern, is the immediate parent language of Carib and Arawak, the two Indian groups who will inhabit the island of Grenada at the time of the Spanish conquest. It is also the parent language of Taino, the peoples who will inhabit the northern Caribbean at the time of the Spanish conquest.

c. 500 B.C.   Ceramic peoples pass through Grenada

The Saladoid peoples begin their northward migrations by first embarking to Trinidad from the South American coast. They slowly filter northward in small bands searching for favorable areas in which to settle, selecting uninhabited areas, intermingling with existing populations, or pushing pre-existing peoples out of the way through conquest. The Saladoids are responsible for introducing the ceramic age—pottery and the techniques of its production—to the Caribbean. As they migrate north, they add to their pottery styles by developing new forms, such as jars, bottles, effigy vessels, and incense burners.

c. A.D. 100   Second migration of ceramic peoples

A second, minor northward migration of Saladoid pottery peoples begins at the Trinidad/South American coast nexus. These second-wave migrants are Arawak speakers and are defined by a new type of pottery incision.

c. A.D. 600   Ceramic peoples arrive at Puerto Rico

The pottery-bearing migrants arrive at the Puerto Rico/Hispaniola frontier. During their one thousand year migration, they have replaced much of the Ortoiroid civilizations that they encountered. At Puerto Rico they divide into at least four new migrant civilizations, collectively called "Ostionoid," which refers to their new cultural styles and adaptations to the local environment. The general difference between the Saladoid pottery style, and the new Ostionoid pottery style is that the former is smaller, finer, and delicately ornamented, while the latter is heavier, bulkier, and more massive. These four groups in turn migrate into the rest of northern Caribbean, including Cuba, Jamaica, Hispaniola, and the Bahamas. They will become the "Tainos," the Indians that Columbus will encounter in 1492. The origin stories of most all Taino groups uniformly identify their homeland as a cave on the island of Hispaniola, which scholars site as evidence of their common ancestry.

A.D. 1200   Final Indian migration into Caribbean

A final northward migration into the Caribbean begins at the Trinidad/South America coast. The migrants are "Carib" speakers who share their linguistic origin with the Tainos and the Arawaks. The Caribs are the last group to migrate through St. Lucia, and they are the inhabitants of the island when the Europeans arrive. Spaniards portray the Carib peoples as warlike and cannibalistic. Scholars contend that this portrayal was overstated in order to justify the suppression of the Indians, but they do acknowledge that Caribs had to be more aggressive to force their way onto islands already inhabited by earlier migrants.

Pre-twentieth Century

1498: August 15   Columbus sights Grenada

During his third voyage to the Americas, Christopher Columbus sails past the island of Grenada and names it Concepción but does not land. Over the next 100 years, European presence on the island will remain nil. The half-hearted attempts at settlement are repulsed by the Carib inhabitants. The name "Grenada" has obscure origins, but it is believed to be a reference to the Spanish city of the same name.

1609   British attempt colonization

A British merchant group attempts to settle the island of Grenada for the purposes of tobacco production. But, like a similar attempt four years earlier on the nearby island of St. Lucia, a combination of sickness and Carib resistance forces them to abandon the project.

1650   French begin colonization

The French purchase Grenada from the English and manage to establish a permanent settlement at the site of modern-day St. George's. The settlers begin a campaign of extermination against the Carib inhabitants.

1654   Last Caribs commit suicide

After more than a century's resistance to European encroachment, the Caribs are pushed to the northern-most cliff on the island. During a final conflict, the last remaining Caribs commit suicide by leaping off a cliff rather than surrender to the French, an event still commemorated in present-day Grenada. The cliff has been named *La Morne des Sauteurs* (Leapers' Hill).

c. 1700   Sugar crops

French planters have introduced sugar onto the island. During the next century, sugar will be the island's most important crop.

1705   Fort Royale built

The French build Fort Royale, a mighty bastion to guard the harbor of St. George's, which was then named Basseterre by the French.

1753   Number of slaves grows

The total number of African slaves on the island reaches 12,000. They have been brought by the French for the purpose of working on the large sugar estates. The number of whites is approximately 1,200 and the number of free "coloreds" only 200.

1762   British gain control

After more than 100 years under French control, Grenada is taken by the British.

1763   Treaty of Paris

France formally cedes Grenada to Great Britain in the Treaty of Paris. Britain begins to implement a vigorous plan to advance sugar production by increasing the number of, and demands upon, slaves.

1763   Evolution of the Big Drum Dance

An African-derived cultural expression of song and dance known as the Big Drum Dance begins to emerge on the island of Grenada. The stronghold of the dance is Carriacou, one of the small Grenadine Islands to the north. Because these northern islands are more isolated and less subject to direct European presence, they serve as a sort of safe-haven for African-based cultural practices. The carriers of the cultural knowledge of the Big Drum Dance are believed to be Cromanti peoples from west Africa whom the British imported as slaves during their push to increase agricultural production. The dance is similar to other African-based singing and dancing throughout the Caribbean, but it holds its own unique Grenadian qualities. The dance involves at least three male drummers and several female singers, and a fairly regularized series of physical motions. The language of the lyrics is the patois of Grenada which combines French and various African languages.

1773   Sugar production and slave population grow

The plan to increase sugar production shows dividends. Production is up at least two times in the last ten years and the slave population has increased three times. Grenada is now the second most valuable British possession in the Caribbean behind Jamaica.

1779   French attack

France recaptures Grenada. All the British ships in the harbor of the capital, St. George's, are sunk, the British governor is jailed, and French soldiers pillage English properties on the island.

1783   Attempts to eradicate French influence

Grenada is restored to the British under the Treaty of Versailles. British planters respond to four years of French rule by trying to suppress Roman Catholicism on the island. All lands and properties of the Church are confiscated, and all religious services have to be presided over by an Anglican minister. Despite their best efforts, the British fail to eradicate Catholicism, and today a majority of Grenadians are Catholic. Many place names on the island are still in French. A common language among the populace is a patois that has French elements.

1794   France abolishes slavery in its possessions

Under the control of the revolutionary Jacobins (the most radical political faction during the French Revolution), the French government abolishes slavery in its Caribbean possessions. France also sends a special representative, Victor Hughes, to Guadeloupe to spread the ideas of the French Revolution throughout the Caribbean.

1795: March   Beginning of the "Brigand" conflict

In the midst of the French Revolution (1789–99) the Caribbean finds itself caught up, even more than usual, in the seemingly perpetual state of conflict between England and France. Britain takes advantage of instability in France caused by the Revolution to seize French territory throughout in the Caribbean. But the British are faced with strong local opposition, both from French nationals, who despise being under the British, and slaves who are anxious to take advantage of France's abolition of slavery. Throughout the tiny islands of the Lesser Antilles, these conflicts take on the dynamic of a guerrilla war against the British, who call the rebels "Brigands." The conflict comes to Grenada in February 1795, and lasts for more than a year. The leader of the rebellion in Grenada is Julian Fédon, a plantation owner of mixed French and African descent. Starting in 1794 he begins organizing armed opposition to the British. He contacts Victor Hughes on Guadeloupe, who provides him with arms and ammunition. Fédon then recruits slaves to his cause. Normally a plantation owner and slaves would have little in common, but under the circumstances, they have a mutual enemy in the British. The first shots are fired on March 1, 1795, when Fédon's rebels attack three British settlements on opposite sides of the island.

1795: April   Fédon gains the upper hand

After a month of bitter fighting between Fédon's rebels and British troops, which had been brought in from neighboring islands, the British governor determines that the rebels cannot be quickly defeated. He pulls his troops back to defensive positions in coastal strongholds, effectively surrendering the interior of the island to the rebels.

1795: December   Rebel offensive

After eight months of guerrilla tactics aimed at weakening the British strongholds, the rebels launch a full offensive to take over the entire island.

1796: February   Fédon pushes British back to St. George's

Under relentless pressure from the rebel offensive, the British are pushed back to a single location, the capital of St. George's. Although Fédon and the rebels cannot break through the British defenses, Grenada can accurately be described, as historian Michael Craton has, "as a black republic under arms." Slaves not directly participating in the rebellion take advantage of their temporary liberty to organize households and plant crops for subsistence.

1796: June   British defeat Fédon

After finally managing to cut the French supply lines, the British break out of St. George's and push Fédon's forces back to a mountaintop near Fédon's own estate, Belvidere. The rebellion is effectively dead. Fédon's fate is unknown, but he is presumed to have drowned trying to escape to Trinidad.

1807   Slave trade abolished

The British government abolishes the slave trade, thereby cutting off the supply of African-born slaves. Current slaves will remain in bondage for nearly two more decades.

1833   Slavery itself is abolished

The British government abolishes slavery in all of its dominions in the West Indies, including Grenada, over the objections of the many and powerful plantation interests. Plantation owners fear financial ruin because of a shortage of cheap and readily available labor. The British financially compensate many owners for their lost slaves and introduce an apprentice system that drags out the Africans' obligations to the plantations another half decade. The number of slaves on the island is approximately 23,000, or eighty-five percent of the population.

1846   Sugar as main export

Sugar exports total 9,100,000 pounds, while cacao is only 375,000 pounds. Over the next four decades, however, the positions of sugar and cacao will reverse.

1849   Yorubas bring African religion

Approximately one thousand Africans of Yoruba origin are settled on Grenada as part of Britain's policy of capturing French and Spanish slaving ships and resettling liberated slaves. The Yoruba from this resettlement introduce the Shango, or Orisha, religion into Grenada.

c. 1850   Nutmeg industry thrives

Diseases in the nutmeg production areas of Asia provide Grenada with an unexpected economic boost from its nutmeg industry. Nutmeg rapidly becomes the island's second most important export crop.

c. 1880   Cacao rises in importance

With the sugar industry on Britain's Caribbean possession in decline, cacao is introduced into Grenada and becomes the island's most valuable export crop.

1881   Cacao replaces sugar

Cacao replaces sugar as the main export. Since 1846 cacao exports have risen to 5,800,000 pounds, while sugar has declined to 2,038,000 pounds.

1885   Grenada is British administrative center

Grenada is established as the headquarters of the British Windward Islands administration. The Windward Islands include St. Lucia, St. Vincent, the Grenadines, as well as Grenada. Grenada remains a under this administration until 1958 with the formation of the West Indian Federation.

Twentieth Century

1915   Publication of *The West Indian*

T. A. Marryshow (1887–1958) and a partner found a newspaper named *The West Indian*. The paper, which Marryshow will edit for the next twenty years, espouses nationalism and equal rights for blacks. Marryshow quickly becomes a popular political activist.

1917   Foundation of RGA

Marryshow founds the Representative Government Association (RGA) to apply pressure on the British government advocating greater popular representation in the political system. At this time all the legislative councils in the Caribbean, with the exception of Jamaica and Barbados, are appointed by representatives of the British Crown.

1919   Butler Returns to Grenada

Tubal Uriah "Buzz" Butler (1897–1977) returns to his native Grenada after having served for the British in World War I (1914–18). Butler will go on to be a major union leader and political figure in Trinidad and Tobago. In the two years that he is on the island, his organizational commitment takes shape. Butler becomes convinced that British colonialism and the control of a few white landowners must eventually cease. He founds the Grenada Representative Government Movement which advocates universal suffrage. He also establishes the Returned Soldiers' Association which demands care for veterans returning from World War I.

1921   Marryshow lobbies for self-government

Marryshow travels to England at his own expense to lobby the British government for greater self-government in Grenada.

1925   Constitution and elections

As a result of reforms by the British government, Grenada is allowed to adopt a new constitution that allows for five of the sixteen-member legislative council to be elected. Marryshow becomes a the council member and holds his seat for the next thirty-three years until his death in 1958.

1931   Protest against tax increases

The British governor raises taxes on the island in an attempt to counter the revenue declines associated with the Great Depression. He soon revokes the increase after a large protest organized by Marryshow.

1933   Legalization of city trade unions

Urban-based trade unions are legalized. Organization among rural laborers remains illegal.

1938   Moyne Commission

In the wake of a series of large and violent labor disputes in Jamaica, the British government dispatches the West India Royal Commission, known as the Moyne Commission, to the Caribbean to investigate ways to prevent future such disturbances.

1939   Moyne Commission in Grenada

The Moyne Commission visits Grenada and reports on a variety of social problems, including a lack of housing, malnourishment, and pitiable wages.

1940   Disappearance of patois

The patois of mixed French and African languages is nearly dead after two centuries of British control over the island.

1944   Nutmeg as main export

Nutmeg accounts for more than fifty percent of the value of all exports. Cacao is second at thirty-two percent.

1946   Trade unions founded

Two trade unions, The General Workers' Union and the St. George's Union, are founded. They are both urban-based.

1950 Political organizations founded

Eric Gairy (b. 1922) founds the Grenada Manual and Mental Workers' Union (GMMWU) and the Grenada People's Party, the name of which will be changed to The Grenada United Labour Party (GULP) in 1951. These organizations begin to make significant inroads among rural workers, although colonial officials reject the GMMWU as a legal bargaining agent for workers. Gairy was born into a poor peasant family outside of Grenville, Grenada's second largest town. As a young man in search of work he left Grenada for Aruba, where he became involved in trade-union organizing. He was expelled from Aruba for his organizational activities in 1949 and returned to Grenada.

1951: February 19 Strike and rural uprising

Gairy and the GMMWU call for a general strike. This is the first general strike in the nation's history, and it quickly turns into a rural rebellion as plantation workers loot and burn homes. On February 21 Gairy leads thousands of workers to the colonial administrative offices in St. George's demanding reforms. The governor declares a state of emergency and exiles Gairy to Carriacou.

1951: March 15 Gairy becomes popular hero

The governor of Grenada is forced to bring Gairy back from exile to help end the continuing violence. In the process, Gairy is able to extract concessions from the plantation owners. The strike catapults Gairy to the status of populist hero among the working masses.

1951: October Gairy wins elections

Elections for the legislative council are held. Gairy and his parties win a resounding victory by receiving seventy-one percent of the vote. Of the seven seats on the council that are elected rather than appointed by the British governor, Gairy's parties win six of them. Gairy himself occupies a seat. He will be a dominant figure in Grenadian politics for the next three decades.

1957 Gairy ousted in elections

Gairy's party loses the election to the Grenada National Party (GNP). Part of the discontent with Gairy is that he is perceived as running his political organizations like a mini-dictatorship. Despite the fact that he came into power on the support of the rural poor, he seems to be trying to ingratiate himself with the establishment. The GNP, led by Herbert Blaize, a landowner from Carriacou, won the election by appealing to the constituents who normally would be strong Gairy supporters. Once in power, however, its policies clearly favor landowners and business. In the next three years planters' incomes increase 170 percent while laborers wages increase only fifteen percent.

1958–62 Grenada in the West Indies Federation

Facing the likelihood that it will soon be granting independence to its colonies in the Caribbean, the British government creates the West Indies Federation. Its purpose is to bring the islands together under a common administrative infrastructure that is more viable than the separate individual islands. The breakup of the federation in 1962 is caused by interregional conflict between the wealthier islands which feel the smaller, poorer islands are an economic drain. Jamaica is the first to pull out in 1961, followed by Trinidad and Tobago. The first wave of independence (1962–66) follows.

1960 High unemployment

Unemployment passes forty percent, putting pressure on Blaize and the GNP government.

1960 New constitution

A new constitution is adopted that allows for the entire legislative council to be elected by popular vote.

1961 GULP wins elections

Able to take advantage of the popular discontent with the GNP government, Gairy and the GULP return to power in the 1961 elections. Once in power, Gairy engages in a variety of corrupt business deals that enrich him and his closest associates.

1962: March Squandermania report

The British colonial administration appoints a special council to look into the rumors of Gairy's corrupt dealings. It issues a report, which comes to be known as the Squandermania Report and accuses Gairy of corruption and running a campaign of intimidation. On the basis of the report, the colonial administration suspends the 1960 constitution, dissolves the GULP government, and places the island under colonial authority once again until elections can be held later in the year.

1962: September GULP returns to power

Gairy and GULP win the election and return to power.

1967: March 3 Self-government

Grenada becomes a sovereign nation associated with the United Kingdom but with full powers of self-government. External affairs and defense issues are the responsibility of the United Kingdom. The position of prime minister is much stronger under the new arrangement.

1967: August Elections

GULP defeats the GNP in national elections, and Eric Gairy remains prime minister. He will remain in power for the next twelve years, largely through corrupt practices of vote rigging and intimidation. Government becomes a veritable dictatorship under Gairy. Even small bureaucratic decisions have to await his approval. He grows quite wealthy by taking advantage of his position in power, while public services such as education and healthcare stagnate.

1970: November   Nurses strike

Approximately thirty nurses from St. George's hospital march the streets demonstrating against the government for the bad conditions in the hospital. Gairy responds by transferring most of them to remote locations in the country.

1970: December   Wider protests

Gairy's response to the nurses' strike triggers wider protests by nurses, teachers, and trade unions. Police beat the demonstrators with clubs. Twenty of the nurses who are arrested are defended by two young lawyers, Maurice Bishop (1944–83) and Kendrick Radix (b. 1941), both fresh out of law school in England and Ireland respectively. Radix and Bishop will emerge as major political opponents to Gairy. Bishop was born to Grenadian parents of humble means on the island of Aruba. After returning to Grenada at the age of six, he emerges as an outstanding student and eventually leaves for England to study law. He intends to use his law practice to defend victims of the Gairy regime.

1973: March 11   New Jewel Movement founded

The New Jewel Movement (NJM) is founded with Maurice Bishop and Unison Whiteman as its co-leaders. The NJM quickly becomes the main opposition movement to Gairy.

1973: November 17   NJM leaders beaten and arrested

The six main leaders of the NJM, including Bishop and Whiteman, are beaten and incarcerated by Gairy's police agents. Once knowledge of the arrests becomes public, a committee is formed by twenty-two of the country's major trade unions and popular organizations. The Committee of twenty-two, as it was called, demands the release of the NJM leaders and calls for a general strike. Gairy responds by releasing the prisoners, and the strike is called off.

1974: February 7   Full independence

Grenada gains full independence from Great Britain. The transition is marked by widespread violence and opposition because Eric Gairy remains prime minister. With the mandate of independence behind him, Gairy abandons restraint and sets up a quasi-dictatorship that crushes opponents and consolidates his power. His government is considered an outcast throughout the Caribbean.

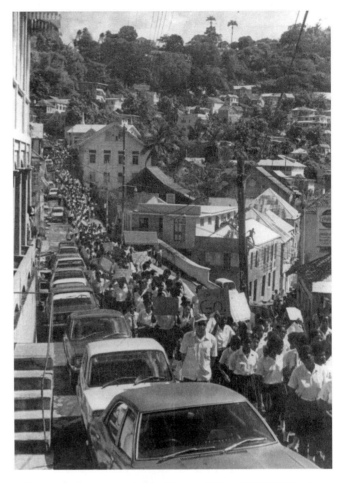

The day before independence, Maurice Bishop (1944–83) leads a demonstration against the government of Gairy (b. 1922). (EPD Photos/CSU Archives)

1976: December   Gairy wins elections

Gairy wins national elections through blatant fraud and intimidation on the part of his police agents. The New Jewell Movement continues to grow in surreptitious popular support.

1979: March 12   Gairy government ousted

NJM supporters who infiltrated Gairy's intelligence services inform their leaders that Gairy has given orders to have them eliminated when he leaves for the United States on a tour. On March 12, Gairy leaves for the United States and on the same day leaders of the NJM, including Bishop, Unison Whiteman, Kendrick Radix, and Bernard Coard (b.1944) launch a bloodless coup that overthrows the government. In its place they declare the People's Revolutionary Government (PRG). The leaders of the new government are strongly nationalistic and socialist-leaning. They also hold a personal admiration for Fidel Castro and not surprisingly, relations with the United States deteriorate as ties to Cuba grow. The Cuban government begins to send aid in the way of construction materials, teachers, and doctors to the new regime. Over the next four years the PRG will institute a variety of social reforms designed to lives of the commoner. But the government will also practice press censorship and refuse to hold democratic elections.

1980: January   Construction begins on a new airport

The PRG begins construction on a new airport to replace the outdated one, which cannot accommodate modern planes. Numerous countries contribute to the high cost of project, including Venezuela and Nigeria, but by far the largest source of aid is Cuba. The PRG says the goal of the new airport is to lay the foundation for an eventual tourist industry on the island. The administration of U.S. President Jimmy Carter is relatively quiet about the project at this point compared to how loud it will be later in accusing Cuba of building the airport as part of a military base. The United States and Grenada regimes are moving further apart.

1981   Mock invasion

The U.S. military carries out a large scale practice invasion on the Puerto Rican island of Vieques. The exercise involves more than 120,000 troops and is a thinly disguised threat to the Grenada government.

1982   Factionalism in the NJM

Bernard Coard resigns from the Central Committee of the NJM, as internal disputes and factionalism within the party are on the rise. The party has developed into a somewhat doctrinaire and secretive apparatus with not more than 200 full members who are in charge of organizing and leading the masses. The membership of the party's Central Committee is never published, nor is Coard's resignation publicized. The main split is between Coard and Bishop. Coard is the more hard-line of the two, and his resignation is due in large part to his conviction that the party lacks proper socialist rigor.

1983: March   Star Wars speech

U.S. President Ronald Reagan delivers his famous "Star Wars" speech, in which he declares Grenada to be a threat to U.S. national security.

1983: July   Caribbean Basin Initiative

The United States Congress approves the Caribbean Basin Initiative (CBI), which goes into effect in January 1984. CBI is a regional assistance package for the Caribbean designed to isolate socialist-leaning governments by encouraging private enterprise through trade and tax incentives. Some of its main components include free trade with the United States and economic incentives to U.S. business investing in the Caribbean. Supporters herald CBI as an economic windfall; detractors describe it as a masked form of U.S. imperialism. Most nations in Central America and the Caribbean are invited to participate, with the exceptions of those deemed to be communist. Conditions of participation include closer ties to the United States in a variety of legal and economic arenas. Amid growing protectionism (the policy of favoring domestically products by applying tariffs on imports) in the United States by 1986, CBI is considered largely a failure.

1983: October   Bishop executed

With army backing, Coard and his supporters arrest Bishop upon his return from a trip to Eastern Europe. Bishop is freed a few days later by supporters, but Coard sends military units after him. Bishop and three of his cabinet ministers are detained and executed on October 19. For the next week Coard and a Revolutionary Military Council attempt to secretly govern the nation.

1983: October 25   United States leads invasion

The United States invades Grenada with 7,000 of its own troops accompanied by a token 500 from other Caribbean islands. U.S. President Ronald Reagan's administration explains the motives behind the invasion as the "rescue" of a group of American medical students, the restoration of democracy, and the eradication of a Soviet-Cuban threat. A modicum of resistance is put forth by a few Grenadians, and approximately 650 modestly armed Cubans working on the airport. Twenty Americans are killed during the invasion, but nearly half of those casualties are not combat related. Approximately twenty-five Cubans and forty-five Grenadians, including eighteen inmates of an insane asylum that is accidentally bombed, are killed. Logistically, the invasion is an embarrassment. The four branches of the U.S. military bicker with one another, and intelligence and communication is sorely lacking. Bernard Coard is arrested to be turned over to a new Grenadian government for trial. The United States will occupy the island for the next eighteen months with the final troops leaving in June 1985. U.S. money will eventually be provided to complete construction of the airport.

1984: December   Elections

Elections held during United States' occupation are won by the conservative New National Party (NNP) and its leader Herbert Blaize.

1986   Faction leaves NNP

Five members of Parliament and two cabinet ministers leave the NNP government and form the National Democratic Congress, which becomes the official opposition party.

1986: December   Coard receives death sentence

Coard and his Jamaican wife Phyliss are sentenced to death by a Grenadian court.

1990: March Elections

The opposition National Democratic Congress defeats the New National Party in national elections. It forms a coalition with the minority GULP, the party once controlled by Eric Gairy, to form a majority coalition. Nicholas Brathwaite (b.1925) becomes prime minister.

1994 Mealy bug found

The first pink mealy bug is found in Grenada. This species of bug has the capacity to destroy foliage at an alarming rate. Its discovery on the island raises concerns over the economic impact of an infestation.

1995: June Elections

The NNP wins a majority, and Dr. Keith Mitchell becomes prime minister. NDC leader George Brizan is the opposition leader.

1997: May Meeting with President Clinton

Prime Minister Mitchell joins President Clinton and fourteen other Caribbean leaders in Barbados in the first-ever United States-regional summit.

1997: August Deputy prime minister dismissed

The Mitchell government dumps Deputy Prime Minister Grace Duncan, causing political tremors to pass through the nation.

1997: December Hotel construction

Plans are laid for a $140 million luxury hotel on the bombed-out headquarters of the regime that was overthrown in the U.S. invasion. The hotel is a cornerstone of the government's new tourism initiative.

1998: June Rastafarians in the UN

A group of Rastafarians from the Caribbean, including a delegation from Grenada, go before the United Nations (UN) to request that their use of marijuana be recognized as a religious practice.

Rastafarianism begins in Jamaica as an Afrocentric movement in response to growing social inequality. The beliefs are based on the teachings of the Jamaican Marcus Garvey (1887–1940), who in the 1920s advocated African racial pride and the eventual return of blacks to Africa. Garvey believed that an African king would rise up and rescue Africans from oppression. In 1930, Ras Tafari (1892–1975) of Ethiopia is crowned Emperor Haile Selassie I (r. 1930–36, 1941–74). Garvey's adherents believe the Ethiopian emperor is the king that he had foreseen. Emperor Haile Selassie claims to be a direct descendent of King David. The Rastafarians reject the European culture that is imposed by colonists and seek a simpler life. By the late 1970s, Rastafarians are known internationally for two of their characteristics: wearing their hair in dreadlocks (long cords of twisted and uncombed hair) and *ganja* (marijuana). The smoking of marijuana is a sacrament in Rastafarianism.

1998: August Castro visits

Fidel Castro visits Grenada, the first time the Cuban leader has been to the island since the U.S. invasion in 1983.

Bibliography

Craton, Michael. *Testing the Chains: Resistance to Slavery in the British West Indies.* Ithaca: Cornell University Press, 1982.

Devas, Raymond. *The Island of Grenada, 1650–1950.* St. George's, Grenada: s.n., 1964.

Emmanuel, Patrick. *Crown Colony Politics in Grenada, 1917–1951.* Cave Hill, Barbados: Institute of Social and Economic Research, University of the West Indies, 1978.

Garvin, Glenn, "Castro, in Grenada, Revisits Cuban Role." *Miami Herald,* August 2, 1998.

Gunson, Phil, et. al., eds. *The Dictionary of Contemporary Politics of Central America and the Caribbean.* New York: Simon and Schuster, 1991.

Luxner, Larry. "Ailing Grenada Struggles to Move Ahead." *Washington Times,* December 9, 1997, p. A15.

McDaniel, Lorna. "The Concept of the Big Drum Dance of Carriacou, Grenada." In *Musical Repercussions of 1492: Encounters in Text and Performance,* ed. Carol Robertson. Washington, D.C.: Smithsonian Institution Press, 1992.

Rouse, Irving. *The Tainos: Rise and Decline of the People Who Greeted Columbus.* New Haven: Yale University Press, 1992.

Schoenhals, Kai, and Richard Melanson. *Revolution and Intervention in Grenada: The New Jewel Movement, the United States and the Caribbean.* Boulder, Colo.: Westview Press, 1985.

Thorndike, Tony. *Grenada: Politics, Economics and Society.* Boulder, Colo.: Lynne Rienner Publishers, 1985.

Weeks, John, and Peter Ferbel. *Ancient Caribbean.* New York: Garland Publishing, 1994.

Guatemala

Introduction

Guatemala is often portrayed in stark, contrasting terms. A nation of aesthetic splendor, ethnic diversity, and rich cultural history, it has also been marked by intense ethnic hatred, economic impoverishment, and brutal authoritarian regimes. Guatemala is distinguished by being part of the historic homeland of the Maya, one of the largest and most advanced Amerindian populations. When the Spaniards arrived to conquer Guatemala in the 1520s, Mayan civilization was in decline. But the last five centuries of Guatemalan history have been dominated by the aftereffects of that clash between Europe and America.

Characteristics and Culture

Guatemala is located on the northern edge of the Central American isthmus (land bridge) connecting the continents of North America and South America. It is bounded on the north by Mexico, on the south by El Salvador and Honduras, on the west by the Pacific Ocean and on the east by Belize and the Caribbean Sea. With an area of approximately 42,000 square miles (109,000 square kilometers), it is roughly the size of the state of Ohio. It is the largest and most populous nation in Central America with approximately 10.5 million people.

Ethnicity is a dominant issue in Guatemala. Slightly over half of Guatemala's population is Indian, descendants of the Maya. The other half is defined as "Ladino," persons of mixed European and Amerindian heritage. Culture separates these two groups. Ladinos speak Spanish and consider themselves "Westernized." Indians speak one of twenty-three Indian dialects as their first language, although many speak some Spanish. They also wear distinctive clothing and have a culture that combines Western and indigenous elements.

Indians dominate the ranks of Guatemala's poorest people. They live predominantly in the rural areas; only twenty percent of Guatemala's urban population is Indian. Most Indians are peasants (*campesinos*), small farmers who survive on small plots of land (*milpas*) on which they grow crops such as corn and beans for personal consumption. Ladinos dominate the political and economic life of Guatemala. Although many

Ladinos are sympathetic to the plight of the Indian majority, there is a strong tendency in Ladino society to consider Indians as inferiors. During the civil war of the last four decades, Indians suffered tremendously at the hands of the Ladino-controlled government and army.

Guatemalan history can be divided into five eras: Preclassic (2,000 B.C. to A.D. 300); Classic (300 to 900); Postclassic (900 to 1530); Colonial (1530 to 1821), and Independence (1821 to the present). The Preclassic period is defined by the transition in Indian society from simple hunting and gathering to the greater complexity of sedentary agriculture, hierarchical social systems, and urban settlements. This period was the foundation upon which Mayan civilization would experience its most productive stage during the Classic period.

The Classic period is known as the Golden Age of the Maya. It is characterized by rapid advances in artistic and intellectual pursuits, highly dense population settlements, and far-reaching economic exchanges. At its height, the Mayan world stretched from Central Mexico to El Salvador and Honduras. There was considerable uniformity throughout this world, especially in culture, but there was also significant regional differentiation, particularly in politics. The Mayans did not have a centralized empire. Rather, loosely affiliated kingdoms ruled over disparate regions. Most studies estimate that the Mayan population was well over one million people at this time.

The Postclassic period marks the decline of the Maya, characterized by the abandonment of major urban areas and the cessation of cultural advancement. Prevailing theories explain this decline as a product of a rapid population growth overtaxing the land. This resulted in a shortage of resources, mainly food, which in turn caused conflict and warfare between the various Mayan groups. The Postclassic Period ended with the arrival of the Spaniards.

The Spaniards arrived in Guatemala in 1524 in the form of a conquering army led by Pedro de Alvarado, the lieutenant of Hernán Cortés, conqueror of the Aztec Empire in Central Mexico. Although the Maya had surpassed their cultural peak, they were still numerous and well organized. Alvarado and his army defeated them in a series of bloody battles between 1524 and 1530. The Spaniards then settled down to

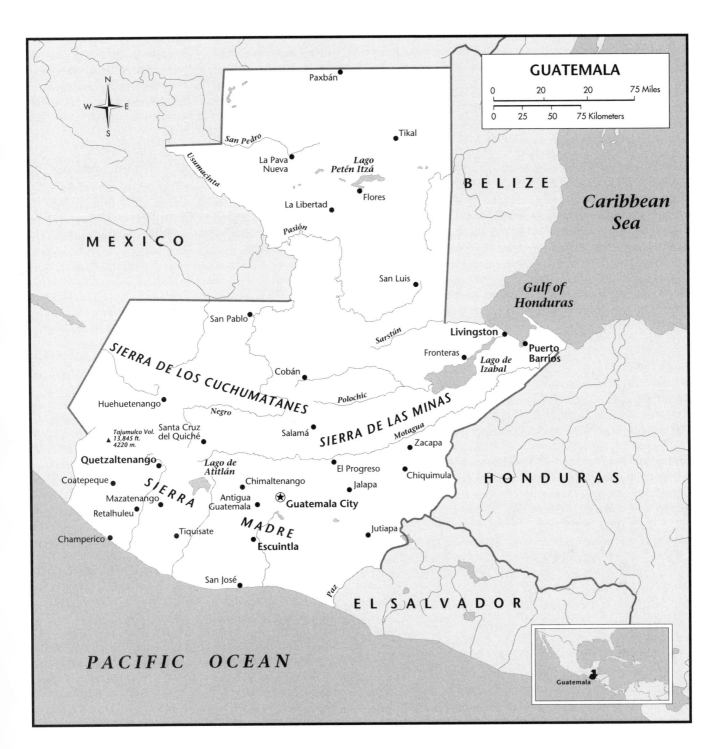

administrating the region on behalf of the Spanish Crown and pursuing economic ventures through the use of Indian labor. Throughout the colonial era, the region of Central America was nominally under the control of New Spain (Mexico), Spain's primary outpost in the Americas. But the region of modern-day Guatemala emerged as the administrative and economic center of Central America. This was a source of conflict with the rest of the region. Never a source of major wealth for the Spanish Crown, Central America did produce a few viable crops such as indigo (a blue dye), cochineal (a red dye), and cacao (chocolate). Merchants in Guatemala controlled the trade in these products.

Central America gained its independence from Spain in 1821 for reasons similar to the rest of Latin America—a growing sense of American identity among the *Creoles* (persons of Spanish descent born in the Americas) and irritation with Crown policies on taxation and administration. The regions of Central America founded a short-lived federation, but colonial-era rivalries predominated and in 1839 the federation collapsed, giving rise to the five separate nations of Central America, including Guatemala.

Nineteenth-century Guatemala was characterized by desperate attempts to create a viable government amidst economic stagnation. The result was incessant political instability and warfare with its Central American neighbors. In the latter part of the nineteenth-century, the world market for primary goods increased dramatically and Guatemala began producing coffee for export. Government centralization followed in tandem. However, the coffee industry had an inherent conflict. The highland region suited for coffee production was an important area of food production for Indian communities. Over the next decades, the Ladino-dominated government enacted land privatization and labor drafts that removed land from the Indian communities and forced Indians to serve as laborers on the new plantations.

Guatemala also turned out to be suitable for banana production. In the first half of the twentieth century, an expansive U.S. company named United Fruit grew bananas for export to markets in the United States. The coffee and banana industries allowed a few people to become very wealthy, but Guatemala's political system remained autocratic, and the majority of its population continued to live in squalid conditions.

A potential respite came in 1944 when a reform-oriented government came into power under Juan José Arévalo. For the next ten years (1944–1954), during a period called "the Decade of Spring," Arévalo and his successor, Jacobo Arbenz, initiated social reforms. But Arbenz was overthrown in a military coup in 1954, aided by the Central Intelligence Agency (CIA) of the United States. Thereafter, Guatemalan politics was dominated by a series of authoritarian military regimes. They overturned the reforms and initiated counterinsurgency campaigns against political opponents. The harshness of these regimes inspired the formation of left-wing guerrilla movements (bands of irregular rebel forces that mount hit-and-run attacks), which the military contended were supported by Cuba and the Soviet Union.

For the next four decades, Guatemala was mired in one of the most devastating civil wars in Latin America. The small guerrilla armies survived in the rugged terrain, but Indian communities were left exposed to military assault. Conservative estimates place the number of Indians killed at more than 100,000 and the number of persons displaced at more than 1,000,000. Despite this large loss of life, the war was a stalemate, with neither the military nor the guerrilla movements able to claim victory.

In the early 1990s, the government underwent a transition in which more reform-minded persons rose to prominence. In 1996, the government and the guerrillas signed a peace accord (agreement) that ended the fighting, but not without serious disagreement as to Guatemala's future. The accord held out promise that an era of peaceful growth would commence. Fortunately for Guatemala, civil war did not devastate the economy and infrastructure to the extent of other war-torn countries such as Angola and Afghanistan. But Guatemala is still a developing country with stark inequalities in wealth and an economy dependent on traditional sectors. Statistics show that during the 1980s the country's poverty rate increased and the extreme poverty rate, defined as those persons unable to meet basic needs, doubled. The government has attempted economic diversification with some success, but to date the fastest growing sector is the service industry, reflecting the growing importance of tourism. With its many Mayan ruins and so much natural beauty, Guatemala has much to offer tourists from wealthier countries in North America and Europe.

Timeline

c. 11,000 B.C.   Earliest people

The first hunting-gathering peoples settle in the highland and lowland regions of the modern Maya.

3114 or 3113 B.C.   Mayan beginning of time

The Maya Long Count calendar, which situates events relative to an arbitrary zero point in the distant past, cites this date as the creation of the world.

2500 B.C.   Proto-Maya

Mayan civilization emerges. It is characterized by a degree of uniformity in culture, including a language now-called "Proto-Mayan," the sole parent language of all modern-day Mayan dialects.

2000 B.C.   Rise of Olmec

This year marks the rise of the Olmec civilization along the Gulf coast plain of modern-day Veracruz and Tobasco, Mexico. It is characterized by social hierarchies, centralized political power, and monumental architecture. Mayan culture will demonstrate Olmec influence.

1500 B.C.   Farming predominates

Settled farming has become the predominant economic pattern throughout Mesoamerica.

900 B.C.   Mayan settlements interact

Characterized by scattered regional settlement until this time, the Mayans begin to exhibit larger circles of interaction such that a "Maya world" can be said to exist. Closer economic ties, political interaction, and greater uniformity of culture is evident.

400 B.C.   Solar calendars in use

Carved in stone, the earliest known solar calendars are in use and may have been in use much earlier.

400 B.C. to A.D. 250   Emergence of aristocracies

Mayan society is characterized by the normalization of social hierarchy and rule by nobles and kings. The widening economic gap between emergent aristocracies and ordinary farmers give the aristocrats the capacity to marshal the labor and economic resources necessary to initiate large-scale building and establish greater political control. The Classic Period of Mayan civilization will be built upon this social foundation.

150 B.C.   Construction begins on Tikal's Great Plaza

Construction begins on the Great Plaza in the Mayan city of Tikal. The plaza consists of four stone floors, one on top of the other. The first floor is designed this year, but construction will continue for decades.

Tikal will eventually include thousands of spectacular buildings, as well. The Temple of the Masks will be built facing the Great Plaza. Along another side of the plaza will be rows of tall spires (stelae). Behind these, the North Acropolis, a multilevel complex of temples with a base that covers over two acres, will be built, as well as the Temple of the Giant Jaguar, a 145-foot (44-meter), nine-sided pyramid.

c. 100 B.C.   Decline of the Olmec civilization

Rise of the Izapan culture is seen along the Pacific coast of modern-day Mexico and Guatemala, and the Pacific slopes emerge as the center of early Mayan culture. Sites like Kaminaljuyú, west of Guatemala City, Izapa in Chiapas, Chalchuapa in El Salvador and Abaj Takalik on Pacific slopes of Guatemala develop into dominant centers. Their prominence is probably based on controlling trade in valuable resources like cacao and obsidian. These areas also initiate the development of the classic Mayan sculpture, iconography, and writing. These in turn show influence from the Olmecs.

This period also maks the rise of the city of Teotihuacan in the highlands of Central Mexico and its emergence as the predominant civilization in Mesoamerica. Teotihuacan civilization will come to influence Mayan culture.

A.D. 300–900   Golden Age of Maya

The Golden Age of the Maya was one of advances in artistic and intellectual pursuits and of highly dense population settlements. The Mayan empire included areas from Central Mexico to El Salvador and Honduras. The culture was fairly uniform, despite the fact that the Mayans did not have a centralized empire. Separate ruled kingdoms were loosely connected over the vast area covered by the empire. The Mayan

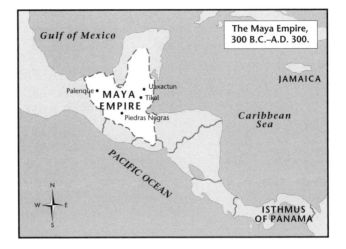

population is believed to have exceeded one million people at this time.

c. A.D. 400   Teotihuacan lords trade with Mayan

Lords from Teotihuacan occupy the city of Kaminaljuyú and establish political and economic exchange with the Maya.

c. A.D. 600   Height of Mayan civilization

The Teotihuacan civilization begins a rapid decline and its outpost at Kaminaljuyú is abandoned. This coincides with some of the main Mayan cities achieving their greatest heights. Palenque, Tikal, and Caracol rise to prominence. Chichén Itzá is founded in the Yucatán region. Copán, located in the foothills of modern-day Honduras, also rises to importance. Copán's rulers claim direct descent from the royal line of Teotihuacan. A Mayan presence is established at Xochicalco and Cacaxtla in central Mexico. Some of these major urban areas have populations of 50,000.

A.D. 682–734   Reign of Ah-Kakaw (or Double-Comb), last of Mayan kings

Ah-Kakaw (or Double-Comb) rules the Mayan empire. Ah-Kakaw reportedly was five feet five inches tall and towered over his subjects. His burial chamber is in the Temple of the Giant Jaguar. (See 1962.)

A.D. 869   Life at Tikal

The Mayans at Tikal, a prominent Mayan city, engage in unique ceremonies. Archaeologists believe that the leaders in Tikal engaged in a deadly ritual ball game—the losers were beheaded. It is also speculated that rulers used thorns to pierce their tongues during a worship ritual.

A.D. 899   Tikal abandoned

Tikal is abandoned. After 650 years and a succession of thirty-nine rulers, the city of Tikal suddenly dies.

Scholars have searched for an explanation, and no single answer has been revealed. However, University of Texas art professor Linda Schele writes in *A Forest of Kings,* "The voiceless remains of the dead, both commoner and noble alike, bear witness to malnutrition, sickness, [and] infection." Many Maya scholars believe overpopulation and environmental problems prevented the society from being able to produce enough food.

c. A.D. 900   Mayan population decline

The core Mayan world in the southern lowlands enters into a rapid decline characterized by the ceasing of most cities from large-scale construction and the carving of sculpture. Population declines and many cities are abandoned.

A.D. 1000   Changes in Mayan empire

As the classic Mayan areas in the south decline, northern areas in the Yucatán, such as Chichén Itzá and Mayapán, rise to prominence. The Quiché kingdom in the west of Guatemala also thrives. All of these are tribute-collecting, conquest states. For the Quiché, the mythic king named Plumed Serpent gains cultural importance.

1244   Chichén Itzá abandoned

Chichén Itzá, a prominent area of Mayan settlement, is abandoned.

1283   Mayapán center of Mayan empire

Mayapán emerges as the dominant center of the Mayan world in the Yucatán.

1441   Rebellion in Mayapán

A rebellion in the city of Mayapán causes it be abandoned. Mayan society in the Yucatán degenerates into more than a dozen rival states, each striving for prominence.

Colonial Era

1519–1521   Epidemic among Indians

An epidemic believed to be smallpox, but possibly measles, breaks out among the Cakchiqueles and the Tzutujiles, two of the three main Mayan groups in Guatemala. The disease spread inland from Spanish sailing expeditions along the Caribbean coast. The epidemic weakens Indian capacity to resist the Spaniards who arrive just a few years later.

1523: December 6   Alvarado embarks for Guatemala

Pedro de Alvarado (1458?–1541), the leader of a Spanish expeditionary force, departs Mexico City for Guatemala. His army consists of 120 cavalry, 300 foot soldiers and a number of Indian allies from Mexico estimated between several hundred and several thousand.

1524: February 24   Alvarado arrives

Alvarado's army enters the territory of the Quiché, the third main Mayan group. Alvarado forms an alliance with the Cakchiqueles. In the last week of February, Alvarado's forces fight and win a heated battle against thousands of Quiché led by Tecúm-Umán (d.1524). Over the next 100 days, the invading force completes the initial conquest of Guatemala and initiates a pacification program that would last nearly six years.

1524   Chichicastenango becomes spiritual center for Quichés

After their defeat by Alvarado, Chichicastenango becomes a spiritual center for the Quiché. Called Santo Tomas by the Spaniards, the town has winding streets leading to a main plaza. (See 1542.)

1524: April   Defeat of the Quiché

In the early days of April, on plains outside the city of Quetzaltenango in the northeastern highlands, Alvarado's forces carry out the final defeat of the Quiché in a large battle. Many of the surviving Indians are branded and enslaved.

1524: July 25   First Spanish capital founded

Alvarado founds the capital city of Santiago de los Caballeros de Guatemala on the Cakchiquel capital of Iximiché and appoints the first *cabildo* (town council). This is the beginning of formal Spanish administration in the region.

1524–1525   Conflict with the Cakchiqueles

Angered by demands for heavy tribute payments in the form of gold, the Cakchiqueles rise up in a prolonged insurrection under the leadership of Sincam. The Quichés and Tzutujiles reject a Cakchiquel offer of unity and ally instead with the Spaniards. By 1525, the insurgents have been largely subdued, but Sincam is not captured.

1526: November 22   Sincam captured

Continuing to resist ongoing Indian rebellions marked by constantly shuffling alliances between the three main Mayan groups, the Spanish capture Sincam, many of his followers, and another rebel leader, Panagel (d. 1526), who is publicly executed.

1527   First Guatemala city founded

Pedro de Alvarado, conqueror of Guatemala (see 1524: April), founds a city on the banks of the Rio Penssativo at the foot of the volcanic mountain. This settlement will come to be known as Ciudad Vieja (Old City), and will later be named Agua (see 1541).

1527: December 18   Alvarado named governor

Emperor Charles V (1500–1558) of the Holy Roman Empire (who was formerly Charles I of Spain) appoints Alvarado as governor and military commander of the region stretching from Chiapas in southern Mexico to Costa Rica—the lands that eventually will constitute the Audiencia of Guatemala. Alvarado serves only infrequently as governor, taking time off to embark on adventures and military conquests designed to bring him greater wealth and fame. Most of these end badly.

1530   End of Indian-Spanish contact

Remaining Indian rebels are defeated or surrender, marking the end of the first phase of Indian-Spanish contact. Despite ongoing and sporadic rebellion, and large areas of Guatamala still outside Spanish control, at this time it can be considered conquered.

1537: April 8   Diocese of Guatemala established

Upon authorization by Pope Paul III, The King of Spain founds the Diocese of Guatemala, formalizing the presence of the Catholic church in the Central American region and beginning the close church-state relationship characteristic of the entire Spanish empire. Father Francisco de Marroquín (1499–1563) is consecrated as the first bishop.

1540: June 10   Congregación begins

The policy of Congregación is begun. This is a policy ordered by the Spanish crown to relocate Indians into towns where they can be more easily administered and taught Christianity. By the 1570s, some 300 of these towns will have been established with approximately 45,000 Indian inhabitants.

1541   Agua, a principal mountain peak, floods and gets its name

Agua destroys the capital of Guatemala, Ciudad Vieja, with a devastating flood from its overflowing crater. The Spanish then name the city Agua, Spanish for "water." Over 1,000 inhabitants are drowned, including the governor, Dona Beatriz de la Cueva. Following the flood, the city of Santiago de Guatemala (modern-day Antigua) is founded to be the center of government. (See 1542)

1541: July 4   Alvarado dies

Alvarado is killed after being crushed by the falling horse of one of his followers. His death occurs during a battle in the Mixtón War in Nueva Galicia in New Spain (Mexico).

1542   Monastery founded at Chichicastenango

A Dominican monastery is founded at Chichicastenango. The famous *Popul-Vuh* manuscript of Maya-Quiché mythology is discovered there (see 1550.)

1543: May 1   New Laws of the Indies

The New Laws of the Indies are established to defend Indians against abuse by Spaniards. Officially proclaimed on May 1, 1543, the laws reflect debate within the Spanish monarchy over the nature of its Indian subjects. While some consider Indians to be sub-human and thus worthy only of slavery, others, such as famed defender Bartolomé de Las Casas (1474–1566), a Dominican priest, argue that Indians have a soul and should be granted legal rights.

At the center of the debate is the institution of *encomienda*, a cornerstone of Spanish conquest practices. An *encomienda* is a grant of Indian labor to a loyal Spanish subject, such as a conqueror. The individual receiving the grant, the *economendero*, has the right to use the Indian labor as he sees fit with the agreement that he will instruct the Indians in Christianity. In reality, *encomienda* becomes the source of manifold abuses against the Indians.

Witnessing the rapid decline in Indian population, and the corresponding fall in wealth and taxes that those Indians represent, the Crown is apt to agree with arguments that Indians should be protected. The result is the passing of the New Laws, which rule that no further *encomiendas* will be granted. Although never rigorously enforced, the New Laws nonetheless represent an important shift in the legal basis of Spanish authority in the Americas.

1542   Capital moved

The capital city is moved for a third time to a site in the valley of Panchoy named Santiago de Guatemala, now the modern-day city of Antigua.

1542: November 20   Audiencia founded

The king of Spain officially creates the Audiencia de los Confines, the administrative unit of Central America. It would be given various names until 1570 when it officially became the Audiencia of Guatemala. It consists of lands stretching from Chiapas, Yucatán, and Tobasco in Southern Mexico to Panama. The initial capital is located at Gracias in Honduras. But the capital is moved to Santiago de Guatemala (Antigua) in 1549 and Guatemala thereafter becomes the administrative and economic center.

c. 1550   *Popul Vuh* transcribed

The ancient book of Quiché tradition, *Popul Vuh,* is transcribed by unknown Indian authors in the western highlands. The *Popul Vuh* contains the laws and the story of origin of the Quiché people. In the pre-Spanish era, Quiché lords consulted the book during council as a foundation for decision making. They referred to it as the "Council Book," "Our Place in the Shadows," or "The Dawn of Life." The authors of the sixteenth-century transcription suggest that their purpose in transcribing the book at the time was to preserve an essence of Quiché culture amidst the Spanish domination and the teachings of Christianity. It is unclear to what extent they were working from an actual document or from their own memory based on oral tradition. This transcribed copy disappears until the turn of the eighteenth century when a friar named Francisco Ximénez (1666–1729) finds it in the city of Chichicastenango. He makes the only surviving copy and adds a Spanish translation.

1570   Territories removed from Audiencia of Guatemala

Tobasco, the Yucatán, and Panama are removed from the administrative authority of the Audiencia of Guatemala.

1594   Cristo Negro commissioned

The Chorti Indian inhabitants of the town of Esquipulas in southeastern Guatemala raise the equivalent of fifty ounces of silver from the sale of cotton grown by their community in order to commission the carving of a sacred religious idol celebrating Nuestro Señor de Esquipulas, commonly referred to as the Cristo Negro (Black Christ). Esquipulas was founded for Chorti Indians in the 1560s near a shrine that the Indians believed to have health-giving qualities. When the Catholic Church outlawed the worship of Indian idols in 1578, the local inhabitants bridged Catholic and indigenous beliefs in the form of the Cristo Negro. The creator of the image was Quirio Cataño (d. 1595), a noted sculptor of Antigua. He completed the image, made of balsam and orangewood, in 1595. Thereafter, the Cristo Negro became a focal point of worship for the townspeople and for pilgrims from all over Central America who believed in its curative qualities. The Catholic Church's later recognition of the symbol signified the blending of Catholicism with indigenous beliefs.

17th century   English explorers mine for gold

Gold washing is carried on by English miners near the coast and along the Motagua river. The region near the mouth of the river becomes known as the Gold Coast.

1676: January 31   University of San Carlos founded

The Spanish king, Carlos II (1661–1700), authorizes the founding of the University of San Carlos in the capital city Santiago de Guatemala (Antigua). Classes begin in January, 1681. Only those of pure Spanish blood who attest their devotion to Catholicism are allowed to enroll. As the premier institute of higher education in Central America, the University becomes a center of Enlightenment, liberal thinking, in the late 1700s, which in turn feeds directly into support for independence from Spain. Over the course of the colonial period, the university will grant 2,415 degrees, including 206 doctorates.

1695   Priest visits surviving Maya

A small number of Mayan communities survive. Father Andres de Avendano y Layola, a Roman Catholic priest, vis-

its Maya king Can-Ek on the island of Tayasal (site of the modern-day city, Flores). Father Avendano converts the king to Christianity. The priest, who had learned to read the Mayan language, uses his understanding of Mayan history in his teachings about the Roman Catholic faith.

1697   Tayasal conquered by Spanish

Tayasal, an island city in Lago Petén Itzá (Lake Petén Itzá), is conquered by the Spanish. They destroy all vestiges of Mayan culture in Tayasal, which is one of the last Mayan cities. The Spanish found the city of Flores on the site.

1730   Cathedral erected on Plaza Major in Guatemala City

A major Roman Catholic cathedral is erected in the center of Guatemala City.

1731   Birth of poet Rafael Landivar

Considered Guatemala's first poet, Rafael Landivar (1731–93) is born in Santiago de los Caballeros (modern-day Antigua Guatemala), the first capital. His father is a supplier of gunpowder to the militia and a manufacturer of fireworks. Rafael is educated at Saint Luke's College run by Jesuits (priests belonging to a Roman Catholic religious order). When the Jesuits are forced to leave Guatemala, Landivar is among them. (see 1767). His epic poem is published in Bologna, Italy, in 1782, and is titled *Rusticatio Mexicana* (A Walk in the Mexican Countryside)

1737   Religious architecture of Santiago de los Caballeros

In an area of about two square miles, there are more than thirty churches, chapels, and oratories in Santiago de los Caballeros, a city with an estimated population of 12,000–15,000. These churches feature ornate plaster designs, with depictions of intertwining vines and leaves on facades and columns. The altars inside are heavily carved and gilded. When Santiago de los Caballeros is destroyed by an earthquake (see 1773), seventeen gilded altarpieces and other treasures from the Church of La Merced are relocated to the Church of La Merced in Guatemala City.

1737   Church recognizes Cristo Negro

The Catholic Church officially recognizes the curative powers of the carved image of the Cristo Negro, in part because Bishop Fray Pedro Pardo de Figueroa claims to have been cured of a contagious disease at the Esquipulas site.

1742   Archdiocese founded

Pope Benedict XIV founds the Archdiocese of Guatemala, giving Guatemala greater church authority in the Central

American region. Bishop Frey Pedro Pardo de Figueroa becomes the first Archbishop in 1743.

1758   Completion of Esquipulas Church

Construction of the Church of Esquipulas is completed. A white structure in baroque style, the church was commissioned by Archbishop Pardo de Figueroa to house the image of the Cristo Negro.

1765–66   Monopoly on tobacco and alcohol

The king of Spain establishes royal monopolies over the production and sale of tobacco and *aguardiente* (alcohol). Seen by the Crown as an administrative reform to enhance control and increase revenues, the monopoly also intensifies local opposition to Spanish authority.

1767: June   Expulsion of the Jesuits

As part of an internal conflict between the Catholic church and the Spanish Crown, the Jesuit order is expelled from all Spanish possessions in the New World. The twenty Jesuit residents in Guatemala are gathered together by colonial officials on the morning of June 26 and sent away on horseback on July 1. Among them is Guatemala's first poet, Rafael Landivar (see 1731).

1773: July 29   Earthquake destroys capital

The capital city of Antigua (first known as Santiago de los Caballeros la Nueva, see 1542) is destroyed in an earthquake. This prompts the relocation of the capital to Ermita, today's Guatemala City. (See 1776.)

1776   Guatemala City founded

When Antigua is destroyed by an earthquake (see 1773: July 29), the site of the small settlement of Ermita, twenty-seven miles (forty-three kilometers) northeast of the former capital of Antigua, is chosen as the new seat of government. The transfer of the capital is completed in 1779. The new capital city is planned carefully, with the streets laid on a simple grid. In order to minimize the potential for damage from earthquakes, (see 1874) ordinances restrict the height of buildings to a maximum of twenty feet (six meters).

1779–82   Military reorganization of Audiencia

A campaign to shore up the military defenses of the Audiencia is initiated under the direction of Governor Matías de Gálvez (1717–83). The campaign is part of a broader reform movement (the Bourbon Reforms) undertaken by the Spanish Crown to increase fiscal and administrative authority in the American possessions. In Central America, the specific goal is to counter threats to Spanish sovereignty by the British and to end contraband trade.

1794   Consulado de Guatemala founded

The king of Spain authorizes the establishment of the Consulado de Guatemala, a merchant guild, to regulate trade and promote economic development throughout the Audiencia of Guatemala. The Consulado will become a powerful political and economic force. As Guatemala is already the administrative and economic center of Central America, the establishment of the Consulado exacerbates regional rivalries.

1794   Publication of *Gazeta*

The first periodical published in Central America, the *Gazeta de Guatemala,* begins publication in Guatemala City under the editorial direction of Ignacio Beteta (1757–1827). Suspended briefly in 1795, it resumes publication in 1797. In 1812, its name is changed to *Gazeta del Gobierno de Guatemala.* The periodical is pro-independence and will continue to be published after independence is declared in 1821.

1794–1795   Sociedad Económica founded

The Sociedad Económica de Amigos del País is chartered and founded in the capital city. With Central America's most prominent liberals as its members, the Sociedad is dedicated to such ideas as free trade, science, and the reformism then current in Spain. It is suppressed for its liberalism on July 14, 1800, but is reinstated in 1811.

1801   Indigo declines

Competition from Venezuela and British India causes the trade in indigo, the main export crop of Guatemala, to enter an extended decline. This contributes to a prolonged period of economic stagnation in the region.

1801–11   González as governor

General Antonio González Mollinedo y Saravia serves as the highest-ranking Spanish authority in Central America, bearing the title of Governor, Captain General (military commander), and President of the Audiencia of Guatemala. His tenure is marked by growing conflict in Spain and corresponding instability in Guatemala as support for independence increases.

1806   Seaport of Livingston founded

Named for the author of Guatemalan laws, the city of Livingston is founded at the mouth of the Rio Dulce on the Gulf of Honduras on the Caribbean Sea.

1807–08   Napoleon invades Spain

Napoleon Bonaparte (1769–1821) and his army invade the Iberian Peninsula, forcing the Spanish monarch Ferdinand VII (1784–1833) to abdicate. Napoleon's brother Joseph Bonaparte (1768–1844) is established as King of Spain, which causes an administrative crisis in Spanish America and lends support to calls for independence.

1811–18   Bustamante as governor

General José de Bustamante y Guerra (1759–1825) replaces Antonio González (see 1801–11) as Governor, Captain General, and President of the Audiencia of Guatemala. His tenure is marked by a power struggle with the *cabildo* (town council) of Guatemala City, which is defending the interests of liberal merchants and their desire for less regulation by the Spanish Crown. Bustamante earns a reputation as being a forceful and sometimes repressive defender of Spain's authority.

1811: November 11   Pro-independence rebellion

A rebellion breaks out in San Salvador (the modern day capital of El Salvador). The rebellion is led by Dr. Matías Delgado (1767–1832) and Don Nicolás Aguilar (1742–1819) against the local Spanish authority, Antonio Gutiérrez Ulloa. Governor Bustamante in Guatemala City sends a military force to restore order and return Crown officials to office.

1812   Cádiz Constitution

Spain's defenders against Napoleon's invasion gather in the town of Cádiz and draft the liberal Cádiz Constitution. It places limits on the power of the Spanish monarch. It will serve as a model for many of the constitutions of the newly independent countries in Spanish America.

1814   Ferdinand VII returns

Ferdinand VII is restored as the Spanish monarch. He immediately abolishes the Cádiz Constitution of 1812, setting off six years of conflict over the future of government in Spain and its American possessions.

1818–1821   Urrutia as governor

General Carlos Urrutia y Montoya replaces Jose Bustamante (see 1811–18) as Governor, with the title having been changed to Jefe Político Superior. His tenure marks the final stages of Spanish administration in Guatemala.

1820: January 1   Restoration of Cádiz Constitution

Colonel Rafael Riego (1785–1832), with troops assembled at Cádiz, revolts against Ferdinand VII. He restores the Constitution of 1812 and secures Ferdinand's declaration of support to it and the idea of a constitutional monarchy.

1821: September 15   Independence declared

Influenced by the Riego revolt in Spain and the declaration of independence in Mexico on April 21, leading citizens gather in the capital city of Guatemala and declare independence from Spain. The declaration comes in the form of the "Act of Independence of the United Provinces of Central America" which consists of nineteen articles.

The region of Chiapas leaves Central America and declares itself part of Mexico.

Independent Era Chronology

1822: January   Annexation to Mexico

Under the leadership of Emperor Agustín Iturbide (1783–1824), the newly independent Mexican Empire announces the annexation of Central America and sends a force of 600 men to Guatemala City to defend its claim.

1823   Monroe Doctrine

United States announces the adoption of the Monroe Doctrine. Technically, the doctrine declares the Western Hemisphere to be free from the colonial designs of European nations, but in actuality it serves as the foundation for the United States's own semi-colonial interests in the region.

1823: February   Mexican political upheaval

Mexican troops crush an opposition movement in San Salvador; Emperor Iturbide is deposed in Mexico, which brings the Empire to an end.

1823: July 1   Independence from Mexico declared

The United Provinces of Central America declare their independence from Mexico.

1824: November 22   Government established for Central America

This date marks the official declaration of the federal government of Central America with a constitutional mandate.

1825–50   Growth of cochineal exports

Cochineal, a red dye, is a primary export until 1850, when competition from chemical dyes erodes the foreign market.

1831   Museum founded in Guatemala City

A national museum is founded in Guatemala City. It is run by the Sociedad Economica, which is led by scholars and historians.

1837–39   Carrera Rebellion

A highland Indian, Rafael Carrera (1814–1865), leads a predominantly Indian rebellion against the government. His troops march triumphantly into Guatemala City for the first time on February 1, 1838, and return again one year later, beginning Carrera's rise to power, a rise that will culminate in his occupation of the presidency for more than two decades. The rebellion is caused in large part by Indian opposition to the government's land and taxation policies, which capture some of the basic tensions in the young federal government. Dominated by liberals, the government had been adopting policies such as land privatization and individual taxation that struck at the Indian communities. The conservatives, as their name implies, oppose such changes and instead support the traditional, albeit paternalistic, protection of Indians and their communal land rights. While no Ladino politicians, whether liberal or conservative, relish an Indian rebellion, the conservatives find Carrera's views more palatable and are quick to cultivate ties with him. Carrera's rebellion initiates the demise of the federation and places Guatemalan politics on a conservative trajectory that will last until the 1870s.

1839   Federation breaks down

The Central American Federation disintegrates amidst years of incessant warfare and regional conflict. The five modern-day nations of Central America emerge: Guatemala, Honduras, El Salvador, Nicaragua, and Costa Rica.

Guatemala claims to have inherited rights to Belize from Spain. Britain counters with its own claim to Belize.

1844   Sevilla la Nueva founded

Sevilla la Nueva, a Spanish settlement near the Gulf of Honduras on the Caribbean Sea, is founded. It will not survive into the twentieth century, however. Livingston will become the main port.

1844–65   Carrera as President

Carrera holds the office of the Presidency for this entire period, with the exception of the years between 1848 and 1851.

1850   Treaty between the United States and Britain

The United States and Britain sign a treaty in which Britain agrees to refrain from occupying any part of Central America, except Belize, because of its claims to prior settlement. Guatemala signs the treaty, claiming that all parties agreed to build a road to the Caribbean coast.

1858   National Theater founded in Guatemala City

The National Theater is founded in Guatemala City and earns a reputation as one of the best theaters in Central America.

1859   British sovereignty over Belize

Guatemala recognizes Britain's sovereignty over Belize.

1870   Coffee leading export

Coffee surpasses cochineal as Guatemala's most valuable export crop and will remain a cornerstone of the Guatemalan economy.

1870–1900   Economic growth

Inspired by the rapid increase in the demand for primary goods as a result of the industrial revolution in Europe and North America, the Guatemalan economy grows. During this thirty-year period, Guatemala's international trade increases twenty times.

1870   Pacaya erupts

A group of volcanic peaks erupt in a series of volcanoes.

1871   Liberal ascendance

Liberals Miguel García Granados (1809–1878) and Justo Rufino Barrios (1835–1885) overthrow the conservative regime and initiate what will come to be called the "Liberal Revolution" in Guatemala. Thereafter, development initiatives are marked by land privatization, coffee cultivation, and reliance on export agriculture. The power of the church declines steadily, as does the land holdings of the Indian communities.

1873–85   Barrios era

Justo Rufino Barrios, considered to be one of the primary liberal leaders and known as "The Reformer," serves as president for more than a decade. He governs in a dictatorial style.

1873   Jesuits expelled

Barrios expels the Jesuits, confiscates all their property, and declares the Roman Catholic Church to be nonexistent.

1874   Earthquake damages Guatemala City and Antigua

A massive earthquake causes extensive damage in Guatemala City, Antigua, and the surrounding region. The ordinances of 1779 restricting the height of buildings to a maximum of twenty feet (six meters) have been ignored over time. Thus, tall steeples and belfries atop churches topple over in the earthquake. (See 1779.)

1877   New land and labor laws affect Indians

After a series of haphazard decrees at both the national and regional levels concerning land tenure, the national government passes a comprehensive land law. Its purpose is to "rationalize common land" by selling off state-owned properties and forcing the privatization of all communally-owned properties. As the two largest possessors of communal land, the Roman Catholic Church and Indian communities suffer the most. Many Indians gain title to former communal plots in the upper highlands, but most of the communal lands in the potentially rich coffee-growing regions on the Pacific slopes will be lost to Ladino speculators.

The government also passes labor laws legalizing debt peonage and *mandamiento*. Debt peonage is a process in which loans are advanced to Indians, who then have to work off payment with labor. Low wages make repayment difficult, and often an individual has to acquire more debt to buy basic necessities. Debts are heritable, so children are bounded by the debt of their parents. *Mandamiento* is a labor system in which an Indian either has to find work voluntarily or be subject to forced labor drafts. A planter requests labor from a regional governor, who then orders local authorities in Indian communities to produce laborers. *Mandamiento* is officially abolished in 1890, but continues to be practiced until the overthrow of Cabrera in 1920. Debt Peonage will be replaced by a Vagrancy Law in 1934.

1878–79   Minor Keith begins banana cultivation

Minor Keith (1848–1929) begins his first banana cultivation in Costa Rica on lands granted to him by the Costa Rican government in exchange for promises to build a railroad.

1879: December   Constitution passed

The constitution creates a national assembly with one deputy (representative) for every 20,000 citizens. The assembly of 1879 includes sixty-nine deputies. They are elected by popular vote for four-year terms. The president is elected to a six-year term. The president has a cabinet of six ministers and a council of state with thirteen members.

1885: February   Barrios announces that he is military chief of Central America

Hoping to unite all the Central American states under his own leadership, President Barrios declares himself to be the military chief and commander of all forces of the five states of Central America (Costa Rica, Honduras, Nicaragua, Salvador, and Guatemala). Salvador, Costa Rica, and Honduras ally to resist his domination.

1885: April 2   Barrios killed invading El Salvador

President Barrios, who wants to become leader of all of Central America, invades El Salvador. He is killed during a battle there.

1885: after April 2   Manuel Barillas becomes president and ends war

Manuel Barillas (1845–1907) succeeds the slain president Barrios and ends the war with El Salvador on April 16. (See also 1906: March.)

1892   General José Maria Reina Barrios elected president

General José Maria Reina Barrios is elected president and serves for five years until the next election. (See 1897)

1893   Majority of population is illiterate

Only eight percent of the population can read and write. Another two percent can read but cannot write.

1894   Land Act

The Land Act divides the undeveloped land of the country, except along the coast, into lots for sale. Limits are placed on the amount of land one person can hold. The maximum acreage one person can purchase is 5,625 acres or 50 *caballerias*. Grants of land, are given free of charge to immigrants or to anyone who promises to build a road or railway through their land to enable travel to the nearest towns.

1895   German archaeologist researches Mayan kingdom of Tikal

German archaeologist Teobert Maler travels to Tikal to research the Mayan ruins there. He sketches and photographs extensively for ten years.

1897   General José Maria Reina Barrios reelected president

General José Maria Reina Barrios is reelected president (see also 1892).

1898: February 8   President Barrios assassinated

General José Maria Reina Barrios is assassinated in 1898. Vice president Morales steps in to fill the position.

1898–1920   Estrada Cabrera era

Another prominent liberal, Manuel Estrada Cabrera (1857–1924), is elected president to fill the term vacated by the assassination of President Barrios (see 1898: February 8) Cabrera serves as president for more than two decades. He and Barrios are considered the architects of Guatemalan liberalism.

1899   United Fruit Company

Minor Keith forms the United Fruit Company by merging his company with that of a main competitor, the Boston Fruit Company.

1900s   Wagon roads and railways provide means of travel

Travel within the country is accomplished by a system of wagon roads, usually by mule. In order to pay for road construction and maintenance, the government charges an annual tax to all men living in the inland towns. In lieu of paying the tax, men may elect to work on the roads for four days each year.

Between Guatemala City, the capital, and Quetzaltenango there is a road suitable for horse-drawn carriages.

The main rail lines include the Southern, owned by a U.S. company and connecting San Jose on the Pacific coast to Guatemala City; the Northern, owned by the government and connecting Guatemala City to Puerto Barrios in the north on the Gulf of Honduras; and the Western, connecting Champerico on the Pacific coast with Quezaltenango.

1900–04 Imports from the United States and the United Kingdom

About one-half of all imports into Guatemala come from the United States and about one-fourth from the United Kingdom. Imports include textiles, machinery, sacks, flour, beer, and wine. Export products are coffee, timber, animal hides, rubber, sugar, bananas, and cocoa.

1901 United Fruit Company contract

The United Fruit Company signs its first contract with the Guatemalan government for banana cultivation.

1902 Santa Maria erupts

The volcano Santa Maria erupts after centuries of no volcanic activity.

1902: April 18 Earthquake destroys settlements

A massive earthquake destroys wealthy cities in the southern agricultural foothills of the Sierra Madre mountains. These include the coffee center, Retalhuleu, which is connected by rail to the port city of Champerica, which is also seriously damaged by the earthquake.

1903 Population approaches two million

The population of Guatemala is reportedly 1,842,132. This represents one-third of all people in Central America. About sixty percent of the population is Indian. About 12,000 are Europeans or North Americans.

1903 Government schools offer primary instruction

There are over 1,000 schools offering instruction for children ages six to thirteen. In addition, any plantation that has more than ten children must provide schooling on their premises.

1904 Railroad deal complete

The United Fruit Company receives a concession from the government of Manuel Estrada Cabrera to complete a railway from Puerto Barrios on the Caribbean coast to Guatemala City. In return, United Fruit receives large tracts of land, a monopoly on rail use, and exemption from taxes. Similar agreements will be negotiated in contracts of 1912, 1924, 1930, and 1936.

1906 Former president Barillas invades Guatemala

Former Guatemalan president Barillas (see 1885) leads an invasion of his homeland. His goal is to establish a stable economy by creating a currency of silver. He has the support of German and British citizens living in Guatemala.

1906: July 20 Marblehead Pact ends conflict

The conflict initiated by Barillas's invasion escalates into a war. The conflict ends when U.S. president Theodore Roosevelt (1858–1919) and Mexican president Diaz intervene. The two negotiate an armistice known as the Marblehead Pact that is signed on the U.S. ship *Marblehead* on July 20.

1906: September 28 Central American states sign a treaty

Four of the five Central American states sign a treaty to bring peace to the region. The treaty specifies rules for trade among the states and between Central America, North America, and Europe. It also creates a system for resolving conflicts between the Central American nations through arbitration by the United States and Mexico. (Nicaragua elects not to join in the treaty.)

1929 United Fruit monopolizes banana production

United Fruit Company buys out its only major competitor, the Cuyamel Banana Company, which is owned by Samuel Zemuray (1877–1961). United Fruit's veritable monopoly over banana production for the U.S. market will bring about an anti trust suit.

1931–44 Ubico era

General Jorge Ubico Casteñeda (1878–1946) serves as President. His era is marked by dictatorial authoritarianism amidst the financial chaos of the Great Depression.

1932 Repression of political opposition and the Great Depression

Ubico launches a campaign of repression against the small Communist Party of Guatemala and other opposition movements.

The impact of the Great Depression reaches its most severe stage. Exports fall to forty percent of their 1929 value.

1934 Debt peonage abolished

The Ubico government abolishes debt peonage and replaces it with Decree 1996, the Vagrancy Law, which requires all landless laborers to work for an employer for 150 days per year. Laborers with subsistence plots have to work for 100 days.

1944: July 1 Ubico Resigns

Amidst widespread protests, marked by a series of work stoppages in the capital city, Ubico resigns the presidency and turns over power to a reformist junta that oversees national elections and the writing of a new Constitution.

1945 Inauguration of Arévalo

Juan José Arévalo (1904–1990) is inaugurated as President.

1945 National Academy of Sciences founded

A National Academy of Medical, Physical, and Natural Sciences is founded at Guatemala City.

1946 Catholic Action founded

Catholic Action is formed by Archbishop Mariano Rossell Arellano (1894–1964). Dedicated to the defense of traditional Catholicism and religious purification, it attempts to counter a growth in Protestantism and the presence of Indian beliefs in religious practices. Although initially conservative, Catholic Action eventually is predominated by more reform-minded priests who stress community action and solidarity. By the 1970s and 1980s, Catholic Action is seen as an important element in promoting popular opposition to the government and military.

1946: April Social Security introduced

Arévalo government passes the Social Security Law, creating the nation's first social security system.

1947 Labor regulations passed

Arévalo government passes the Labor Code which establishes a minimum wage, collective bargaining, and the right to strike.

1949: July 18 Arana assassinated

The chief of staff of the army, Colonel Francisco Arana (1905–49), is assassinated. He is considered a front-runner for the presidential election of 1950. Another main candidate, minister of defense Jacobo Arbenz (1913–71) is strongly suspected. Both Arana and Arbenz had been heavily involved in the ouster of Ubico in 1944. As a powerful figure in the military, Arana has strongly defended Arévalo by forestalling various military conspiracies and coups. But Arana is a lukewarm supporter of Arévalo's reforms and has used his power to badger Arévalo. Arbenz is much more supportive of Arévalo's reforms, but his support is more mass-based, centering on such elements as the students and urban trade unions. The death of Arana allows Arbenz to win the election of 1950.

1952: June 17 Agrarian reform

The most ambitious and controversial of Arbenz's reforms, Decree 900, the Agrarian Reform Law, is passed. It allows for the expropriation (confiscation) of private properties above a certain minimum size. Compensation is based upon the owners's self-declared value of the property for tax purposes. The Land Reform is the subject of intense debate. Its most vehement critics, including large landowners, the Church, and the Eisenhower (1890–1969) administration in the United States, accuse Arbenz of being a communist. In actuality, the Arbenz administration considers land reform to enhance capitalist relations of production. The government bases the need for

reform on the nation's maldistribution of land. Statistics from the 1950 census show that two percent of the population owns seventy-four percent of the arable (fit for growing crops land while seventy-six percent of the nation's farming units account for only nine percent of arable land. The goal of the land reform is to create a greater number of capitalist-oriented farms. By 1954, nearly one million acres of land from 1,000 of the nation's largest plantations have been expropriated and redistributed to approximately 100,000 landless families. Ultimately, this represents a significant but far from wholesale transformation in agricultural production.

1953: February Government expropriates banana plantation land

The Arbenz government expropriates 200,000 acres from one of United Fruit's plantations at Tiquisate in the Pacific lowlands. The government argues that United Fruit owns nearly 550 million acres of land, only fifteen percent of which is cultivated. United Fruit responds that the potential for banana disease makes large tracts of uncultivated land necessary.

1953: August Operation Success

The administration of U.S. president Dwight D. Eisenhower gives the Central Intelligence Agency (CIA) permission to begin "Operation Success," the overthrow of the Arbenz government. The CIA eventually recruits Colonel Castillo Armas (1914–57) as the leader of the movement. Training is carried out in Honduras and Nicaragua. Armas's force consists of approximately 100 men.

1954: February Plantation land expropriations

The government expropriates 172,000 acres from United Fruit's holdings in the eastern lowlands.

1954: June Armas invades and Arbenz resigns

Armas's small force crosses the Guatemalan/Honduran border. U.S. mercenaries fly reconnaissance and bombing missions. CIA operatives use radio as propaganda to announce an "invasion." Less than one week later, Arbenz resigns and goes into exile. Crucial to Arbenz's resignation is the refusal of the military to support the government by defending it against the invasion.

1955 Armas as president

The final provisional junta after Arbenz's departure declares Armas to be president of the country. Armas's tenure is marked by a strongly counter revolutionary trend. Nearly ninety-nine percent of the land expropriated under the Land Reform is returned to its original owners. Arbenz's supporters and other opponents of Armas are jailed, exiled, or killed.

Despite efforts at modernization, much of Guatemalan society remains traditional. The weekly market day remains a place for trading and gossip. (EPD Photos/CSU Archives)

1957 Armas is killed

Armas is killed by one of his bodyguards as part of an internal party conflict. One year of political instability occurs. Presidential elections are marked by riots. The military temporarily assumes control of the government.

1958 Ydigoras becomes president

The conservative Manuel Ydigoras (1895–1982) wins new elections and becomes president.

1960 Common Market established

A Central American Common Market is established, which creates an economic trade zone for all of Central America with the exception of Costa Rica, which joins three years later.

1962 Archaeologists discover evidence of last Mayan king

Archaeologists discover a skeleton and artifacts at the tomb of Mayan ruler Ah-Kakaw (or Double-Comb; see A.D. 682–

734). The remains of a burial chamber are unearthed beneath the Temple of the Giant Jaguar. Over sixteen pounds of jade jewelry and ninety engraved animal and human bones are among the items found.

1962   Rebel Armed Forces begins operations

The main guerrilla army, FAR (Rebel Armed Forces) comes into existence. It is headed by Yon Sosa (1932–1970) and Turcios Lima (1941–66) and is centered in the eastern highlands.

1963   Army assumes control of government

The army assumes control of the government after Ydigoras is ousted by Minister of Defense Peralta Azurdia (b. 1908). This begins an extended period of direct military control over government.

1965: March   FAR splits into factions

Ideological differences split the FAR. Lima remains head of the FAR, but Yon Sosa leaves to form a second guerrilla front, MR-13.

1966   Death squads begin operations

This year marks the appearance of White Hand (*Mano Blanco*) and other right-wing paramilitary death squads, believed to be responsible for tens of thousands of deaths over the next decade.

1966–70   Guerilla activity suppressed

An intensified counter-insurgency campaign on the part of the Guatemala military takes place, suppressing guerrilla activity for the time being. In 1970, Yon Sosa is killed in a clash at the Mexico border.

1967   Asturias wins Nobel Prize

Miguel Angel Asturias (1899–1974) wins the Nobel Prize for Literature. Asturias's writings combine Indian themes with social protest to capture some of Guatemala's most pressing social dilemmas. One of his best-known novels, *El Señor Presidente* (*The President*), first published in Mexico in 1946, is a biting indictment of the dictatorship of Manuel Estrada Cabrera.

1968   Ambassador assassinated

U.S. Ambassador John Mein is assassinated by guerrillas. The guerrillas view Ambassador Mein as a legitimate target since he represents U.S. interests—the chief target of guerrilla animosity.

Miguel Asturias (1899–1974) wins the Nobel Prize for Literature in 1967. His works are best-known for their social protest. (EPD Photos/CSU Archives)

1969   Common Market collapses

The "Soccer War" between Honduras and El Salvador causes the Central American Common Market to collapse.

1970: November   State of siege

The army declares the country to be in a state of siege. An estimated 1,000 political killings occur in the next six months.

1971: January   Organization of People in Arms forms

The guerrilla organization OPRA (Organization of People in Arms) forms and begins activity in the Pacific lowland region. It eventually moves its operation to the highlands and becomes predominantly Indian in membership.

1972: January   Guerrilla Army of the Poor forms

Another guerrilla organization, EGP (Guerrilla Army of the Poor), forms in the western highlands. By 1975, EGP is active in the Quiché region and shows recruitment among Indians.

1976   Coca-Cola locks out workers

A lockout of workers at a Coca-Coca plant results in the formation of the National Committee of Trade Union Unity (CNUS).

1977   U.S. military aid suspended

Under president Jimmy Carter (b. 1924), the U.S. administration suspends all military aid to Guatemala for human rights violations. Israel and South Africa become primary weapons suppliers.

1978–82   Intensified counterinsurgency

The most intense period of government repression takes place in the highland zones. As part of a broad counterinsurgency campaign, the army employs a series of so-called civic action tactics. These include strategic hamlets and civil patrols, where entire villages are relocated to a determined "safe" zone and all other areas are considered free-fire. Loyalist villagers are organized into civil defense patrols to serve as the local eyes and ears of the military. Tens of thousands are killed and as many as 1,000,000 are displaced, many of them ending up in refugee camps across the Mexican border.

1980: January 31   Attack on Spanish embassy

Thirty-nine Indian peasants and their supporters are killed by the army in the Spanish embassy. The protesters had peacefully occupied the embassy as part of a campaign to draw international attention to army abuses in the highlands. Ignoring the protests of the Spanish government, the army units attack and burn the embassy, killing all but one of the occupants. Spain breaks off diplomatic relations with Guatemala.

1982: January   Guerrilla alliance

Four main guerrilla movements (FAR; EGP; OPRA, and the PGT, Guatemalan Worker's Party) announce the formation of the Guatemalan National Revolutionary Union to coordinate campaigns against the government. One month later, they are joined by the Democratic Front Against Repression, a coalition of 170 organizations. Highest estimates of guerrilla armies are still at less than 5,000.

1992: October   Menchú wins Nobel Peace Prize

Rigoberta Menchú (b. 1959), a Mayan Indian who lost most of her family to military repression, is awarded the Nobel Peace Prize for her work to aid Indians against abuse. Her life story is widely circulated in a subsequent book, *Me llamo Rigoberta Menchú* (I, Rigoberta Menchu).

1993: May   President forced to resign

President Jorge Serrano (b. 1945) is forced to resign after widespread condemnation of his attempt to remain in power through a "self coup." Congress chooses Ramiro de León Carpio, previously the human rights ombudsman, to complete the remainder of Serrano's term. Carpio makes the end of civil war a primary goal.

1994: March   Human Rights Accord

In Mexico City, representatives from the government and the guerrilla armies sign a "Human Rights Accord," which includes provisions for the government to defend human rights, dismantle paramilitary bands, and not protect human rights violators. The accord also establishes a United Nations peacekeeping mission in Guatemala. This is the first of five accords, signed over the next two and a half years, that culminate in the Peace Treaty of December, 1996.

1995   Harbury draws attention to U.S. role

As a result of her five-year campaign to learn the whereabouts of her husband, Jennifer Harbury focuses attention on the participation of the U.S. government in the civil conflict in Guatemala. A lawyer and U.S. citizen, Harbury married Efrain Baranca Velásquez, a Guatemalan guerrilla commander, in 1991. Baranca was captured by the Guatemalan army during a mission in the spring of 1992. The army reported him killed in combat. Counterintelligence evidence suggested that he was held, tortured, and probably killed many months later.

Harbury holds a thirty-two-day hunger strike in Guatemala to demand the release of her husband. Later information, originating in part from then-Representative Robert Torricelli in the United States, confirms Baranca's death in captivity. The new evidence also indicts Colonel Julio Roberto Alpírez, a paid informant of the CIA, in the capture and killing of Baranca. Alpírez is also suspected in the 1990 killing of an American innkeeper named Michael DeVine in Guatemala. In 1996, Harbury files suit against the CIA for information relating to her case.

1995   Roman Catholic bishops launch Historic Memory Project

Catholic bishops create the Historic Memory Project (known by its Spanish acronym REMHI). As part of the project, religious leaders conduct 6,500 interviews, almost two-thirds of which are in one of thirty-five Mayan languages. Bishop Juan José Gerardo directs the project, which issues a 1,400-page, four-volume report in 1998. The report analyzes the struggle between leftist guerrillas and a series of U.S.-backed military governments. It provides details of thousands of acts of violence, including over 400 massacres, most of which have been carried out by the army or paramilitary death squads. After the report is issued, Bishop Gerardo is murdered. (See 1998: April)

1996: January   Arzú elected president in run-off

Alvaro Arzú, a conservative businessman and former mayor of Guatemala City, is elected president in a run-off election.

1996: October 16   Fans killed in crowd at soccer match at Mateo Flores stadium

Eighty-two people die in the crush of soccer fans at Mateo Flores Stadium in Guatemala City. Over 100 more are seriously injured. The stadium had been overcrowded with fans hoping to see a World Cup qualifying match. When spectators began pushing out of the stadium into its tunnels, injuries and deaths occurred.

1996: December   Peace Accords

In Guatemala City, the government and the guerrilla armies sign the Peace Accords, formally ending the nearly four-decade long civil war. The guerrillas are represented by four main commanders, Rolando Morán: Pablo Monsanto, Carlos González and Jorge Rosas. The accords spell out the demobilization of the guerrilla armies, estimated at less than 3,000 combatants, and their reinsertion into civil society. The Guatemalan military is supposed to reduce its 46,000-solider force by at least one-third. The accords include provisions to end discrimination against Indians, reform the judicial and law enforcement systems, and identify those who committed atrocities during the war.

1998: April   Assassination of human rights figure

Bishop Juan José Gerardo, author of a monumental report on human rights abuses committed by the army and paramilitary groups during the war (Historic Memory Project's "Guatemala Never Again"), is murdered. (See 1995.)

1998: October   Hurricane Mitch

Guatemalans and others throughout Central America suffer hardships following Hurricane Mitch. Many are isolated when roads and bridges are washed out by torrential rains. Crops, especially bananas and coffee, are destroyed.

Bibliography

Barry, Tom. *Inside Guatemala: The Essential Guide to Its Politics, Economy, Society and Environment.* Albuquerque: Inter-Hemispheric Education Resource Center, 1992.

Carmack, Robert. *Rebels of Highland Guatemala: The Quiché Mayas of Momestenango.* Norman: University of Oklahoma Press, 1995.

Culbert, Patrick, ed. *Classic Maya Political Chronology: Hieroglyphic and Archaeological Evidence.* Cambridge: Cambridge University Press, 1991.

Delli Sante, Angela. *Nightmare or Reality: Guatemala in the 1980s.* Amsterdam: Thela Publications, 1996.

Dosal, Paul. *Doing Business with the Dictators: A Political History of United Fruit in Guatemala, 1899–1944.* Wilmington: SR Books, 1993.

Handy, Jim. *Gift of the Devil: A History of Guatemala.* Toronto: Between the Lines Press, 1984.

———. *Revolution in the Countryside: Rural Conflict and Agrarian Reform in Guatemala, 1944–1954.* Chapel Hill: University of North Carolina Press, 1994.

Henderson, John. *The World of the Ancient Maya,* 2nd ed. Ithaca: Cornell University Press, 1997.

Jeffrey, Paul. "Killing Draws Notice to Violent History." *National Catholic Reporter,* vol. 34, no. 27, May 8, 1998, p. 3+.

Jones, Oakah. *Guatemala in the Spanish Colonial Period.* Norman: University of Oklahoma Press, 1994.

Kramer, Wendy, *Encomienda Politics in Early Colonial Guatemala, 1524–1544: Dividing the Spoils.* Boulder: Westview Press, 1994.

Leder, Dennis. "Memory Medicine and the Laureate Landivar." *America,* 1994, vol. 170, no. 10, March 19, p. 20+.

Lovell, George. *Conquest and Survival in Colonial Guatemala: A Historical Geography of the Cuchumatán Highlands, 1500–1821,* 2nd ed. Montreal: McGill-Queen's University Press, 1992.

Manegold, Catherine, "The Rebel and the Lawyer: Unlikely Love in Guatemala," *New York Times,* March 27, 1995, p. A1.

McCreery, David. *Rural Guatemala, 1760–1940.* Stanford: Stanford University Press, 1994.

Nyrop, Richard, ed. *Guatemala: A Country Study.* Washington DC: Department of the Army, 1983.

Perera, Victor. *Unfinished Conquest: The Guatemalan Tragedy.* Berkeley: University of California Press, 1993.

Popul Vuh: The Mayan Book of the Dawn of Life. New York: Simon and Schuster, 1996.

Risen, James, "Mission Impossible?" *Los Angeles Times,* June 7, 1995, p. E1.

Rother, Larry. "Bishop's Death Shakes Hope for Guatemala Peace," *New York Times,* May 9, 1998, p. A3.

Rother, Larry. "Guatemalans Formally End Their 36-Year Civil War," *New York Times,* December 30, 1996, p. A4.

Schele, Linda. *A Forest of Kings: The Untold Story of the Ancient Maya.* New York: Morrow, 1990.

Sharer, Robert. *Daily Life in Maya Civilization.* Westport: Greenwood Press, 1996.

South America, Central America and the Caribbean, 6th ed. London: Europa Publication, 1997.

Tedeschi, Ted. "Natural Attractions of Mundo Maya." *Audubon,* vol. 98, no. 5, September–October 1996, p. 83+.

Weiner, Tim. "Records Tie CIA Informer to Two Guatemala Killings," *New York Times,* May 7, 1996, p. A1.

Woodward, Ralph Lee. *Rafael Carrera and the Emergence of the Republic of Guatemala, 1821–1871.* Athens: University of Georgia Press, 1993.

Guyana

Introduction

Cooperative Republic of Guyana

In December 1997 a seventy-seven-year-old American expatriate became president of Guyana and promptly promised to be "president of all of the people." In most nations that phrase would have been part of a typical victory speech, but in Guyana, it carried deeper meaning.

President Janet Jagan (b. 1920), a U.S.-born white politician from Chicago, faced the difficult task of ruling a young nation where the descendants of East Indians and Africans had historically competed for power with one another.

The racial divisions are the remnants of Guyana's colonial past, when the Dutch, and later the English, brought African slaves to work in the sugar cane fields. When the British abolished slavery in the mid-nineteenth century, they brought indentured Portuguese and Chinese laborers to take the place of the freed slaves. But the bulk of the workforce came from northern India—nearly 240,000 indentured workers between 1838 and 1917.

The legacy of the plantation culture was a segregated society, with minimal contact between the different ethnic groups—including the surviving tribal peoples of Guyana, who were pushed deep into the Guyanese wilderness by European expansion.

Geography and Population

Guyana, slightly smaller than the state of Idaho, is the third-smallest country in South America with 83,000 square miles (214,970 square kilometers) of territory. It is bounded on the north by the Atlantic Ocean, on the east by Suriname, on the south and southwest by Brazil, and on the northwest by Venezuela. Border issues remain unresolved between Guyana and two of its neighbors, Venezuela and Suriname. Venezuela claims an area over three-fifths the size of Guyana, while Suriname claims a largely uninhabited area of 5,800 square miles in the southeast.

Guyana has three main natural regions. Much of the low-lying coastal plain is below high-tide level, protected by sea-walls and drainage canals. It extends for about 270 miles and ranges from 10 to 40 miles in width. A region that accounts for five-sixths of Guyana's land area is heavily forested, with rolling, hilly land that contains most of the nation's mineral wealth. Savannas and mountains are in the south and west. There are many large rivers, but few are navigable for any distance above the plains because of rapids and falls. Guyana's climate is subtropical and rainy, with average temperatures of 81°F and about eighty-five percent humidity.

Ethnic Diversity

Ethnicity has been one of the defining characteristics of Guyanese society and politics, even though all of the immigrant groups eventually adapted to the dominant British culture and the official language, which is English. Their ancestors worked in the plantations—the Africans as slaves and the groups that followed as indentured workers. They did the same work and lived in the same kind of housing, had the same bosses, and the same regulations.

Today fifty-one percent of the popoulation is Asian Indian and forty-three percent is African.

Spanish, Dutch, and English Influence

Explorer Christopher Columbus (1451–1506) sailed past the northeastern coast of South America on his third trip to the New World in 1498. Known as *Guiana* for "land of water," the region was inhabited by Amerindians of the *Arawak*, *Carib*, and *Warrau* language groups. While they claimed it for the Spanish Kingdom, the Spaniards showed little interest in the area. The muddy, flooded coastal region did not seem to have much to offer.

By the mid-1700s, the Dutch had established settlements on the Essequibo, Demerara, and Berbice Rivers to trade with the indigenous groups and fought English and French colonization efforts. In time the Dutch moved from trading to agriculture and built, with slave labor, a massive dam system to reclaim the coastal areas from the sea.

By the early 1800s England, in a series of wars fought between European nations, occupied the Dutch settlements.

By 1831 the former Dutch colonies officially became the Colony of British Guiana.

The English, as the Dutch had done before them, relied on slavery to maintain their plantations. In 1838, when slavery came to an end in the British colonies, the plantations remained a powerful symbol of oppression for the freed slaves, who moved on to other work or started their own farms.

Faced with a severe labor shortage, the British began to import indentured workers into their Caribbean colonies. Among the first to arrive were Portuguese, followed by East Indians, and later Chinese laborers.

The political system established by the Dutch, and later modified by the English, was designed to maintain power in the hands of the elite, white plantation owners. By the late nineteenth century, new generations of Africans, East Indians, and Portuguese began to demand more representation in a government that could not claim to represent them. Political and social changes quickened in the early twentieth century. Early worker strikes in 1905 led to a strong, organized labor movement. By the end of World War I, a new elite began to challenge the supremacy of the sugar barons.

World War II forced the English to rethink their imperial policy. Slowly, England began to divest itself of its colonies. In British Guiana they approved a new constitution that provided for universal suffrage and increased home rule in 1953 in preparation for full independence.

The Jagans and the Formation of the People's Progressive Party

Cheddi Jagan (1918–97), whose grandparents had migrated to British Guiana as indentured laborers, emerged as the new chief minister of the colony in 1953. Jagan and his American wife, the former Janet Rosenberg, founded the People's Progressive Party (PPP), the colony's first political party, in 1950. The Jagans would become two of the most powerful political figures in the country, with both eventually ruling the young nation as president in the 1990s.

Jagan called for independence and sweeping social and economic reforms. His leftist rhetoric angered Britain's prime minister, Winston Churchill (1874–1965), who considered Jagan a puppet of the communists. Six months into Jagan's first ministerial term, the British sent ships and troops to depose him.

The Jagans embarked on a civil disobedience campaign modeled on the one led by Mohandas K. Gandhi (1869–1948) against British rule in India. They were arrested and jailed for six months. They emerged from prison as heroes, both winning cabinet posts in the 1957 elections.

By 1961 the PPP, with wide support from the Indo-Guyanese community, had consolidated its power, and Jagan was elected prime minister of the British colony. He continued to press for independence and promised to install a socialist

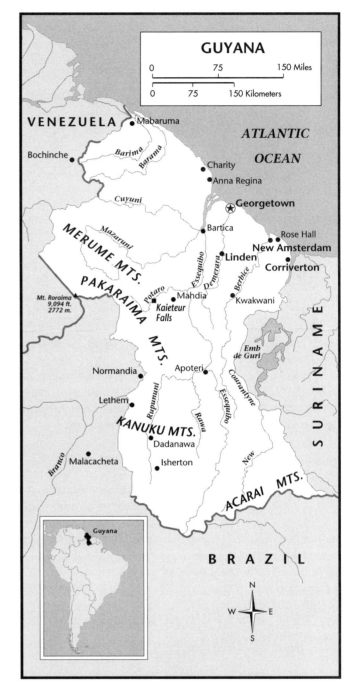

economy, raising the concern of U.S. policy makers, who were determined to stop any other nation in the Western Hemisphere from becoming another like Cuba.

In an interview with the *New York Times* in 1994, President Jagan remembered his visit with President Kennedy (1917–63) in 1961. "I went to see President Kennedy to seek the help of the United States, and to seek his support for our independence from the British. [Kennedy] was very charming and jovial. Now, the United States feared that I would give Guyana to the Russians. I said if this is your fear, fear not. We will not have a Soviet base."

Under President Kennedy's orders, the Central Intelligence Agency (CIA) began a destabilization campaign against Jagan. In 1964 the Americans convinced England to delay independence and change the electoral system to keep Jagan out of office. Through manipulation of Guyanese labor unions, United States intervention created serious racial disturbances in British Guiana between 1961 and 1964. Mobs twice destroyed sections of Georgetown, and as many as 100 people died in race riots between Indo-Guyanese and Afro-Guyanese.

Burnham and Independence

American intervention led to the election of Forbes Burnham (1923–85) in 1964. Burnham had been an ally of the Jagans in the 1950s and a member of the PPP, but philosophical differences drove him away, and he started his own moderate socialist party, the People's National Congress (PNC). The party became the predominant political voice for blacks, while a majority of Indo-Guyanese remained loyal to the PPP.

The rise of the two parties in the 1950s was the beginning of a racial politics that polarized the country well into the 1990s.

British Guiana became the independent nation of Guyana in 1966, with Burnham at the helm. By 1970 he declared Guyana a "cooperative republic," and installed a socialist economy. Claims of electoral fraud were made against him, human rights abuses increased during his tenure, and some major political opponents were assassinated. To support his economic policies, he incurred a foreign debt of more than two billion dollars, a sum Guyana was still trying to pay back at great cost in the 1990s.

From Socialism to Free Market

Hugh Desmond Hoyte (b. 1930) became president after Burnham's death in 1985 and slowly dismantled Burnham's policies, moving from state socialism to a free market economy. His austerity measures shocked the country, and thousands of Guyanese left in the late 1980s. In October 1992 Jagan once again became leader of the country, this time as president. With President Bill Clinton (b. 1946) in office, and backed by the United States, Jagan signaled his willingness to abandon racial politics by naming a multiethnic cabinet.

When Jagan died after heart surgery at Walter Reed Army Hospital in Washington, D.C., on March 6, 1997, his wife Janet was elected president of Guyana in December of that year. With an Afro-Guyanese, Sam Hinds, as her vice-president, the PPP captured more than fifty-six percent of the votes. But violent events in the streets after the elections and the unsuccessful attempts by the PNC to keep Jagan from taking office showed that Guyana was far from abandoning ethnic politics.

"It will take years to deal with that issue in a really constructive manner," Reverend Malcolm Rodrigues, a Jesuit priest of Portuguese ancestry and chairman of Guyana's non-governmental Electoral Assistance Board, remarked in an interview with the *Miami Herald* before the 1997 elections.

"For too long, we have allowed the political culture to set the tone on ethnicity," he told the *Herald*. "We have to move away from the political parties to set the tone . . . to get people involved together in activities other than politics. It will take a long time. We have to begin at the bottom and build it up."

In 1999 Guyana celebrated its thirty-third birthday as a free nation and its people continued to mend the wounds left by racial conflict. With a new century on the way, they faced the challenge of accepting what some Guyanese called a nation of six peoples: the descendants of the original native tribes, Europeans, Africans, Chinese, Portuguese, and East Indians.

Timeline

Pre-1492   The Pre-Columbian Era

Before Christopher Columbus (1451–1506) arrives in the Western Hemisphere, Guyana's inhabitants are divided into two groups, the Arawak along the coast and the Carib in the interior. The northeastern region of South America, which now encompasses modern Guyana, Suriname (former Dutch Guiana), and French Guiana, is collectively known as Guiana. The word means "land of waters" and is used by the indigenous peoples to describe an area rich in rivers. Europeans come to call the entire region the Guianas.

Historians and anthropologists believe the Arawak and Carib originate deep in the South American jungles and migrate northward, first to the present-day Guianas and then to the Caribbean islands.

The peaceful Arawak hunt, fish, and cultivate the land. Over time they migrate to the Caribbean islands. The warlike and violent Carib follow them, disrupting their lives and displacing them throughout the Caribbean.

The Arrival of the Europeans and Dutch Control

1498   Columbus sails along the Guyanese coast

Columbus sights the Guyanese coast in 1498 during his third voyage to the Americas. Although they claim it on behalf of the Spanish kingdom, the Spaniards take little interest in the area. The Caribs, who fight hard to maintain their freedom, deter early European occupation of the Guianas. And so does the muddy, mangrove-covered coastal region, which is not suitable for farming.

1616   Dutch begin to colonize the Guianas

The Dutch are the first Europeans to settle what is now Guyana. The Netherlands, which obtains independence from Spain in the late 1500s, emerges as a major commercial power by the 1600s.

The Dutch begin trading with the English and French colonies in a group of islands known as the Lesser Antilles off the coast of Guyana. By 1616 they establish a trading post several miles upstream from the mouth of the Essequibo River in Guyana to trade with the indigenous peoples. The Dutch establish other settlements along other rivers in the area.

1648   Dutch declare ownership of the Guianas

Dutch sovereignty of the Guianas is officially recognized with the signing of the Treaty of Munster in 1648. By then the focus of the colony has changed from trading to growing crops.

As agriculture grows in importance, the Dutch face a labor shortage. The indigenous populations are not adapted for work on plantations, and many of them die from diseases introduced by the Europeans. The Dutch West India Company, which in 1621 receives complete control over the trading post on the Essequibo River from the Dutch government, begins to import African slaves to work the land.

Initially the Dutch settle far from the coastline, which is covered in water during high tide. But with greater profits for agricultural products, especially sugar, they need more land. By the 1700s the Dutch begin to reclaim coastal lands.

The Dutch have ample experience in land reclamation. Back home, they have created the polder system, a technique that creates usable land by damming and then draining a water-covered area. The Dutch rely on slave labor to build the polder system and reclaim the coastal plains, which historically remain one of Guyana's most productive plantation areas.

The polder system relies on the use of a front dam, or facade, along the shorefront. The dam is supported by a back dam of the same length and two connecting side dams, which form a rectangular tract of land known as a polder. The dams keep the salt water out, and fresh water is managed by a network of canals that provide drainage, irrigation, and a system of transportation.

1660   Indigenous peoples retreat deep into Guyana

By the 1660s the slave population has grown to about 2,500. In the meantime, the indigenous people, estimated at about 50,000, retreat deep into the wilderness of the Guianas.

By the 1990s the Amerindians of Guyana have gone through extensive acculturation. They speak English or Portuguese as a first or second language, integrate into the national

Georgetown, 1784.

economy (usually at the lowest levels), and often convert to Christianity.

In modern Guyana Amerindians are broadly grouped into coastal and interior tribes (the term *tribe* is a linguistic and cultural classification rather than a political one).

The coastal Amerindians are the *Carib, Arawak,* and *Warao.* The interior Amerindians are classified into seven tribes: *Akawaio, Arekuna, Baruma River Carib, Macusi, Patamona, Waiwai,* and *Wapisiana.*

1663: February   Brutal working conditions give rise to slave rebellions

Much like other colonies in the Western Hemisphere, slaves are treated brutally by the Dutch. On February 1663 slaves on two plantations on the Canje River in Berbice rebel and take control of the region.

As many as three thousand rebels are led by Cuffy, the national hero of Guyana. As plantations fall to the slaves, many Europeans flee the colony. Cuffy and his troops are finally defeated with the assistance of troops from neighboring French and British colonies.

1746   Dutch allow British to immigrate to Guyana

Guyana remains sparsely populated, and the Dutch begin to encourage British immigration from the nearby islands of the Lesser Antilles to stimulate the economy.

British plantation owners are lured to the Dutch colonies by richer soil and the promise of land ownership. By 1760 the English become a majority in the colony of Demerara. By 1786 the internal affairs of this Dutch colony are effectively under British control.

1781–84 European powers struggle to control the Guianas

In 1781 war breaks out between England and the Netherlands. The English briefly occupy the colonies of Berbice, Essequibo, and Demerara. France, allied with the Dutch, captures the colonies and governs for two years. During that time, the French build the town of Longchamps at the mouth of the Demerara River.

The Dutch regain power in the Guianas in 1784 and move the colonial capital to Longchamps, which they rename Stabroeck. The colony eventually becomes known as Georgetown, the capital of Guyana.

England Takes Over

1795–1814 Colony becomes British Guiana

The French Revolution and the resulting Napoleonic Wars eventually lead to the British takeover of Guyana.

In 1795 the French occupy the Netherlands, and the British declare war on France. In 1796 the British send an expeditionary force from Barbados to occupy the Dutch colony of Berbice and the United Colony of Demerara and Essequibo. The British takeover is bloodless, and they allow the Dutch to continue administering the colonies.

The British return the colonies to the Dutch in 1802, but peace is short-lived. By 1803 the French and English are at war, and Britain again seizes the Dutch colonies.

At the London Convention of 1814, the colony of Berbice and the United Colony of Demerara and Essequibo are formally ceded to Britain. In 1831 the colonies are unified as British Guiana. The colony remains under British control until independence in 1966.

The early social structure of British Guiana is similar to England's other colonial holdings. A small class of white plantation owners, known collectively as the European Planter class, controls government and the economy. They have links with commercial interests in London and enjoy close relations with the governor, who is appointed by the monarch. This elite own the slaves, control exports, and employ the majority of the population.

There are some freed slaves, most of mixed African and European heritage, who enjoy limited freedom. The indigenous peoples remain unconnected to colonial life.

1808–38 Slavery slowly ends in British Guiana

The international slave trade is abolished in the British Empire in 1807, but slavery continues until 1838, when slaves are freed. The end of slavery significantly alters life in British Guiana.

Many former African slaves leave the plantations, and some of them move to towns and villages. Many Africans consider plantations a symbol of their slavery and consider fieldwork degrading. Other former slaves pool their resources and purchase the abandoned estates of their former masters. They create self-sustained communities and grow and sell food. In time, their independence threatens the traditional planter class, which no longer holds a near-monopoly on the colony's economy.

The exodus of the former African slaves from the sugar plantations creates a labor shortage in British Guiana. As they have done in their other colonies in the Western Hemisphere, the British begin to bring indentured workers (workers contracted for a certain period of time prior to gaining their freedom) to their South American colony to keep up production in the plantations.

Descendants of the Africans, the Afro-Guyanese come to see themselves as the true people of British Guiana. That belief is based on their long history in British Guiana and a sense of superiority over recent arrivals based on their British colonial values, their literacy, and Christianity.

By the early twentieth century, Afro-Guyanese hold most of the professional jobs in the urban centers, where they are a majority. Yet, by the 1930s, the Indo-Guyanese begin to catch up and enter the middle class in large numbers.

The New Immigrants

1835 Portuguese among new immigrants in British Guiana

The Portuguese, who come from the North Atlantic island of Madeira, are among the first indentured workers the British bring to Guyana, starting in 1835 and ending in 1882.

The Portuguese spend little time in plantation work and quickly move into other parts of the economy, especially retail business. The Portuguese enjoy economic success, but they are treated as socially inferior by the British plantation owners and colony officials because of their indentured past and Roman Catholic religion.

Yet despite discrimination, the Portuguese become an important part of the colony's middle class by the end of the nineteenth century and become competitors with the new Afro-Guyanese middle class.

1838 English bring East Indians to work to the land

Beginning in 1838 and lasting until 1917 almost 240,000 East Indian indentured workers are brought to British Guiana to help relieve the labor shortage. The East Indians can ask to be repatriated at the end of their contracts, but most of them choose to settle in British Guiana instead of going back to India with their savings.

The majority of the East Indians are from northern India, with variations in caste and religion. About thirty percent of the East Indians are from agricultural castes and thirty-one percent are from low castes or are *untouchables*. Brahmans, the highest caste, constitute fourteen percent of the East Indian immigrants. Another sixteen percent are Muslims.

Their British bosses house them together and put them to work together without consideration to caste or religion.

Unlike the African slaves, the East Indians are allowed to retain their cultural traditions.

And unlike the Africans, who are denied land, the planters make it available for East Indians late in the nineteenth century. Yet the process of assimilation makes the culture of the modern Indo-Guyanese more homogeneous than that of their caste-conscious immigrant ancestors.

1853   Chinese are brought to the plantations

The first indentured Chinese workers come to British Guiana from the south coast of China. Because almost all of the Chinese are men, they tend to intermarry with both East Indians and Africans. The Chinese language and most Chinese customs, including religion, disappear after a few generations.

The Chinese seek other work as soon as their indenture contracts at the plantations end. Many of them, like the Portuguese, enter the retail trade, while others become farmers.

1887   British Guiana and Venezuela break relations over disputed territory

Throughout the nineteenth century England and neighboring Venezuela dispute territory that England considers part of British Guiana.

In 1835 German explorer Robert Hermann Schomburgk (1804–65) at the request of the British government, had mapped British Guiana. Under British orders, he drew the western boundary with Venezuela at the mouth of the Orinoco River.

When England published the map in 1840, the Venezuelans protested, claiming the entire area west of the Essequibo River—three-fifths of British Guiana's territory—belonged to them.

In 1850 both nations had agreed not to occupy the disputed zone, but the British violated the agreement when gold was discovered in the area. British settlers moved into the contested area, and the British Mining Company began to mine the deposits.

Venezuela breaks diplomatic relations with Britain in 1887 and appeals to the United States for help. The British at first reject U.S. mediation, but agree to let an international tribunal mediate the dispute when President Grover Cleveland (1837–1908) threatens to intervene by citing the Monroe Doctrine. Created in 1823, the Monroe Doctrine had warned European countries not to intervene in the newly independent nations of the Western Hemisphere.

In a three to two decision, two Britons, two Americans, and a Russian award ninety-four percent of the disputed territory to British Guiana in 1899. Venezuela receives only the mouth of the Orinoco River and a short stretch of the Atlantic coastline just to the east. More than half a century will pass before Venezuela presses its case again.

1889   Race shapes politics and power structure

For most of the nineteenth century British and Dutch sugar planters maintain their privileged status by controlling the colony's political structure and denying access to other groups.

While some reforms are made to increase participation, the planters refuse to accept the Portuguese as equals and seek to prevent them from voting. The political tensions lead the Portuguese to establish the Reform Association. After anti-Portuguese riots in 1889, the Portuguese begin to work with the Afro-Guyanese and other disenfranchised groups to demand greater participation in the colony's affairs.

The Reform Association and a similar group known as the Reform Club represent a small and articulate emerging middle class. While they are sympathetic to the working class, these groups do not represent a national, political, or social movement.

1905: November   Bloody labor dispute leads to organized labor movement

Georgetown stevedores (laborers who load and unload ships) go on strike demanding higher wages, and they are joined by sympathetic workers from other parts of British Guiana. Together, they create the first urban-rural worker alliance.

1905: December 1   Black Friday

On what is now called Black Friday, a large group of porters protesting at the Plantation Ruimveldt refuse to disperse under orders by a police patrol and a detachment of artillery. They open fire and injure four workers.

News of the shootings quickly spread through Georgetown, and many residents take to the streets in protest, taking over several buildings. Seven people die during the disturbances and seventeen more are injured. British troops finally bring an end to the uprising. The stevedores' strike fails, but the riots and worker alliance lead to an organized trade union movement.

1914–18   World War I affects British Guiana

World War I does not have a direct effect on the colony, but it is the cause of many social changes. Many Afro-Guyanese join the British military. Upon their return, they become the nucleus of an elite Afro-Guyanese community. During the war the British end indentured labor because Indian nationalists criticize the program as a form of human bondage.

After World War I, the sugar planters begin to lose power as world sugar prices drop, and the colony's economy diversifies. Producers of other commodities, including rice and minerals, begin to press for political changes and a more representative political system.

1917   Workers organize to form labor union

Workers become more organized. Facing widespread business opposition, a group forms the colony's first trade union in 1917. The British Guiana Labour Union (BGLU), under the leadership of H.N. Critchlow, mostly represents Afro-Guyanese dockworkers, as many as thirteen thousand by 1920. While the union is legally recognized a year later, other unions have no official standing until 1939.

1928   British Guiana becomes a crown colony

The Combined Court and the Court of Policy had ruled British Guiana during colonial times. This system of government established by the Dutch in the late seventeenth century, with a few changes, was later adopted by the British. Historically the power structure benefited the white plantation owners.

By the 1920s with sugar prices falling, the planters had petitioned the Combined Court to fund new irrigation and drainage programs. But other sectors of the economy also needed funding. The Combined Court had become paralyzed by the demands of so many competing groups. In 1928 the British Colonial Office, in order to end the fighting, announces a new constitution for British Guiana and places the colony under the tight control of a handpicked governor.

A Legislative Council replaces both the Combined Court and the Court of Policy. A majority of its members are appointed. Despite the changes, the rising middle-class, workers, and political activists believe the government continues to favor the planters.

1938   Commission uncovers deep divisions in the colony

By the 1930s the Great Depression hurts British Guiana when the price of its major exports drops dramatically. As unemployment soars, violent demonstrations break out throughout the colony. The same problems plague other British colonies in the Caribbean.

A royal commission under Lord Moyne is established to find the cause of the riots and to make recommendations to improve the crisis. The commission discovers deep divisions between the country's two largest ethnic groups, the Afro-Guyanese and the Indo-Guyanese. The Afro-Guyanese are largely urban or bauxite miners and have adopted European culture. By the turn of the century, they become a majority of the eligible voters and dominate national politics. The Indo-Guyanese are mostly rice producers or merchants and have maintained their traditional culture. Generally, they do not participate in politics.

The Moyne Commission calls for social reforms and a more democratic form of government. It favors giving women and people who do not own land the right to vote. And it supports the trade union movement. Changes are stalled by the outbreak of World War II.

British influence in Guyana is widespread. This is Town Hall in Georgetown, the country's capital. (EPD Photos/CSU Archives)

The End of British Colonial Rule

1950   Indo-Guyanese form powerful political party

British imperial policy changes after World War II, and the British slowly move toward granting full independence to their colonies. In British Guiana demands for independence are matched with political reforms. The decade of the 1950s also sees the founding of Guyana's major political parties.

Cheddi Berret Jagan, Jr. (1918–97) and his American wife, the former Janet Rosenberg (b. 1920), organize the People's Progressive Party (PPP), which claims to speak for the lower social classes without consideration of race. The PPP seeks an end to British rule.

Jagan's parents are Indian immigrants of modest means. His father, a driver, managed to save enough money to send Jagan to Queen's College in Georgetown and later to the United States, where he studied dentistry at Northwestern University in Evanston, Illinois.

Jagan returned in 1943 and opened his own dentistry clinic. He quickly became involved in politics, first as treasurer for a group of sugar workers who were mostly Indo-Guyanese. In 1946 he joined a group called the Political

Affairs Committee, which promoted Marxist ideology and an end to British rule.

Jagan's political standing within the Indo-Guyanese population climbs after he helps organize a peaceful demonstration to protest the police shootings of five Indo-Guyanese workers in June 1948.

1953   England makes political reforms in British Guiana

The British introduce a new constitution, with a bicameral legislature and universal voting rights. The PPP, which has been branded as a communist organization by conservative leaders, captures eighteen of twenty-four elected seats during the first legislative elections.

1953: October 9   Constitution suspended

The United Kingdom suspends the constitution and sends troops to the colony, charging communist subversion of the government. An interim government controls the country until 1957.

1955   A PPP leader breaks away to start his own party

When the Jagans organize their party, they bring Linden Forbes Sampson Burnham (1923–85) into the party. They believe the London-educated lawyer can appeal to the Afro-Guyanese community.

Burnham, whose father is headmaster of Kitty Methodist Primary School near Georgetown, belongs to the colony's educated class. But as an Afro-Guyanese, he is aware of racial discrimination. In the mid-twentieth century a wide gap separates the different racial groups in British Guiana. Those divisions are also found within each group. In the Afro-Guyanese community, there's a mulatto elite, a black professional middle class, and a black working class.

By 1955 the PPP is leaning far to the left politically, and the more conservative Burnham decides to break away and start his own faction of the PPP. Two years later he creates the People's National Congress (PNC). The party quickly becomes the predominant political voice for Guyanese blacks, while a majority of Asian Indians remain with the PPP.

The rise of the two parties is the beginning of a long struggle for power between Jagan and Burnham, who contribute to a legacy of racially polarized politics that remains well into the 1990s.

1961   PPP victory leads to U.S. concern

A new constitution introduces a bicameral system and gives the colony full internal self-government. The constitution introduces a thirty-five-member Legislative Assembly and a thirteen-member Senate. All the legislative seats are elected, while a British-appointed governor chooses the senators. A prime minister post is to be filled by the majority party in the Legislative Assembly.

The PPP, which won the elections in 1957, wins again in 1961 by a substantial margin, with twenty seats in the Legislative Assembly. Burnham's party only wins eleven seats. Jagan, by now firmly a Marxist-Leninist, is appointed prime minister by the PPP.

Jagan creates nervousness in U.S. and British policy makers by calling for nationalization of foreign holdings, especially in the sugar industry.

1961: October   Jagan travels to U.S. to ask for economic aid

Despite his leftist rhetoric, Jagan's admiration for Fidel Castro (b. 1927), and his increasing ties to Eastern European Communist nations, Jagan travels to the United States to ask President Kennedy (1917–63) for economic aid. But Kennedy is not convinced by Jagan's assurances that he is not seeking to establish a Marxist state. Kennedy orders the CIA to destabilize Jagan's government.

1961–64   Jagan's government rocked by protests

Riots and demonstrations against the PPP administration increase in frequency. In 1962 and 1963 mobs destroy part of Georgetown.

Classified documents released in the 1990s show that the CIA, through covert operations, incites a general labor strike and racial violence between Jagan's Asian Indian followers and his opponents, mainly Burnham's Afro-Guyanese supporters.

President Kennedy urges Britain to delay full independence and change the colony's electoral system in 1963 to keep Jagan from winning another term.

British troops are sent to restore order. By the end of the crisis, 160 people are dead and more than one thousand homes have been destroyed.

1962   Venezuela again presses for Guyanese territory

Decades after an international tribunal has sided with British Guiana in a territorial dispute, the posthumous notes of a lawyer involved in the arbitration claim the tribunal president coerced several members into assenting to the final decision.

In 1962 Venezuela declares it will not abide by the 1899 arbitration because of this new information. In 1966 Venezuela seizes the Guyanese half of Ankoko Island in the Cuyuni River and later claims a strip of sea along Guyana's western coast. The border dispute has not been settled late in the 1990s.

1964: October   U.S. undermines Jagan's re-election

The British, who don't want the PPP back in power, once again tinker with the constitution to provide for more proportional representation and a fifty-three-member unicameral legislature.

Even though Jagan's election is undermined by United States intervention, his party wins twenty-four of the fifty-three seats. But under new parliamentary rules, the PPP is unable to form a government.

Burnham's leftist party and the United Force Party, a conservative alliance of big business and the Roman Catholic Church, form a coalition and select Burnham as prime minister.

Until his death in 1985, Burnham rules Guyana in an increasingly autocratic manner, first as prime minister and later, after the adoption of a new constitution in 1980, as executive president.

His party increases its parliamentary majority at the polls, but the elections are viewed as rigged, both in Guyana and abroad. Burnham's government suppresses human rights and civil liberties, and according to some accounts, eliminates powerful dissenters of his regime.

Two major political assassinations occur during his regime. Bernard Darke, a Jesuit priest and journalist, is killed in July 1979, and Walter Rodney, a distinguished historian and leader of the small multi-ethnic Working People's Alliance, in June 1980. Burnham's agents are believed to have been responsible for both murders.

The Beginning of a New Nation

1966: May 23   New nation named Guyana

An independence conference held in November 1965 approved a new constitution granting sovereignty to British Guiana. On May 23, 1966, the colony officially becomes the independent nation of Guyana.

1968   Burnham solidifies his power with victory at the polls

In the 1968 elections Burnham's party wins thirty seats and doesn't need the United Front Party to rule the country. Burnham announces that Guyana will become a Socialist nation. With the support of the Afro-Guyanese community, Burnham begins preparing the nation for a major shift in economic policy.

1970: February 23   Guyana declares itself a "cooperative republic"

In 1970 Guyana proclaims itself a "cooperative republic" and cuts all ties to the British monarchy. Under the proclamation the government embarks on a policy of cooperative socialism by nationalizing the bauxite industry, seeking a redistribution of national wealth, encouraging self-reliance, and establishing cooperative enterprises for most sectors of the economy. Within a decade, eighty percent of the economy is in the public sector.

Burnham improves relations with Cuba and turns his country into an influential voice in the Nonaligned Movement of nations. The nonaligned countries do not side with the United States or the Soviet Union.

In 1972 he welcomes the leaders of nonaligned nations during a conference in Georgetown, where he attacks imperialism and shows his support for the African liberation movements in South Africa. In the mid-1970s he allows Cuban troops to use Guyana as a transit point on their way to the war in Angola.

1978: November 18   The Jonestown Massacre

The government of Guyana had allowed American James Warren "Jim" Jones to establish the People's Temple commune in a wilderness area. The commune is one of many government attempts to colonize its sparsely populated wilderness.

When many Americans in the United States became concerned with allegations of abuse and coercion at the hands of Jones in the commune, which was known as Jonestown, U.S. Representative Leo J. Ryan traveled to Jonestown to investigate. He and four other U.S. citizens were murdered at a nearby airstrip by Jones's followers. On November 18 Jones and more than 900 of his followers commit suicide by drinking poisoned punch.

1980–85   Political situation deteriorates

The PNC claims seventy-seven percent of the vote and forty-one seats of the popularly elected seats in the December 15, 1980, election. The opposition claims Burnham has stolen the elections again. A team of international observers upholds the charge of electoral fraud.

As the PNC continues to tighten its control of Guyana, the opposition becomes more vocal. The government responds with arrests and intimidation. The assassination in 1980 of Dr. Walter Rodney, a leading opposition figure, escalates the conflict. During Burnham's last five years in office, there is a steady increase in human rights violations.

1985: August 6   Guyana's only leader since independence, dies

Burnham undergoes surgery for a throat ailment and dies on August 6. Vice President Desmond Hoyte (b. 1930) becomes the new executive president and leader of the PNC.

Hoyte slowly breaks away from Burnham's policies and seeks to improve Guyana's relations with non-socialist nations, particularly the United States. In September 1988 he visits the United States and a month later, he publicly breaks away from Burnham's legacy by announcing the liberalization of the economy.

1989   The Economic Recovery Program

Beginning in the 1970s Guyana's economy had suffered a severe decline, mostly because of the high costs of imported oil and petroleum products. At the same time, price and pro-

In 1978, over 900 people commit suicide by drinking poisoned punch at the People's Temple in Jonestown, Guyana. The deceased are members of a cult led by American Jim Jones. (Corbis Corporation)

duction of the country's exports dropped. By 1982 there were serious shortages of basic commodities, and the country could not meet its debt obligations.

In 1989 Hoyte's government launches the Economic Recovery Program with the aid of international lending institutions. The program marks a drastic reversal in government policy.

What is a state-controlled, socialist economy moves to a more open, free market system. The government eliminates price controls, removes import restrictions, promotes foreign investment, and sells many state-owned enterprises. Inflation, which runs at an annual average of up to 100 percent between 1989–91 drops to fourteen percent by 1992.

1990 Government approves Equal Rights Act

The Equal Rights Act is supposed to end sex discrimination, but it proves difficult to enforce because it lacks a clear definition of discrimination. There is no legal protection against sexual harassment in the workplace or laws that prevent dis-

missal on the grounds of pregnancy. In 1990 the Legislature strengthens women's property rights in cases of divorce.

1992 Jagan returns to power

Despite Hoyte's attempted reforms, Guyana grows tired of the PNC and in 1992 voters elect Jagan as the country's new president and give his party, the PPP, thirty-six seats in the National Assembly. It is considered the first free election since 1965.

Jagan's leftist politics mellow with age, and this time the United States supports his election. The seventy-eight-year-old Jagan dies in March 1997 while still in office. Prime Minister Samuel Hinds is appointed interim president until elections are held and proclaims six days of official mourning for "the greatest son and patriot that has ever walked this land."

Jagan attempts to minimize the racial divisions that have plagued his nation by diversifying his cabinet. He appoints a prime minister and a cabinet consisting of eight Indo-Guyanese, four Afro-Guyanese, and two Guyanese of Portuguese descent, one of Chinese descent, and one of Amerindian descent. Two members of the cabinet and thirteen members of the National Assembly are women.

During his term, Jagan consolidates Guyana's massive foreign debts and leads the country to sustained economic growth.

1995 Illiteracy rate among the lowest in South America

Educational standards are high in Guyana, but the schools have suffered chronic shortages of teachers and materials during the 1980s and 1990s. Yet the adult illiteracy rate remains at about 1.9 percent. School attendance is free and mandatory for ten years for children between the ages of five and fourteen. All schools in Guyana are public, church and private schools having been taken over by the government in 1976.

Many Guyanese leave the country to obtain a university education. The country's first university, the University of Guyana, is established in 1963 and graduates its first class in 1967.

1997: December American expatriate becomes Guyana's president

Janet Jagan, an American expatriate who, along with her Indo-Guyanese husband, Cheddi Jagan, founded the PPP, becomes the country's new president.

Writing for the *Washington Post* on December 21, one week after the election, freelancer Judy Flander, a cousin of the president, criticizes the American media for their portrayal of Janet Jagan as a docile grandmother.

"To kiss her off as a grandmother!" she writes. "This champion of plantation workers, native Amerindians and women's rights. This co-founder of the frankly Marxist People's Progressive Party (PPP). This veteran of riots, bombings, jail and house arrest. This leftist firebrand who, with her husband, led the drive for Guyana's independence from Great Britain in 1966. This editor of the *Mirror* newspaper in Georgetown from 1972 to 1997. This founder of two art museums, longtime member of Parliament and former acting ambassador to the United Nations. This 1997 recipient of UNESCO's Gandhi Gold Medal for Peace, Democracy and Women's Rights.''

Flander said that Janet's marriage to an Indo-Guyanese had caused great consternation in her family. She had voted in the 1947 elections in the colony, and effectively renounced her U.S. citizenship by doing so. From that point on, her future was tied to the future of the colony. She became her husband's chief political partner and adviser. A student nurse active in political causes, she encouraged her husband to study philosophy, political science, economics, and sociology.

1999: February All Internet restrictions lifted

Guyana's minister of information, Moses Nagamootoo, announces that the government has removed all restrictions on access to the Internet.

Nagamootoo says the government doesn't believe the indiscriminate restricting of specific sites is useful, even though some of the restricted sites deal with pornography. The decision, he says, is consistent with the government's stance on freedom of expression and free access to information.

The government of Guyana has generally respected constitutional provisions for freedom of expression and the press, although during Burnham's administration, there had been many reports of government harassment of independent and opposition media. One of the more established newspapers in the country, the *Guyana Chronicle*, is state-owned, while one of its main competitors, the *Stabroek News,* is independent.

Bibliography

Braveboy-Wagner, Jacqueline Anne. *The Venezuela-Guyana Border Dispute: Britain's Colonial Legacy in Latin America.* Boulder, Colo.: Westview Press, 1984.

Burrowes, Reynold A. *The Wild Coast: An Account of Politics in Guyana.* Cambridge, Mass.: Schenkman Publishing Co., 1984.

Daly, Vere T. *A Short History of the Guyanese People.* London: Macmillan Education, 1975.

Mecklenburg, Kurt K. *Guyana Gold.* New York: Carlton Press, 1990.

Naipaul, V.S. *The Middle Passage.* New York: Macmillan, 1963.

Riviera, Peter. *Individual and Society in Guyana*: *A Comparative Study of Amerindian Social Organization.* Cambridge: Cambridge University Press, 1984.

Rodney, Walter. *A History of the Guyanese Working People, 1881-1905.* Baltimore: Johns Hopkins University Press, 1981.

Singh, Chaitram. *Guyana: Politics in a Plantation Society.* New York: Praeger Publishers, 1988.

Smith, Raymond T. *Kinship and Class in the West Indies: A Genealogical Study of Jamaica and Guyana.* Cambridge: Cambridge University Press, 1990.

Spinner, Thomas J., Jr. *A Political and Social History of Guyana, 1945-1983.* Boulder, Colo.: Westview Press, 1984.

Williams, Brackette F. *Stains on my Name, War in my Veins: Guyana and the Politics of Cultural Struggle.* Durham, N.C.: Duke University Press, 1991.

Haiti

Introduction

When Jean-Bertrand Aristide (b. 1953) became President of Haiti in 1990, hundreds of thousands of Haitians took to the streets to dance in celebration. It seemed impossible that a Catholic priest who championed the rights of the poor could win a fair election in a nation twisted into obedience by a series of brutal dictatorships. Fledgling democracy was unable to bear the weight of history. Aristide was deposed a year later, and the dream of freedom became a nightmare for thousands of his supporters.

The invasion of Haiti by twenty thousand American troops and pressure from the international community brought Aristide back to power in October of 1994. To a large extent, the history of Haiti has been shaped by foreign powers: in colonial times by Spain and France, and later the United States. Strategically located in the Caribbean Sea, Haiti held a great deal of value for the world's powers.

Geography and Environment

Haiti occupies the western third of the island of Hispaniola and lies approximately 700 miles southeast of Florida. It is located between the islands of Cuba, Jamaica and Puerto Rico. Haiti, with an area of 10,714 square miles (27,750 square kilometers), is slightly larger than the state of Maryland.

The republic includes the islands of La Tortuga (La Tortue), Gonave, Les Cayemites, and Vache. Haiti is bounded on the north by the Atlantic Ocean, on the east by the Dominican Republic, on the south by the Caribbean Sea and on the west by the Windward Passage and the Gulf of Gonave. Haiti's capital city, Port-au-Prince, is located on the west coast.

Haiti's climate is generally tropical, hot and humid, with two rainy seasons. Once covered with virgin forests, Haiti is in danger of becoming a desert. Much of the natural vegetation has been razed by agriculture, grazing, and harvesting of timber. Only eleven percent of the land is arable and the country faces environmental devastation due to deforestation and soil erosion.

About ninety-five percent of Haitians are descendants of the approximately half million enslaved West Africans who

won their freedom from France in 1804. Only five percent are mulatto (of mixed black and white descent). Roman Catholicism is the official religion of Haiti, but nearly all Haitians practice at least some aspect of voodoo (West African mysticism) or are affected by the religion. French and Haitian Creole are the two national languages, but fluency in French is held in higher esteem.

Haiti is one of the world's poorest countries, mostly a nation of small farmers who practice subsistence agriculture on their meager plots of land. More than two-thirds of the population live in rural areas. Since the 1970s deforestation and erosion have driven more people to the cities, where light export industries, mainly involved in manufacturing clothes for U.S. markets, held the promise of better jobs. Haitian factory workers earned about $2.34 per day in 1997–98, making them the lowest-paid labor force in the Western Hemisphere.

Early Slave-holding Nation Created Racial Distinctions

Exploitation of cheap labor has been a common thread in Haitian history. The Spaniards were the first to do so when they arrived on the island of Hispaniola, now the countries of Haiti and the Dominican Republic. In 1492 when Columbus (1446?–1506) sailed to the island, as many as 400,000 Tainos lived there. By 1550 all but 500 or so had died from inhumane forced labor, violence, and exposure to European diseases. West African slaves replaced the Tainos in the fields. Spain ceded the western third of the island of Hispaniola to the French in 1697. The colony, known as Saint-Domingue, quickly rose to become one of the wealthiest in the world through the backbreaking labor of slaves. By 1791 a half million slaves worked in the fields there.

Modern Haitian society owes its origins to this slaveholding system. It would be difficult to understand present-day Haiti without considering the role of race and class. The color of their skin, the language they speak, and their work have historically defined Haitians and their place in society.

Under French rule, mulattos (a person of mixed black and white ancestry) were treated as second-class citizens, but they had some rights, unlike black slaves, who had none. Their place in society fostered a sense of superiority among

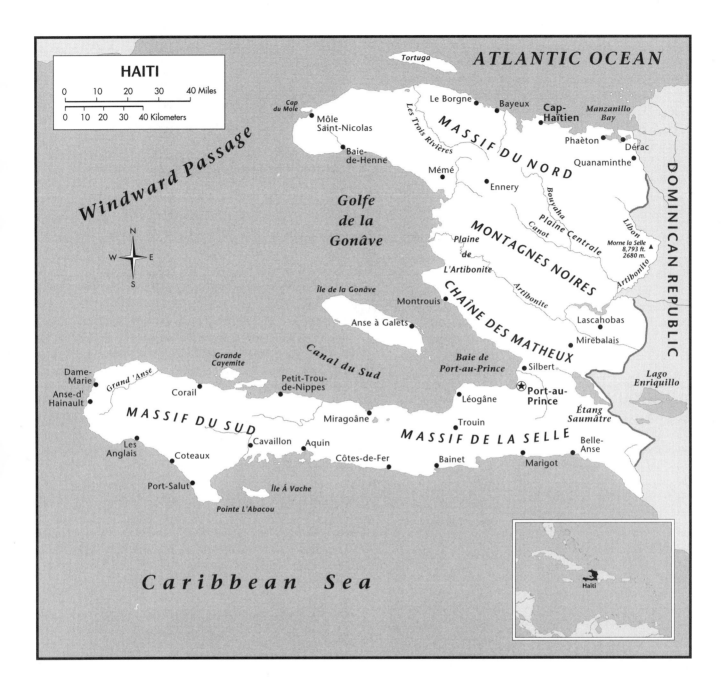

HAITI

0 10 20 30 40 Miles

0 10 20 30 40 Kilometers

Windward Passage

ATLANTIC OCEAN

Tortuga

Le Borgne

Bayeux

Cap-Haïtien

Manzanillo Bay

Cap du Môle

Môle Saint-Nicolas

Baie-de-Henne

MASSIF DU NORD

Phaèton

Dérac

Quanaminthe

Les Trois Rivières

Mémé

Ennery

Golfe de la Gonâve

Bouyaha

Plaine Centrale

Canot

MONTAGNES NOIRES

Libon

Morne la Selle 8,793 ft. 2680 m.

Plaine de L'Artibonite

Artibonite

DOMINICAN REPUBLIC

Île de la Gonâve

Montrouis

CHAÎNE DES MATHEUX

Artibonite

Lascahobas

Anse à Galets

Mirebalais

Baie de Port-au-Prince

Silbert

Lago Enriquillo

Dame-Marie

Grand 'Anse

Grande Cayemite

Corail

Petit-Trou-de-Nippes

Canal du Sud

★ **Port-au-Prince**

Léogâne

Étang Saumâtre

Anse-d' Hainault

MASSIF DU SUD

Miragoâne

Trouin

MASSIF DE LA SELLE

Belle-Anse

Les Anglais

Cavaillon

Aquin

Côtes-de-Fer

Bainet

Marigot

Coteaux

Port-Salut

Île À Vache

Pointe L'Abacou

Caribbean Sea

Haiti

mulattos, who placed themselves above blacks. These distinctions have continued to haunt Haiti into the waning days of the twentieth century. A Creole proverb is a typical reminder of the relationship: *Milat pov se neg, neg rich se milat* (A poor mulatto is a black man, a rich black man is a mulatto).

Uprisings Led to Independence

In 1791 black slaves and mulattos began a series of uprisings against French rule which ultimately led to Haiti's independence in 1804, only the second country in the Western Hemisphere–after the United States–to declare sovereignty from a European power. Such a momentous victory may have united many other countries, but not Haiti. Mulattos and blacks had

been divided and pitted against one another during a long period of enslavement. The euphoria of independence quickly disintegrated into a power struggle between the two groups. In 1806, the country, nearly destroyed by war, broke into a northern monarchy led by a former black slave and a southern republic led by a mulatto. The country was reunited in 1820.

Haiti was ostracized after declaring independence. The world's powers relied on slavery to drive their economies, and they were not ready to accept a free black nation. The United States benefited from Haiti's struggle for freedom. Americans eagerly helped the blacks and mulattos in their fight against the French, theorizing that France could never exploit or defend the Louisiana territory without using Haiti as a jumping point. Instead, occupied by events in Europe and the Haitian insurrection, Napoleon Bonaparte (1769–1821) was forced to sell Louisiana to the United States.

The legacy of French colonial rule left Haiti woefully unprepared for independence. Most Haitians were illiterate and divided by race and class. The elite mulatto lived in towns and controlled trade, the military, and government. They mimicked European culture and used the French language to conduct all of their affairs. The peasants, blacks who made up the majority of the population, lived on tiny farms, spoke Creole, and were excluded from power.

For most of the nineteenth century, dictatorial governments did little to improve the lives of Haitians. By the late 1800s, Haiti faced the recurring fear of outside intervention. The Germans had invested heavily on the island and wielded a great deal of power. Their presence did not escape notice. The United States on the other hand, desired that Europe stay out of the Western Hemisphere. The Americans feared that the German presence in Haiti could ultimately pose a security threat.

U.S. Invaded Haiti

In 1915, the murder of a Haitian president by an angry mob gave the United States a green light to invade Haiti. But Haiti's social and political instability were not the only reasons for intervening. The island was strategically located near the U.S.-administered Panama Canal. The United States was also concerned that Germany was preparing to open a military base in Haiti. Even while President Woodrow Wilson (1856–1924) advocated for the sovereignty of nations to the rest of the world, U.S. Marines were fighting and killing Haitians in their own land. Many Americans opposed the occupation, and the United States finally withdrew in 1934. Haiti quickly unraveled, suffering through a series of dictatorial governments. Two of the most notorious dictators of the twentieth century, the Duvaliers, ruled Haiti from 1957 to 1986.

Duvalier Regimes

Americans began to understand the atrocities of the François "Papa Doc" Duvalier (1907–71) regime in the 1960s, when refugees started pouring into the United States. Then, between 1972 and 1991, at least 55,000 and perhaps as many as 100,000 Haitians escaped to Florida, surviving a dangerous water crossing in small crowded boats. These "boat people" were fleeing the repressive government of Jean-Claude "Baby Doc" Duvalier (b. 1951), who succeeded his father as dictator-for-life.

The Duvaliers, especially "Papa Doc," sought to control every aspect of Haitian society. They left a legacy that has been difficult to shake as Haiti heads into the twenty-first century. In early 1999 President Rene Preval was engaged in a power struggle with Parliament that threatened to spin the country into yet another crisis. With Preval and legislators

exchanging accusations, and political murders on the rise, Haitians prepared themselves for the prospect of another dictatorship.

Timeline

Pre-1492   The Pre-Columbian Era

When Columbus (1446?–1506) reaches the island that he names Hispaniola–now home to the Dominican Republic and Haiti–indigenous peoples known as the Tainos have already been living there for several hundred years. The Tainos belong to the language group known as Arawak, which means "meal" or "cassava eater." The Arawaks originally come from South America and over a long period of time, they move up the chain of Caribbean islands from Trinidad to Cuba.

On the island of Hispaniola, as many as 400,000 Tainos live in five different provinces, each with its own *cacique* (leader). The native people call the island *Ayti*, or *Hayti* (mountainous). The eastern part of the island is known as *Quisqueya* (mother of all lands), a name sometimes used by contemporary Dominicans to refer to their country.

Columbus describes the Tainos as kind and good-natured. Their diet consists of yucca, sweet potatoes, chilis, peanuts, and corn. They supplement their meals with fish, shellfish, turtles, and even manatee.

Their farm tools are made of sharpened sticks and ax heads of polished stone. Land is owned and worked communally. They live in simple huts covered with woven straw and palm leaves. They have many gods and practice polygamy. They are skilled in sculpture, ceramics, and basket weaving. They wear little or no clothes at all, but they wear jewelry, often made of gold.

The Tainos share the island with a small group of Caribs. The Caribs are said to be cannibalistic, and prey on the Tainos.

Columbus and Conquest

1492   Tainos welcome Columbus

Christopher Columbus sights the island of Hispaniola on his first voyage to the "Indies." On Christmas Day, 1492, Columbus' flagship, the *Santa Maria,* runs aground in the vicinity of what is now Cap Haïtien, in Haiti, on the northern coast. Taino natives welcome Columbus, who names the place Navidad (Christmas) and establishes a makeshift settlement.

The island is fertile, but Columbus and his fellow explorers are more interested in gold. The Spaniards discover that they can barter for gold with the Tainos, who wear golden jewelry. The Spaniards also obtain gold from alluvial deposits on the island.

The friendly *cacique* allows thirty of Columbus's men to remain in the area. Columbus returns a year later to find all his men dead, killed by the Tainos for abusing their hospitality.

1494   Tainos rebel against invading Spaniards

Columbus establishes a second colony on Hispaniola, named Isabela. The settlers plant melons, wheat, and sugarcane, but spend most of their time searching for gold.

The presence of gold, the base for the new mercantilist system in Europe, attracts many Spaniards to the "New World." Instead of settling the land, most of them are initially interested in getting rich quickly.

As soon as they settle in Hispaniola, the Spanish begin to abuse the Taino, taking their food, abusing their women, and forcing them into labor. The formerly peaceful Indians rebel, but they are quickly crushed by the Spaniards.

1496   Columbus's brother founds oldest city in the New World

Bartholomew Columbus (1445?–?1514) succeeds in founding a permanent settlement on the island. He names it Santo Domingo, the oldest city in the New World. It is here that the New World's first cathedral, first university, first hospital and first insane asylum are built.

Before the turn of the century, the surviving Tainos have been subdued. Thousands have died, killed by the Spaniards or diseases brought by the Europeans. Countless others commit suicide to prevent their capture. Many others attempt to flee. By the 1550s the island's Taino population has virtually disappeared.

Hispaniola becomes a logistical base for the conquest of most of the Western Hemisphere and remains an important way station for the next thirty years.

1499   Columbus makes changes before giving up governorship

Columbus rules the colony as royal governor until 1499, when he attempts to end the more serious abuses against the Taino Indians. He prohibits foraging expeditions against them and regulates the informal taxation imposed by the settlers. The Spanish settlers resent Columbus's new rules and force him to create the *repartimiento* system. Under the system, a settler is granted in perpetuity (forever) a large tract of land and the services of the Indians living on it. This system gives Spain's royalty little control over the settlers and the lands they occupy.

1500   Columbus accused of poor management

Columbus and his brother fall out of favor with the majority of the colony's settlers, mostly because of jealousy and avarice. The crown accuses them of failing to maintain order. A royal investigator orders both to be imprisoned briefly in a Spanish prison.

1503   Spanish Crown takes control

Feeling that they have too little control over the settlers, The Spanish Crown replaces the *repartimiento* system with the *encomienda* in 1503. Under this new system, all land in the New World becomes property of the Crown, and the native people living on it are considered tenants on royal land.

The settlers are granted encomiendas, land holdings that allow them to use a number of indigenous serfs to work the land. In exchange, the conquistadors teach them the Christian faith. In essence, the encomienda is a slave system. And many of those who rely on it abuse the indigenous population.

In theory, settlers can lose their rights to the land if they abuse the native people under the encomienda system, but it simply serves to strengthen local authority and centralize power.

The new system does not curb the death of native peoples, and the Spaniards begin to import African slaves into Hispaniola and other holdings in the New World. By the 1520s Spain relies almost exclusively on slave labor. Only 500 Taino out of several hundred thousand who lived on the island before Columbus' arrival are alive by the late 1540s.

1509   Columbus's son becomes Hispaniola's new governor

Columbus' oldest son, Diego (1480?–1526), becomes the new governor and serves until 1515. He also holds the important position of viceroy of the Indies from 1520 to 1524. Diego marries into Spanish aristocracy and builds a magnificent palace in Hispaniola, the Alcazar.

His ambitions arouse the suspicions of the crown, which establishes the *audiencia* in 1511 to diminish the governor's power. The *audiencia* is a tribunal composed of three judges whose jurisdiction extends over all of the West Indies. It is the highest court of appeals. The audiencia eventually spreads throughout the New World under Spanish rule.

1524   Audiencia's power grows

The tribunal's influence grows as the Spaniards begin to conquer lands in the Americas. Its name is changed to the Royal Audiencia of Santo Domingo, which has jurisdiction in the Caribbean and Mexico, and along the Atlantic coast of Central America and the northern coast of South America.

The audiencia represents the crown, with expanded powers in administration, legislation, and other functions. Its decisions in criminal cases are final but civil suits are appealed to the Royal and Supreme Council of the Indies in Spain.

Charles V (1500–58) creates the Council of the Indies in 1524 to direct colonial affairs. During most of its existence, the council exercises almost absolute power in making laws, administering justice, controlling finance and trade, supervising the church, and directing armies.

Despite these changes, Santo Domingo's influence begins to decline, first with the conquest of Mexico by Hernán Cortés (1485–1547) in 1521 and the discovery of gold in large quantities both in Mexico and later in Peru. At this time, alluvial deposits of gold have been nearly exhausted in Hispaniola, and many settlers leave for Mexico and Peru. New settlers bypass Santo Domingo, and the island's population begins to fall.

1564   Earthquake destroys two cities

An earthquake destroys Santiago de los Caballeros and Concepcion de la Vega, the island's two main inland cities.

1586   Santo Domingo is attacked

The Western Hemisphere has been parceled by papal decree between Spain and Portugal, a decision that is not accepted by the other European powers.

The Spaniards attempt to maintain a monopoly on trade and colonization in the Americas. But Britain, Holland, and France begin to challenge Spanish supremacy, first resorting to smuggling, piracy, and war, especially in the Caribbean region.

The English Admiral Sir Francis Drake (1545?–1596) sacks, burns, and leaves Santo Domingo in ruins in 1586, but the Spaniards are more concerned about the French, who try to establish several colonies along Hispaniola's northern coast.

1655   English fleet bypasses Santo Domingo and captures Jamaica

An English fleet commanded by Sir William Penn (1644–1718) tries to take Santo Domingo. After meeting heavy resistance, the English sail farther west and take Jamaica. The English use Jamaica as a base to support other settlements along the Caribbean Coast. Gradually English, French, and Dutch presence grows in the region.

1659   French establish permanent settlement near Hispaniola

The first settlers of Tortuga Island, just off the northern coast of Hispaniola, are reportedly expelled by the Spanish from Saint Christopher (Saint Kitts). English and French begin to inhabit the island of Tortuga as early as 1625.

The French settlers sustain themselves by curing the meat and tanning the hides of wild game and attacking Spanish ships. Soon the French acquire the name buccaneers, derived from the Arawak word for the smoking of meat.

As Spain weakens, the buccaneers grow in strength and numbers, and Tortuga Island is officially declared a permanent French settlement by 1659 under the commission of King Louis XIV (1638–1715).

1664   French send clear indications they plan to take Hispaniola

The Spaniards withdraw from the northern coastal region of Hispaniola and concentrate their power in Santo Domingo. In the meantime, French settlers have been moving into the northwestern coast of Hispaniola.

The Spaniards destroy French settlements several times, but they continue to return. In 1664 the creation of the French West India Company to direct trade between the colony and France signals that it intends to colonize the western end of the island. For the next three decades, Spain and France fight for control of western Hispaniola.

French Colony Prospers with Slave Labor

1670   French establish first major settlement in Hispaniola

The French establish their first major community, Cap François (later Cap Français, now Cap-Haïtien), in Hispaniola. During this period, the western part of the island is commonly called Saint-Domingue.

1697   Treaty gives western third of island to France

In 1697 under the Treaty of Ryswick, Spain cedes the western third of the island to France. The exact boundary of this territory (Saint-Domingue–now Haiti) is not established at the time of cession and remains in question until 1929.

1750   France builds its wealth with slave labor

While the Spanish side of the island languishes, the French turn the western side into one of the richest colonies in the world by the mid-eighteenth century. By this time the colony produces about sixty percent of the world's coffee and about forty percent of the sugar imported by France. The colony accounts for almost two-thirds of French commercial interests abroad and forty percent of foreign trade.

France's emerging economy is built with forced labor extracted from African slaves, who will number as many as 500,000 by 1791. Voracious in its appetite for wealth and cruel in its treatment of slaves, the French colony continually brings new slaves from West Africa to keep its economy moving.

Many of the white French masters keep African women as concubines. Their union produces a small, elite mulatto population that sets itself apart and above the impoverished black slaves.

Saint-Domingue's colonial society consists of three classes: *les blancs*, or white colonists; *les affranchis*, or free blacks (usually mulattos, called *gens de couleur*), and black slaves.

The *affranchis* are not restricted from buying land or lending money and many of them become wealthy, but they have a subservient role in colonial society. The white landowners, or *grands blancs*, discriminate against them through legislation. Statutes forbid *gens de couleur* from marrying whites, practicing certain professions, wearing European clothes, carrying weapons, or even socializing with whites. The laws in essence create a caste system.

By the mid–1750s many slaves escape to the mountains and dense forests and begin to fight the white colonists. The runaway slaves become known as *marrons* (maroons). Their attacks plant the seed of rebellion against the French colony.

1758   Maroon leader draws on religion to oppose French colonialists

From their hideouts maroons begin to attack the white-owned plantations, securing provisions and weapons to avenge themselves against their oppressors. Their numbers grow, and bands of several thousands carry out hit-and-run attacks throughout the French colony.

While the maroons lack centralized organization and leadership, one maroon leader stands out. François Macandal leads a six-year rebellion (1751–57) that leaves as many as 6,000 dead. Reportedly a *boko*, or *voodoo* sorcerer, Macandal draws from African traditions and religions to motivate his followers. He is captured by the French and burned at the stake in Cap Français in 1758. Popular accounts of his execution say the stake snaps during his execution, enhancing his legendary stature.

Voodoo derives from many African religious beliefs and is considered the national religion of Haiti. The word *voodoo* means "spirit" and comes from the *Fon* language of Benin (formerly Dahomey) in West Africa. Popular misconceptions have created a negative stereotype of the religion. Books and movies depict *voodoo* as a cult of sorcerers and followers who practice "black magic."

In Haiti the *voodoo* religion takes shape over many years as African slaves from many different tribes bring their beliefs with them. They believe in spirits who act as intermediaries with a single God. Some of these spirits are ancestors of the living, while others represent human emotions and forces of nature. A system of beliefs and spirits unique to the slaves in Haiti develops over time. The spirits are known as *loas*. They are inherited or can be purchased by families. *Loas* are paid tributes to bring good fortune, protect, and attack enemies. Payment is usually in the form of food, drink, or other gifts offered during rituals.

Colonists attempt many times to Christianize the slaves, who refuse to yield to spiritual enslavement. In time the *voodoo* religion borrows freely from Roman Catholicism. Many Catholic symbols and prayers blend with *voodoo* rituals and traditions, creating a unique Haitian religion. Modern Haiti is about eighty percent Roman Catholic, while Protestants of various denominations number twenty percent. But most Haitians practice some aspect of *voodoo*.

1789   French Revolution outlaws slavery in France

By the late eighteenth century, the slaveholding system in Saint-Domingue is unstable. The system has immense economic benefits for whites, some for the *gens de couleur,* and none for blacks. But mulattos are not content with their status as second-class citizens and seek equality with whites. The events set in motion in 1789 by the French Revolution shake up and eventually shatter the social structure of the colony.

France's new government requires the white Colonial Assembly to grant *affranchis* the right to vote. The white colonists, who have rejected any measures to improve the lives of the black and mulatto populations, refuse to let mulattos participate in government and press for the continuation of slavery. Their intransigence leads to the first mulatto rebellion in the colony in 1790.

Vincent Ogé (1755–91) leads the rebellion, which fails when the white militia reinforces itself with a corps of black volunteers. The French constantly use racial tension between the *affranchis* and black slaves to their advantage.

Ogé and another freeman leader, Jean-Batiste Chavannes, are tortured. Many others are hanged, and others are sentenced to various terms in prison.

From Rebellion to Independence: A Bloody Struggle

1791: August 14   Blacks meet to plan rebellion

By 1791 the political and social situation in Haiti deteriorates rapidly. While whites appear to relent to mulatto demands for equality, black slaves are unable to change their status peacefully. Slave leaders begin to prepare for an insurrection against all white slave owners. The rebellion of August 22 begins a bloodbath that will claim thousands of lives on all sides and leads to a twelve year struggle for independence. In time, all of Hispaniola will be at war, and France, England and Spain will fight to control the island.

According to accounts of the rebellion, on August 14, black leaders meet to adopt a plan for an uprising against slave owners. The leaders seal their agreement with a voodoo ceremony conducted by Boukman, a maroon and voodoo priest.

Among the leaders of the rebellion are Pierre Dominique Toussaint L'Ouverture (1743–1803), Georges Biassou, who later makes Toussaint his aide; Jean-François, who will command forces, along with Biassou and Toussaint under the Spanish flag; and Jeannot, who is particularly noted for his savagery.

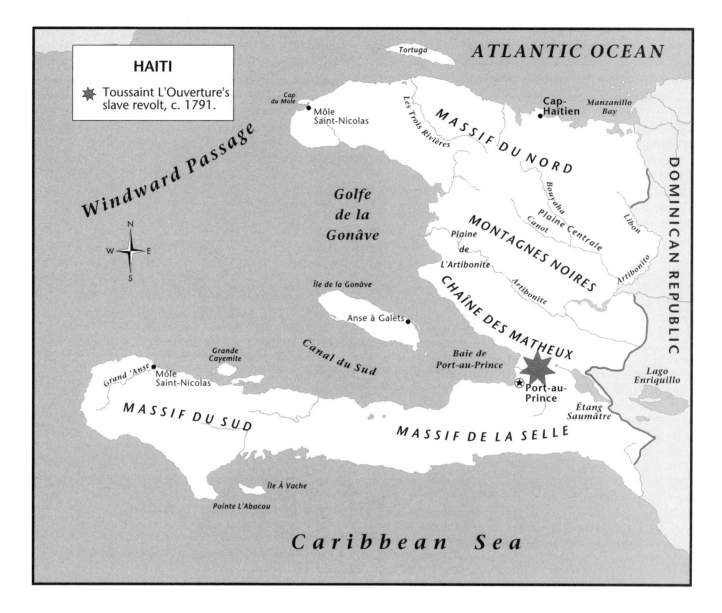

HAITI

⭐ Toussaint L'Ouverture's slave revolt, c. 1791.

ATLANTIC OCEAN

Tortuga

Cap du Mole

Môle Saint-Nicolas

Windward Passage

Les Trois Rivières

MASSIF DU NORD

Cap-Haïtien

Manzanillo Bay

DOMINICAN REPUBLIC

Bouyaha

Plaine Centrale

Canot

Libon

N W E S

Golfe de la Gonâve

Plaine de L'Artibonite

MONTAGNES NOIRES

Artibonito

Île de la Gonâve

Anse à Galets

Artibonite

Artibonite

CHAÎNE DES MATHEUX

Baie de Port-au-Prince

Lago Enriquillo

Canal du Sud

Grand 'Anse

Grande Cayemite

Môle Saint-Nicolas

Port-au-Prince

Étang Saumâtre

MASSIF DU SUD

MASSIF DE LA SELLE

Île Á Vache

Pointe L'Abacou

Caribbean Sea

1791: August 22   Rebellion claims thousands of lives

The revolt begins on August 22, and by the end of the month it spreads throughout the northern plains of Haiti. Whites are massacred and their houses and plantations are burned. The whites respond just as violently against the slaves. The rebellion leaves an estimated 10,000 blacks and 2,000 whites dead and more than 1,000 plantations sacked and razed.

The mulattos, under the leadership of Alexandre Petion (1770–1818) and others, also begin to mount attacks. The mulattos fight the white colonists (Royalists) but not the whites of the new French Republic, who favor mulatto enfranchisement and human rights.

The black (slave) forces also are split. Some fight against the white colonists, while others fight both the whites and the mulattos. Spain and Britain, involved in a power struggle with France, are poised to take Saint-Dominque away from the French.

1792   French try to assuage mulattos in Haiti

Civil commissioners from the French Republic are dispatched to Haiti to bring order, but they alienate white slaveholders when they try to put into effect a French decree granting mulattos full equality.

1793   English, Spanish involvement worsens Haiti's situation

By April French forces fighting under the republican banner take control of Cap Haitien with the help of thousands of blacks. Blacks have joined the French against the Royalists on the promise of freedom. In August Commissioner Léger-Félicité Sonthonax abolishes slavery in the colony. Some of the slaveholders appeal to the governor of Jamaica, a British colony. The British send troops to end the rebellion and take possession of Haiti.

Two black leaders refuse to commit their forces to France. Jean-François and Biassou believe allegiance to a king would be more secure than allegiance to a republic, and they accept commissions from Spain.

The Spaniards deploy troops in coordination with Jean-François and Biassou to take the north of Saint-Domingue. Toussaint, who joins the Spaniards in February 1793, comes to command his own forces independently of Biassou's army.

By the end of 1793 Toussaint cuts a swath through the north, swings south to Gonaïves, and effectively controls north-central Saint-Domingue.

1794: May 6   Toussaint pledges his support for France

Toussaint, a former slave, has been fighting on behalf of Spain, which has promised emancipation. But the Spaniards show no signs of keeping their word, and the British reinstate slavery in the areas they control. On May 6 Toussaint pledges his support to France, driven by France's decision on February 4, 1794, to abolish slavery. It is a crucial decision that seals Haiti's fate.

Toussaint turns his troops against his former Spanish allies and scores several victories.

By this time, Toussaint L'Ouverture has become one of the most important black leaders following the insurrection of 1791. Born in Saint-Domingue, Toussaint belongs to a small class of slaves who work as personal servants. Before the rebellion, he serves as house servant and coachman and receives an education, becoming one of the few literate black revolutionary leaders. He reads works by Julius Caesar (100 B.C.–44 B.C.) and others, which helps him with strategic and tactical planning when he joins Biassou's forces, and later, when he commands his own troops.

1794: June   British capture key positions

Some historians believe that Spain and Britain reach an informal arrangement to divide the French colony between them. Britain would take the south and Spain, the north.

British forces land at Jérémie and Môle Saint-Nicolas (the Môle), and in June they capture Port-au-Prince.

In the meantime, the Spaniards have launched a two-pronged offensive from the east, but French forces slow them down. Unable to advance towards Port-au-Prince in the south, the Spanish push through the north and occupy most of it by 1794.

Spain and Britain are on the brink of taking Saint-Domingue, but several factors keep them from attaining their goal. Tropical diseases thin British ranks far more quickly than battle against the French. Southern forces led by Rigaud (1761–1811) and northern forces led by another mulatto commander, Villatte, also prevent a complete victory by the foreign forces.

Toussaint comes to believe that only black leadership can assure the continuation of an autonomous Saint-Domingue. He does not trust the intentions of the world's powers regarding the future of slavery and he is weary of the mulattos and their intentions towards blacks. As he sets out to consolidate his political and military positions, he undercuts the positions of the French and the resentful *gens de couleur*.

1794: July 22   France and Spain agree to end hostilities

France and Spain agree to end their conflict. The agreement is finalized a year later with the signing of the Treaty of Basel. The accord directs Spain to cede its holdings on Hispaniola to France.

The armies of Jean-François and Biassou disband, and many of their former soldiers join Toussaint's army.

1796: March   Toussaint consolidates his power

Toussaint rescues the French commander, General Étienne-Maynard Laveaux, from a mulatto-led effort to depose him as the primary colonial authority.

To express his gratitude, Laveaux appoints Toussaint lieutenant governor of Saint-Domingue, giving the former slave a great deal of power over the affairs of his homeland.

A year later Toussaint becomes commander-in-chief of all French forces on the island and moves to establish an autonomous state under black rule. He expels Léger-Félicité Sonthonax, the French commissioner who earlier proclaims the abolition of slavery. And he negotiates an agreement to end hostilities with Britain.

1800   Blacks and mulattos go to war

Rigaud remains a powerful mulatto leader, and Toussaint, who wants to incorporate the majority of mulattos into his national project, seeks his allegiance. But the French consider Rigaud as their last opportunity to retain control of Saint-Domingue. They have used racial animosity before to remain in control, and they enlist Rigaud to regain Saint-Domingue.

Toussaint's predominantly black forces clash with Rigaud's mulatto army in what is sometimes called the War of the Castes. Toussaint seeks help from the United States, promising President John Adams (1735–1826) that France will not be allowed to use Saint-Domingue as a base to attack North America.

Adams is convinced France can't take Louisiana without first restoring sovereignty over Saint-Domingue, and he allows American merchants to send weapons, ammunition, and supplies to help Toussaint's forces defeat Rigaud.

1800: May   Toussaint becomes leader of all of Saint-Domingue

After securing all of Hispaniola, Toussaint becomes a military dictator and concentrates on restoring domestic order and the economy. He abolishes slavery but reinstates the plantation system, using enforced contract labor to produce sugar, coffee, and other commodities. While Toussaint never formally declares independence from France, his autonomy angers French leaders. Slave-holding nations, such as Britain and the United States, also become concerned that slaves in their countries may follow Saint-Domingue's example.

1802: January   French troops land in Hispaniola

Napoléon Bonaparte (1769–1821), French first consul, (his position in the French leadership prior to his becoming emperor) dispatches troops to Saint-Domingue. Bonaparte resents the former slaves' attempts to break away from France and considers Saint-Domingue as essential to French exploitation of the Louisiana Territory.

Bonaparte's brother-in-law, General Charles Victor Emmanuel Leclerc (1772–1802), lands with 16,000 to 20,000 French troops on the north coast of the island on January 2.

Toussaint commands an army of similar size, but the French, with help from white colonists and mulatto forces commanded by Alexandre Pétion and others, eventually out-match his troops.

Two of Toussaint's chief lieutenants, Jean Jacques Dessalines (1758–1806) and Henri Christophe (1767–1820), agree to transfer their allegiance to France when they realize they cannot win.

1802: May 5   Toussaint surrenders

Toussaint surrenders to Leclerc after receiving assurances that he will be allowed to retire quietly. But a month later he is seized and transported to France, where he dies of neglect in the frigid dungeon of Fort de Joux in the Jura Mountains on April 7, 1803. Toussaint is considered a national hero.

The betrayal of Toussaint and Bonaparte's restoration of slavery in Martinique undermines the collaboration of Dessalines, Christophe, and Pétion. Blacks and mulattos become convinced that the same fate awaits them and begin to battle against Leclerc's forces.

Leclerc dies of yellow fever in November 1802, about two months after requesting reinforcements to end renewed resistance. Leclerc's replacement, General Donatien Rochambeau (1725–1807), wages a bloody campaign against the insurgents. Resistance and events outside Saint-Domingue doom France's attempts to hold on to the island.

1803   War resumes between France and Britain

By 1803 war resumes between France and Britain. Bonaparte concentrates his energies on the struggle in Europe but is unable to look after his interests in Saint-Domingue and Louisiana and signs a treaty selling the North American territory to the United States.

Rochambeau's reinforcements and supplies never arrive in sufficient numbers. He flees to Jamaica in November 1803 and surrenders to British authorities. The era of French colonial rule in Haiti comes to an end.

Independent Haiti: A Divided Nation

1804: January 1   Haiti proclaims its independence

In 1804 Haiti declares its independence, becoming the second nation in the Western Hemisphere, after the United States, to declare sovereignty from European domination. More importantly, Haiti becomes the first free black republic in the world. It is a unique position for this tiny, impoverished nation.

The world's nations refuse to acknowledge Haiti's independence, some fearing that the slave insurrection will spread to their borders. Other nations, which remain silent on slavery, are shocked by the brutality against white colonists. For enslaved peoples around the globe, the tiny new nation is a symbol of freedom.

Yet Haiti is in no position to inspire or support slave rebellions in other parts of the world. Its leaders have plenty of domestic problems to keep them occupied, and they fear continued intervention by foreign powers.

1804: October 8   Dessalines declares himself Emperor of Haiti

Dessalines, an ex-slave who commands black and mulatto forces during the final days of the revolution, becomes Haiti's leader under independence and rules under the dictatorial 1801 constitution. Independent Haiti has been devastated by years of war, most whites have been killed or forced to leave, and the rift between blacks and mulattos has not been mended. About ninety-five percent of Haitians remain illiterate.

Dessalines at first attempts to narrow the differences between blacks and mulattos. But it is a difficult undertaking for a man who does not trust mulattos and who is seen by mulattos as a crude, illiterate leader.

To improve agricultural production, Dessalines reestablishes the hated plantation system and institutes harsh measures to keep laborers in their assigned work places. Penalties are imposed on runaways and on those who protect them. The dictatorial and military nature of Dessalines' government sets a precedent for the new nation. In the next 150 years the armed forces play a major role in politics.

Dessalines' unpopular politics briefly unite black and mulattos in an insurrection. Dessalines is assassinated on October 17, 1806, splitting the nation into two rival enclaves.

1806: November   Racial politics divide country

Following the death of Dessalines, army officers and *anciens libres* (pre-independence freedmen) landowners elect a constituent assembly to establish a new government. They draft a constitution that creates a weak presidency and a strong legislature. The assembly selects Henri Christophe (1767–1820), a

Jean Jacques Dessalines (1758–1806), a Haitian nationalist, becomes the first ruler of an independent Haiti in 1804. (The Library of Congress)

former black commander under Toussaint, as president and Pétion, a powerful mulatto leader, as head of the legislature.

Historians see these appointments as the earliest attempts in Haiti to establish what later becomes known as the *politique de doublure* (politics by understudies). Under this system, a black leader serves as figurehead for mulatto elitist rule.

Christophe rejects the arrangement and attempts to take Port-au-Prince by force. When he fails, he retreats and establishes his own government in the northern part of the country, ruling from Cap Haitien.

1809 Land reforms have profound effect on Haiti

Pétion undertakes the first large-scale distribution of land in the Western Hemisphere, granting fifteen acres of land in perpetuity to every soldier. Larger grants are awarded to officers. He later authorizes the sale of public land at low prices. Some historians claim that large plantations are parceled into smaller plots as a backlash to the colonial slave-holding sys-

tem. The policy gives more land to more people, but it contributes to lower output and fewer exports.

Historians point out that former slaves are unwilling to work for wages in the larger plantations. That forces the owners to parcel out their land on a sharecropping basis. Haitian inheritance laws cause further divisions of the original plots. In time, the plots are only good for subsistence farming.

1811 Christophe declares himself king

Christophe declares himself King Henry I of Haiti, dividing the country into a northern monarchy and a southern republic led by Pétion. Other former revolutionaries attempt to establish their own governments Rigaud establishes a separate government in the southwest in 1810, but it collapses following his death a year later.

Christophe, a strict leader, installs a nobility of mainly black supporters who are given the titles of earls, counts, and barons. He surrounds himself with African warriors from Dahomey to enforce his laws.

1820: October Haiti unified after King Henry commits suicide

After Pétion's death, the republican senate selects General Jean-Pierre Boyer (1776–1850), the mulatto secretary and commander of the Presidential Guard, as the new president. He rules from 1818 to 1843.

Boyer, upon hearing that King Henry has committed suicide after losing control of his army, captures Cap Haïtien, once again unifying Haiti into a single nation. In 1822 he invades the Spanish side of Hispaniola, which in time sparks a Dominican rebellion.

1825 France recognizes Haiti's independence

France becomes the first nation to recognize Haiti's sovereignty, but at a high price for the small island nation. Boyer empties the treasury to settle claims with France. As a result, Haiti is broke and at the financial mercy of French banks.

1843: February Haiti spins out of control as Boyer is deposed

Haiti makes little progress during Boyer's presidency, and relations worsen between blacks and mulattos. By the late 1830s, a poet and liberal thinker, Hérard Dumesle, and his followers begin to decry Boyer's administration for its poor economy, lack of freedom, and dependence on imported goods.

They also criticize the country's elite for adhering to French culture instead of developing a national identity. This opposition group comes to be known as the Society for the Rights of Man and of the Citizen. But Boyer will not listen to them and his intransigence against change triggers violent clashes in the southern part of the country.

Charles Rivière-Hérard, a cousin of Dumesle, and his troops sweep through the southern peninsula toward the capital, where Boyer is informed that most of his army has joined the rebel forces. Boyer sails to Jamaica and Rivière-Hérard replaces him as president.

1843–1915   Twenty-two heads of state rule chaotic nation

During several decades, corrupt leaders compete in bloody struggles to rule this small nation. Between 1843 and 1915, only one head of state serves his prescribed term of office. Three of them die while in office, one is blown up in his palace, a second is hacked to pieces by a mob, and another one is possibly poisoned. Fourteen are deposed and one resigns before suffering the same fate. Among them is President Faustin Elie Soulouque (1785–1867), who in 1849 declares himself Emperor Faustin I. He is overthrown by Nicholas Fabre Geffrard (1806–79), who reestablishes the republic.

Overthrowing governments practically becomes a business venture. Foreign merchants–usually Germans–fund the rebellions with the expectation of substantial returns. The merchants are attracted by the potential profits they can gain from supporting a corrupt leader. The government does little to improve the nation, which remains deeply impoverished.

1844   Dominican Republic proclaims its independence from Haiti

Nationalist forces led by Juan Pablo Duarte seize control of Santo Domingo on February 27. The untrained Haitian forces quickly capitulate to the rebels. In March Rivière-Hérard attempts to recapture the eastern half of the island, but Dominicans put up fierce resistance. Haitian forces attempt to take the Dominican Republic in 1849, 1850, 1855, and 1856.

1860   Catholicism becomes national religion

Haiti signs an agreement with the Vatican in 1860 that expands the presence of the Roman Catholic Church in Haiti. While Haiti retains freedom of religion, education falls under the control of the Catholic Church. French religious orders establish and maintain Catholic schools, which in time become non-secular public schools. The new teachers are mostly French clergy who promote an attachment to French culture at the expense of Haitian culture. Few priests attempt to educate the peasantry, widening the social gap between the urban elite and lower classes.

1862   United States recognizes Haiti's sovereignty

Early in the nineteenth century, the United States helps or refuses to help Haiti in order to further its own interests against the European powers.

While it allows commerce, the United States withholds recognition of Haiti for nearly six decades. One of the reasons is prejudice against blacks and fear that recognizing a nation

of former slaves will inspire American slaves to launch their own rebellion. With the United States deeply divided by Civil War, President Lincoln (1809–65) recognizes Haiti in 1862.

The United States decides to recognize Haiti for various reasons: chiefly to keep other nations from taking it at a time when Americans have a greater economic stake in the Caribbean. American statesmen also believe they can resettle former American slaves in countries like Haiti or send them back to Africa. Congress appropriates $600,000 for resettlement purposes, but the plan eventually fails.

1870   U.S. sets its eyes on island of Hispaniola

Several times during the 1800s, the United States considers the annexation of the island of Hispaniola. For the Dominican Republic, annexation by the U.S. means protection from other nations, including Haiti. In 1870 President Ulysses S. Grant (1822–85) proposes annexation of the Dominican Republic, but the U.S. Senate rejects the idea.

By the late nineteenth century, the United States its on its way to becoming a world power, especially after winning the Spanish-American War of 1898. One of the spoils of war is Puerto Rico, but some policy makers in the United States favor a base in Haiti to protect its growing economic interests in the Caribbean and the Panama Canal, which is completed in 1914. While Haiti refuses to yield any territory to a foreign power, Germans begin to wield greater economic influence on the island.

1897: December   Germans wield great economic power in Haiti

German merchants become a major economic power in Haiti by the late nineteenth century and rely on military might to protect their interests. In December 1897 a German commodore in charge of two warships forces Haiti to pay an indemnity (repayment for loss or damage) to a German national who has been deported from the island. Five years later another German warship intervenes in a Haitian uprising.

1910   German influence in Haiti worries U.S.

In 1823 the United States adopts the Monroe Doctrine, which warns European countries not to intervene in the newly independent nations of the Western Hemisphere. In 1904 the United States, with the Roosevelt Corollary to the Monroe Doctrine, declares itself the policeman of the Western Hemisphere.

Despite these warnings, Germans appear set to widen their presence in the Caribbean. In 1902–03 Germany bombards a Venezuelan port to force that country to pay its debts. Americans remain concerned that Germany is poised to buy the Danish West Indies, now the U.S. Virgin Islands. By 1910 the German community in Haiti grows to about 200 people and is said to control eighty percent of Haiti's international trade.

The growing German influence in Haiti worries U.S. policy makers. The United States believes Germany wants to establish a military base on the island, threatening American interests and the Panama Canal.

1912–15 Bloody confrontations sets the stage for U.S. intervention

In August of 1912 President Cincinnatus Leconte dies in an explosion in the National Palace. In the next three years, five Haitians contend for the presidency.

In March of 1915, General Vilbrun Guillaume Sam, who had earlier helped Leconte reach the presidency, assumes power under hostile opposition. On July 27 Guillaume Sam executes 167 political prisoners, provoking violence in the streets of Port-au-Prince.

Guillaume Sam seeks refuge at the French Embassy, but he is captured and torn to pieces by angry protesters. The dismembered corpse is paraded through the streets.

Haiti's social and political instability, American trade, investments and growing concerns over German influence on Haiti, and the island's strategic importance lead to a U.S. invasion. Shortly after Guillaume Sam's body is dragged through the streets, U.S. Marines land in Port-au-Prince.

United States Occupation of Haiti

1915–34 U.S. Marines occupy Haiti

U.S. troops land in Haiti in 1915 and occupy neighboring Dominican Republica year later.

Americans bring some degree of stability and order to Haiti, which in its previous seventy-two year history has experienced 102 revolts, revolutions, civil wars, and coups. Yet while the United States brings needed improvements to Haiti–building new roads, hospitals, and schools–it does little to improve the lives of Haitians. When Americans finally leave the island in 1934, illiteracy and poverty rates remain unchanged.

The leaders of the military occupation, shaped by the racial prejudice of American society, have deep disdain for Haitians. The Americans discriminate equally against blacks and mulattos, excluding them from real positions of power in government and the new, U.S.-trained military force.

Immediately after the invasion, the occupying forces set out to reorganize the government. Under the command of Admiral William Caperton (1855–1941), a government with Haitian figureheads is installed. While Haitians continue to run local institutions, American representatives wield veto power over all governmental decisions and Marine Corps commanders become administrators in the provinces. Caperton declares martial law, which lasts until 1929. A new constitution allows the purchase of land by foreigners, a practice that had been outlawed by Dessalines in 1804.

Americans also establish the Gendarmerie d'Haïti (Haitian Constabulary), the country's first professional military force. Over time, the constabulary, which is later called the Garde d'Haïti (the National Guard) becomes a major player in national politics.

1918–20 Thousands of Haitians rebel against U.S. occupation

By 1918 many Haitians fear the Americans want to reinstate slavery. Admiral Caperton seems to be doing little to persuade Haitians that this is not the case. Caperton reintroduces forced labor to build roads, and workers are treated like prisoners. American forces are accused of committing atrocities against Haitians.

The same year thousands of Haitians begin an insurrection against U.S. forces. A Haitian leader, Charlemagne Peralte, organizes as many as 15,000 men to fight against U.S. Marines.

As many as 3,000 and perhaps many more Haitians are killed by U.S. troops during the conflict. One of the casualties is Peralte, whose half-naked body is lashed to a door and his photo distributed throughout the country to demoralize his followers.

An American general reports 2,250 Haitian casualties and only 13 American dead during a five-year period. Some Haitian historians say as many as 15,000 Haitians die during American occupation.

1929 Killing of Haitian peasants raises concerns over U.S. occupation

In December U.S. Marines kill at least ten Haitian peasants during a march to protest local economic conditions. President Herbert Hoover (1874–1964), who is concerned about the effects of the occupation, appoints two commissions to study the situation.

One of the commissions, headed by former governor general of the Philippines, W. Cameron Forbes (1870–1959), praises the physical improvements brought by the United States, but criticizes the exclusion of Haitians from positions of power. The commission points out that occupation has failed to curb the social forces that create instability, chief among them poverty, illiteracy, and the tradition or desire for an orderly, free government.

1934 The United States leaves Haiti

By 1932 the withdrawal of U.S. troops is well under way. Newly elected President Franklin D. Roosevelt (1882–1945) visits Haiti in July, 1934, and reconfirms the intent to fully withdraw. The United States leaves the National Guard in charge.

The nineteen year occupation leaves a legacy of anti-American feeling. The percieved intolerance of white Americans angers Haitians. A new generation of Haitian writers, historians, and artists begins to reflect racial pride in their work. Yet, while Haitians briefly unite against the racism of the occupying forces, racial and social divisions persist between blacks and mulattos.

From End of Occupation to Brutal Dictatorship

1937: October   Dictator orders death of thousands of Haitians

In 1937 Dominican Republic dictator Rafael Leónidas Trujillo Molina (1891–1961), seeking to expand his power over all of Hispaniola, sends his army to butcher an estimated 15,000 to 20,000 Haitians on the Dominican side of the Massacre River.

The massacre has great political implications on the Haitian side, where Trujillo may have supported an abortive coup attempt. Haitian President Stenio Vincent purges the officer corps of all members suspected of disloyalty, and he dismisses his top commander, Colonel Démosthènes Pétrus Calixte.

1950   Haiti celebrates bicentennial

Haiti marks its two hundredth year with a Bicentennial Exposition in Port-Au-Prince.

1950–56   Chaos returns to Haiti

After the U.S. withdrawal, leaders come and go under the watchful eye of the military, which holds virtual veto power over election results. In 1950 General Paul Magloire takes power in a military coup, ending a long period of political instability. But his economic policies lead to a serious depression. In December 1956 a strike by business, labor and professional leaders forces Magloire into exile. After his departure, seven governments attempt to reestablish control of the country.

1957: September   Elected president becomes brutal dictator

François Duvalier, (1907–71), a middle-class black physician known to his followers as "Papa Doc," maneuvers himself into the presidency, scoring a major victory in 1957. His followers take two-thirds of the legislature's lower house and all of the seats in the Senate.

From the beginning his supporters portray Duvalier as the savior of the nation, an honest and humanitarian public health administrator who only wants to serve his country. His propaganda machine portrays him as "one of the greatest leaders of contemporary times." Duvalier, whose earliest experience in government is as minister of labor, quickly sets the stage to take complete control of the country.

Haitians celebrate the opening of their country's Bicentennial Exposition in Port-au-Prince. (EPD Photos/CSU Archives)

A year following his election, Duvalier begins to rule by decree and in 1961 has himself elected for another six years. On June 22, 1964, Duvalier declares himself president for life in an election that is universally accepted as rigged in his favor.

In his acceptance speech, Duvalier shows why he is a man to be feared. Some of his words: "He wants to lead as a master. He wants to lead as a true autocrat," he says of himself. "That is to say, I repeat, he does not accept anybody else before him but his own person ...Other great citizens who lead their countries with firmness and with all the necessary savagery know what they are doing. Duvalier also, ever since he was practicing the profession of doctor, knew what he was doing."

Knowing that a powerful and independent army can replace him as easily as it has done with past presidents, Duvalier purges U.S.-trained army officers and replaces them with loyal military leaders. He creates an elite presidential guard and consolidates his power throughout Haiti by ruling largely through his private security force, the *tonton makouts* (derived from the Creole term for a mythological bogeyman) and other thugs known as *cagoulards*.

In time the *tonton makouts* surpass the power of the army, providing Duvalier with a feared network of paramilitary activists who keep Haitians in line. With extreme brutality, they extinguish all opposition to Duvalier. According to some estimates, as many as 50,000 Haitians are killed during his fourteen year rule.

Thousands of Haitians leave the country and settle in Puerto Rico and Venezuela. Duvalier opponents attempt to invade Haiti in 1964, 1969, and 1970. Many Haitians migrate to the United States, where they mount anti-Duvalier campaigns and attempt to invade Haiti in 1964, 1969, and 1970.

1964   Duvalier declares himself president for life

Following a pattern set by past presidents, Duvalier names himself President-for-life. His regime is marked by terror, corruption, and extremes of wealth and poverty. While his country languishes in poverty, Duvalier's associates become wealthy. At one point during his administration, the blatant misappropriation of U.S. aid money angers the Kennedy administration. The U.S. suspends aid, but after Kennedy's (1917–63) death, the United States follows a policy of support for Duvalier because of his anti-Communist stand and opposition to Cuba.

1971   Duvalier leaves his son in charge

On January 22, the elder Duvalier names his son Jean-Claude (b. 1951) as his successor. Papa Doc dies on April 21, and Jean-Claude, known as "Baby Doc," becomes president-for-life a day later. A playboy who has never shown an interest in politics, he assumes power at the age of nineteen.

Early in his administration, his mother, Simone Ovid Duvalier, manages important administrative, tasks while he attends ceremonial functions and lives a decadent lifestyle.

Baby Doc is not considered as brutal or as politically astute as his father. Although he seeks to ease political tension, his administration continues the practice of imprisonment and torture of opponents.

1978   Haitians attempt to reform educational system

In the 1970s Haitians attempt to reform the school system to make education more accessible to the poor. Before reforms, the school system is modeled on the French system, following a classical curriculum that emphasizes literature.

The 1978 reforms change the curriculum and materials and allow the use of Haitian Creole as the language of instruction in the first four grades. Despite reforms, education remains elusive for most Haitians. In 1982 more than sixty-five percent of the population over the age of ten receives no formal education at all, and only eight percent receive more than a primary education. While schools are free, they remain beyond the means of most Haitians, who can't afford the uniforms, supplies, and supplemental fees.

Statistics for 1990 show that only twenty-six percent of primary school aged children are enrolled in school. The same year, adult literacy rate is estimated at fifty-three percent.

1982   Haiti unfairly targeted as source of AIDS

The United States Centers for Disease Control mistakenly classify Haitians as a high-risk group for AIDS. The decision is based on earlier studies that wrongly suggest the disease originated in Haiti. Others trace AIDS to tourism in the 1970s, when cheap vacation packages brought visitors to Haiti, stimulating the growth of prostitution and sexually transmitted diseases. The Centers for Disease Control drop the high-risk classification in 1985.

While Haiti has one of the highest HIV infection rates in the Americas, other serious health problems affect the small nation. Haiti has the highest infant mortality rate and the highest maternal mortality rate in the Americas. Malnutrition and gastrointestinal diseases are responsible for more than half the deaths. In 1992 there is only one doctor per 10,855 people. During 1995, the average life expectancy was 55.5. In 1994–95 only twenty-four percent of Haitians have access to adequate sanitation.

1983   Pope John Paul II visits Haiti

Declaring that "something must change here," Pope John Paul II (b. 1920) calls for a more equitable distribution of income and more freedom for Haitians. His message revitalizes the opposition and leads to social and political activism.

Nuns and priests are angered by the poverty and suffering. Along with young Haitians, they help lead the opposition against Baby Doc.

1985–86   Haitians rise against Baby Doc

Demonstrations and raids on food-distribution centers quickly spread to several cities, including Cap Haitien. Jean-Claude attempts to quell the protests with a ten percent cut in staple food prices. He closes independent radio stations speaking against him and reshuffles his cabinet. Police and army units attempt to put down the rebellion.

1986   United States cuts aid to Haiti

In January U.S. President Ronald Reagan's (b. 1911) administration begins to pressure Duvalier to give up the presidency and leave the country. The United States declines to give him political asylum but offers to help him depart Haiti.

At first Baby Doc agrees to leave but changes his mind, provoking more violence in the streets. On January 31, the United States announces a cutback in aid to Haiti, symbolically isolating Duvalier.

Loss of military support finally convinces Duvalier to leave the country on February 7 and he and his family depart for France.

The Duvalier legacy leaves Haiti in shambles, with inconsequential political institutions, a ravished economy, rampant poverty, and the military in control.

1987   Creole earns status as official language

For most of its history, the French and Creole languages of Haiti have punctuated the social and racial divisions of the nation. It is not until 1987 that the constitution gives official status to Creole.

Some scholars believe Creole is derived from a pidgin that develops between French colonists and African slaves in the colonies. Others believe it is already a language before it is used in Saint-Domingue. Despite its origins, Creole is linguistically a separate language and not just a corrupt French dialect.

Some intellectuals use the term *ayisyen* (Haitian) for the language to distinguish it from the term "creole," which refers to many languages, and as a symbol of national pride. This wasn't always the case.

French and Creole serve different functions during the history of the nation. Haitian Creole is the informal everyday language of all Haitians, with no regard to social class. French language is used in government, the courts, schools and newspapers. While all Haitians speak Creole, only ten percent speak French.

Since only whites and mulattos spoke French in colonial times, the French language became an important distinguishing feature between those who were emancipated before the revolution (the *anciens libres*) and those who achieved freedom through the revolution.

The *anciens libres* ensured their superior status with their use of French, which is also associated with culture and refinement. Some of the most nationalist Haitians of the nineteenth century placed little value on Creole.

Attitudes toward Creole began to change in the twentieth century, especially after the United States occupation. Haitian intellectuals began to consider Creole as the authentic language of the country.

The first Creole newspaper appeared in 1943 and in the 1950s, some Haitians began a movement to give Creole official status. Their efforts seem doomed when the constitution of 1957 reaffirmed French as the official language. In 1969, Creole was given limited legal status and in 1979, authorities allowed the use of Creole in the classrooms. The constitution of 1983 declares that Creole and French are the national languages but French remains the official language.

An example of a Haitian proverb in Creole: *Bouch manje tout manje, men li pa pale tout pawol* (The mouth may eat any food but should not speak on any subject.) The proverb implies that discretion is important.

1987   Chaos engulfs nation

Before leaving the country in 1986, Duvalier had created the National Council of Government (Conseil National de Gouvernement or CNG). It is made up of two military officers and three civilians. Taking several steps that are well received by Haitians in 1987 the CNG releases political prisoners and disbands the *tontons makouts*. And it prepares the nation for a democratically elected government.

But hopes for democracy fade when the presidential election scheduled for November is postponed because at least thirty-seven people at the polls are killed by gangs of thugs and soldiers. Two presidential candidates are assassinated. The armed forces effectively control the country for the next three years, installing governments that have no credibility at home or abroad.

1990   New leader symbolizes Haiti's struggle for democracy

Jean-Bertrand Aristide (b. 1953), a priest at the St.-Jean-Bosco Church in Port-au-Prince, is one of many young and progressive Roman Catholic priests and nuns who together are called Ti Legliz or the Little Church. They have been organizing peasants and slum dwellers since the 1970s and openly challenge the country's rulers.

Aristide preaches a brand of liberation theology, which rises in the 1960s in Latin America to help defend the poor from oppressive governments. Liberation theologians use Jesus' teachings to raise political consciousness among the poor. Liberation Theology is not widely accepted by the Catholic Church and has been denounced by the Vatican.

By 1990 Aristide has grown into one of the most powerful priests in the country, with a vast following. Preaching from a small church in the heart of a slum, he talks about the need for a *lavalas*, a flood to cleanse the country of corruption. In retaliation for his tough words, his church is burned down.

1990: December 16   Aristide becomes president in historic elections

In October Aristide announces he will seek the presidency. With the help of a political coalition of fifteen parties, known as the National Front for Change and Democracy, Aristide wins the December election with more than sixty-seven percent of the vote.

Opponents attempt a coup a month before his inauguration, but can't stop Aristide from becoming the first democratically elected president in the history of Haiti.

In his acceptance speech Aristide proposes "a marriage between the army and the people." Aristide understands that bringing the armed forces under control is key to staying in power. Six of the top seven military commanders turn in their resignations, and Aristide names Colonel Raoul Cedras his chief-of-staff.

1991: September   Aristide is forced to give up the presidency

Aristide faces insurmountable problems, and he is unable to bring the country or the military under control. As human rights violations drop dramatically during his tenure, Aristide contributes to the climate of uncertainty with his strong rhetoric, which is interpreted as encouraging violence against the rich and the opposition.

With mounting political assassinations and disorder, signs point to an imminent military takeover. Cedras takes power and with the intervention of the French Embassy, he allows Aristide to board a plane for Venezuela.

A Haitian rights group announces that as many as 1,000 people are murdered within the first few weeks following the coup. Aristide supporters are beaten and human rights violations are documented daily by national and international organizations. In October the United Nations General Assembly condemns the coup and announces that it does not recognize the regime. Thousands of Haitian refugees board small vessels and begin to make their way to the United States.

1991: November 5   President George Bush rejects help for new regime

President Bush (b. 1924) signs a commercial embargo against Haiti on all products except for humanitarian aid. In Haiti the military regime tries to drum up support, but Haitians continue to defy the new government.

In the same month, the United States sends back 538 fleeing Haitian refugees, the first boatload to reach the United States since the coup. In the next two years the United States repatriates more than 30,000 Haitians.

1991–94   Aristide seeks help from international community

From exile, Aristide attempts to gain support for his return to Haiti. He appeals to international organizations and the United States. The United Nations and the Organization of American States forge an agreement with Cedras that would return Aristide to power in October, 1993. The military refuses to accept the deal, and Aristide pleads with U.S. President Bill Clinton (b. 1946) to help him.

1994: September   U.S. prepares to invade Haiti

The Clinton administration secures international support for a military invasion to force Cedras from power. Just before U.S. troops invade the island, President Clinton negotiates a peaceful solution to end the crisis.

1994: October   Aristide returns to Haiti

More than 22,000 American troops peacefully take control of Haiti. Aristide returns to power. Former military leaders go into exile.

Aristide, as he did when he was elected president, faces severe challenges. After a three-year absence, he is expected to rescue the country's ravaged economy and curb the violent street crime. The United Nations sends a peacekeeping force in March 1995 to help Aristide reestablish order.

1994   Tourism falls dramatically

While Haiti has not been a favored tourist destination in the Caribbean, visitors contribute millions of dollars to the economy each year. Starting in the 1980s, tourism steadily began to decline. In 1990, 143,700 tourists arrive in Haiti. But political problems, the military coup, and the alleged link between Haitians and AIDS dramatically affect the number of visitors. In 1994 only 70,200 tourists come to the island.

1995: December 17   New president elected to succeed Aristide

In March 1995, Haitians go to the polls to elect local and national leaders. The elections are generally peaceful and regarded as a success for an emerging and fledging democracy.

Many Haitians want Aristide to remain president. Aristide, who feels he has been cheated of the presidency by the coup, wavers about honoring a provision in the 1987 constitution that bars the president from seeking a second consecutive term. Ultimately, he agrees to step down and endorses Rene Preval, a close associate, for the presidency.

Preval receives ninety percent of the vote and takes office in February 1996, becoming Haiti's second democratically elected president in the country's 191-year history as an independent nation. He inherits a corrupt and inefficient bureaucracy with almost no money in the treasury.

1996   Preval agrees to reform economy

Despite stable political conditions, Haiti remains one of the world's poorest nations, with unemployment estimated at seventy to eighty percent. In May Preval agrees to economic reforms demanded by the International Monetary Fund. Haiti agrees to privatize state-owned enterprises.

Austerity measures provoke nation-wide protests, and by January 1997 Haitians burn tires and throw stones demanding a suspension of negotiations with international lending institutions.

1998   International aid to Haiti decreases

The United Nations Development program announces that international support for Haiti drops sharply over the two previous years. Yet at $47 per person in 1997, Haitians receive four times more aid than citizens of other developing countries.

In 1995 Haiti receives $540 million from foreign governments and institutions. The aid drops to $420 million in 1996 and $351 million in 1997. The United States, which remains the top contributor, withholds some aid because Haiti has been unable to resolve a political stalemate that leaves the government without a prime minister for more than a year.

1998: March 24   Haitian immigrants press for amnesty

Thousands of Haitians protest in front of the U.S. Capitol in favor of equal treatment under immigration laws. Haitians are alarmed by turmoil in their country and concerned they will be forced to leave the United States. They want Congress to create a general amnesty for Haitians who left the island before December 1995. An estimated 100,000 Haitians face potential deportation unless Congress acts on their behalf.

1999: January   Power struggle threatens fragile democracy

A power struggle between President Preval and Parliament reaches a crisis on January 11, when Preval announces he will dissolve Parliament and appoint a Prime Minister by decree.

Preval's move is seen as a powerful blow against Haiti's weak democracy, and his opponents claim the president has placed Haiti on the road to dictatorship again. Parliament has rejected at least five presidential nominees for prime minister since June 1997, when Prime Minister Rosny Smarth resigns, accusing the president of complicity in election fraud.

Two days after making his announcement, Preval's sister, Marie-Claude Calvin, is shot, and her driver is killed in an attack. She is not a leading political figure but her husband, Serge, is an executive-committee member of the Lavalas Family Political Party founded by Aristide.

The opposition has accused Preval and Aristide of conspiring to establish a dictatorship. Aristide is campaigning to return to the presidency in the 2000 election. He remains a powerful figure in Haiti, even though he has been accused of shady dealings.

"We are living in very difficult moments," Edgard LeBlanc Fils, who heads the Haitian Senate, tells the *Sun-Sentinel* of South Florida. "We are ashamed to be seen by foreign leaders in this way. But there are no relations between the branches of government."

Bibliography

Abbott, Elizabeth. *Haiti: The Duvaliers and their Legacy.* New York: Simon and Schuster, 1991.

Aristide, Jean-Bertrand. *Dignity.* Charlottesville and London: University Press of Virginia, 1996.

Dayan, Joan. *Haiti, History, and the Gods.* Berkeley: University of California Press, 1995.

Dupuy, Alex. *Haiti in the New World Order: The Limits of the Democratic Revolution.* Boulder, Colo.: Westview Press, 1997.

Lawless, Robert. *Haiti's Bad Press.* Rochester, Vermont: Schenkman Books, 1992.

Logan, Rayford W. *Haiti and the Dominican Republic,* New York: Oxford University Press, 1968.

McFadyen, Deidre, and Pierre LaRamee, eds. *Haiti: Dangerous Crossroads.* Boston: South End Press, 1995.

Nichols, David. *From Dessalines to Duvalier: Race, Colour, and National Independence in Haiti.* New Brunswick, N.J.: Rutgers University Press, 1996.

Plummer, Brenda Gayle. *Haiti and the United States: The Psychological Moment.* Athens: University of Georgia Press, 1992.

Ridgeway, James, ed. *The Haiti Files: Decoding the Crisis.* Washington, D.C.: Essential Books, Azul Editions, 1994.

Honduras

Introduction

In Tegucigalpa, the capital city of Honduras, life is difficult for new arrivals. Peasants with no land of their own come here looking for work, but the city has little to offer them. Most of them settle in the surrounding hills of Tegucigalpa, in the sickly shantytowns that surround the city, and try to survive.

The peasants are a reflection of this small nation's painful history. Historically, the country's leaders have been quite willing to sell the nation's resources, and the presidency has been a revolving door for dictators who have done little to develop the social structures of the nation.

Honduras has been exploited by its neighbors and by foreign companies who have often proven more powerful than the government. The American-owned banana companies that prospered in the early 1900s virtually ran the country. They manipulated elections, bribed government officials to avoid taxes and toppled presidencies to protect their interests. In time, Honduras became known as a "Banana Republic," a derogatory term widely used to describe Latin American countries economically controlled by the United States.

Geography

Honduras, with a total area of 43,278 square miles (112,090 square kilometers) is slightly larger than the state of Tennessee. It is bounded on the north and east by the Caribbean Sea, and on the south by Nicaragua and the Gulf of Fonseca, which opens to the Pacific Ocean. El Salvador is on the western border, and Guatemala lies on the northwest. The capital city of Tegucigalpa is located in the central highlands, home to seventy percent of the population.

Mountains dominate the country's landscape, with peaks reaching 10,000 feet. Rain forests, covered by constant mist, are found in the higher altitudes. The northern Caribbean area and the southern coastal plain have a wet, tropical climate. The country is rich in flora and fauna, with many varieties of tropical trees, birds, and mammals.

Spanish Colonialism and Independence

Well before the arrival of Christopher Columbus (1451–1506) in the Americas in 1492, the Mayas built one of their greatest cities—Copan—in Honduras. Copan prospered, becoming a major cultural and trade center for more than three centuries. But sometime in A.D. 800, Copan and the other Mayan cities were inexplicably abandoned. After the abandonment of the cities by the ruling class and priests, only Mayan peasants remained in the area, and other Amerindian indigenous groups, from as far as South America and Mexico, moved into Honduras.

The arrival of the Spaniards in present-day Honduras was fraught with dissent from the start. Four Spanish expeditions descended on Honduras at about the same time, claiming rights to the area's wealth. Colonial struggles for power set a precedent that would last for centuries.

During the colonial period disease and slavery decimated the indigenous populations of Honduras, and there was little wealth to sustain much Spanish attention. Isolated by its rugged terrain, the country remained poor and undeveloped during Spanish rule and was mostly ignored by its more powerful neighbors.

Independence of the Central American colonies from Spain in 1821 worsened Honduras's situation. Its two major cities, Tegucigalpa and Comayagua, fought for control of the newly independent province. But Honduras also faced outside pressures coming from the stronger colonies of Guatemala, Nicaragua, and El Salvador, which fought for supremacy in Central America. Honduras became a fully independent nation in 1838, but its troubles were not over.

National rivalries, political divisions, and civil wars in other nations affected Honduras deeply. For the next 150 years, the country was ruled by dictators and threatened by coups and coup attempts.

United States Involvement

Starting in 1889 Honduras increasingly came under the growing sphere of influence of the United States. In that year the first boatload of bananas was shipped from Honduras to the

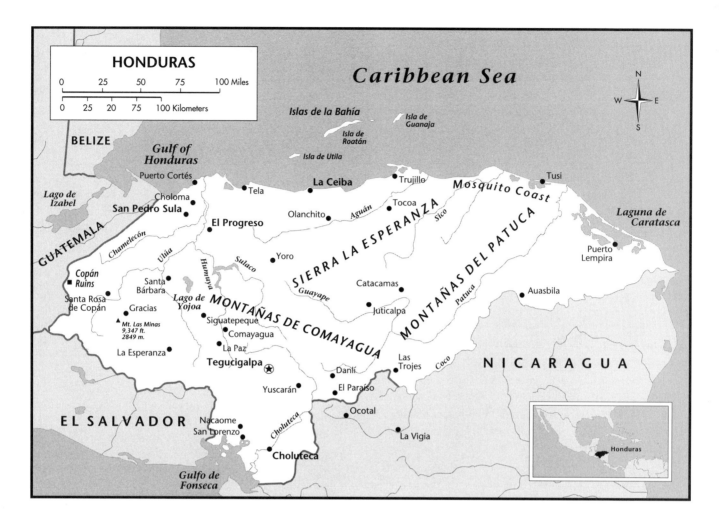

HONDURAS

0   25   50   75   100 Miles

0   25   20   75   100 Kilometers

Caribbean Sea

BELIZE

Gulf of Honduras

Islas de la Bahía

Isla de Guanaja

Isla de Roatán

Isla de Utila

Puerto Cortés

Tela

La Ceiba

Trujillo

Tusi

Mosquito Coast

Lago de Izabel

Choloma

San Pedro Sula

Olanchito

Aguán

Tocoa

Sico

Laguna de Caratasca

El Progreso

Chamelecón

Yoro

SIERRA LA ESPERANZA

Puerto Lempira

GUATEMALA

Ulúa

Humuya

Sulaco

MONTAÑAS DEL PATUCA

■ Copán Ruins

Santa Bárbara

Guayape

Catacamas

Patuca

Auasbila

Santa Rosa de Copán

Gracias

Lago de Yojoa

MONTAÑAS DE COMAYAGUA

Juticalpa

▲ Mt. Las Minas 9,347 ft. 2849 m.

Siguatepeque

Comayagua

La Esperanza

La Paz

Tegucigalpa ✪

Danlí

Las Trojes

Coco

NICARAGUA

Yuscarán

El Paraíso

EL SALVADOR

Nacaome

San Lorenzo

Ocotal

Choluteca

La Vigia

Honduras

Choluteca

Gulfo de Fonseca

United States. Within a couple of decades, as bananas became the mainstay of the Honduran economy, the American banana companies grew very powerful and began to dominate the politics of the country. With the growing investment, the United States government became concerned about Honduras's political instability. The United States made it clear that they would intervene if U.S. interests were threatened and dispatched warships as a demonstration of its commitment. The United States succeeded in maintaining stability but at the cost of democracy. For most of the twentieth century strong U.S. supported military men kept Honduras in check.

Honduras Base for Contras

In the early 1980s, U.S. foreign policy practically turned Honduras into a military base for the U.S.-supported *contras,* a guerrilla group fighting the leftist Sandinista government of Nicaragua. The Honduran military (which ruled in the background), with unprecedented economic aid from the United States, wielded enormous power in the small nation, committing numerous atrocities against its citizens in the name of fighting communism. For a while, it seemed democracy for this small nation was a vanishing dream.

In 1987 the five Central American nations signed an historic peace agreement, and began to look at their own internal problems. In Honduras, beginning in 1982, presidents sought to improve the economy and curb the power of the military.

In the last decade of the twentieth century, Hondurans have seen glimpses of a better future. In the early 1990s a vigorous president and human rights activist, Carlos Roberto Reina (b. 1926), pursued a tough agenda to bring the military under civilian control. His administration filed fraud charges against past officials who used their offices to enrich themselves. The November 1997 elections marked the fifth consecutive presidential election held peacefully since the country returned to democracy in 1982.

Today Honduras remains a poor nation, with an overwhelming number of people below the poverty line. It has one of the highest illiteracy rates in the Americas, and distribution of land remains one of the most contentious issues.

Continued struggle for democracy

The indigenous peoples of Honduras, long ignored, have found a voice in the emerging democracy and have pressured the government to uphold their rights. But not without a price. Many of their leaders have been murdered. The government also has come under pressure by faceless enemies. In recent years, frequent bombings have made it clear that the nation's struggles for a free and democratic society are far from over.

Timeline

A.D. 500   Mayas begin to build one of their great cities

A complex mixture of indigenous people live in pre-Columbian (before Christopher Columbus) Honduras. The most advanced are related to the Maya of the Yucatan (in present-day Mexico) and Guatemala. The Mayas reach western Honduras about A.D. 500 and begin to build Copán, a major ceremonial center. For the next 300 years, the Mayas develop Copán into one of the principle centers of their culture. From here they develop extensive trade routes spanning as far as central Mexico.

The Mayas excel in astronomical studies and art, and Copan becomes a leading center for both. Over a thousand years later, one of the longest Mayan hieroglyphic inscriptions is discovered at Copán.

A.D. 800   Mayas abandon Copán

At the height of the Mayan civilization, Copán—as well as other Mayan cities—is abandoned. The educated class—priests and rulers who build the temples and inscribe the hieroglyphs, study mathematics and astronomy, and do the art work—simply vanish. It appears that much of the population scatters throughout the area, but they have no memory of the meaning of the hieroglyphs and why the city is abandoned. The decline of the Mayas continues to puzzle contemporary experts. The last dated hieroglyph in Copán is A.D. 800.

With the breakup of the Maya empire, other indigenous peoples begin to inhabit what is now Honduras. Groups related to indigenous peoples of Mexico move into western and southern Honduras. The Chorotegas, who arrive from central Mexico, settle near the present-day city of Choluteca, while some migrate and settle further south in Costa Rica. The Nahua-speaking people, whose language is related to that of the Aztec, settle along the Caribbean coast and other parts of Central America.

There is also a northward movement of people. Indigenous groups with languages related to those of the Chibcha of Colombia establish themselves in northeastern Honduras. The Lenca, who also are believed to have migrated from Colombia, settle in central Honduras.

With so many different indigenous peoples in a small area, hostilities are frequent. Yet there is considerable trade within the region and with groups as far as Mexico and Panama. There appears to be a population of 500,000 in Honduras when the Spaniards arrive in the New World.

From Columbus to Conquest

1502: September 18   Columbus arrives in present-day Honduras

Christopher Columbus (1451–1506), the great explorer at the service of Spain, arrives in Honduras on his fourth and last voyage to the New World. He sails past the Islas de la Bahía (Bay Islands) and then reaches the mainland of Central America. At one of the islands, he captures a large canoe loaded with many goods. The paddlers appear to be Mayan traders.

Columbus spends little time in *Honduras* ("depths"), a name he gives the area apparently for the deep waters off the northern coast.

1523   Colonization gets under way

The Spanish do little exploration of Honduras until the 1520s. The expedition of Hernan Cortez (1485–1547) into Mexico and his conquest of the mighty Aztec Empire revive interest in the mainland. Expeditions from Mexico, Panama, and the Caribbean move into Central America.

Explorer Gil Gonzales Davila discovers the Golfo de Fonseca on the Pacific coast and one year later, four separate land expeditions move into Honduras.

The Spaniards fight one another for control of Honduras. Cortez, who arrives in the area in 1525, restores some order, but power struggles continue after his departure.

1534   Colony on the verge of collapse

Spaniards continue to fight for control of Honduras, with the native populations caught in the middle of the power struggle. The native peoples are quickly decimated through disease, mistreatment, and exportation of large numbers to the Caribbean islands as slave labor.

Pedro de Alvarado (c.1485–1541), the Spanish governor of Guatemala, intervenes in 1536 and keeps the town of Higueras from being abandoned. Alvarado earns a reputation as a cruel man who has no qualms against using violence to subjugate the native peoples of Central America.

Alvarado is attracted to Honduras by the prospect of gold and develops a profitable gold-mining industry. The discovery of gold and silver increases the demand for indigenous labor, which leads to renewed resistance by the native people.

1538   Spaniards murder Amerindian chief to end resistance

The cruelty of the Spanish conquistadors leads to serious revolt. The leader of the uprising is a young Lenca chieftain named Lempira. To his people, he is known as Lord of the Mountains. Lempira establishes his base on a fortified hill known as the Penol de Cerquin. The Spanish cannot subdue him and, inspired by his example, other native people begin revolting.

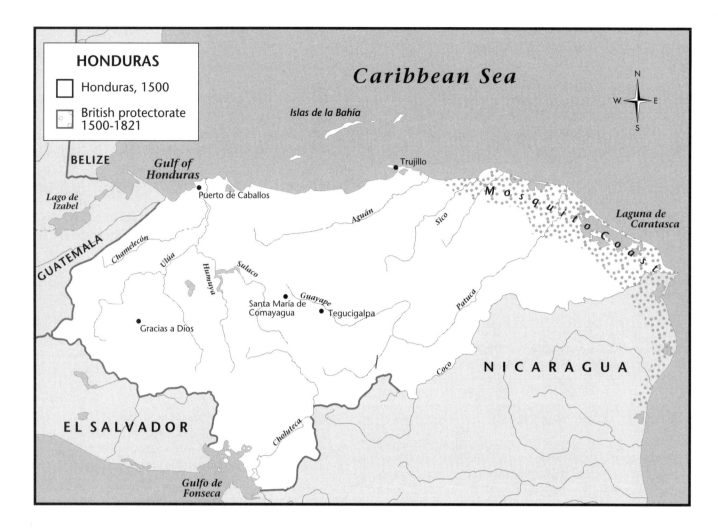

HONDURAS

☐ Honduras, 1500

☐ British protectorate 1500-1821

Lempira commands a large army and at one time rules as many as 200 small towns. He is considered a deity by many of his followers. The Spanish, however send an army of 800 Spaniards and Indian allies to deal with Lempira, who is known to have killed 120 men in battle. Lempira is murdered while negotiating with the Spaniards. His death demoralizes his people, quickly bringing an end to the war.

Lempira is declared a national hero after Honduras gains independence from Spain. The country's monetary unit is the Lempira, and his face appears on some coins.

1539 Decimation of native population follows Lempira's death

After Lempira's death, Honduras falls under the jurisdiction of Guatemala and is divided into two provinces: Comayagua and Tegucigalpa.

By 1541, almost half of the 15,000 native people under Spanish control have died. Those who remain become victims of the Spanish *encomienda* system. Spaniards rely heavily on their soldiers for the economic and social development of their provinces. The soldiers are granted *encomiendas*, land holdings that allow them to use a number of indigenous serfs

to work the land. In exchange, the conquistadors teach them the Christian faith. In essence, the encomienda is a slave system, and many of those who rely on it abuse the indigenous population. With declining native populations, the Spaniards treat the remaining people even more ruthlessly.

1542 Catholic priest defends native people

The exploitation leads to serious clashes between Spanish settlers and their leaders against the Roman Catholic Church. Father Cristóbal de Pedraza becomes the first bishop of Honduras. He tries to stop the abuses of the native people with little success.

1643 English-Spanish rivalry turns violent

An English expedition destroys the town of Trujillo, at the time the major port for Honduras. Late in the sixteenth century, Dutch and English *corsairs* (pirates) attack the Caribbean Coast and threaten Spanish rule. In time the English attempt to colonize the Caribbean Coast, placing them in direct conflict with the Spaniards.

In Honduras the English get help from two groups known as the Sambo and the Miskito, racially mixed people of

Native American and African ancestry. The British manage to establish settlements in Honduras, but they soon come under attack by the Spanish.

1545 Slave labor fuels economic growth

Conflicts among the Spaniards decrease after Lempira's death, and they begin to develop cattle ranching and large-scale agriculture. Mining gold and silver drives the growing economy.

With the rapidly decreasing native populations, the Spanish bring African slaves into Honduras—as many as 2,000 by 1545.

Mining gives Honduras political weight in the new Spanish colonies and in 1544 it becomes the seat of government for the Central American provinces. In 1549 the capital is moved to the more populated Antigua, Guatemala. When gold and silver production decline in the 1560s, Honduras quickly loses its importance.

1569 City of Tegucigalpa grows in importance

New silver strikes near Tegucigalpa revive the economy. Tegucigalpa begins to rival Comayagua as the most important town in the province. But the silver boom peaks in 1584, and economic depression returns. By the seventeenth century Honduras is neglected and practically forgotten by the Spanish colonial empire.

1780s Spanish drive out British

During the 1780s, the Spaniards manage to regain control over the Islas de la Bahía and drive the majority of the British out of Honduras. In 1786, the Anglo-Spanish Convention gives recognition of Spanish sovereignty over the Caribbean Coast. But the British manage to hold on to British Honduras, which gains independence as the nation of Belize in 1981.

Building a Nation

1808 Overthrow of Spanish king leads to independence

In the early nineteenth century, Spain goes into rapid decline and in 1808, Napoleon Bonaparte forces the Spanish king to abdicate. Spanish people revolt, setting off a chain of uprisings throughout Latin America. The idea of independence has been circulating for many years in the Americas, and the power struggle in Spain provides the perfect opportunity to rise against the Spanish. Independence is quickly declared in many parts of Latin America, including powerful Mexico.

The present nations of Guatemala, Honduras, El Salvador, Nicaragua, and Costa Rica and the state of Chiapas in Mexico have been functioning as colonial provinces during Spanish rule. Antigua, Guatemala, is the capital city, but each of the provinces operates with different degrees of autonomy.

While they have a shared history and culture, the provinces develop in different ways during colonial rule. Isolation, the makeup of native and Spanish populations, local economies and their leaders, among many other factors, help shape the provinces.

1821: September 15 Central America declares independence

Shortly after Mexico declares independence from Spain, the Spanish colonies of Central America also proclaim independence. The news of independence is not received warmly in Honduras, where there is no strong popular movement to break away from Spain.

The provinces of Comayagua and Tegucigalpa are caught in their own struggle and react differently to the news. Tegucigalpa wants to join a union of Central American states, while Comayagua favors union with Mexico, which is ruled by Agustin Iturbide (1783–1824). Their differences almost lead to war, which is averted when Guatemala decides to join Mexico, carrying all the provinces along.

Iturbide rules Mexico and Central America for about a year, but geographical isolation, colonial heritage, and social fragmentation stand in the way of a unified empire. The union collapses when Iturbide is forced from the throne, and Mexico is declared a republic.

1823: August 20 Former Spanish colonies break away from Mexico

Mexico, which holds on to Chiapas, recognizes Central American independence. Honduras joins the United Provinces of Central America, along with Guatemala, El Salvador, Nicaragua, and Costa Rica. The provinces set up their capital city in Guatemala.

In Honduras, the competing provinces of Tegucigalpa and Comayagua agree to enter the federation as one province, but the union is short-lived. Spanish rule has created an atmosphere of divisiveness and distrust, and the provinces agree on very little while also suspecting that Guatemala wants to dominate the union. Yet the greatest threats are the political differences between liberals and conservatives.

Generally in the Americas, "liberals" believe in the separation of church and state, with the state in a dominant position. Liberals believe formal education can improve society's problems, and they favor public education, freedom of the press, and religious tolerance. They seek to incorporate the native people into society.

"Conservatives" tend to align themselves more closely with the church, often belonging to the upper class, and do not generally favor social legislation. They prefer a centralized government with monarchical leanings and a church monopoly of education. They want to maintain native people in traditional, subservient roles.

Honduras produces two of the most prominent leaders of the union, the liberal Francisco Morazán and the conservative José Cecilio del Valle (1780–1834).

1826–42   A Honduran general goes to war to save the union

Francisco Morazán (1792–1842) is often called the George Washington of Central America. Morazán fights and is later executed for his attempts to save the union.

The United Provinces of Central America is doomed from the start as disputes between liberals and conservatives surface almost immediately. Morazán is a member of a wealthy colonial family, an educated Honduran who is attracted to the arts and sciences. He develops an interest in judicial and administrative posts and enters public life at the age of 29, when he helps defend Tegucigalpa during conflicts leading to the creation of the Central American union.

Morazán, like many of his contemporary liberals, is greatly influenced by the American and French Revolutions. During his tenure as president of the united provinces, Morazán seeks to establish democratic principles in a united Central America.

1829: April 13   Morazán and his troops enter Guatemala

Morazán and his troops, the "Allied Army for Protection of the Law," take over the federal capital in Guatemala, and he is declared president of the union. Morazán has little time for legislation because conflicts continue to break out throughout Central America.

Morazán moves the federal capital to the city of San Salvador, in El Salvador, in 1833, but rebellions continue. Slowly, the situation deteriorates, with a cholera epidemic worsening social stability. Wealthy landowners and the clergy push for a dictatorship or a monarchy to bring order. Morazán is asked to rule dictatorially several times, but he refuses.

1838: May 30   Morazán removed from office

By 1838 there appears to be no hope to save the United Provinces of Central America. Costa Rica withdraws and proclaims complete independence. In Guatemala civil war breaks out, and Honduras and Nicaragua attack El Salvador.

On May 30, the Central American Congress removes Morazán from office and declares that the individual states can establish their own governments. Later the same body recognizes the provinces as "sovereign, free, and independent political bodies."

Morazán holds on to El Salvador and later seeks refuge in South America.

1838: November 15   Honduras declares independence

The union proves to be a disaster for Honduras. Ideological disputes, chaos, and economic problems batter the small province. Even the British take advantage of the situation, retaking the Islas de la Bahía. Honduras formally secedes from the union in 1838 and adopts a constitution the following year.

In 1839 a Honduran army twice attacks El Salvador, which is ruled by Morazán. After both invasions fail, Honduras focuses on its internal problems. Through the rest of the nineteenth century, Honduras is in turmoil, with liberals and conservatives trying to gain control.

Honduras is also affected deeply by regional conflicts and political allegiances that transcend borders. Exiled opposition figures seek refuge from sympathetic states and use them as bases to launch attacks against their own governments. Honduras' neighbors constantly interfere in the country's internal politics well into the twentieth century.

1842   Morazán seeks to rebuild the federation

Morazán returns to Central America, toppling the repressive Costa Rican government of Braulio Carrillo, who has been declared president for life. He attempts to revive the Central American federation from Costa Rica but is overthrown and publicly shot in San Jose.

1845   Prominent priest encourages higher education

Father Jose Trinidad Reyes (1797–1855) founds an institute that later becomes the National University. Because of his humble origins, he is not allowed to attend a prestigious school in Honduras. He travels to Nicaragua, where he receives a degree in philosophy, theology, and canon law. He becomes a priest in 1822 and in 1840 is named bishop of Honduras. But Honduran leaders tell Rome church officials that he is dead. In 1844–45 he serves prison time, accused of opposing the regime.

In the next ten years, Reyes is named the poet laureate of Honduras and deputy to a Central American Congress. For his work as poet, educator, politician and priest, he is considered the father of higher education.

1855–57   American invades Nicaragua, threatens Central America

The liberals and conservatives briefly set aside their differences to battle William Walker, an American adventurer who invades Nicaragua with a private army.

Walker is a soldier of fortune, a racist who believes, like many Americans of his time, that it is his nation's "manifest destiny" to control other peoples. Born in Nashville, Tennessee, Walker first tries to provoke a secessionist movement in Sonora, Mexico, and fails. In June 1855—at the invitation of liberals engaged in a fight for power with conservatives—he arrives in Nicaragua with a private army.

Walker declares himself president of Nicaragua after defeating the conservatives and then sets his sights on Costa Rica. Walker's power grows when southern slaveholders offer

The former Presidential Palace in Tegucigalpa reflects Spanish influence. The building is currently a museum. (EPD Photos/CSU Archives)

him men, weapons, and money. In return, Walker promises to institute slavery in Nicaragua. He dreams of conquering the weak and divided Central American nations and turning them over to his supporters as part of a Confederacy of Southern States.

Walker is finally defeated in April 1857 and forced to leave Central America. In 1860 he attempts a landing in Honduras but is forced to surrender to a British sea captain who turns him over to Honduran authorities. He is promptly executed.

In the next two decades, the presidency changes hands many times, with liberals and conservatives caught in a bloody struggle for power. During this time, General Jose Maria Medina serves as president or dictator eleven times. In 1876, with help from Guatemala, Medina and his conservative supporters are driven from power.

1873–74   Small school budget reflects country's problems

In 1873 the government sets aside the equivalent of $720 for education, all of it designated for the National University. Honduras provides few opportunities for education. In the mid-nineteenth century, there are no libraries. Newspapers are not published on a regular basis, and there's little effort to promote culture. In the mid-1870s there are only 9,000 students attending school.

1875   One of Honduras's great writers is born

Froilan Turcios (1875–1943) is considered one of Honduras's best writers. His works deal with the supernatural, legend, death, love, and nostalgia. Among his best-known works are *Cuentos de Amor y de la Muerte* (Tales of Love and Death), *Fantasma Blanca* (White Ghost), *El Vampiro* (The Vampire) and *Paginas de Ayer* (Pages from the Past).

Turcios and Juan Ramon Molina (1875–1908) are the most prominent Hondurans in the *Modernismo* ("modernism") literary movement at the beginning of the twentieth century. The foremost modernist is Ruben Dario, the world-renowned Nicaraguan poet (see Nicaragua entry). The movement is a rebellion against the rigid academic standards of the day, challenging conformity and a pervasive naturalism. *Modernismo* emphasizes free verse, individualism, and creativity.

Honduras never develops a strong literary movement. The country has high illiteracy rates, poor educational institutions, and little money for literature. Most Hondurans write for journals or migrate to other countries to continue their work. However, poetry is a strong fixture of Honduran literature. Clementina Suarez (b. 1906), the first lady of arts and letters, is one of the leading poets.

1876: August   Popular president rules Honduras

Liberal president Marco Aurelio Soto (1846–1908) rules provisionally from August 27, 1876, to May 30, 1877, with support from Guatemalan General Justo Rufino Barrios. Soto rules through 1883, when he falls into disfavor with Barrios and is forced to resign.

Soto restores order to the chaotic nation and implements many basic reforms in education, finance, and public administration during his tenure. He is considered tolerant of political opposition and is credited with starting a postal and telegraph service, opening a national library and mint, and setting up free primary-grade schooling. Under him an effective separation of church and state takes place.

After Soto, Honduras again falls prey to foreign companies and governments. Power struggles are the norm, with the presidency a revolving door for dictators and ineffectual leaders.

The "Banana Republic"

1889   The rise of a "Banana Republic"

The Vaccaro Brothers of New Orleans ship their first boatload of bananas from Honduras to New Orleans in 1889. They are the founders of the Standard Fruit and Steamship Company (which later becomes Standard Fruit Company).

In 1902 railroad lines are built to handle an increase in banana exports. The Honduran government extends favorable exemptions from taxes. The banana companies are given free reign to develop roads, piers, and other infrastructure to support the industry. These companies quickly become powerful economic empires and influence government policy. Rivalries also develop among the banana companies, which control some of the most desirable coastal land in the country.

The presence of the American banana companies opens the door to U.S. intervention in Honduras's internal affairs. Soon the government of Honduras is simply a puppet whose strings are pulled by the United States government and the banana companies. Countries like Honduras give rise to the term "Banana Republic," a negative term often used to designate poor Latin American countries economically controlled by the United States.

Until the twentieth century, the United States plays a minimal role in Central America. But the country becomes heavily involved in the region after defeating Spain and capturing its Caribbean colonies in the Spanish-American War of 1898. During this time the Americans are building the Panama Canal (see Panama entry) and opening many business ventures in Central America.

The United States takes on the role of arbiter and active participant in Central American events, landing troops anytime its interests are threatened. It wields enormous power, even toppling governments that fall into disfavor. In October of 1913, President Woodrow Wilson delivers a speech in the name of democracy, warning European powers to stay out of the American hemisphere. But many Latin American leaders are wary of American intervention. United States interference in Central American affairs continues into the 1990s.

1917   Banana companies nearly spark a war between neighbors

The Cuyamel Fruit Company, supported by the Honduran government, begins extending its rail lines into disputed territory along the Guatemalan border. Guatemala is supported by the United Fruit Company and sends its troops to the border, threatening a war. The United States mediates an end to the conflict.

Appalling working conditions and low pay, among other factors, give rise to organized labor in Honduras in 1917. Workers strike for the first time against the Cuyamel Fruit Co., but the Honduran military intervenes to end the strike. Workers begin to rebel at other banana plantations.

1920   General strike hits Caribbean Coast

Workers call for a general strike against the banana companies. The United States sends one of its warships to the area, and the Honduran government arrests strike leaders. The strike collapses when Standard Fruit offers a new wage equivalent to $1.75 per day. Standard Fruit reports earnings of $2.5 million for the year.

1930   Honduras becomes one of world's leading producer of bananas

Banana exports peak in 1930, when Honduras produces one-third of the world's supply of bananas. The country also enjoys a brief period of peace. A year earlier United Fruit had acquired one of its major competitors, the Cuyamel Fruit Co.

But the onset of the Great Depression hits Honduras hard. Banana exports decline rapidly during the 1930s, and thousands of workers lose their jobs. Pay is reduced for those who remain employed. The government, which in 1931 borrows $250,000 from the banana companies to pay its army, violently suppresses workers' strikes.

1932–48   General dominates country's politics

A peaceful transition of power brings General Tiburcio Carias Andino (b. 1876) to the presidency. The elections stand in contrast to what is going on in other parts of Latin America, where dictators take over in many countries as the depression deepens.

General Carias doesn't waste any time consolidating his power and almost immediately begins to maneuver for reelection. His first order of business is to bring weapons from El Salvador to crush an attempted uprising. He strengthens the military and the secret police, founds the Military Aviation School, and names an American colonel to run it.

Carias outlaws the Communist Party and suppresses the labor movement, earning the backing of the banana companies. He reintroduces the death penalty and takes away women's right to vote. Many of his opponents are exiled or imprisoned. Carias cultivates close relations with other dictators.

Firmly in control of government, Honduras enjoys a period of relative peace and order. The country's economy improves slightly. Yet democracy suffers a setback, and Carias sacrifices national interests to bolster foreign interests, his supporters, and relatives.

At the same time, Carias introduces modest social reforms, including an eight-hour workday, paid holidays, and more money for education.

1952   United States threatens Guatemalan government

The new Guatemalan government of President Jacobo Arbenz Guzman expropriates lands owned by United Fruit Company, angering the United States. American policy-makers also are concerned that Arbenz's left-leaning government is encouraging banana workers to rise against the banana producers. The United States begins to consider overthrowing the Arbenz government. The developments in Guatemala eventually involve Honduras.

1954: May   More than 30,000 Hondurans go on strike

Thousands of banana workers go on strike to ask for better working conditions, medical benefits, overtime pay and the right to collective bargaining.

1954: May   United States and Honduras sign military agreement

The United States pressures Honduras and Nicaragua to stand against Arbenz. It sends large quantities of weapons to Honduras. Most of the weapons end up in the hands of Guatemalan exiles opposed to the Arbenz regime. The U.S. Central Intelligence Agency (CIA) has been planning a major covert operation with the assistance of the Honduran and Nicaraguan governments. Exiles launch an invasion from Honduras, forcing Arbenz into exile.

1954: July   Workers reach compromise with banana companies

The end of the Guatemala crisis also brings an end to the strikes in Honduras. Labor leaders accused of having ties to Guatemalan activists are jailed. The final settlement gives workers few gains.

1955: January 24   Women gain the right to vote

Honduran women win the right to vote for the first time.

From Puppet to Holder of the Strings: The Military Years

1956   Military coup removes contentious president

The military removes Julio Lozano Diaz from the presidency. Lozano, who is credited with making some positive social reforms, is widely popular at first, but his support vanishes when he attempts to hold on to power.

The coup marks a turning point for the military. For the first time in the country's history, military forces act as an institution. In the past, they had simply acted as servants of a political party or individual leaders. The military hands power back to a civil government in 1957, but it remains the most powerful institution in the country, the final arbiter in national affairs through the second half of the twentieth century.

1958   President gets social reforms under way

Honduras lacks a national education system until the late 1950s, when the government of Ramon Villeda Morales (1957–63) creates a public education system and begins to build schools. Before the reforms, few parents can afford to send their children to private schools.

President Morales also introduces a new labor code, establishes a social security system, and begins some agrarian reforms, which meet with mounting opposition from rich landowners and conservatives opposed to social legislation.

1963: October 3   Military coup topples Villeda

General Oswaldo Lopez Arellano, an air force officer, leads a military coup against Villeda's government ten days before presidential elections are to be held. Lopez immediately works to consolidate his power, even though the United States has broken relations with Honduras in opposition to the coup. Lopez suppresses leftist politicians and begins to dismantle Villeda's agrarian reforms. He is elected to the presidency in 1965 for a six-year term.

His early years in power are relatively quiet. But in the late 1960s, Honduras begins to experience economic decline along with severe social problems and growing opposition to his regime.

1969   Hondurans blame bad economy on Salvadoran immigrants

The government and private groups begin to blame economic problems on the 300,000 undocumented Salvadorans living in Honduras. Illegal land invasion is one of the alleged charges. Tensions mount between the two nations as some Salvadorans are forced to return to their small, overpopulated country.

1969: June   Soccer matches lead to war

In June Honduras and El Salvador are competing for a spot in World Cup for soccer. There are serious disturbances in the first game in Tegucigalpa. In the second game in El Salvador, Salvadorans desecrate the Honduran flag and beat up many of the visiting fans. In response, many Salvadorans are killed or beaten in Honduras. Thousands of Salvadorans leave Honduras.

1969: June 27   Honduras and El Salvador break diplomatic relations

Political and economic tensions between the two countries result in a break in relations.

1969: July 14   El Salvador attacks Honduras

The Salvadoran Air Force hits several targets inside Honduras, and the army launches a major offensive along the border, moving its troops about five miles into Honduran territory.

Honduras responds by blowing up oil storage facilities in El Salvador and crippling its air force. The Organization of American States demands a cease-fire. The war lasts about 100 hours, but peace is not accomplished for another ten years.

The war is devastating for both nations. Between 60,000 and 130,000 Salvadorans are displaced and more than 2,000 people, mostly Hondurans, lose their lives. In Honduras, support for the army plummets among the civilian population.

In June 1970, the two countries accept a seven-point peace plan, creating a "no-man's land" demilitarized zone running along the border. In 1973 they begin bilateral talks to end their disagreements. Finally, in 1980, El Salvador and Honduras sign a treaty settling their dispute.

1971   General Lopez steps aside for free elections

Lopez allows free elections in 1971 and Ramon Ernesto Cruz rises to the presidency. Lopez remains chief of the armed forces. Cruz doesn't last long. After a period of political instability, Lopez overthrows Cruz and returns to power on Dec. 14, 1972. He suspends the National Congress and all political activity.

1974   Banana tax scandal rocks the country

Honduras and several banana-growing nations agree to set a tax on banana exports. Honduras, which begins collecting the tax in April, suddenly cancels it four months later.

Soon, in what becomes known as "bananagate" in the United States, it is discovered that high-ranking Honduran officials have accepted $1.25 million in bribes from United Brands Co. in exchange for a fifty percent reduction in the tax.

1974: September   Hurricane devastates Honduras

Hurricane Fifi claims the lives of 10,000 people and causes widespread destruction throughout the country, hitting the banana plantations hard.

1975: March 31   Army officers remove Lopez from power

Lopez is implicated in the banana tax scandal, and the powerful Supreme Council of the Armed Forces strips Lopez of the presidency. The council is composed of twenty to twenty-five colonels who wield enormous power. They name Colonel Juan Alberto Melgar Castro to replace Lopez.

1978   Melgar's presidency short-lived

Melgar manages to hold on to power until 1978, when charges of corruption and military links with the illegal drug trade, contraband, and emerald traffic become widespread. Opponents accuse Melgar of failing to protect the nation, and he quickly loses support among the large landowners.

Melgar had promised free elections, but appears to make little progress in this regard. Eventually military leaders oust him from power replacing him with a three-man *junta,* which promises to hold elections.

The Long Road to Democracy

1982: November   Latest constitution seeks to balance power

Loss of faith in the armed forces helps ignite democratic reforms that lead to an elected president. A new constitution, which keeps the basic form of government Honduras has had under its fifteen previous constitutions, is drafted. Under the constitution, a strong president is elected by direct popular vote every four years. A key change deprives the president of the title of commander-in-chief of the armed forces. That responsibility is transferred to the army chief-of-staff.

It also provides for the popular election of deputies to the unicameral (having a single legislative chamber) National Assembly. They are elected to four-year terms concurrent with the president. Judicial power is exercised by the nine-member Supreme Court and is theoretically independent of the other two.

1982   New president assumes power

Roberto Suazo Cordova becomes the country's new president, but the armed forces retain broad powers, including veto power over cabinet appointments and responsibility for national security.

1982–87   Honduras caught in Central American conflict

In 1979 a leftist revolution topples the Nicaraguan dictator Anastasio Somoza Debayle. The United States has supported Somoza's repressive regime, and President Ronald Reagan fears other Central American nations will fall to leftist governments.

The Reagan administration begins to give millions of dollars in military aid to Nicaragua's neighbors and seeks to build an alliance against the Frente Sandinista de Liberacion Nacional (FSLN), the revolutionary group that topples Somoza.

Honduras is a major recipient of aid. From 1975–80, it receives $16.3 million in aid from the United States. From

1981–85, it receives $169 million. The country's military budget also skyrockets, from seven percent in U.S. funds in 1980 to more than seventy-five percent in 1985.

By 1983 several thousand anti-Sandinista guerrillas (popularly known as "contras") operate from Honduras. At the same time, Honduras, backed by the United States, helps the Salvadoran government in their fight against leftist guerrillas.

President Suazo Cordova works closely with the United States on domestic and foreign policy, and U.S. military presence in Honduras grows rapidly. The U.S. Central Intelligence Agency begins to use Honduras as a base for covert operations against the Sandinista regime.

Brigadier General Gustavo Álvarez Martínez, commander of the armed forces, orchestrates Honduras's support for U.S. policy. His opponents also accuse him of operating a personal death squad to hunt down leftist Hondurans. Violent repression of government opponents becomes common. As many as 184 left-wing activists may have been murdered by Alvarez's forces (see entry for 1998: May).

In time Hondurans begin to resent the contra presence and the close alliance to the United States. Military leaders force Alvarez out of power and exile him to Miami.

1984 U.S. military aid to Honduras reaches all-time high

Honduras armed forces receive $77.5 million from the United States. This represents an all-time high in U.S. military aid to the country.

1985: November New president seeks independent foreign policy

Jose Simon Azcona Hoyo (governed 1985–89) becomes the country's new president in the first peaceful transfer of power between elected executives in half a century. Azcona attempts to distance himself from U.S. foreign policy and criticizes U.S. support for the contras.

1987 Costa Rican president engineers Central American peace plan

Costa Rican President Oscar Arias Sanchez (b. 1940) receives the Nobel Peace Prize for structuring the Guatemala Peace Plan. On August 7, 1987, five Central American nations agree to end armed conflict gradually.

His peace plan, signed by Guatemala, Nicaragua, Costa Rica, El Salvador, and Honduras provides for free elections in all countries, a cease-fire by contra and government forces in Nicaragua, an end to outside aid to the contras, and amnesty for the contras. It also calls for repatriation or resettlement of all war refugees and an eventual reduction in the armed forces.

The plan significantly eases tensions in Central America and leads to free elections in Nicaragua. The Sandinistas lose in that election.

1989 Peaceful transition of power continues

With the threat of continued war in neighboring Nicaragua, Hondurans turn their attention to domestic issues. The new president, Rafael Leonardo Callejas (r. 1989–93), focuses on the economy. Callejas moves to reduce the deficit, but has little success in repairing the economy. During his administration, the lempira, the country's currency, is devalued by 350 percent.

1989 An activist's life story becomes a book

One of the most prominent political activists in the emerging democracy is Elvia Alvarado, who is trained by the Roman Catholic Church to organize women's groups to combat malnutrition. She is the grandmother of eleven children and only reaches the second grade. Under her leadership, a group of women farmers become part of one of the most prominent peasant groups in the country. In her book *Don't Be Afraid Gringo: A Honduran Woman Speaks from the Heart*, Alvarado talks about her imprisonment and torture and her country's plight.

1993 U.S. military aid slashed

U.S. military aid to Honduras is reduced to $2.7 million. It falls again in 1996, to $400,000.

1994 President Reina seeks to strengthen democracy

Carlos Roberto Reina, known for his support of human rights, clean government, and integrity, calls for a "moral revolution" to fight crime, poverty, and widespread corruption in government and private sectors. Reina is the former president of the Inter-American Court of Human Rights.

President Reina struggles to bring the military under civilian control. He abolishes the draft, including the notorious press-gang conscription of young men who are seized and forced into military service. His administration also dismantles the military-controlled Public Security Forces and replaces them with a new civilian force.

But Reina struggles with the country's economic problems. A drought in 1994 affects production of hydroelectric power, driving up food and fuel prices and causing chronic power outages.

In the meantime, the international financial community demands tough structural reforms, but poor Hondurans are unwilling to shoulder the sacrifices. Reina falls into disfavor with Hondurans, who blame the seventy-year-old president for high inflation, slow economic recovery, and an increase in violent crime. Still Reina manages to bring the country out of recession by 1995.

1995 Corruption charges pile up

By late 1994 Reina's aggressive attorney general's office begins to file formal charges of corruption against Callejas and other top government officials.

In August of 1995, Ernesto Paz Aguilar, who briefly serves as Callejas's foreign minister, is arrested on charges of selling Honduran passports at $25,000 each to thousands of Hong Kong Chinese citizens without authorization.

Attempts to indict Callejas, who also is implicated in the passport scandal, as well as in the illegal distribution of a $23-million oil fund and the illegal sale of state-owned machinery to a group of former employees have been unsuccessful.

1995 Illiteracy rate amongst highest in the Americas

The constitution of 1982 states that primary education is free and mandatory for children between seven and fourteen, yet few children receive proper education. Honduras has a small number of schools, and those that exist are poor and understaffed.

In 1990 the average classroom in primary schools has 37.2 students per teacher. In 1995 the rate of illiteracy among adults is estimated at nearly twenty-eight percent, with rates for women and men about the same.

1995 Bananas and coffee keep country going

Bananas and coffee account for fifty-one percent of total Honduran export revenues. The country remains one of the poorest in Latin America, with a per capita gross domestic product of less than $600. About seventy percent of Hondurans live below the poverty line.

1996 Government fortifies sugar with Vitamin A

From 1989–95, nearly twenty percent of children under the age of five are considered malnourished. In 1996 the government begins to fortify sugar with vitamin A. Health conditions are among the worst in the Americas, with few doctors and hospitals to treat the population.

1996 Emerging seafood industry helps Honduran economy

Commercial shrimp farming in the Pacific Coast province of Choluteca continues to grow. Shrimp and lobster exports bring $139 million in 1993 and $165 million in 1994. A virus affects shrimp production in 1995, but it bounces back with shellfish revenues of $178 million in 1996.

1996 The Armed Forces continue to shrink

Reina's military reforms help reduce the size of the armed forces from a high of 26,000 in the 1980s to about 12,000 by 1996. Only half of them are military personnel. The rest are members of the national police. With army recruits earning less than five dollars per month, there are few volunteers.

But Reina is careful not to push too hard and avoids unnecessary confrontations with the armed forces. His limited influence on the army is reflected by the inability of the government to bring active or retired military officers to trial for human rights abuses committed in the 1980s.

Military officers angry with the Reina government are suspected of setting off bombs targeting the president, human rights activists, Congress, and the Supreme Court. Several judges handling human rights cases have received death threats.

1996 Tourism increases

With relative social stability, tourism has become an important industry for Honduras. Tourism revenues grow from $29 million in 1990 to $156 million in 1996. The Mayan ruins of Copan and the country's natural beauty attract visitors.

1997 Honduras lowers banana tax

As part of economic restructuring demanded by the International Monetary Fund, Honduras agrees to drop the banana export tax from fifty cents for a forty-pound box to four cents by 1999.

1997: February Remains of ancient civilization discovered

Archaeologists discover a network of hidden caves with the remains of an unrecognized jungle civilization that lived in the shadows of the Maya and Olmec (ancient civilization in southeastern Mexico) empires.

James Brady, the archaeology team leader from George Washington University, tells the *Los Angeles Times:* "All of a sudden, we are showing that there was a tremendous early population" before the arrival of Christopher Columbus.

1997: March Honduras becomes a drug transit point

The Economist reports in its March 29, 1997, issue that Honduras has become a major transit point for illicit drugs going to the United States. More than 300 tons of drugs pass through Honduras, according to statistics by the U.S. Drug Enforcement Agency. Most of it is cocaine. The article reports that drug traffickers use dozens of airstrips once used by the contras in the early 1980s.

1997 Protest by indigenous groups

Indigenous groups organize massive demonstrations demanding land rights and investigations into the murder of their leaders.

1997: November Fifth consecutive presidential elections held

An unprecedented fifth consecutive election for the presidency is held, with the conservative Carlos Roberto Flores, capturing 52.8 percent of the vote.

His opponent is Nora Melgar, who served as mayor of Tegucigalpa during the Callejas administration. She is better known for her three years as first lady, when she was married to General Juan Melgar, one of the last Honduran dictators to rule the country (see entries under 1975 and 1978). Yet her candidacy is seen as a breakthrough for women in Honduras. She is the first woman to run for the presidency since women gained the right to vote in 1955.

Honduras has never experienced such an extended period of peaceful electoral politics leading to the 1997 elections. Part of the reason lies in declining political interference by the military and the willingness of politicians to accept electoral results. But Honduran democracy faces many challenges. Chief among them is the loss of faith among voters in electoral democracy.

There are no ideological differences between the two major parties, and the 1997 election, as in past elections, becomes a popularity contest with candidates exchanging personal attacks.

"Trapped in poverty, disgusted by corruption, and fearful of rising crime, many Hondurans have already concluded, perhaps too hastily, that democratic politics has little to do with them," writes J. Mark Ruhl in the February 1997 edition of *Current History, A Journal of Contemporary World Affairs.*

1998   International human rights group criticizes Honduras

Amnesty International, a human rights organization, reports continuous violations in Honduras. It accuses the Public Security Forces of attacking and sometimes raping young street children. It also reports the murder of several leaders representing indigenous groups.

1998: May   Mass grave discovered in Honduras

A human rights group announces the discovery of a mass grave with the remains of ninety-eight people, including the body of an American, James Francis Carney, a priest who had joined a guerrilla group in 1982 to fight the Honduran military.

James Francis Carney came to Honduras as a missionary in 1968 and was expelled from the country ten years later for his political activism in favor of peasants. He quit the Jesuit religious order in 1982 and traveled to Nicaragua, where he joined a Honduran guerrilla group opposed to the Honduran right-wing government. Honduran troops captured about 100 guerrillas, including Carney, in 1982, but it was not known what happened to them after their capture. Carney was among 184 people missing, believed to have died in the hands of the military.

1998   Hurricane Mitch

Hurricane Mitch strikes Honduras on October 30–31, 1998 and leaves a trail of death and destruction in its wake. The hurricane, the strongest North Atlantic storm in two centuries, renders approximately 7,000 Hondurans dead, and 400,000 homeless. Most deaths and dislocations result from floods and mudslides caused by the hurricane's torrential rains. The destruction includes nearly seventy percent of all roads and bridges as well as $800 million in crops.

Although much work remains before normal conditions return, emergency aid is having an effect. People are returning to their homes and nearly all bridges have been repaired. Much of the assistance is from USAID/OFDA which has provided over $2.3 million in assistance.

Bibliography

Alvarado, Elvia. *Don't Be Afraid Gringo: A Honduran Woman Speaks from the Heart.* Ed. Benjamin Media. San Francisco: Institute for Food & Development Policy, 1988.

Cline, William R. (ed.) *Economic Integration in Central America.* Washington, D.C.: Brookings Institution, 1978.

Durham, William H. *Scarcity and Survival in Central America: Ecological Origins of the Soccer War.* Stanford, Calif.: Stanford University Press, 1979.

Euraque, Dar'io A. *Reinterpreting the Banana Republic: Region and State in Honduras, 1870-1972.* Chapel Hill: University of North Carolina, 1996.

Karnes, Thomas L. *The Failure of Union: Central America: 1824-1975.* Rev. ed. Tempe: Arizona State University, Center for Latin American Studies, 1976.

Kelly, Joyce. *An Archeological Guide to Northern Central America: Belize, Guatemala, Honduras and El Salvador.* Norman, Okla.: University of Oklahoma Press, 1996.

Ruhl, Mark J. "Doubting Democracy in Honduras." *Current History: A Journal of Contemporary World Affairs.* vol. 96 (Feb. 1997): 81-86.

Schulz, Donald E. and Deborah S. Schulz. *The United States, Honduras and the Crisis in Central America.* Boulder, Colo.: Westview Press, 1994.

Todorov, Tzvetan. *The Conquest of America.* Trans. Richard Howard New York: Harper & Row, 1984.

Torres Rivas, Edelberto. *History and Society in Central America.* Austin: University of Texas Press, 1993.

Weinberg, Bill. *War on the Land: Ecology and Politics in Central America.* London: Zed Books Ltd., 1991.

Jamaica

Introduction

Jamaica is the third largest island in the West Indies, and the largest English-speaking Caribbean island. With an area of 4,244 square miles (10,991 square kilometers), Jamaica is slightly smaller than the state of Connecticut. The island's interior terrain is mostly mountainous, with the highest peaks belonging to the Blue Mountains in the eastern third of the island. The John Crow Mountains, composed of a limestone plateau, lie to the north. To the west lie rolling plateau land. The coastline is varied and includes beaches that attract the tourist trade, a mainstay of the Jamaican economy. Jamaica has a population of 2.5 million, with over 700,000 (nearly a third of the island's population) living in Kingston, the capital located on the island's southeastern coast.

Christopher Columbus (1451–1506) landed on Jamaica during his second transatlantic voyage in 1494 and claimed the island for Spain. Over the succeeding decades, the Spanish began to settle the island; the indigenous Arawak population was largely eliminated within a hundred years due to increased exposure to non-native disease and the hardships of enslavement. Over the next century and a half, the Spanish used the island mostly as a supply base, and it became a popular target for pirate attacks. In 1655, the British invaded and claimed Jamaica, which was to remain under the control of Britain for three hundred years.

Soon after the British takeover, Jamaica became a major sugar producer, an economic development that had long-lasting social consequences. To supply the vast amounts of labor required by the sugar industry, large numbers of slaves were imported from Africa, with Jamaica eventually becoming a major center of the slave trade in the Western Hemisphere. In the course of the eighteenth century, over 600,000 slaves were brought into Jamaica. By the end of the century, the island's blacks far outnumbered its white population, with 300,000 black slaves to 20,000 whites. As a result of the racial composition created by the institution of slavery, today over ninety-five percent of the Jamaican people are of African descent.

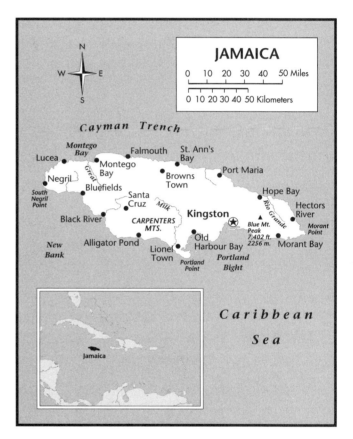

The course of Jamaican history was changed once again when first the slave trade and then (in 1834) slavery itself were outlawed in the British empire. The sugar industry declined, and Jamaican blacks began the long struggle for decent living conditions and full political rights as free citizens. The denial of political and land ownership rights to blacks sparked the Morant Bay Rebellion of 1865, in which over 1,000 blacks lost their lives.

In the first decades of the twentieth century, Jamaicans of all races developed a growing political consciousness, leading to the formation of the island's two major political parties, the People's National Party (1938) and the Jamaica Labour Party (1943). In 1944, Jamaica was granted political autonomy (self-rule) by the British Empire in internal matters, including the right to form its own parliament. Its first prime minister, elected the same year, was JLP leader Alexander Bustamente (1884–1977). During the 1950s Jamaica moved toward full independence, first becoming a member of the West Indies Federation (1958–1961) and then a fully independent nation (1962).

Since then, political rule has alternated between the PNP and the JLP. The JLP won the first elections held after independence, remaining in power for ten years. From 1972 to 1980, Michael Manley (b. 1923) (the son of PNP founder Norman Manley) presided over a socialist-style PNP government that implemented important reforms but alienated other Western governments by its leftist stance, which placed the nation at an economic disadvantage in its trade relations. During this period, Jamaican elections, while still democratic, became noted for the accompanying violence committed by partisans on both sides, which climaxed in 1980 when at least 600 lives were lost to election-related violence.

In the 1980s, the government turned to the right under the JLP and its leader, Edward Seaga (b. 1930). When Seaga called local elections a year ahead of schedule in 1983, the PNP withdrew from participation in the political system for much of the decade, until it regained power in the 1989 elections. Since then, three consecutive PNP governments have been elected, and Percival Patterson has served as prime minister since taking over from Michael Manley in 1992.

"Out of Many, One People"

Today Jamaica's national motto—"Out of many, one people"—is, on one hand, an apt descriptive phrase and, on the other, a goal still to be fulfilled. As a description, it applies to the racial mingling that has occurred on the island over time, between not only whites and blacks, but also encompassing the Asians who have emigrated to Jamaica since the nineteenth century, including East Indians, Chinese, and Lebanese. Culturally, the motto evokes the intermingling of the legacy left by the 300-year British presence on the island with the African roots of most of its inhabitants.

In the twentieth century, the discovery of these roots and their transformation into a source of black identity and pride were most memorably linked to one Jamaican: Marcus Garvey (1887–1940), who promoted black self-awareness and a worldwide sense of connection between blacks and their African homeland. Through his United Negro Improvement Association and his newsletter, the *Negro World*, Garvey gave hope and a new sense of dignity to blacks both in Jamaica and around the world.

A uniquely Jamaican offshoot of Garvey's ideas was the birth of the Rastafarian movement in the 1930s. Based on their connection to Africa—specifically Ethiopia and the newly crowned Haile Selassie (1892–1975) (see Ethiopia)—the Rastafarians developed a religion that provided them with

a set of principles and beliefs to live by, faith in a better future, and a unified community of believers. The Rastafarian movement has continued to grow and has spread to other countries around the world.

In the twentieth century, Jamaica's multicultural society has given rise to distinctively Jamaican achievements in the arts, and Jamaica has become one of the most important cultural centers in the Caribbean. Since the 1920s, visual artists have moved away from the older Eurocentric schools and emphasized themes derived from the black experience. Important pioneers in this regard were sculptor Edna Manley (1900–87) and painters John Dunkley (1881–1947) and Kapo (1911–89), who was also a sculptor. In Jamaican literature, poet and novelist Claude McKay (1890–1948) is celebrated for his works describing turn-of-the-century Jamaican life, especially the 1933 novel *Banana Bottom*, which some have hailed as the first true West Indian novel. Other important voices in Jamaican literature are those of Roger Mais (1905–55), Una Marson (1905–65), and, more recently, Anthony Winkler (b. 1942). Musically, Jamaica attained worldwide recognition with the rise of reggae, pioneered in the 1970s by Bob Marley (1945–81). Derived from Jamaican "ska" music and strongly rooted in Rastafarianism, reggae is a quintessentially Jamaican art form, and Marley, its most famous exponent, was known for his dedication to bringing together the diverse Jamaican people into a harmonious society.

While Jamaican culture has approached the ideal of "Out of many, one people," social conditions on the island suggest that there is still much progress to be made in overcoming the barriers and divisions that separate different segments of society. Poverty is widespread on the island, and unemployment is high. Many Jamaicans perform unskilled labor for extremely low wages or farm small patches of land, barely earning enough to survive. While the houses of the poor often lack amenities such as indoor plumbing and electricity, a small number of Jamaicans, mostly in the Kingston area and on the north coast, live in great wealth. Another element of disunity is the increasing violence that has plagued the country since the 1970s, some of it political, some drug- or gang-related, and some the result of excesses by the police themselves.

Nevertheless, in spite of their problems, Jamaicans can point with pride to their nation's political stability. Unlike other countries in the region, and in spite of strong divisions between its two parties, the political process in Jamaica since independence has, on the whole, remained democratic and orderly, with no military coups, assassinations, or other challenges to constitutionally mandated rule. In 1997, the PNP was returned to power for a third term, under the leadership of Percival Patterson, who, since 1993, has been the first black Jamaican to serve as elected prime minister of the country. Through cooperation with other countries in the region, such as that represented by the 1997 Caribbean Community and Common Market (CARICOM) meeting held in Montego Bay, Jamaica has the opportunity to unite with its neighbors to help realize the dreams of both its own people and the larger multicultural community of which it is a part.

Timeline

c. 200 B.C. to A.D. 1200   Arawak Indians settle in Jamaica

The peaceful Arawak Indians arrive on the island of Jamaica from the northern coast of South America, an area that today comprises Guyana, Venezuela, and northern Brazil. They call the island Xaymaca, today interpreted as either "land of springs" or "land of wood and water." ("Jamaica" is later derived from this name.) The Arawaks eventually settle throughout the island.

Living in villages of up to 3,000 people, they subsist as farmers and fishermen. They grow a variety of crops, including tobacco, cotton (from which they weave cloth), maize (corn), cassava (a plant whose roots are made into a starch), and other fruits and vegetables. English words derived from the Arawak language include "canoe," "guava" (a fruit), "barbecue," "hurricane," and "manatee" (a marine mammal).

1494: May 5   Columbus arrives in Jamaica

Christopher Columbus (1451–1506) lands on the north coast of Jamaica during his second voyage across the Atlantic Ocean. He describes Jamaica as "the fairest isle that eyes have beheld." Columbus and his men receive a hostile reception from the Arawak Indians they encounter and eventually subdue them in combat aided by crossbows, armor, and dogs. Columbus claims the island for Spain, naming it Santiago.

1503–04   Columbus returns to Jamaica

On his fourth transatlantic voyage, Columbus's two remaining ships (of four) run aground on Jamaica, and his expedition is marooned on the island for over a year. Eventually Arawak Indians transport them to the island of Hispaniola, from which he returns home on a rescue ship sent by Spain. Columbus dies on May 20, 1506, less than two years after his return to Spain.

1509: December 9   The Spanish begin to settle Jamaica

Christopher Columbus's son, Diego Columbus (c. 1480–1526), is appointed viceroy of Jamaica and sponsors the first party of Spanish settlers in Jamaica. They found the settlement of Sevilla la Nueva (New Seville) on the north coast of the island. The Spanish enslave the Arawaks, who are virtually eliminated from the island within decades due to harsh treatment and the diseases introduced by the Spaniards.

Spain makes few efforts to develop the island as a colonial center, using it mostly as a shipping base. The colonial government is corrupt and torn by internal strife, and the island falls prey to pirate attacks, often supported by rival

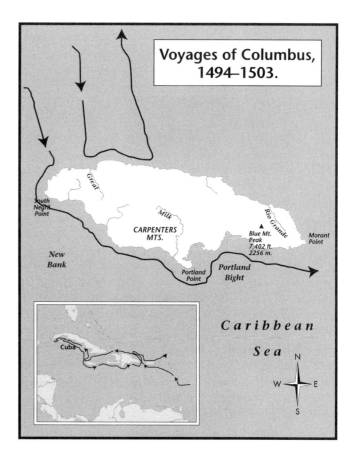

Voyages of Columbus, 1494–1503.

European governments seeking to gain a foothold in the Caribbean region.

1517   First African slaves are brought to Jamaica

African slave labor is introduced. As the Arawak population is decimated, Indian slaves are replaced by Africans. Slavery will become central to the Jamaican sugar economy in the seventeenth and eighteenth centuries.

1534   Villa de la Vega is founded

The Spanish abandon New Seville due to its unhealthy marsh environment (the marsh is humid and disease spreads easily there) and settle farther inland, founding the settlement of Villa de la Vega "Village of the Fertile Valley".

1640   Sugar is introduced into Jamaica

The first sugar is grown in Jamaica. By the eighteenth century, it will be the staple of the Jamaican economy.

1655: May 10   British invade Jamaica

After a humiliating defeat in their efforts to capture Santo Domingo on the Caribbean island of Hispaniola, a British fleet sent by British leader Oliver Cromwell (1599–1658) sails into Kingston Harbor and lands at Caguaya, quickly winning control of Jamaica from the Spanish. At the time of

Columbus describes Jamaica as "the fairest isle that eyes have beheld." (EPD Photos/CSU Archives)

the British takeover, Jamaica's population, composed of Spaniards and black slaves, numbers about 3,000.

The Spanish free many of their slaves in hopes that they will help them in their attempts to retake the island. The freed slaves (and their descendants) come to be known as Maroons, from the Spanish word *cimarron*, which means "wild" or "untamed." They remain at large and launch guerrilla attacks of their own against the British.

1656   British colonists arrive in Jamaica

The first British settlers—1,600 altogether—come to Jamaica from the island of Nevis under the leadership of Governor Luke Stokes. More settlers—about 250—follow within the year, from Bermuda. However, they are not prepared for the unwholesome marsh environment in the Port Morant area, and three-fourths die of yellow fever and dysentery within a few months. The settling of Jamaica is part of a larger program of Caribbean colonization overseen by Cromwell called the "Western Design."

1658   Spanish attempt to retake Jamaica

The Spanish make sporadic attempts to regain control of the island. The most famous occurs when Spanish troops land on

the island's north coast and attempt to unseat the British. The British and Spanish fight a land battle in which close to 300 Spanish and twenty-eight British lose their lives.

1660s   Pirates cooperate with British colonial government

In exchange for royal protection, pirates make trouble for the Spanish remaining in Jamaica, attacking their ships and settlements.

1660: May 9   Last Spanish troops leave Jamaica

The last remaining soldiers from the 1658 military attack leave for Cuba.

1664: January 20   Assembly meets for the first time

Jamaica's first assembly of elected representatives meets at St. Jago de la Vega. Allocation of seats depends on the wealth and political influence of each region.

1670s   Growth of the sugar industry begins

In the 1670s, Jamaica's primary crop shifts from cocoa to sugar. This shift also marks the replacement of small farm holdings with large estates and the inauguration of the labor-intensive plantation economy for which large numbers of African slaves are needed. Increasing numbers of slaves are brought in to replace the large number who die prematurely from harsh living conditions and overwork. There are 57 sugar plantations in Jamaica in 1673; by 1740 there are 430 or more.

1670   British take formal possession of Jamaica

The Treaty of Madrid gives the British official control of Jamaica, and ends the necessity of defending the island against Spanish attempts to retake it. Spanish Town is declared Jamaica's capital.

1677   Assembly's power is revoked

The British crown takes away the Jamaican assembly's right to pass laws, decreeing that Jamaica's laws are to be enacted in England, with Jamaican colonists given only the right to ratify or reject them. This move arouses much opposition among the colonists.

1690   The First Maroon War begins

Beginning in the Parish of Clarendon and spreading throughout the island, the First Maroon War is the first major slave rebellion in Jamaica. Escaped slaves from Clarendon flee to the interior of the island and join with Maroons (former slaves freed by the Spanish) to launch guerrilla attacks on plantations, setting fires and stealing animals. The revolt lasts for decades. The British hire Indians from the Mosquito

Coast (in present-day Honduras and Nicaragua) and freed slaves to fight the rebels (see 1739).

1692: June 7   Port Royal is decimated by an earthquake

The worst earthquake yet to occur in the western hemisphere destroys nearly the entire city of Port Royal. The second shock of the quake, lasting a full minute, crushes buildings and opens crevices in the earth as long as some streets. Landslides and a tsunami (giant tidal wave) push the northern part of the city into the sea and drown all its inhabitants. Close to 2000 people are killed immediately, and about as many die in epidemics that resulted from the disaster. A freak rescue occurs when a ship grounded for repairs is lifted by the tsunami, sails over submerged buildings, and some people are carried to safety by grabbing onto its lifelines before it runs aground on top of a half-submerged building.

1694   French launch attack on Jamaica

A French fleet lands on the island and pillages over fifty sugar plantations. They are defeated by the British at Carlisle Bay.

1739   End of the First Maroon War

British and Indian leaders negotiate a peace agreement ending the decades-long uprising. The Maroons are officially freed, this time by the British and land as well as hunting rights are allocated to them. They are also granted the right to self-government in their own communities. These include Moore Town, Nanny Town, Charles Town, Scotts Hall, and Accompong Town. The Maroons also agree not to free any more current slaves and to help suppress further slave rebellions and aid in hunting down rebel slaves.

1760   Runaway slave leads major rebellion

A runaway slave named Tacky instigates a revolt beginning with an attack on a British fort in Port Maria, which touches off a series of uprisings. The rebellion is put down after several months with the aid of Maroons.

1777   Hurricanes lead to famine; thousands of slaves die

A series of hurricanes devastates Jamaican farmland including its sugar plantations, resulting in widespread famine by 1778. The owners of the sugar plantations, having no work for their slaves, refuse to feed them, provoking rioting in which some whites are killed. Within a six-month period, two-thirds of the island's 25,000 slaves are dead from starvation and weakness to disease.

1795   Second Maroon War begins

A public whipping of two Maroons sparks a five-month rebellion. Several Maroon chiefs are taken captive by the British, and Trelawney Town is burned down. Ultimately, the British

use bloodhounds to help them track down the rebels and put down the rebellion. Over 600 Maroons are deported from the island, sent to Canada at first and eventually to Sierra Leone. They become the first black slaves ever to return to Africa.

1808   Britain outlaws the slave trade

The British outlaw the slave trade, an act that is to have a profound effect on Jamaica, whose labor-intensive sugar industry is heavily dependent on slave labor, thereby making it a center of the slave trade. Between 1700 and 1807, more than 600,000 slaves have been brought to Jamaica, but the living conditions of slaves is so harsh that six die for each one that is born in Jamaica. By the late eighteenth century, there are about 300,000 black slaves on the island.

1831: December   Jamaica's last slave revolt

Slaves on an estate near Montego Bay set fire to work houses on their plantation. Under the leadership of Sam Sharpe, a black Baptist preacher, 20,000 slaves desert and band together, eventually gaining control of all the plantations on the island and setting fire to the homes of many of the owners.

Once the British have fooled Sharpe's forces into laying down their arms through fraudulent promises, they exact a terrible vengeance from them, killing thousands of slaves. The atrocities committed by the Jamaican authorities are an important factor in motivating the British to abolish slavery throughout their empire three years later.

1834: August 1   Slavery is abolished in the British empire

The British abolish slavery throughout the lands they rule. However, the newly emancipated slaves are required to serve four-year "apprenticeships" with their former masters before truly gaining their freedom.

Losing its large pool of slave labor, the Jamaican sugar industry declines in the decades following the abolition of slavery. There are attempts to replace the African slaves with immigrant labor from other countries, but during the middle of the century this disruption of labor, coupled with other economic problems, helps to create friction between white planters and black laborers.

1834–65   Indentured laborers arrive from India

Over 6,000 indentured workers are brought in from India to work on Jamaica's sugar plantations. These immigrants are thought to be linked to a nationwide cholera epidemic (see 1850).

1834   Newspaper The Gleaner is founded

The Gleaner—still Jamaica's major daily at the end of the twentieth century—is established by Jewish immigrants.

1836   First commercial bank opens

An office of the British Barclays Bank is established in Jamaica, becoming the island's first commercial bank.

1843   Railway system is established

Jamaica's railroad—the world's first colonial railroad—begins operations. Built by the Jamaican Railway Company, the railroad is later sold to the Jamaican government.

1846   Sugar Duties Act weakens ailing sugar industry

Britain's Sugar Duties Act, ending trade preferences for the import of Jamaican sugar by England, deals yet another blow to the island's struggling sugar producers, already devastated after the abolition of slavery in the British empire.

1850   Cholera epidemic kills 32,000

A massive cholera outbreak throughout Jamaica results in 32,000 deaths. Many mistakenly blame the epidemic on indentured laborers brought from India to work the sugar plantations.

1860–61   Great Revival religious movement sweeps Jamaica

The Great Revival draws thousands of converts to Christianity but, at the same time, blends Christianity with practices derived from native African religions.

1865   Paul Bogle leads the Morant Bay rebellion

In the period following the abolition of slavery, blacks are still denied many political and land rights by white authorities. During the trial of a supporter of Paul Bogle, a freed slave and spokesman for the black community, rioting occurs and a police station is attacked. The government harshly suppresses the unrest, killing over 400 blacks and hanging Bogle and another influential blacks accused of instigating the rioting. A thousand blacks have their houses torched. The incident leads the British government to overhaul the colonial administration of Jamaica, making it a crown colony. (See 1865.)

1865   Jamaica becomes a crown colony

In response to abuses of power used to quash the Morant Bay rebellion, the British government recalls Govenor Edward John Eyre (1815–1901) and makes Jamaica a full crown colony under direct British rule. Its parliament is disbanded and its traditional elite loses power. Under this new administrative structure, the British implement major social welfare reforms and undertake the improvement of the island's physical infrastructure.

1872   Kingston becomes the capital of Jamaica

The capital is moved from Spanish Town to Kingston.

1881   Artist John Dunkley is born

Dunkley is a self-taught artist and a prominent member of the early-twentieth-century "Intuitive" school of Jamaican painting, which emphasizes primitive, native themes. Dunkley is known for his use of Biblical and folkloric imagery. A practicing barber in Kingston, he covers the entire interior (including walls and furniture) of his shop in decorative painting.

1887: August 17   Birth of black nationalist leader Marcus Garvey

Marcus Garvey plays an important role in shaping the self-image and self-awareness of blacks in his native Jamaica, the Western Hemisphere as a whole, and throughout the world. Garvey is born into a large family in St. Ann's Bay in northern Jamaica. At a young age, he is apprenticed to a printer and becomes proficient in the printing trade, a skill he will later draw upon in disseminating his social and political views.

In 1914, Garvey founds the Universal Negro Improvement Association (UNIA) to improve the lot of blacks throughout the world. Through this organization, Garvey hopes to establish a new independent black state in Africa. In 1918, Garvey, by then living in the United States, publishes a newspaper, the *Negro World*, which is read by blacks throughout the world. The following year he establishes the Black Star shipping line, also as a part of his "Back to Africa" vision. As Garvey becomes increasingly famous, his speeches advocating black pride and independence turn many—including both whites and blacks—against him. He survives a serious assault in 1919 and is imprisoned in the United States on mail-fraud charges in 1925. In 1927 he is deported to Jamaica and settles in London in 1934. He dies of a heart attack there in 1940 and is buried in Jamaica, where he is regarded as a national hero.

1890: September 15   Birth of poet and novelist Claude McKay

McKay is born and spends his youth in Jamaica. In 1912 he leaves for the United States, attending college at the Tuskegee Institute and moving to New York in 1914. In the 1920s he becomes involved in the Harlem Renaissance, publishing two volumes of poetry, *Spring in New Hampshire* (1920) and *Harlem Shadows* (1922). Later, McKay lives in France, Spain, Morocco, and the former Soviet Union.

McKay's *Home to Harlem* (1928) is the most acclaimed novel written by a black up to that time. Although he spends most of his life in the United States and abroad, McKay continues to base many of his works on his early experiences in Jamaica. *Banana Bottom* (1933) which depicts rural Jamaican society in the early 1900s, is considered by many to be the first West Indian novel. McKay returns to the United States in 1934 and writes articles for a variety of publications. He also writes *A Long Way from Home*, an autobiography. McKay dies in 1948, and a book of his selected poems is published posthumously in 1953.

1892: July 4   Birth of statesman Norman Manley

Manley grows up in modest circumstances, receiving an education through scholarships. In 1914 he wins a Rhodes Scholarship to study at Oxford University, where he receives a law degree after taking time off to serve in the British armed forces during World War I (1914–1918). In 1922 Manley returns to Jamaica, establishing himself in the legal profession and also getting involved in public service to improve the lives of ordinary Jamaicans. He organizes the Jamaica Banana Growers Association, which obtains a favorable contract from the U.S.-owned United Fruit company.

In 1938 Manley organizes Jamaica's first political party, the People's National Party (PNP). When Jamaica is granted the right of self-government in 1944, the PNP contests the first election against the Jamaica Labour Party (JLP), formed by Alexander Bustamente. The JLP becomes the dominant party until 1953, when the PNP gains power, and Manley becomes prime minister. He holds that office through the period when Jamaica belongs to the West Indies Federation, and oversees the referendum that calls for secession from the federation. The JLP wins Jamaica's first elections as an independent nation in 1962, and Manley serves as leader of the opposition party. Manley retires from politics shortly before his death in 1969.

1900   Birth of sculptor Edna Manley

Edna Manley, Jamaica's most famous sculptor, is born in England to a Jamaican mother. She moves to Jamaica in 1922 after marrying her first cousin, Norman Manley, who will later found the People's National Party. Their son, Michael Manley (1924–1997), becomes the head of the PNP and also serves as Jamaican prime minister in 1972–1976 and 1989–1992.

Edna Manley leads the struggle to free Jamaican art from English aesthetic standards and does much to encourage the success of self-taught artists. Her first solo exhibit is held in 1937. She teaches art classes at the Institute of Jamaica and is instrumental in the formation of the Jamaican School of Art and in the opening of commercial galleries that encourage local art. Manley remains one of the foremost figures in Jamaican art until her death in 1987.

1905   Poet Una Marson is born

Marson is a poet, playwright, journalist, and feminist who works for racial equality and Jamaican independence from Britain. After founding a magazine in Jamaica, she works for the League of Colored Peoples in England and then as private secretary to Ethiopian Emperor Haile Selassie (1892–1975). Marson founds the Save the Children Fund, and a progressive newspaper, *Public Opinion*. She also writes the first play to be performed in London by black actors, *At What a Price* (1932). Other plays include *London Calling* (1937) and *Pocamania* (1938). She publishes four volumes of poetry between 1930 and 1945, and her writing has a strong influence on other Jamaican authors, especially in the 1940s. During World War II she hosts a popular program on the British Broadcasting Service (BBC) called "Caribbean Voices." Marson dies in 1965.

1905   Birth of author Roger Mais

Mais is a journalist, painter, and novelist who writes about life in the slums of Kingston. He is regarded as having written some of the best Jamaican poetry of the 1930s. Toward the end of his life he writes three novels, *The Hills Were Joyful Together*, (1981) *Brother Man (1974),* and *Black Lightning (1983)*. Mais dies in 1955.

1907: January 14   Earthquake devastates Kingston

Kingston is rocked by the worst earthquake to hit Jamaica since 1692. Much of the city is destroyed as giant cracks open in roads and sidewalks and buildings topple. In addition, the quake sets off a tsunami (giant tidal wave) that roars over Anotta Bay, sweeping hundreds of houses into the sea. Live wires from the city's powerhouse snap, setting off fires that cannot be extinguished because the water mains are broken. Over twenty-five blocks of houses and shops are destroyed. Over 1,400 people are killed.

1911–21   Hurricanes ravage Jamaican banana crop

Four hurricanes in a ten-year period cause widespread destruction on Jamaica's banana plantations.

1911   Birth of sculptor and painter Kapo

Kapo (Mallica Reynolds) is a leading artist of Jamaica's twentieth-century "primitivist" school. He begins his career as a wood carver, producing one of the finest bodies of work of any Caribbean artist. Much of his art is inspired by his Revivalist religion, which combines African and Christian religious elements. He is also known for the incorporation of Jamaican history and folklore in his work.

In the last three decades of his life, Kapo becomes one of Jamaica's most accomplished painters, known especially for his depiction of the Jamaican landscape. In 1981 the Jamaican government presents one of his works, *A New Spring*, to Prince Charles and Princess Diana as a wedding gift. Kapo dies in Kingston in 1989.

1914–18   World War I benefits Jamaican economy

During World War I, sugar prices go up and the demand for logwood rises, aiding Jamaica's economy.

1914   Birth of writer Victor Reid

Victor Reid pursues a career as a novelist, historian, and journalist. He is the author of *New Day* (1949), a historical novel that uses Jamaican folk dialect extensively, and *The Leopard* (1958), a novel set in Africa. Reid dies in 1987.

1920–1940   Women leaders pave the way for feminism

A series of women social activists and writers, of both European and African descent, reach across racial and class lines to work for social justice, launching a flourishing women's movement by the 1930s. Included among these pioneers are Amy Bailey, Edith Dalton James, Ethlyn Rhodd, Lillie Mae Burke, Una Marson (1905–65), and Carmen Lusan.

1924: December 10   Birth of political leader Michael Manley

Manley is the son of one of Jamaica's prime ministers, Norman Manley, and the country's leading sculptor, Edna Manley. Norman Manley is also the founder of one of Jamaica's two main political parties, the People's National Party (PNP).

The younger Manley is first elected to Parliament in 1967, becoming the head of the PNP after his father's death in 1969. He becomes prime minister himself in 1972 and adopts socialist-style policies to deal with Jamaica's social and economic problems. These policies and his close relations with Cuban leader Fidel Castro, alienate the United States and other Western nations, resulting in a decline in foreign aid. Manley's PNP loses the 1980 elections but is returned to power nine years later. During his second term as prime minister (1989–92), Manley moderates his policies, adopting a free-market approach. Manley resigns the prime ministership for health reasons in 1992, before his term is over. He dies on March 6, 1997, at his home near Kingston.

1930s   Emergence of the Rastafarian movement

Rastafarianism is a social and religious movement that originates primarily among the rural poor in Jamaica. It is based on the belief in a special, biblically-mandated relationship between all black people and the nation of Ethiopia, especially in relation to its newly anointed king, Haile Selassie I (1892–1975), also called Ras Tafari (the origin of the term "Rastafarians"). The notion of a connection between Jamaican blacks and Ethiopia comes at least partly from the writing and speeches of black nationalist leader Marcus Garvey (1887–1940; see 1887).

The Rastafarians regard Haile Selassie as their god (a belief that does not suffer when he eventually dies in 1975). They believe that through his power, the poor will someday be delivered from oppression and led back to Ethiopia, which they regard as heaven. This belief provides hope to many of the poor in the Depression-stricken 1930s. However, Rastafarianism endures beyond this period and continues to grow, particularly after the late 1950s.

Rastafarians are most readily identified by their long hair styled in dreadlocks, a custom that grows out of a religious prohibition on cutting their hair. In addition, they observe the Hebrew dietary laws specified in the Old Testament and regulate their diet in other ways as well. *Ganja* (marijuana) is used in their rituals, playing a role similar to that of the bread and wine in the Christian sacrament.

By the 1960s Rastafarianism comes to attract educated young people in addition to the rural poor that have been its traditional adherents. Reggae music, pioneered by Bob Marley (see 1945) in the early 1970s, is profoundly influenced by Rastafarianism.

1930s   Institute of Jamaica is catalyst for artistic development

The Institute of Jamaica, founded by Sir Anthony Musgrave, spurs the development of a Jamaican art movement, bringing together such diverse artists as Albert Huie, John Dunkley (see 1881), Edna Manley (see 1900), Ralph Campbell, Henry Daley, and Carl Abrahams. These artists all become interested in portraying the historical experience of Jamaicans of African descent. Landmark artworks in this movement include Manley's *Negro Aroused, Young Negro*, and *The Diggers*.

1933   Novel *Banana Bottom* is published

Banana Bottom, a landmark in Jamaican literature regarded by many as the first West Indian novel, is published by Jamaican-born poet and novelist Claude McKay. It depicts rural Jamaican society in the early 1900s. (See 1890.)

1935   Bustamente forms first labor union

Alexander Bustamante (see 1884) establishes Jamaica's first labor union, the Jamaica Trade Workers and Tradesmen Union (JTWTU).

1936   Founding of Jamaica Welfare Ltd.

Jamaica's first non-government organization devoted to community development is launched by activist Norman Manley (1893–1969) and Samuel Zemurray, president of United Fruit. It becomes a model for similar organizations in other developing nations in Latin America and Africa.

1938   Labor riots spread throughout Jamaica

Armed police kill several workers and wound a number of others during a strike at the West Indies Sugar Company in Westmoreland, spurring protests, civil unrest, and looting throughout Jamaica. These events are often viewed as the point when modern Jamaican nationalism is born.

1938: September 18   People's National Party is formed

Norman W. Manley (see 1892) establishes the People's National Party (PNP), which remains a major force in Jamaican politics throughout the twentieth century. Begun as a nationalist movement, the PNP moves toward socialism in the 1940s.

1939   Artists storm meeting to declare cultural independence

Forty respected artists break into the annual meeting of the Institute of Jamaica, an important national cultural institution, demanding greater Jamaican independence from European artistic traditions. One of them declares that the portraits of English governors displayed on the walls should be replaced by Jamaican paintings.

1940: June 10   Black nationalist leader Marcus Garvey dies

Black nationalist leader Marcus Garvey dies in London. (See 1887.)

1941   Founding of the Little Theatre Movement

The Little Theatre Movement (LTM) is established by Henry and Greta Fowler. It is the venue for the annual national pantomime, which has taken place every year for over fifty years. The pantomime is an annual Jamaican folk musical with traditional songs and dances, bright sets, costumes and commentary on current affairs.

1942   Birth of contemporary novelist Anthony Winkler

Winkler, born in Jamaica, divides his time between Jamaica and the United States. A college professor, he writes literary criticism and textbooks as well as novels, including *The Painted Canoe* (1983) and *The Lunatic* (1987). *The Great Yacht Race* (1990) is a satire of life in Jamaica before independence from Great Britain. *Going Home to Teach* (1995) is an autobiographical account of Winkler's experiences teaching at a rural teacher-training college in Jamaica. His novel *The Lunatic* is made into a movie in 1991.

1943   Bustamente forms the Jamaica Labour Party

Labor leader and unionist Alexander Bustamente (see 1884; 1935) forms Jamaica's second major political party, the Jamaica Labour Party, or JLP.

1944: November 20   Jamaica adopts a new constitution

The British grant Jamaica a new constitution, giving the island political autonomy (self-government) modeled on the British system and establishing universal adult suffrage.

In the 1970s Bob Marley (1945–81) and the Wailers help popularize reggae music throughout the world. (EPD Photos/CSU Archives)

1944: December 14   Alexander Bustamente wins first Jamaican elections

Jamaica's first elections as a self-governing political entity are won by the Jamaica Labour Party (JLP), headed by Alexander Bustamente. Bustamente is one of the giants of twentieth-century Jamaican politics, becomes the nation's first prime minister. Born February 24, 1884 to a poor plantation overseer, the future statesman lived abroad from 1903 to 1932 and changed his last name from Clarke to Bustamente when he was adopted at the age of fifteen by a Spanish seaman of that name. He first came to public attention speaking out against the substandard living conditions of most of the Jamaican population during the colonial era. In 1938 he was imprisoned by the British for organizing the Bustamente Industrial Trade Union (BITU).

Bustamente remains Jamaica's prime minister for the next decade. He is prime minister again when the island secedes from the short-lived West Indies Federation in 1962, after gaining full independence from Britain. He remains influential even after his formal retirement from politics in 1967. Bustamente is knighted by Queen Elizabeth II in 1954 and named a national hero of Jamaica shortly before his death in 1977.

1945: February 6   Birth of reggae superstar Bob Marley

Robert Nesta Marley is born in St. Ann Parish to a Jamaican mother and an English-born father, who deserts the family while Marley is still a child. In his teens, Marley, then living in Kingston's Trench Town ghetto, forms a musical group with three friends: Junior Braithwaite, Peter Macintosh, and Neville Livingstone. They call themselves the Wailing Rude Boys (the name is later shortened to The Wailers.) Marley marries, and his wife, Rita, becomes the leader of a performing group of her own.

In the late 1960s, Marley lives briefly in the United States with his mother, then returns to Jamaica, where he tries his hand at farming and becomes a Rastafarian. He develops a new musical style which will become known as reggae and begins performing again. In 1973 he releases two albums (*Catch a Fire* and *Burnin'*) that bring him worldwide recognition and success. More hit albums follow throughout the 1970s.

In addition to his success as an entertainer, Marley is widely known for his public stand against social injustice and as a voice for peace. His political involvement nearly costs him his life during the election campaign of 1976, when armed assassins invade his Kingston home, injuring him, his wife, and others. Two days after the incident, Marley gets political rivals Michael Manley and Edward Seaga (b. 1930) to join hands onstage at a concert as a gesture toward ending the political violence shaking the nation. Marley dies at the age of 36, on May 11, 1981, of brain cancer. Just before his death he is awarded Jamaica's highest honor, the Order of Merit. His body lies in state in the National Arena in Kingston.

1947–48   Jamaican Herb McKenley wins 400-meter world title

Herb McKenley sets world records for the 400-meter dash in 1947 and 1948. McKenley also wins one gold and two silver medals at the Helsinki Olympics (see 1952). The 1947 record is the first time-measured record set by an English-speaking West Indian.

1947   Death of artist John Dunkley

Dunkley, one of Jamaica's foremost twentieth-century painters, dies. (See 1881.)

1948   Arthur Wint wins Olympic gold in London

Arthur Wint wins an Olympic gold medal in the 400-meter track event at the London Olympics, edging out fellow Jamaican Herb McKenley, who takes home the silver.

1948: May 22   Death of poet and novelist Claude McKay

Leading black author Claude McKay dies in Chicago. (See 1890.)

1949   PNP wins Jamaica's second national elections

Norman Manley's People's National Party (PNP) wins the second Jamaican elections.

1951: August 17   Hurricane Charlie kills fifty-four people

Hurricane Charlie, which is originally expected to bypass Jamaica, instead veers into it, with winds reaching 125 miles (201 kilometers) per hour. The storm causes massive destruction, first in Morant Bay and then in Port Royal. Trees are uprooted, houses smashed, and crops washed into the ocean. After destroying the entire Port Royal waterfront, the storm sweeps inland, ruining ninety percent of the banana crop as well as much of the coconut yield. When it is over, 54 people are dead, 2,000 injured, and 50,000 are left homeless.

1952   Bauxite production begins

Reynolds Metals, an American company, inaugurates bauxite production in Jamaica. The limestone plateau in the island's interior has abundant deposits of this mineral (a claylike ore that is the source of aluminum), and Jamaica soon becomes the world's leading bauxite producer, which spurs rapid economic growth.

1952   Jamaican runners capture gold at the Helsinki Olympics

Two Jamaican runners win honors at the summer Olympics in Helsinki, Finland. George Rhoden is the gold medalist in the 400-meter event, while Herb McKenley wins the gold medal for the 400-meter relay and silver medals for the 100- and 400-meter events.

1955   Louise Bennett joins staff of Jamaica Welfare Ltd.

Bennett, a folklorist and charismatic performer, begins a long collaboration with this prominent cultural organization. She travels throughout Jamaica performing original, folklore-inspired material, and her performances help legitimize the use of Jamaican dialect in the arts.

1955   Author Roger Mais dies

Mais, author of poems, journalism, and novels, dies. (See 1905.)

1955   The first Jamaica National Festival of Arts is held

This year-long festival marks the tercentenary of the British presence in Jamaica. (See 1655.) Performances take place throughout Jamaica, bringing many local talents into the spotlight. The festival eventually becomes an annual event (see 1965).

1957   Jamaica's worst train wreck kills 175

A twelve-car train carrying 1,500 passengers returning from a holiday trip to Montego Bay derails outside Kendal while attempting to negotiate a dangerous S-curve. The accident—the worst train wreck in Jamaica's history—kills 175 persons and injures more than 750.

1958: January 3   The West Indian Federation is established

Believing that the small island nations of the British West Indies can best promote their interests by banding together, their leaders join together to form the West Indian Federation, which includes Jamaica, Trinidad and Tobago, Barbados, and several other islands in the region.

Ultimately, the unity of the federation is undermined by the inequality of its members, notably the economic dominance of Jamaica and Trinidad, and also by the rivalry between these two islands.

1961: September 19   Jamaicans vote to withdraw from the federation

In a special referendum, the Jamaican people overwhelmingly vote "no" to continued participation in the West Indian Federation. The federation is soon dissolved, and most of its former members eventually gain independence on their own, with others remaining British possessions.

1962   National Dance Theatre Company is formed

Jamaica's National Dance Theatre Company (NDTC) is formed, under the directorship of Professor Rex Nettleford. A former Rhodes Scholar and a leading Caribbean intellectual, Nettleford has published several books on issues of political development, race, ethnicity and cultural identity, including the Smithsonian Institution publication: *Race, Discourse and the Origins of the Americas.*

1962: August 5   Jamaica declares independence

With British consent, Jamaica proclaims its independence from Great Britain. There is a large public celebration in the capital city of Kingston, where the Union Jack (the British flag) is lowered for the last time and the new Jamaican flag is raised. Alexander Bustamente becomes the new nation's first prime minister. The British monarch remains Jamaica's ceremonial head of state and is represented by a governor-general.

1963: April   Government makes arts festival an annual event

The government announces that the Jamaica National Festival of Arts, which has been held in 1955 and 1958, will become an annual event. Minister of Development and Welfare (and later prime minister) Edward Seaga (b. 1930) calls the festival "a national stage where Jamaicans from all walks of life [will] have the opportunity to create their own brand of artistic expression, reflecting their life history and their life styles."

1965   Death of feminist writer Una Marson

Marson, a playwright, journalist, and radio personality, dies. (See 1905.)

1969   Jamaican Stock Exchange opens

The Jamaican Stock Exchange is launched by the Bank of Jamaica, becoming the first stock exchange in the Caribbean region.

1969: September 2   Political leader Norman Manley dies

One of the chief framers of Jamaican independence, former Prime Minister Norman Manley dies. (See 1892.)

1972   PNP wins control of the government in national elections

The leftist People's National Party defeats the rival Jamaican Labour Party, winning control of the government and inaugurating a period of economic reform. Its prime minister is Michael Manley (see 1924), son of party founder Norman Manley (see 1892).

1972–76   Manley implements "People's Projects"

A series of reforms, popularly known as "People's Projects," are carried out during Michael Manley's first term as prime minister. These include healthcare, education, labor, and distribution measures, the introduction of a statutory minimum wage, and the construction of public-sector housing.

1973   Jamaica takes part in formation of CARICOM

Jamaica is a founding member of the Caribbean Common Market (CARICOM), designed to further economic integration among nations in the region through collaboration in trade, foreign policy, and other areas.

1974   Government literacy program begins

The Jamaica Movement for the Advancement of Literacy is launched, with the goal of providing adult education programs to 100,000 people a year. By the late 1970s, Jamaican literacy is eighty-five percent.

1976   PNP wins its second national election

The People's National Party wins its second consecutive national election, increasing its majority. However, the campaign is characterized by political violence, especially in the Kingston slum districts, where armed supporters of the two major parties attack each other. Prime Minister Manley responds to the unrest by declaring a state of emergency and establishing a special court to deal with the offenders.

1980   Jamaica Labour Party regains power

The JLP wins the national elections, and Edward Seaga becomes prime minister. The election campaign is marred by the worst violence in Jamaica since the 1865 Morant Bay Rebellion. An estimated 600 people are killed in street fighting between supporters of the JLP and the PNP.

1980: May 20   Election-related arson rumored in nursing home fire

The Eventide Home housing 204 elderly, indigent women burns down, killing 157 of its occupants. Minutes after the fire department arrives, the building collapses, trapping the women and burning many alive. Each side in the ongoing election campaign accuses the other of setting the fire. However, fire officials determine that the cause of the fire is a short circuit in one of the walls, combined with the highly flammable pinewood of the building.

1981: May 11   Reggae legend Bob Marley dies

Marley dies at age 36 of brain cancer. (See 1945.)

1981: May 21   Bob Marley's funeral is held

The funeral of Reggae star Bob Marley, dead of a brain tumor at age 36, is as lavish as that of a head of state—his funeral cortege extends for fifty miles. He is buried near his childhood home, a one-room cottage in St. Ann's Parish.

1983: October   Jamaican troops participate in U.S. invasion of Grenada

Jamaican troops support the U.S. effort to oust the Marxist regime that has seized power in Grenada, murdering its moderate socialist prime minister, Maurice Bishop (1944–83).

1983   Seaga holds local elections early

Attempting to capitalize on the military victory over Grenada, Prime Minister Seaga calls local elections a year ahead of schedule. The opposition PNP boycotts the elections, and thus the JLP rules Jamaica single-handedly from 1984 to 1989. However, the PNP continues to carry out peaceful opposition activities outside the framework of government.

1987   Author Victor Reid dies

Novelist Victor Reid dies. (See 1914.)

1987   Sculptor Edna Manley dies

Manley, a leader in the movement to throw off European artistic traditions and restrictions, dies at the age of 86. (See 1900.)

1988   Jamaican bobsled team qualifies for the Winter Olympics

A bobsled team from Jamaica beats the odds to make it to the Winter Olympics in Calgary, Canada. Never having seen snow before in their lives, they make their first actual run on ice only days before qualifying. The movie *Cool Runnings* (see 1993) is based on their experiences.

1988: September 12   Hurricane Gilbert wreaks unprecedented destruction

After pounding the Dominican Republic and Haiti, Hurricane Gilbert reaches Jamaica, where the eye of the storm traces a path across the island, with wind speeds of 174 to 186 miles (280 to 300 kilometers) per hour. For a major disaster, the number of dead is relatively low (forty-five), but infrastructure (roads, bridges, etc.), property, and crop damage disrupts daily life on the island for months. Some of the hurricane's effects on the country's economy last for years.

1989   Jamaica's first national parks are established

Two national parks are set up as pilot projects under the Protected Areas Resource Conservation project. One is the Blue Mountain/John Crow Mountain National Park, consisting of 200,000 acres. The other is the Montego Bay Marine Park, intended to protect offshore coral reefs from erosion, pollution, and over-fishing.

1989: February   Artist Kapo dies

Kapo (Mallica Reynolds), one of the most renowned twentieth-century Jamaican artists, dies. (See 1911.)

1989   The PNP is returned to power

The People's National Party wins the national elections. The type of political violence that marred the 1976 and 1980 elections is largely avoided through an agreement between the two major parties.

Michael Manley again becomes prime minister. He pursues more moderate policies than those adopted during his previous term in office. Modifying his former pro-Communist stance, he establishes contact with U.S. President George Bush (b. 1924). Manley also fosters stronger ties with other countries in the region, including non-English-speaking nations such as Mexico and Venezuela.

1992   Manley steps down; Percival Patterson takes over as president

When Michael Manley is obliged to retire before the end of his term for health reasons, Percival J. Patterson (b. 1940) assumes the prime ministership, gaining growing support with his moderate political approach and willingness to cooperate with the opposition party. One source of popular appeal is the fact that Patterson is one of very few highly placed Jamaican politicians who is black.

1993 Movie *Cool Runnings* celebrates exploits of bobsled team

The Hollywood movie, *Cool Runnings,* is loosely based on the events surrounding the Jamaican bobsled team's participation in the 1988 Winter Olympics in Calgary. Through the film, the Jamaicans' story becomes widely known and public recognition of bobsledding also spreads. (See 1988.)

1993: March Patterson calls early elections to prove support

In early elections scheduled by the new president, Patterson's party, the PNP, gains an impressive legislative victory, assuring Patterson of a full term in office. However, violence flares as the election date approaches. A dozen people are killed, and many more are injured.

1994 Bobsled team finishes fourteenth in Albertville, France

Jamaican bobsledders participate in the Winter Olympics for the second time, placing ahead of more experienced competitors including those from the United States, Italy, and France.

1997: March 6 Death of former Prime Minister Michael Manley

Manley, longtime leader of the People's National Party and son of the party's founder, dies at his home near Kingston. (See 1924.)

1997: June 30 Jamaica hosts CARICOM meeting

The Caribbean Community and Common Market (CARICOM) holds its annual summit meeting in Montego Bay. The delegates agree to admit Haiti to the organization, bringing the total number of members to fifteen and doubling the combined total population of its member states. There is also new progress toward the establishment of a Caribbean common market by 1999.

1997: December 18 PNP wins third consecutive term in power

The People's National Party (PNP) wins fifty-six percent of the vote in national elections, the first time since Jamaican independence that one party has stayed in power for three consecutive terms. The victory returns Percival Patterson to the prime ministership. Of Jamaica's sixty parliamentary seats, forty-nine are won by PNP candidates.

Compared with the 1976 and 1980 elections, the 1997 vote is relatively peaceful. The election is monitored by former U.S. President Jimmy Carter (b. 1934; president 1976–81) and sixty international observers from the Carter Center, including former U.S. Chief of Staff General Colin Powell (b. 1937), who is the son of Jamaican immigrants.

Bibliography

Bakan, Abigail B. *Ideology and Class Conflict in Jamaica: The Politics of Rebellion.* Montreal: McGill-Queen's University Press, 1990.

Barrett, Leonard E. *The Rastafarians:Sounds of Cultural Dissonance.* Boston: Beacon Press, 1997

Bayer, Marcel. *Jamaica: A Guide to the People, Politics, and Culture.* Trans. John Smith. London, Eng.: Latin American Bureau, 1993.

Davis, Stephen. *Reggae Bloodlines: In Search of the Music and Culture of Jamaica.* New York: Da Capo Press, 1992.

Douglass, Lisa. *The Power of Sentiment: Love, Hierarchy, and the Jamaican Family Elite.* Boulder, Colo.: Westview Press, 1992.

Looney, Robert E. *The Jamaican Economy in the 1980s: Economic Decline and Structural Adjustment.* Boulder, Colo.: Westview Press, 1987.

McKay, Claude. *Banana Bottom.* New York: Harper and Row, 1961.

Nettleford, Rex. *Caribbean Cultural Identity: The Case of Jamaica.* Kingston: Institute of Jamaica, 1978.

Sherlock, Philip, and Hazel Bennett. *The Story of the Jamaican People.* Princeton, NJ: Markus Wiener Publishers, 1998.

Stone, Carl. *Class, State, and Democracy in Jamaica.* New York: Praeger, 1986.

Mexico

Introduction

Mexico, known more formally as *Estados Unidos Mexicanos* (United States of Mexico), is the third largest country in Latin America and the only one situated in North America. It is a nation of many contrasts that capture the imagination of the foreign visitor.

Mexico covers 756,066 square miles of territory. From north to south it measures approximately 1,800 miles and from east to west, 1,200 miles at its widest point and roughly 130 miles at its narrowest. It has coasts on three major bodies of water: the Pacific Ocean, the Gulf of Mexico, and the Caribbean Sea. Mexico has two major rivers, the Río Grande which shares a common border with the United States, and the Papaloapan. Its other rivers are much shorter.

Modern Mexico is a mixture of Amerindian, and Spanish culture. The first inhabitants of Mexico were descendants of Asians who entered North America over 50,000 years ago via a land bridge across the Bering Straits connecting Siberia with Alaska. The earliest record of human settlement in the region dates back to Tepexpan Man (12,000 B.C.), the oldest human remains found in Mexico about 1500 years after the beginning of corn cultivation (c. 1500 B.C.), the Toltecs emerged as the first major civilization of Mexico.

The Aztecs supplanted the Toltecs around A.D. 1100 and established one of the most advanced civilizations in the Americas. The ultimate dominance of the Aztecs over the Toltecs lay not only in their intrepid, highly skilled society, but also in the Aztec's systematic, sacrificial method of dealing with the enemy. Significant achievements of Aztec civilization included the establishment of a canal system, public buildings, and wide roads. The wealthy and vast capital of Tenochtitlán (on which today's Mexico City stands) exemplified the accomplishments of the Aztecs. In its day, Tenochtitlán was a stunning achievement not only in its mammoth population density, but also in its architectural engineering feats.

Spanish Conquest

Today's Mexico emerged as a direct result of events ushered in during the Spanish conquest that began in 1519. In that year, Spanish *conquistador* (one who participated in the Spanish conquest of Mexico and Peru in the 16th century) Hernán Cortés led an expedition to Mexico charged with the task of conquering the area for the Spanish crown. The Aztecs, under the leadership of emperor Montezuma II—who believed the Spaniards represented the Aztec god Quetzalcoatl—initially did not oppose the conquerors. Soon, however, the Spanish (with the help of some of the Aztecs' subject groups) overwhelmed Montezuma's forces and established control over the empire. An Aztec revolt followed under the leadership of Montezuma's nephew, Guatemotzin, who succeeded in forcing the Spainish out of Tenochtitlán on June 30, 1520; that same night Montezuma died. The Aztec victory proved short-lived, however. The following year Cortés returned to Tenochtitlán, defeated the forces of Guatemotzin, and destroyed the once-magnificent Aztec capital. Upon its ruins, the *conquistadores* established what came to be known as Mexico City.

Once the Spanish settled in Mexico, they set out to reorganize the adminstration of the country and introduce Spanish culture. Mexico became a Spanish colonial territory dominated by wealthy Spanish landowners who introduced the *encomienda* system to govern relations with the impoverished Amerindian peasantry. Spaniards used force to keep the native population in check and exploited their labor.

The most important cultural influences brought by the Spanish were their Roman Catholic faith and the Spanish language. Beginning in 1524, Franciscan monks set out converting the native inhabitants of Mexico. Apart from its missionary work, the Catholic Church also tried to improve relations with, and respect for the Amerindians. Under the leadership of Fray Bartolomé de las Casas, the Church became a leading voice for better treatment of the natives. Over time, relations between the Spanish and Mexicans improved and the *Mestizos* (those of mixed Spanish-Amerindian descent) become a majority of Mexico's population.

As new economic and social patterns emerged in Mexico in the decades after the Spanish conquest, and Spain contin-

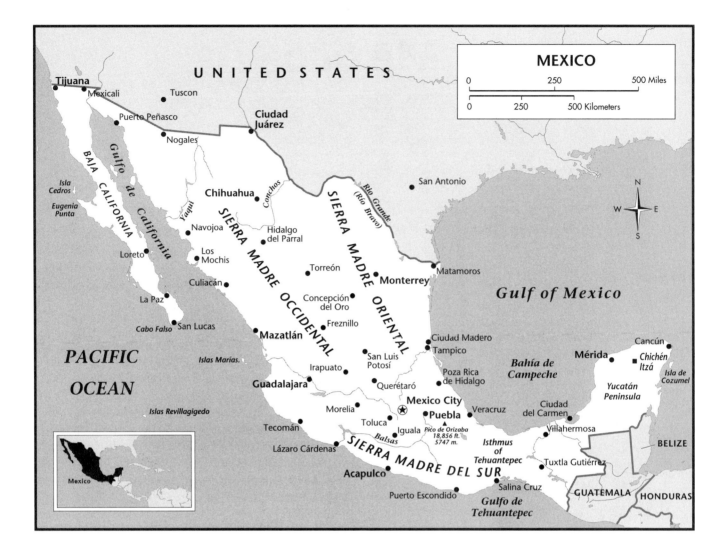

ued to try to exercise centralized authority over its colony, Mexicans felt resentment toward continued Spanish rule. As early as 1556 the Spaniards in Mexico revolted against the imperial Spanish government. Yet despite several uprisings, Spanish rule in Mexico continued until the early nineteenth century.

Fight for Independence

In 1810, a parish priest, Miguel Hidalgo y Costilla, called for an uprising against the Spanish. Hidalgo planned an uprising for December that leaked to Spanish authorities and his arrest was ordered. In September, Father Hidalgo was forced to prematurely distribute the *Grito de Delores,* an appeal for social and economic reform, to his parishoners and nearby residents. With little organization and no training, a revolutionary army of thousands (primarily composed of Indians and mestizos) overwhelmed royal forces in Guanajuato, and proceeded on a systematic plan of murder and pillage. The force continued to Mexico City and defeated royalists on the outskirts, but due to their somewhat vulnerable position, the revolutionary army

did not enter the city, and instead returned home. Although his revolt failed (and Hidalgo lost his life), the following year, José María Morelos y Pavón—another priest—took up the standard of revolt. In 1813 Morelos proclaimed Mexico an independent republic. Although Morelos' action also failed, (in 1815, after his defeat he was shot) popular support for independence remained. The resistence of Hidalgo and Morelos, coupled with an emergence of liberalism in Spain allowed the conservative oligarchy of Mexico to favor independence as a way of holding on to their own power. In 1820 Agustín de Iturbe (1783–1824) began yet another revolution against the Spanish. This time, the revolt succeeded and in 1821 Spain recognized Mexican independence.

However, independence did not bring stability. Since gaining freedom from Spain, Mexico has been saddled with chronic socio-political unrest, economic debt, and often uneasy relations with the United States, relations which several times have deteriorated into open hostilities.

Iturbide, a soldier and politician, quickly declared himself emperor of Mexico in 1822, but was ousted a year later by Antonio López de Santa Anna and Guadalupe Victoria. In 1824 Victoria promulgated a new constitution and became president. During his five-year term, Victoria ousted Spaniard opponents of his regime, and in 1827 he abolished slavery. The 1830s and 1840s witnessed the increasing influence of Santa Anna along with intermittent hostitilities with the United States. Santa Anna attempted to centralize Mexican adminstration and preserve Mexican sovereignty over the outlying regions of the country. While some successes were had in his centralization campaign, Mexico's foreign policy suffered grave defeats. In the 1830s, Texas won its independence and joined the United States in 1845. This helped spark a war with the latter in 1846 in which Mexico suffered a disastrous defeat. The Treaty of Guadalupe Hidalgo in 1848, reaffirmed the loss of Texas, forced Mexico to recognize the Rio Grande River as the southern boundary of the United States, and to cede nearly the entire area of the present Southwestern United States, including Arizona, New Mexico, Nevada, Utah, and parts of California and Colorado. In return for the territory ceded, Mexico received $45 million from the United States. In 1853, the United States purchased the Gila River valley in Arizona and New Mexico for $10 million and fixed the modern United States border with Mexico.

The 1850s and 1860s brought renewed civil strife as Santa Anna was ousted (due to his intrigue into Archduke Maximillian's rule) and new reforms attempted. These efforts culminated in a new constitution in 1857. This constitution raised the ire of conservatives, as the constitution abolished military and clerical immunities, from which the conservatives had much to lose. The conservatives consequently revolted in 1858, thus starting a three-year civil war. Pitted against the leadership of President Benito Juárez, the main proponent of the reforms, the conservative forces lost. Juárez's victory also upheld the separation of church and state as enshrined in the *Leyes de Reforma* (Laws of Reform) of 1859. However, the end of the civil war did not end Mexico's strife. In 1861 French forces landed in Mexico with the stated aim of collecting its debts. After initial Mexican successes, French forces proved overwhelming and they occupied the country. In 1864, the French proclaimed Archduke Maximillian of Austria Emperor of Mexico. Maximillian's rule was totally dependent upon foreign and conservative Mexican backing. Thus, when the United States demanded and received an end to the foreign occupation of Mexico in 1866, Maximillian's days were numbered.

The following year Juárez and José de la Cruz Porfirio Díaz ousted Maximillian who was then executed. The United States had demanded an end to foreign military occupation, but did not stipulate on political leadership alone. Consequently, Maximilian remained in power while the French troops pulled out in order to not cross the United States. It was this vulnerability that eventually led to his demise. Juárez, who served as president until his death in 1872, continued his earlier reform program. The ascendancy of Díaz in

1876, however, brought in over three decades of one-man rule that lasted until 1911. Under his leadership, Mexico stressed modernization although without any concern for the welfare of the great peasant majority.

The Twentieth Century

By the early twentieth century, Díaz's rule had dissaffected the middle classes as well as the peasantry; his reelection in 1910 set the stage for a revolution that forced the long-serving president into exile. Opposition leader Francisco Indalecio Madero succeeded Díaz as president, but was soon beset by troubles of his own as the revolutionary forces became radicalized. Revolutionary peasant leader Emiliano Zapata gained a large following through his *Plan de Ayala*, a program of land reform which called for the occupation and expropriation of large estates. In 1913, rebel forces under the command of Victoriano Huerta, ousted Madero and assassinated him. Huerta's regime proved divisive and a new rebellion led by Venustiao Carranza and Alvaro Obregón seized power in 1914. Mexico descended into civil war that lasted until 1917. As president, Carranza successfully suppressed revolts by Zapata and Pancho Villa (who prompted United States military intervention after he raided a border town in New Mexico).

The most significant development of the revolutionary years was the constitution of 1917. This document implemented many reforms including separation of powers among the three brances of government, a bicameral legislature, a single term for the president, public ownership of land and resources, labor recognition, removal of the church from education, and social security (although this last program did not become a reality until the 1940s). This constitution remains in force today.

The establishment of a new constitution failed to bring immediate political stability to Mexico. Obregón overthrew Carranza in 1920, and in 1928, after tinkering with the constitution to allow for his own reelection, the former was himself assassinated prior to taking office once again. Only under the presidency of Emilio Portes Gil (1928–34) did stability return to Mexico.

The presidency of Gil coincided with the establishment of the *Partido Nacional Revolucionario* (National Revolutionary Party), a grouping that—despite two name changes—remained in power uninterrupted until the end of the century.

Since that time, Mexico has been relatively stable. During the Second World War, Mexico contributed a token force to the Allied cause. Following that conflict, successive governments stressed economic modernization and the growth of the middle class. Efforts have also been made to exploit the country's vast oil reserves. Mexico was in the global spotlight as a host for the XIX Summer Olympiad in Mexico City in 1968 and the World Cup in 1970 and 1986. In 1994 Mexico entered into the North American Free Trade Agreement

(NAFTA), a free trade zone with the United States and Canada.

However, widespread government corruption and chronic economic problems beginning with the recession of the early 1980s continue to plague the country and have begun to undermine governmental stability. In 1994, rebels from the impoverished state of Chiapas calling themselves the Zapatista Army of National Liberation took up arms against the government and demanded land reform. That same year the ruling Institutional Revolutionary Party's (or PRI) presidential candidate and secretary general were assassinated. In 1997, another rebel group appeared in the state of Guerrrero, the Popular Revolutionary Army. The Mexican government responded to both uprisings by sending in the armed forces to restore order. In addition to the open rebellions in the provinces, the PRI has seen its support dwindle at the polls. In 1997, the PRI lost its majority in the Chamber of Deputies, as well as the mayoralty of Mexico City to the rival *Partido de la Revolución Democratica* (PRD). Chances for an opposition victory in the 2000 presidential elections appeared good.

Timeline

50,000 B.C.   First inhabitants arrive

The first inhabitants of North America arrive from Asia by way of the Bering Straits.

12,000 B.C.   Age of the Tepexpan Man

Archaeologists believe the evidence of the Tepexpan man indicates that he is the oldest found in Mexico.

1500 B.C.   Cultivation of corn

Itinerant tribes in central Mexico cultivate corn, according to theories of archaeologists.

A.D. 900–1100   Emergence of the Toltecs

Predecessors of the Aztecs, the Toltecs attain control of the central plateau of Mexico and remain in charge until about 1100.

900–1521   Postclassic era

The postclassic era is characterized by the militarization of theocratic societies.

1325   Construction of Tenochtitlán begins

The name Aztec applies to the American Indian people living in the area known as the valley of Mexico, and making up a mighty empire that thrived during the 1300s, 1400s, and early 1500s. The people who built the spectacular city of Tenochti-

tlán (pronunciation: tay notch TEE tlohn) are also sometime called Aztecs, but they called themselves by other names, such as Cohua-Mexica and Tenochca.

According to legend, the Tenochca (Aztecs) were to build a city at the spot where they would find an eagle perched on a cactus while consuming a snake. They find this omen on an island in a lake. This omen becomes and remains the Mexican national symbol. The name Tenochtitlán means "place of the high priest Tenoch." By the time the Spanish explorers and conquerers arrive in the early 1500s, Tenochtitlán is a complex city, with a system of canals, vast public buildings, and wide raised roadways. Aztec society includes architects, engineers, astronomers, artists, potters, and metalworkers. The Aztec method of cultivating maize (corn) is advanced; maize provides the primary staple of the Aztec diet.

1474?   Birth of Fray Bartolomé de las Casas

Fray Bartolomé de las Casas (1474?–1566), defender of Indians, is born and educated in Seville, Spain. Intrigued by the tales he hears of the New World, de las Casas comes to Mexico as an *encomendero* (one granted land in the Americas by Spain). He joins the Dominican religious order He observes the way the Spaniards treat the natives, is saddened by what he sees and he speaks out in their defense.

As an early historian of the period of the Conquest (Spanish conquest of Native American societies), he writes many books. Among his important works is *Historia de las Indias*.

1485   Birth of Hernán Cortés

Hernán Cortés (1485–1547) is born in Medellín, Spain. Before his twentieth birthday he comes to America. When Diego Velázquez, his brother-in-law, becomes governor of Cuba he accompanies him to the island, eventually becoming mayor of Santiago de Cuba. At thirty-three, named head of the expedition charged with the conquest of Mexico, he realizes this goal within three years. He is an effusive correspondent writing to the Emperor of his deeds. Of these the *Cartas de relación*, of 1519, 1520, 1522, 1524, and 1526 are considered of historic and literary merit.

1502   Montezuma becomes emperor

Montezuma II (Moctezuma or Motecuhzoma, 1466–1520), ruler of the Aztecs, becomes emperor of the Aztecs. He is the ninth Aztec emperor to rule Mexico. When Hernán Cortés arrives in Mexico (see 1519), Montezuma believes he may represent the Aztec white god, Quetzalcoatl. Based on this belief, Montezuma gives Cortés and his party gifts, and even invites them to his palace at Tenochtitlán.

The Spaniards overpower him and declare themselves rulers. (See 1521.)

Aztecs

The Aztecs, a group of tribes who spoke the Nahuatl language, were the last great native civilization to rule Mexico before its conquest by the Spanish under Hernando Cortez in 1519. Originating in northern Mexico as a nomadic people, they migrated to Mexico's central basin, where they conquered the existing city-states and forced their people to pay tribute in the form of food and other goods. Emerging as the dominant group by the middle of the fifteenth century, the Aztecs forged an empire that extended over much of Mexico and was second in size only to that of the Incas in Peru.

The Aztecs were ruled by a king whose throne did not pass automatically to his sons upon his death, although it did usually pass to a brother or other relative, who was chosen by a special council of warriors and priests. The Aztec religion included a number of different gods, all of whom were believed to live in heaven, with the most powerful gods found on its upper levels. Most Aztec religious festivals were related to aspects of the agricultural cycle, such as the harvest or the need for rain.

Although their civilization was an advanced one for its time, the Aztecs were a violent and warlike people—their most important god was Huitzilopochtli, the god of war. Their armies, organized into companies of between 200 and 400 men, consisted of career warriors as well as men recruited from other walks of life—all Aztecs except merchants and slaves had to perform military service. Aztec warriors carried spears and javelins, and they wore padded cotton armor soaked in salt water to stiffen it.

The Aztecs enslaved members of the tribes they had defeated in war, and thousands of these slaves were used as human sacrifices in religious rituals that involved feeding their blood or their hearts to statues of the Aztec gods. As many as 50,000 people are thought to have been sacrificed in a single four-day ritual dedication of a new temple. Nevertheless, the Aztecs also developed an advanced agricultural economy and a complex legal code; built island cities connected by an intricate network of canals; created two separate calendars, one for the farming year and another for religious and astrological use; and produced sculptures that are still considered great works of art.

The Aztec empire came to an end with the Spanish conquest of Mexico begun by Cortez in 1519. In addition to their superior weapons, the Spaniards were aided by the fear that their appearance engendered in the Aztecs, who had never seen white men, horses, or guns before and thought that the conquerors were gods. This belief also accorded with an Aztec legend prophesying the destruction of their empire by a god named Quetzalcoatl who would arrive by ship.

1518   European exploration of coast from Cozumel to Cabo Roxo

Spanish explorer Juan de Grijala voyages along the Mexican coast from Cozumel to Cabo Roxo.

1519   Conquistadores sent to conquer Mexico

Diego Velázquez de Cuéllar, conqueror of Cuba, initiates conquest of Mexico by sending Francisco Fernández de Córdoba and Hernán Cortés to survey the country, the former to the Yucatán and the latter to Mexico. He appoints Cortés to head an expedition of over 500 men to attempt to conquer Mexico. (See 1521.)

1519: April   Cortés establishes first Spanish settlement

Hernán Cortés establishes a settlement at Villa Rica de la Vera Cruz. According to legend, prior to beginning his exploration of the interior, he orders the sinking of all the Spanish ships.

1520   Arrival of Pánfilo de Narváez in Mexico

Cortés goes out to confront Pánfilo de Narváez, believing Narváez has come to Mexico to replace him. Cortés leaves Pedro de Alvarado in charge of Tenochtitlán. Pedro de Alvarado permits a religious service for the Spaniards. During the service, Indians surround and kill many of them. The attack is commemorated as *La Matanza del Templo Maya.*

1520: June 30   Death of Montezuma II

After eight months of Spanish occupation, the Aztecs revolt against the Spaniards and stone to death their own leader, Montezuma.

Montezuma is succeeded by Cuitlahuac and eighty days later by his nephew, the last Aztec ruler, Guatemozín (also known as: Cuauhtémoc Cuauhtemoctzín, Guatémoc, or Quauhtémoc, 1495?–1525) who takes up the anti-Spanish battle after his uncle's death. He drives the Spanish from

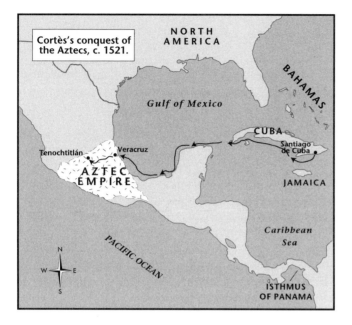

Cortès's conquest of the Aztecs, c. 1521.

NORTH AMERICA

BAHAMAS

Gulf of Mexico

CUBA

Santiago de Cuba

Tenochtitlán

Veracruz

AZTEC EMPIRE

JAMAICA

Caribbean Sea

PACIFIC OCEAN

N W E S

ISTHMUS OF PANAMA

Tenochtitlán during a battle known as *"La noche triste"* (the sad night).

1521   Conquest of Tenochtitlán by Hernán Cortés

Hernán Cortés, leading a large expedition of men (see 1519), wins victory over the city of Tenochtitlán, completing his conquest of Mexico. The Spanish conquerors destroy the city of Tenochtitlán, signaling the end of the Aztec empire. On its ashes they begin the construction of Mexico City, destined to become the largest city in the world.

1521   *Encomienda* system instituted

The Spaniards enact an *encomienda* system which grants landowners authority to extort labor and money from the Indians.

1524   Arrival of the Franciscans

The Franciscans (members of a Roman Catholic religious order named after St. Francis of Assisi), the first of the Spanish missionaries, arrive in Mexico.

1531   Appearance of the Virgin

According to legend, the Virgin Mary appears to Juan Diego, a humble Indian youth. The vision occurs on the hill of Tepeyac, north of the capital. A basilica in honor of the Virgin of Guadalupe (as this vision of the Virgin Mary becomes known), Mexico's patroness, is built on this site. December 12, the feast day of the Virgin of Guadalupe, is a major Mexican religious festival.

1541   Mixtón War erupts

A short-lived insurrection among the Chichimen Indians of western Mexico occurs. This is known as the Mixtón War.

1551   Founding of the national university

Universidad Nacional Autónoma de México (National Autonomous University), known more popularly by its acronym UNAM, is founded. It is the oldest university in North America.

By the late 1990s, over 70,000 students study at UNAM in a modern educational complex decorated with murals by some of Mexico's greatest artists.

1651   Birth of Sor Juana Inés de la Cruz (Juana de Asbaje)

Sor Juana Inés de la Cruz (known as Juana de Asbaje, 1651–95) is born, the illegitimate daughter of a *mestiza* (mixed Indian and Spanish descent) and a Spaniard. She learns to read at age three and begins to write poetry at age eight. She becomes a nun, and is considered an activist for women's rights. An early feminist, she authors plays, verse, and philosophical, theological, and scientific treatises.

1720   Abolition of the *encomienda*

Though officially obliterated, the evils of the *encomienda* system persist. (See 1521.)

1763   Fray Teresa de Mier born

Fray Teresa de Mier (1763–1827) is born in Monterrey to a rich and powerful family. He becomes a priest who gains fame as a great preacher. He finds fault with untrue historical myths, and sets out to dispel them. On December 12, 1794, in the presence of the Archbishop and the Viceroy and other important citizens, he alters the Church's account of the appearance of the Virgin of Guadalupe (see 1531); in addition, he insists that Saint Thomas had preceded the conquistadors in the New World, where he preached to the natives. Fray Teresa de Mier is the author of numerous sermons and studies including *Apología y relaciones de su vida*. His most notable work is *Profecía del Doctor Mier sobre la Federación Mexicana*. He dies in 1827.

1776   Birth of José Joaquín de Lizardi

Writer José Joaquín de Lizardi (1776–1827) is born in Mexico City. He is ranked as the "father" of the Spanish American novel and the most influential writer in nineteenth century Mexico. He founds seven newspapers and authors numerous pamphlets where he sets forth his wide-ranging ideas. He publishes *El periquillo sarniento* in 1816, an adventuresome work categorized as the first Spanish American novel. He dies of tuberculosis in 1827 in Mexico City.

1785   Establishment of the *Real Academia de Bellas Artes*

Under the sponsorship of King Charles III of Spain (1716–88), the *Real Academia de las Bellas Artes de México,* (Royal Academy of Fine Arts of Mexico) is established, commonly known as the "Academia de San Carlos." To staff the new school, the academy brings in teachers from Europe, among whom are the sculptor and architect Manuel Tolsá, the painter Rafael Ximeno y Planes and the engraver Jerónimo Antonio Gil.

1790   Alvarez Juan Concepción de Atoyac is born

Alvarez Juan Concepción de Atoyac (1790–1867) is born. A rebellious political leader prior to and after independence is won from Spain (see 1821), Alvarez Juan Concepción de Atoyac serves as provisional president in 1855 (see 1855). He dies August 21, 1867 in Acapulco.

1790   Discovery of the Aztec Calendar Stone

The Aztec "Calendar Stone" is discovered. It is a flat stone disc, 12 feet (3.7 meters) in diameter, and is used in ceremonies honoring the sun god Tonotiuh. Tonotiuh's face is at the center of the stone. Rows of carved religious symbols and symbols for days of the month surround his image. This stone is preserved in the National Museum of Anthropology in Mexico City (see 1825).

1806   Birth of Benito (Pablo) Juárez

Benito (Pablo) Juárez (1806–72), a Zapotec Indian, is born. He becomes a national hero during the civil war (see 1858), when he assumes the presidency. He is elected president in 1861 and serves until his death in 1872.

1810: September 16   Miguel Hidalgo y Costilla leads first revolt against Spain

Parish priest Miguel Hidalgo y Costilla (1753–1811) of Dolores, unable to tolerate Spanish colonialism, starts to plan with some fellow liberals to separate from Spain. He calls his parishioners to Mass where, instead of the customary homily, he urges them to rise up against Spain. One of the most famous orations in Mexican history, this sermon is more commonly known as the "Grito de Dolores" (the Shouting of Dolores). The rebellion fails and Hidalgo is executed.

1813   José María Morelos y Pavón fights for freedom

José María Morelos y Pavón (1765–1815) resumes the rebellion interrupted by the death of his fellow priest Miguel Hidalgo y Costilla (see 1810: September 16), and convenes the first National Congress which proclaims the nation's freedom from Spain. During 1813 he speaks for the first time about absolute independence for Mexico. This then leads to the Constitutional Decree for the freedom of Mexican America, given at Apatzingán the 22nd of October, 1814. This

Decree serves to foster the further growth of Mexico's distinct political image. Morelos obtains an army, but his military power would be broken through his untrained soldiers. His soldiers are basically untrained guerrillas, and though they have great patriotic fervor for Mexico, they fight without discipline or coordination.

1815   Execution of Morelos

José María Morelos y Pavón is executed in the ruins of the palace of San Cristóbal Ecatepec, on December 22, 1815. This results in the end of revolutionary activities in Mexico.

1820   Iturbide starts revolution for independence

The first emperor of Mexico, Augustín de Iturbide was born in 1783 in Valladolid (Morelia); Iturbide joined the army in 1800, by 1810 he was fighting with the royalists, and by 1820 held the rank of colonel.

1821   Iturbide proclaims the Plan de Iguala

Along with Vicente Guerrero, Iturbide proclaims the Plan de Iguala, or the Plan of Equals.

This proclamation is also known as the Plan de Las Tres Garantías, or the Plan of the Three Guarantees. The Plan de Iguala has three goals: one religion, unification of all social groups, and Mexico's independence.

1821   Mexico declares independence

In August, Iturbide and the viceroy Juan de O'Donojú sign the Treaty of Córdoba, ratifying the Plan de Iguala and confirming Mexico's independence.

1822   Iturbide is declared Emperor of Mexico

Iturbide masterminds his coronation as Agustín I, emperor of Mexico. After a year of lavish spending, he finds himself broke.

1823   Antonio López de Santa Anna and Guadalupe Victoria lead an uprising

Antonio López de Santa Anna and Guadalupe Victoria head an uprising that results in Iturbide's abdication and exile.

1824   Federal constitution proclaimed and Iturbide executed

Mexico proclaims a new federal constitution. Guadalupe Victoria, born Manuel Félix Fernández, becomes president. As president (1824–29), he banishes Spaniards who oppose his government and policies. In April 1824 the congress, having already declared Iturbide's administration void, revoke his pension and declare Iturbide a traitor. When he returns from exile in Italy to Mexico in July 1824 he is apprehended on arrival in Tamaulipas and executed the next day, July 19, 1824, in Padilla, Tamaulipas.

1825   National Museum of Anthropology founded

The National Museum of Anthropology is founded in Mexico City. It houses artifacts and includes a library of over 300,000 volumes. Among its treasures are the Aztec Calendar Stone (see 1790) and a 137-ton statue of the Aztec god of rain, Tlaloc.

1827   Slavery abolished

President Guadalupe Victoria abolishes slavery.

1833   Santa Anna takes over as dictator

Santa Anna (1794–1876) a figure with a long military career had gained prominence through his role in installing Iturbide to power in 1821, and in thwarting the Spanish plan of re-taking Mexico in 1829. In 1833 he takes over as dictator and sells southern areas of modern-day New Mexico and Arizona to the United States.

1833–47   Presidency of Valentín Gómez Fareos

Valentín Gómez Fareos becomes president. He occupies the presidency five times between 1833–47, the longest period between December 24, 1846 to March 20, 1847.

1844–51   Presidency of José Joaquín Herrera

José Joaquín Herrera serves as president five times, the longest period being from June 2, 1848 to January 14, 1851.

1846: May   U.S. officially declares war on Mexico, and the Battle of Palo Alto

In the first skirmish of the Mexican-American War, Mexican troops cross the Río Grande and attack Fort Brown, jeopardizing U.S. General Zachary Taylor's supply line. The encounter ends in a definitive victory for the United States.

1848: February 2   Treaty of Guadalupe Hidalgo

The Treaty of Guadalupe Hidalgo is signed. It brings to a close war between the United States and Mexico. Modern-day Texas and territory including Arizona, New Mexico, Nevada, Utah, and part of Colorado, as well as California, are ceded to the United States.

1853   Gadsen Purchase

Though the Pierce administration wants to purchase considerable property in the northern provinces from Mexico, the U.S. emissary, James Gadsen, succeeds in acquiring only the Gila River valley. This $10 million purchase finalizes the modern-day northern Mexican border with the United States.

1854: December 23   Birth of Victoriano Huerta, future president

Victoriano Huerta (1854–1916) is born in Colotlán. Later he joins the army and advances to the rank of general. When president Porfirio Diaz goes into exile, Huerta assumes the presidency (see 1913–14).

1855   Interim president named

Political leader Alvarez Juan Concepción de Atoyac (1790–1867) serves as provisional president (see 1790).

1857   New constitution announced

A new constitution is proclaimed as the basis of Juárez's La *Reforma*. Its intent, is to eradicate Spanish colonialism from Mexico.

1858–61   Civil war breaks out

In this civil war, known as the War of Reform, forces led by Pablo Juárez (1806–72) defeat the conservative forces.

1858   Juárez becomes President

Benito (Pablo) Juárez becomes president; he is electcd to the office in 1861, and holds the position until his death, except for two days (see 1860: August 13–14 and 1867–72). He advocates for separation of church and state.

1859   *Leyes de Reforma* ratified

Leyes de Reforma (laws of reform) pass that allow for freedom of religion; nationalization of Church properties; recognition of marriage as a civil contract; and establishment of a public registry.

1860: August 13–14   Presidency of José Ignacio Pavón

José Ignacio Pavón serves the shortest of all Mexican presidencies—his terms lasts just two days.

1861   France invades Mexico

France, supported by England and Spain, sends a fleet to Vera Cruz to attempt to pressure Mexico into paying its debts to these European countries.

1862: May 5   National Army defeats French forces

The National Army defeats the French forces at Puebla. The anniversary of this victory is commemorated both in Mexico and the United States and is known popularly in both countries as *El cinco de mayo* (The Fifth of May).

1864–67   Maximilian of Austria declared emperor

Through French military intervention directed by Napoleon III, (allegedly because Mexico had not paid her debts) Ferdinand Joseph Maximilian (1832–67) of Austria is installed and declared the emperor of Mexico. His wife, Carlota (1840–1927), is declared empress. He proves an ineffectual leader whose liberal leanings upset his conservative backers. Maximilian plans much of the *Paseo de la Reforma* in Mexico City. He reigns from April 10, 1864 to June 14, 1867.

1866   United States pressures France to withdraw troops

Diplomatic messages sent from the United States pressure France to withdraw troops from Mexico. Troops are withdrawn, leaving Maximilian vulnerable.

1867   Juárez works to unseat Maximilian

With U.S. aid, Pablo Juárez succeeds in toppling Maximilian who surrenders to Juárez and is executed by Juárez's firing squad. Maximilian's wife, Carlota, returns to Europe where she dies insane in 1927.

1867–72   Pablo Juárez becomes president

Benito (Pablo) Juárez becomes president once again (see 1858). He initiates reforms to bring economic stability to Mexico, and takes preliminary steps to establish a public school system. He dies in office.

1873: January 1   Novelist Mariano Azuela born

Mariano Azuela (1873–1952) is born in Lagos de Moreno. He becomes a physician, but in 1911, joins up with the revolutionary forces. He publishes *Los de abajo,* the first of his novels dealing with the Mexican Revolution of 1910, in installments from October to December 1915 in the newspaper *El Paso del Norte.* He is often referred to as the "father" of the novel of the Mexican Revolution. He dies in 1952.

1875   Birth of Doctor Atl

Gerardo Murillo is born in Guadalajara. He is known in art circles as Doctor Atl. He is a painter and author regarded as one of the pacesetters in the Mexican movement for artistic nationalism.

1877   First term as president for Porfirio Díaz

One of Juárez's generals and a mestizo born in Oaxaca, Porfirio Díaz (1830–1915) becomes president. Except for a four-year period (1880–84) when Díaz appoints a Gonzalez (essentially a puppet president) to the presidency to get around the law prohibiting reelection, Díaz governs Mexico until 1911, longer than any other person. (See 1881–1911.)

1878: June 5   Birth of Pancho Villa

Revolutionary leader Pancho (Francisco) Villa (1878–1923) is born Doroteo Arangol in Hacienda de Río Grande, San Juan del Río. He is a vigorous revolutionary fighter in the Mexican Revolution (see 1910–20). He and Emiliano Zapata (1879–1919) are defeated by Venustiano Carranza (see 1915). The two withdraw to northern Mexico, where they continue to lead guerrilla attacks. In 1916 Villa's deadly attack on some U.S. citizens in Santa Isabel result in U.S. President

Woodrow Wilson's decision to impose an arms embargo against Mexico. Villa is assassinated on June 20, 1923.

1881–1911   Dictatorship of Porfirio Díaz

Rather than conform to the law, Porfirio Díaz changes the law and rules without regard to moral or legal mandates until 1911, believing in ruling from the top down. His first name has passed into the Mexican vocabulary. *Porfirianismo* refers to totalitarianism in government accompanied by conservatism in politics and excess in the decorative arts.

1882   José Vasconcelos is born

Essayist, philosopher, and educator José Vasconcelos (1882–1959) is born in Oaxaca. He is an active participant in the Mexican Revolution of 1910 (see 1910–20). His five-volume autobiography, *Ulises criollo,* and his analysis of the racial future of Latin America, *La raza cósmica,* are among his greatest works. He serves as secretary of education twice, encouraging public education and establishing public libraries. He dies in 1959.

1883   Birth of José Clemente Orozco

Leading muralist and painter José Clemente Orozco (1883–1949) is born. He studies engineering and architecture in Mexico City, and continues his education at the Academia San Carlos, where he studies art. His physical handicaps prohibit his active participation in the Revolution of 1910. Nonetheless he accompanies the troops, gathering the harsh images which would inform his later drawings and paintings. His first exhibition is in Paris, in 1925. In Mexico City a major retrospective of his work is staged in 1947. He creates urban murals in Mexico and the United States, and works in the United States at the New School for Social Research in New York and at Pomona College in Claremont, California. He dies in 1949.

1886?   Birth of Diego Rivera, leading muralist and painter

Diego Rivera (1886?–1957), leading muralist and painter, is born in Guanajuato. He wins a scholarship to study painting in Madrid, and Paris. He begins painting murals in public buildings in 1921. His subjects include the history of Mexico and the life of the Mexican people. His depictions of common people's uprisings earn him a reputation as a revolutionary artist who wants to bring art to the masses. He is part of a trio of internationally famous Mexican muralists early in the twentieth century (see 1898). He works in the United States, primarily at the Institute of Fine Arts in Detroit, Michigan, from 1930–34. During that period, he creates murals depicting the life of factory workers. His mural, *Man at the Crossroads* (1933) was moved from Rockefeller Center to Mexico City. In the 1940s and 1950s Rivera focuses on his antiwar and atheist views. He is credited with starting the Mexican school of mural painting. He marries artist Frida Kahlo (see 1907: July 6). Rivera dies in 1957.

1887: October 6   Birth of novelist Martín Luis Guzmán

Novelist Martín Luis Guzmán is born in Chihuahua. He becomes a novelist, historian, and journalist. One of the best writers of the revolutionary period, he is editor and publisher of *El Tiempo,* a magazine similar to *Time.* He writes *El águila y la serpiente,* chronicling his experiences in the Mexican Revolution (see 1910) and *Memorias de Pancho Villa.* He dies December 22, 1976 in Mexico City.

1888   Birth of writer Ramón López Velarde

Poet and writer Ramón López Velarde (1888–1921) is born in Zacatecas. He writes patriotic poems that are full of irony and moving love poems. His complete poems are compiled in *Poesias completas,* published in 1957.

1889   Alfonso Reyes born in Monterrey

Noted poet, essayist, and short story author Alfonso Reyes (1889–1959) is born in Monterrey. He graduates from UNAM as an attorney and begins his career as a professor there. He holds diplomatic positions in Europe (notably Spain) and Latin America. In 1939 Reyes retires from his diplomatic career to devote time to writing. His *Diario 1911–1930,* published in 1960, offers his perspective on Mexican literature in that era.

His works are widely translated into many languages. Among these are *Selected Essays* (1964), and two books about Mexico, *The Position of America* (1950) and *Mexico in a Nutshell* (1964). His fame rests primarily on his essays. In 1956 he is nominated for the Nobel Prize in Literature. He dies in 1959.

1896   Painter David Alfaro Siqueiros is born

David Alfaro Siqueiros (1896–1974) is born in Chihuahua. He participates in the Mexican Revolution (see 1910–20) by joining the forces of Carranza in 1913, as well as in the Spanish Civil War as an officer in the Republican army. With Diego Rivera (see 1886?) he launches the review magazine, *El Machete,* in 1922. He is considered responsible for reviving the fresco in Mexican art and he creates frescoes for the National Preparatory School in Mexico City, among others. His art and life are characterized by social protest. He founds the Center of Realist Art in Mexico City in 1944. His most celebrated works include murals on the UNAM campus: *From Porfirio's Dictatorship in the Revolution* (at the National History Museum) and *March of Humanity* (at the Hotel de Mexico). He dies January 6, 1974 in Cuernavaca.

1899   Birth of Rufino Tamayo

Artist Rufino Tamayo (1899–1991) is born in Oaxaca, the son of Zapotec Indians. He is orphaned at an early age and taken by an aunt to live in Mexico City. He studies at the Academy of San Carlos there (1917–21), and becomes a curator at the National Museum of Anthropology (1921–26). He lives in New York (where he teaches at the Dalton School) and Paris, for about ten years, but works for most of his career in his native Mexico. His knowledge of pre-Columbian and indigenous art, developed while a curator, is reflected in his work. His murals are found in the Palacio de Bellas Artes (National Palace of Fine Arts) in Mexico City and in UNESCO (United Nations Educational, Scientific, and Cultural Organization) headquarters in Paris. Other works include the painting *Women of Tehuantepec* (1939, at the Albright-Knox Gallery in Buffalo, New York). He dies in 1991.

1899: June 13   Carlos Chávez is born

Composer Carlos Chávez (1899–1978) is born Carlos Antonio de Padua Chávez y Ramírez in Mexico City. Chávez is a world-renowned musician and composer who incorporates elements of traditional popular music into his compositions, utilizing modern techniques. He forms the Mexican Symphony Orchestra in 1928. He is also a founder and director of the Palacio de Bellas Artes from 1947–52. His works—ballets, symphonies, concertos—are influenced by folk music themes. He dies August 2, 1978, in Mexico City.

1900?   U. S. consul in Yucatán purchases abandoned hacienda

The exact year is uncertain, but very early in the 1900s Edward H. Thompson acquires a deserted hacienda for seventy-five dollars. To his surprise his acquisition encompasses Chichén Itzá, the ancient Mayan city—a treasure trove of Mayan artifacts.

1900: February 2   Birth of filmmaker Luis Buñuel

Surrealist filmmaker and director Luis Buñuel (1900–83) is born in Calanda, Spain. He makes films with Spanish artist Salvador Dali (see Spain), including *Un Chien Andalou* (An Andalusian Dog, 1928) and *L'Age d'or* (The Golden Age, 1930). Without Dali, his first film is a documentary about poverty, *Las Hurdes* (Land Without Bread, 1932). It is banned in Spain and Luis Buñuel is forced to leave the country. He settles in Mexico in 1947. In 1950 he makes the award-winning *Los Olvidados* (The Young and the Damned), a documentary about juvenile delinquency. His later films use images that are often erotic, and employ dark, ironic humor to express his hatred of the Roman Catholic Church. He dies on July 29, 1983, in Mexico City.

1902   Birth of poet Jaime Torres Bodet

Internationally recognized Mexican poet and essayist Jaime Torres Bodet is born. During his career, he represents Mexico on several diplomatic missions and serves as Mexico's minister of education. He inaugurates President Manuel Avila

Camacho's campaign against illiteracy (see 1944). In 1958, again as minister of education, Jaime Torres Bodet begins a national program to supply Mexico with primary schools and teachers to assure every Mexican an elementary-level education.

1904   Birth of Agustín Yáñez

Author Agustín Yáñez is born in Guadalajara. Yáñez is considered an avant-garde author for his time. He successfully combines professorial and political careers. He is elected governor of his native state; he heads the Mexican delegation to UNESCO in 1960; and he serves as minister of education in 1964. His novels include: *Al filo del agua, La tierra pródiga,* and *Las tierras flacas.*

1905: July 6   Juan O'Gorman born

World-renowned architect and muralist Juan O'Gorman (1904–82) is born in Coyoacán. He is director of the town-planning administration, beginning work as an independent architect in 1934. He collaborates on the Library of the National University of Mexico (1952) in Mexico City and creates many mosaic designs, murals, and frescoes for the exteriors of buildings. He is found dead in Mexico City, January 18, 1982.

1907: April 24   Birth of cinematographer Gabriel Figueroa Mateos

Gabriel Figueroa Mateos (1907–97) is born in Mexico City. As a cinematographer, he is internationally renowned for his striking use of Mexican landscape in approximately 200 films. He works with many prominent film directors, among them American John Ford and Mexican Luis Buñuel (see 1900: February 2). He dies April 27, 1997, in Mexico City.

1907: July 6   Frida Kahlo (de Rivera) born

Artist Frida Kahlo (1907–54) is born in Coyoacán, Mexico City. Her given name is Magadalena Carmen Frida Kahlo y Calderón, but she is known simply as Frida Kahlo. Her father is a German Jewish photographer who had immigrated to Mexico; her mother is a Mexican Roman Catholic. Frida is seriously injured in an accident at the age of fifteen. During her recovery, she begins to paint. She shares her work with Diego Rivera (see 1886?), the famed Mexican muralist, whom she eventually marries. Her dreamlike portraits often focus on pain and the challenges faced by women, and frequently feature herself as the subject because as she says, "I am the subject I know the best."

At the 1940 International Surrealist Exhibition, she and her husband Rivera both display their art. In 1946 Frida wins a prize at the Annual National Exhibition at the Palace of Fine Arts. She dies July 13, 1954. In 1958 the Frida Kahlo museum opens in her childhood home in Coyoacán.

1908   Madero publishes *The Presidential Succession in 1910*

Francisco Madero (1873–1913) publishes a controversial book, *The Presidential Succession in 1910.* It calls for genuine political freedom, and results in the formation of an opposition party that opposes reelection; Madero becomes their presidential candidate. This action is the spark to set off forces which overthrow Díaz.

1910–20   Mexican Revolution of 1910

The Mexican Revolution of 1910 is credited with bringing the nation into the twentieth century. Unlike uprisings in other parts of the world, this one does not have one major leader or one principal cause. Rival leaders keep the country in a state of constant turmoil for a decade.

1911   Zapata proclaims his *Plan de Ayala*

Revolutionary leader Emiliano Zapata (1879–1919) puts forth his *Plan de Ayala* dealing with the agrarian problem. He occupies large estates, demanding the return of land to the native Indians. At first he supports Madero, but he becomes impatient when reforms are delayed. The motto for his revolution is *Tierra y Libertad* (Earth and Freedom). He operates a commission to distribute land, and sets up a Rural Loan Bank.

1911: March 6   United States mobilizes troops against Mexico

The turmoil and confusion growing in Mexico frightens the United States into taking measures to prevent revolution from spreading to America. U. S. President William Taft orders U.S. soldiers to the Mexican border. He also sends naval units to the Pacific as well as the Gulf of Mexico.

1911: March 20   Alfonso García Robles is born

Activist Alfonso García Robles (1911–91) is born in Zamora. He is an early proponent of nuclear disarmament and shares the 1982 Nobel Peace Prize with Alva Reimer Myrdal (1902–86; see Europe: Sweden) of Sweden. Robles dies September 2, 1991 in Mexico City.

1911: May 21   Díaz steps down as president

Porfirio Díaz resigns as president; on May 31 he embarks from Veracruz for Europe.

1911: June 7   Madero becomes president

Francisco Indalecio Madero enters Mexico City triumphantly from his headquarters in Ciudad Juarez and succeeds Díaz as president (see 1908). His official election as president comes in October 1911. His moderate positions do not satisfy any factions, and he faces revolts led by Emiliano Zapata (1879–1919). His presidency lasts less than two years (see 1913: February).

Mexicans burn the town of Nuevo Laredo in order to prevent its capture by American forces during the upheavals of 1914. (EPD Photos/ CSU Archives)

Madero is the son of wealthy landowners, and is educated in Paris. He is a spiritualist, vegetarian, and practices alternative healing techniques. He is assassinated in 1913.

1911: August 12   Birth of Cantinflas

Comedian and film star Cantinflas (1911–93) is born as Mario Moreno Reyes in Mexico City. Frequently likened to Charlie Chaplin, Cantinflas appears in forty-nine films, including *Around the World in 80 Days.* He is idolized by the Mexican people for both his comic gifts and his philanthropic efforts. He dies April 20, 1993, in Mexico City.

1911: November 27   Reelection of officials prohibited

A decree limiting the president, vice president, and state governors to one term is proclaimed.

1913–14   Presidency of Victoriano Huerta

Victoriano Huerta (1854–1916; see 1854) becomes president, a post he holds until 1914. His despotic administration polarizes diverse revolutionary groups against him and forces him out of office. His stance toward the United States is defiant. He dies January 13, 1916, in El Paso, Texas.

1913: February   Francisco Madero cedes the presidency

Betrayed by Victoriano Huerta (1854–1916), Madero is ousted from office. He and his vice president Pino Suárez are taken prisoners. Huerta assumes military and civil control. While being transported to jail, Madero and Suárez are assassinated.

1913: November 17   Nellie Campobello is born

Nellie Campobello is born in Villa Ocampo. She gains fame as a dancer and teacher of dance. She becomes director of the *Escuela Nacional de Danza* (National School of the Dance). She is the only woman to write novels of the Mexican Revolution of 1910 (see 1910–20): *Cartucho* and *Las manos de mamá.*

1914   Birth of writer Octavio Paz

Poet and essayist Octavio Paz (1914–98) is born in Mexico City. He attends the National University of Mexico, and serves as ambassador to India in the 1960s. Paz publishes his first book at the age of twenty-nine. His prolific production of poetry places him among the most outstanding Spanish-language poets. In 1943 he receives a Guggenheim fellowship to study in the United States. His verse is characterized by strong existential themes (existentialism being the philosophic method of defining an individual's existence: often through the plight of sadness, despair and the individual's relationship to God and the universe). His volumes of poetry include: *Piedra de sol* (Sun Stone, ten volumes); *Raíz del hombre, La estación violenta,* and *Viento entero.* Collections of essays include *El laberinto de la soledad* (The Labyrinth of Solitude, 1962); *Postdata* (The Other Mexico: Critique of the Pyramid, 1970); *El arco y la lira* (The Bow and the Lyre, 1973); and *Vislumbres de la India* (Glimpses of India, 1995). In 1990 he is the recipient of the Nobel Prize in literature. He dies in 1998.

1914   Architect Carlos Lazo born

Renowned architect Carlos Lazo (1914–55) is born. He utilizes the talents of Mexican architects such as O'Gorman (see 1915) and Mario Pani, among others. He supervises the construction of Ciudad Universitaria, 1949–54. He dies in 1955.

1914: February 3   U.S. arms embargo lifted

U.S. President Woodrow Wilson lifts the arms embargo against Mexico.

1914: April 21   U.S. naval forces occupy the port city of Veracruz

In an effort to stop the German boat, *Ipiranga,* from unloading military supplies for President Huerta's forces, U.S. Navy personnel take over the port of Veracruz. This action results in a break in diplomatic relations between the United States and Mexico three days later.

1914: November 23   U.S. military forces return Veracruz to Mexico

U.S. military personnel hand over control of the port city of Veracruz to Mexican authorities.

1915   Venustiano Carranza becomes president

Venustiano Carranza (1859–1920) overcomes other revolutionary leaders to seize control of the country and become president. He is a moderate, but fails to enforce the new constitution (see 1917), and is eventually overthrown (see 1920).

1915: April   Battle of Celaya

One of the goriest battles in the Mexican Revolution (see 1910–20) is fought at Celaya, Guanajuato. Troops under the command of Alvaro Obregón defeat Pancho Villa.

1915: July 2   Díaz dies in Paris

Former dictator Porfirio Díaz dies.

1915: October 19   U.S. president Wilson proclaims arms embargo

U.S. President Woodrow Wilson declares an embargo on all arms destined for Mexico except those destined for Carranza.

1916: March 9   Pancho Villa attacks Americans

Pancho Villa enters U.S. territory and attacks the town of Columbus, New Mexico, killing fourteen Americans—seven military and seven civilians.

1916   American troops enter Mexico in retaliation

Under the command of General John J. Pershing, U.S. troops enter Mexico to capture and punish Pancho Villa.

1917   Social security comes to Mexico

The concept of social security is incorporated into the constitution of 1917. It will not be until 1943 that legislation is passed to make social security a reality.

1917: January   New constitution approved

A new constitution, Mexico's fifth, is promulgated. It provides for separation of legislative, judicial, and executive branches. Legislative power is vested in a bicameral legislature. The Chamber of Deputies is enlarged to 500 legislators, and the Senate expanded to 128 members.

1917: January 5   U.S. troops leave Mexico

Negotiations between the United States and Mexico conclude concerning the immediate and unconditional withdrawal of U.S. troops from Mexico.

1917: May 1   Carranza becomes Mexican president

Carranza enters Mexico City and takes over as constitutional president of the country.

1919: April 20   Zapata assassinated

Emiliano Zapata head of the agrarian revolution, is betrayed and killed by government troops in Chinameca.

1920: April 24   The Plan de Agua Prieta promulgated

Generals dissatisfied with Carranza put forth the *Plan de Agua Prieta,* in which they refuse to recognize him as president.

1920: September 5   Alvaro Obregón unseats Carranza to become president

Alvaro Obregón (1880–1928) overthrows Venustiano Carranza (see 1915). The Cámara de Diputados declares Alvaro Obregón president and holds the post until 1924. He wins reelection in 1928 (see 1928).

Alvaro Obregón is born near Alamos. He fights with Francisco Madero against the revolutionaries and aids Venustiano Carranza in putting down the revolt led by Pancho Villa and Emiliano Zapata, after which Carranza becomes president (see 1915).

1920: December 11   Elena Garro is born

Journalist and playwright Elena Garro (1920–98) is born in Puebla. She starts her literary career writing one-act plays. Despite positive reception of her work for the theater, Garro turns to the novel. Her first one, *Los recuerdos del porvenir* (Recollections of Things to Come) wins for her the Xavier Villaurrutia Prize. She utilizes "magical realism" in her narrative before it becomes widespread in Latin America. Garro is designated one of the most important women writers in Mexico. She dies August 22, 1998.

1921: September   Mexico celebrates anniversary of independence

Mexico commemorates the centenary of its independence. Many nations send official delegations to the festivities.

1922   Birth of playwright Carlos Solórzano

Carlos Solórzano is born in Guatemala but lives a good part of his life in Mexico and is identified with Mexican literature and culture from the age of seventeen. With degrees in architecture and literature from UNAM, he is director of the Teatro

Universitario, founder of the Grupos Teatrales Estudiantiles, director of the Museo Nacional de Teatro, and a professor at his alma mater. His dramas, noted for their utilization of the latest theater techniques, include: *Las manos de Dios; Los fantoches; Doña Beatriz;* and *El sueño del ángel.* He also distinguishes himself as a historian of Mexican theater and a drama critic.

1922: June 16 American bankers extend financial aid to Mexico

An agreement is signed in New York by Mexico's secretary of state and U.S. bankers outlining a plan for the payment of Mexico's foreign debt.

1923: January 11 Apostolic delegate challenges government order

The Most Reverend Philippi, Apostolic Delegate to Mexico, sets the first stone in what is to be a monument to Christ the King in the state of Guanajuato. Since this action is prohibited in article twenty-four of the constitution, it is an act of deliberate challenge to the government. He is given three days in which to leave the country.

1924 Mexican Airlines founded

Mexicana de Aviación, S.A. and the oldest airline in North America, is established. Mexican Airlines, more formally known as CIA, is founded in Tampico. Current headquarters are in Mexico City.

1924 Presidency of Plutarco Elias Calles

Plutarco Elias Calles (1877–1945) becomes president, a post he holds until 1928. He is fanatically opposed to foreign influence over the Mexican oil industry. During his administration, he founds a national bank. In an effort to establish economic stability, he balances the national budget and exerts a major effort to pay Mexico's debts. A year after leaving office, he is a founding member of the National Revolutionary Party (PRN—see 1929) and thus exercises considerable influence over subsequent presidents. He openly opposes the policies of his former protégé, Lázaro Cárdenas. Calles is consequently sent into exile in 1936 at the loss of this contest. However, he is allowed to return under an amnesty in 1941.

He was born in Guaymas, Sonora, and began his career as head of a school. He served as governor of Sonora (1917–19) and secretary of the interior (1920–24).

1925 Birth of Rosario Castellanos

Poet and prose writer Rosario Castellanos (1925–74) is born in Comitán, Chiapas where she spends her youth. Land reform forces the family to resettle in Mexico City, where she explores several career possibilities. She travels abroad extensively, but returns to her native Mexico where she establishes

Government reform efforts that curbed the power of the Roman Catholic Church have widespread popular support in the 1930s. Over 150,000 Mexicans marched in support of the government in Mexico City in 1934. (EPD Photos/CSU Archives)

the *Instituto de Ciencias y Artes de Chiapas.* Her writing includes novels, short stories, and poetry; she is generally regarded as among the preeminent Hispanic poets. On August 7, 1974, Castellanos is accidentally electrocuted while serving as Mexican ambassador to Israel. With Sor Inés de la Cruz (see 1651) and Elena Garro (see 1920), Castellanos is deemed one of the three most significant women writers in Mexico's literary history.

1926–29 Cristero Revolution

A revolution erupts, mainly centered in the Mexican states of Colima, Michoacán, and Jalisco. It is inspired by opposition to restrictions placed upon the Roman Catholic Church by the new amendments to the Mexican constitution (see 1917 and

1926). The revolution becomes violent at times, including the dynamiting of a train.

1926   Mexican constitution amended

The Mexican government enacts the Regulating Law of 1926 which expands the anticlerical emphasis in the constitution of 1917 by designating specific punishments for violations concerning religious matters.

1926   Archbishop of Mexico rejects new Mexican constitution

The archbishop protests legislation in the new constitution limiting religious activities. The government responds by deporting foreign-born religious personnel and forcing all others to register with civil authorities.

1927   Change in presidential term

The Senate approves lengthening the term of the president to six years; Mexico's Chamber of Deputies follows suit on November 21, 1927.

1928   Birth of Carlos Fuentes

Novelist and playwright Carlos Fuentes is born in Panama City, Panama. As the son of a diplomat, Fuentes travels extensively as a child. He serves as secretary of the Mexican delegation of the International Labor Organization (ILO), an agency of the United Nations. He serves at the Mexican Embassy in Geneva, Switzerland (1950–52), and as press secretary at the United Nations Information Centre in Mexico City. From 1975–77, he is Mexico's ambassador to France. In 1954 he publishes a collection of short stories inspired by myths, *Los dias enmascarados* (The Masked Days). Other works include: *La muerte de Artemio Cruz, Cambio de Piel,* and *Terra Nostra.*

1928   Presidency of Emilio Portes Gil

Following the assassination of Alvaro Obregón (see 1928: July 17), Emilio Portes Gil (1891–1978) becomes interim president. He wins election to a full term (see 1930).

Emilio Portes Gil was born October 3, 1891, in Ciudad, Victoria; he dies December 10, 1978, in Mexico City.

1928: July 17   Obregón assassinated before taking office

After a constitutional amendment to circumvent the ban against reelection, Alvaro Obregón wins a second term as president. Before he can assume office, he is killed by José de León Toral, a religious fanatic.

1929   Founding of the Partido Nacional Revolucionario

The Partido Nacional Revolucionario is founded. The party name is changed twice—first to Partido de la Revolución Mexicana and later to Partido Revolucionario Institucional. It is known by the acronym PRI, and dominates national and local politics for over fifty years.

1930   Interim president Gil elected to a full term

Emilio Portes Gil (1891–1978), serving as interim president (see 1928), is elected to a full term as president.

1930: September 6   Birth of future billionaire

Emilio Azcárrraga Milmo (1930–97) is born. As one of Mexico's most successful businessmen, he creates what is known as the "Spanish-speaking world's largest media empire" by building up his family's radio and television network. He dies April 16, 1997.

1932   Presidency of Abelardo Rodríguez

Abelardo Rodríguez serves as president for two years, from 1932–34.

1934   Minimum wage laws enacted

A minimum wage law goes into effect. Two factors determine the wage rate: the type of work and the regional cost of living. The highest minimum wages are paid in the capital, Mexico City, and in towns along the U.S. border, such as Tijuana and Ciudad Juárez.

1934   Mexico enacts a law forbidding export of relics

To preserve historical artifacts, a law is enacted to prohibit the shipment out of the country of pre-Columbian relics except by those specifically licensed to do so.

1934   Presidency of Lázaro Cárdenas

Lázaro Cárdenas del Rio (1895–1970) becomes president, a post he holds until 1940. Unlike his predecessors, he does not acquire wealth or power from his position. Cárdenas sends Plutarco Calles (see 1924) and his labor bosses out of the country to live in exile in the United States in 1936. Champion of the little man, he does not live in the presidential palace in Chapultepec. He hastens land distribution, nationalizes the railways, and seizes foreign oil and land interests. He founds the Instituto Politécnico Nacional.

1937   IPN founded

As the technical counterpart to UNAM, Instituto Politécnico Nacional (IPN) is established in 1937. Prior to the 1950s both institutions account for over seventy percent of Mexico's enrollment in institutions of higher learning.

1938   Mexico expropriates U.S. and British oil facilities

President Franklin D. Roosevelt does not interfere when Mexico seizes U.S. and British petroleum works. In appreciation of this action, Mexico cooperates with the U.S. in World War II (1939–45) and contributes a squadron of airmen to the war effort.

1940–46 Manuel Avila Camacho president

Manuel Avila Camacho (1897–1955) becomes president, a post he holds for six years. He creates the Instituto Mexicano del Seguro Social and proclaims the Ley del Seguro Social.

1943 Eruption of Paracutín

In a farmer's field, the volcano Paracutín explodes, spewing lava in all directions, inundating the nearby town of San Juan Parangaricutira. By the time it becomes dormant in 1952, the volcano reaches a height of 1700 feet.

1944 National program to reduce illiteracy in Mexico

President Manuel Avila Camacho and Jaime Torres Bodet, minister of education, mandate and implement a plan where each literate Mexican has to teach one illiterate person to read and write. Within a year, more than 700,000 learn to read and write.

1946–52 Presidency of Miguel Alemán Valdés

President Miguel Alemán Valdés is responsible for undertaking the construction of the Ciudad Universitaria, the new campus of UNAM. He favors foreign investments in Mexico.

1946 Final resting place of Cortés discovered

Scholars open a wall in the chapel of the Hospital of Jesus in Mexico City and uncover the tomb containing the remains of Hernan Cortés (1485–1547). Scholars maintain that Cortés had ordered the construction of the chapel where his crypt is found.

1952 Residents of Oaxaca rebel

When the inhabitants of Oaxaca organize a general protest against the governor forced upon them by PRI, the party rescinds the appointment and puts a more acceptable candidate in the position.

1953 Women receive the right to vote

Women's right to vote in Mexico is formally acknowledged by an amendment of Article 34 of the constitution.

1954 Dedication of the new campus of UNAM

After four years of work by the best Mexican architects and engineers, Mexico dedicates the new campus of UNAM built on over five hundred acres of volcanic wasteland, just outside of the capital.

1957 Earthquake hits Mexico City

During the summer of 1957 an earthquake hits Mexico City killing approximately sixty people and resulting in property damage in excess of $25 million.

1958–64 Presidency of Adolfo López Mateos

López Mateos nationalizes the electrical industry.

1960 Birth of Fernando Valenzuela

Fernando Valenzuela is born. A gifted baseball pitcher, he joins the U.S. team, the Los Angeles Dodgers. He wins the Cy Young Award his rookie season (1981).

1964 United States and Mexico sign the Chamizal border treaty

U.S. President Lyndon B. Johnson and Mexican President Adolfo López Mateos sign the Chamizal treaty. Under its terms, the United States returns 437 acres of borderland to Mexico.

1968 Riots in Mexico City cast shadow on Olympic Games

A series of political riots in Mexico City on the eve of the Olympic Games jeopardizes the international competition. Order is restored with no threat to political stability.

1968 Summer Olympic games in Mexico City

The XIX Olympiad is held in Mexico City. Thirty-four world and thirty-eight Olympic records are set.

Seventeen-year-old Felipe Muñoz wins Mexico's first gold medal in these games when he takes the 200-meter breaststroke competition. Mexico wins a total of two gold, two silver, and two bronze medals.

1970–76 Presidency of Luis Echeverría Alvarez

Luis Echeverría Alvarez (b. 1922) becomes president, a post he holds for six years. During his administration, he attempts to initiate reforms such as land redistribution and expansion of social security. His efforts are hampered by high inflation and unemployment. He founds the *Universidad Autónoma Metropolitana* (UAM, Autonomous University of Mexico City—see 1974).

Luis Echeverría Alvarez was born in Mexico City on January 17, 1922.

1970 Mexico hosts World Cup

Mexico hosts the World Cup soccer championship, held in Mexico City.

1974 Women receive legal equality

With the reform of Article 4 of the constitution, women are guaranteed legal equality with men.

1974 UAM established

The *Universidad Autónoma Metropolitana* (UAM) opens. It is created in part to relieve the heavy enrollments in Universidad Nacional Autónoma (de) México (UNAM) and

Well into the twentieth century much of Mexico remains rural and underdeveloped. (EPD Photos/CSU Archives)

Instituto Politécnico Nacional (IPN), and in part to counter the influence of those two institutions.

1974: February Mexican president visits Pope Paul VI

Ignoring Article 130 of the Constitution of 1917 which requires separation of church and state, Mexican president Luis Echeverría visits Pope Paul VI (1897–1978) in the Vatican. He also appoints a Mexican chargé d'affaires to the Vatican.

1975 Mexico puts into effect the National Water Plan

With the assistance of the World Bank and the United Nations Development Program, Mexico strives to implement a national water plan. The plan focuses on management of existing water resources and the development of new ones.

1976 Presidency of José López Portillo y Pacheco

José López Portillo y Pacheco (b. 1920) becomes president of Mexico, a post he holds until 1982. He attempts to stimulate the economy through various programs (see 1982), but his administration is corrupt, and the economy continues to struggle.

José López Portillo y Pacheco was born in Mexico City in 1920. He entered government in 1959 after a career in law. Before becoming president, he served as finance minister (1971–75).

1978   Mexico adopts National Human Settlements Law

Mexico ratifies the *Ley General de Asentamientos Humanos* (National Human Settlements Law) which sets forth guidelines regulating both urban and regional growth and development.

1979   Pope visits Mexico

Pope John Paul II (b. 1920) inaugurates his pontificate with a trip to Mexico (see also 1991: May). Public outpouring of affection for the Pontiff underscores the vitality of the Roman Catholic Church in Mexico.

1980   Mexico introduces VAT

Mexico introduces a value-added tax (VAT) on goods and services.

1981: May   Mexico agrees to eliminate discrimination against women

Mexico ratifies the international Convention on the Elimination of All Forms of Discrimination Against Women.

1982–88   Presidency of Miguel de la Madrid Hurtado

President Miguel de la Madrid Hurtado is chosen as the candidate of the Institutional Revolutionary Party in 1981 and assumes the presidency in 1982. In response to the nation's economic crisis, he institutes an austerity program.

De la Madrid was born December 12, 1934, in Colima. He studied law in Mexico City; he continued his education at Harvard University in Cambridge, Massachusetts, where he studied public administration. He served as an advisor to the Bank of Mexico and became minister of planning and budget under José López Portillo.

1982   López Portillo nationalizes banks

President López Portillo (see 1976) nationalizes Mexican banks. To assure the permanency of this decision he makes it part of the national constitution.

1983   Government recognizes value of work in the home

Article 288 of the Civil Code of the Federal District puts a monetary value on work done in the home, thus easing the economic straits in which divorced partners (mostly women) find themselves.

1984   Establishment of SNI

The National System of Researchers (SNI) is established to encourage and reward research and publication activity by university professors. In 1992 it is extended to recognize outstanding teaching by university faculty members.

1985: September 19   Earthquake hits Mexico City

At 7:18 a.m. a major earthquake hits Mexico City. Though the center of the earthquake lies under the Pacific coast of Mexico, it causes considerable damage in the capital and kills an estimated ten thousand people.

1985: October   Baker debt relief plan announced

U.S. Treasury Secretary James Baker announces a three-part plan to rectify Mexico's debt predicament: the country would implement "new macroeconomic policies"; the International Monetary Fund would increase its attention to Mexico's debt management; and commercial banks would fund these endeavors with addcd lcnding. This proposal is commonly known as the Baker Plan.

1986   Mexico joins GATT

Mexico becomes a signatory to the General Agreement on Tariffs and Trade, commonly referred to as GATT. Gatt is an anti-inflationary policy that sets an initial maximum import tax of twenty percent, which is later reduced to less than ten percent.

1986   Mexican legislature enlarged

The de la Madrid administration increases the size of the Chamber of Deputies to 500.

1986   Forestry Law approved

A new forestry law is passed. It is designed to implement more environmentally and socially sensitive forestry policies. Austerity measures eventually change the government's Forestry Subsecretariat into the National Forestry Commission. The National Forestry Commission is a weaker agency and implements fewer, less rigorous forestry policies.

1986   Mexico hosts World Cup

Mexico hosts the World Cup soccer championship for the second time (see 1970).

1987: December   President de la Madrid and labor sign pact

In an effort to stabilize Mexican economic conditions via the imposition of fiscal restraints and wage and price controls, President de la Madrid and representatives of all economic sectors, including agriculture, sign the Pact for Economic Solidarity.

1988–94   Presidency of Carlos Salinas de Gortari

President Carlos Salinas de Gortari is elected to office with only fifty percent of the votes, due to the maverick candidacy of Cuauhténoc Cárdenas. The latter's challenge marks the first time since 1952 that a candidate not endorsed by the PRI has mounted such a campaign.

1988   Economic pact extended

President Salinas de Gortari's administration approves the Pact for Stability and Economic Growth, an extension of the de la Madrid measure. Despite several revisions, the two are popularly mentioned together as the Pact.

1989   Mexico City fights pollution

In an effort to decrease air pollution in the capital, Mexico City's government inaugurates "Hoy no Circula" (One Day Without a Car) program. Despite its goals, the program has been plagued by corruption as well as inadequate public transportation.

1989   President Salinas de Gortari creates PRONASOL

Salinas announces establishment of PRONASOL (Programa Nacional de Solidaridad) designed to aid the deprived in Mexican society. Among its successes is Mujeres en Solidaridad (Women in Solidarity), which trains women and allows implementation of projects by women.

1989   Creation of PRD

The Partido de la Revolution Democrática, a political party with strong leftist inclinations is founded.

1989   Educational reform projects

The Mexican Ministry of Public Education initiates a number of programs designed to assess and improve universities in Mexico. These include, among others, the submission of self-evaluation reports and peer review.

1989: February   Water commission established

In response to increasing problems with the amount and purity of water resources, the Mexican government sets up the Comisión Nacional del Agua, designated as the only authorized agency of the federal government to deal with problems concerning water.

1990   Octavio Paz awarded Nobel Prize

Octavio Paz (1914–98), writer, is awarded the Nobel Prize for Literature by the Swedish Academy. Paz is recognized as "one of the most influential voices in late twentieth-century letters." (See 1914.)

1991   Salinas administration implements pollution-control measure in Mexico City

In an effort to decrease air pollution in the capital, the Salinas administration implements Programa Integral Contra la Contaminación Atmosférica (PICCA) in Mexico City. This measure mandates the use of unleaded gasoline and catalytic converters.

1991   Demonstrations against proposed highway

The Comité Nacional para la Defensa de las Chimalpas (CNDCHIM), a national environmental group, objects to the construction of a new highway. The highway is scheduled to be built through the Ocote Reserve, a critical biodiverse area.

1991: May   Second visit by Pope John Paul II

The Pontiff, Pope John Paul II, returns to Mexico (see 1979). The pope is the head of the Roman Catholic Church. President Salinas de Gortari, who exchanged personal representatives with the Pope in 1990, greets him at the airport. By doing so, he signals a more receptive attitude in Mexico's relations with the Vatican.

1991: October   LANIA established

Laboratorio Nacional de Informática Avanzada (LANIA) is founded in Xalapa in the state of Veracruz. LANIA is a national research center in advanced computer techniques.

1992: February 26   Mexico implements constitutional reforms

The new "Ley Agraria," a constitutional reform, goes into effect. Initiated by President Salinas, this revision of Article 27 of the constitution has an impact on land tenure and serves as part of the motivation for the 1994 uprising in Chiapas. (See 1994: January)

1992: June   Chamber of Deputies modifies constitution

The Chamber of Deputies modifies articles of the constitution concerning religion in Mexico. The new Law of Religious Associations and Public Worship modifies constitutional restrictions on the practice of religion in Mexico.

1992: December   Mexican congress approves National Water Law

The Mexican legislature approves a new National Water Law, built upon and expanding a 1975 law.

1993: September   Electoral reform enacted

Electoral reforms, putting limits on spending during political campaigns, are enacted.

1994: January Indians rebel in Chiapas

In Chiapas, one of Mexico's poorest provinces, Indians calling themselves the Zapatista Army of National Liberation, stage a brutal uprising. Motivated in part by the unrealized land reform promised in the constitution of 1917 and inspired by their leader, "subcomandante Marcos," *campesinos* (farmers) seize many of the ranches in the province.

1994: January 1 North American Free Trade Agreement takes effect

Mexico enters into an economic agreement, known as the North American Free Trade Agreement (NAFTA), with Canada and the United States to facilitate free trade among the three countries.

1994: March 23 Presidential candidate assassinated

The PRI presidential candidate Luis Donaldo Colosio, is murdered while campaigning in Tijuana.

1994: July 14 Toxic waste ban passed

The Chamber of Deputies passes a measure to stop the importation of toxic waste into Mexico. The measure is based on a plan advanced by Greenpeace-Mexico.

1994: September PRI secretary general assassinated

José Francisco Ruiz Massieu, secretary general of the PRI, is assassinated.

1994: October "Operation Gatekeeper" initiated by the United States

The U.S. Immigration and Naturalization Service implements "Operation Gatekeeper" along the United States border with Mexico. This program is designed to stop illegal immigration along the border.

1994: November Mexican president condemns approval of Proposition 187

In his State of the Nation address, President Salinas Gortari speaks against the approval of Proposition 187. The proposition, passed by California voters, prohibits illegal immigrants from receiving public benefits such as education and health benefits.

1994: December 1 Inauguration of Ernesto Zedillo

Ernesto Zedillo is inaugurated as president of Mexico, the chief executive who will lead Mexico into the twenty-first century.

1994: December 20 Mexico devalues peso

Mexico devalues its currency, the peso, to stabilize its economic status. Contrary to expectations, within three months the peso has fallen fifty percent in its exchange with the dollar.

1995 Peso devaluation results in inflation

Devaluation of the peso triggers inflation at a rate of slightly over fifty percent.

1995: January 3 Government responds to growing economic crisis

President Zedillo, in an effort to counter an economic crisis, initiates a sweeping program designed to limit prices and wages, place a cap on governmental spending, and extend privatization.

1995: January 17 Government negotiates with rebels

Three groups—PRI, PAN (Partido Acción Nacional), and PRD (Partido de la Revolución Democrática)—agree to arbitrate issues concerning rebels in Chiapas.

1995: February 9 Rebel leaders arrested

The truce ends when the Mexican government sends troops to arrest rebel leaders.

1996 Revision of Mexican electoral procedures

Mexico amends its electoral ordinances. The new procedures are designed to ensure the political rights of its citizens, establish criteria for the certification of political groups as national political parties, and guarantee free and peaceful participation in political affairs.

1996 Government implements austerity program

The Mexican government implements an International Monetary Fund austerity program to assure economic stability.

1997 Porfirio Muñoz Leda heads Chamber of Deputies

Porfirio Muñoz Leda of PRD becomes the first non-PRI member to head the Chamber of Deputies.

1997 New rebel group, Popular Revolutionary Army (EPR), appears

Popular Revolutionary Army (EPR), a new rebel group, appears in Aguas Blancas in the state of Guerrero. The group brands the Zedillo presidency as illegal.

1997: July 6 Cárdenas elected mayor of Mexico City

In what is viewed as a blow to PRI, Cuauhtémoc Cárdenas Solórzano, candidate of the *Partido de la Revolución Democrática (PRD)* takes forty-eight percent of the vote to become the first elected mayor of Mexico City.

1997: July 6   PRI loses majority in the Chamber of Deputies

In the mid-term elections (similar to off-year elections in the United States, when voter interest and turnout is often light), PRI fails to win the number of seats necessary to claim a majority, with PRD and PAN each receiving slightly more than twenty-five percent of the total votes.

1997: September   PRI approves resolution

The PRI approves a resolution prohibiting those with less than ten years as party members from being presidential candidates.

1998   "Subcomandante Marcos" alive

Putting an end to speculation that his four-month silence resulted from his death and/or a serious illness, the leader of the Zapatista National Liberation Army, which sparked the 1994 uprising in Chiapas, sent the following message to federal troops in the region:

"¡Yepa! ¡Yepa! ¡Andale! ¡Andale! ¡Arriba! ¡Arriba!".

He signed the message "Speedy González," the name of the famed cartoon figure.

Bibliography

Collier, Simon, Harold Blakemore, and Thomas E. Skidmore, eds. *The Cambridge Encyclopedia of Latin America and the Caribbean.* 2nd edition. New York: Cambridge University Press, 1992.

Johnson, William Weber and the editors of Life. *Mexico.* New York: Time, Inc., 1961.

Kaplan, Robert D. "Foreign Affairs: History Moving North—Mexico Seems to Be Evolving Backward. Has Nationhood Been Merely an Interlude?" *The Atlantic.* 279:2 (1997):21.

Krauzem, Enrique. "Biography of Power: A History of Modern Mexico, 1810–1996." Trans. Hank Heifetz. *The New Republic,* Sept. 8, 1997: 38.

Mellin, Maribeth. "The Magic of Mexico City." *San Diego Magazine,* 47:1 (Nov. 1994):84.

National Geographic, 190:2 (Aug. 1996). "Emerging Mexico". Entire isssue devoted to Mexico.

Randall, Laura, ed. *Changing Structure of Mexico.* Armonk, N.Y.: M.E. Sharpe: 1966.

Randall, Stephen J. "NAFTA in Transition: The United States and Mexico." *Canadian Review of American Studies,* 1997.

Semo, Enrique, Lidia Lozano, Richard J. Salvucci "The History of Capaitalism in Mexico: Its Origins, 1521-1763." *The Hispanic American Historical Review,* 75:3(1995):471.

Skidmore, Thomas E., and Peter H. Smith. *Modern Latin America.* 4th edition. New York: Oxford University Press, 1997.

Stern, Steve J., and Richard Boyer. "The Secret History of Gender: Women, Men and Power in Late Colonial Mexico." *The Hispanic American Historical Review,* 77:2 (1997):314.

Williamson, Edwin. *The Penguin History of Latin America.* New York: Penguin Books, 1992.

Nicaragua

Introduction

Nicaragua is a country struggling to overcome a historical legacy of political turbulence, foreign intervention, and widespread poverty. In the past twenty years, it has emerged from a forty-two-year period of dictatorship only to suffer a decade-long guerrilla war. Although its has had three democratically elected presidents within this period, deep political divisions continue to hamper the nation's progress toward economic stability.

Nicaragua is the largest country in Central America. With an area of slightly more than 50,000 square miles (80,450 square kilometers), it is about the same size as New York state. It is bordered on the east by the Caribbean Sea, on the south by Costa Rica, on the west by the Pacific Ocean, and on the north by Honduras. Nicaragua's three main geographic regions are the Pacific lowlands in the west, the central highlands, and the Caribbean lowlands in the east (also called the Mosquito Coast, named for the Miskito Indians who live there). Nicaragua's population is unequally distributed, with the majority of Nicaraguans living in the western part of the country. Nicaragua is home to nearly 4.5 million people. Managua, the capital, is the largest city, with a population of nearly 700,000.

Hundreds of years before the first Europeans set foot in present-day Nicaragua, its native Indian peoples were raising corn and speaking a language similar to the ones spoken by the Aztecs and Mayas of Mexico. In 1522, Spaniards from Panama became the first Europeans to explore Nicaragua, and settlements were begun soon afterward. However, Spain, more interested in the lands of Mexico and Peru, due to their rich resources, left Nicaragua largely undeveloped until the eighteenth century. During this period, the British first laid claim to the Mosquito Coast area, which would remain a source of contention for over two hundred years. In the eighteenth century, Nicaragua began to develop an active export trade in agricultural goods, (such as coffee, cotton, and sugarcane) and a long-lived rivalry began between the political conservatives (pro-clerical and aristocratic), based in Granada and the liberals (anti-clerical and merchant-supported) based in León.

Joining the revolutionary tide rolling through Central and South America at the beginning of the nineteenth century, the Captaincy General of Guatemala, and with it the Province of Nicaragua, declared independence from Spain in 1821, to become part of Mexico. Two years later, Nicaragua and four other former Spanish provinces banded together to form the United Provinces of Central America federation, but the shaky and contentious union dissolved within fifteen years. Nicaragua became self-governing but was still torn by strife between the liberals and conservatives. Finally, in 1854, the Republic of Nicaragua was created, and the young nation soon banded together to overthrow a self-proclaimed and unwanted president, the American adventurer William Walker (see 1855).

After a long period of political stability in the late nineteenth century, the country was once again torn by divisions between political factions at the beginning of the twentieth century, and the resulting instability introduced a new element to Nicaraguan politics—intervention by the United States, which stationed marines in the country in 1912 in response to a request for aid by conservatives. This was the beginning of a twenty-year U.S. military presence in Nicaragua, which gave rise to a peasant rebellion and a lasting distrust of U.S. interference on the part of many Nicaraguans. The rebellion, which took place between 1927 and 1933, was led by Augusto Sandino, and both its method (guerrilla warfare) and the name of its fighters (Sandinistas) would be taken up half a century later in another struggle against a government supported by the United States.

The most prominent feature of politics in twentieth-century Nicaragua is its domination by a single family, the Somozas. Anastasio Somoza García, head of the National Guard, gained the presidency in 1937 and retained power, either as president or through hand-picked successors, until his assassination in 1956. He was succeeded as president by his sons, first Luis Somoza Debayle (1956–1967) and then Anastasio Somoza Debayle (1967–1979). The last Somoza regime was overthrown by the leftist Sandinista National Liberation Front (FSLN), which then governed the country for a decade, first under a *junta*, or controlling council and then under president Daniel Ortega (see 1984: November 4). For most of the Sandinistas' period in power, they faced military opposition from

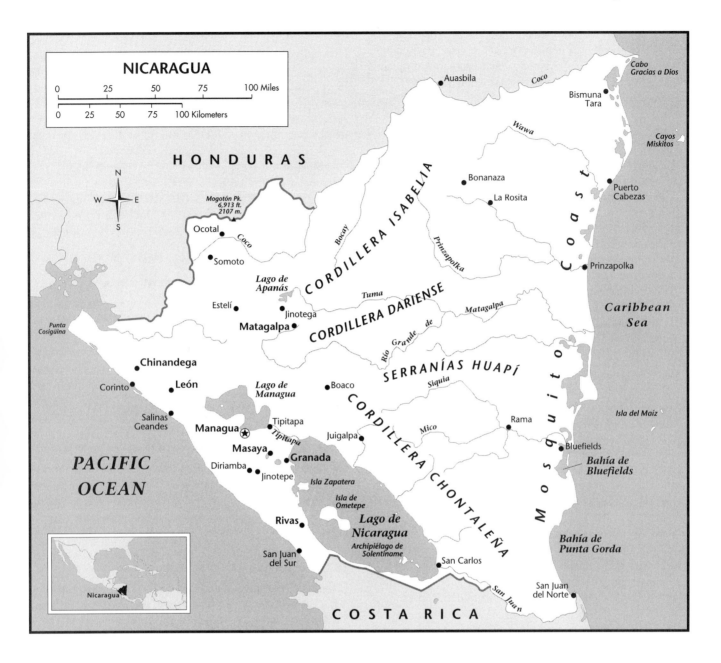

NICARAGUA

0   25   50   75   100 Miles
0   25   50   75   100 Kilometers

U.S.-financed Contra (counter-revolutionaries) forces. In 1990, Ortega was unseated by Violeta Barrios de Chamorro, the widow of a newspaper publisher slain by the Somoza regime. Barrios de Chamorro's government was unable to deal effectively with the country's accumulated economic and social problems. In 1996, Arnoldo Alemán, a former right-wing mayor of Managua, was elected president, defeating Daniel Ortega, who had sought to regain his former office.

"Everybody Is Considered a Poet"

With its turbulent history, Nicaragua is a country in which cultural life is closely intertwined with social and political concerns. Some of its leading poets—like prominent literary figures throughout Latin America—have been actively

involved in the political arena, and many have worked as journalists. Nicaragua's most renowned poet, Rubén Darío, wrote newspaper articles about the Spanish-American War, and his poetry reflected his increasing concern with his country's political future and its ties to Hispanic culture worldwide. The *vanguardista* literary movement of the 1920s was inspired by the peasant rebellion led by Augusto Sandino. Paralleling Sandino's call for an end to foreign domination in the political sphere, the *vanguardistas* advocated artistic freedom from European traditions. Another major poet, Pablo Antonio Cuadra, joined Sandino's rebel army in the 1920s. He was imprisoned by the Somoza government in the 1950s and took over as editor of the country's leading newspaper, *La Prensa*, in 1978 after the political assassination of its publisher, Pedro Chamorro whose widow later became president of Nicaragua. Ernesto Cardenal, who pioneered the *exteriorista* style of poetry, was minister of culture under the Sandinista government of the 1980s and created ambitious programs to make all Nicaraguans active participants in the arts.

During the Contra War (1981–90), the National Theater Workshop traveled to the nation's war zones and produced plays there. Members of the theater troupe would spend time living with the people of each region and learning about their concerns and their way of life. They would then create plays based on themes central to the lives of the community. During the same period, poet and culture minister Ernesto Cardenal set up thirty-two centers throughout the country for the study and production of popular culture. At these centers, people could take classes in drawing, painting, crafts, and other forms of artistic expression to help them become active participants in the arts.

Although Cardenal's programs suffered from budget cuts both during and after the Sandinista regime, a new and important form of popular art emerged that combines artistic and political expression: the murals that cover the walls of many public spaces throughout Nicaragua. These bold and colorful art works, often created by groups of people, characteristically center around the political developments of recent decades or express nationalistic themes based on historical events from Nicaragua's past.

If Nicaragua can be said to have a national art form, it is poetry. It was a Nicaraguan—Rubén Darío—who revolutionized Spanish poetry in the twentieth century, not just in Nicaragua but throughout the Hispanic world. His poetry changed the whole of Spanish syntax and metrics, emphasizing perfection of form, musical expression, and overwhelming sadness. Importantly, Nicaragua has produced more poets than almost any other Spanish-speaking country. Poetry is part of everyday life for many Nicaraguans. It is widely read and quoted, and works by contemporary Nicaraguan poets are widely available in the nation's bookstores. In addition, Nicaraguans from all walks of life write poetry themselves. Former president Daniel Ortega wrote poems during the years he spent in prison between 1967 and 1975, and one of them, "I Never Saw Managua When the Miniskirt Was in Style," gained widespread popularity throughout the country. And it was Ortega who described not only his people's love of the arts but also an important aspect of their national character in the following much-quoted words: "In Nicaragua everybody is considered a poet until he proves otherwise."

Timeline

10,000–5,000 B.C . First inhabitants people Nicaragua

The first Indian tribes are thought to have settled in present-day Nicaragua 6,500 to 11,000 years before the arrival of the first Europeans.

300 B.C.–A.D. 300 Indians develop farming villages

Originally cave dwellers, the native inhabitants of Nicaragua form small villages and begin to practice agriculture. Those in the central highlands and Pacific lowlands are culturally related to the indigenous groups of Mexico, including the Aztecs and Mayas. Like these groups, their primary crop is maize (corn). Most speak some dialect of the Pipil language, which is similar to the Aztec language, Nahuatl.

Most of the native people living on the east coast (in the Caribbean lowlands) have migrated from the region that today is Colombia and speak in dialects of the Chibcha language. Another distinguishing feature of these groups is their use of round, thatched huts, sometimes called *bohios*, as dwellings.

1500s Three main tribes inhabit Nicaragua

The first European explorers encounter three principal Indian tribes: the Niquirano, Chorotegano, and Chontal. Each lives in a different region and has its own laws, cultures, and ruler, but all have systems of government similar to monarchies. The chief of the Niquirano is known as Nicarao or Nicaragua.

The Colonial Period

1522 Avila leads first European expedition to Nicaragua

Gil González Avila leads a party north from Panama through present-day Costa Rica. Racked by illness and beset by heavy rain on their journey, Avila and his men are welcomed by the Indians they encounter, many of whom convert to Catholicism.

1523 Córdoba explores Nicaragua

Francisco Hernández Córdoba (c. 1475–1526) explores Nicaragua, sent by Panama's governor, Pedro Arias (Pedrarias) Dávila (c. 1440–1531). Córdoba founds the settlements of Granada and León. He tries to make the new region independent of Panama, but his claims are denied and he is executed.

1524 First Catholic church is built

Nicaragua's first Catholic church is established in the city of Granada.

1528 Dávila is appointed the first governor of Nicaragua

The Spanish crown grants Pedro Arias Dávila the title of governor of Nicaragua. He holds the post until his death in 1531.

1530–1700 Nicaragua remains largely undeveloped

After their initial exploration and settlement, Spain's interest in Nicaragua wanes as the Spanish concentrate on claiming the wealth of Mexico and Peru. Many settlers depart for South America, and the Indians who have been enslaved are transported there as well. By the end of the sixteenth century, only the initial settlements of León and Granada remain.

1538 Nicaragua becomes part of the *audiencia* of Panama

Nicaragua is officially placed under the jurisdiction of Panama by the Spanish government.

1544 Nicaragua is assigned to the *audiencia* of Guatemala

Nicaragua becomes part of the newly created audiencia of Guatemala, which includes southern Mexico as well as present-day Panama, Honduras, and Guatemala.

1648–63 Earthquakes cause widespread destruction

The Province of Nicaragua suffers heavy damage from a series of major earthquakes.

1651–89 Nicaragua becomes a target for pirate raids

French, English, and Dutch pirates carry out violent raids on Nicaragua, facilitated by the lack of Spanish fortifications at the mouth of the San Juan River. Toward the end of this period, they plunder and destroy the city of Granada, carrying off gold, indigo (blue dye), and other treasures. The raids are eventually reduced as a result of treaties between Spain and Britain.

1687 British lay claim to Nicaraguan coastal area

The British governor of Jamaica claims Nicaragua's Caribbean coastal area (the "Mosquito Coast") for England. The Spanish had constructed forts along the coast to defend their claim on the land but are unable to keep the British at bay. The British, however, keep their settlements along the coast, leaving the interior under Spanish control. A member of the native Miskito population is crowned Jeremy I, "King of Mosquitia" by the British. Control of the area will remain in dispute until the end of the nineteenth century.

18th century Nicaragua thrives with growth of exports

After a long period of relative neglect, Spain begins to develop Nicaragua as a source of agricultural exports, primarily animal products including beef, hides, and tallow, which remain the colony's primary exports until coffee becomes dominant in the mid-nineteenth century. Economic progress is especially evident for a period after 1741 and again after 1778. However, both trade and wealth become sources of contention as political liberals and conservatives divide over the freer trade policies of the Bourbon dynasty that has taken over the Spanish throne following the War of the Spanish Succession.

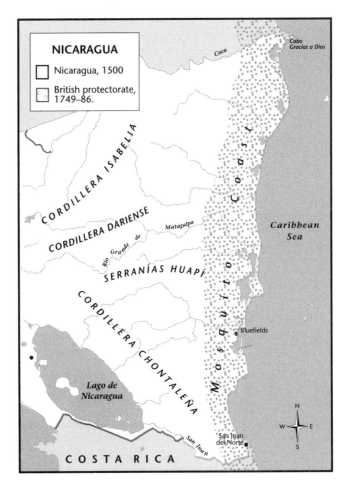

NICARAGUA

☐ Nicaragua, 1500

☐ British protectorate, 1749–86.

CORDILLERA ISABELIA

CORDILLERA DARIENSE

Matagalpa

Río Grande de

SERRANÍAS HUAPÍ

CORDILLERA CHONTALEÑA

Lago de Nicaragua

San Juan

San Juan del Norte

COSTA RICA

Coco

Cabo Gracias a Dios

Mosquito Coast

Caribbean Sea

Bluefields

1701–14 Rivalry between Granada and Léon

León becomes a center for liberals, who favor the new trade policies, while Granada is the stronghold of the conservatives, who want to retain the old system. This division fuels the existing rivalry between the two cities, which will continue beyond Nicaraguan independence.

1749–86 British protectorate on the Mosquito Coast

Through the governors of Jamaica, the British oversee the Mosquito Coast as a protectorate. By maintaining a presence in the area, they hope to protect their timber interests in Belize from the Spanish.

1767 Jesuits are expelled from Spanish America

In line with the secular sympathies of the Bourbon monarchy, the Jesuit order is expelled from all of Spain's colonial possessions in the New World.

1786: July 14 British subjects leave the Mosquito Coast

By signing the Convention of London, Britain agrees to withdraw its subjects from the Mosquito Coast. However, it does not completely relinquish its claims to the territory.

1812 National University of Nicaragua is founded

Nicaragua's oldest and largest university was founded in León. In the 1990s, it has over 22,000 students and campuses in both Managua and León.

Independence

1821: September 15 Guatemala declares independence from Spain

The Captaincy General of Guatemala, of which Nicaragua is a part, formally declares independence from Spain and becomes part of the Mexican Empire, together with Honduras and Guatemala. This date is still celebrated as independence day in Nicaragua.

1823: July United Provinces of Central America is formed

Nicaragua is one of five Spanish colonies—the others are Costa Rica, El Salvador, Guatemala, and Honduras—that oppose control by Mexico and form their own government. Each retains its own internal administration.

1826–29 Civil war within the United Provinces

Nicaraguans clash with Guatemala over the construction of a proposed canal through Nicaragua. Differences between conservatives (who defend the interests of the aristocrats and the Catholic church) and the liberals (who favor separation of church and state) lead to warfare between members of the United Provinces of Central America. Unity is temporarily achieved under Honduran liberal leader Francisco Morazán (1799–1842).

1835: January 22 Nicaragua is shaken by volcanic eruption

Coseguina, Nicaragua, is the site of the most violent eruption to occur in the Western Hemisphere. The volcano, which has a crater a mile in diameter and is 2,000 feet deep, shows its first sign of activity on January 20, producing a white cloud that later turns gray, and, eventually, yellow and crimson. Ash showers begin to fall, with some ashes landing up to 100 miles away. Darkness over the area lasts for three days.

On January 22, the volcano erupts with an explosion that can be heard 400 miles away and is mistaken for cannon fire. Pumice spews from the volcano's cone for over seven hours, with darkness covering everything for 50 miles. Nearly 800 residents of the area lose their lives in the disaster.

1838 United Provinces is dissolved; Nicaragua is declared independent

The United Provinces of Central America is dissolved, and an assembly of elected representatives proclaims Nicaragua independent and sovereign (self-ruling). Formation of an effective central government is stalled by rivalry between the liberals in León and the conservatives in Granada.

1840s Coffee is first grown commercially

As coffee consumption grows in North America and Europe, coffee, introduced to Nicaragua early in the century, is first grown commercially. Over the coming decades, increasing amounts of land will be cleared for the crop, beginning a coffee boom that lasts into the twentieth century.

1842 New Central American Federation is formed

Nicaragua, Honduras, and El Salvador attempt to establish a Central American federation, but it fails after two years.

1848 Gold rush in the U.S. raises interest in building a canal through Nicaragua

The California gold rush and the rapid development of the U.S. western frontier kindle interest in building a canal through Nicaragua to link the Atlantic and Pacific oceans. The U.S. and Britain spar over the rights to this canal—which is never built—through the nineteenth century and into the beginning of the twentieth. The canal is eventually built in Panama instead.

1848 British retake control of southern Caribbean coast

British forces seize San Juan del Norte at the mouth of the Río San Juan and drive out Nicaraguan government officials, claiming sovereignty over Nicaragua's southeastern coastal land.

1849: August 26 Vanderbilt wins canal rights

The Nicaraguan government signs an agreement granting U.S. business tycoon Cornelius Vanderbilt (1794–1877) exclusive rights to develop a canal through Nicaragua linking the Atlantic and Pacific oceans. Anticipating the construction of a canal, Vanderbilt sets up steamship service from the U.S. to Nicaragua and establishes overland transport as well.

1850: April 19 Clayton-Bulwer Treaty governing canal rights is signed

Tensions erupt between the U.S. and Britain over canal rights through Nicaragua. Officials of the two governments hold talks and draw up the Clayton-Bulwer Treaty, by which both sides agree to refrain from purchasing territory in Central America or blocking each other's canal-building efforts.

1852 Accessory Transit Company begins operating

Cornelius Vanderbilt's Accessory Transit Company provides overland transportation across Nicaragua to would-be gold prospectors en route to California. The route across Central America is deemed safer and more convenient than crossing the continental United States. Moreover, the Nicaraguan route is attractive to Vanderbilt because the San Juan River, which follows the border between Nicaragua and Costa Rica and Lake Nicaragua form a natural waterway for steamships. Overland transportation takes travelers from the western bank of Lake Nicaragua to the Pacific coast, where a steamship continues the journey to California. By the end of 1852, thousands of North Americans have used Vanderbilt's service.

The Nicaraguan Republic

1854: February 28   Republic of Nicaragua is formed

Under Conservative leader Fruto Chamorro, the Nicaraguan republic is established. However, civil war is still widespread among the young nation's competing political factions. In addition, the U.S. and Britain continue to compete for the rights to develop a canal linking the Atlantic and Pacific oceans by way of Nicaragua.

1855   William Walker arrives in Nicaragua

William Walker (1824–60), a mercenary (soldier for hire), is brought in from the U.S. by liberals in León to help them defeat the nation's conservative government. Arriving in Nicaragua with a force of fifty-seven men, he seizes control of the country, proclaims himself president, and rules as a dictator. Thousands of U.S. residents are enticed to the country with promises of free or cheap land. Walker institutes slavery, makes English the official language, gives land to American companies, and attempts to get Nicaragua admitted to the United States.

1857: May 1   Central Americans join forces to defeat Walker

Central Americans, aided by the British, unite in the "National War" (1856–57) to fight Walker. He is decisively defeated in battle in the town of Rivas. A truce is arranged, and Walker and his forces are deported. Between 1857 and 1860, Walker makes four subsequent attempts to regain power. In 1860, he is taken prisoner by the British Navy and executed by a firing squad.

1858   A new constitution is adopted

Nicaragua adopts a new constitution as the nation begins a period of reconstruction following the National War and the defeat of William Walker.

1858   Managua becomes the capital of Nicaragua

The city of Managua is designated as the nation's new capital in hopes of defusing tensions between the competing cities of Granada and León.

1860   Treaty of Managua ends British protectorate over the Mosquito Coast

The British protectorate over the Mosquito Coast is formally dissolved. The region remains autonomous until the 1890s, and the British maintain a presence there until the beginning of the twentieth century.

1863–93   Los Treinta Años (The Thirty Years)

Nicaragua enters a period of political stability, as the conservatives rule the nation for thirty years. The city of Managua becomes increasingly powerful, superseding León and Granada, and the historical rivalry between these two cities becomes less important.

The coffee, banana, and timber industries grow, gold mining thrives, and railroads are developed.

1867: January 18   Birth of internationally renowned poet Rubén Darío

Darío, Nicaragua's most renowned writer, is born Félix Rubén García Sarmiento (1867–1916). He begins writing poetry at a young age and has already adopted his pen name by the age of fourteen. In 1886 he moves to Chile, where, two years later, he publishes his first important volume, *Azul*, a collection of poems, sketches, and short stories influenced by French Parnassian poetry. The goal of the Parnassians is producing poetry exact and faultless in form, using rigidity of style and emotional detachment to the subject. In 1893, Darío is appointed Colombian consul to Buenos Aires, Argentina. He becomes a leading literary figure and the leader of the Modernismo poetry movement that spreads throughout Latin America. Though borrowing many of its ideals from many sources, Modernismo is spontaneous, and in the poetry often creates an array of exotic landscapes with symbolic characters within them. His next major works, *Prosas profanas y otros poemas* (Profane Hymns and Other Poems, 1896), applies the style of French Symbolist poetry to Spanish.

In 1898 Darío journeys to Europe to report for the newspaper *La nación* and becomes increasingly concerned with social and political issues. His growing interest in solidarity and identity in the Spanish-speaking world is evident in his next book, *Cantos de vida y esperanza* (Songs of Life and Hope, 1905) which is generally regarded as his masterpiece. In 1914, Darío, in need of money, embarks on a lecture tour in the United States. While in the U.S. he develops pneumonia and returns to Nicaragua, where he dies on February 6, 1916.

Darío's work is acclaimed for transforming and rejuvenating Spanish poetry and for breaking away from nineteenth-century conventions. From the publication of *Azul* in 1888 to his death in 1916, Darío is the world's major Spanish-language poet.

1876   Telegraph service begins

The first telegraph service is inaugurated in Nicaragua.

1881   War of the Comuneros

The Nicaraguan military crushes a rebellion by Indians protesting seizure of their lands by coffee growers.

1885   Earthquake hits Managua

Nicaragua's capital is nearly leveled by a major earthquake, whose tremors and resulting fires kill thousands.

1887   Birth of painter and crafts artist Asilia Guillén

Born to a well-off family in Granada, Guillén learns embroidery as part of the standard aristocratic education for girls. However, her native talent enables her to turn this feminine "pastime" into art. Eventually she replaces the conventional embroidery patterns with images specific to Nicaraguan history and to the country's natural landscape, and she widens the range of colors that are traditionally employed.

In 1951 Guillén takes up painting, transferring the style of her embroidery to her oil paintings. In 1962 her paintings are exhibited at the gallery of the Pan American Union in Washington, D.C. They are an important influence on the primitivist painting that becomes popular in Nicaragua during the Sandinista regime of the 1980s. Guillén dies in 1964.

1887: August 26   Birth of composer Luis Delgadillo

Nicaragua's foremost twentieth-century composer is born in Managua and studies at Italy's Milan Conservatory at a young age. After spending five years in Europe, he returns to Nicaragua to teach, conduct, and compose. He serves as director of the Escuela Nacional de Música (National School of Music) in Managua from 1950 to 1962 and conducts the Orquesta Nacional. He travels on extensive tours of Latin America in the 1920s, conducting programs featuring his compositions, and his works are performed at Carnegie Hall in 1930.

Altogether, Delgadillo writes over four hundred compositions, which include sixteen symphonies, seven string quartets, five masses, an assortment of vocal works, and more than fifty piano compositions. Delgadillo dies in 1962.

1888   Groundbreaking poem "Azul" is published by Darío

Poet Rubén Darío publishes "Azul" (Blue), a pioneering work that is generally regarded as the beginning of the Modernismo movement in Nicaragua.

1890–1910   Baseball is introduced to Nicaragua

Baseball is introduced to Nicaragua by Nicaraguans who have studied in the United States and bring the game to their hometowns. Among them are Juan Deshón and David Arellano, who also introduces baseball to Granada.

1891   First baseball series is played

A series of games between Granada and the Recreation Society of Managua is inaugurated.

1893–1909   Zelaya regime

Taking advantage of divisions among the conservatives, General José Santos Zelaya (1853–1919) seizes control of the government and remains in power until 1909. Zelaya rules as a dictator and alienates foreign countries, especially the United States, by opposing their involvement in the Nicaraguan economy. Under Zelaya, Nicaragua adopts a new constitution implementing traditional liberal policies, including separation of church and state and state-run education. Zelaya promotes export production, strengthens the Nicaraguan military, and fosters the growth of nationalism and anti-foreign sentiment in the country.

1894   Nicaragua takes control of the Mosquito Coast

Under President Zelaya, the Mosquito Coast comes under direct Nicaraguan administration for the first time in its history.

1895   Birth of national hero Augusto Sandino

Augusto César Sandino (1895–1934) is born into poverty as the illegitimate son of a businessman and a coffee picker. As a child he picks coffee alongside his mother. Later, he lives with his father but is treated like a servant. He is forced to leave school and work in his father's business. Sandino lives for a time in Costa Rica and later in Mexico, where his exposure to the variety of political philosophies espoused by workers there begins his political education. Returning to Nicaragua in 1926, he becomes active in the liberal opposition to the conservative regime then in power. In 1927 he organizes his own military resistance against both the conservative government and the U.S. marines sent to support it.

For the next six years, Sandino leads his peasant army—known as Sandinistas— in guerrilla attacks on government and U.S. military targets from their base in the village of El Chipote. Sandino's forces grow to at least 5,000 by 1930 and are legendary for their ability to elude capture. In 1932 liberal candidate Juan Batista Sacasa (1874–1946) is elected president and the U.S. begins withdrawing its forces from the country. In 1934 Sandino meets with Sacasa in Managua to forge an agreement. As Sandino and his associates are leaving the presidential palace, they are abducted by National Guard troops and murdered. Sandino becomes a national hero, and the name of his army is adopted by the guerrilla forces organized a generation later to overthrow the son of Anastasio Somoza (1896–1956), who as head of the National Guard engineered Sandino's assassination.

For most of Nicaragua's history, the military is a bastion of political conservatism. Here, the Nicaraguan army passes for review in 1912. (EPD Photos/CSU Archives)

1896　National Museum of Nicaragua is founded

The National Museum of Nicaragua is established to house archaeological, zoological, botanical, and geological exhibits.

1896: February 1　Birth of dictator Anastasio Somoza García

Anastasio Somoza García (1896–1956), who sires a dynasty of political dictators that rules Nicaragua for forty-two years, is born in San Marco and educated in Philadelphia. His excellent command of English helps Garcia ingratiate himself with U.S. leaders later on. When Somoza returns to Nicaragua, Garcia enters the military and marries the niece of President Juan Batista Sacasa. He rises rapidly within the military and wins the attention of top U.S. officials. In 1927 the U.S. forms the Nicaraguan National Guard to keep order in the country so the U.S. can withdraw its own military presence from the country. Anastasio Somoza is named head of the Guard, mak-

ing him one of Nicaragua's most powerful men. In 1934 he oversees the assassination of rebel leader Augusto Sandino following peace talks between Sandino and President Sacasa. In 1936 Somoza ousts Sacasa, his own relative, from the presidency and seizes power himself.

Somoza assumes the presidency in 1937 and retains it for nearly twenty years through a combination of charm, guile, and ruthlessness. He remains very popular with the U.S., which strongly supports his regime. During his time in power, Somoza amasses a personal fortune estimated at over $60 million, becoming the country's largest private landowner and holding investments in many industries. On September 21, 1956, Somoza is assassinated in León by a young poet, Rigoberto López Pérez. The U.S. ambassador has Somoza airlifted to a U.S. military hospital in Panama, where he dies eight days later. He is succeeded in office by his son Luis Anastasio Somoza Debayle (1925–80).

1908   Birth of painter Rodrigo Peñalba

Peñalba, the foremost figure in modern Nicaraguan painting, studies in Chicago, Mexico City, Madrid, and Rome. After a sojourn in Italy, he returns to Nicaragua in 1947. There he becomes a famed art instructor and director of the Escuela Nacional de Bellas Artes (National School of Fine Arts) in Managua. His works include murals, figure studies (including miniatures), and experimental works painted during the last two decades of his life. His adoption of modernism has a major influence on twentieth-century Nicaraguan art. Former students of Peñalba form the country's first avant-garde movement, Praxis (see 1963). Peñalba dies in 1979.

1911–12   Organized league baseball games are launched

The first organized baseball league play is introduced in Managua, with five teams participating. The winning club, Bóer, is still in existence today.

1912   U.S. marines are stationed in Nicaragua

The conservative government that takes over following the ouster of Zelaya in 1909 calls for the U.S.'s help in defeating a liberal rebellion. The United States sends in 2,700 marines, and from then until 1933 maintains an almost continuous military presence in Nicaragua. As in Cuba, Panama, and the Dominican Republic, the U.S. becomes heavily involved in Nicaraguan affairs and U.S. investors gain increasing control over the Nicaraguan economy. The U.S. takes over running the country's railroads and posts troops in major cities; U.S. forces also monitor Nicaraguan elections.

1912: November 4   Birth of poet Pablo Antonio Cuadra

Cuadra (b. 1912), a poet, journalist, and playwright, combines literature and political commitment throughout his career. Cuadra is active in the original Sandinista rebel movement (led by Augusto Sandino in the 1920s and early 1930s) and begins writing professionally at the age of nineteen. His first volume of poetry, *Poemas Nicaragüenses* is published in 1934. During the 1930s, Cuadra also edits several literary and political publications. After a period of travel in the 1940s, during which he also writes several major essays, Cuadra returns to Nicaragua. When Anastasio Somoza García is assassinated in 1956, Cuadra, who is identified with the political opposition, is imprisoned. He later describes this experience in *América o el purgatorio* (1955).

In the late 1950s and early 1960s, Cuadra wins major literary awards in both Spain and Central America. Among his most important volumes from this period are *El jaguar y la luna* and *Poesía: Selección* (1929–62). Among Cuadra's later works are *Cantos de Cifar y del Mar Dulce* (1971), *Esos rostros que asoman en la multitud* (1976), and *La ronda del año: Poemas para un calendario* (1988). After Pedro Barrios de

Chamorro is assassinated in 1978, Cuadra takes over editorship of the daily newspaper *La Prensa*. In spite of his long-standing sympathy with the Sandinista cause, the FSLN regime closes down the newspaper in 1986 due to Cuadra's criticism of the regime's human rights record. Cuadra continues to pursue his literary career through the late 1980s and into the 1990s, winning Fulbright and Guggenheim fellowships in the United States, publishing his complete works, and participating in literary symposiums.

1915   First nationwide league baseball play takes place

The first nationwide games between baseball leagues are played. Teams from Managua, Granada, León, Masaya, and Chinandega compete in the games.

1916   Chamorro-Bryan Treaty restores canal rights to U.S.

Although construction has recently been completed on the Panama Canal, the United States continues to maintain an interest in building a similar canal in Nicaragua. The Chamorro-Bryan Treaty gives the U.S. exclusive canal rights as well as rights to build a naval base on the Gulf of Fonseca and a ninety-nine-year lease on the Corn Islands off the country's eastern coast (for which the U.S. pays $3 million).

1916: February 6   Acclaimed poet Rubén Darío dies

Internationally renowned poet Rubén Darío dies in León at the age of 49 (see 1867).

1925–26   U.S. marines withdraw but quickly return

Feeling secure in its political control of Nicaragua, the U.S. withdraws its marines from the country in 1925. However, following leftist revolts, U.S. President Calvin Coolidge (1872–1933) sends marines back to Nicaragua in 1926 to protect U.S. interests there by assuring the continuation of governments favorable to the United States.

1925: January 20   Birth of poet and social activist Ernesto Cardenal

Cardenal is an influential poet, essayist, and minister of culture in the Sandinista government of the 1980s. In his youth, Cardenal receives a Catholic education. He studies philosophy in Mexico (1942–47) and literature at Columbia University in New York City (1947–49), where he is introduced to the poetry of Ezra Pound and T. S. Eliot. Their works inspire his *exteriorista* poetic style, which emphasizes concrete detail over metaphor. Early works include *La ciudad deshabitada* (1946) and *Proclama del conquistador* (1947). After studying in Spain, Cardenal returns to Nicaragua, where he operates a bookstore and a small publishing company. He participates in the April 1954 revolt against the Somoza government, which forms the subject of *La hora cero* (1957).

In the late 1950s Cardenal turns to the Catholicism of his youth, undertaking religious studies in Mexico and Colombia as well as studying with the Trappist monk Thomas Merton (1915–68) in the United States. His religious conversion during this period is reflected in works such as *Gethsamani, Kentucky* (1960), and *Salmos* (1964), a modern-day version of the Biblical Psalms. In 1966 Cardenal forms an experimental Christian colony on the island of Solentiname, where he works with the mostly illiterate rural peasantry. He also maintains his political activism there, and the community is destroyed by the National Guard in 1977. After serving as a cultural ambassador for the Sandinistas, he becomes the minister of culture after their victory in 1979. However, the nation's economic problems and the Contra war block implementation of many of his projects. The Ministry of Culture is dissolved as an independent agency in 1988 and many of Cardenal's programs are discontinued under the Chamorro government.

1925: December 5   Anastasio "Tachito" Somoza is born

"Tachito," the younger son of Anastasio Somoza García "Tacho," will become the final heir of the political dynasty that controls Nicaragua for nearly half the twentieth century. A military man, Somoza graduates from West Point in 1948 and holds several important positions in Nicaragua's National Guard, eventually becoming its head. In this capacity, he provides the "muscle" supporting the administration of his elder brother Luis, who serves as president from 1956 to 1963.

When Luis Somoza dies in 1967, the final voice restraining Tachito's ambitions is silenced, and he launches a political regime infamous for its repression and brutality. Like his father and brother, he courts U.S. favor with free-enterprise, anti-Communist policies. Opposition to Somoza's regime grows in the 1970s, spearheaded by the guerrilla operations of the Sandinista National Liberation Front (FSLN). In spite of a heart attack, in 1977, Somoza retains power, even in the face of widespread opposition, which escalates after the assassination of moderate opposition leader Pedro Chamorro in 1978. Ultimately, however, Samoza is driven out of Nicaragua on July 17, 1979 and takes refuge in Paraguay, under the protection of President Alfredo Stroessner. He is killed by a car bomb on September 17, 1980.

1927–33   Sandino leads armed revolt

General Augusto César Sandino (1895–1934) leads a guerrilla rebellion to drive U.S. marines out of Nicaragua. His followers, mostly peasants, are known as Sandinistas. By 1930 Sandino's forces number at least 5,000. Bombing and other military responses fail to defeat the Sandinistas.

1927   Vanguard literary movement established

The Vanguard is a Nicaraguan writers' movement inspired by Augusto Sandino's guerrilla campaign to end the U.S. military presence in Nicaragua. They adopt the parallel goal of ending the dominance of European literary traditions over Nicaraguan literature. Its members, known as *vanguardistas*, include well-known poets Joaquín Pasos, Pablo Antonio Cuadra (b. 1912), and Ernesto Cardenal (b. 1925).

1928   Liberal candidate wins presidential election

In spite of U.S. supervision of the election (and support for the conservatives), liberal candidate General José María Moncada wins the presidency and implements policies less favorable to the U.S. than those of his conservative predecessors. However, the revolt led by his former ally Sandino continues. Sandino charges Moncada of selling out to the Americans.

1929   Violeta Barrios de Chamorro born

Barrios de Chamorro, the daughter of wealthy landowners in Nicaragua's southern Rivas province, marries Pedro Joaquín Chamorro Cardenal in 1950. Barrios de Chamorro becomes the editor of the daily newspaper *La Prensa* and an outspoken critic of the ruling Somoza family. The public outrage provoked by his assassination in 1978 helps fuel the drive to remove Anastasio "Tachito" Somoza (1925–80) from power, and he is ousted by the following year. Violeta Barrios de Chamorro (popularly known as "Doña Violeta") serves briefly as part of the Sandinistas' ruling *junta*. However, she resigns in less than a year.

For the remainder of the decade that the FSLN is in power, Barrios de Chamorro, who has taken over the newspaper *La Prensa* from her late husband, is a critic of their regime and a supporter of the contra rebels seeking to unseat them. In 1990 she runs for president against FSLN leader and incumbent Daniel Ortega and wins with fifty-five percent of the vote, supported by the fourteen-party UNO (United Nicaraguan Opposition) coalition. Once she is elected, however, the coalition proves unstable and the nation's economic problems remain difficult to resolve. Barrios de Chamorro does not run for reelection in 1996 when right-wing candidate Arnoldo Alemán defeats former president Daniel Ortega.

1930s   First professional baseball is first played in Nicaragua

Professional baseball is launched in Nicaragua. The country's official team, sponsored by political leader Anastasio Somoza, is Cinco Estrellas (Five Stars).

1931   U.S. begins withdrawal of its troops

United States marines begin a gradual withdrawal from Nicaragua over a two-year period.

1931   Earthquake wreaks destruction in Managua

Nicaragua's capital suffers heavy damage from a major earthquake and subsequent fires.

1932   Sacasa elected president

Former rebel Juan Batista Sacasa (1874–1946) is elected president in the last U.S.-supervised election.

1934   Sandino assassinated following meeting with Sacasa

Sandino and Sacasa meet in Managua for peace negotiations, but after the meeting Anastasio Somoza (see 1896), head of the National Guard, has Sandino and his top military officials assassinated when negotiations break down.

The Somoza Era

1937   Somoza becomes president

After forcing the resignation of Sacasa, Anastasio Somoza García (see 1896) becomes president of Nicaragua in an election in which the National Guard counts the ballots. The Somoza family rules Nicaragua for the next forty-two years, and Somoza amasses a personal fortune of millions of dollars for himself and his family. Somoza maintains close ties with the United States, encouraging U.S. investment in Nicaragua.

1941–45   Somoza serves second consecutive term as president

Somoza's authority over the National Guard gives him full control of the country, and he continues in the presidency for a second term when his first one expires.

1945   Labor Code goes into effect

Nicaragua's first major labor law is adopted. It provides for vacation days and places limits on overtime.

1945: November 11   Birth of Sandinista leader Daniel Ortega

Daniel Ortega Saavedra is born in the department (province) of Chontales, with a family history of political activism. His parents, supporters of the rebel general Augusto César Sandino in the 1920s, were both imprisoned by Anastasio Somoza (see 1896) in the 1940s. Ortega is active in political organizations from adolescence. He briefly attends law school, where he organizes student protests. In 1963 he joins the Sandinista National Liberation Front (FSLN), a recently formed anti-government guerrilla group with the goal of overthrowing the Somoza regime. From 1963 to 1967 Ortega takes part in FSLN operations. In 1964 he is imprisoned and tortured. In 1967 he is again imprisoned, this time for eight years. He is released in 1974 as part of a prisoner-hostage exchange following a Sandinista raid and becomes one of the highest-ranking FSLN leaders.

After Somoza is removed from power, Ortega is a member of the Sandinistas' governing junta and serves as the FSLN's major representative to foreign nations. In 1984 he is elected president of Nicaragua with over two-thirds of the popular vote. As president, he confronts the nation's ailing economy and the security threat posed by the U.S.-backed Contra (short for *contrarevolucionarios*, or counter-revolutionaries) rebels. Ortega participates in the creation of the Arias peace plan that finally ends a decade of fighting. He runs for reelection at the end of his six-year presidential term but loses to Violeta Barrios de Chamorro (see 1929), widow of slain political leader and publisher Pedro Chamorro. Ortega remains active in the FSLN and in Nicaraguan politics. In 1996 he makes another run for president, this time against right-winger Arnoldo Alemán, but is once again unsuccessful.

1947   Somoza chooses his own successors

After Somoza's second term, he hand-picks two successors, Leonardo Argüello and Benjamin Sacasa, quickly replacing them at will when they prove too independent. Finally, he settles on a relative, Victor Reyes, as president.

1948   Nicaragua hosts amateur baseball championship games

Nicaragua hosts the World Amateur Championship for the first time, in the newly completed National Stadium.

1950s   Cotton exports grow in importance

Following increased demand during the Korean War (1950–53), more cotton is produced. It becomes the nation's second most important agricultural export, surpassed only by coffee.

1950   Somoza returns to the presidency

Somoza once again becomes president, single-handedly controlling virtually all government institutions and functions, including the judiciary, police, and military as well as taxation, finance, broadcasting, and other government functions.

1950   First woman serves in Nicaraguan cabinet

Olga Nuñez de Sassalow becomes the first woman to serve in a cabinet-level position when she is appointed minister of public education. She is also the first woman to receive a law degree in Nicaragua.

1955   Voting rights for women extended

Women get the right to vote on the same basis as men.

1956: September 21   Somoza is fatally shot

Rigoberto López Pérez, a young poet, shoots Somoza in León (see 1896).

Anastasio Somoza (1896–1956) becomes the president of Nicaragua in 1937, and is the founder of the Somoza family dynasty which dominates the country for forty-two years.

1956: September 29   Somoza dies

Anastasio Somoza García (see 1896) dies in Panama in a U.S. military hospital. The Nicaraguan government declares a state of siege and the National Assembly elects Somoza's older son, Luis Somoza Debayle (1922–67), to the presidency. A younger son, Anastasio Somoza Debayle (1925–80), is chosen to head the National Guard, leaving government control firmly in the hands of the Somoza family.

1957   Social security program adopted

Nicaragua's first social security legislation is passed. It provides for medical, maternity, death, and survivors' benefits, pensions, and disability compensation.

1961   Sandinista National Liberation Front is formed

Carlos Fonseca Amador, Tomás Borge Martínez, and Silvio Mayorga form the Sandinista National Liberation Front (FSLN), a guerrilla resistance force which begins operations

two years later. Within its first few years, most of its leaders are killed by government forces or jailed, but the group reorganizes and continues its struggle, drawing members from among the ranks of peasants, students, and other middle- and upper-class youths. By 1978 the FSLN will number about 3,000.

1961   Central American University opens

The Central American University, operated by the Roman Catholic church, opens in Managua.

1961: April   Nicaragua is staging area for Bay of Pigs invasion

Nicaragua serves as a staging area for the United States' aborted CIA-backed attempt by Cuban exiles to unseat Cuban Communist leader Fidel Castro (b. 1926).

1962   Death of Luis Delgadillo

Nicaragua's foremost twentieth-century composer dies in Managua (see 1887).

1963   Schick is elected president

René Schick Gutiérrez, a Somoza supporter, is elected to the presidency, essentially extending the period of Somoza family rule. National Guard head Anastasio "Tachito" Somoza carries out a purge of alleged subversives, arresting opposition leaders both in the military and among the civilian population.

1963   Artists launch avant-garde movement

Praxis, Nicaragua's first avant-garde art movement, holds its first exhibition in Managua. The exhibit consists of thirty-seven abstract oil paintings by fifteen artists. The group consists of painters and sculptors, many of whom are former students of painter Rodrigo Peñalba (see 1908). "Praxis" is the name of a cooperative art gallery that serves as the central base for the group as well as a gathering place for leading intellectuals of all kinds. The members of Praxis seek to combine elements of Western modernism with uniquely Nicaraguan traditions, especially those involving the influence of indigenous art. The group is also concerned with the relationship of politics and the arts and supports the cause of the recently formed Sandinista National Liberation Front.

Members of Praxis include Alejandro Aróstegui (b. 1935), Arnoldo Guillén (b. 1941), Omar de León (b. 1929), César Izquierdo (b. 1937), Genaro Lugo (b. 1946), and others.

1964   Nicaraguan team wins baseball championship

The Nicaraguan baseball team, Cinco Estrellas, wins the Inter-American Series.

1964   Death of painter Asilia Guillén

Guillén, a pioneering female embroiderer and painter, dies in Granada (see 1887).

1966: August   Schick dies in office and is succeeded by his vice president

President Schick dies and is succeeded by Vice President Lorenzo Guerrero Gutiérrez.

1967: January   Mass demonstrations brutally crushed

In response to the presidential candidacy of Tachito Somoza, head of the National Guard, mass demonstrations are staged outside the presidential palace. They are broken up by the National Guard, which kills hundreds of demonstrators.

1967: February   Anastasio Somoza Debayle becomes president

Allegedly defeating the opposition candidate, Anastasio Somoza Debayle "Tachito" is elected president of Nicaragua.

1971   Congress is dissolved

Confronted with an alliance between opposition parties, pro-Somoza forces dissolve Congress and transfer its powers to the president.

1971   U.S. ambassador brokers agreement to appoint successor to Somoza

U.S. ambassador Turner Shelton presides over a pact calling for Somoza to be succeeded by a triumvirate of liberal and conservative leaders. However, Somoza remains head of the National Guard.

1972: December 21–22   Major earthquakes destroy Managua

A series of earthquakes levels three-fourths of Nicaragua's capital city, leaving 200,000 homeless. Estimates of the death toll range from 7,000 to 20,000. The city's population is cut by more than half, from 325,000 to 118,000. The third, and strongest, tremor measures 6.25 on the Richter scale and triggers the aftershocks that cause most of the damage. The fires that follow cannot be controlled because of damage to water mains. Radio transmitters are also destroyed, temporarily knocking out communication between the beleaguered city and the outside.

In the aftermath of the quakes, survivors are evacuated to prevent an epidemic, and many of the remaining buildings are torn down, including the U.S. embassy. About 2,000 of the victims are buried in a mass grave outside the city. The bodies of many others are burned. In the chaos that ensues, widespread looting takes place. International aid arrives from all over the world, including $3 million from the United States. Somoza and the National Guard are accused of confiscating

food, medicine, and supplies and selling them to those in need. Somoza also profits from the disaster by investing in the businesses that will rebuild homes and roads and shuttle people to and from their temporary shelters. Furthermore, he is thought to have directly pocketed foreign aid money.

Scientists note that Managua is directly on a fault line and that locations a mere twenty-five miles away sustain little or no damage. Yet even after suffering multiple earthquakes, the capital city is rebuilt on the same site.

1974   Somoza elected president

Amid widespread repression and election fraud, Somoza, who has been head of the National Guard, regains the presidency. However, the dire condition of the country—high unemployment, widespread poverty, illiteracy, and malnutrition—gives rise to growing opposition to the Somoza regime throughout the 1970s.

1974: December   FSLN raid wins release of prisoners

The Sandinista National Liberation Front (FSLN) wins nationwide attention by raiding the Christmas party of Somoza associates and winning the release of imprisoned FSLN members, including future Nicaraguan president Daniel Ortega.

1977   Coffee growers' cooperative is formed

The Eastern Regional Coffee-Growers Cooperative enables members to purchase farm supplies at reduced prices and helps them market their crops. During the 1980s, its membership grows to 400.

1978   FSLN occupies presidential palace

Led by Edén Pastora (b. 1937), the FSLN occupies the presidential palace, demanding the release of political prisoners as well as a ransom payment and the right to publicly broadcast a message. FSLN popularity and membership grows with the success of this and similar widely-publicized operations.

1978: January 10   Opposition leader Pedro Barrios de Chamorro assassinated

Pedro Joaquín Barrios de Chamorro, leader of the opposition Democratic Liberation Union (UDEL), is assassinated. Three days of protests and a general strike by business and labor groups follow the killing. The protests are violently put down by the National Guard. UDEL demands Somoza's resignation. Students and the Catholic church openly join the opposition to Somoza.

1978: September   FSLN launches major offensive

The FSLN set off a large-scale popular revolt against the Somoza government. At least 2,000 people are killed by government tanks and bombs.

1979 Death of painter Rodrigo Peñalba

Modernist pioneer Rodrigo Peñalba dies in Managua (see 1908.)

The Sandinista Government and the Contra War

1979: July 19   FSLN wins control of Nicaragua

Throughout the spring of 1979, Somoza desperately hangs on to power, ordering the bombing of civilian areas. An estimated 50,000 Nicaraguans die during this period. Finally, Somoza escapes to the United States and then to Paraguay, and the FSLN takes over the government, setting up a three-member ruling junta.

The FSLN successfully implements literacy and health programs, setting up clinics and schools throughout the country. Two thousand Cuban teachers are brought in for the literacy program. An agrarian reform program redistributes land to poor peasants, as the FSLN maps out a socialist agenda. Large agricultural estates, banks, and some factories are nationalized (taken over by the government).

1979: October   Environmental agency is formed

The Nicaraguan Institute of Natural Resources and Environment (Instituto Nicaragüense de Recursos Naturales y del Ambiente—IRENA) is established and takes charge of the nation's resource conservation.

1980s   National Theater Workshop offers "people's theater"

During the years of the Sandinista regime, the National Theater Workshop, directed by well known playwright Alan Bolt, travels to towns and villages throughout Nicaragua, improvising plays based on the lives of the people who live there.

1980   Nicaragua wins UNESCO literacy award

The Sandinista government wins the grand prize for its literacy programs from UNESCO (United Nations Educational, Scientific, and Cultural Organization).

1980: September 17   Anastasio Somoza dies in Paraguay

Anastasio Somoza Debayle "Tachito", president of Nicaragua from 1967 to 1979, is killed by a car bomb in Asunción, Paraguay, where has he gone into exile following his ouster as president. (See 1925)

1981–90   The Contra War

Throughout the 1980s, the FSLN battles destabilizing opposition from the U.S., including the cutting off of foreign aid and the funding of "Contra" forces formed to resist the Sandinista government. The Contras are mostly made up of former members of Anastasio Somoza's National Guard, eventually joined by peasants from the north and others, including the Miskito Indians living in Nicaragua's Caribbean coastal lands. These indigenous people, who are both culturally and racially different from the majority population to the west, are alienated by the Sandinista government's attempts to relocate them and administer their land. Under the leadership of President Ronald Reagan (b. 1911), the United States supports the Contra efforts because the FSLN is a Marxist (communist) and authoritarian regime; the Reagan administration believes that the Contras will provide more stability in the region and that they are more in sync with American democratic ideology.

The Contras operate from bases in Honduras, carrying out raids in northern Nicaragua. In late 1981, the Reagan administration authorizes an initial budget of $19 million to support their anti-Sandinista campaign. Eventually their numbers grow from 500 to 12,000.

The Contras' activities eventually draw international censure because of human rights abuses, including attacks on civilians. The U.S. Congress votes to cut off aid (see 1984), at which point the administration secretly continues funding the Contras through the National Security Council, using money from the oil-producing nations of the Middle East, in what will eventually become known as the "Iran-Contra Affair." Because of the Contras' association with Somoza's National Guard, the group never gains widespread popularity within Nicaragua itself. Once Sandinista leader Daniel Ortega loses the presidency to Violeta Barrios de Chamorro in 1990, the Contras disband. However, some will later rearm in response to actions of the Chamorro government which they perceive as pro-Sandinista.

1981: January 23   U.S. cuts off aid to Nicaragua

Newly inaugurated president Ronald Reagan suspends all foreign aid to Nicaragua on grounds that the Sandinista government, aided by the U.S.S.R. and Cuba, is providing arms to left-wing guerrillas in El Salvador, a claim that the Nicaraguan government denies.

1981: December   Reagan approves contra funding

The Reagan administration authorizes the CIA to spend $19.8 million to support opponents of Nicaragua's FSLN government. The CIA forms and trains the counter-revolutionary "contras," based in camps in neighboring Costa Rica and Honduras, to carry out military operations against the Nicaraguan government.

1982   Family allowances are introduced

Family allowance legislation provides employees who qualify with benefits for children under the age of fifteen.

1983: August   Geothermal energy plant opens

A geothermal generating plant with a capacity of 70,000 kw begins operations at the foot of the Momotombo volcano. By the 1990s, it is providing about a third of Nicaragua's electricity.

1984   International court condemns U.S. role in Nicaragua

The International Court of Justice rules against U.S. actions designed to destabilize the Nicaraguan Sandinista government. The U.S. Congress also bans further support for these actions.

1984   U.S. Congress votes to cut off aid to the Contras

The U.S. Congress approves the Boland Amendment, cutting off funding for the Contras.

1984: November 4   FSLN victorious in national elections

The FSLN wins nationwide elections by a wide margin, even though two major opposition parties boycott the election. Votes come from more than eighty percent of the electorate and Daniel Ortega Saavedra (see 1945: November 11) is elected president with a majority vote of sixty-eight percent.

1985: May   Reagan declares trade embargo on Nicaragua

President Reagan announces a total embargo on U.S. trade with Nicaragua. The embargo lasts five years.

1986   Bishop is deported for supporting contras

The government expels Bishop Pablo Antonio Vega for supporting President Reagan's plan to provide $100 million in aid to the contra rebels.

1987   Arias proposes Guatemala Accord

The Guatemala Accord, proposed by Costa Rican president Oscar Arias Sánchez (b. 1941), calls for a cease-fire between government forces and the contra rebels and provides for open elections. At first, the plan is rejected by both sides. However, after the U.S. Congress cuts off aid to the contras the following year, they are forced to reconsider the agreement.

1989   Central American peace plan approved by all sides

The peace plan sponsored by Costa Rican president Oscar Arias Sánchez is approved by all parties in the contra war. Between 1983 and 1987, 40,000 lives have been lost in the fighting. Arias wins the Nobel Peace Prize for drawing up the plan.

The 1990s

1990: February   Violeta Chamorro elected president

In the first national elections following approval of the Arias peace plan, Violeta Barrios de Chamorro (b. 1929), widow of slain Somoza opposition leader and newspaper publisher Pedro Chamorro, is elected to the presidency. She is backed by United Nicaraguan Opposition (UNO)—a broad-spectrum coalition of fourteen parties—and by the Bush administration in the U.S.

1990: April   Chamorro takes office

Newly elected president Violeta Chamorro is inaugurated. In order to assure a smooth transition of power, she takes a conciliatory approach toward the Sandinistas, promising to maintain some of their policies, such as land reform measures and the 1987 constitution. Her government faces serious economic problems. Over half of all government expenditures have gone into funding the contra war, leaving social programs stalled, and inflation has risen rapidly. The economy does not respond quickly to Chamorro's programs, and her political coalition, the UNO, begins to dissolve. In addition, the contras, unhappy with the government role played by the FSLN, refuse to disarm completely.

1991   "Re-contras" cause new violence

Refusing to accept the terms of the 1989 peace agreement, an estimated one thousand former contras calling themselves "Re-contras," launch a new wave of violence. The Re-Contras also object to President Violeta Chamorro's decision to retain the Sandinistas' Minister of Defense, General Humberto Ortega, as head of the army.

1992   Government defuses "Re-Contra" crisis

The Chamorro government works out an agreement to conciliate the Re-Contras and prevent further violence, promising them funds, houses, and land.

1992: September 1   Earthquake and tidal waves pummel Pacific Coast

An earthquake under the Pacific Ocean floor causes tsunamis (giant tidal waves) that strike a 150-mile (241-kilometers) stretch of Nicaragua's Pacific Coast, reaching heights of fifty feet (fifty meters). In the aftermath of the disaster, 116 people are dead and 150 are reported missing. More than 16,000 people are left homeless as a result of the quake, which measures 7.0 on the Richter scale. Countries in Latin America, Europe, and Asia send relief aid. The United States contributes $5 million in emergency aid.

1994 Right-wing opponents try to oust Chamorro

The "Group of Three" (Vice President Virgilio Godoy, Alfredo César, and Managua's Mayor, Arnoldo Alemán) attempt to oust Chamorro, undermining her political position by calling illegal congressional sessions.

1995 New Sandinista party is formed

In the 1990s, factional divisions occur within the FSLN based on the issue of how much to support the Chamorro government. A new group, the Sandinista Renovation Movement (MRS), is founded by former FSLN leaders. Along with the traditional nationalist goals of the Sandinistas, it advocates greater internal democracy. FSLN leader and former president Daniel Ortega attempts to negotiate a truce with the MRS in time for the scheduled presidential elections in 1996, but he is not successful.

1996: January 25 Attempt is made to assassinate presidential candidate

An attempt is made on the life of right-wing presidential candidate Arnoldo Alemán while he is campaigning in the town of Quilali. Alemán is unhurt, but one of his bodyguards is killed and three other people are injured. It is suspected that the would-be assassins are military personnel of the former FSLN government.

1996: February Pope John Paul II visits Nicaragua

Pope John Paul II travels to Nicaragua. He holds a mass in Managua, attended by hundreds of thousands of Nicaraguans. President Violeta Barrios de Chamorro offers a public apology to the pope for harassment by Sandinista supporters during a mass he conducted on his last visit to the country in 1983. Sandinista leader Daniel Ortega also publicly apologizes for the incident in a full-page newspaper advertisement.

1996: October Alemán is elected president

José Arnoldo Alemán Lacayo, candidate of the right-wing Liberal Alliance party, wins the presidential election, defeating former Sandinista president Daniel Ortega Saavedra. Ortega at first challenges the results, but former U.S. president Jimmy Carter (b. 1924), monitoring the election, claims it was fair. Ortega concedes defeat. Voter turnout for Alemán, a former mayor of Managua, shows support for his policies of free-market reform, private investment, and closer ties with the U.S.

1997 Drought attributed to El Niño

A drought in the latter half of 1997 is blamed on the effects of the warm Pacific Ocean current known as El Niño.

1997: January 10 President Alemán is inaugurated

Newly elected president Arnoldo Alemán Lacayo takes office, replacing Violeta Barrios de Chamorro. Alemán entreats Nicaraguans to transcend their differences, which have been evident in the months following the October elections.

Alemán's Liberal Alliance party holds forty-two seats in the ninety-two-seat assembly, while the rival FSLN holds thirty-six, and minor parties hold the rest.

1997: November 26 Land distribution law is passed

A law is passed resolving the long-standing controversy over the treatment of lands seized by the Sandinista government in the 1980s. Persons occupying small plots of this land may occupy it and receive legal title. Those with plots larger than a specified size must either return it to its original owners or pay for it. However, the original owners must officially prove that the land is theirs.

Bibliography

Borge, Tomás, et al. *The Sandinistas Speak: Speeches and Writings of Nicaragua's Leaders.* New York: Pathfinder, 1982.

Brentlinger, John. *The Best of What We Are: Reflections on the Nicaraguan Revolution.* Amherst, Mass.: University of Massachusetts Press, 1995.

Gambone, Michael D. *Eisenhower, Somoza, and the Cold War in Nicaragua, 1953–1961.* Westport, Conn.: Praeger, 1997.

Hale, Charles R. *Resistance and Contradiction: Miskitu Indians and the Nicaraguan State, 1894–1987.* Stanford, Calif.: Stanford University Press, 1994.

Kagan, Robert. *A Twilight Struggle: American Power and Nicaragua, 1977–1990.* New York: Free Press, 1996.

Luciak, Ilja A. *The Sandinista Legacy: Lessons from a Political Economy in Transition.* Gainesville, Fla.: University of Florida, 1995.

Morley, Morris H. *Washington, Somoza, and the Sandinistas: State and Regime in U.S. Policy Toward Nicaragua, 1969–1981.* New York: Cambridge University Press, 1994.

Rudolph, James D., ed. *Nicaragua: A Country Study.* 2d ed. Washington, D.C.: Government Printing Office, 1994.

Sabia, Debra. *Contradiction and Conflict : The Popular Church in Nicaragua.* Tuscaloosa: University of Alabama Press, 1997.

Stanislawski, Dan. *The Transformation of Nicaragua.* Berkeley: University of California Press, 1983.

Walker, Thomas W., ed. *Revolution and Counterrevolution in Nicaragua.* Boulder, Colo.: Westview Press, 1991.

Whisnant, David E. *Rascally Signs in Sacred Places: The Politics of Culture in Nicaragua.* Chapel Hill: University of North Carolina Press, 1995.

Panama

Introduction

Panama's geography has been instrumental in determining its history and culture. As the narrowest point of passage between the Atlantic and Pacific oceans, it has attracted the attention of a succession of governments and trading concerns since it was claimed for Spain in the early sixteenth century as a transport route between the Atlantic and Pacific Oceans. This attention, in turn, has made Panama a cosmopolitan society with a colorful variety of people and traditions. In the twentieth century, Panama became the site of an engineering feat that permanently changed the world of maritime commerce: the Panama Canal.

Panama is an isthmus, a narrow strip of land connecting two larger land areas (in this case, Central America and South America). At its thinnest, Panama is only about 50 miles (80 kilometers) across, making it the narrowest point separating the Pacific Ocean from the Atlantic (via the Caribbean Sea). With an area of 29,762 square miles (75,649 square kilometers), Panama is about the size of the state of South Carolina. It contains a variety of topographic features, including rain forests, lowland plains, several mountain ranges, and thousands of small islands along both of its coasts. Panama's estimated population total in 1997 was 2.7 million, with over 600,000 people living in the capital, Panama City. Though Spanish is the official language of Panama, approximately twenty-five percent of the population also speaks English.

The people of Panama are a diverse group made up of mestizos (persons of mixed European and Indian descent), mulattos (persons of European and African descent), whites, Antillean Blacks, and tribal indigenous groups. The most numerous are mestizos and mulattos, who represent approximately seventy percent of the population; Antillean Blacks represent fourteen percent, whites ten percent, and indigenous peoples approximately six percent. Immigrants of East Asian, South Asian, European, North American, and Middle Eastern origins settled in Panama to take advantage of the commercial opportunities presented by the Panama Canal. The Black Antilleans in modern Panama are descendents of laborers originally brought from the British West Indies to help construct the Canal. They are the largest minority group, and are set apart from the predominantly Roman Catholic Spanish-speaking mestizo population in their English language and Protestant faith. The majority of Panama's population (ninety-three percent) is Roman Catholic. The main Indian groups in Panama are the Cuna (Kuna) of the San Blas Islands off the Caribbean coast, the Chocó, and the Ngobe-Buglé. More than one-half of the indigenous groups in Panama live in the northwest region of the country. While some tribes have maintained their traditional language and belief systems, others have assimilated more into general Panamanian society and adopted the Spanish language.

Panama's rich folklore is expressed in its traditional dances and costumes. Women typically wear a colorful *pollera* (dress), and *templeques* (hair ornaments) for traditional dances. Men wear an embroidered, long-sleeved shirt, calf-high trousers, and a straw *montuno* hat. Traditional dances and celebrations are common during a town's Patron Saint festivals, and in February and March during Carnaval, a four-day celebration. Other large celebrations include Easter week, or Semana Santa, and the Festival of the Black Christ in October.

A large percentage of Panama's public budget is spent on education. The education system is divided into three levels: primary school (six years), secondary school (six years), and university/higher education. Education is mandatory between the ages of six and fifteen and provided free by the government through the university level. According to the World Bank (1998), ninety-one percent of Panamanians aged fifteen and over can read and write.

Gateway to the World

People have been crossing Panama to travel between the Atlantic and Pacific Oceans for centuries. The first person in recorded history to make the trip was the Spanish explorer Vasco Núñez de Balboa (1475–1519), who journeyed westward across the isthmus and sighted the Pacific Ocean on September 25, 1513. The discovery of this body of water opened up new trade possibilities for Europe. Goods could be shipped east to west without the long trek around South America. By the middle of the sixteenth century, Panama had become a major trading center. Its port city of Nombre de

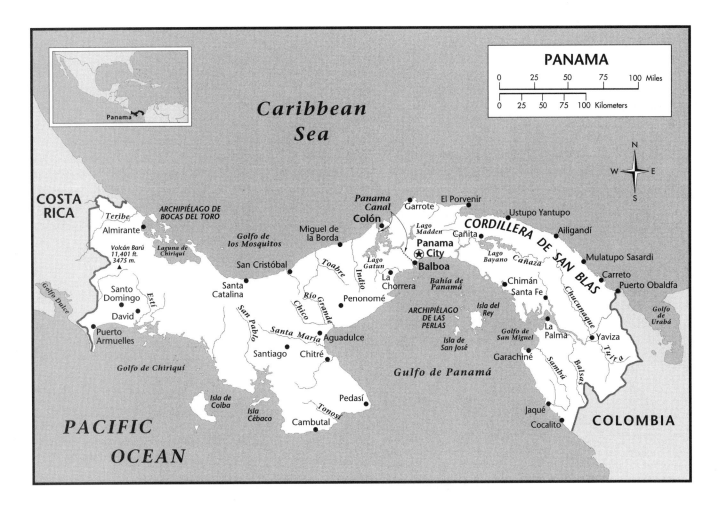

Dios was one of only three ports officially approved for trade between Spain and the New World. Gold and other minerals from Peru were transported up the Pacific coast of South America and carried overland across Panama to the Caribbean coast for shipment to Europe, while goods from Europe were brought in at the same spot for distribution in the Americas. By the end of the century, the trade center had been shifted to Portobelo, which became famous for its massive fairs, where a lavish assortment of goods was available.

By the end of the seventeenth century, Panama's prominence as a trading center declined. Raids by buccaneers who had discovered Panama's accessibility to the sea, the depletion of Peruvian gold and silver, and the passage of new laws limiting Panama's monopoly on transatlantic trade added to the decline. Panama rose to prominence again as a transport center during the California gold rush of 1849. Many preferred making the journey by way of Panama rather than attempting the treacherous overland journey across the continental United States. They were soon accommodated by the construction of a railway across the isthmus, completed in 1855. Travel along this route boomed for over a decade but then declined with the completion of the Transcontinental Railroad in the United States. In Panama, attention then

shifted to the most ambitious transport project of all: the construction of a canal.

Having just overseen the construction of Egypt's Suez Canal, in 1869, the Frenchman Ferdinand de Lesseps (1805–1894) seemed like the perfect person to take charge of a canal project in Panama. He organized a company in 1879 and construction began a year later. However, engineering and management difficulties, combined with the loss of workers from diseases that flourished in the swampy environment, ruined the French canal team, and the project was abandoned in 1889 with two-fifths of the excavation completed.

After paving the way with the necessary diplomatic and political maneuvers, the United States Army Corps of Engineers began construction of a canal in 1904. A crucial first step was the elimination of yellow fever, malaria, and other diseases that had killed thousands of laborers during the ill-fated French project. By installing modern sewage systems in the urban areas around the Canal Zone, they were able to reduce the region's disease-carrying mosquito population. Finally, after ten years, the $387 million project, which required the largest excavation in history, was completed and the first ship passed through the canal in August of 1914.

In the middle of the twentieth century, Panama's reputation as an international trading center was further enhanced by the establishment of the largest free trade zone in the Western Hemisphere, located in the city of Colón. Here, companies from countries throughout the world could bring goods to be processed, stored, or shipped elsewhere without payment of duties or taxes. By the 1990s, the free trade zone employed thousands of people and handled goods worth billions of dollars annually.

Claiming a National Identity

In spite of (in some cases, because of) their unique geographic position, the Panamanians have experienced a long struggle to establish recognition of their unique identity as a people and gain full political and economic control of their land.

The beginning of the sixteenth century ushered in three hundred years of Spanish colonial government. The first Spanish explorer, Rodrigo de Bastidas (1460–1526) arrived in Panama in 1501, and Spaniards began to colonize the region by 1510. During the Spanish colonial period, Panama was governed as part of the Spanish viceroyalty of Peru and, later, as part of Colombia (then called New Granada). Even after 1821, when it declared independence from Spain, Panama still remained part of Colombia for eighty-three years.

When Panama achieved independence from Colombia in 1903 it was with the support of the United States. That same year, Panama signed the Hay-Bunau-Varilla Treaty, guaranteeing the United States exclusive rights within a Canal Zone that cut a ten-mile-wide swath through the middle of the country. The treaty, which remained in effect until 1936, also granted the United States power to intervene in Panama's internal politics, making Panama, in effect, a U.S. protectorate during this period.

Throughout the twentieth century, the U.S. intervention in Panama's affairs has been a catalyst for civil unrest in Panama, culminating in the Torrijos-Carter Treaties of 1977, which provided for full Panamanian control of the Canal Zone by the year 2000 following a gradual U.S. withdrawal. The U.S. withdrawal from the Canal Zone proceeded after 1978, when the treaty was ratified by the U.S. Senate. Panamanians still faced political oppression, but this time it came from within their own borders. Beginning in 1983, General Manuel Noriega exercised political control of the country, imposing a harsh and corrupt military dictatorship. The United States, at first a supporter of Noriega, eventually reversed its position and pressured him to resign, finally ousting the dictator in a military invasion at the end of 1989. In the 1990s, Panama faced the dual tasks of preparing for the U.S. departure from the Canal Zone while reestablishing democracy within their country. In the first election following the removal of Noriega, Panamanians elected Ernesto Pérez Balladares (b. 1947) president in 1994.

In 1999, Panama's relationship with the United States remains characteristically complex. In spite of a history of anti-United States sentiment, polls show that a majority of Panamanians favor some form of continued U.S. military presence to insure political stability. In addition, many fear the loss of income and employment that will result from the pullout of U.S. military personnel and their families. In late 1995, President Balladares and U.S. President Clinton met to discuss the issue and scheduled further exploratory talks on the future role of the United States in Panama. A new law ratified in 1997 created the Canal Authority to administer the Canal after the United States relinquishes its control over the Canal on December 31, 1999.

Presidential elections are to take place in May 1999. Three presidential candidates are campaigning. The leading candidate, Marin Torrijos Espino, is supported by the large governing body, the Democratic Revolutionary Party (PRD).

Timeline

15th century A.D. Panama is occupied by over 50 Indian tribes

By some estimates, there are as many as 500,000 to 800,000 Indians from more than fifty different tribal groups living in Panama at the end of the fifteenth century. Most of these Indians come from the Mayas of Guatemala and the Chibchas of Colombia. The most numerous are the Cuna on the San Blas islands, the Chocó in the jungles of the Darién region, and the Guaymí in western Panama.

Spanish Colonial Rule

1501 First European arrives in Panama

The Spanish explorer Rodrigo de Bastidas (1460–1526), one of Columbus' captains on his second voyage, is the first documented European to set foot on Panama. Setting sail from Venezuela, he arrives in Panama and travels about 93 miles (150 kilometers) in search of gold. He then continues on to the West Indies.

1502 Columbus visits Panama

Christopher Columbus (1451–1506) explores part of Panama's coast on his fourth and final transatlantic voyage. He gives the name Puerto Bello (beautiful port) to one of the spots he visits. It is later renamed Portobelo.

1510 First Spanish settlement is established

Panama's first Spanish settlement, Antigua, is founded in the Darién region. The settlers elect two co-mayors: Vasco Núñez de Balboa (1475–1519), a former member of Rodrigo Bastidas's crew, and Martin Zamudio. Balboa encourages agriculture in the new settlement, and it thrives.

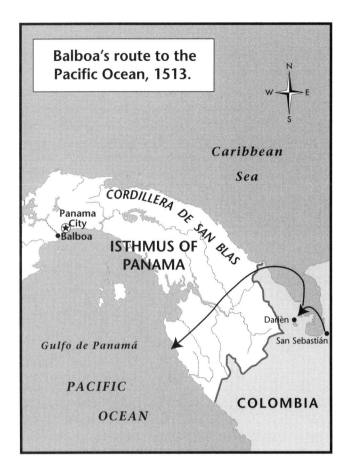

Balboa's route to the Pacific Ocean, 1513.

1513: September 25 Balboa is the first European to sight the Pacific Ocean

In search of gold and a fabled sea described by local Indians, Balboa sets out across the isthmus of Panama with a band of 190 Spaniards, a large number of Indian slaves, and a pack of dogs. The party crosses the Darién jungle and reaches the mountains. On September 25, atop the peak of Pirre, Balboa sees a huge body of water, which he names the Southern Sea. He descends the mountain, wades into the ocean, and claims it and all the shores it touches for Spain.

1514 Pedro Arias Dávila is appointed governor of Panama

The Spanish monarchy appoints Pedro Arias Dávila (1442–1531) who is also known as Pedrarias, to be the new governor of Panama. Pedrarias, known for his cruelty, later charges Balboa with treason and has him beheaded.

1519 Arias Dávila founds Panama City

Moving his base from the jungle of Darién westward to the Pacific coast, Arias Dávila founds a new village called Panama, an Indian name meaning "plenty of fish." It will later become Panama City.

1538 Audiencia established in Panama

Spain sets up an audiencia (court) in Panama to administer Spanish possessions extending from Nicaragua to Cape Horn at the southernmost tip of South America.

Mid-16th century Panama becomes a major trading center

Early in the sixteenth century, Spain designates the Panamanian port of Nombre de Dios as one of only three ports authorized to handle trade between itself and its American colonies. Gold, silver, and other raw materials from Peru and other locations enter Panama en route to Europe, and European goods are made available to the colonists.

1563 New, more limited, audiencia is created

Because the first audiencia had encompassed too vast an area, a new audiencia is established which more nearly coincides with the present-day boundaries of Panama.

1567 Panama becomes part of the Viceroyalty of Peru

Panama is attached to the Viceroyalty of Peru but retains its own audiencia.

1597 Trade center moved to Portobelo

After raids by the British pirate Sir Francis Drake (c. 1540–96) demonstrate the vulnerability of the port of Nombre de Dios, Panama's Atlantic trade is moved to Portobelo, which has a well-protected natural harbor. In the coming decades, great fairs are held here. Portobelo becomes the terminus for supplying much of Spanish America with European goods.

Late 17th century Portobelo trade declines

During the latter part of the seventeenth century, the importance of Portobelo as a trading center declines due to a combination of factors including the liberalization of trade under the Bourbon monarchy, the exhaustion of the Peruvian gold and silver mines, raids by buccaneers, and increased smuggling.

1671: January 29 British pirate sacks Panama City

British buccaneer Henry Morgan (1635–88) sails up the Río Chagres to Panama, crosses the isthmus, and lays siege to Panama City, leading a force of 1,400 men. Many of its residents gather whatever treasure and other possessions they can and flee, either to the countryside or to Peru. The city is looted and destroyed by fire for the second time (see 1644). It is later rebuilt at a new site (see 1673). Morgan and his men leave Panama with 600 prisoners and 175 mules carrying the accumulated spoils from their looting.

1673: January 21　Panama City is founded at a new site

Antonio Fernández de Córdoba y Mendoza re-establishes Panama City at a new location roughly seven miles (eleven kilometers) from its original site. It is strongly fortified to defend against pirate attacks such as that which destroyed the old city. The city has about 900 residents at its founding. Its population grows to 1,600 by 1675 and 20,000 by the turn of the century.

1698　British entrepreneur attempts to found a colony

Toward the end of 1698, William Paterson (1658–1719), a British financier, has 1,200 would-be colonists transported to the San Blas area in an attempt to form a settlement. They give up and return to England after six months, unaware that 1,600 reinforcements have been sent. The new colonists stay until early 1700 and then leave, when many are ravaged by illness.

Early 18th century　First play written in Panama

Victor de la Guardia (1772–1824) writes the first Panamanian play. A tragedy in verse, *La política del mundo,* is the first dramatic work composed and performed in Panama.

1717　Panama becomes part of the Viceroyalty of New Granada

Spain decrees that Panama will be administered by the Viceroyalty of New Granada, along with Colombia, Venezuela, and Ecuador.

1737　Fire erupts in Panama City

A major fire ("the big fire") causes heavy damage in Panama City.

1739　British capture and destroy Portobelo

Forces led by British admiral Edward Vernon (1684–1757) seize Portobelo and demolish it. Afterward, this Panamanian city declines in importance as a trading center. Much of South America's Spanish trade is routed around Cape Horn.

1749　Jesuit university is founded

The Jesuits (a Catholic order) establish the first university in Panama. However, it closes when the Bourbon monarchy of Spain expels the Jesuits from Spanish America in 1767.

1819: August 7　Victory at Boyacá liberates Colombia

Colombia (then called New Granada) is liberated by Simón Bolívar's forces at Boyacá, and the Spanish viceroy flees to Panama, where he rules until his death in 1921.

1820　First printing press in Panama

Printing is brought to Panama.

Colombian Province

1821: November 28　Panama declares independence from Spain

At a meeting in Panama City, Panama declares its independence from Spain. It becomes part of the newly formed Gran Colombia, which also includes Ecuador and Venezuela.

1826: June　Congress of Panama is held

Four liberated Latin American states—Mexico, Peru, Gran Colombia, and the United Provinces of Central America—attend the Congress of Panama to discuss regional cooperation. A mutual defense treaty is drawn up, but Colombia is the only country to sign it, and it is eventually abandoned. Simon Bolívar's lifelong dream of a unified Latin America goes unrealized.

1830–40　Three Panamanian secession attempts are made

Three different attempts are made by Panama's leaders to secede from Gran Colombia. The first attempt is brought about by the acting governor of Panama and his opposition of the president's policies. The second attempt is a plot by an unpopular dictator, who is later deposed and executed. The final attempt to secede is in response to a civil war in Colombia and declared by a popular assembly. None of the secession attempts are successful.

1848　California gold rush makes Panama a popular passage to the west

With the discovery of gold in California, Panama becomes a strategic location for both people and supplies moving westward. Prospectors reach the west by sailing to Panama, crossing the isthmus by an overland route, and continuing westward via the Pacific Ocean. This journey is easier and safer than the long trip west by covered wagon. Within twenty years, 375,000 people will make the westward crossing through Panama, and another 225,000 will use the same route to travel eastward.

1855: January 28　Railroad across Panama is completed

The New York-based Panama Railroad Company completes construction of a railroad across the isthmus of Panama linking the Caribbean Sea and the Pacific Ocean. The forty-seven-mile-long railroad running from the port city of Colón on the Caribbean to Panama City on the Pacific takes five years to construct. Some 6,000 lives are lost in the process, many from diseases carried by mosquitoes breeding in the swampy terrain. After the completion of the Transcontinental Railroad in the United States, in 1869, the Panamanian route westward declines in popularity.

1879   French company is formed to build a canal in Panama

Throughout the nineteenth century, both Panama and Nicaragua attract interest as possible sites for a canal linking the Atlantic and Pacific oceans. Ferdinand de Lesseps, the Frenchman who oversaw construction of the Suez Canal in Egypt, forms a company to construct a canal in Panama.

1880s   Baseball is introduced to Panama

Even before its independence from Colombia, baseball games are played in Panama by U.S. and British residents and by Panamanians who learn the game while attending school in the United States.

1880: January 1   Work officially begins on de Lesseps canal

An official ceremony launches de Lesseps' canal project, but construction doesn't actually start until the following year. Progress is slowed by worker illness and disagreements between de Lesseps and his engineers. By the time half the projected construction time has passed, only about ten percent of the project has been completed, and four-fifths of the workers are too ill to work. Actual construction of the first canal lock does not begin for eight years.

1888   Birth of poet Ricardo Miró

Miró (1888–1940), one of Panama's foremost twentieth-century poets, is born. His verse collections *Segundos preludios* (1916) and *Caminos silenciosos* (1929) help him achieve the honor of being named Panama's national poet. Miró also founds the review *Nuevos Ritos.*

1889   French canal project is abandoned

The canal company formed by Ferdinand de Lesseps goes bankrupt and all work on the project is stopped, leaving thousands of laborers unemployed. However, nearly half the excavation necessary for construction of a canal has been completed, and numerous buildings have been constructed, conditions that later canal-builders can use to their advantage.

1892   First baseball club is established

Panama's first baseball club is formed in Colón and participates in matches with players from other cities.

1892   National library is founded

The National Library of Panama is established in Panama City under the name Biblioteca Colón. In 1942 it is reorganized and renamed the National Library.

1901: December   United States and Britain sign treaty over canal rights

The Hay-Pauncefote Treaty invalidates the provisions of the 1850 Clayton-Bulwer Treaty. The Clayton-Bulwer Treaty arises out of a rivalry between Great Britain and the United States and states that neither country will seek exclusive control over any part of Central America. Britain approves construction of a Central American canal solely by the United States, as long as it remains a politically neutral zone.

Independence

1903: November 3   Panama proclaims independence from Colombia

When the Colombian government refuses to grant canal rights to the United States, the U.S. backs Panamanian revolutionary José Augustín Arango, who ousts the Colombians and declares Panama's independence. U.S. President Theodore Roosevelt recognizes Panama as an independent U.S. protectorate.

1903   Panama's first national flag

Painter Manuel Amador (1869–1952), a pioneer of modern art in Panama, designs Panama's first national flag. Amador holds a number of government posts during the course of his career, including minister of finance and consul in Hamburg, Germany, and New York City. Amador's paintings reflect the influence of German expressionism and of American artist Robert Henri (1865–1929). Most of his work is produced between 1910 and 1914 and after 1940. He is known particularly for his portraits, including *Cabeza de Estudio* (1910) and *Rabbi* (1948). There is a major collection of Amador's work at the University of Panama.

The Panama Canal

1903   U.S. and Panama sign Hay-Bunau-Varilla Treaty

Following the Colombian Congress rejection of Hay-Herran Treaty, the people of Panama declare independence on November 3, 1903. The United States recognizes independence three days later and on November 18, 1903 the Hay-Bunau-Varilla Treaty is signed in Washington, D.C. The treaty gives the United States the right to build a canal five miles wide and grants them control over a zone ten miles wide through the middle of Panama. In return, the United States agrees to guarantee Panamanian freedom and assumes the task of maintaining the law and also agrees to pay $10 million, or $28 per acre.

The construction of the Panama Canal is one of the engineering feats of the twentieth century. Here construction materials are hauled by railroad car. (EPD Photos/CSU Archives)

1904 Construction of the Panama Canal begins

The U.S. senate ratifies the Hay-Bunau-Varilla Treaty. The Hay-Bunau-Varilla Treaty establishes the Canal Zone, a strip of land five miles wide on either side of the canal. The administration of the Zone is carried out by the Panama Canal Company, a self-sustaining corporate organization of the U.S. Department of Defense. The U.S. President serves as governor of the Canal Zone.

President Roosevelt appoints a commission to supervise canal construction. Roosevelt puts the Army Corps of Engineers in charge of the project, with Colonel George Washington Goethal (1858–1928) as head commander. William C. Gorgas (1854–1920), an army doctor, is hired to take care of sanitation in the region.

Gorgas oversees an extensive program to rid the area of the malaria, yellow fever, and other diseases that claimed the lives of at least 20,000 workers during the de Lesseps project (see 1879). The principle task is to eliminate the mosquitoes that carry the deadly diseases, and Gorgas's staff oversees measures that drastically lower the mosquito count, including fumigation, the draining of standing water, and the establishment of a fresh water supply, ultimately saving an estimated 70,000 lives and millions of sick days in the course of the ten-year-long project.

1904: February 1 Panamanian constitution is adopted

Panama adopts a constitution similar to that of the United States. It also authorizes U.S. intervention in case of civil disorder.

1905 John Stevens replaces Wallace as chief canal engineer

John Stevens (1853–1943) takes over supervision of the canal project and determines that a lock canal must be built, rather than one at sea level as had originally been planned. Stevens is later replaced by Lieutenant Colonel George Goethals (1858–1928), who remains with the project until it is completed. Three sets of locks are built to raise canal-going ships above sea level to the level of the artificially created Gatun Lake and then lower them back down. Over 150,000 West Indian and European immigrants come to Panama to work on the canal during its ten-year construction period.

1908 Teatro Nacional is completed

Construction of the Teatro Nacional (National Theater), designed by Italian architect Genaro Ruggieri, is completed. The inaugural performance is the opera *Aida* by Verdi, followed by Bizet's *Carmen* and *Lucía.*

1912 Liberal government inaugurates period of rapid development

A twenty-year period of liberal rule is launched as the liberal party, headed by Belisario Porras (1856–1942), wins control of the government. Major development efforts include the construction of railroads, highways, schools, and hospitals. Business opportunities expand as the banana, logging, and cattle industries thrive.

1913 Academia Nacional de Pintura is established

Panama's first art school, the Academia Nacional de Pintura, is founded by painter Roberto Lewis, who serves as its first director. Its name is changed to Escuela Nacional de Pintura in 1939, and to Escuela Nacional de Artes Plásticas in 1952.

Born in Panama City, Lewis (1874–1949) is a leading Panamanian painter, sculptor, and art educator. He studies in Paris in the 1890s and is later named Panamanian consul to the city from 1904 to 1912. Between 1905–07, he paints the ceiling, curtains and murals in the foyer of the Teatro Nacional, the Palacio de Gobierno, and also paints many murals and a gallery of portraits of presidents in the official Presidential Palace. From 1913 to 1938, he is a teacher of drawing at Instituto Nacional, Escuela Normal, Escuela de Artes y Oficios, and Director of Museo Nacional and Academia de Pintura. Lewis's neoclassical paintings adorn the interiors of many public buildings in Panama, including the Teatro Nacional, the Palacio de Gobierno, and the Presidencia. By contrast, his landscapes, influenced by post-impressionism, are known for their bright colors and bold brush strokes.

1914: August 15 First ship passes through the Panama Canal

The Panama Canal is officially opened and the first ship, the *S.S. Ancón* with U.S. captain John Constantine at the helm, passes through it.

1923 Smithsonian launches research institute in Panama

The Smithsonian Institution opens a tropical research institute in Balboa.

The Growth Of Nationalism

1923: August 19 Acción Comunal is formed

Middle-class Panamanians who fear that U.S. influence will overwhelm Panama's national identity establish the nationalist group *Acción Comunal* (Community Action). The group also opposes corruption and inefficiency in the Panamanian government. It successfully opposes ratification of a new treaty with the United States in 1926.

Prison laborers, such as this "ball and chain gang" are used in the construction of the Panama Canal. (EPD Photos/CSU Archives)

1925 Cuna Indian rebellion is suppressed

Cuna Indians, led by U.S. citizen Richard O. Marsh, carry out a revolt against Panama's government, creating their own republic, which they name the Republic of Tule. The uprising is quickly contained with help from the United States and the new "republic" dismantled. The United States brokers a treaty that gives the Cunas' homeland, the San Blas Islands, the status of a semiautonomous (partly self-governing) territory.

1925: October 10–13 Tenants' strike results in rioting

Tenants in Panama City organize a strike to protest substandard apartment conditions and high rents. After renters clash with police, resulting in four deaths, the strike spreads, virtually shutting down the city. U.S. military forces are called in at the request of the government and remain stationed in Panama until October 23. Two more people are subsequently killed in confrontations with U.S. soldiers.

1929 National stadium is inaugurated

Construction is completed on Panama's baseball stadium.

1929: February 13 Birth of national leader Omar Torrijos

Omar Torrijos Herrera (1929–81), Panama's most celebrated twentieth-century leader, is born. Pursuing a military education and career, Torrijos joins Panama's National Guard, gradually advancing in rank. In 1968, he is among a group of colonels who lead a *coup d'état* that ousts Arnulfo Arias from the presidency only ten days after Arias has taken office. Torrijos emerges as the strongest leader of the group and becomes the sole head of the National Guard and political leader of the country. Although criticized for his dictatorial methods, he is also lauded for his many accomplishments in both the domestic and international arenas. Internationally, Torrijos is best remembered for his role in negotiating the 1977 Torrijos-Carter Treaty that mandates U.S. withdrawal from the Canal Zone by the year 2000. Control of the canal has long been a goal of Panamanian nationalists, as well as a catalyst for civil unrest, and Torrijos gains great popularity by helping make it a reality.

Domestically, Torrijos implements a number of measures to improve living conditions in Panama, including a new labor code, government-subsidized public housing, an improved transportation system, expansion of the banking industry, increased government spending in rural areas, and expanded political participation for members of the middle and working classes. Aside from the canal treaty, Torrijos steers a skillful diplomatic course, managing to balance security cooperation with the United States with opposition to the U.S. blockade of communist Cuba and aid to the Sandinista rebels in Nicaragua.

1932 Nationalist Harmodio Arias Madrid is elected president

During the 1920s, nationalist, anti-United States sentiment grows in Panama in reaction to the intervention rights granted by the 1903 Hay-Bunau-Varilla Treaty. After a coup by the nationalist group Acción Comunal (Community Action), Harmodio Arias Madrid assumes the presidency. A mestizo from a poor family, Arias is the first Panamanian president to execute relief efforts for the poor countryside.

1935 The University of Panama is founded

President Arias founds the University of Panama.

1935 Baseball team places second in Central American Games

Panama's national team wins second place in the Central American Games. Teams from Cuba and Nicaragua visit Panama in 1934–35 to help the Panamanians prepare for the games.

1936: March 2 New United States–Panama treaty is signed

The Hull-Alfaro treaty ends Panama's status as a U.S. protectorate and with it the U.S. right of intervention in Panama's internal affairs. U.S. special privileges in Panama are radically reduced. The new treaty is signed by U.S. Secretary of State Cordell Hull, but it is not ratified by the Senate until three years later.

1939–45 World War II

During World War II (1939–45), the United States uses 134 sites in Panama for military facilities, including landing fields, warning stations, antiaircraft batteries, radar stations, and a naval base.

1939 Museo Nacional opens

Panama's first permanent public museum, the Museo Nacional, is established.

1940: June Arnulfo Arias Madrid is elected president

Arnulfo Arias (1901–88), brother of former president Harmodio Arias (see 1932) is elected president on a nationalist platform known as Panameñismo, based not only on anti–United States sentiment, but also on hostility toward all non-Hispanics in Panama, including Chinese, West Indians, and Jews. In the political climate of World War II, Arias's views correspond with those of the European fascists and Nazis.

1941 Women's suffrage is granted

Women win the right to vote on the same basis as men.

1941 Arias draws up a new constitution

The Constitution of 1941 rescinds the citizenship of nonwhite immigrants and extends Arias's presidential term from four to six years.

1941 Social security fund is set up

A social security program is established to provide old age and disability pensions, maternity care, hospitalization and medical care, and funeral benefits. The retirement age is fifty-seven for women and sixty-two for men.

1941: October Arias is ousted in coup

Arias' dictatorial methods and his pro-Nazi beliefs, result in his overthrow by the National Police while he is out of the country. He is succeeded in office by his minister of justice, Ricardo de la Guardia (1899–1970). Although jailed and then exiled from Panama, Arias continues to build a following using his anti-U.S. political stance.

1942   National Academy of Sciences is established

The National Academy of Sciences of Panama is founded to advise the government on scientific issues.

1943   Trans-Isthmian Highway opens

The Trans-Isthmian Highway linking the Atlantic and Pacific oceans is inaugurated, and Panama switches from driving on the left-hand side of the road to driving on the right.

1945: November   Panama joins the United Nations

Panama becomes a charter member of the United Nations.

1946   New constitution is adopted

A new democratic constitution replaces the one implemented by Arnulfo Arias.

1946   Panama extends U.S. lease on military installations

Panamanian officials agree to give the United States an additional twenty-year lease on the sites of thirteen military installations. However, ten thousands Panamanians protest the lease in a violent mass demonstration instigated by university students, launching a period of animosity between students and the National Police. Both students and police are killed in the hostilities, and the National Assembly votes to reject the lease extension. The United States vacates all military installations except for those within the Canal Zone within two years.

1946   Professional baseball is launched

Panama's first professional baseball league is established. In its first year, it consists of thirty-two players from Panama, sixteen professional players from the United States, twenty Cubans, four Mexicans, three Dominicans, and two Nicaraguans. The New York Yankees visit Panama for exhibition games with professional all-star teams fielded by Panama and by the Canal Zone. In addition to the Panamanians playing in their country's national league, players from Panama compete on Venezuelan and Mexican teams and in the U.S. Negro Leagues.

1948   Arias is elected president again

Arnulfo Arias is pronounced retroactive winner of the presidency. He announces the impeding suspension of the 1946 constitution and is met with widespread opposition. Antonio Remón, head of the national police, delays his installment in office.

1950   Juan Manuel Cedeño paints well-known work

The painter, Juan Manuel Cedeño (b. 1914), paints *Octavio Méndez Pereira,* one of his best-known works. Born December 28, 1914, Cedeño studies with Roberto Lewis in Panama at the Escuela Nacional de Pintura and later at the Chicago Art Institute, graduating in 1948. From the 1950s to the 1970s, he is Panama's leading portrait painter. He is also a leading art educator, serving as director of his alma mater, the Escuela Nacional de Pintura (renamed the Escuela Nacional de Artes Plásticas in 1952), from 1948–67, and as a professor at the University of Panama from 1967–78. Cedeño is also known for his experiments with geometrization (incorporation of geometric shapes) in working with folk themes of his native land. (An example is *Domingo de Ramos,* 1955.) His works also include landscapes and still lifes.

1951   Arias is overthrown in coup

As impeachment charges against him are pending in the legislature, President Arias is once again ousted in a coup.

1952   José Antonio Remón is elected president

Head of the national polic, José Antonio Remón (1908–55) seeks and wins the presidency. In spite of the questionable methods he employs to win the office, he implements a number of important measures as president, including social reforms and efforts to promote economic development. However, Remón's presidency is beset by rumors of corruption.

1953   Art exhibition celebrates independence from Colombia

The fiftieth anniversary of Panamanian independence from Colombia is commemorated in Panama's first major art exhibition, which is sponsored by the government.

1953   Colón Free Zone opens

The Zona Libre de Colón, or Colón Free Zone, is the largest trading zone in the Americas and the second largest in the world, surpassed only by Hong Kong. By the 1990s, over 700 companies operate in the zone, which takes up about a fourth of the city of Colón. Goods enter the zone and are warehoused, processed, assembled, or exhibited, and can then leave the country without the payment of duties or taxes. Thousands of Panamanians are employed in the free trade zone, and merchandise worth billions of dollars passes through it every year.

1955   Remón is assassinated

President Remón is killed by assassins.

1958   Bridge of the Americas is completed

The Bridge of the Americas replaces the Thatcher ferry as the link in the Pan-American Highway at Panama City.

Anti–U.S. Protests and New Treaties

1958: May State of siege is declared in response to student riots

Anti-United States demonstrations by university students lead to nine deaths and the imposition of a state of siege by the government.

1959: November Anti–U.S. violence escalates

U.S. troops are alerted in response to threats that Panamanians may overrun the Canal Zone and plant their national flag there. Hundreds of Panamanians clash with the National Police and U.S. military. Mobs damage U.S. buildings outside the Canal Zone and tear down the American flag outside the residence of the U.S. ambassador. The United States puts up a fence around the Canal Zone, and U.S. citizens implement a boycott against Panamanian merchants, who depend heavily on their business.

1962 Instituto Panameño de Arte is founded

Panama's first institution devoted to private patronage of the arts is established.

1964: January 9 "Flag riots" result in multiple injuries and deaths

A confrontation over raising the Panamanian flag at a U.S.-run high school in the Canal Zone leads 200 Panamanian students to storm the Canal Zone with their flag and confront residents there. In the scuffle, the Panamanian flag is torn and thousands of Panamanians mob the Canal Zone in three days of rioting. Two dozen people are killed and many more are injured. Diplomatic relations between the United States and Panama are temporarily broken off.

Later that year, Panamanian president Marco Aurelio Robles (1905–90) and U.S. President Lyndon Johnson (1908–73) negotiate several new treaties governing the operation and defense of the canal, and additional U.S. foreign aid is authorized for Panama. However, the treaties are ultimately rejected as offering little real advantage for Panamanians.

1965 University of Santa María la Antigua is founded

The University of Santa María la Antigua, a private Roman Catholic college, is established in Panama City.

1966 Rubén Blades releases first album

Born July 16, 1948 in Panama City, salsa singer, songwriter, and actor Rubén Blades releases his first album in 1966. Blades' mother is a pianist and singer; his father is a bongo player. He attends law school in Panama but sets his sights on a career in music. After working as a lawyer for the National Bank of Panama, he emigrates to New York in 1974, getting a mail-room job at Fania Records, a leading Latin American record label. By July 1974, Blades appears with Ray Barretto's band at Madison Square Garden.

Blades writes songs for a number of Latin bands in the 1970s and begins collaborating with trombonist Willie Colón. Their 1978 album *Siembra* (Seed) becomes one of salsa's most significant and biggest-selling records. Blades introduces innovations to traditional salsa, adding rock elements and synthesizers, as well as lyrics that incorporate social commentary. After spending a year getting a master's degree from Harvard Law School (1984–85), Blades moves to Los Angeles and begins a film career, starring in the low-budget movie *Crossover Dreams*. He subsequently appears in numerous films including *The Milagro Beanfield War* (1988), *The Two Jakes* (1989), and *Dead Man Out* (1989).

In 1992, Blades becomes active in Panamanian politics, forming the populist party Papa Egoro ("Mother Earth" in the Indian language, Embera) and becoming its presidential candidate in the 1994 election. After leading briefly in the polls, he comes in third in a field of seven candidates. Blades remains involved in the politics of his homeland while also juggling his music and acting careers. In early 1998 he has a starring role on Broadway in Paul Simon's doomed musical, *Capeman*.

1968: May 12 Arnulfo Arias returns to the presidency

Arnulfo Arias, who last served as president in the early 1950s, is elected once again.

He assumes office on October 1. His immediate removal of two senior National Guard officers antagonizes the military, and they plan his ouster.

1968: October 11 Arias is ousted in coup; Torrijos comes to power

The military overthrows President Arias. After a period of turmoil, Omar Torrijos Herrera (see 1929) gains control of the government. Although he projects the image of a populist (a ruler who stands for the common people), he rules as a dictator and maintains power through harshly repressive measures, including censorship, torture, assassination, and exiling his political opponents from the country. However, Torrijos gains great popularity among Panamanians and implements significant reforms in labor legislation, education, health care, and social welfare programs. He also gains the support of the rural poor by addressing agrarian problems and initiates talks with the United States on a new canal treaty.

1970s First art galleries are opened

Panama's first art galleries appear, furthering the careers of contemporary artists including Aguilar Ponce (b. 1943), Tabo Toral, Antonio Madrid (b. 1949), and Emilio Torres (b. 1944).

1970   Banking law leads to growth of financial sector

A new banking law removes many restrictions on the banking industry, leading to a rapid growth in international banking and earning Panama the nickname of "the Switzerland of Latin America." Taxation on transactions is reduced, funds are allowed to move into and out of the country without limitations, and secret accounts are legalized.

1972   Boxer Roberto Durán wins lightweight world championship

Roberto Durán (b. 1951) becomes the first Panamanian boxer to win the lightweight world title in 1972, a title he holds for seven years. Born in Panama on June 16, 1951, Roberto Durán becomes Panama's foremost sports hero. He begins boxing as an amateur, and at the age of sixteen is chosen to represent his country at the Pan-American Games. However, already exhibiting the temperamental personality that will be his hallmark throughout his career, he forfeits his place in the games because of conflicts over his personal behavior. In 1968, he gives up his amateur status to become a professional boxer. In 1980, he wins the welterweight title, becoming the first boxer to defeat U.S. boxer Sugar Ray Leonard in a professional match. This victory is celebrated so wildly in Panama that a number of people are injured. After the welterweight title, Durán goes on to become the junior middleweight champion in 1982.

Following defeats in 1983 and 1984, the boxer suffers a period of depression and gains over fifty pounds. However, after a hiatus, he returns to the ring. At thirty-seven, he wins the middleweight championship (1989), becoming one of only four men in boxing history to hold four different titles in the course of his career. Durán remains an active boxer in the 1990s.

1972   Torrijos regime convenes assembly and adopts constitution

The military government of Omar Torrijos enacts a new constitution expanding the National Assembly to 505 members.

1972   New law implements labor reforms

The newly enacted Labor Code introduces a minimum wage and compulsory arbitration of labor disputes.

1972   Professional baseball is discontinued

Panama's professional baseball league is dissolved.

1975   Museum of Natural Sciences opens

Panama's Museum of Natural Sciences showcases the country's natural history and plant life, as well as the fauna of the surrounding region.

1977: September 7   New canal treaties are signed

Two new canal treaties, negotiated by Panamanian president Omar Torrijos and U.S. president Jimmy Carter (b. 1924), are signed. (The U.S. Senate ratifies the treaties in 1978.) Panama wins full control of the canal on December 31, 1999, with the United States gradually phasing out its operations there. Both countries pledge to keep the canal a neutral zone, open to the peaceful passage of ships by all nations. The Panama Canal Company will be replaced by a Panama Canal Commission, with an administrator to be chosen by the U.S. president.

1978   Torrijos resigns but remains head of the National Guard

Omar Torrijos resigns from the presidency but continues to head Panama's National Guard. Aristides Royo (b. 1940) becomes the president of Panama.

1979   Shah of Iran seeks to live in exile in Panama

The Shah of Iran, overthrown by Islamic revolutionaries under the leadership of Ayatollah Khomeini, lives in exile in Panama.

1980   Two oil strikes are made in Panama

Oil is discovered off the San Blas Islands, and a 50-million-barrel deposit is found east of Panama City.

1980   Boxer Roberto Durán wins welterweight championship

Boxer Roberto Durán (see 1951: June 16) wins the world welterweight title, becoming the first boxer to defeat U.S. boxer Sugar Ray Leonard in a professional match.

1981   President Torrijos is killed in plane crash

President Omar Torrijos is killed in a plane crash in a mountainous region of western Panama. Many believe that General Manuel Noriega, who later seizes control of the government, is responsible for the crash. (See 1983.)

1982   Trans-Panama oil pipeline goes into operation

A fifty-mile (eighty-one-kilometer) pipeline to transport Alaskan crude oil to the eastern part of the United States is placed in operation by Petroterminal de Panama. It has a capacity of 850,000 barrels a day. Next to the Panama Canal, the pipeline is Panama's greatest generator of income.

1983: April   Election reforms are adopted

The system of electing presidents by a vote of the National Assembly is replaced by direct popular vote.

The Noriega Regime

1983: August 12   Manuel Noriega becomes head of the military

General Manuel Noriega becomes commander-in-chief of the National Guard, a position from which he exercises political control of the country. Although not serving as president himself, he controls the presidency, picking and ousting presidents at will. Noriega remains in power through the 1980s. In Panama, he removes all opposition through measures including torture and murder. Internationally, he is involved in a wide range of unethical and illegal activities, including deals with Colombian drug cartels, illegal supply of arms to both the Nicaraguan Contra rebels and guerrilla fighters in El Salvador, and selling intelligence information to several governments. Subsequent presidents all have his backing, and those he disapproves of are removed from office. Noriega continues Torrijos's build-up of the National Guard, which he renames the Panamanian Defense Force (PDF). Eventually, Panama will have one soldier for every eighty civilians.

Born into a family of modest means, Noriega is educated in public schools and at the Peruvian Military Academy, from which he graduates in 1962. He also receives counterintelligence training at the U.S. School of the Americas in Panama. Noriega becomes a National Guard officer upon completing his education, and rises through the ranks, with significant promotions granted by Omar Torrijos (see 1929) in the late 1960s.

Noriega is eventually found to have sold intelligence secrets to several countries including both the United States and Cuba. He is also involved in smuggling drugs for Colombia's Medellín cartel and in supplying arms to both the Sandinistas and the Contras in Nicaragua. After putting up a lengthy resistance, Noriega is finally ousted by the December 1989 U.S. invasion of Panama. He surrenders to U.S. forces on January 3, 1990, and is deported to Miami to stand trial on charges of drug trafficking and racketeering. On July 10, 1992, he is sentenced to forty years in prison by a Florida court.

1984: May 6   Noriega-supported candidate wins presidential election

Nicolas Ardito Barletta (b. 1938) defeats Arnulfo Arias in an allegedly rigged election. The United States supports Ardito in hopes that the U.S.–educated politician can effect a separation between the government and the PDF. However, Ardito is forced out of office the following year (see 1985: September 28).

1985   National Association for the Conservation of Nature founded

A member of a wealthy Panamanian family raises money from businesses and individuals to launch the National Asso-

ciation for the Conservation of Nature (ANCON). Based in Panama City, ANCON has thousands of volunteers and is privately funded, with no government funds in its budget. It aggressively lobbies the Panamanian government on environmental issues, such as for the creation of five national parks.

1985   Government opponent is found decapitated

Dr. Hugo Spadafora, a prominent critic of the Noriega government, is found murdered and decapitated, arousing widespread outrage throughout the country.

1985: September 28   Ardito is forced to resign

Due to the outcry raised by the murder of Dr. Spadafora, President Ardito is forced out of office.

1986   United States pressures Noriega to step down

Revealing that Noriega has sold classified information to both the United States and Cuba and also been involved in drug smuggling, the U.S. government begins pressuring the military strongman to give up his position.

1986   Oil spill

An oil storage tank located at a refinery twelve kilometers northeast of Colón ruptures, spilling almost 100,000 barrels of crude oil. Oil from the spill, carried by winds, is spread along an eighty-five-kilometer stretch of coast.

1987: June   Ousted colonel reveals damaging information about Noriega

After Noriega ousts Army Chief of Staff Colonel Roberto Díaz Herrera, Díaz Herrera publicly attests to Noriega's involvement in election fraud, drug trafficking, and arms smuggling, and his responsibility for the murder of government critic Hugo Spadafora (see 1985).

1987: July   Mass opposition to Noriega is mobilized

Panamanian citizens launch a massive nonviolent protest campaign against Noriega involving large peaceful demonstrations.

1988   University opens a school of fine arts

A school of fine arts is established at the University of Panama in Panama City.

1988: February   Noriega indicted by U.S. court on drug charges

The state of Florida indicts Noriega on drug and racketeering charges for conspiring with Colombia's Medellín drug cartel to import cocaine into the United States.

1988: February 25   President tries to oust Noriega

Panama's president, Eric Arturo Del Valle (b. 1937), attempts to remove Noriega from his position as head of the PDF but is fired by the Congress. Manuel Solis Palma becomes interim president but is controlled by Noriega.

1988: March 4   The United States freezes Panamanian assets

To put pressure on Noriega and bring him down, the United States announces that it will impose economic sanctions on Panama and freeze $56 million in Panamanian assets currently deposited in U.S. banks.

1988: March 16   U.S.-backed coup attempt fails

A U.S.-supported military coup is foiled by the National Guard.

1989: May 7   Endara wins presidential election; results are annulled

Guillermo Endara (b. 1936) wins Panama's promised free elections with seventy percent of the vote, but Noriega orders the election voided on the pretext of interference by international observers. The observers charge election fraud.

1989: May 10   National Guard disperses opposition protest using violence

Noriega's "Dignity Battalions" attack opposition members protesting the election annulment with lead pipes and other objects.

1989: May   United States sends troops and recalls its ambassador

U.S. President George Bush sends 2,000 troop reinforcements to Panama and recalls the U.S. ambassador.

1989: December 15   Panama declares "state of war" against the United States

Panama's Congress proclaims a "state of war with the United States" and appoints Noriega as official head of the government. The following day, an off-duty U.S. lieutenant who has wandered into the wrong area is killed by Panamanian officers.

1989: December 20   United States invades Panama

The United States launches a full-scale invasion of Panama named "Operation Just Cause." Some 12,000 troops land in Panama. Panamanian armed resistance soon dwindles except for Noriega's "Dignity Battalions," and, within three days, peace is achieved. Noriega eludes U.S. forces, who destroy the headquarters of the PDF in their attempt to capture him.

Democracy is Restored

1989: late December   Guillermo Endara becomes president

Following the U.S. invasion of Panama, Guillermo Endara is installed as president based on his victory in the annulled May 1989 elections. He governs with the help of two vice-presidents, the cooperation of U.S. authorities, and a billion dollars in U.S. aid. Nevertheless, scandals surround his administration.

1990   Panamanian stock exchange open

Panama's first stock exchange, the Bolsa de Valores de Panama, enhances the country's position as an international financial center.

1990: January 3   Noriega is taken into custody

Manuel Noriega surrenders to U.S. officials and is transported to Miami, where he is arraigned on drug trafficking charges. He is incarcerated in the Metropolitan Center prison.

1990: March   President Endara goes on a hunger strike

President Guillermo Endara fasts for thirteen days in the Metropolitan Cathedral to protest delays in the arrival of U.S. financial aid promised in the wake of the 1989 invasion. In April, the U.S revokes the economic restrictions, and in May, the U.S. Congress approves $420 million in financial assistance for Panama.

1991: September   Noriega trial opens in Florida

The trial of former dictator Manuel Noriega on drug trafficking charges opens in Florida.

1992: July 10   Noriega is convicted on drug and racketeering charges

Manuel Noriega is convicted on eight counts of drug trafficking and racketeering. He is sentenced to forty years in prison.

1992: Sepbember 1   Ernesto Perez Balladeres becomes president

Ernesto Perez Balldderes (b. 1947) wins the presidential election and assumes the presidency September 1.

1993   Jorge Serrano is exiled from Guatemala

The president of Guatemala, Jorge Serrano (b. 1945) is forced out of his country. He enters Panama to live in exile (see Guatemala).

1994   Raoul Cedras leaves Haiti for exile in Panama

Raoul Cedras (b. 1950) flees Haiti with his wife and children for exile in Panama. His departure is part of a plan to restore democracy to Haiti (see Haiti).

1995   Tensions over Indian rights claims lead to violence

The Ngobe-Bugle Indians charge encroachment on their tribal lands by private landowners. Conflicts over their claims result in violence.

1995   Communal assets recognized in marriages

For the first time, Panamanian law recognizes communal assets in marriages, helping to end the destitution faced by many divorced women.

1995: August 4   Four people are killed at pro-union demonstration

In Panama City, a pro-union demonstration intended to kick off an eight-day general strike leads to a violent confrontation between demonstrators and police in which four people are killed, dozens are hurt, and hundreds are arrested. The strike is a protest against proposed labor legislation,which is intended to attract foreign investment, but would reduce the power of Panama's labor unions. After an initial day of violence, it continues peacefully for eight days.

1996   Government pledges crackdown on drug smuggling activity continues

President Ernesto Perez Balladares says that his government's fight against illegal drug trafficking will continue. New anti-smuggling programs include closer surveillance at the Colón free zone located at the eastern entrance to the Panama Canal. However, because banking regulations in Panama are lax, laundering of drug money is difficult to track and prosecute. The U.S. government grants Panama $400,000 to improve monitoring of financial transactions.

A Panamanian newspaper investigation reveals corruption among customs officials.

1996: December 22   Leftist rebels in Peru release Panamanian ambassador

Panama's ambassador to Peru is one of roughly 400 hostages taken by members of the Tupac Amaru Revolutionary Movement at a party in the Japanese ambassador's residence in Lima, Peru. He is among the 225 hostages released on the sixth day of the siege.

1997   Ecuadoran president Abdala Bucaram seeks refuge

Abdala Bucaram, forced by the Ecuadoran congress to give up the presidency of that country, arrives in Panama where he plans to live in exile.

1997: June   Canal Authority created

A new law creates the autonomous Canal Authority to administer the Panama Canal after the United States relinquishes its control on the waterway on December 31, 1999.

1997: July   Government sends troops to contain Colombians in Darién

Panama's government sends 1,200 armed police to the Darién region to contain violence generated by tensions between rival groups of Colombians. Rebel groups in Colombia have used villages near the Panama border as bases, leading to attacks on innocent villagers by paramilitary forces hunting the rebels. All three groups—rebels, villagers, and paramilitaries—have been entering Panama, in some cases endangering the lives of Panamanians who have accidentally gotten caught up in hostilities.

1999   Presidential elections

Three main candidates run for the presidency, campaigning for elections to take place in May 1999. The leading candidate, Martin Torrijos Espino, is from the large governing party, Democratic Revolutionary Party (PRD). Mireya Moscoso, widow of former president Arnulfo Arias, is running under the Arnulfista Party. The third candidate, Alberto Vallarino, is a former member of the Arnulfista Party who shifted his allegiance to the Christian Democrats. The reversion of the Canal lands and the Canal will be a benchmark by which Panama will be watched internationally over the next few years.

Bibliography

Conniff, Michael. *Black Labor on a White Canal: Panama.* Pittsburgh: University of Pittsburgh Press, 1985.

Flanagan, E. M. *Battle for Panama: Inside Operation Just Cause.* Washington, D.C.: Brassey's, Inc., 1993.

Greene, Graham. *Getting to Know the General.* New York: Simon & Schuster, 1984.

Guevara Mann, Carlos. *Panamanian Militarism: A Historical Interpretation.* Athens: Ohio University Center for International Studies, 1996.

Hedrick, Basil C. and Anne K. *Historical Dictionary of Panama.* New Jersey: The Scarecrow Press, Inc.. 1970.

LaFeber, Walter. *The Panama Canal: The Crisis in Historical Perspective.* New York: Oxford University Press, 1989.

Major, John. *Prize possession: The United States and the Panama Canal, 1903–1979.* New York: Cambridge University Press, 1993.

McCullough, David G. *The Path Between the Seas: The Creation of the Panama Canal, 1870–1914.* New York: Simon & Schuster, 1977.

Meditz, Sandra W., and Dennis M. Hanratty, eds. *Panama: A Country Study, 4th ed.* Washington, D.C.: Library of Congress, 1989.

Moffet, George D. *The Limits of Victory: The Ratification of the Panama Canal Treaties.* Ithaca, N.Y.: Cornell University Press, 1985.

Noriega, Manuel Antonio. *America's Prisoner: The Memoirs of Manuel Noriega.* 1st ed. New York : Random House, 1997.

Pearcy, Thomas L. *We Answer Only to God: Politics and the Military in Panama, 1903–1947.* 1st ed. Albuquerque: University of New Mexico Press, 1998.

Richard, Alfred Charles. *The Panama Canal in American National Consciousness, 1870–1990.* New York: Garland, 1990.

Watson, Bruce W., and Peter G. Tsouras, eds. *Operation Just Cause: The U.S. Intervention in Panama.* Boulder, Colo.: Westview Press, 1991.

The World Bank Group, Annual Report, Washington, D.C., 1997.

Zimbalist, Andrew S. *Panama at the Crossroads: Economic Development and Political Change in the Twentieth Century.* Berkeley, Calif.: University of California Press, 1991.

Paraguay

Introduction

Paraguay, or "land with a great river" as it is translated from the Indian language, Guarani, lies in the heart of South America. With its 157,047 landlocked square miles, Paraguay is comparable to the size of California. The Paraguay River divides the nation into two physically distinct regions. The eastern region is home to two-thirds of the 4.8 million Paraguayans as well as the capital city of Asuncion. The western region, Chaco, named for the different species of animals (*chacu*) inhabiting the area, is hot and barren. The Paraguayans, ninety-five percent of whom are a mixture of Spanish and Guarani Indian, are one of the most ethnically pure nations in South America. Spanish is the official language for politics, business, and education. However, Guarani is the language of choice among family and friends.

The earliest civilization in Paraguay dates back to the Guarani Amerindians. Shortly after the first Indian nations arrived in South America, the Guarani settled into the region surrounding the Paraguay River. The Guarani were known to be fierce warriors, practicing cannibalism. They were also recognized as a very religious tribe, with mystical beliefs and colorful superstitions. Paraguayan culture is largely derived from the traditions of the Guarani. For example, the Guarani art of embroidery or *aho poi*, is believed to be a mixture of Guarani and sixteenth-century European art.

The first recorded history in Paraguay dates back to 1516, with the failed expedition of Juan Diaz de Solis (1470?–1516) to the New World. Solis was commissioned by the King of Spain to find a waterway which would connect the Atlantic and Pacific oceans. Unfortunately, Solis's expedition ended when he was captured and killed by savages in Uruguay. One of Solis's men, Aleixo Garcia, on his return to Spain wrecked his ship off the coast of Buenos Aires. For several years, Garcia studied and eventually mastered the Guarani language. Garcia was intrigued by stories of powerful kings and lavish kingdoms lying west into the Andes. Gathering some 2,000 Guarani warriors, the young adventurer set out to conquer *El Rey Blanco*. On his journey Garcia amassed vast amounts of silver, but was eventually captured and killed by Indian allies of the Incas.

Garcia's stories of wealth made it back to Europe, sending a surge of European explorers into South America. The first permanent Spanish settlement was founded at Asuncion on the Feast of the Assumption in 1537.

The next two centuries brought domination by Jesuit missionaries, who began one of the most amazing social experiments in the New World by protecting the Amerindians from Portuguese slave trading and Spanish colonists. For shortly after Asuncion's founding, rigourous missionary works began. The Jesuits organized Amerindian families in special mission villages (named *reducciones*), which were designed as communes. The Amerindians were taught improved cultivation methods, trades and the fine arts along with religion. Yet most importantly, for awhile these communes protected them from exploitation by colonists and slave traders. However, as the communes became increasingly successful without benefit to the Spanish colonists, a firestorm was ignited against the Jesuits. The King of Spain even interceded, when he was convinced that the Jesuits were attempting to create a sovereign kingdom or republic within the New World. In 1767, he completely expelled the Jesuits from all Spanish posessions in the New World. The communes disappeared and even Asuncion became merely an outpost by 1776.

Independence

In 1811, Paraguay, following the lead of Buenos Aires, declared its independence from Spain and deposed the last royal governor in the country. After rejecting the leadership of Buenos Aires, Dr. Jose Gaspar Rodriguez de Francia (1761?–1840) and General Fulgencio Yegros were elected to serve on a consulate which ruled Paraguay. Francia emerged as the dominant leader in power, and was appointed dictator for life in 1816, the first in a long line of authoritarian leaders.

Francia, or *El Supremo,* ruled with an iron fist. He virtually closed off Paraguay to the outside world by eliminating all commerce, except for heavily regulated trade with Brazil and Argentina. Internally, Francia attempted to create a classless society. He forbade the ethnic elite from marrying within their own clique, stripped the Roman Catholic Church of all property and authority, and prohibited any sort of political

activity. After Francia's death in 1840, Paraguay was a prosperous nation, but politically aimless.

In 1844, Antonio Lopez (1760–1842), a lawyer and landowner was appointed president by Congress. Lopez continued with Francia's authoritarian rule although he attempted to modernize Paraguay by trading with neighboring countries.

Lopez died in 1862 and was succeeded by his son Francisco Solano Lopez (1827–70). Solano Lopez aspired to be the "Napoleon of South America." With the Brazilian invasion of Uruguay in 1864, Lopez looked as if he might live his

dream, and declared war on Brazil and Argentina. Shortly thereafter, a peace treaty was signed between the puppet government of Uruguay, Brazil, and Argentina. The three allies turned their aggressions towards Lopez, igniting the war of the Triple Alliance. The war ended with the death of Lopez at the battle of Cerro Cora. Paraguay had been devastated by violence disease, and poverty. Over half of its population was killed in the war and only 28,000 Paraguayan males were left alive.

After the war a provisional government was established by the occupying forces. Crime and corruption ran high in the nation. Out of the confusion of Paraguay's reconstruction, two new political parties emerged, the Colorados (Reds) and the Liberals (Blue). Ideologically the two parties were not much different. The Colorados, nationalistic and conservative, claimed to be the descendants of Francisco Solano Lopez. The Liberals were labeled the "new wave" of politicians in Paraguay. The Colorados, with more experience in politics, gained control of the government and appointed General Bernardino Caballero (1831–85) president in 1882.

For more than a decade the Colorados controlled power in Paraguay. However, in 1904, Liberals launched an attack against the Colorado government. After enduring civil war for four months, the Colorados finally relinquished control of the government, marking the beginning of Liberal rule that lasted for the next thirty years.

The 1920s brought tension and violent encounters between the Paraguayans and Bolivians which was heightened after the discovery of oil in the Chaco. Coinciding with similar movements in Europe, a new movement of nationalism in Paraguay began to take root. In 1922, political parties clashed as the Colorados and Schaeristas (followers of the former President Eduardo Schaerer) called for the resignation of Liberal President Eusebio Ayala (1875–1942). The two parties finally went to battle in another civil war and the Liberals were defeated by the Schaeristas.

In 1932, tensions over the control of the Chaco broke out into war after Bolivian soldiers stormed a Paraguayan fort. Although Bolivia's troops were more numerous, they were no match for the highly motivated Paraguayans. Just as a Paraguayan victory seemed imminent in the Chaco war, President Ayala agreed to a cease-fire. This unfortunate decision gave the Bolivians time to regroup, and the war continued on for another two years. With this blunder, the Liberals lost what legitimacy they had left among the people of Paraguay.

In 1935 a cease-fire in the Chaco war was signed. However, shortly thereafter, violence erupted once more as the Liberals were forced from their thirty-year-reign in what came to be called the Febrerista revolt. The revolt was supported, and to a certain extent organized, by a set of Paraguayan heroes of the Chaco War. The Febreristas wanted sweeping social changes, including the expropriation of large estates and land allotments for the poor. In 1947 the Febreristas and Colorados waged a bloody civil war that culminated in the Colorados regaining power. Through a series of coups, a hero of the Chaco War, General Higinio Morínigo, emerged as president in 1940. Amazingly, through World War II, Morínigo received massive amounts of aid from the United States, while permitting widespread Axis activity in Paraguay. When Morínigo retired in 1948, after successfully silencing critics of his policies, the political vaccum he left was filled a year later by Federico Chavez, who remained in power until 1954. In that year, with the support of the Colora-

dos and military, General Alfredo Stroessner (b. 1912) rose to power. Stroessner's regime, one of the longest running in Latin America, brought a long era of repression to Paraguay. Stroessner ruled in a state of siege, using the military as his means to control the nation, outlawing any opposition to his regime. It was only under pressure from the international community in the 1960s and 1970s that Stroessner allowed political opposition and authorized the release of political prisoners.

Stroessner was reelected for the last time in 1988. Under the leadership of General Andres Rodriguez, Stroessner was overthrown in a February 1989 coup. Shortly thereafter, Rodriguez was elected president. Rodriguez led the way for liberalization and reform. In 1992, a new democratic constitution, which severely limited the powers of the president, was ratified. In May 1993 Carlos Wasmosy (b. 1938) was elected the first civilian president in forty years. Today, Paraguay continues to progress both economically and politically.

Timeline

Pre-1492   The Guarani Civilization

As the earliest Indians make their way down through North America, into Central America, and finally into South America, one particular tribe, the Tupi-Guarani family settles into the Amazon and Plata basins. It is along the Parana River system that the Guarani-Paraguayan culture develops. The Guarani are fierce warriors, practicing cannibalism. They are also known as mystics, believing in enchanted cities and colorful superstitions. The Guarani family is comprised of several smaller tribes, the Cairos, Tapes, Itatines, and Chiriguanos among them. The Cairos are the most advanced tribe of the Guarani. They work as farmers, and are skilled in weaving cotton and making household products such as utensils and pottery. The Spanish are particularly intrigued by their art. It is the Cairos who establish themselves in the area of Asuncion, which later becomes the capital of Paraguay.

The First Europeans in Paraguay

1515: October   Solis departs for the New World

Commissioned by King Ferdinand of Spain to find a waterway which would connect the Atlantic and the Pacific oceans, Juan Diaz de Solis sets sail for the New World in 1515.

1516: January 1   Solis establishes the bay of Rio de Janeiro

The Bay of Rio de Janeiro is established by Solis.

Mid-1516   Solis killed

Spotted by the Charrua tribe off the coast of Uruguay, Solis is captured and eaten by them.

1524   Aleixo Garcia sets out to raid *"El Rey Blanco"*

On his return to Spain, Alexio Garcia, a Portuguese adventurer, and a member of the Solis expedition, wrecks his ship off the coast of Brazil at Santa Catarina Island. While in Santa Catarina, Garcia, develops a working knowledge of Guarani. Intrigued by the Guarani myth of the immensely wealthy "White King," Garcia gathers 2,000 Guarani warriors and sets out to invade *"El Rey Blanco."* By the time Garcia has made his way to the boundaries of the Inca Empire, he has accumulated vast amounts of silver. Eventually, Garcia is captured and killed by Indian allies of the Incas. The stories of his travels and wealth bring many European explorers across the sea in search of their own fortunes.

1526–30   John Cabot explores the Rio Paraguay

Having heard of the wealth amassed by Garcia, Sebastian Cabot (1476?–1557), the son of the Italian explorer, John Cabot (1450–98) sets out to investigate the Rio Paraguay. Cabot is the first European to explore the estuary, renaming it "Rio de la Plata."

1537: May 15   Spain establishes its first permanent settlement

Spanish explorer, Juan de Salaz founds the first permanent Spanish settlement, Asuncion (*Nuestra Senora de la Asuncion*), on the Feast of the Assumption. The population of Asuncion is comprised of Spaniards, Frenchmen, Germans, the English, and the Portuguese. Because there are no female colonists, the settlers unite with Guarani women.

1539–40   Irala becomes governor

Domingo Martinez de Irala (1487–1557) is appointed governor of the Rio de la Plata province.

1542   Paraguay becomes part of the viceroyalty of Peru

For administrative purposes, Paraguay is incorporated into the viceroyalty of Peru.

1556: April 2   Roman Catholic Church is established in Paraguay

The Roman Catholic Church is established in Paraguay with the arrival of Father Pedro Fernandez de la Torre.

Early 1600s   Jesuits begin missionary efforts

In an attempt to protect the Amerindians from exploitation by the Portuguese slave traders and Spanish colonists, the Jesuit priests group the Guaranis into *reducciones* or self-sufficient communities. In these communities, the Amerindians, in addition to learning new trades and new methods of cultivation, study the fine arts and religion. An estimated 100,000 Amerindians are organized into these settlements.

1720s–30s   Colonists rebel

The Paraguayans protest against the Jesuits' economic success with the *reducciones*. This is the first revolt against Spain in the region.

1767   Spain expels Jesuits from the New World

The Jesuits cooperative experiment proves financially fruitful. This provokes the poor settlers and farmers to protest against the *reducciones*. Fearful that the Jesuits are trying to build a new kingdom, the King of Spain banishes the Jesuits from the region.

1776   Buenos Aires becomes the capital of La Plata

With the founding of Buenos Aires as the new capital of the viceroyalty of La Plata, Asuncion loses some of its vitality.

Independence and Authoritarian Rule

1811: May 11   Paraguay declares independence

Following the example set by Buenos Aires, Paraguay declares complete independence from Spain. However, Paraguay rejects the leadership imposed by Buenos Aires. Following the orders of the Buenos Aires *junta,* Manuel Belgrano (1770–1820) attempts to take Paraguay by force. Belgrano and his army are turned back by a Paraguayan militia.

1811: May 14–15   Last royal governor is deposed

The last Spanish royal governor is ousted from Paraguay under the leadership of captains Pedro Juan Cabalero and Fulgencio Yegros.

1811: October 12   Paraguay declares independence

Buenos Aires sends Nicolas de Herrera to Asuncion in an attempt to pressure Paraguay into a military union. In response, former theologian turned civilian lawyer, Dr. Jose Gaspar Rodriguez de Francia (1761?–1840), calls on Congress to formally declare Paraguay an independent republic. In exchange for de facto recognition of Paraguayan independence, Paraguay enters into a military alliance with Buenos Aires.

1811–12   Argentina cuts off waterways to Paraguayan goods

Paraguay refuses to lend its troops to Buenos Aires for the resolution of the country's internal squabbles. Argentina retaliates by blockading waterways to Paraguayan goods

1813: October   Consulate governs Paraguay

Congress establishes a consulate of two men, Yegros and Francia, to govern Paraguay.

1814   Congress appoints Francia as dictator

Francia emerges as the dominant figure in the consulate, convincing Congress and Paraguay that only one dictator is necessary. Francia is appointed to a five-year term as the "supreme dictator of the republic" or *El Supremo.*"

1816   Francia becomes dictator for life

Francia is named dictator for life by Congress. Internationally, Francia practices an ideology of isolationism, transforming Paraguay into a virtually self-sufficient nation. Francia cuts off all commerce, except for severely limited and regulated trade with Argentina and Brazil. No one is allowed to leave or enter the country without his permission. Undesirable foreigners are often incarcerated and detained for years. Domestically, Francia seeks to establish a "classless society." Free speech is forbidden. The Roman Catholic Church is made powerless, all religious orders are barred and church property is collected by the state. Land is confiscated from the wealthy and redistributed to the poor. Additionally, the ethnic elite are forbidden to marry within their own clique, and forced to marry mestizos, Indians, or mulattos.

1820   "The Good Friday Conspiracy"

Francia discovers a conspiracy to unseat him from power. Hundreds from Paraguay's elite are arrested and thrown into prison. Sixty-eight men, including Francia's former colleague, Fulgencio Yegros are put to death.

1840: September 20   Death of Francia

Francia leaves Paraguay a prosperous nation. However, he also leaves Paraguay with a legacy of authoritarian rule, and no political direction.

1841: March   Congress elects Carlos Antonio Lopez first consul

A Congress of five hundred elects landowner and lawyer, Carlos Antonio Lopez (1790–1862) as first consul.

1844   Lopez becomes president

The 1844 constitution gives extreme powers to the president. Although many of Francia's policies continue under Antonio Lopez (i.e., free speech is still prohibited), he initiates the modernization of Paraguay. He constructs roads and builds hundreds of new elementary schools. Lopez also reestablishes trade relations with neighboring countries.

1852   Argentina opens waterways to Paraguay

After the fall of Argentinean dictator, Juan Manual de Rosas (1793–1877), Argentina recognizes Paraguay's sovereignty. Argentina allows Paraguay to utilize the region's main waterways once again, stimulating commerce, and tripling Paraguay's trade over the following decade.

1853–54   Lopez sends son to meet with European leaders

Francisco Solano Lopez (1827–70) is sent to Europe as a special representative of his father. Solano Lopez attempts to reestablish relations with the continent and open up armament trade. He holds audiences with such leaders as Napoleon III (1808–73) and Victor Emanuel II (1820–78). While in Europe, Lopez commissions engineers to build a national theater and a state palace in Paraguay, as well as South America's first railroad, and new ports in Asuncion.

1862: September   Congress names Francisco Solano Lopez president

Congress appoints the thirty-six year old son of Antonio Lopez as president. Lopez (1827–70), an extreme nationalist, longs to be the "Napoleon of South America."

1864: September   Brazil invades Uruguay

Brazil sends troops into Uruguay. Lopez views Uruguay's sovereignty as imperative to Paraguay's independence.

1864: November 12   Solano Lopez declares war on Brazil

Having ignored Lopez's demands to leave Uruguay as an independent nation, Lopez declares war on Brazil. Lopez orders the capture of the Brazilian war steamer, *Marques de Olinda*. In addition, he deploys a portion of his navy northward to invade Mato Grosso, Brazil. In the south, Lopez mobilizes his army for an attack on Brazilian forces in Uruguay.

1865: March   Solano Lopez declares war on Argentina

Paraguayan troops are prohibited from marching through Argentina to attack Brazilian forces in Uruguay. Lopez declares war on Argentina.

1865: May   Triple Alliance declares war on Paraguay

The puppet government of Uruguay, along with Brazil and Argentina, sign the Treaty of the Triple Alliance. Under the treaty, all three countries are committed to the destruction of Solano Lopez's government.

1866–68   Cholera disables Paraguay's population

Paraguay's troops and civilian population are devastated by the spread of Asiatic Cholera.

1866: May 2   Battle at Tuyuty

Triple Alliance invade Paraguay. Solano Lopez orders suicidal attacks on allied forces at Tuyuty, and loses 20,000 of his best men, including most of the remaining Spanish males in Paraguay.

1869   National Library and Archives

Paraguay founds its National Library and Archives.

1869: January   Allied forces take Asuncion

Solano Lopez retreats northward through Paraguay with allied forces following close behind. Asuncion is captured by allied troops. Solano Lopez begins to act irrationally, ordering the execution of hundreds of his officers, including his two brothers and two of his brothers-in-law. His own mother barely escapes being executed.

Occupation and Reconstruction

1870: March 1   Triple Alliance War ends

The Brazilian cavalry destroys Solano Lopez's last camp at Cerro Cora. Lopez is killed in combat. 450, 000 Paraguayans perish over the course of the war. Only 28,000 Paraguayan males remain alive. The Triple Alliance War is considered the bloodiest war in South America's history.

1870   Provisional government is established

A provisional government is established by the Triple Alliance. A constitution of laissez-faire liberalism, similar to the constitutions of Argentina and the United States, is ratified.

1870   Paraguay's economy plunges into debt

Devastation from the war and interference in Paraguay's internal affairs by occupying forces lead to an ever-increasing economic debt. At this time many Paraguayans emigrate to neighboring Argentina.

1876   Occupation troops leave Paraguay

Six years after its disastrous defeat, Paraguay is free of foreign occupation troops.

1878: November 12   Rutherford B. Hayes decides boundary issue

U.S. President Rutherford B. Hayes (1822–93) awards Paraguay most of Argentina's claims in the Chaco.

1880s   Bolivia begins to move into the Chaco region

After the loss of its seacoast in the War of the Pacific with Chile, Bolivia looks to the Chaco region and the Parana river as a prospective waterway to the ocean. Bolivian soldiers and colonists begin to settle in the region.

1885   Two political parties emerge

Two new political parties emerge from the confusion of reconstruction, the National Republican Association or the Colorados (Reds), and the Liberal Party (Blues). It is difficult to distinguish between the two parties ideologically. The Colorados, conservative and nationalistic, claim to be the descendants of Francisco Solano Lopez. Labeled as reactionaries by the Colorados, the Liberals are representative of a "new" generation of politicians. One's political preference is not based on a party's platform, but instead on family tradition. General Bernardino Caballero (1831–85) a Colorado and war veteran from the time of Solano Lopez, is named president.

1889   National University of Asuncion is established

Paraguay establishes a school of higher eduction, the National University of Asuncion.

1904: December 12   Civil War: Colorados relinquish powers to the Liberals

In response to over a decade of political corruption, economic debt and disunity within the Colorado party, Liberals launch an attack on the Colorado government. Four months of fighting precedes the Pact of Pilcomayo, which forces President Juan Antonio Ezcurra, to relinquish his power to the Liberals.

1904–22   Instability arises within the Liberal Party

Warring within the Liberal Party contributes to the election of fifteen different presidents from 1904 to 1922.

1917   Novelist Augusto Roa Bastos is born

The Paraguayan novelist, Augusto Roa Bastos is born. Bastos authors the biography of Jose Gaspar Rodgriguez de Francia, Paraguay's first dictator, *Yo El Supremo.*

1920s   Paraguay moves into the Chaco region

As Paraguay forces itself into the Chaco region, hostile encounters with the Bolivians increase.

1920s   National Independent League founded

The National Independent League is the result of a stark nationalist movement. Supported by university students, laborers, intellectuals, and military officials, the league calls for a "New Paraguay."

1920: August   Liberal Manuel Gondra is elected president

The liberals take office with the election of Manuel Gondra (1872–1927) to the presidency.

Relations between Paraguay and Bolivia deteriorate during the 1920s and lead to war in 1932. A contingent of Paraguayan troops stands at attention during a drill following the 1928 crisis with Paraguay. (EPD Photos/CSU Archives)

1921: November   Squabbling factions force Gondra from office

Internal squabbling within the Liberal party forces Gondra from the presidency. Eusebio Ayala (1875–1942) is appointed provisional president of Paraguay.

1922   Civil War

In an attempt to oust Ayala from power, former president, Eduardo Schaerer, his followers, the Schaeristas, and the Colorados demand new presidential elections. A law is passed in Congress calling for new elections, however, Ayala vetoes the law. In response, the Schaeristas seek the support of the army. President Ayala, backed by the party youth and loyal Liberal military officers, forms the Constitutionalist "Army."

1922: June 9   Constitutionalists attack the capital

Constitutionalists attack the capital forcing the Schaeristas back.

1922: July 10   Schaeistas defeat the Constitutionalists

In a violent counterattack, the Constitutionalists are defeated by the Schaeristas, bringing the war to a close.

1925   Newspaper is founded

A leading newspaper, *La Tribuna* is founded.

1928: December   Rafael Franco burns down Fortin Vanguardia

Rafael Franco, a major in the Paraguayan army torches Bolivia's Fortin Vanguardia. Fortin Vanguardia is one of

Bolivia's newly constructed forts along the Paraguay river in the Chaco region.

1930s   Social conditions deteriorate

The Great Depression reveals Paraguay's desperate need for social reform, particularly, in education, the work environment, and public services.

1931: October 23   Troops open fire on students

Paraguayan soldiers kill ten people when they open fire on a mob of students attempting to storm the presidential palace. The students are part of the New Paraguay movement, which is based on devout nationalism. After the Liberal government reluctantly agrees to reconstruct Bolivia's Fortin Vanguardia, the nationalists protest the Liberals as a party of traitors.

1932: June 15   Start of the Chaco War

A Paraguayan fort is stormed by Bolivian troops. President Eusebio Ayala orders General Jose Felix Estigarribia (1888–1940) to push Bolivian forces out of the Chaco.

1933: December   Victory at Campo Via

Although Bolivia's troops outnumber Paraguayan forces, Bolivia cannot compete with the Paraguayan drive to defend the homeland. Additionally, the Paraguayans are more familiar with the terrain of the Chaco, giving their forces a strategic advantage. Just as a Paraguayan victory seems imminent at Campo Via, President Ayala agrees to a truce. This action proves detrimental to Ayala and Paraguay; Bolivia reorganizes, forcing the war to last another two years. This poor leadership on the part of the Liberals is the last straw for the opposition movement in Paraguay.

1935: June 12   Cease-fire in Chaco War

Diplomatic intervention by the United States, Argentina, Brazil, Chile, Peru and Uruguay leads to a cease-fire in the Chaco War.

1936: February 17   Febrerista revolt

The war leaves a bad taste in the mouths of the Paraguayans. Poor leadership shown by the Liberals during the war, refusal by the government to pay disabled war veterans a pension while filling the pockets of top military officials, and the expulsion of Colonel Rafael Franco (the leader of the Febrerista Revolutionary Party) from Paraguay for speaking out against Ayala, all lead to the Paraguayan army's invasion of the Presidential Palace. Ayala is forced to step down from his post, ending over thirty years of Liberal party rule. Franco is named president.

1937: August   Government overthrown

Because Franco does not deliver promised social reforms, and in view of his totalitarian ideology, Franco is deposed by the army. The Liberals return to power.

1938: July 21   Peace treaty with Bolivia ratified

Under the treaty, Paraguay is permitted to retain its formerly conquered territory. This area amounts to three-fourths of the Chaco.

1939   General Estigarribia elected president

The Liberals need a hero to gain public support, therefore General Estigarribia (1888–1940), a Chaco war hero, is named president.

1940: August   Estigarribia's constitution ratified

Estigarribia's constitution is ratified by a plebiscite, (a vote, taken by the public by ballot, which supposedly reflects the will of the people) and remains in effect until 1967. Estigarribia balances the budget, lowers the public debt, and develops plans to build highways and public utilities while in office.

1940: September 7   Estigarribia replaced by Higinio Morinigo

The cabinet names War Minister General Higinio Morinigo (1897–1985) president after the death of Estigarribia in a plane crash. Morinigo, somewhat of an authoritarian leader, places all social and economic institutions under state control.

1942   Morinigo bans Liberals

Morinigo deposes all Liberals in the government, and bans the party in Paraguay.

1943   Social insurance legally instituted

With the creation of the Social Security Institute under the Ministry of Public Health, social insurance is legally instituted.

1943   Morinigo nominates himself for president

Morinigo nominates himself for the presidency, and wins the single-candidate election.

1943: August   Morinigo meets with President Roosevelt

Morinigo travels to the United States to meet with President Franklin Delano Roosevelt (1882–1945). He reassures President Roosevelt of his loyalty to the Allies during World War II, despite his openly expressed sympathy for the Axis powers.

1945   Merchant marine established

Paraguy establishes a merchant marine.

1945: February   Morinigo declares war on the Axis

In order to insure Paraguay's membership in the United Nations, Morinigo declares war on the Axis powers.

1945: October 24   Member of United Nations

Paraguay becomes one of the first members of the United Nations.

1946: July   Colorado-Febrerista coalition government

Morinigo founds a Colorado-Febrerista coalition government. However, all of the Paraguayan political parties unite at the end of July to overthrow the dictator, Morinigo.

Early 1947   Civil war ignited

The Colorados and Febrerista parties clash over the number of posts each party holds in the cabinet. Morinigo and the Colorados form a coalition to fight the Febrerista.

1947: January 27   Colorados gain control of the government

The Morinigo-Colorado coalition imprisons the Febrerista ministers. The Colorados regain the power they had lost during the civil war of 1904.

1948   Colorados oust Morinigo

In the 1948 election, Morinigo is again the only candidate for president. Fearful that Morinigo will not step-down peacefully, the Colorados successfully convince him to retire. Over the next six years, Paraguay is governed by six different presidents.

1950s   Colorados split into two factions

The "officialist" Colorados are supporters of the Stroessner dictatorship. The second faction, the People's Colorado Movement (MOPOCO), are supporters of representative democracy.

The Stroessner Era

1954: May 4   Frederico Chaves overthrown

President Chavez is overthrown by the commander-in-chief of the army, General Alfredo Stroessner (b. 1912).

1954: August 15   Stroessner inaugurated president

The key to Stroessner's long regime is repression. Stroessner's main support is the Colorado party and his main instrument is the military, which he uses to control the country. Throughout his eight terms as president, Stroessner consistently institutes a state of siege, as deemed legal by the consti-tution. Critics of the Stroessner regime are imprisoned, tortured, or murdered.

1957   IMF and U.S. government provide aid

To relieve Paraguay of smothering inflation, the International Monetary Fund (IMF) and the United States provide funds for the creation of a free exchange system which is expected to encourage public investment.

1959: April   Stroessner removes state of siege

Under pressure for reform, Stroessner lifts the state of siege, allows opposition exiles to return to Paraguay, frees political prisoners, and institutes freedom of the press. The Liberal party also reemerges in Paraguayan politics, proving once more to be the Colorados' largest opposition. Unfortunately, it does not take long before Stroessner reinstates his siege.

1960s–1970s   Meat-packing industry develops

Meat becomes Paraguay's single most important export.

1961   Women given suffrage

Paraguay becomes the last nation in the Americas to give women the right to vote.

1961   British sell Paraguay railroad system

Great Britain sells the Paraguayan Central Railroad to Paraguay for 560,000 dollars. Paraguay renames the railroad for the dictator, President Carlos Antonio Lopez.

1961   Minimum wage introduced

The minimum wage is introduced by the Ministry of Labor.

1967: August 25   Constitution ratified

To Stroessner's liking, the new constitution of 1967 concentrates power in the executive office.

1968   Paraguayan troops sent to fight in Vietnam

Stroessner sends Paraguayan troops to aid the United States in Vietnam.

1968   First turbine inaugurated

The first turbine engine goes into operation at the intersection of the Acaray and Monday rivers.

1970s   International community is critical of Stroessner

Human rights violations of the Stroessner regime, such as genocide against the Amerindians, are no secret to the world. In the 1970s, the international community, including the Roman Catholic Church, Amnesty International, and the United States, speak out against the Stroessner regime, and

PARAGUAY

Land gained from Gran Chaco War, 1932–35.

demand domestic reform in Paraguay. In 1977, U.S. President Jimmy Carter (b. 1924) cuts off armaments to Stroessner.

1970s–80s Forestland decreases dramatically

Forestry is a lucrative industry for Paraguay. Therefore, the government is hesitant to institute any sort of deforestation policy. From the 1970s to 1980s forest land decreases from forty-five percent of the land to thirty percent.

1973 A pension plan established

Paraguay's social security institute draws up a new pension plan for the state's citizens.

1973 Construction of the Itaipus power plant begins

The joint venture between Brazil and Paraguay begins construction. The Itaipus power plant is to be the world's largest hydroelectric project.

1979 Stroessner releases political prisoners

Under pressure from the United States, Stroessner releases hundreds of political prisoners.

1980s Economic slowdown

The decline in international market prices for Paraguay's agricultural products, and poor weather conditions in Paraguay lead to economic slowdown.

1980s Colorados divide into three factions

The Colorados are weakened, as three distinct factions within the party arise. The "militants" support Stroessner, the "traditionalists" call for the resignation of Stroessner, and the reformist "ethical" faction calls for an end to government corruption.

1980 Somoza assassinated in Asuncion

Paraguay severs relations with Nicaragua after former Nicaraguan dictator, Anastasio Somoza Debayle (1925–80), is assassinated in Asuncion.

1982 Itaipu hydroelectric project finished

Itaipu project is finished, supplying Paraguay with seventy percent of its electricity. The project also stimulates Paraguay's economic expansion.

1982 Liberals split into three factions

The Liberals split into three factions: the Authentic Radical Liberal Party (PLRA), the Liberal Teete Party (PLT), and the Radical Liberal Party (PLR). This factionalism discourages participation in the Liberal Party.

1986 Construction of an international airport begins

Construction on an international airport at Civdaddel Erste is launched.

1988 Stroessner reelected for an eighth time

Stroessner, in office for thirty-four years, wins reelection a record eighth time.

Overthrow and Democratization

1989: February 3   Coup led by General Andres Rodriguez

The military attempts to force the resignation of General Andres Rodriguez as commander of the Paraguayan First Army Corps, in order to free the position for Stroessners's son, Gustavo Stroessner. On Februay 3, Rodriguez retaliates with an attack on the presidential escort battalion. Stroessner is overthrown. As a member of the traditionalist faction of the Colorados, Rodriguez pledges political liberalization and reform.

1989: May 1   Congressional elections held

Rodriguez is chosen as the Colorados' candidate for president. This election is Paraguay's first free municipal election.

1989: May 15   Rodriguez inaugurated president

Rodriguez's presidency is marked by democratic reform and the liberalization of the economy. Such reforms include the easing of restrictions on freedom of speech and tax incentives for investment.

1990s Paraguay's leading cash and export crops

Cotton, tobacco, and sugarcane are Paraguay's leading cash and export crops.

1991 Paraguay maintains the lowest foreign debt in Latin America

In stark contrast to much of Latin America, Paraguay maintains a low foreign debt.

1991 Paraguay joins the GATT

Paraguay applies for membership in the General Agreement on Tariffs and Trade (GATT).

1991: March 26   Paraguay joins MERCOSUR

Paraguay becomes a member of MERCOSUR (The Common Market of the South). MERCOSUR aims at establishing a common market between the countries of Argentina, Brazil, Uruguay and Paraguay by the beginning of 1995.

1992: June 20   New democratic constitution ratified

The new democratic constitution keeps most of the structure of the 1967 constitution. However, the new document calls for a division of powers which severely limits the authority of the president. Also included in the new constitution is protection of basic worker rights, i.e. severance pay and the right to strike.

1993 Five women elected to congress

In a major victory for women's rights, Paraguayan voters elect five women to serve in congress.

1993 First general labor strike in thirty-five years

The protesters are met with some resistance from the government. Further strikes in Paraguay are conducted with little government interference.

1993: May 9 Carlos Wasmosy elected president

Running as a member of the Colorado party, Carlos Wasmosy (b. 1938) is elected the first civilian president in forty years. The election has been closely monitored by international observers, who rule that the elections have been conducted in both fair and free fashion. Wasmosy continues with many of his predecessor General Andres Rodriguez's policies, including economic, political and judicial reform, as well as the transfer of control of the military over to the civilian sector.

1994 Paraguay exports ninety-two percent of its domestic electrical production

The export of domestic electrical production in Paraguay reaches a phenomenal ninety-two percent.

1995 Government bails out banks

The government intervenes in a 1995 banking crisis, which results in a severe decrease in commercial sales.

1996 Eighty-two Internet hosts in Paraguay

In yet another sign of modernization, Paraguay claims eighty-two Internet hosts.

1996 AIDS in Paraguay

Paraguay reports only 206 cases of Acquired Immuno-Deficiency Syndrome (AIDS).

1996 Maria Argana elected president of Colorados

A pro-Stroessner faction leader, Maria Argana is elected president of the Colorados. The Colorados remain severely divided ideologically. The PLRA are still the largest opposition party.

1996: April Oviedo attempts overthrow

Army Chief General Lino Oviedo attempts to remove President Wasmosy from power with a small rebellion. However, without the support of the people, he fails. Oviedo is eventually acquitted of all the charges brought against him.

1997 General Oviedo a contender for president

General Lino Oviedo wins the Colorado Party presidential primary, making him a presidential hopeful for the 1998 election. However, Oviedo, one of the generals who oversaw Stroessner's demise becomes locked in a complex political struggle, leading to his imprisonment.

1998: May 10 Next [residential election

Raul Cubas Grau wins the presidential election. Grau is an ally of Oviedo and vows to have Oviedo share power with him, even if it is from prison. Only three days after taking power, Grau releases Oviedo, which causes a disgruntled outburst from Congressional leaders.

Bibliography

Chronology of Women Worldwide. Supplied by Eastword Publications Development.

Department of State, Bureau of Inter-American Affairs. Office of Public Communication. *Background Notes: Paraguay.* Washington: GPO, 1997.

Department of State, Bureau of Inter-American Affairs. Office of Public Communication. *Post Report: Paraguay.* Washington: GPO, March 1994.

Department of State. *Country Reports on Human Rights Practices for 1997.* Washington: GPO, 1997.

Historic World Leaders. Supplied by Eastword Publications Development.

Lewis, Paul H. *Paraguay Under Stroessner.* Chapel Hill: University of North Carolina Press, 1980

Paraguay. Supplied by Eastword Publications Development.

Roett, Riordan, and Scott Sacks. *Paraguay. The Personalist Legacy.* Boulder: Westview Press, Inc., 1991.

United States Library of Congress. *Paraguay: A Country Study.* Washington: GPO, 1990.

Warren, Harris Gaylord. *Paraguay: An Informal History.* Norman: University of Oklahoma Press, 1949.

Wiarda, Howard J. (ed.), and Harvey F. Kline (ed.). *Latin American Politics and Development.* Boulder: Westview Press, Inc., 1996.

Peru

Introduction

Peru is South America's third-largest country in size, after Brazil and Argentina, and its fourth-largest in population. It has a diverse landscape ranging from the rugged slopes of the Andes to the humid jungles of the Amazon. Its three primary geographical areas are the Andean highland (or Sierra), the eastern foothills and surrounding plains, and the coast and coastal plains.

The people of Peru, like its landscape, are diverse, with sharp social and political divisions. A small wealthy elite in Lima, descended from Peru's European conquerors, have historically dominated the native population living in the interior and the mixed-race, or *mestizo,* population, many of whom live in Peru's other coastal cities. In 1996 Peru's population was over 24.5 million people, over one-fifth of whom lived in the capital city of Lima.

Pre-Incan and Incan Civilizations

Peru's first inhabitants were nomadic tribes who crossed over the Bering Strait from Asia thousands of years ago and continued moving south from North America. Settled, agricultural communities emerged after about 5000 B.C., including the pre-Incan Chavín, Mochica, Nazca, Paraca, and Chimú groups that flourished between 1000 B.C. and A.D. 1000. Hundreds of years ago Peru was the center of the vast, powerful Inca empire, the most advanced native civilization in the Americas. The great expansion, which eventually covered a third of South America, occurred between the thirteenth and fifteenth centuries. The Incas united this vast territory with an extensive network of paved roads radiating outward from their central city of Cuzco. They fed their populace using advanced agricultural techniques, including terracing, crop rotation, and irrigation. They developed a calendar and a decimal number system, but did not have money, the wheel, or written records.

Spanish Conquest

Already weakened by civil war, the Incas easily fell prey to the Spanish conquistadors who arrived on their shores in 1532 under the leadership of Francisco Pizarro. By the following year, their leader, Atahualpa, had been killed, and their capital, Cuzco, overrun by the Spanish. The native population was subjected to a system of forced labor, toiling in the gold and silver mines of their native land to enrich the European conquerors. The city of Lima, founded in 1835, became the capital of the Viceroyalty of Peru, the seat of Spanish power in South America. Peru's original conquerors became embroiled in factional fighting for years, and Pizarro himself was assassinated by rival forces in 1541.

The Viceroyalty of Peru was unified and stabilized under the rule of its fifth viceroy, Francisco de Toledo, between 1569 and 1581. Under Toledo's rule, the remaining Inca stronghold was captured in 1571, and the last Inca leader, Tupac Amaru, was beheaded. The colonial system established by Toledo remained largely the same for over 200 years. In that time, the only major revolt by the native population occurred in 1780 under the leadership of a *mestizo* (of mixed Eurpean and Amerindian ancestry) who adopted the name Tupac Amaru II. The uprising was put down within two years and its leader killed.

Peruvian Independence

For Peru, as for its South American neighbors, independence from Spain came early in the nineteenth century. The Peruvian campaign for independence was actually led by liberators generally associated with neighboring countries: José de San Martín of Argentina and Símon Bolívar of Venezuela. (Bolívar also served as Peru's first president.) Peru proclaimed its independence on July 28, 1821, but the Spanish were not driven out until 1824. The early years of Peruvian independence were marked by political turmoil. Individual governments were short-lived, with numerous revolts. Stability came with the administration of Ramón Castilla (1844–50). Castilla introduced social reforms, such as ending the forced labor of the Indian population, and also brought economic progress to his country. During his presidency, Peru

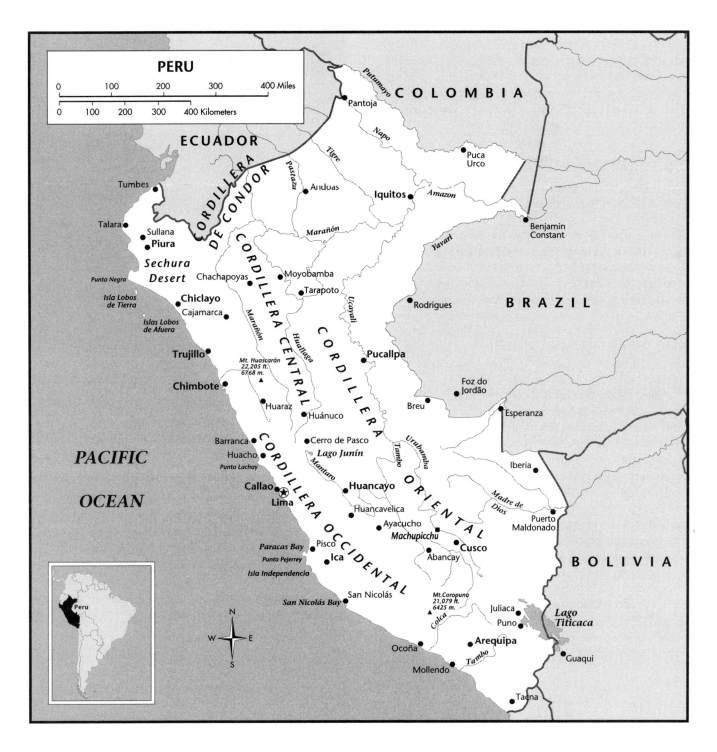

began mining and exporting its large reserves of nitrate-rich bird droppings called guano, an important source of fertilizer that could be exported to nations overseas.

War of the Pacific and Leguiá's Presidency

The latter part of the nineteenth century was marked by international tensions. In the 1860s Peru was joined by Ecua-dor and Chile in driving the Spanish from the Chincha Islands, the site of the nation's guano reserves. Between 1879 and 1884, Peru became involved in the War of the Pacific over Chile's seizure of Bolivian nitrate reserves. At home, a fourteen-year period of military rule restored the political stability that had been lost after the presidential tenure of reformist president Ramón Castilla ended in 1862.

The outstanding figure in the political landscape of early-twentieth-century Peru was Augusto Leguiá y Salcedo, who served intermittently as president between 1908 and 1930, including eleven consecutive years from 1919 to 1930. Leguiá's style of governing was authoritarian, but he instituted important social reforms and encouraged United States investment in Peru. The twentieth century's most significant opposition party, a left-wing group known as APRA, was founded in 1924 by Victor Raúl de la Torre.

Civilian and Military Governments

Peru was mostly under dictatorial rule from the departure of Leguiá in 1930 to the election of José Luis Bustamante, backed by a liberal coalition, in 1945. Bustamante was overthrown by the military in 1948. In the years since then, Peru has alternated between civilian and military governments. This political instability has been fueled by serious economic problems that have plagued the country in recent decades, most notably a massive foreign debt and astronomical levels of inflation. However, migration from the country's interior to the coast since 1950 has begun to weaken the country's traditional division between the wealthy elites in Lima and the impoverished Indian population of the Andes, as the middle class has grown and gained power.

Since the 1980s Peru has also been rocked by the terrorist acts of the Maoist Shining Path guerrilla group. The 1995 reelection of Alberto Fujimori to the presidency was partly motivated by his perceived success in reducing terrorism, even though it had entailed a suspension of constitutional government since 1992 (the "autogolpe"). Fujimori's reputation for fighting terrorism was further enhanced by the final outcome of the four-month-long siege in which members of the MRTA terrorist group held hostages in the residence of the Japanese ambassador. In April 1997, with 73 of the initial 600 hostages still being held, government commandos stormed the residence and freed those imprisoned inside. One hostage died of a heart attack after receiving a bullet wound, and all fourteen terrorists were killed. In 1996, over the protests of the opposition party, Peru's legislature amended the constitution to allow Fujimori to run for a third term as president in the year 2000.

Timeline

Pre-conquest Societies: Before the Incas

900–350 B.C.   The Chavín culture flourishes

The Chavín are the first group to control both the Andean highlands and the coastal areas of present-day Peru. They grow maize (corn) over large cultivated areas, a notable achievement because it assures a continued food supply and

The Nazca attach mummy masks, such as this, to nobles for religious purposes. (EPD Photos/CSU Archives)

the accumulation of a surplus. The Chavín are also known for their stone masonry and metalwork.

A.D. 100–1000   Rise of the Moche and other groups along the coast

The Moche live in the Moche Valley and surrounding areas near the northern part of Peru's coastline, where they flourish between A.D. 100 and 700. Eventually they have some 100,000 acres (40,000 hectares) of land under cultivation. They develop an elaborate irrigation system, building their towns on terraced hillsides and using guano (bird droppings) to fertilize their crops. The Moche are known for their advanced metalwork and ceramic pottery and their great Pyramid of the Sun, south of Trujillo, composed of 140 million mud bricks and nearly as extensive as the Great Pyramid in Egypt. The Moche practice a complex system of ritual burial. It is thought that drought and earthquakes bring about the end of the Moche civilization.

Other coastal groups that live along the Peruvian coast in the first thousand years after Christ include the Nazca, Paracas, and Chimú. The Paracas are known for their beautiful textiles, and the Chimú for the ruins of their capital at Chan Chan.

A.D. 100–600   Nazca create giant drawings in the Peruvian desert

The "Nazca Lines" are huge line drawings created in the soil of the southern Peruvian desert over a thousand years ago by the Nazca people. The representations, which are so large they are only visible from the air, are of abstract geometric shapes and animals, including a 300-foot (90-meter) monkey and a 600-foot (180-meter) lizard. The Nazca create the lines by removing darker topsoil so that the lighter soil underneath showed. The Nazca Lines are widely thought to have been drawn for religious purposes, though some still contend that they are the work of extraterrestrials. They have been studied intensively by one woman, Maria Reiche, since 1946. A German by birth, Reiche fled Nazi Germany to find a home in Peru. Though she had a degree in mathematics, she found interest in an archeological dig near where she was staying. It was at this dig that she overheard about the Nazca Lines. Drawn to them, she studied them for years without notoriety, until she published her book, *Secret of the Pampa*. Today the giant drawings are one of Peru's main tourist attractions and have been declared a "world cultural heritage site" by the United Nations Educational, Scientific and Cultural Organization (UNESCO).

A.D. 500–1000   The Tihuanaco and the Huari

Both these pre-Inca groups arise near Lake Titicaca, in present-day Bolivia. The Tihuanaco conquer other groups and at one time dominate southern Peru. The large stone structures of their central settlement south of Lake Titicaca include the famous still standing Sun Gate that has an image of their sun god engraved on it.

Preconquest Societies: the Inca Empire

1200s   The Incas begin to expand their territory

The Incas are originally a small Quechua-speaking tribe centered in Cuzco, in the southern Andean highlands. They begin slowly expanding their territory in the thirteenth century A.D., assimilating the peoples they conquer. Expansion continues at a slow pace until the fifteenth century.

1438–71   Reign of Pachacuti

The Incas' expansion increases rapidly under the rule of their ninth emperor, Pachacuti Inca Yupanqui. After first conquering the neighboring Chancas, he continues expanding the Incas' territory, both to the north and the south. He also rebuilds the capital city of Cuzco. The Incas call their empire Tuhuantinsuyo ("land of four sections") because it is divided into four provinces. Agriculture—using crop rotation, terracing, and a sophisticated irrigation system—is the basis of its economy. People work the land collectively, living in small communities called *ayllus*. The Incas are an advanced civilization without written records, money, or the wheel. Records are kept using knotted strings called *quipus*.

1471–93   Reign of Pachacuti's son, Topa Inca Yupanqui

The expansion of the Inca empire continues under Pachacuti's son. Eventually it stretches from present-day southern Colombia to central Chile, with an area of some 380,000 square miles (1 million square kilometers) and a coastline of 2,500 miles (4,000 kilometers). The Inca lands are linked by an extensive system of roads, with one major thoroughfare leading from Cuzco to each of the empire's four provinces.

Late 1400s–1532   The Inca empire begins to decline

By the late fifteenth century, overexpansion has begun to weaken the Incan empire. The capital city has been moved from Cuzco to the newly built city of Quito (in what is today known as Ecuador). The emperor at the time, Huayna Capac (r. 1493–1524), attempts to divide the empire between his two sons, Atahualpa (c. 1500–33), who lives with his father in Quito, and Huáscar (c. 1495–1533), whose home is in Cuzco.

1528   Civil war ravages the Incan empire

After Huayna Capac's sudden death in 1524, civil war breaks out between his sons, Atahualpa and Huáscar, over the division of the Inca empire. Atahualpa emerges from the five-year conflict victorious, but the war weakens the Incas just as the Spanish are arriving in Peru.

The Spanish Conquest

1532: November 16   Pizarro's forces ambush Atahualpa at Cajamarca

Spanish explorer Francisco Pizarro (c. 1478–1541) arrives on the northern coast of Peru with a band of around 180 men and twenty-seven horses late in 1532. He arranges for Atahualpa to meet him at the city of Cajamarca, which has been virtually abandoned during the recent Incan civil war. Atahualpa arrives with over five thousand troops, but his forces are ambushed by the Spanish, who are hiding in vacant buildings.

As part of the ambush plan, a priest, Vicente de Valverde, approaches Atahualpa, who is riding on a litter carried by eighty men, formally asks him to submit to the authority of the pope, and hands him a prayer book. The Inca, having never seen a book, throws it on the ground, and Pizarro's soldiers use this "outrage" as a pretext to charge on the unsuspecting Incas, cornering them in the town square. The Incas, armed only with axes and slings, are terrified and outmatched by the Spaniards' firearms, horses, and steel swords and armor. Thousands are killed, and Atahualpa is captured.

1532–33   The Incas attempt to ransom Atahualpa

Atahualpa offers to pay the Spanish a lavish ransom consisting of gold and silver—reportedly a roomful of each—collected from the four corners of the Incan empire, and they accept, holding him prisoner for months as his subjects gather the ransom from their vast empire. Several powerful Inca armies—any of which could destroy the small Spanish detachment—are still encamped in various regions. The Spanish fool Chalcuchima, one of the Inca generals, into visiting Atahualpa in captivity and murder him, thus reducing a portion of the threat that hangs over them. In April 1533 the

The ruins of the Inca city of Macchu Picchu are located in central Peru. (EPD Photos/CSU Archives)

Spaniard Diego de Almagro (c. 1475–1538) arrives from Panama with 150 more men.

1533: August   Atahualpa is executed

The treasure is delivered, but the Spaniards still bring Atahualpa to trial and execute him for fear that his subjects may stage a rebellion. They offer him a more humane death if he first converts to Christianity, and he accepts. Many Spaniards condemn the execution.

1533: November 15   Cuzco falls to Pizarro's forces

Once Atahualpa is slain, the Spanish move to complete their conquest of the Incas, taking advantages of the political divisions engendered by the Incas' civil war. Atahualpa's rival, Huáscar, aids the Spanish in hopes of regaining his throne in return. Pizarro appoints Manco Capac II (c. 1500–44), a relative of Huáscar, to rule as a puppet emperor under Spanish control.

On November 15 Pizarro leads his forces into Cuzco, the birthplace of the Incan empire, and they loot its treasures.

1534   Quito is conquered by the Spanish

Having conquered Cuzco and the southern part of the Incan territories, the Spanish still need to overcome Incan resistance in the provinces to the north, surrounding Quito. This area falls to Spanish forces under the command of Sebastián Benalcázar (c. 1495–1550) and Diego de Almagro in 1534. Pedro de Alvarado (c. 1495–1541), a conquistador from Mexico, contends for control of this region, but he is paid off with a share of the gold gathered to ransom Atahualpa.

trative districts called *audiencias*, and is divided into provinces.

1537–46 Civil war erupts between the followers of Pizarro and Almagro

In Cuzco Almagro and his followers challenge the Spanish forces under the command of Francisco Pizarro's brothers and then try to wrest Lima itself from Pizarro, launching a lengthy civil war between Pizarro supporters and those who support Almagro ("Almagristas").

1537 First Roman Catholic diocese in Peru founded

The diocese of Cuzco is founded.

1538 First bullfight held in Peru

Pizarro brings the first bulls to the New World, and with them the popular Spanish sport of bullfighting. Lima's first permanent bullring, built in 1768, is the world's third-oldest ring. Today bullfighting is among Peru's most popular sports.

1538: April 26 Almagro is defeated at battle of Las Salinas

Hernando Pizarro (c. 1475–1578) and forces under his command defeat Almagro and take him prisoner. Almagro is tried and killed by strangling, further exacerbating the rivalry between the two factions. Three years later, on June 26, 1641, the Almagristas take their revenge: twenty men break into Francisco's palace in Lima and knife him to death.

c. 1539 Author Garcilaso de la Vega born

Born to a Spanish soldier and an Incan princess, Garcilaso de la Vega (also called "the Inca") is the first great writer of the Americas. His *Royal Commentaries of Peru* records in detail the traditions of the Incas and the history of their homeland. He dies in 1616.

1542 Spain issues new laws governing policy in Peru

In order to centralize control of Peru, Spain issues the "New Laws for the Indies" that curb the *encomienda* system, by which Spanish colonists are induced to settle in newly captured territories by awards of land and the right to tribute (forced payment to an overlod) and free labor from the native population. This system makes the local leader landowners, or *encomenderos,* too independent to be readily controlled by the Spanish government. Instead, Spain divides the land into administrative districts called *corregiemientos* governed by Spanish-appointed officials.

1542: September 16 Almagristas defeated at the Battle of Chupas

The factional rivalry between the Pizarro and Almagro supporters (both of whose leaders are dead) ends only through

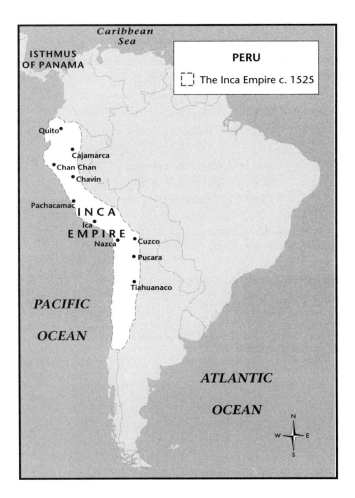

Spanish Colonial Rule

1535 Indian rebellion is led by Manco Capac

Manco Capac realizes that he will never return Peru to Inca rule and attacks the Spanish forces in Cuzco. A similar revolt is staged at Lima. Aid arrives from other Spanish possessions in the Western Hemisphere, and the revolts are crushed. In 1537 Manco Capac's men are decisively defeated at Cuzco by the forces of Diego de Almagro, newly returned from a grueling and fruitless two-year expedition to Chile.

1535: January 6 Pizarro founds city of Lima is founded by Francisco

Pizarro founds Lima in order to have a more centrally located capital city than Cuzco. Its official name is "City of the Kings," but it soon becomes known as Lima, a corruption of the name of the Rimac River, near whose mouth it is located. In 1542 Lima becomes the seat of the Viceroyalty of Peru, the center of Spanish power and wealth in South America and the administrative center for all of Spain's South American territories except Venezuela. The viceroyalty consists of adminis-

Spanish intervention. In 1542 a Spanish official, Cristóbal Vaca de Castro (d. 1558), leads an army of Pizarro's forces against the Almagristas, whom he defeats at the battle of Chupas on September 16, 1542.

1544: May   Spain appoints first viceroy

Blasco Nuñez Vela is appointed by Spain's king, Charles I, as the first viceroy of Peru. He enforces the new administrative system and sets up a high court. The *encomenderos,* rebel and forcibly oust Nuñez Vela from Lima in October 1544. He is killed two years later.

1544–48   Pizarro's brother leads revolt against New Laws

Francisco Pizarro's brother, Gonzalo (c. 1506–48), leads a rebellion against Spain's "New Laws for the Indies." In the course of this revolt, Blasco Nuñez Vela, Spain's first viceroy of Peru, is killed, and Pizarro attempts to have himself proclaimed king. On April 9, 1548, the rebellion is put down by Spanish forces under the command of Pedro de la Gasca (1485–1567), and Gonzalo Pizarro is executed for treason, ending the bloody reign of the Pizarro brothers in Peru. However, the New Laws are not enforced, and the exploitative *encomienda* system continues.

1546   Archdiocese is established at Lima

A Roman Catholic archdiocese, with its center at Lima, is placed in charge of nearly all church activities of the Spanish territories in the Andes Mountains.

1548   Italian painter Bernardo Bitti emigrates to Peru

Bernardo Bitti is one of the first European painters to settle in Peru. European artists introduce oil painting on canvas and fresco painting to the newly colonized territory. Church paintings and other decorations are often imported directly from Spain.

1551   University of San Marcos is founded by Dominican missionaries

Establishment of the National Autonomous University of San Marcos (Universidad Nacional Autónoma de San Marcos)—the oldest institution of higher learning in the New World—is decreed. The university does not open its doors, however, until two decades later.

1553   Birth of composer Gutierre Fernández Hidalgo

Hidalgo (1553–1620), a composer of sacred music and resident of Lima, is the foremost sixteenth-century composer in Latin America.

1560   Quechua grammar book is written

Domingo de Santo Tomas, a Dominican friar, produces a grammar and vocabulary text of the native Quechua language of the Incas.

1569–81   Spanish consolide rule under Toledo

Francisco de Toledo y Figueroa (c. 1515–84), Spain's fifth viceroy, sets the course of Spanish policy in Peru. Many of his laws involve controlling the remaining Indian population. He consolidates their settlements, taxes them, and inaugurates a system of forced labor called the *mita.* Toledo also places the Huancavelica silver mines under government control. The system created by Toledo's policies largely remains in place until the early nineteenth century.

1572   Túpac Amaru, last Inca leader, is killed

The last Inca stronghold at Vilcabamba is overrun by the Spanish, at the direction of Viceroy Toledo. The Inca leader, Túpac Amaru (c. 1544–72), is captured and beheaded in Cuzco.

1579   Birth of St. Martín de Porres, first black saint

One of two Peruvian saints canonized during the Spanish colonial period, St. Martín de Porres is the first person of African descent to be canonized. He is born in Lima to a Spanish nobleman and an African woman and becomes a Dominican monk in 1610. He founds a home for youth in Lima, which still stands today and is regarded as his monument. He is celebrated for his kindness to all people, especially the unfortunate. His feast day is November 5. The civil rights movement in the United States regards him as the patron saint of justice and harmony for all races.

1584   First printing press is brought to Peru

The arrival of the first printing press stimulates the colony's literary output. Much writing is religious in nature. Also common is autobiographical writing by Spanish priests, government officials, and soldiers. Works of fiction are strictly censored.

1586   Saint Rose of Lima is born

Saint Rose of Lima, the first person born in the Americas to be canonized, lives from 1586 to 1617. She is a nun born in Lima who lives a secluded life devoted to God. She is considered the patron saint of the Americas and the Philippines. Her feast day is celebrated on August 30.

1599   Jivaro Indians stage a revolt

The Jivaro Indians of northern Peru carry out a major revolt against the Spanish, attacking and setting fire to cities. They are thought to have killed between twenty and thirty thousand civilians. Among these is the Spanish governor of the region, whom they torture by forcing him to swallow melted gold.

1615   Waman Puma completes account of Peruvian history

Felipe Waman Puma de Ayala, a Peruvian Indian, completes *Nueva Cronica y Buen Gobierno*, a detailed history of Peru. Waman Puma labors from roughly 1576 to 1615 to produce the massive work, which includes details about all facets of Peruvian life under both the Incas and Spanish colonial rule.

1650: March 31   Cuzco is struck by major earthquake

This event gives birth to one of Cuzco's oldest festivals. After the earthquake occurs, a small statue of Christ, known as *Nuestro Señor de los Temblores* (Our Lord of the Earth-quakes), is paraded through the streets of Cuzco. This action, believed to have saved the city from further damage, inaugurates a tradition that is still observed every year on Monday during Holy Week. The statue, which local people call *Taitacha* (little father), is carried on a three-hour parade of the city streets.

1687   Earthquake decimates central Peru

A major earthquake causes large-scale damage in Lima and Callao, destroying much of the irrigation system in the region. Lima's economy sustains long-term harm because the region is no longer self-sufficient agriculturally and has to import food from other areas, most notably wheat from Chile.

1700s   Bourbon monarchy attacks colonial corruption

The Bourbons, who have replaced the Hapsburgs in Spain, stir up unrest by trying to curb corruption in the colonial regime. At the time, public offices are purchased from the crown, and embezzlement is common. Accountability by the colonial treasury is minimal, and commercial regulations often go unenforced. However, many of the native-born Peruvians, or *criollos,* are pleased with the ineptitude of the Spanish colonial regime because it allows them greater freedom.

1701   First opera performed in the Americas is staged in Peru

La púrpura de la rosa by Peruvian music director Tomás de Torrejón y Velasco (1644–1728) is performed in Lima.

1717–20   Epidemics kill many in native population

By 1570 the pre-conquest Indian population, thought to number as many as 9 million people, has been reduced to about 1.3 million by warfare, disease, and deprivation. By 1620 it has been slashed to 600,000, and falls to its lowest level during an onslaught of epidemics a hundred years later.

1718   New colonies reduce size of Viceroyalty of Peru

The Viceroyalty of New Granada is created, partially on territory formerly belonging to the Viceroyalty of Peru. The creation of the Viceroyalty of the Río de la Plata in 1776 once again reduces the extent of the lands belonging to Peru.

1777–88   Botanical expedition records native species

One of the first botanical expeditions in South America is organized under a special Spanish scientific commission, led by Spanish scientists Hipólito Ruiz and José Antonio Pavón, studies the quinine-producing cinchona (tropical South America).

1780–82   Indian uprising is led by Túpac Amaru II

José Gabriel Condorcanqui (1742–81), a *mestizo* who claims descent from the last Inca ruler Túpac Amaru II (c. 1544–71), adopts his name and leads the most extensive Indian revolt in two centuries to protest the Spanish exploitation of the native Indian population. He executes a Spanish official and recruits an army of thousands of Indians. Condorcanqui is captured in 1781. Although he and many of his followers are tortured and killed, the uprising continues until the following year, led by his brother Diego Cristóbal Túpac Amaru. Eventually, the Spanish actually implement some of the reforms sought by the rebels.

1782   Native Quechua language is outlawed following uprising

Quechua, the native tongue of the Incas, is outlawed by the Peruvian government following the Túpac Amaru uprising. Nearly two hundred years later it is declared an official language of Peru, along with Spanish.

1791   Early Peruvian newspaper is established

El Mercurio Peruano is one of the first newspapers published in Latin America.

Liberation from Spain

1808   Napoleon Bonaparte (1769–1821), French Emperor, invades Spain

Bonaparte's overthrow of the Spanish king, Ferdinand VII (1784–1833), leads to revolutionary turmoil in Spain's South American colonies. As the colonial territory with the largest and strongest Spanish army, Peru is the most conservative and the slowest to revolt. Rebellions by the native population and by *criollos* (Peruvian-born Spanish) between 1808 and 1815 are crushed by forces loyal to Spain. Ultimately, it takes leaders from other colonies to liberate Peru.

1810   Private art schools are founded in Lima

Javier Cortés opens private art schools in Peru's colonial capital.

1820: September   San Martín lands in Peru

Argentinian general, José de San Martín (1788–1850), fights for Argentine liberation in 1814 and liberates Chile in 1817. In 1820 he lands in Pisco, on Peru's southern coast, with about five thousand men. Victory comes rapidly in coastal cities, but the capital city of Lima is not liberated until the following year.

1821: July 28   Peruvian independence is proclaimed

San Martín proclaims Peru an independent republic and is named as its protector. He proposes a constitution that would guarantee civil rights to the native population and abolish slavery in Peru. He also wants to recognize the native Peruvian language of Quechua as an official language. Some Peruvians however, view San Martín's activities as foreign interference and refuse to cooperate with him. San Martín then seeks help from another leading South American liberator, Simón Bolívar Palacios (1783–1830).

1822   San Martín and Bolívar meet in Guayaquil

The two liberators meet to discuss working together to complete the task of freeing Peru from Spanish rule. They are unable to agree on a joint course of action, and San Martín leaves for Chile in 1822, turning over leadership of the Peruvian independence effort to Bolívar.

1824: December 9   Spanish forces are defeated at Ayacucho

Led by Antonio José de Sucre Alcalá (1795–1830), Bolívar's forces decisively defeat Spanish troops at the Battle of Ayacucho near Huamanga in the southern highlands. This defeat signals the end of Spanish colonial rule not only in Peru but in all of South America.

The Republic of Peru

1824–26   Bolívar serves as president

Liberator Simón Bolívar is Peru's first president. Peru is ill prepared for the transition from colonial to independent rule. The new country is beset by factional rivalries, especially between the military and the aristocracy. The social inequalities created during the colonial era remain, but power is transferred from aristocrats appointed by the Spanish monarch to native-born Peruvians of Spanish ancestry. Over the next two decades, the new nation undergoes repeated armed revolt. There are a number of military governments and six different constitutions. Altogether there are some twenty-four different

governments, or about one per year, during this chaotic period.

1833   Birth of author Ricardo Palma

Palma (1833–1919), noted literary critic and historian, begins writing at a young age. In 1860 he participates in efforts to overthrow President Ramón Castilla (c. 1797–1867) and is exiled to Chile for two years (1861–63). He travels to Europe in 1864. In 1865 he returns to Peru, filling a number of government positions, including that of senator (he is elected to three terms). He also writes for the Buenos Aires (Argentina) newspaper *La Prensa* and works as a librarian. Palma enlists in the reserves at the beginning of the War of the Pacific (1879–84). In 1883 he becomes head of Peru's National Library and works to restore its collection, following damage inflicted by the Chilean army during the war.

Palma is known for introducing the romantic movement to Peru and for combining literature and history in his ten-volume short story collection, *Tradiciones Peruanas* (Peruvian Traditions; 1893–1896). Many of the stories are about both colonial Peru and Inca history and culture. For furthering interest in Peruvian history and culture, both at home and abroad, Palma is regarded by many as Peru's greatest literary figure.

1839   *El Comercio*, nation's oldest newspaper, is founded

El Comercio, the oldest newspaper still in circulation in Peru, is established.

1844   Castilla assumes the presidency

The election of Ramón Castilla (c. 1797–1867) to the presidency brings stability and progress to Peru. Castilla serves as president from 1844–50 and again from 1855–62. He implements numerous social reforms. He inaugurates public education, abolishes slavery, ends the *mita* system of forced labor by the native Indian population, and encourages immigration. Under his leadership Peru begins exploiting its reserves of nitrates (natural salts/chemicals used as fertilizers and in manufacturing) and guano (bird droppings used as fertilizer).

1845   Catholicism becomes Peru's official religion

Roman Catholicism is made the official state religion of Peru. Although members of other religions are permitted to conduct services, Peruvians are legally forbidden to attend them. Religious freedom is subsequently granted by the 1920 constitution.

1849   Nitrate content of guano discovered

Scientists verify the rich nitrate content of guano, the thick beds of droppings that have been deposited for centuries by birds feeding on schools of anchovies off of Peru's desert coastlands. Guano has been used as fertilizer by native peoples of Peru even before the Incas. In the second half of the nineteenth century, the export of guano to Europe becomes Peru's most important source of income.

1854   Easter Island raided for slave labor

Traders raid Easter Island off the coast of Chile and kidnap its inhabitants to mine Peru's guano deposits.

1860s–70s   Chinese immigrants enter Peru

Chinese immigrants are brought to Peru to mine the guano deposits on the Chincha Islands. They also work on the railroads and on cotton plantations. This is Peru's greatest wave of foreign immigration.

1864–66   War with Spain over guano deposits

Spain seize Peru's uninhabited Chincha Islands, the source of the nation's wealth from exporting guano. The Spanish claim they are entitled to the islands as payment of outstanding debts owed by Peru. Ecuador, Bolivia, and Chile join forces with Peru, and the Spanish are expelled by 1866.

1868: August 13   Earthquake on Peru/Ecuador border

On the afternoon of August 13, a gigantic earthquake on the border between Peru and Ecuador kills a total of twenty-five thousand people from both countries.

1868–74   Andean railways are built

Profits from the guano industry fund the construction of a network of railways in the Andes, under the supervision of an engineer from the United States, Henry Meiggs.

1877   Italian writes opera about Inca chief

Carlo Enrico Pasta (1855–98) writes *Atahualpa*, an opera about the Inca king defeated by the Spanish conquistadors. The work inspires an interest in opera among native Peruvians.

1879–84   War of the Pacific

In 1879 Chile seizes Bolivia's nitrate interests in the Atacama Desert, and Peru is obligated to join Bolivia in fighting Chile because of a mutual defense agreement it has previously made with Bolivia. Lima itself is occupied by Chilean forces from 1881 until the war's end. Chile emerges victorious in the Treaty of Ancón that resolves the dispute. Peru loses its nitrate-rich province of Tarapacá, and Chile wins temporary control of Tacna and the port of Arica. After prolonged disputes over these provinces, Chile returns the province of Tacna to Peru under a 1929 agreement mediated by the United States.

1884   First literary guide for women is published

The first literary guide for Peruvian women is published by Clorinda Matto de Turner (1854–1909), a feminist and advo-

cate for Indian rights. Turner is also a novelist and founder of *El Recreo*, a journal for women.

1886–95   Military rule restores political and economic order to Peru

Peruvian politics is plagued by instability and corruption after the tenure of President Ramón Castilla ends in 1862. Most disastrously, succeeding governments have allowed Peru's foreign debt to grow to unmanageable proportions. The government of General Andrés Aveline Cácares (1836–1923), who serves as president from 1886–90 and 1894–95, restores political and economic order. It is under Cácares's leadership that Peru satisfies its foreign creditors through its arrangement with the Peruvian Corporation (see, 1888–90).

Late 1800s   Artists turn to native culture for themes

Peruvian artists begin turning to their country's native traditions in their works. The *costumbrista* movement advocates art based on the daily life of Peru's native peoples.

1888–90   Foreign investors assume Peru's debts

Peru is economically devastated following the disastrous War of the Pacific. Its economy is rescued by a British-based group of investors who come together to form the London-based Peruvian Corporation, which assumes responsibility for Peru's large foreign debt in return for a substantial economic stake in the country. They gain control of the nation's railways, ports, remaining guano deposits, and Amazon Basin rubber reserves.

1888: March 11   Birth of artist José Sabogal

Sabogal (1888–1956) is a twentieth-century Peruvian artist who works with native themes, and a leading member of the artists' movement known as the "Generation of 1919." He studies and teaches art in Argentina and is influenced by the work of Mexican mural painters. After returning to Peru, Sabogal teaches at the Escuela de Bellas Artes (School of Fine Arts) in Lima.

1889   Nuevo Teatro Principal opens

Lima's first opera theater opens to the public. The Teatro Municipal (Municipal Theater) is opened five years later.

1892   First soccer game played in Peru

Soccer, brought to Peru by British immigrants, becomes Peru's national sport (called *fútbol*). The first soccer club is formed in 1897, and league play is begun in 1912 (see 1912).

1895   Birth of poet César Vallejo

Internationally recognized Peruvian poet César Vallejo (1895–1938) is born into a middle-class family and studies literature at the National University of Trujillo, Peru, receiving a bachelor's degree in 1915. He begins working as a schoolteacher in 1917. In 1920 and 1921 Vallejo is jailed on charges of political violence in Santiago de Chuco, his hometown. In 1923 he leaves for Europe, where he will spend the rest of his life. He mostly lives in France, spending periods of time in Spain and Russia as well. In the 1930s he works for the Republican cause in the Spanish Civil War.

Vallejo writes novels, short stories, journalism, and dramas, but he is best known for three volumes of poetry: *Los heraldos negros* (The Black Heralds; 1918), the groundbreaking *Trilce* (1922), and *Poemas humanos* (1939). In *Trilce*, considered a major poetic achievement, the fragmented language of the poems mirrors the breakdown of faith in the twentieth century. In spite of the time he has spent living in Europe, Vallejo is celebrated for creating a uniquely Peruvian poetry distinct from European and modernist influences. He dies in Paris in 1938.

1895   Piérola unseats Cácares as president

The charismatic political leader Nicolás de Piérola (1839–1913), who has led popular revolts against the government in 1874 and 1877, succeeds in overthrowing the military regime of Cácares. Piérola, who serves as president from 1895 to 1899, institutes a less centralized government. Actions taken during his term in office include legalization of civil marriage and adoption of the gold standard. Piérola's presidency inaugurates a period of relative political stability and economic progress that lasts until World War I.

1902   Copper smelter is built

The Cerro de Pasco Corporation, formed by North American investors, builds a copper smelter, which refines metal by melting it to extract pure copper. Copper soon overtakes silver as Peru's most profitable metal.

1905   Government assumes responsibility for education

The Peruvian government takes over responsibility for public education.

The Leguiá Presidency

1908   Leguiá begins first presidential term

Augusto Leguiá y Salcedo (1863–1932), the dominant Peruvian political figure of the early twentieth century, serves his first term as president from 1908 to 1912 (he is also president from 1919 to 1930). In spite of his dictatorial rule, Leguiá brings economic and social progress to Peru, instituting educational and labor reforms. He also encourages U.S. companies including W. R. Grace and Standard Oil to invest in Peru and expand their operations there.

1911   Ancient city of Machu Picchu is discovered

While seeking the ruins of the Inca city of Vilcabamba, American archaeologist (and later historian and statesman) Hiram Bingham (1875–1956) comes upon the "lost city" of Machu Picchu, an Inca enclave that was never discovered by the Spanish. Located above the Urubamba Valley, its ancient houses and temples, surrounded by mountain terraces, are found nearly intact and provide an unparalleled glimpse into the architecture and civilization of the Incas.

1912   First soccer league is established

Peru's soccer clubs form their first league.

1914   Standard Oil purchases oil rights

A Canadian subsidiary of the Standard Oil company purchases rights to develop Peru's main oil fields (La Brea y Pariñas).

1918   Natural history museum established

The Natural History Museum of the University of San Marcos is founded.

1919   Leguía returns to power

Former president Augusto Leguía, in exile outside the country, resumes the presidency after arranging a coup to unseat the current regime. During his second term in office, he carries out major housing, education, banking, and infrastructure programs.

1920s   Peruvian researcher probes health effects of high altitudes

Carlos Monge Medrano carries out the first major scientific studies investigating how living at high altitudes affects health. He finds that the native Peruvian habit of chewing coca leaves has a positive effect on metabolism at high altitudes.

1920   Highway Conscription Act is implemented

In a move that some compare to the *mita* forced labor system of the colonial period, President Leguía's government forces rural villagers into mandatory service to provide the labor needed for infrastructure projects. All males between the ages of eighteen and sixty are required to work between six and twelve days per year on public construction projects.

1924   APRA reform party founded

Víctor Raúl Haya de la Torre (1895–1979), in exile in Mexico, founds the *Alianza Popular Revolucionaria Americana* (APRA), a left-wing political party representing the interests of the working class and the native population. APRA opposes the policies of President Leguía, including the growing role of the U.S. in the Peruvian economy. Twenty years after it was founded, APRA comes to power in national elections. Its founder remains active in national politics until his death in 1979.

1925   Artist Fernando de Szyszlo is born

Fernando de Szyszlo (b. 1925), the best known of Peru's contemporary artists, is born to a Polish father and a mother of Peruvian Indian ancestry. He studies art in Lima and becomes part of a group of artists (called *Espacio*) who use themes and symbols from indigenous cultures in their works. One of Szyszlo's works is a series of paintings based on an elegy on the death of Inca leader Atahualpa. After traveling to Europe in 1949, Szyszlo turns from figurative to abstract painting. In 1953 he travels to the United States, where he serves as a visual arts consultant to the Organization of American States (OAS) between 1957 and 1960. He is an artist in residence at Cornell University in 1962 and a visiting professor at Yale in 1966. Szyszlo returns to Lima permanently in 1970. Szyszlo's work has had a major influence on other Peruvian painters.

1927   Peru hosts the eleventh South American soccer championship games

Peru makes its international soccer debut as host of the eleventh South American Championship. However, its team is decisively defeated by reigning soccer heavyweights Argentina (5–1) and Brazil (4–0). Peru does take third place over Bolivia.

1927   Birth of singer Yma Sumac

Internationally renowned singer Yma Sumac (b. 1927) is born in a small village in the Peruvian Andes. (Her real name is Emperatriz Chavarri.) She begins performing in concerts while still a child. In Lima she meets her husband, composer Moises Vivanco, director of the Peruvian National Board of Broadcasting. Vivanco helps advance Sumac's career, and she becomes well known in Latin America.

In 1947 Sumac and Vivanco emigrate to the United States, where she achieves success with the release of her first recording, *Voice of Xtabay*, issued by Capitol Records. More records follow, and Sumac's total record sales total over one million. She also appears in theater, films, and on television, winning fame for her unparalleled four-octave vocal range and ability to reproduce the sounds of Peru's native animals. Much of the music performed by Sumac is written by her husband and modeled after Peruvian folk music. She also performs internationally in operas, including *The Magic Flute* and *La Traviata*. Sumac becomes a U.S. citizen in 1955.

1929   Socialist party founded by Mariátegui

José Carlos Mariátegui founds a new socialist party, which later becomes affiliated with the Communist Party. Mariátegui, a journalist, is known for the analysis of Peru's social and political problems found in his volume of essays, *Siete ensayos de interpretación de la realidad peruana* (Seven Interpretive Essays on Peruvian Reality; 1928). He is also the founder of a journal, *Amauta*. His ideas are part of the *indigenismo* of the 1920s: a current of Latin American political ideology that envisions a return to the values and culture of the region's Indian communities. Víctor Raúl Haya de la Torre is also associated with *indigenismo*.

Thinkers like Mariátegui link the collective nature of ancient Indian civilizations to socialist ideals. Mariátegui argues that Peru's Indian peasants are its true revolutionary class rather than the industrial workers of classical Marxism. Abimael Guzmán Reynoso, founder of the *Sendero Luminoso* (Shining Path) terrorist group later in the century, will draw on the ideas of Mariátegui (see 1960s). The name "shining path" is, in fact, taken from Mariátegui's writings.

1930s   Peruvian painters espouse nationalism

A movement toward nationalism in painting, especially as expressed in Indian themes, is led by José Sabogal and Julia Codesido.

1930   Peru enters the first World Cup soccer championship

Peru qualifies for the first-ever World Cup games. Its team is eliminated in the first round however, after losses to Romania (3–1) and Uruguay (1–0).

The Military Takes Over

1930   Leguiá is ousted in coup d'état

Longtime president Augusto Leguiá is overthrown and jailed by Colonel Luis Sánchez Cerro's forces. Cerro claims victory in the 1931 presidential election, over the protests of Haya de la Torre, leader of the APRA party, which claims election fraud. Sánchez Cerro remains in office until 1933. However, his tenure in office is marked by strikes and popular rebellions.

1931   Peru's Central Reserve Bank founded

The Central Reserve Bank, the only bank to issue currency, is founded in Lima.

1932   Conflict with Colombia over disputed Letícia border territory

Cerro's government nearly goes to war with Colombia after attempting to reclaim an area in the Amazon jungle that Peru has ceded to Colombia in a treaty in 1927. The dispute is settled with international intervention.

1932: July   APRA revolt against the Cerro regime

The APRA opposition party stages an armed revolt in Trujillo. About sixty military officers are killed. The government retaliates, killing a thousand or more APRA members (known as *apristas*). In addition to ground fighting, APRA supporters are attacked by government bomber planes.

1933   Peruvian soccer club tours Europe

The Universatorio de Deportes soccer club embarks on a six-month, thirty-game tour of Europe that is important in advancing the status of Peruvian soccer. Unfortunately, much of the goodwill created by this tour is destroyed by an incident that occurs at the 1936 Olympics in Berlin. Peruvian fans riot during overtime in a game with Austria, and the Peruvian team takes the opportunity to score two goals. The Peruvians refuse to take part in the replay that is ordered and leave the game, and Austria is awarded the two points in question.

1933: April 30   Sánchez Cerro is assassinated

President Luis Sánchez Cerro is assassinated by a militant member of the APRA party.

Former president Oscar Benavides completes Sánchez Cerro's five-year presidential term. The results of the 1936 election are again disputed by APRA. Benavides remains in office until 1939, ruling as a dictator. Manuel Prado y Ugarteche (1889–1967) is installed as president after Benavides and serves until 1945.

1936   Birth of acclaimed novelist Mario Vargas Llosa

Mario Vargas Llosa (b. 1936) is Peru's most famous living literary figure. He is the first Latin American to serve as president of the international writers' organization, PEN, and he wins Spain's Cervantes prize in 1984. Acclaimed for novels including *The Time of the Hero, Conversation in the Cathedral,* and *Aunt Julia and the Scriptwriter,* Vargas Llosa has also written essays, plays, and an autobiographical work entitled *A Fish in the Water.*

Vargas Llosa is born in Arequipa and is enrolled in a military school in Lima when he is ten years old. (He later describes this experience in his first novel, *The Time of the Hero,* which is banned by the Peruvian military.) He lives abroad for much of the time between 1959 and 1974, mostly in Paris. He maintains a strong interest and involvement in his country's social and political affairs, however, beginning as a radical left sympathizer in the 1960s and gradually moving toward the right. In 1990 Vargas Llosa contends unsuccessfully against Alberto Fujimori for the presidency. Considered the front-runner, he loses to Fujimori on the second ballot. Dissatisfied with the political direction his country is taking, Vargas Llosa moves to London shortly after the 1990 election and becomes a Spanish citizen.

1938   Establishment of orchestra in Lima

Austrian émigré Theo Buchwald establishes Lima's symphony orchestra, which contributes to the advancement of Peruvian musicand performs works by native composers.

1939–45   World War II bolsters Peruvian economy

During World War II Peru benefits from U.S. demands for minerals, cotton, sugar, rubber, and quinine. Politically, the Allied victory in the war helps move Peru toward democratic government. In addition, the international cooperation fostered during the war leads the opposition APRA party to soften its previous anti-American stance. Peru does not directly take part in the war. Its president does formally declare war on Germany in February 1945, but by then the war is nearly over.

1939   Science academy is founded

The Lima Academy of Exact, Physical, and Natural Sciences is founded in Lima.

1943   Trans-Andean Highway is completed

The Trans-Andean Highway, the only road directly connecting eastern and western Peru, is completed.

1944   Workers' Confederation of Peru is formed

A national federation of labor groups is begun. At first most of the power is wielded by communists. In 1956 the group is reorganized and renamed the Peruvian Revolutionary Workers' Center (Central de Trabajadores de la Revolución Peruana, or CTRP). Still later, it is assimilated into the larger Democratic Trade Union Front.

1945: October 31   Peru joins the United Nations

Peru becomes a charter member of the United Nations.

Civilian and Military Governments

1945   Liberal coalition candidate is elected president

José Luis Bustamante y Rivero, (1894–1989) the candidate of the APRA-supported National Democratic Front, is elected to the presidency. The National Democratic Front also wins control of both houses of the legislature. The new government introduces democratic and social reforms, lifting censorship, restoring civil rights, and instituting labor and educational reforms. APRA's economic development plans are also set into motion.

1946   Secondary schooling becomes free

For the first time, secondary education is free, although facilities are not adequate to meet the demand.

1946 Non-governmental agency launches peasant literacy

The Peruvian North American Cooperative Service (SEC-PANE) launches a literacy project in the Andean highlands, where one-third or more of the local population do not read or even speak Spanish.

1948: October Bustamante is overthrown in military coup

The military overthrows President Bustamante in a right-wing coup led by General Manuel Odría, who is installed as provisional president. Odría imposes strict dictatorial rule, outlawing dissent, censoring the opposition, and dissolving the country's labor unions. Odría's government institutes a major transportation program that builds roads in eastern Peru. Odría is elected president in 1950, but he is the only candidate.

1950s Peruvians migrate from the interior to the coast

Peruvians begin migrating in large numbers from the central mountains to the cities of the coast, especially Lima. This rural-urban migration trend has continued to the present. Almost a third of Peru's population live in Lima by the mid-1990s.

1950 Women's suffrage

Women win the right to vote. In 1956, they also win the right to hold elected office.

1956 Prado chosen as president in free elections

Peru returns to civilian government with the election of former president Manuel Prado (1889–1967). Previously president during the Second World War, Prado now restores greater freedom to the people and authorizes public works programs. Pedro Beltrán, appointed as prime minister and finance minister in 1959, implements an economic program that brings inflation under control within three years. He also organizes a housing program, *Techo y Tierro* (Roof and Land), to improve conditions in urban shantytowns, redistribute land belonging to large estates, and resettle families to areas east of the Andes.

1960s Shining Path leftist group is formed

Abimael Guzmán Reynoso, a philosophy professor at the National University of San Cristobal de Huamanga in the Ayacucho department, forms the *Sendero Luminoso* (Shining Path, or SL), a left-wing Maoist splinter group. (His name within the group is "President Gonzalo.") The SL dedicates itself to a "prolonged popular war" with the goal of destroying the nation's capitalist economy and creating a new society. Women play a prominent role, both in the ranks and among the group's leaders. Shining Path supporters will eventually come to control about a fifth of the country. In the 1980s the group emerges as a major terrorist threat (see 1980s).

1962: January 10 Shattered glacier causes an avalanche, killing 3,500

Large amounts of sunshine cause a glacier atop Mt. Huascarán in the Andes to rupture, leading to an avalanche that roars down the mountainside and destroys four villages and one city in the valley below, all within seven minutes. After accumulating boulders and other debris, it sweeps into the town of Yanamachico and three other towns, killing all 800 of their residents. With the added ruins of this destruction, and traveling at approximately sixty miles per hour, the avalanche crashes into Ranrahirca, near Yungay, killing almost all of its 2,700 residents.

1962: July 18 Military overthrows government

The 1962 election is a three-way race between Odría, APRA leader Haya de la Torre, and Fernando Belaúnde Terry, who is favored by the military. Because no candidate gains enough votes to win, the election is to be decided by congress. The military, fearing that de la Torre would emerge as either president or as vice president under Odría, takes control of the government. They establish a junta but rule with restraint and promise free elections in 1963.

1962: November 27 Jet crashes over the Andes

A Boeing 707 en route from Río de Janeiro to Los Angeles experiences problems shortly before a scheduled landing in Lima and crashes over the Andes, killing all ninety-seven persons aboard. The reason for the crash remains unknown.

1963: June 19 Belaúnde is chosen as president in free elections

Belaúnde (b. 1913) wins the presidency in free and open elections. He attempts to implement social reform and economic development, but his administration's programs produce budget deficits and a new round of inflation. Belaúnde's government grows increasingly unstable—there are five different cabinets between 1967 and 1968.

1964 Site of lost Inca capital is discovered

Gene Savoy, an American, finds the site of Vilcabamba, the capital established in the jungle by Inca leader Manco Capac after the Spanish took over Cuzco in the sixteenth century.

1964: May Soccer riot in Lima

Rioting occurs at a soccer game in Lima when the Argentinean team beats Peru by scoring a last-minute goal. Nearly three hundred people are killed.

Military Rule

1968: October 3   Military ousts Belaúnde

Displeased with Belaúnde's policies and the state of the nation's economy, the military removes him from office and exiles him to Argentina. General Juan Velasco Alvarado (1910–77) becomes the head of yet another military dictatorship, which carries out radical land reforms as well as social and cultural reform. Ties with the United States are strained by government seizure of U.S. fishing boats and property belonging to a subsidiary of Standard Oil.

1968: December   Five-year reform plan is announced

The military government announces a five-year plan to transform Peru from an agricultural to an industrial society. Large estates are to be taken over by the government, and the land will be redistributed to the peasants.

1969   Radical land reform law is enacted

The Agrarian Reform Law is designed to remedy the monopoly on landowning by a small elite. The government transforms large private estates into cooperative farms. By 1973 land reform is being carried out virtually throughout the country, and by 1980 the program has largely been completed.

1969: June 24   Government launches land program on "Day of the Indian"

Albarado's government announces sweeping land reform measures designed to give large amounts of land to Indian peasants.

1970: May 31   Earthquake, flooding, and landslides kill 70,000

In the most destructive natural disaster ever to strike the Western Hemisphere, an earthquake measuring 7.75 on the Richter scale strikes the department of Ancash (an area in northwestern Peru) and surrounding areas, killing 70,000, injuring 140,000, and leaving over 500,000 homeless. It can be felt in Ecuador to the north, in the Amazon jungle to the east, and 300 miles south of Lima, in the village of Nazca. The city of Chimbote, located twelve miles from the quake's epicenter, suffers tremendous damage; three-fourths of its houses are totally destroyed, and over 200 people are killed. The neighboring resort city of Huaras is completed decimated.

The most catastrophic destruction, however, occurrs in Yungay, which is flattened by glacial ice dislodged from the top of El Huascarán, Peru's tallest mountain. Out of 20,000 residents, only 2,500 survive the earthquake. In addition to the fissures and tremors of the quake itself, added damage is caused by avalanches, and flooding rivers and dams. Rescue efforts a seriously hampered for the first three days by heavy fog trapped between the Andean peaks and rising to heights of 18,000 feet over the region. The fog makes it impossible to airlift supplies to the survivors, who are left in the cold without food or shelter for three days, resulting in additional fatalities.

1971   Peru hosts South American baseball championship

The South American baseball championship games are held in Lima.

1971: March 19   Earthquake causes avalanche that kills 400–600 people

When an earthquake causes a mountaintop in the Andes to crumble and plunge into a nearby lake, the overflow from the lake creates an avalanche as it sweeps boulders, trees, and soil along with it down the mountainside. The avalanche destroys the isolated mining camp of Chungar, located 10,000 feet up in the Andes and an eight-hour walk from the nearest town. When help arrives, only about a third of the camp's 1,000 inhabitants are found alive. The rest are buried under the debris carried by the avalanche. Fifty injured persons are flown out to medical facilities.

1972   Education Act expands Ministry of Education

The government assumes control over the hiring of all teachers in the public schools and also increases its authority over appointments in private schools. The native Quechua and Aymará languages are made languages of instruction for the non-Spanish-speaking native population.

1973   National airline, Aeroperú, is established

Aeroperú is created to provide both domestic and international service. The government-run airline is privatized in the 1990s.

1973   Peruvian publishes influential book on liberation theology

Gustavo Gutiérrez's *A Theology of Liberation* has a strong influence on the developing trend toward social activism in the church, both in Peru and in other countries.

1974   Government silences last independent political magazine

Caretas, the remaining magazine that expresses independent views, is shut down by the government, which also seizes control of all private national newspapers that have more than twenty thousand readers.

1975: August 29   Bermúdez replaces Alvarado

Widespread strikes and antigovernment demonstrations lead to the replacement of President Alvarado with a more moderate general, Francisco Morales Bermúdez (b. 1921), who promises to move the country toward a civilian government.

Inflation and unemployment, however, are both high, and the national debt is extremely large. Peru receives a loan from the International Monetary Fund in 1978, but the economic conditions imposed on the country in return make life even harder for the average Peruvian.

1978   Peru qualifies for the World Cup soccer championship

The Peruvian soccer team qualifies for the World Cup and advances to the second round after defeating Scotland and drawing with Holland in the first. But Peru loses to Brazil, Argentina, and Poland in the semifinals.

1978: May   Civil unrest follows economic austerity measures

Demonstrations and rioting follow an increase in food prices resulting from government measures to stabilize the economy.

Late 1970s   Peru enters the international cocaine trade

Peruvians begin to export both fully refined cocaine and partially processed coca paste to drug lords in Colombia and Bolivia.

Return of Civilian Government

1980s   Shining Path leftist group carries out terrorist attacks

The left-wing *Sendero Luminoso* (SL), whose influence now extends over a thousand-mile stretch along the Andes, inaugurates a reign of terror that will last throughout the decade and into the 1990s. The group carries out attacks in the department of Ayacucho and surrounding areas, destroying public buildings, blowing up trains, and killing landlords, public officials, and other government supporters. The populations of towns in the region fall by as much as two-thirds, as residents migrate to other areas. In the SL's first major operation, it dynamites a polling center in the village of Chuschi during the 1980 presidential election. Eventually the SL gathers thousands of followers among poor laborers in Indian farming communities.

By the late 1980s and early 1990s the SL turns to urban terrorism, bombing hydroelectric plants and a variety of public and private buildings in cities, including the capital. The SL's tight organizational structure makes it difficult for government agents to infiltrate the group. By the end of the decade, the SL's guerrilla activities, which include kidnapping, robberies, and attacks on soldiers and police, result in about twenty thousand deaths and drive as many as two hundred thousand people from their homes in rural areas.

1980s   Sugar industry declines

Peru's sugar industry suffers setbacks from bad weather, inadequate financing and equipment, heavy indebtedness, and ill-conceived pricing policies.

1980   Civilian government returns with election of Belaúnde

Former president Belaúnde is chosen as president in free elections (Peru's first ever with universal suffrage), and civilian government is restored after twelve years of military rule. His five-year term in office is troubled by strikes, an economic depression, and dramatic levels of inflation. Peru is forced to stop paying off its enormous foreign debt. The international community pressures the nation to halt the rising coca plant production that has resulted from increasing international demand for cocaine.

1981: February   Border dispute with Ecuador erupts into violence

A five-day series of border clashes with Ecuadoran armed forces ends when the Organization of American States intervenes and brokers a cease-fire. Peru and Ecuador agree to have the dispute placed under international arbitration.

1983   Strongest recorded El Niño brings flooding to the northwest

During the first half of the year, the most powerful El Niño current ever recorded was followed by 100 inches (250 cm) of rain on the northwestern plains of Peru. A strong El Niño current in the Pacific Ocean leads to unusual weather patterns in North and South America.

1985   Pope John Paul II visits Peru

A visit by Pope John Paul II (b. 1920) is the occasion for massive public rallies.

1985: April   Belaúnde is unseated by APRA candidate

APRA wins Peru's national elections by a large margin. Alan García Perez (b. 1949) becomes the first APRA candidate to win the presidency. Under Perez the government takes over banks and insurance companies and assumes stricter control over the economy as a whole. However, his policies fail to rescue Peru from its economic woes. The nation remains unable to get international loans, and its inflation rate reaches three thousand percent by 1990. Population growth and unemployment remain high, and Peru's cities are falling into decay.

1986: June   Shining Path prisoners revolt

SL members held in Lima prisons stage a mass rebellion. Altogether 254 of the prisoners are killed.

1987 Nighttime vandals stumble upon 2,000-year-old treasures

Rural villagers vandalizing an archaeological site near the village of Sipán, north of Trujillo, unknowingly lead archaeologists to a royal Moche Indian tomb and burial ground containing priceless gold, silver, and turquoise jewelry, ceramics, and sculpture.

1989–90 Peruvians protest nation's economic situation

Wage increases are demanded in a series of strikes, as Peru's populace becomes increasingly distraught over the nation's runaway inflation and other economic problems.

1990 Fujimori is elected president

Dissatisfied with García's efforts to revive their economy and curb the threat from the Shining Path guerrillas, Peruvian voters turn the APRA politician out of office in 1990. The primary contenders for the presidency that year are both unconventional candidates. Internationally renowned novelist Mario Vargas Llosa (see 1936), who is supported by Peru's governing classes and the U.S., advocates free-market economic reforms. Alberto Fujimori, (b. 1938) the son of Japanese immigrants, stresses social welfare programs and rural development. Fujimori unexpectedly wins the election by a wide majority. He becomes the first president of Asian ancestry in the Western Hemisphere.

1990–95 Fujimori takes aggressive action to rescue Peru's economy

Upon taking office, Fujimori enacts major free-market economic reforms, including privatization of state industries, cuts in public spending and government employment, and tax increases. Ironically, many of his policies are similar to those put forth by his defeated opponent, Mario Vargas Llosa. The effects of the far-reaching and, in some cases, drastic measures are referred to by many as "Fuji-shock." They drive up unemployment, and, with growing inflation, real salaries drop. By 1992 however, economic growth is rising and by 1994 it is the highest in South America. Direct foreign investment rises to $400 million by 1993.

1991: January Cholera epidemic strikes Peru

Just one year after newly elected president Alberto Fujimori takes office, Peru is hit by a major outbreak of cholera, which first appears in the fishing port of Chimbote. The epidemic is especially devastating because it occurs just as Fujimori's policies are reducing Peru's social welfare programs and forcing an economic austerity program on the nation. In addition to the heavy toll in lives lost, the outbreak also cuts revenues from tourism and food exports.

1992: April 5 Fujimori suspends government and rules by decree

Thwarted by congress in his campaign for additional powers to combat terrorism, President Fujimori, with the aid of the military, stages a coup against his own government (which he calls an *autogolpe,* or "self coup"). He suspends Peru's constitution and dissolves the legislature, and imposes censorship on the press, granting himself the power to rule by decree. There is little international reaction, and many Peruvians hope that the extra powers seized by their president will enable him to deal more decisively with terrorism, the economy, and government corruption. The United States signals its disapproval of Fujimori's actions by suspending all foreign aid to Peru. A new constitution is adopted by the legislature in 1993.

1992: July 16 Shining Path bombs major tourist facilities

Shining Path terrorists set off a car bomb in the Miraflores district, causing over one million dollars (U.S.) in damage to six large hotels and seriously damaging Peru's tourist industry. In the same month the government issues a law guaranteeing amnesty for terrorists who turn themselves in and cooperate with authorities by providing information about their fellow guerrillas.

1992: September Shining Path leader captured by police

Abimael Guzmán Reynoso, the SL leader, is arrested by government forces and sentenced to life in prison. Captured together with him, in a house in suburban Lima, are three other high-ranking SL officials. SL membership declines dramatically after Guzmán's capture, although isolated terrorist acts still occur.

1995: April Fujimori is elected to second term

President Fujimori is reelected by a wide margin in national elections, in spite of a chaotic preliminary election in which his own wife, Susana Higuchi, runs against him and alleges corruption in the legislature. Higuchi's candidacy is declared illegal in connection with fraud charges, and she and her husband begin divorce proceedings. In what is expected to be a close presidential election between Fujimori and former UN Secretary General Javier Peréz de Cuellar (b. 1920), Fujimori triumphs, winning a nearly two-thirds majority of the popular vote. In addition, his political party (Cambio '90 Nueva Mayoria) wins a small majority in congress.

1996: August 23 Third term created for Fujimori

The congress, controlled by Fujimori's party, approves a constitutional amendment that will allow Fujimori to run for president in the year 2000. The current constitution, approved in 1993, limits the president to two terms. The amendment, however, justifies the exemption on the grounds that Fujimori's first term was served under the old constitution. Members of the opposition party leave the chamber in protest before the vote.

1996: December 17   Guerrillas storm Japanese ambassador's house, take hostages

Members of the Marxist Túpac Amaru Revolutionary Movement (MRTA) storm the residence of the Japanese ambassador to Peru during a party, taking more than 600 people hostage. The group demands that the Peruvian government release hundreds of MRTA members jailed under President Fujimori's strict anti-terrorism laws. They also demand safe passage to a remote jungle location, the enactment of various government policy measures, and threaten to kill the hostages one at a time until their demands are met.

This hostage-taking is considered the most serious terrorist act in Peru's history. The MRTA, founded in 1984, is smaller than the SL and has previously received less attention.

1996: December 22   MRTA rebels release 225 hostages

MRTA members release 225 of the hostages they have been holding since December 17 at the home of the Japanese ambassador. Of the original 600 persons taken hostage, 200 women are released on the first night of the crisis, and 38 more hostages are allowed to leave on December 20. After the December 22 release, there are still about 100 high-ranking government officials among the hostages, including several congressmen and the president of the supreme court. The brother of President Alberto Fujimori is also among the hostages.

1997: February 18   Mudslide kills 250 people

Some 250 people are killed when a mudslide destroys the villages of Ccocha and Pumaranra in the southern Andes. The Tamburco district, where the disaster takes place, experiences mudslides on an annual basis. This occurrence is the worst in a decade.

1997: April 22   Peruvian troops free hostages

The four-month-long terrorist siege by members of the MRTA ends when Peruvian troops storm the Japanese ambassador's residence and free the remaining hostages. One hostage dies after being struck by a stray bullet and then suffering a heart attack. All fourteen terrorists are killed, as well as two Peruvian soldiers. Negotiations to free the hostages had been stalled since mid-March.

The raid is accomplished with the help of tunnels dug under the ambassador's residence. It is begun with a bomb set off underneath the living room of the house, where eight of the guerrillas are playing soccer. The hostages, who are on the second floor, have secretly been in contact with intelligence officials throughout much of the ordeal, and they are warned of the raid ten minutes before it begins. After the bomb goes off, government commandos rush the house through holes in the floor and roof, kill the remaining rebels, and lead the hostages to safety.

International reaction to the raid is mixed, with some leaders calling the move "highly risky," while others call the Peruvian government's handling of the crisis "responsible." Within Peru, President Fujimori is lauded for ending the 126-day-long ordeal, the longest terrorist siege ever to occur in Latin America.

Bibliography

Degregori, Carlos Ivan, Robin Kirk, and Orin Starn, eds. *The Peru Reader: History, Culture, Politics.* Durham. N.C.: Duke University Press, 1995.

Dobyns, Henry F., and Paul L. Doughty. *Peru: A Cultural History.* New York: Oxford University Press, 1976.

Gootenberg, Paul. *Imagining Development: Economic Ideas in Peru's "Fictitious Prosperity" of Guano, 1840–1880.* Berkeley, Calif. : University of California Press, 1993.

Hemming, John. *The Conquest of the Incas.* New York: Harcourt Brace Jovanovich, 1970.

Herzog, Kristin. *Finding Their Voice: Peruvian Women's Testimonies of War.* Valley Forge, Pa.: Trinity Press International, 1993.

Hudson, Rex A., ed. *Peru in Pictures.* Minneapolis: Lerner, 1987., ed. *Peru: A Country Study.* 4th ed. Washington, D.C.: Library of Congress, Federal Research Division, 1993.

Keatinge, Richard., ed. *Peruvian Prehistory: An Overview of Pre-Inca and Inca Society.* New York: Cambridge University Press, 1988.

Moseley, Michael Edward. *The Incas and their Ancestors: The Archaeology of Peru.* New York: Thames and Hudson, 1992.

Mörner, Magnus. *The Andean Past: Land, Societies, and Conflicts.* New York: Columbia University Press, 1985.

Olson, Richard S., with Bruno Podesta and Joanne M. Nigg. *The Politics of Earthquake Prediction.* Princeton, N.J. : Princeton University Press, 1989.

Quiroz, Alfonso W. *Domestic and Foreign Finance in Modern Peru, 1850–1950: Financing Visions of Development.* Pittsburgh: University of Pittsburgh Press, 1993.

Stokes, Susan Carol. *Cultures in Conflict: Social Movements and the State in Peru.* Berkeley, Calif.: Univ. of California Press, 1995.

Strong, Simon. *Shining Path: Terror and Revolution in Peru.* New York: Times Books, 1992.

Puerto Rico

Introduction

On July 25, 1998, Puerto Ricans observed the one-hundredth anniversary of the landing of U.S. troops on the island. This invasion, during the Spanish-American War (1898), ended 400 years of Spanish occupation.

But for many Puerto Ricans, the invasion simply meant that a declining colonial power had been replaced by a more powerful imperialist nation; the arrival of U.S. troops squelched dreams of an independent country. For other Puerto Ricans, the invasion signaled an opportunity for improving the poverty-stricken island, which had been mostly neglected by the ruling Spaniards.

For the next 100 years, Puerto Ricans could not seem to resolve one of the most pressing issues facing them. They remained deeply divided on the political status of the island; should they seek independence, remain a commonwealth under the United States or become the fifty-first state? By the 1990s, only a small minority favored full independence. The rest were almost equally divided in support of the commonwealth or becoming the fifty-first U.S. state.

Puerto Rico, about 1,000 miles from the U.S. mainland, is the smallest and most easterly of the islands that make up the Greater Antilles. These islands screen the Caribbean Sea from the Atlantic Ocean. Roughly rectangular, Puerto Rico has a total area of 3,435 square miles.

The main island extends 111 miles east-west and 36 miles north-south. It is crossed from east to west by mountain ranges, with the Cordillera Central rising to nearly 4,400 feet. About fifty short rivers descend from the mountains into the coastal plain, which widens to fifteen miles at its broadest point. Four other islands, Culebra and Vieques to the east and Mona and Desecheo to the west, are part of Puerto Rico.

Puerto Rico, "rich port," was the name given to the island by Spaniards who arrived with Christopher Columbus in 1493 hoping to find gold. To its native inhabitants, the Taino, the island was known as Borinquen. The Spaniards did more than just change the name of the island. As they had done through-

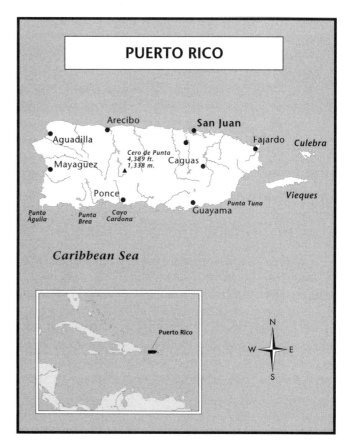

out the New World, they enslaved the native people, decimating the Taino population as they sought to extract wealth from the small island. As the gold supplies were quickly exhausted, the Spanish conquistadors turned to sugarcane growing and began to import African slaves to do the heavy work.

By the 1820s, Spain had lost most of its colonies in the New World, but held on to Puerto Rico and Cuba. In Puerto Rico, the Spaniards continued importing slaves to prop up the failing economy at a time when most nations had abolished slavery.

By the 1860s, many Puerto Rican intellectuals and the working classes fought to end slavery and to gain more independence from Spain. Yet, they could not agree on the extent of that independence. Some sought local independence while maintaining political connections with Spain. Others wanted full independence.

But events in Cuba would have a profound effect on Puerto Rico's future. Cubans, most notably Jose Marti (1853–95), a writer and revolutionary in favor of independence, rebelled against Spanish rule.

Many of the Cuban and Puerto Rican revolutionaries escaped to New York and helped turn U.S. public opinion against Spanish rule. By 1898, the United States, fast becoming one of the most powerful nations in the world, had sided with the Cuban revolutionaries.

On February 15, 1898, with relations between the United States and Spain rapidly worsening, the battleship U.S.S. *Maine,* sent to protect Havana harbor, exploded. The United States blamed Spain for the explosion and declared war, attacking Spanish defenses in Cuba and Puerto Rico.

After the war, Puerto Rico was ceded to the United States and the Americans quickly established their authority there. The military, which ruled the island for two years, abruptly changed its name to Porto Rico and changed the language of public-school instruction from Spanish to English.

Military rule ended in 1900, but Puerto Ricans remained wary of American intentions. Islanders pressed for political changes, even for autonomy, but were rebuffed by the U.S. government.

By the 1930s, the United States had banned the Puerto Rican flag and closely monitored and suppressed the island's independence movement, which had become violent. During those years, under the leadership of Pedro Albizu Campos, independence had widespread support among Puerto Ricans.

In 1950, Albizu was charged with planning the attempted assassination of President Harry Truman and in 1954, Puerto Rican nationalists fired shots inside the chambers of the U.S. House of Representatives, injuring several legislators. Pro-independence forces continued their attacks in the 1970s and 1980s, killing American marines and destroying eight aircraft at a National Guard installation in Puerto Rico.

Yet in 1993, only 4.4 percent of Puerto Ricans voted for independence during a plebiscite—a vote by which a population can accept or reject a political proposal—to decide the island's future. The vote to keep the commonwealth won by a slim margin over the drive for statehood.

Puerto Ricans have not yet decided the final fate of their island. They are expected to vote once again on its political status. Yet, any decision will need the approval of U.S. Congress, which in 1998 remained highly divided on the issue of statehood. One of the obstacles to Puerto Rican statehood is that many Americans believe Puerto Rico should declare English its official language before it is accepted into the union.

Timeline

400 B.C. Records show first settlements

The first inhabitants of Puerto Rico descend from cultures that migrated from South America and to the islands of the Caribbean. Archeological records show that at least three Amerindian cultures exist on the island long before the arrival of Christopher Columbus in 1493. The first group belongs to the Archaic Culture, believed to have migrated from Florida. They have no knowledge of agriculture or pottery and rely on fishing for food. Their remains have been found buried in caves.

The Igneri come from northern South America and are believed to be descendants of the Arawak people, also of South America. The Igneri bring agriculture and pottery to the island. Their remains have been found along the coast.

The Taino, also of Arawak ancestry, settle in Borinquen, the name they give to present-day Puerto Rico. They flourish by combining fishing with agriculture and develop a religion. They appear to lead a peaceful, sedentary and spiritual life, building ceremonial centers used frequently for religious activities. They use loud maracas and other religious artifacts made from stone and clay to protect the island from Juracan, the evil god of wind and rain.

Hurricanes and storms frequently hit the island, but the Taino also face hostile attacks from other Caribbean tribes, who arrive by canoe and kidnap women and children for use as slaves.

Columbus and Conquest of Puerto Rico

1493: November 19 Christopher Columbus arrives in Puerto Rico

Columbus lands on the northwestern side of the island during his second trip to the New World. He claims the island for the Crown of Spain and names it San Juan Bautista (Saint John the Baptist).

At this time the island is populated by as many as 30,000 Taino living in small villages.

1508: August 12 First Spanish settlers arrive in Puerto Rico

Juan Ponce de Leon (1460–1521), a young and ambitious nobleman, shapes the first Spanish settlement on the island. He seeks to transplant and adapt Spanish civilization to Puerto Rico's tropical habitat and begins to turn the island into one of the most important Spanish colonies in the New World. In doing so, Ponce de Leon decimates the Taino popu-

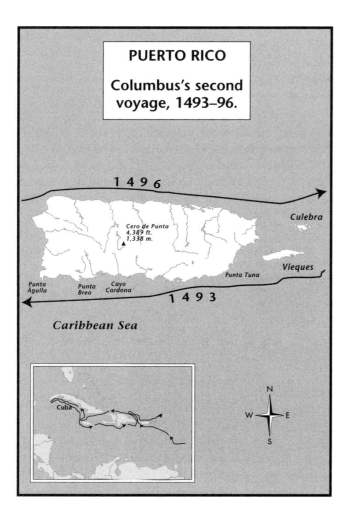

PUERTO RICO

Columbus's second voyage, 1493–96.

1 4 9 6

Cero de Punta
4,389 ft.
1,338 m.

Culebra

Vieques

Punta Tuna

Punta
Águila

Punta
Brea

Cayo
Cardona

1 4 9 3

Caribbean Sea

Cuba

N
W E
S

lation. He burns their villages, massacres those who oppose him and turns the survivors into slaves.

While the conquerors discover little gold and precious stones in Puerto Rico, the island becomes a hub for wealth moving from the New World to Europe. The port city of San Juan Bautista becomes a final stop for travelers returning to Europe.

1514   Only 4,000 Taino people survive conquest

After their arrival in 1508, the well-armed Spaniards began to subjugate the Taino and force thousands of them to mine for gold.

In 1511, Ponce de Leon orders the execution of 6,000 Taino natives following a major uprising. Many survivors fled to the mountains or left the island.

By 1514, the rigors of gold mining and losses from rebellion reduce the Taino population to about 4,000. By this time, gold supplies have been nearly exhausted and the economy shifts to growing sugarcane.

With the demise of the Taino, the Spaniards begin to import slaves from Africa to work in the sugar fields. For the

next 400 years, Puerto Rico's economy depends almost entirely on sugar.

1521   Puerto Rican settlements strengthened against invasion

Spain begins to build massive defenses around Puerto Rico to protect it from European invaders. One of those defenses is El Morro, a castle with eighteen-foot-thick walls in San Juan. Other forts are built and the Spanish surround San Juan with a protective wall.

1528   French sack and burn the small town of San German

The Spaniards attempt to maintain a monopoly on trade and colonization in the Americas. But Britain, Holland and France begin to challenge Spanish supremacy, first resorting to smuggling, piracy and war, especially in the Caribbean region.

Sir Francis Drake (1540?–1596) attacks Puerto Rico in 1595, the British invade the island in 1598 and the Dutch burn San Juan in 1625.

1538   Monks build Church of Porta Coeli

Spanish monks build the Church of Porta Coeli in San German. It is believed to be the oldest church in the Western Hemisphere.

1809   Artist paints more than 400 paintings before his death

Jose Campeche (1752–1809), the son of a freed slave, becomes one of the first great painters in Latin America, producing more than 400 paintings. He learns to draw and paint from his father. Later, Campeche is trained by a court painter from Spain. Campeche, a devout Catholic, paints for churches, but also becomes a noted portrait painter.

1840   A new style of dancing appears on the island

Starting in the mid-nineteenth century, a new style of dance unique to Puerto Rico begins to capture the attention of the islanders. The *Danza,* somewhat similar to European classical music, is considered the national dance of Puerto Rico. *Danzas* are of two types: romantic and festive, each with its own style.

Manuel G. Tavarez (1843–83) is a young pianist who raises the level of sophistication of the *danza.* He composes many *danzas* but is remembered for "*Margarita*" and "*Tu ausencia*" (Your absence). His disciple, Juan Morel Campos (1857–96) takes the musical form to a higher level and composes more than 300 *danzas.* While *Danza* loses its allure in the twentieth century, Puerto Ricans still listen to Campos' "*Felices Dias*" (Happy Days) and "Laura and Georgina." Puerto Rico's national anthem, "*La Borinquena*,'' is based on a romantic *danza.*

Struggle for Independence and the Spanish-American War

1867 Census shows population has grown to more than 650,000 people

According to the census, 346,437 whites and 309,891 people "of color" live on the island. People of color are classified as blacks, mulatto (of European and African descent), and mestizos of (European and Amerindian descent).

1868: September 23   Rebellion shows Puerto Rican discontent

By the 1860s, Spaniards face increasing challenges from Puerto Ricans seeking an independent state. The Spaniards prohibit any meeting that doesn't have their prior approval and enforce a 9 p.m. curfew. They imprison or exile those who challenge their rules. By this time, other nations have abolished slavery, but Spain, which has lost nearly all of its colonies in the Western Hemisphere, refuses to end slavery on the island.

Ramon Emeterio Betances, who publishes a document titled "The Ten Commandments of free men," is one of several well-educated Puerto Ricans leading the struggle for independence. Some of their demands include the abolition of slavery, the right to determine taxes and the right to elect their own leaders. Many of the independence leaders are writers and poets, yet the majority of the population remains poor and illiterate.

On September 23 about 600 rebels, most of them laborers, slaves and small farmers, rise against the Spaniards. The attempted revolution becomes known as *Grito de Lares* (Shout of Lares). The Spanish suppress the revolt when the revolutionaries are unable to get weapons and support from other islanders.

1870 Early political parties debate autonomy

The new Liberal Reform Party and the Liberal Conservative Party debate whether to remain part of Spain or seek independence. The Liberal Conservatives oppose any movement for reform. The Liberal Reformers are split between those who want Puerto Rico to be as much like Spain as possible and those who want autonomy from Spain.

1870 Puerto Rican travels through Latin America pressing for social changes

Eugenio María de Hostos y Bonilla (1839–1903) becomes one of the most prominent voices for Puerto Rican independence and earns an international reputation for upholding human rights.

He joins the pro-independence Cuban Revolutionary Junta in 1869 and a year later, begins a four-year trip through

Spanish monks build the Church of Porta Coeli in San German in 1538. It is believed to be the oldest church in the Western Hemisphere. (EPD Photos/CSU Archives)

Also unique to Puerto Rico is *Bomba* and *plena*, two musical styles with African roots. *Bomba* is first performed in the early 1800s by black laborers working in the sugar plantations. *Bomba* dancing and singing is often accompanied only by percussion instruments and danced by mixed couples.

Plena is more melodic than *bomba* and sometimes the music style is called "*el periodico cantado*" (the sung newspaper) because lyrics deal with everyday events. Couples also dance to *plena* music, but the dance is not an integral part of the performance.

Panderetas (hand-held drums), accordion, harmonica, *requinto* (solo drum), tambourines, *guiro* (scraped gourd once played by the Taino natives) and other instruments are used to perform the plena.

While *plena* is revived in the late twentieth century, it is the *salsa* rhythm that moves the islanders. While *salsa* is performed by many Latin American groups, it owes its roots to the Puerto Rican community of New York. *Salsa,* which develops after World War II, draws heavily from Cuban and African-Caribbean rhythms. The music is popular throughout the world. One of the great *salsa* musicians is Tito Puente (b. 1920), a New Yorker of Puerto Rican heritage.

Latin America speaking against slavery and pushing for support of a federation of Caribbean nations. In Peru, he comes to the defense of maltreated Chinese laborers and in Chile, he helps women gain admittance to professional schools. In Argentina, he advocates for a trans-Andean railway. In the 1870s and 1880s, he helps reform the educational systems of Chile and the Dominican Republic.

He writes fifty books and many essays. His most important work is *La Peregrinación de Bayoán*, which promotes Cuban independence. He writes his own epitaph: "I wish that they will say: In that island (Puerto Rico) a man was born who loved truth, desired justice, and worked for the good of men."

1873: March 22   The Spanish Crown abolishes slavery in Puerto Rico

Under intense political pressure, Spain abolishes slavery. One of the foremost abolitionists and political figures in Puerto Rico is Ramon Emeterio Betances (1827–98), who founds a clandestine society to help liberate the slaves.

Receiving a medical degree in Paris in 1855, Betances returned to Puerto Rico to fight a cholera epidemic. The Spanish government exiled him several times for his work against slavery and for promulgating independence.

In 1867, he fled to present-day Dominican Republic and founded the Revolutionary Committee of Puerto Rico. One year later, he organized the Grito de Lares insurrection. When it failed, he returned to Paris to continue working for Puerto Rican independence.

The French government rewards him with the Legion of Honor for his contributions to literature. He dies in Paris.

1876   Spanish create forest reserve

El Yunque, a 28,000-acre rain forest, is designated a forest reserve, one of the oldest in the Western Hemisphere.

1887: March   New party seeks political changes in Puerto Rico

Roman Baldorioty de Castro (1822–89) is one of the founders of the Autonomous Party. Its platform seeks home government for Puerto Rico and representation in the Spanish parliament. Some of his colleagues criticize him for espousing reforms instead of revolution.

Yet Baldorioty argues against slavery and presses for establishing a constitution that guarantees the rights of islanders.

He had studied in Europe and returned to teach in Puerto Rico in 1853. He founded and edited a magazine and in 1878, started a political weekly, *La Cronica*, as a vehicle for his ideology. After founding the Autonomous Party, he is arrested on charges of publishing seditious propaganda.

United States Intervention in the Caribbean

1890   U.S. foreign policy seeks to increase influence in the Caribbean

Alfred T. Mahan (1840–1914) writes *The Influence of Sea Power upon History,* which advocates the taking of the Caribbean Islands, Hawaii, and the Philippine Islands for bases to protect U.S. commerce. He advocates the building of a canal in Central America to enable fleet movement from ocean to ocean as well as building a large navy with steam-driven armor-plated battleships. His work has great influence on U.S. foreign policy.

While many Americans advocate the imperial ambitions of the United States, many others denounce it. Among those against U.S. imperialism is author Mark Twain (1835–1910).

1895: June 12   Residents rise against Spain in Cuba

Spain, which lost most of its New World colonies by the 1820s, still holds on to Cuba and Puerto Rico. Toward the late nineteenth century Cubans and Puerto Ricans begin a more organized campaign for independence. Often, Cuban and Puerto Rican liberators work together and events in Cuba eventually affect Puerto Rico. As Spain loses its grip in Latin America, the United States emerges as a world power that seeks to expand its role in the Americas.

Cubans challenge Spanish rule with insurrections. At first, U.S. President Grover Cleveland proclaims U.S. neutrality.

1896: February 28   United States calls for Cuban independence

Clashes in Cuba between independence forces and the Spanish government continue and the United States approves the joint John T. Morgan/Donald Cameron resolution calling for recognition of Cuban belligerency and Cuban independence. The House of Representatives also approves its own version of the Morgan/Cameron resolution.

The resolutions pressure the White House to pay more attention to the Cuban crisis. In the meantime, Spain, thanks to Great Britain's opposition, is unable to muster support for its policies in Cuba.

1896: December 7   United States threatens to get involved in Cuban crisis

U.S. President Grover Cleveland declares that the United States may take action in Cuba if Spain fails to resolve the crisis.

1897: November 25 Spain gives Puerto Rico some autonomy

Under United States pressure, Spain agrees to an autonomous constitution for Puerto Rico. It allows the island to retain its representation in the Spanish government while providing for a bicameral legislature. This legislature consists of a Council of Administration with eight elected and seven appointed members, and a Chamber of Representatives with one member for every 25,000 inhabitants.

1898: February 9 New Puerto Rican government is inaugurated

Governor General Manuel Macías inaugurates the new government of Puerto Rico under the Autonomous Charter, which gives town councils complete autonomy in local matters. Under the arrangement, the governor has no authority to intervene in civil and political matters unless authorized to do so by the Cabinet.

1898: February 15 Explosion sinks the U.S.S. *Maine* in Havana harbor

The explosion that sinks the U.S.S. *Maine* turns public opinion in favor of going to war against Spain and on March 9, the U.S. Congress approves $50 million for this purpose. On April 19, Spain and the United States suspend all diplomatic relations and President McKinley orders a blockade of Cuba.

U.S. Forces Attack Puerto Rico

1898: May 10 Americans and Spanish exchange fire

Spanish forces in the fortress of San Cristóbal in San Juan exchange fire with the U.S.S. *Yale* under the command of Captain William Clinton Wise.

1898: July 21 U.S. sends forces to Puerto Rico

A convoy of 3,300 soldiers and nine transports escorted by the U.S.S. *Massachusetts* sail for Puerto Rico from Guantánamo, Cuba. Already U.S. ships have bombarded San Juan and blockaded the harbor there.

1898: July 25 U.S. troops land in Puerto Rico

U.S. Troops under the command of General Nelson Miles disembark in Guánica on the southern coast of Puerto Rico. A day later, Brigadier General George Garretson and Guy V. Henry arrive at Yauco and gain control of the key railroad line connecting it with Ponce, the largest city on the island.

Miles and his troops arrive in Ponce. From here, Miles presides over civil and military affairs on the island through early August. He later proclaims that the purpose of the U.S. invasion is to bring Puerto Rico a "banner of freedom."

American forces continue to arrive in the island and on August 9, they inflict heavy losses on the Spanish garrison in Coamo. American troops encounter heavy resistance from Spanish troops in the mountains.

1898: August 12 U.S. and Spain sign armistice

U.S. President William McKinley and French Ambassador Jules Cambon, acting on behalf of the Spanish government, sign an armistice with Spain relinquishing its sovereignty over the territories of Cuba, Puerto Rico, and the Philippines. The fate of these territories will be decided during peace talks.

American troops end their attacks on the Spanish in Puerto Rico on August 13. On September 9, U.S. and Spanish representatives meet to discuss the withdrawal of Spanish troops and the cession of the island to the United States.

1898: September 29 Puerto Rico is officially ceded to the United States

By October 18, the Spanish withdraw all their troops from Puerto Rico and General John Rutter Brooke becomes head of the U.S. military government on the island. General Guy V. Henry later succeeds General Brooke as military governor of Puerto Rico.

A New Era for Puerto Rico: U.S. Colonialism

1898: December 10 Spain loses last remnants of once mighty empire

The Treaty of Paris ends the Spanish-American War of 1898. As a result of this treaty, Spain loses the last of its empire in the New World. The United States receives Puerto Rico and Guam, liquidates its possessions in the West Indies and agrees to pay $20 million for the Philippines. Cuba becomes an independent nation.

1899: August 8 Devastating hurricane hits Puerto Rico

Hurricane San Ciriaco hits Puerto Rico, killing more than 3,000 people. Thousands lose their homes and the sugar and coffee industries are devastated. The hurricane provokes a major economic crisis.

1899 Puerto Rican pushes for statehood

Jose Barbosa (1857–1921) forms the pro-statehood Republican Party on July 4. He is a prominent politician and medical doctor who believes that statehood will benefit Puerto Rico because the United States grants its citizens more rights than many independent Latin American countries. He establishes the *El Tiempo* newspaper in 1907 and holds a seat in the Puerto Rican Senate.

1900: April 12   Congressional Act ends military rule in Puerto Rico

The U.S. Congress passes the Foraker Act, establishing a civilian government in Puerto Rico under U.S. control. The Act provides for an elected House of Representatives on the island, but not for a vote in Washington. On May 1, Governor Charles H. Allen is sworn into office, the start of U.S. civilian government in Puerto Rico.

On June 5, President McKinley names an executive cabinet under Allen that includes five Puerto Ricans and six U.S. members.

1903   Rain forest becomes part of national forest system

The United States officially designates the Luquillo Forest Reserve in Puerto Rico as the only tropical rain forest in the U.S. National Forest System.

1904   New party gains a following

Luis Muñoz Rivera (1859–1916) and José de Diego (1867–1918) found the Unionist Party of Puerto Rico to fight against the colonial government established under the Foraker Act.

De Diego is as well known for his poetry as his politics. One of his poems, "To Laura," becomes a favorite among his contemporaries. He writes the poem after an unhappy love affair.

De Diego advocates Puerto Rican independence and the establishment of a confederation of Spanish-speaking islands in the Caribbean, including the Dominican Republic. He publishes several poetry books. After the United States annexes the island, de Diego champions the use of Spanish in public schools and argues for an end to U.S. colonialism.

1906: November 6   Unionist Party gains votes

A new electoral law gives the vote to all males twenty-one and older. The Unionist Party wins the elections to the Legislative Assembly and sends Tulio Larrinaga to Washington as Resident Commissioner.

1906: December 11   U.S. President visits Puerto Rico

During a visit to Puerto Rico, U.S. President Theodore Roosevelt addresses the Puerto Rican Congress and recommends that Puerto Ricans become United States citizens.

1912   New Party demands full independence

Rosendo Matienzo Cintrón, Manuel Zeno Gandía, Luis Llorens Torres, Eugenio Benítez Castaño, and Pedro Franceschi found the Independence party. It is the first party in the history of the island to exclusively want Puerto Rican independence. Though short-lived, it establishes a precedent for future organizations with similar ideologies.

1915   Puerto Ricans demand more autonomy

A delegation from Puerto Rico, accompanied by Governor Arthur Yager, travels to Washington to ask Congress to grant the island more autonomy.

1916: December 5   U.S. President pushes for citizenship

President Woodrow Wilson urges Congress to pass the Jones Act which would allow Puerto Ricans to become U.S. citizens. Yet many Puerto Ricans oppose U.S. citizenship. They see it as a ploy to end independence sentiments on the island and to make Puerto Rico a permanent possession of the United States.

1917   One of Puerto Rico's most respected painters dies

Francisco Oller (1859–1917) is best known for *El Velorio* (The Wake), a painting he exhibits at the Paris Salon in 1895. Painted in a realistic style, it portrays the ceremonial wake held for infants who die on the island. Oller, who studies in Spain and Paris, becomes one of the first Latin American impressionists. His work has been exhibited at some of the most important museums around the world.

1917: March 2   Puerto Ricans become U.S. citizens

President Woodrow Wilson signs the Jones Act, giving Puerto Ricans U.S. citizenship and a bill of rights. The Act also establishes a locally elected Senate and House of Representatives. However, the Foraker Act still determines economic and fiscal aspects of government.

1922   Court case declares that Puerto Rico is a territory

A U.S. Supreme Court decision declares that Puerto Rico is a territory rather than a part of the Union. The decision states the U.S. constitution does not apply in Puerto Rico.

1924   Cuba and Puerto Rico mourn death of writer

Lola Rodríguez de Tió (1843–1924), one of the most prominent women in the independence movement, has been an important literary figure until her death in Cuba in 1924.

In 1868, inspired by the "Grito de Lares" insurrection, she wrote patriotic lyrics to the tune of "La Borinqueña," a native dance. The song became popular but caused her trouble with the Spanish government. She was exiled several times for her writings and her support of revolutionary sentiments against Spain, both in Puerto Rico and Cuba. Involved in Cuba's struggles for independence, she returned to that country to help found the Cuban Academy of Arts and Letters in 1910.

One of the foremost feminists, she publishes several essays dealing with women and society including "The Influence of Women on Civilization." She was once recognized for her suggestion that the Puerto Rican flag be fashioned after the Cuban flag, with its colors reversed. One of her most famous poems is "Cuba Y Puerto Rico Son" (Cuba and Puerto Rico Are).

1932: May 17   Puerto Rico gets its name back

During U.S. military rule of Puerto Rico shortly after the Spanish-American war, the name of the island is changed to Porto Rico. Now the U.S. Congress approves a law to change the name back to its original—Puerto Rico.

1946   Woman fights sexism to gain mayoral seat

Felisa Rincón de Gautier (1897–1994) becomes the first woman mayor of San Juan in 1946, the highest political post held by a woman in the island's history. To achieve this Rincón challenged the male hierarchy, including going against her father's wishes when, in 1932 she registered to vote and quickly became involved in politics. The Popular Democratic Party nominated her as a candidate for mayor of San Juan in 1944, but she declined when her husband opposed the nomination.

Now, as mayor, she is very popular declaring city hall the "house of the people." She remains active in national politics until her death in 1994.

1947–52   Political changes lead to commonwealth status

In 1947, Congress approves a law that allows for popular election of Puerto Rico's governor and in 1948, Luis Muñoz Marín (1898–1980) is elected. Educated at Georgetown University, he campaigns for land ownership reform and better economic conditions.

While he becomes one of the most influential political figures of Puerto Rico in the twentieth century, his ideas on Puerto Rican sovereignty remain ambiguous. Early on, he criticizes U.S. policy in Puerto Rico, but later seems to claim the island is not ready for independence. Instead, he pushes for commonwealth status, which gives Puerto Rico more independence but keeps it tied to the United States.

During his administration, he begins "Operation Bootstrap" to attract U.S. investment to the island. He helps draft the island's constitution and oversees its transformation into a self-governing Commonwealth.

In 1950, the U.S. allows Puerto Rico to create its own constitution. Voters approve the measure on June 4, 1951. The new constitution is ratified by popular referendum on March 3, 1952.

Puerto Rico officially becomes a commonwealth freely associated with the United States on July 25, 1952. Yet there are no fundamental changes in the relationship between the United States and Puerto Rico.

The Commonwealth of Puerto Rico enjoys nearly complete internal autonomy. The chief executive is the governor, elected by a popular vote to a four-year term. The legislature consists of a twenty-seven-member Senate and fifty-one-member House of Representatives elected by popular vote to four-year terms. The Supreme Court and lower courts are tied

Felisa Rincón de Gautier (1897–1994) makes history when she becomes San Juan's first female mayor in 1946. (AP/Wide World Photos)

to the U.S. federal court system. Appeals from Puerto Rican courts can be heard by the U.S. Supreme Court.

1950   Puerto Rican wins Oscar and Tony awards

Jose Ferrer (1912–92) becomes the first actor to win a Tony and an Oscar in the same year. Ferrer is considered one of the best character actors of his generation and his role in Shakespeare's *Othello* earned him praise and recognition in 1942. He earns an Oscar in 1950 for *Cyrano de Bergerac*. Ferrer also plays supporting roles in *Lawrence of Arabia, Ship of Fools, The Caine Mutiny* and *A Midsummer Night's Sex Comedy.*

1953   One of Puerto Rico's leading poets dies in New York

Julia de Burgos (1914–53) is one of the leading members of the literary Vanguard movement in San Juan in the late 1930s. Influenced by the Chilean Nobel laureate Pablo Neruda

(1964–73), de Burgos' work places her among the greatest poets in Latin America. She publishes several books. One of her best-known poems is "Rio Grande de Loiza."

She is a member of the Nationalist Party, which seeks independence from the United States and many of her poems are dedicated to the nationalist leader Pedro Albizu Campos. Her poems celebrate the Puerto Rican landscape and empathize with women and the poor.

1954: March 1   U.S. Representatives shot by Puerto Rican extremists

Puerto Rican extremists open fire from the visitors' galley of the U.S. House of Representatives, injuring five members. Representatives wounded in the attack include Alvin M. Bently, Michigan; Ben F. Jensen, Iowa; Clifford Davis, Tennessee.; George H. Fallon, Maryland.; and Kenneth A. Roberts, Alabama. Four suspects are arrested and sent to prison.

One of the most prominent nationalists of the twentieth century is Pedro Albizu Campos (1891–1964). Educated at Harvard, he advocated the legal or illegal overthrow of U.S. colonial authority. He joined the Nationalist Party in 1924 and quickly became a leader of the independence movement. In 1930, he became president of the Nationalist Party and in 1936, he was sentenced to ten years in prison in Atlanta, Georgia, for seditious conspiracy (treason).

In October of 1950, he was accused of masterminding a nationalist uprising and planning an assassination attempt on President Truman. He was sentenced to fifty-three years in prison and was offered a conditional pardon in 1953. But Governor Luis Munoz Marin withdraws the offer after the 1954 attack on the U.S. House of Representatives.

1967   Plebiscite upholds Commonwealth status

Puerto Ricans go to the polls and favor the continuation of the commonwealth, with 60.5 percent of voters in favor of it and 38.9 percent in favor of statehood.

1973   Puerto Rican inducted into the Hall of Fame

Roberto Clemente (1934–72), becomes one of baseball's legendary figures. Born into a poor family in Carolina, he played for the Pittsburgh Pirates from 1955 to 1972, compiling a lifetime batting average of .317. He won four National League batting titles and was the league's most valuable player in 1966. He was selected for twelve All-Star teams and earned twelve Gold Glove awards.

Outside baseball, Clemente organizes many social and humanitarian causes on the island and sponsors many athletic programs for disadvantaged youth. In 1972, he organized a relief effort for earthquake victims in Managua, Nicaragua and died in an airplane crash later while carrying supplies to that nation.

1976   New U.S. tax law leads to more jobs in Puerto Rico

A new tax code allows American companies to operate in Puerto Rico without paying taxes. It is a boost for the island, which has depended heavily on sugar, coffee and other agricultural products. In 1952, there are only 82 labor-intensive plants in Puerto Rico. In 1992, there are 2,000.

1979   U.S. President Jimmy Carter releases four Puerto Rican nationalists

President Carter, through executive clemency, frees Lolita Lebrón, Andrés Figueroa Cordero, Rafael Cancel Miranda and Irving Flores. The four nationalists had been involved in the attack on the U.S. House of Representatives in 1954.

Since their arrest supporters have claimed that the four are political prisoners and have demanded their release. The White House cites humanitarian concerns for releasing them.

1985   Puerto Rican actor stars in hit movie

Raul Julia (1940–94) plays a major role in the film *Kiss of the Spider Woman,* launching him into worldwide fame. Julia, born in San Juan, is a versatile actor who plays many roles in television, theater and cinema. He dies in 1994 after a brief illness and is buried in Puerto Rico.

1991   "Greatest living guitarist" wins sixth Emmy

Jose Feliciano (b. 1940), often called a guitar virtuoso, wins his sixth Emmy award. Born in the town of Lares, Feliciano demonstrated great musical talent at an early age. During his career, he has been prominent in English and Spanish pop music and well-received as a classical guitarist. Nominated for eleven Grammy awards, he has thousands of fans throughout the world. Two of his better known songs are "Light My Fire" and "Feliz Navidad," a Christmas song.

1993   Puerto Ricans vote to keep commonwealth

By a slight margin Puerto Ricans vote to remain a commonwealth rather than seek independence or statehood. The plebiscite draws 1.7 million voters, about 73.6 percent of the electorate. For commonwealth: 48.4 percent; for statehood: 46.2 percent; for independence: 4.4 percent.

1993   Puerto Rican wins international award

Ricardo Alegria (b. 1921), considered the driving force behind the creation of the influential Institute of Puerto Rican Culture, wins the United Nations Educational, Scientific and Cultural Organization's Picasso Medal for his efforts to preserve historical monuments. The same year, President Clinton awards him the Charles Frankel Award of the Humanities.

Alegria is a distinguished archeologist and writer. He advocates the preservation of Puerto Rican culture and has conducted important excavations on the island.

1998: March   House to let Puerto Ricans decide fate

The U.S. House votes 209–208 to let Puerto Ricans decide whether to become the fifty-first state, remain a commonwealth or seek full independence. It remains a commonwealth.

1998: June   Cruise line agrees to pay fine for polluting

Royal Caribbean Cruises agrees to pay $9 million in fines for dumping oily bilge waste in Puerto Rican waters. The fine sets aside $1 million for conservation projects in Puerto Rico and Florida.

1998   Hurricane devastates Puerto Rico

Hurricane Georges, with 110 mph winds, strikes the island, killing several people and leaving thousands without homes.

Bibliography

Berbusse, Edward J. *The United States in Puerto Rico, 1898-1900*. Chapel Hill: University of North Carolina Press, 1982.

Betances, Samuel. (ed. By Maria Teresa Babin and Stan Steiner), *Race and the Search for Identity: An Anthology of Puerto Rican Literature*. New York: Vintage Books, 1981.

Fernandez, Ronald. *The Disenchanted Island: Puerto Rico and the United States in the Twentieth Century*. 2d ed. Westport, Conn.: Praeger, 1996.

Figueroa, Loida. *History of Puerto Rico*. New York: Anaya, 1974.

Morales, Carrion, Arturo. *Puerto Rico: A Political and Cultural History*. New York: Norton, 1983.

Morris, Nancy. *Puerto Rico: Culture, Politics, Indentity*. Westport, Conn.: Praeger, 1995.

Rouse, Irving. *The Tainos*. New Haven, Conn.: Yale University Press, 1992.

St. Kitts and Nevis

Introduction

In August of 1998, St. Kitts and Nevis, already among the world's smallest nations, seemed poised to break into two nations that would be even smaller. Historically, residents on the island of Nevis felt neglected and even ignored by the bigger island—and greater population—of St. Kitts.

Many of the other island nations in the Caribbean were weary of Nevis' attempts to secede from the two-island federation. Nearing the end of the twentieth century, the post-colonial island nations were seeking unity to compete in the global economy. The possible break up of St. Kitts and Nevis was seen as a step back.

Ultimately, the secessionists failed to gain a two-thirds majority—even though sixty-one percent of the votes cast on the island of Nevis called for secession—and Nevis remained, for the moment, a part of the federation.

One of the smallest nations in the world, the Federation of St. Kitts and Nevis is slightly larger than the area occupied by Washington D.C. With the Caribbean Sea on one side and the Atlantic Ocean on the other, St. Kitts and Nevis are part of the Leeward Islands. They lie approximately 200 miles southeast of Puerto Rico.

St. Kitts, the larger of the two-island nation, is twenty-three miles long and five miles across at its widest point, with a total area of 104 square miles. Nevis, separated from St. Kitts by a two-mile channel, is eight miles long and 6 miles wide, with a land area of 36 square miles. The capital city, Basseterre, is located on St. Kitts.

"Mother Colony of the West Indies"

The lack of any visible resources may have kept the Spanish off the islands, sighted by Christopher Columbus in 1493. They remained free from European colonization for the next 130 years.

St. Kitts has been referred to as the "mother colony of the West Indies" to reflect its status as the first English colony in the Caribbean. The British established a small settlement

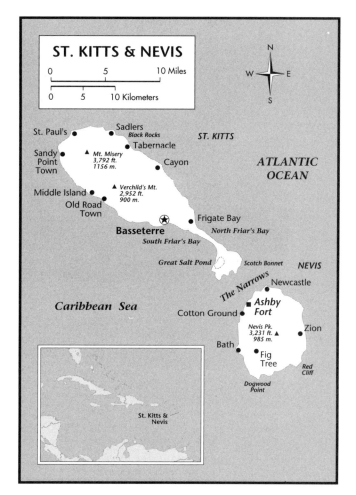

there in 1623, and the French arrived within a year. They soon joined forces to battle the native SCarib population, which they decimated through violence, enslavement, and disease. From St. Kitts, the English and the French colonized other islands and some passed through here before moving on to colonize North America. One of them was Captain John Smith, who established the colony of Virginia. U.S. statesman Alexander Hamilton was born in St. Kitts in 1757.

The first white settlers of St. Kitts and Nevis attempted to recreate European society on the islands. They planted indigo and tobacco, but their product could not match the quality of the tobacco produced in such North American colonies as Virginia. Over time, the settlers began to cultivate sugarcane.

Sugar became a major crop in the Caribbean, slowly spreading from island to island and then to the rest of the Americas. When the European colonies could not meet the world demand for sugar, they began importing African slaves to cultivate sugarcane and work in the sugar factories. Between 1518 and 1870, most new arrivals to the Caribbean were African slaves. By the early nineteenth century, thousands arrived each year—55,000 in 1810. Those who survived the horrible and inhumane transatlantic crossing, stuffed like cargo in miserable, unhealthy ships, were put to work under brutal conditions.

By the late eighteenth century, St. Kitts and Nevis became important sugar producers. Although African slaves were a numerical majority in the islands, the plantation owners remained firmly in charge. But the English and the French, both with colonies on the island, did not even want to share the wealth between themselves. For more than a century, they fought one another for control of St. Kitts and Nevis. The English ultimately prevailed and took formal possession of St. Kitts and Nevis in 1783.

Modern St. Kitts and Nevis society owes its origins to the sugar plantation system. As in other British possessions, white plantation owners generally held power in the islands using slave-labor to create wealth for the Europeans. When the slave system was abolished in the mid-nineteenth century, the freed slaves soon discovered that they faced new problems. They could not afford to buy land and began to compete as wage laborers. By this time, sugar prices were declining in the world market and the economy, dependent on sugarcane, began spinning out of control throwing St. Kitts and Nevis into chronic economic problems.

A desire for independence grew among the British colonies throughout the Caribbean in the mid-twentieth century. Some envisioned one nation in the Caribbean: indeed, the colonies had a common past, but little to hold them together as one nation. St. Kitts and Nevis, and the neighboring island of Anguilla had been governed as one unit for much of their colonial history. But even for the three islands, unity remained elusive.

A West Indies Federation was formed between islands in the Caribbean in 1958 but it was short-lived, disbanding four years later, when Jamaica withdrew its membership. In 1967, St. Kitts, Nevis and the island of Anguilla became an associated state with Britain. Although this status gave them the ability to run their internal affairs in accordance with their own constitution, Britain retained jurisdiction over foreign relations and other matters. But the relationship among the three islands remained strained and Anguilla rebelled in 1969. Britain intervened and allowed Anguilla to secede in 1971.

In 1983, St. Kitts and Nevis became an independent federated state within the British Commonwealth. However that

didn't improve the relationship between the two islands. The new constitution gave Nevis its own legislature and the power to secede from the federation. A vote to secede was placed on the ballot in August 1998, and failed. However, the issues that divide the two islands remain.

Timeline

c. A.D. 600   Arawak and Carib Indians inhabit islands

Before explorer Christopher Columbus arrives in the Western Hemisphere in 1492, the earliest known inhabitants of St. Kitts and Nevis are the *Arawak* and the *Carib* Indians. Historians and anthropologists believe the Arawak and Carib originate in the South American jungles and migrate northward to the Caribbean islands.

The peaceful Arawak hunt, fish and cultivate the land. The Arawak flourish throughout the Caribbean, developing religion and living from fishing and agriculture. They appear to lead a peaceful, sedentary, and spiritual life, building ceremonial centers used frequently for religious activities. Remnants of glossy red Arawak pottery survive to the present day, and are sometimes found by beachcombing vacationers.

The Caribs are highly mobile, using their canoes to wage war on the Arawaks whom they displace throughout the Caribbean. The Carib are not as socially organized as the Arawaks and their villages are small, often only consisting of members of an extended family. Carib males are notorious for their ferocity. They use bows, poisoned arrows, javelins, and clubs to attack Arawak villages. The main purpose of the attack is to capture Arawak women for marriage to Carib men.

The Arrival of the Europeans and English Control

1493   Columbus sights St. Kitts and Nevis

Columbus sights the two islands in 1493. He is given credit for naming the islands St. Christopher and Nevis, which comes from *Nuestra Senora de las Nieves* (Our Lady of the Snows). To Columbus, the white clouds that surround Mt. Nevis, a mountain in the middle of Nevis, resemble snow. The Spaniards don't establish any settlements.

1607   John Smith stops at Nevis

Captain John Smith, on his way to establish the colony of Virginia, stops in Nevis to enjoy the hot sulfur baths in 1607.

1623 St. Kitts becomes first English colony in the Caribbean

St. Christopher fails to attract any Europeans until 1623, when a small group of English colonists led by Sir Thomas Warner (d. 1649) settle on Sandy Bay, at Old Road Bay. St. Kitts, as the English call the island, becomes known as the "mother colony of the West Indies" for being the first English colony in the Caribbean, although the French establish a settlement on the island the following year.

1626 Caribs slaughtered in final battle

The English and French join forces to vanquish the Caribs, who are slaughtered at a place that becomes known as Bloody Point. Three years later, the Europeans withstand a Spanish attack.

Without the Caribs or significant Spanish opposition, the English proceed to claim the islands of Nevis, Antigua, Barbuda, Tortuga, and Montserrat and the French claim Martinique and Guadeloupe. Both nations will fight for control of St. Kitts and Nevis well into the eighteenth century.

1628 The English colonize neighboring Nevis

The island is settled by a group of eighty English residents of St. Kitts, headed by the tobacco planter Anthony Hilton. More settlers come from England and the land is cleared for tobacco planting.

1632 Expedition to Antigua

After establishing a settlement on Nevis (see 1628), Warner leads an expedition to Antigua.

1660 Slavery fuels profitable sugar industry

By the mid-1600s, some 4,000 Europeans have settled in St. Kitts and are engaged in the sugar trade.

Although some of the settlers are experienced tobacco growers (see 1628), the tobacco grown on Nevis can't match the quality of tobacco being produced by Britain's North American colonies, chiefly Virginia. The islanders turn to sugarcane production and by the early eighteenth century, Nevis becomes one of the most important sugar producers in the English colonies. Earning the nickname "Queen of the Caribbees," Nevis becomes one of the wealthiest colonies in the British Empire.

The plantations in St. Kitts and Nevis are similar to those held by the British in other Caribbean islands. Absentee landowners own large tracts of land and grow cash crops for export. Indentured laborers and later African slaves do all of the hard labor.

1671 Caribbean islands join forces

For greater joint security, St. Kitts and Nevis join Antigua (with Barbuda and Redonda) and Montserrat as part of the Leeward Caribbees Islands Government under a British governor.

1690 Earthquake shakes the islands

A large earthquake causes heavy damage in the European colonies and a tidal wave destroys a small community in Nevis.

1713 Treaty gives St. Kitts to Britain

Although France officially holds St. Kitts in 1664, this is merely a formality, for by 1713 Britain is granted St. Kitts under the Treaty of Utrecht. The French continue to press for control and the British are unable to secure the island until the late eighteenth century.

1778 Nisbet Plantation established

The Nisbet Plantation is established on Nevis. It is the home of Frances Nisbet, who marries Lord Horatio Nelson (1785–1805) there. By the 1990s, the plantation has been converted to a thirty-eight room resort known as the Nisbet Plantation Beach Club.

1782: January 11 to February 12 French invade

More than 8,000 French soldiers invade Basseterre and attack the British garrison on Brimstone Hill, defended by about 1,000 British soldiers, and armed slaves. On February 12, "The siege of Brimstone Hill" ends when the British surrender and France takes control of the island.

1783 The Versailles Treaty of 1783 returns St. Kitts to Britain

The Treaty effectively ends French control of St. Kitts.

Late 1700s Nevis attracts wealthy visitors

By the late eighteenth century, Nevis becomes a playground for thousands of wealthy international visitors. Thermal baths at Charlestown and lavish entertainment at the Bath House Spa and Hotel attract visitors. Many grand state houses are built during this time.

1816 St. Kitts and Nevis become part of new colony

The British reorganize their Caribbean colonies and incorporate St. Kitts, Nevis, and Anguilla to form a single colony: the British Virgin Islands. The official colonial designation changes, yet it means essentially the same thing to the people of St. Kitts and Nevis. In 1871 this designation changes again to become the Leeward Islands Federation. Though the official standing of the islands changes, the legal standing of St. Kitts and Nevis is still colonial in makeup.

1834   England takes first step to abolish slavery in its colonies

The English agree to abolish slavery over a five-year period. The act includes two measures that benefit the slave owners: it creates a system of "apprenticeship" that forces former slaves to continue working for their masters without pay, while compensating the former owners for their loss of property.

1932   Labor unrest gives rise to powerful political leader

Sugar prices collapse during the Great Depression, fueling labor unrest in St. Kitts and Nevis. Robert Bradshaw (1916–78) organizes the Workers League in 1932 and builds a large following. By 1940, his union establishes a political arm, the St. Kitts and Nevis Labour Party. Bradshaw's party will dominate politics in the islands for the next three decades.

Bradshaw was born and raised in St. Kitts. By 1932, he is an apprentice at the St. Kitts sugar factory machine shop. He helps stage a walkout during a wage dispute with management. When the company refuses to re-employ him, Bradshaw and several colleagues form the Workers League. Bradshaw quickly begins to play a major role in politics, serving in several national posts before becoming the first premier (similar to prime minister) of St. Kitts, Nevis, and Anguilla.

1958–62   Caribbean islands can't hold together

By the mid-twentieth century, Britain and its Caribbean colonies begin to negotiate for the eventual independence of the region. In 1958, St. Kitts, Nevis, and Anguilla join Jamaica, Trinidad and Tobago, Barbados, Grenada, Antigua and Barbuda, St. Lucia, St. Vincent and the Grenadines, Dominica, and Montserrat in the Federation of the West Indies.

But the federation is doomed from the start. The islands cannot compromise on an equitable way to share power and the federation collapses when Jamaica opts to leave in 1961.

1958   Sugar production ends

Sugar production ends and the mills are abandoned. Some sugar mills are later restored as tourist hotels. For example, the Golden Rock Plantation, the Hermitage, and the Monpelier Plantation Inn are all hotels today. Guests have included international celebrities such as British Princess Diana, and her sons, Princes William and Harry.

1961   First radio broadcasts get under way

Local radio transmission begins, but the government has a near-monopoly on the dissemination of information since it owns the only radio stations on the islands (see also 1972).

1967   St. Kitts, Nevis and Anguilla form one government

In preparation for full independence, the three islands of St. Kitts, Nevis, and Anguilla become an associated state with Britain, with full internal autonomy under a new constitution.

Robert Bradshaw (see 1932) becomes the first premier of the associated state and he establishes local councils in Nevis and Anguilla to give them more authority to deal with local affairs. Still the islanders of Anguilla rebel against rule by St. Kitts. British paratroopers intervene and Anguilla is allowed to secede from the associated state in 1971.

1970   Nevis presses for equal status

Nevis' new political party, the Nevis Reformation Party, advocates secession from St. Kitts as the only solution to the island's lack of autonomy. In 1975, the party receives eighty percent of the vote on Nevis and its candidates win seats in the Nevis legislature.

1972   Television broadcasts begin

Television is first broadcast on the islands by government-owned television stations. The government essentially controls dissemination of information on the islands since it owns the radio and television stations (see also 1961).

1975   Government takes control of all sugarcane fields

The government nationalizes the sugarcane fields and assumes ownership of the central sugar factory in Basseterre. This move by the Labour government is not widely accepted. In Nevis, residents increasingly feel antagonized by St. Kitts' power structure and Bradshaw's government.

1978   Death of powerful leader gives way to political changes

The death of Bradshaw in 1978 (see 1932) followed by the death of his replacement, C. Paul Southwell a year later are serious blows to the strength of the Labour Party. Disorganized and weakened by the loss of their leaders, Labour Party supporters begin to look to new leaders.

1980   Two-party coalition holds power

A two-party coalition holds a slim majority over the weakened Labour Party (see 1978). The Nevis Reformation Party and the People's Action Movement (a middle-class party organized in 1965) begin to push for full independence from Britain.

1983: September 19   St. Kitts and Nevis become an independent federation

St. Kitts and Nevis become an independent nation. The 1983 Constitution provides the islands with a Parliament headed officially by Queen Elizabeth II, who is represented by a governor-general.

The Constitution allows Nevis to have its own legislature, premier, deputy governor-general and cabinet members and guaranteed central government representation.

Eleven of the fourteen members of the House of Assembly are elected, while three are nominated (two on the advise of the prime minister and one on the advise of the opposition leader). The prime minister, as leader of the majority party in the House, leads a cabinet of four other ministers and an attorney general.

1991   Government seeks education reforms

A National Committee on Education is established to consider changes to the country's educational system, both at the primary and secondary level. While the nation has a literacy rate of more than ninety percent, the government considers introducing technical and vocational education and training. It also looks at raising the qualifications of teachers and improving their working conditions. Nearly eighty percent of primary school teachers are women. At the secondary level, fifty-nine percent of teachers are women.

May 1998   Caribbean leaders once more discuss unification

Caribbean island leaders, fearing that they will become economically disadvantaged in the world market, begin to talk again about a confederation of island nations. One proposal would allow each nation to keep a national government and internal laws, but the islands would create joint foreign and economic policies, share defense, judicial and higher education goals. Among the possible members are St. Kitts and Nevis, Antigua and Barbuda, St. Vincent and the Grenadines, Montserrat, Barbados, St. Lucia, Grenada, and Dominica.

1998: August   Suspected drug dealer threatens U.S. students

In the early 1990s, the United States cites St. Kitts as a major drug-smuggling transit point. The disclosure is a shock to many on St. Kitts, who know little to nothing of its existence previously. The U.S. State Department announces that a local businessman known as Charles "Little Nut" Miller has threatened to kill American students enrolled at a St. Kitts veterinary school if the United States is successful in extraditing him. The announcement underscores the region's increased cocaine trade.

1998: August   Breakup of nation fails by narrow margin

Nevisians who have long complained that St. Kitts has failed to distribute resources equitably, go to the polls to vote for secession. But Nevis remains part of the union when secessionists fail to gain a two-thirds majority. Denzil Douglas, prime minister, opposes secession for Nevis and pledges to develop a plan for greater autonomy for Nevis.

1998: September   Hurricane Georges causes damage

St. Kitts suffers extensive damage from Hurricane Georges, but resort hotels pledge to repair in time for winter travel season.

Bibliography

Brown, Whitman, T. *From Commoner to King: Robert L. Bradshaw—Crusader for Dignity and Justice in the Caribbean.* Lanham, MD: University Press of America, 1992.

Cox, Edward L. *Free Coloreds in the Slave Societies of St. Kitts and Grenada, 1763–1833.* Knoxville: University of Tennessee Press, 1984.

Hamshere, Cyril. *The British in the Caribbean.* Cambridge, MA: Harvard University Press, 1972.

Meinig, D.W. *The Shaping of America, A Geographical Perspective on 500 years of History.* New Haven and London: Yale University Press, 1986.

Moll, V.P. *St. Kitts-Nevis.* Santa Barbara, CA: Clio, 1994.

Olwig, Karen Fog. *Global Culture, Island Identity: Continuity and Change in the Afro-Caribbean Community of Nevis.* Philadelphia: Harwood, 1993.

St. Lucia

Introduction

St. Lucia is a small island in the eastern Caribbean located in the middle of the Lesser Antilles, the chain of small, volcanic islands running north-south between Venezuela and the Virgin Islands. At 238 square miles (616 square kilometers), St. Lucia is the second largest of the Windward Islands, the southern half of the Lesser Antilles, and is more than three times as large as Washington, D.C. The island is shaped like a pear with the narrow end pointing north. It is roughly twenty-seven miles long and fourteen miles wide at its broadest point.

The island is bisected by a ridge of mountains running north-south, the tallest of which is Mount Gimie at 3,145 feet (958 meters). The landscape is traversed by numerous small rivers flowing down from the mountains. The soil is fertile and capable of sustaining diverse crops, but much of St. Lucia's history has been dictated by a search for competitive agricultural commodities. Currently the nation is heavily dependent on banana production. The southwest portion of the island is a highly active geothermal area, particularly the boiling sulfur springs near the town of Soufriére. This region has enormous potential for geothermal energy and recently the government has explored the feasibility of extraction. If this resource could be tapped, the nation would be freed from the expense and burden of petroleum imports.

Language and Religion

The ethnic composition of St. Lucia reflects the nation's slave history. Nearly ninety percent of the roughly 160,00 people are of African descent. English is the official language, but most people, as much as ninety percent, speak a form of patois that combines French, English, and African languages. There is a strong class dimension to language in St. Lucia. English is the language of the educated and economic elite. Patois is the language of the majority underclass. Twenty percent of the population speaks no English. Religious practices reflect the historic presence of the French. Perhaps as much as ninety percent of the population is Catholic. But African-based religious practices have a strong presence on the island and intermingle to varying degrees with Christianity. The capital city, Castries, contains more than 55,000 of the nation's inhabitants.

Pre-Columbian Era

Given its short period of independence (since 1979), St. Lucia's history is perhaps best examined in only two periods, before and after the arrival of the Europeans. In the era prior to the arrival of Europeans, St. Lucia's role in the history of the Caribbean was that of a stepping stone in the migratory path of the northward moving Indians. With the exception of one migration from the north, all the Indian peoples of the Caribbean arrived from the south, specifically the interior of modern-day Venezuela and Brazil. As these migrants moved northward, they passed through St. Lucia. However, it was only the final group of migrants, the Caribs, who inhabited St. Lucia by the time of European contact. The Caribs left the South American coast in approximately A.D. 1200 and went only as far north as Guadeloupe.

English and French Struggled over St. Lucia

Although Spanish explorers sighted and identified St. Lucia in the early 1500s, the island remained unsettled by Europeans for more than 150 years. This is partly because European powers did not make concerted efforts to occupy the island, but also the Carib inhabitants vigorously repulsed the few attempts that were made. Finally in 1650 the French established the first permanent settlement and in the process pushed the Caribs into extinction. Over the next 150 years the island changed hands between the French and the British at least fourteen times, and probably more. Settlers from both France and England wanted to turn the island into a productive economic enterprise and began importing slaves and experimenting with tropical agricultural commodities in hopes of finding one that would bring steady, long-lasting revenues. Sugar was one of these products, but there was also tobacco, ginger, and cotton.

The ongoing conflicts between England and France led to a popular insurrection on the island in 1795. The revolutionary government of France abolished slavery throughout its Caribbean possessions in 1794, and slaves on English-controlled islands were willing to go to extreme measures to gain their freedom. When this hope for freedom combined with the ire of French nationals, the result was protracted guerrilla warfare against British authority. The British defined these rebels as "Brigands," and on St. Lucia they succeeded in pushing the British back to solely defensive positions during a year of fighting in 1796. The conflict was finally resolved when the British made a variety of concessions to the rebels.

British Colony and Independence

St. Lucia was a British colony from 1814 until independence in 1979. It remained throughout that period an agricultural island dedicated to producing tropical commodities. Unfortunately for St. Lucia, none of these crops, nor the policies of economic administrators, has allowed the island to move beyond its status as a primary producer for the fickle international market. St. Lucia's current dependence on bananas reflects a long historical pattern of dependency on one or more crops. Unlike other banana producing regions throughout Central America and the Caribbean, much of St. Lucia's banana production is carried out by small-holders, usually family-run units working from a small plot of land. This has the potential of creating a society with a relatively equitable distribution of income, which is a significant step towards achieving economic diversification. In the 1990s U.S. pressure on the European Union to abandon its subsidies for Caribbean banana producers caused great concern on St. Lucia. The island's producers face the prospect of competing on the open market against massive U.S. fruit corporations. This scenario has been St. Lucia's basic dilemma throughout history, and although it is not a unique problem, it does not make the prospect of major economic transition any easier. St. Lucia's per capita income remains very low, and agriculture is still the island's main economic enterprise. Like many other nations in the Lesser Antilles, St. Lucia is looking to expand its tourist industry to fulfill an economic need.

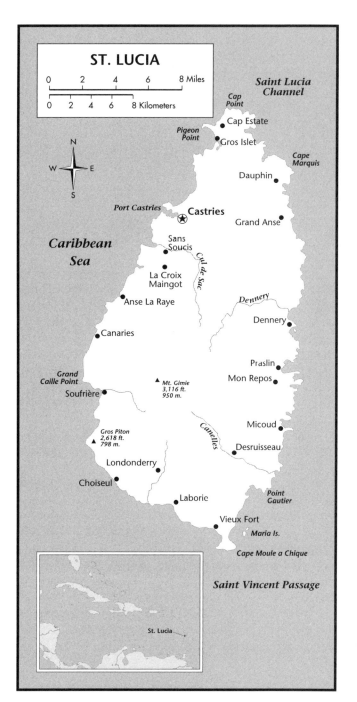

Timeline

c. 4000 B.C.   First major migration into Caribbean

The first inhabitants of the Caribbean begin migrating into the region from the north, probably from the Yucatán peninsula. They are Lithic peoples and are termed "Casimiroid" by archaeologists. They settle on the islands of Cuba, Hispaniola, and Puerto Rico. Although they had to travel by water during their migrations, they are a land-oriented, rather than a water-oriented people, which explains why they did not migrate further south. Around 400 B.C. these Casimiroids will be replaced by migrants from the south, but a handful will remain in western Cuba by the time of Columbus's arrival and will be known as the Guanahatabey ethnic group.

c. 3500 B.C.   Proto-Arawakan

The language known as Proto-Arawak emerges deep in the Amazon basin of modern-day Brazil. This is the parent language of Indian groups who will eventually inhabit the Caribbean, including St. Lucia, at the time of Columbus's arrival.

c. 2000 B.C.   Second major migration into Caribbean

"Ortoiroid" peoples begin migrating north from the region of Trinidad and the South American coast. These people are "Archaic," meaning that they have technological capacities beyond stone chipping such as the production of stone, bone, and shell artifacts. Contrary to their Casimiroid counterparts to the north, they are sea-oriented.

c. 2000 B.C.   Emergence of pottery civilizations

In the upper reaches of the Orinoco River in modern-day Venezuela, a civilization known as "Saladoid" emerges. Its distinct characteristic is the development of pottery. Migrants carrying this new innovation will follow in the footsteps of their Ortoiroid predecessors and migrate northward into the Caribbean through the island of St. Lucia. These peoples are the ancestors of most of the Caribbean Indians that the Europeans will encounter. They are also part of a language family known as Proto-Maipuran which is directly descended from Proto-Arawak.

c. 1000 B.C.   Frontier with Casimiroids

The northward migrating Ortoiroid peoples arrive on the island of Puerto Rico, come into contact with the Casimiroids, and establish a frontier region between Puerto Rico and Hispaniola.

c. 1000 B.C.   Ceramic peoples arrive at the coast

From their homeland in the interior of South America, the pottery-bearing Saladoid migrants arrive at the coast, near Trinidad, and establish a foothold that will serve as their stepping off point to the Caribbean. The language of these migrants is now known as Proto-Northern, to distinguish it from the language of the Proto-Maipuran speakers who remained behind. Proto-Northern is the immediate parent language of Carib and Arawak, the two Indian groups that will occupy the southern Caribbean at the time of the European arrival. It is also the parent language of Taino, the peoples of the northern Caribbean at the time of the Spanish arrival.

c. 500 B.C.   Ceramic peoples pass through St. Lucia

The Saladoid peoples begin their northward migrations by first embarking to Trinidad from the South American coast. This migration does not occur as a large wave of conquerors. Rather the new migrants slowly filter northward in small bands searching for favorable areas in which to settle. They either select uninhabited areas, intermingle with existing populations, or push pre-existing peoples out of the way through conquest. These Saladoid peoples are responsible for introducing the ceramic age—pottery and the techniques of its production—to the Caribbean. As they begin their migration northward, they add to their pottery styles by developing new forms, such as jars, bottles, effigy vessels, and incense burners.

c. A.D. 100   Second migration of ceramic peoples

A second, minor, northward migration of Saladoid pottery peoples begins at the Trinidad/South American coast nexus. These second-wave migrants are Arawak speakers and are defined by a new type of pottery incision. They do not migrate beyond Guadeloupe.

c. A.D. 600   Ceramic peoples arrive at Puerto Rico

The pottery-bearing migrants arrive at the Puerto Rico/Hispaniola frontier. During their one thousand year migration, they have replaced much of the Ortoiroid civilizations that they encountered. Upon arriving at Puerto Rico, they divide into at least four new migrant civilizations, collectively called "Ostionoid," which refers to their new cultural styles and adaptations to the local environment. The general difference between the Saladoid pottery style and the new Ostionoid pottery style is that the former is smaller, finer, and delicately ornamented, while the latter is heavier, bulkier, and more massive. These four groups in turn migrate into the rest of the northern Caribbean, including Cuba, Jamaica, Hispaniola, and the Bahamas. They will become the "Tainos," the Indians who Columbus will encounter in 1492. The origin stories of most all Taino groups uniformly identify their homeland as a cave on the island of Hispaniola, which scholars site as evidence of their common ancestry.

A.D. 1200   Final Indian migration into Caribbean

A final northward migration into the Caribbean begins at the Trinidad/South America coast. The migrants are "Carib" speakers who share their linguistic origin with the Tainos and the Arawaks. The Caribs are the last group to migrate through St. Lucia, and they are the inhabitants of the island when the Europeans arrive. The Carib migration only goes as far north as Guadeloupe. Spaniards portray the Carib peoples as warlike and cannibalistic. Scholars contend that this portrayal was overstated in order to justify the suppression of the Indians, but they do acknowledge that the Caribs had to be more aggressive to force their way onto islands already inhabited by earlier migrants. The oral traditions of the Caribs can be traced to the island of Trinidad and the Guyanas on the South American mainland.

Pre-Twentieth Century

1498   Columbus possibly sights St. Lucia

According to folklore, Columbus sights St. Lucia on St. Lucy's Day in 1498. There is no evidence to corroborate this belief, but December 13 is still celebrated as the date of the island's discovery by Europeans.

1504   Spanish navigator observes St. Lucia

If Columbus did not sight St. Lucia in 1498, then the first European observer is likely to have been Juan de la Cosa in 1504. He is a map maker and navigator who explores the Windward Islands (the southern half of the Lesser Antilles) in the early 1500s.

c. 1550   Pirates use St. Lucia as base

Although Europeans have still not settled on St. Lucia, French pirates establish a small outpost on the island and use it to harass passing Spanish galleons.

1605   First attempt at colonization

The English engage in one of the first attempts at colonizing and establishing settlement on the island. The goal is to establish tobacco production. They are rebuffed by the combined influence of disease and resistance from the Carib Indians.

1638   Second attempt at colonization

The English make a second attempt at colonization with much the same results as the first effort; they are defeated by disease and strong Indian resistance. The settlement actually survives for eighteen months amid amicable relations with the Caribs. But when the English kidnap some Indians, the English are attacked and driven out.

1650   First European settlement

The French finally succeed in establishing a permanent settlement on the island. This is due largely to the fact that the settlement's leader, Rousselan, has married a Carib woman.

1660   Treaty with the Caribs

The French sign a treaty with the Carib Indians on the island allowing for the continuation of French settlement.

1664   English regain the island

Under the leadership of Thomas Warner, son of the governor of Saint Kitts, the English retake St. Lucia from the French.

1667   French take control

The Treaty of Breda restores French control over St. Lucia.

1674   Administration from Martinique

The French place St. Lucia under the control of the French administration on the island of Martinique.

1718   St. Lucia becomes grant from monarchy

The French monarchy grants the island of St. Lucia to the Marshal d'Estrées.

1722   Island is neutral ground

The British monarch George I allows for another attempt at British colonization on St. Lucia by granting the initiative to the Duke of Montague. France manages to prevent the settlement, but is forced to declare the island neutral territory.

1743   France takes possession

France resumes possession of St. Lucia.

1748   Neutral status again

The Treaty of Aix-la-Chapelle between France and England again establishes St. Lucia as a neutral island.

c. 1750   Slavery brings African culture

Under increasing importation of African slaves, the island of St. Lucia begins to show greater amounts of African cultural influences. In the arena of language, the slaves create the patois language that combines French and African languages, as well as some English words. In religion the Catholic overlay mixes with African-based religions to create a single belief system that merges the two together. The specific religious influence is the "Shango" cult, or Orisha, from the region of Nigeria. Belief in "obeah," or witchcraft, is a central component of the belief system. The names and symbols of Catholicism, especially the pantheon of saints, is appropriated to fit the meanings and identities of African deities.

1762   British capture island

The British, under the guidance of Admiral George Rodney and General Robert Monchton capture the island from the French.

1763   French gain back St. Lucia

St. Lucia is returned to the French with the Treaty of Paris.

c. 1772   Governor resists using St. Lucia as a refuge

The French governor of St. Lucia refuses to provide refuge for the "Black Caribs" of neighboring St. Vincent. The Black Caribs are an ethnic mix of Caribs and Africans living on the Atlantic side of St. Vincent. When the British authorities of St. Vincent attempt to extend agricultural cultivation and settlement into their region, the Black Caribs resist, partly in alliance with the French, who are always looking for opportunities to counter British expansionism in the Caribbean. After nearly a decade of conflict, the British declare victory over the Black Caribs and begin entertaining ideas of removing them from the island or limiting them to a reservation-style area.

1778   British recapture island

The British retake St. Lucia once again and use its harbors as a naval base.

1782–1803   St. Lucia changes hands many times

England and France exchange control over St. Lucia numerous times.

c. 1782   French construct road

During one of their periods of occupation of St. Lucia, the French proceed to build the *Chemin Royal*, a road that essentially circles the entire island along the coast. The current round-the-island road corresponds closely to the French precursor.

1794   France abolishes slavery in its possessions

Under the control of revolutionary forces, the French government abolishes slavery in its Caribbean possessions.

1795–97   Brigand's War

As part of the ongoing fighting between France and England, the English begin seizing many of France's colonial possessions in the Lesser Antilles, the string of small islands stretching south from Puerto Rico, including St. Lucia in June 1795. The British land as many as five thousand troops on St. Lucia and take Morne Fortuné, just to the south of Castries, after a month of fighting in May 1796, suffering over five hundred casualties. But gaining control over the remaining portions of the islands proves nearly impossible as the British are faced with a guerrilla campaign on the part of French holdouts and former slaves desperate to maintain their emancipated status as granted by the French in 1794. This conflict is known as the Brigand's War. For more than a year, British troops endure illness and sabotage as they try to pursue the rebels through the nearly impenetrable forests and mountains of St. Lucia's interior. The war comes to an end in the middle of 1797, when the British commander finally makes concessions and convinces the rebels to put down their arms. An amnesty is granted, the Africans are not re-enslaved, and all rebels are allowed to either settle on available land or join the British military.

1803   England prevails over St. Lucia

Following a series of hard-fought naval battles with the French, the British control St. Lucia for the final time. Although the British will control the administrative affairs of the island hereafter, and English will become the official language, Patois remains the principle language of the masses.

1807   Slave trade abolished

The British government abolishes the slave trade, thereby cutting off the supply of African-born slaves. Current slaves will remain in bondage for nearly two more decades.

1814   Island becomes Crown colony

France recognizes Britain's sovereignty over St. Lucia in the Treaty of Paris. The island becomes a Crown colony.

1833   Slavery is abolished

The British government abolishes slavery in all of its dominions in the West Indies over the objections of the powerful plantation interests. Plantation owners fear financial ruin because of a shortage of cheap and readily available labor. The British financially compensate many owners for their lost

slaves and introduce an apprentice system that maintains the the Africans' obligations to the plantations another half decade. In search of cheap labor, St. Lucia's planters will look to East Indian migrants. Just prior to abolition, St. Lucia has 13,000 African slaves, 2,600 free blacks, and 2,300 whites. With abolition, many former slaves leave the plantations to rent, buy, or squat (to settle without right or title to do so) on available land. They form the nucleus of an independent peasantry that grows crops primarily for personal consumption but also for sale at market. Throughout the Lesser Antilles, this peasantry is responsible for introducing and trying out new export crops. In St. Lucia the growing of cacao is attempted, but limes prove to be more profitable.

1838   Barbados adminsiters St. Lucia

St. Lucia becomes part of the British Windward Islands administration and is placed under the control of Barbados. It will remain as such until 1885.

c. 1848   Free-holding farmers

The number of free-holding farmers is estimated at 2,500. Many of these farmers are former slaves or descendants of slaves who left the plantations at abolition and set up small agricultural plots on whatever land they could find.

c. 1883   Malaria plagues the island

St. Lucia is declared to be one of the most malaria-ridden regions in the Caribbean. This may be the result of African-born slaves carrying the malaria pathogen in years before. Though slavery has been abolished for years, malaria thrives in St. Lucia's hot, humid climate, where there is no shortage of water or mosquitos.

1883   Castries as coal port

Castries emerges as one of the main coal ports in the British West Indies.

1885   Adminstrative duties transferred to Grenada

Administrative control of St. Lucia is transferred from Barbados to Grenada. The local political council in St. Lucia remains responsible for police forces, judicial systems, and treasuries.

Twentieth-Century

1921   Wood Commission

The British colonial government sends the Wood Commission, under the direction of Major Wood, to discover ways to alleviate the social tensions that resulted in a recent surge of worker discontent throughout the British West Indies. The Commission concludes that a modicum of political and economic reform would do much to reduce social conflict. The extent to which the commission's modest proposals are followed fails to significantly undermine labor discontent. Strikes will continue to erupt over the next two decades.

1924 Constitution drafted

The by-product of the Woods Commission is a new constitution that allows for three additional seats on the Legislative Council to be determined by popular vote. The majority of the seats are still picked by the British colonial officials.

1936 Workers strike

Stokers in the naval yards in Castries strike. They are joined by unemployed urban and rural workers. Although the strike is conducted peacefully, the governor responds by calling out the militia, bringing a warship to Castries harbor, and having marines patrol the streets to arrest suspected demonstrators.

1938 Barbados Settlement Company

A scheme is introduced to advance the sugar industry, whereby Barbadians control the sugarcane factories in the south of the island. The company constructs a hospital.

1939–46 The U.S. period

St. Lucia is one of the Caribbean islands ceded to the United States by the British government as part of the "Lend-Lease Agreement" during World War II (1939–45). In exchange for fifty old destroyers, Britain grants the United States ninty-nine year leases for military bases on Antigua, St. Lucia, Jamaica, Guyana, and Trinidad. On St. Lucia the U.S. military begins a rapid construction campaign that includes roads, a sea base at Gros Islet, and an airstrip at Vieux Fort. Many local residents are employed to help in the construction projects.

1948 Fire in Castries

A great fire sweeps through the city of Castries destroying many buildings.

1950 SLP founded

The St. Lucia Labour Party (SLP) is founded by George Charles (b. 1916) and Allen Lewis (b. 1909). Charles is the leader of the St. Lucia Workers' Co-operative Union, but Allen becomes the SLP's president.

1951 Universal male suffrage

Universal adult male suffrage is adopted, and elected members become a majority of the Legislative Council.

1958–62 St. Lucia in the West Indies Federation

Facing the likelihood that it will soon be granting independence to its colonies in the Caribbean, the British government creates the West Indies Federation. Its purpose is to bring the islands together under a common administrative infrastructure that is more viable than the separate individual islands. The breakup of the federation in 1962 is caused by interregional conflict between the wealthier islands which feel the smaller, poorer islands represent an economic drain. Jamaica is the first to pull out in 1961, followed by Trinidad and Tobago.

1960 New constitution

With the adoption of a new constitution, the office of governor of the Windward Islands is abolished, and St. Lucia becomes an autonomous island in the West Indies Federation. This is accompanied by a greater degree of self-government.

1964 National elections

The SLP suffers defeat to the new United Worker's Party (UWP) which was founded in part by SLP dissidents, including the new premier, John Compton. Compton is a wealthy planter who once worked at an oil refinery in Curaçao and later studied at the London School of Economics. He joined the SLP in 1956 and held the position of Chief of Trade and Industry (1958–61) under Chief Minister George Charles. He resigned his position and left the SLP in 1961 to form the National Labour Movement, which merged with the People's Progressive Party in 1964 to from the UWP. Compton will remain chief minister (which is changed to chief premier in 1967) for fifteen years until independence in 1979. Compton is a fiscal conservative stressing tourism and foreign investment.

1964 Sugarcane production ceases

Sugarcane, once the primary crop of the island, largely ceases, as sugar plantations are converted over to banana production. Bananas emerge as the principal cash crop.

c. 1965 Inter-island migration

As throughout much of the Caribbean in the mid-1960s, many workers from St. Lucia migrate off the island in search of employment. Trinidad is a common destination given its larger size and more vigorous economy.

1967 SLP changes leadership

George Charles steps down as the leader of the SLP in favor of Kenneth Foster.

1967: March 1 West Indies Act

With the passage of the West Indies Act, St. Lucia becomes a sovereign nation associated with the United Kingdom but with full powers of self-government. External affairs and defense issues are left to the responsibility of the United Kingdom.

1973   Work stoppage among banana laborers

The St. Lucia Action Movement (SLAM) leads a large strike among the banana workers.

1973   Political alliance

SLAM forms a political alliance with the SLP. George Odlum, a co-founder of SLAM, is a proponent of the merger. An economist by training and a leftist, Odlum worked in the Ministry of Trade and Industry and then in the Commonwealth Secretariat in London.

1977   New leader for SLP

The retired judge, Allan Louisy (b. 1916), is elected leader of the SLP over George Odlum, who becomes Louisy's deputy. Odlum is hampered by his being a proponent of "black power."

1979: February 22   Independence

St. Lucia achieves independence from Great Britain. It remains a parliamentary democracy in the British Commonwealth. Queen Elizabeth II remains the titular head of state.

1979: July   First elections

In the first national elections of the newly independent nation, the left-leaning SLP wins, and Louisy becomes minister. The liberal and charismatic George Odlum becomes foreign minister. John Compton and the UWP are pushed from power after fifteen years. The success of the SLP is owed largely to the unity represented by the alliance six years ago with members of SLAM. The SLP draws much of its support among banana laborers. Its nationalist stance in opposition to a U.S.-owned firm's plan to build an oil terminal on the island contributes to its populist image. The victory by the SLP is part of a shift in the Eastern Caribbean as a whole toward more leftist and populist governments.

1979: December   Factionalism in SLP

The alliance between the SLP and SLAM begins to break apart. Six months after the election, Louisy is being pressured from within his party to hand over the position of prime minister to George Odlum. With the United States backing him, Louisy refuses and accuses Odlum and others of being too leftist and linked to Cuba.

1980   Hurricane

Hurricane Allen devastates the banana crop on the island. The impact of such storms is especially hard on banana producers because it takes many years to bring banana plantations back into production after a "blowdown." The storm causes unemployment to increase rapidly.

1981   Organization of East Caribbean States

The SLP government of St. Lucia helps to form and then joins the Organization of East Caribbean States (OECS). The OECS, though is stresses community amoung St. Lucia and other nearby island nations, is primarily a monetary union between them.

1981   Louisy resigns

Prime Minister Louisy resigns but continues to rebuff Odlum and his opponents in the SLP by naming Attorney-General Winston Cenac to be his replacement.

1981   PLP formed

Odlum breaks with the SLP and forms the Progressive Labour Party (PLP).

1982   Cenac resigns; UWP comes to power

Amid growing party factionalism within the SLP, Winston Cenac resigns as prime minister. Under heavy pressure from the opposition accompanied by street demonstrations, an interim government is formed under PLP deputy Michael Pilgrim with the agreement that elections will be held. In the elections the UWP wins a strong victory by taking fourteen of the seventeen seats in the House of Assembly. John Compton becomes prime minister. The conservative UWP comes to power amid a wave of victories by similar conservative parties throughout the Caribbean. Some analysts associate this shift in politics to U.S. anti-communist initiatives in the Caribbean, which include promises of economic aid for cooperative regimes.

1983   Compton supports U.S. invasion

The government of John Compton comes out as a strong supporter of the U.S. invasion of Grenada.

1983   Compton government considers wage cuts

A Tripartite Commission is established by the Compton government to study the feasibility of wage cuts for workers nationwide. The government argues that in order for St. Lucia to deal with economic hardship and become more competitive, it must explore the possibilities of difficult and unpopular options such as these.

1983: July   Caribbean Basin Initiative

The United States Congress approves the Caribbean Basin Initiative (CBI), which goes into effect in January 1984. CBI is a regional assistance package for the Caribbean designed to isolate socialist-leaning governments by encouraging private enterprise through trade and tax incentives. Some of its main components include free trade with the United States and economic incentives to U.S. business investing in the Caribbean. Supporters herald CBI as an economic windfall; detractors describe it as a masked form of U.S. imperialism. Most nations in Central America and the Caribbean are invited to participate, with the exceptions of those deemed to be communist. Conditions of participation include closer ties to the United States in a variety of legal and economic arenas. One important area of exchange is in the arena of labor organization. U.S.-based labor organizations, funded by the U.S. government, make attempts to train and gain ties with labor unions in the Caribbean. The challenge before these unions is to convince workers to back free trade and U.S. investment, which can negatively impact wages. In St. Lucia, for instance, the U.S.-trained union leaders support Compton's Tripartite Commission on wage cuts despite heavy opposition from workers. By 1986, with U.S. protectionism on the rise, CBI is considered largely a failure.

1984   St. Lucia delegation at womens' conference

The First Caribbean Womens' Encounter is held in Antigua and a delegation of women from St. Lucia attend. The purpose of the conference is to encourage pan-Caribbean women's organization. The sponsor is the Antigua Women's Movement, a wing of a leftist political movement, the Antigua Caribbean Liberation Movement. Because of its political affiliation, the conference is heavily monitored by Antigua's police units.

1984   Hunte becomes SLP Party leader

Julian Hunte (b. 1940), who is ironically the brother-in-law of UWP leader John Compton, becomes the party chairman of the SLP by winning an intra-party electoral contest over left-leaning Peter Josie (b. 1941). The contest between Josie and Hunte captures the ideological differences within the SLP. Hunte, an insurance and property agent by profession, tends to be more conservative. He joined the UWP in 1971 and also served as mayor of Castries that same year. He resigned from that position to join the left-wing St. Lucia Action Movement, which he subsequently left to join the SLP, becoming its deputy chairman in 1982. Josie is the one-time leader of the Seamen, Waterfront, and General Workers Union. He too joined the St. Lucia Action Movement in the early 1970s and left it in 1973 to join the SLP. Josie became party chairman and was the losing candidate to Compton in the 1982 election.

1984: March   U.S. military units arrive

U.S. military trainers arrive on St. Lucia and a number of other nations in the Eastern Caribbean for the purpose of training Special Service Units. The training program is supposed to be secret, but news of it is released. One of the challenges before the U.S. in implementing the program is that, with the exception of Jamaica and Barbados, the other nations involved have only police units and no military forces. U.S. law prohibits the training of foreign police units. The Reagan government redefines the police units as "paramilitaries."

1984: October 28   Sesenne named Queen of Culture

A folk singer of poor origins known as Sesenne, whose real name is Marie Clepha Descartes, is declared the Queen of Culture in St. Lucia. She once received the British Empire Medal (BEM) in 1972.

1985   Calabash declared the national tree

Recognizing its historic significance to St. Lucia's economy and culture, the government declares the calabash to be the national tree. The Calabash produces a large pear-shaped fruit known locally as a "kalba." It has been a staple commodity of St. Lucia's peasant society. When hollowed out, kalbas are transformed into a variety of implements. Dishes made from them are known as "kwi" and are used to serve food. The kalbas have also been used as carrying gourds, musical instruments, and decorative masks used in religious practices.

1986: January   CDU formed

The Caribbean Democratic Union (CDU) is formed in Kingston, Jamaica. It is a regional organization joined by numerous governments throughout the Caribbean, including that of St. Lucia. Its purpose is to bring unity to pro-American governments in the region. It is soon revealed that the CDU is an outgrowth of the National Endowment for Democracy (NED) a U.S.-funded organization to promote U.S. interests abroad.

1987   Tourism boost

A complex for cruise boats is completed and opened near Castries. The construction of this site is part of the island's attempt to expand its tourist industry.

1987   National elections

The UWP manages to stay in power as the majority party, but not without slipping significantly from its dominating victory in 1982. It wins only nine of the seventeen seats in the Assembly, representing a narrow one-seat majority. The UWP's victory is largely the result of the factionalism in the opposition SLP as its two leaders, Hunte and Josie fail to agree on electoral strategy. Deputy Chairman Josie cannot convince the more conservative Chairman Hunte to form an alliance with the left-leaning George Odlum and his Progressive Labour Party, which drew off nine percent of the vote. Prime Minister Compton calls for a new election three weeks later in hopes of increasing the UWP's control in the House, but the results are identical. Compton wins over an important member of the SLP, Neville Cenac, by appointing him foreign minister.

c. 1988   Geothermal energy

Dependent on antiquated diesel plants for energy, the government is exploring the feasibility of exploiting the islands substantial geothermal reserves. If these initiatives can be brought to fruition, the impact on the economy is potentially very large, as dependence on petroleum imports will decline.

1992   National elections

The UWP increases its majority in the House of Assembly to elenen of seventeen seats.

1992   Walcott wins Nobel Prize

St. Lucian-born poet and dramatist Derek Walcott (b.1930) wins the Nobel Prize for Literature. His work is characterized by a syncretic style that blends the classical western tradition with distinct Caribbean and African influences. Walcott's capstone work, the epic poem "Omeros," (1990) is the epitome of this style. The sixty-chapter poem is written like the classic epics, such as *Aeneid* and *The Iliad*, but its characters, plots, and contexts are Caribbean, and the characters tend to have a dispossessed quality to them, reflecting Walcott's thematic search for West Indian identity. He published his first poem at the age of eighteen and was educated at the University of the West Indies in Jamaica. In 1981 he received a MacArthur Foundation "genius" award and that same year he joined the faculty at Boston University. He founded the Trinidad Theater Workshop in 1959, the Boston Playrights' Theater at Boston University in 1981, and the Rat Island Foundation in the early 1990s, an international writers' retreat on a small island off the coast of St. Lucia.

1993   Banana preference for Europe

The European Union, which has the highest per capita consumption of bananas in the world, adopts a system of preferences that guarantees access to Europe's banana market for former British and French colonial possessions. A limit is subsequently placed on the number of bananas imported from Central and South America which are produced primarily by U.S.-owned firms. The preferences are scheduled to be abolished in 2002.

1994   Tropical storm

A tropical storm destroys as much as sixty-five percent of St. Lucia's banana crop.

1995   Tropical Storm

Another tropical storm, Iris, destroys up to twenty percent of St. Lucia's banana crop.

1996: October   Pink mealy bug

The government confirms that the crop-eating pink mealy bug has established itself on St. Lucia. There is concern throughout the Eastern Caribbean about the pink mealy bugs' spread, since it is an insect with an incredible capacity to destroy foliage and crops. The government says that the pink mealy's

presence on St. Lucia is limited to the area around the capital Castries.

1997: April   National elections

After fifteen years out of power, the SLP wins a landslide victory over the UWP. Kenneth Anthony becomes prime minister. The UWP had been in power for the last fifteen years and had ruled for all but two and one-half years since 1964. Corruption on the part of the UWP government and a recent economic downturn are the most commonly cited reasons for the UWP downfall.

1997: May   Americans pressure Europeans on bananas

Under urging from American banana producers, Chiquita Brands and Dole Foods, the Clinton administration begins to pressure the European Union to abandon its system of preferences for Caribbean banana producers. Concern grows on St. Lucia among the banana planters, most of whom are small landholders, over the likely impact of having to compete on the open market with giant U.S. fruit companies.

1997: September 30   Diplomatic conflict with Taiwan

Taiwan breaks off diplomatic relations with St. Lucia when the latter recognizes the government of the People's Republic of China. Relations between St. Lucia and Taiwan were cordial under the UWP-led government. But the new SLP government has solid ties with the Chinese government, which has promised to provide St. Lucia with financial aid for a number of capital projects on the island. At the center of the dispute is St. Lucia's vote in the General Assembly of the United Nations (UN) as to whether Taiwan should be admitted to the UN.

Bibliography

Auguste, Joyce, ed. *Oral and Folk Traditions of Saint Lucia.* St. Lucia: Lithographic Press, 1986.

Claypole, William, and John Robottom. *Caribbean Story, Book Two: The Inheritors.* Essex: Longman, 1986.

Craton, Michael. *Testing the Chains: Resistance to Slavery in the British West Indies.* Ithaca, N.Y.: Cornell University Press, 1982.

Gunson, Phil, et. al., eds., *The Dictionary of Contemporary Politics of Central America and the Caribbean.* New York: Simon and Schuster, 1991.

Hartigan, Patti. "Poet Walcott of BU Wins Nobel Prize," *The Boston Globe.* 9 October, 1992.

Jesse, C. *The Amerindians in St. Lucia.* Castries, St. Lucia: St. Lucia Archaeological and Historical Society, 1968.

Paquette, Robert, and Stanley Engerman, eds. *The Lesser Antilles in the Age of European Expansion.* Gainesville, Fla: University of Florida Press, 1996.

Rohter, Larry. "Trade Storm Imperils Caribbean Banana Crops," *The New York Times.* 9 May, 1997.

Rouse, Irving. *The Tainos: Rise and Decline of the People Who Greeted Columbus.* New Haven: Yale University Press, 1992.

"St. Lucia and Taiwan waging 'War of Words' Over Break in Diplomatic Ties," CANA News Agency, Bridgetown, BBC Monitoring International Reports, as provided by BBC Worldwide Monitoring, September 12, 1997.

Simpson, George. "The Kele (Chango) Cult in St. Lucia." *Caribbean Studies* 13 (1973): 110–116.

Smikle, Patrick. "A New Era in Politics?" Inter Press Service, January 23, 1996.

Sunshine, Catherine, *The Caribbean: Survival, Struggle and Sovereignty* (Washington, D.C.: EPICA, 1988).

Taylor, Robert. "Derek Walcott's Poetry." *The Boston Globe*, October 9, 1992.

Weeks, John, and Peter Ferbel. *Ancient Caribbean.* New York: Garland Publishing, 1994.

St. Vincent and the Grenadines

Introduction

St. Vincent and the Grenadines is located twenty-one miles southwest of St. Lucia and about 100 miles west of Barbados in the Caribbean Sea. Scattered between St. Vincent and Grenada are more than 100 small islands called the Grenadines, half of which belong to St. Vincent and the other half to Grenada. Only about a dozen of the country's 120 islands are populated. The Grenadines belonging to St. Vincent include Union Island, Mayreau, Canouan, Mustique, Bequia, and many other uninhabited cays, rocks, and reefs. St. Vincent has an area of 134 square miles, with a coastline of fifty-two miles. Bequia, the largest of the Grenadines, has an area of 7 square miles. St. Vincent and the Grenadines has a population of about 118,000, with over ninety percent living on the island of St. Vincent. Kingstown, the capital, is located on the southwestern coast of that island and has a population of about 27,000. St. Vincent is a rugged island with dark volcanic sand beaches. Its highest point is Soufrière, an active volcano that rises 4,048 feet above sea level. Only five percent of St. Vincent's surface has slopes of less than five degrees. The low-lying Grenadines have wide beaches and shallow bays and harbors. The islands have a pleasant tropical climate throughout the year, with the average temperature ranging from 77°F in January to 81°F in September. The islands lie in the Caribbean hurricane belt and were devastated in 1780, 1898, and 1980.

About sixty-five percent of the islanders are descendants of slaves brought from Africa. About twenty percent is of mixed origins and about 3.5 percent is of European descent. Some 5.5 percent of the islanders are descendants of nineteenth-century East Indian indentured laborers (workers contracted for a fixed amount of time). About 2 percent of the people are indigenous Caribs. The mixture of Africans and the native Caribs created an ethnicity known as the Black Caribs. Some scholars believe that Black Caribs descended from escaped slaves, while others think they may have been descendants of a stranded group of thirteenth century West African explorers.

The Ciboney, a branch of the Arawaks, first settled on the island around 5000 BC. The Arawaks were the largest group of ancient people that lived in South America. By the fourteenth century, Caribs from the Orinoco Basin of South America had driven out the Arawaks. The Spanish raided islands across the Lesser Antilles and abducted Caribs for use as laborers on the early sugar plantations. The remaining Caribs retreated to the mountainous Windward Islands (St. Lucia, St. Vincent, Grenada, and Dominica) and continued resisting the Spanish, which delayed European colonial development.

In 1627, King James I of England granted control of St. Vincent to the Earl of Carlisle, but the French also wanted control of the island. In 1660, the two sides agreed to leave the island to the native Caribs, but the agreement was soon broken. Shipwrecked slaves arrived on the island of Bequia in 1675 and began mixing with the native Yellow Caribs to create the new racial group that the colonists called the Black Caribs. By the early 1700s, there were tensions between the Yellow and Black Caribs, and the French divided the island into Yellow and Black Carib districts. The Yellow Caribs allowed the French to construct a permanent settlement on the island, and the French settlers began plantation agriculture with labor supplied by African slaves. During the mid-eighteenth century Britain and France competed for control over the island. Britain finally won control in 1762, but lost it to France during the American War of Independence (1775–83). In 1779, France captured the island and held possession until 1783. By 1797, all the Carib rebels had either been killed or deported.

St. Vincent became a crown colony in 1833 and the following year slavery was abolished throughout Britain's territories in the West Indies. Soon indentured workers from India were brought in to work on the plantations. The sugar cane industry went into decline in the second half of the nineteenth century as European countries began cultivating sugar beets. With Europe producing their own supply of sugar, the demand for importing it from other areas sharply decreased. By the beginning of the twentieth century, arrowroot and bananas had become the main cash crops.

In 1902, St. Vincent's volcano, Soufrière, erupted killing some two thousand people. The first elections using a limited franchise were conducted in 1925, and in 1951 universal adult suffrage was granted. During the 1960s, the infrastructure was greatly improved, as several islands were developed for tourists or as exclusive resorts. Soufrière erupted again in 1979, forcing the evacuation of St. Vincent. In October, 1979, St. Vincent and the Grenadines became an independent nation. It was a steady progression from being a colony to being part of the British Commonwealth. In 1989, a fire devastated the center of Kingstown.

Timeline

5000 B.C.   Earliest inhabitants of St. Vincent

St. Vincent may have been inhabited as early as 5000 B.C. The Ciboney, a branch of the Arawaks, first settle on the island at about that time.

The Arawaks, Caribs, and Spanish

Agrarian Arawaks, who call themselves the Lokono, first immigrate to St. Vincent from South America. They are believed to be the largest group of ancient people that lived in South America. Between the tenth and fourteenth centuries, Caribs from the Orinoco Basin of South America migrate north and begin to drive out the Arawaks. The Caribs eventually conquer the Arawaks, killing all the men and assimilating the women into their own culture. The Arawaks are skilled artisans, while the Caribs are experienced canoe builders. The Caribs are also politically organized and have a more sophisticated knowledge of cultivation and fishing. (Recent discoveries indicate that there may have been several waves of Amerindian migration to the Lesser Antilles and that the differences between the Arawaks and Caribs may not have been as great as once thought.) The Caribs settle on St. Vincent (which they called Iouloumain) and other islands throughout the Caribbean Sea. The Caribs are named by the Spanish after the word caribal (cannibal), although it is unclear whether the Caribs actually eat human flesh.

1498: January 22   Columbus allegedly sights St. Vincent

Christopher Columbus allegedly sights St. Vincent on the feast day of the island's namesake. Some historians now dispute that the sighting actually occurred. European settlers later arrive with African slaves.

Early 1500s   Spanish introduce bananas

The yellow banana, originally from Malaya (a British colony on the southern end of the Malay peninsula, now Malaysia), is cultivated by early Spanish explorers. The Spanish also

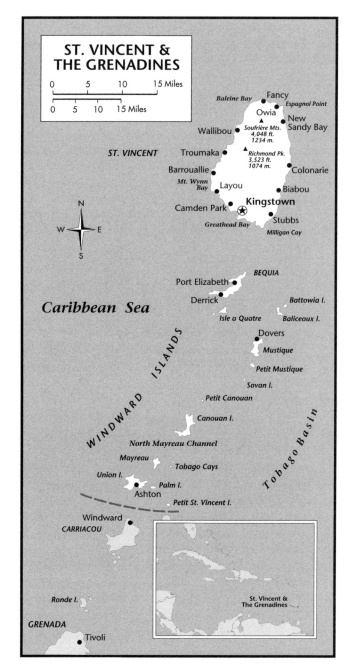

introduce plantains (a tropical banana plant with coarse fruit) from India and red bananas. Bananas later are to become an important cash crop throughout the Caribbean.

Late 1500s   Caribs resist Spanish conquistadors

The Spanish regularly raid islands across the Lesser Antilles and abduct Caribs to use as laborers on the early sugar plantations. These raids decrease the native population on many of the islands, and the remaining Caribs frequently go into hiding in the mountainous Windward Islands (St. Lucia, St. Vincent, Grenada, and Dominica). From these islands, the Caribs assemble a military force to resist the Spanish. Their opposition stalls the development of colonization by European powers.

Britain and France Vie for Control

1627: July 2   Earl of Carlisle given title to several islands in the Lesser Antilles

Several English merchants arrange to make their patron, James Hay (d. 1636), Earl of Carlisle in Scotland, owner of St. Vincent and most of the Lesser Antilles. The earl is well-liked by King James I and owes the merchants a lot of money. The king issues a royal patent to the earl, making him "Lord Proprietor of the English Caribbee Islands."

1660   British and French agree to leave island alone

St. Vincent is one of the last islands of the West Indies to be settled by Europeans. The British and French agree to leave the island to the native Caribs, but the agreement only lasts a few years. The island has a sizable Carib population until the 1720s.

1675   Shipwrecked slaves arrive on Bequia

Slaves from a Dutch shipwreck make it to the island of Bequia and are given shelter by the native, or Yellow, Caribs. Slaves from Barbados and St. Lucia later manage to escape to the island. The mixture of slaves and Yellow Caribs creates a new group known as the Black Caribs. The Caribs ardently resist European settlement on St. Vincent until the eighteenth century.

1700   French divide St. Vincent

The French make an administrative division of St. Vincent, with the west going to the Yellow Caribs and the east going to the Black Caribs. Tensions between the two groups lead to conflict and territorial division. The perpetual hostilities delay colonial development of the island, while several nearby islands already have an advanced sugar industry.

1718   Soufrière erupts

A passing ship logs the first recorded eruption of La Soufrière, St. Vincent's volcanic mountain. Scientists now believe that an earlier eruption occurred sometime between A.D. 160 and 350.

1719   French begin plantation agriculture

In order to appease the French, the Yellow Caribs permit some French colonists to build a settlement on St. Vincent. The Yellow Caribs fear the growing influence of the Black Caribs. The French settlers begin cultivating coffee, tobacco, indigo, cotton, and sugar on plantations worked by African slaves. The Black Caribs head to the inland hilly forests and continue resisting French rule.

1722   British colonization intensifies

King George I (1660–1727; r. 1714–27) gives the islands of St. Vincent and St. Lucia to the Duke of Montague. The British send Captain Braithwaite to build a settlement on St. Vincent.

1748   Treaty of Aix-la-Chapelle

French control over the island is very tenuous. Under the Treaty of Aix-la-Chapelle, the island is officially declared neutral by Britain and France. St. Lucia and Dominica are also declared neutral. The peace, however, is fleeting and the two colonial powers continue to fight over possession.

1762   British take control of island

During the Seven Years' War, Britain captures St. Vincent under the leadership of General Robert Monckton.

1763   Treaty of Paris

The French officially cede the island to Great Britain. The British rule thereafter except during 1779–83, when the French temporarily gain control.

1765   St. Vincent Botanic Gardens founded

The St. Vincent Botanic Gardens on the outskirts of Kingstown is the most ornate garden in the Caribbean and the oldest of its kind in the Western Hemisphere. General Robert Melville, then Governor-in-Chief of the Windward Federation, founds the gardens as a nursery for plants useful to medicine and commerce. The French introduce exotic Asian spices (such as cinnamon) to the gardens. Cloves are brought from Martinique and nutmeg and black pepper plants from French Guiana.

1767   Government restricts slave participation in the economy

Though slaves participate in the economy, the government makes it illegal for slaves to cultivate any crop that the island exports.

1773   British make treaty with Caribs

Several Black Caribs sign a peace treaty, which places many restrictions upon them under British law. The treaty is written in English, but the Black Caribs communicate in French, and so it is likely that the Black Carib leaders do not fully understand the document.

1779–83   French rule restored

During the American War of Independence, trade between Britain and its Caribbean colonies is blocked. As a result, supplies and arms are often in short supply in the Britain's Caribbean possessions. The French support the Americans in their fight against the British, and take advantage of the opportunity by seizing St. Vincent. The Black Caribs prefer the French over the British, and ask for help from the French on Martinique. With the Treaty of Versailles, British rule is restored in 1783.

1793: January 23   Breadfruit introduced

Breadfruit trees are first brought from Tahiti in the South Pacific by Captain Bligh. The 1,200 trees are cultivated in St. Vincent and Jamaica before being sent to other islands in the Caribbean. Breadfruit becomes a staple of the slave diet during the early nineteenth century.

1795–96   Second Carib War

The Black Caribs and the remaining Yellow Caribs start an uprising against the British, at French instigation. The British put down the rebellion after bringing in reserve troops.

1796: March 14   Death of Chatoyér, Carib chief

Joseph Chatoyér (also known as Chatayer), a Carib chief, is instrumental in organizing a rebellion to drive the British down the western coast of the island to Kingstown. In the hills overlooking Kingstown, Chatoyér is killed in battle. The few remaining Yellow Caribs surrender and withdraw to the remote northern region of St. Vincent called Sandy Bay, where their descendants still live. The Black Caribs fight on for another year. Chatoyér is today considered a national hero of St. Vincent.

1797   Caribs deported to Roatán

The remaining 5,000 Black Carib insurgents have their villages destroyed and their crops obliterated. They are then deported to the island of Roatán, which lies off the northern coast of present-day Honduras in the western Caribbean Sea. In the early 1800s, they go on to settle in what is now Honduras and Belize.

1806   Fort Charlotte completed

West of Kingstown, Ft. Charlotte sits atop a ridge 600 feet above sea level. With its elevation and sturdy construction, it was one of the most formidible forts constructed in the region.

1812: April 30   Soufrière erupts

The eruption of Soufrière kills fifty-six people. The Carib community is devastated by the volcano's damage. The eruption, which lasts for three days, is heard as far away as the coast of Guyana.

1815   Black Point Tunnel opens

The Black Point Tunnel connects the north and south parts of the island, which are separated by mountains. The 300-foot tunnel is hand-dug by slave and Carib laborers, and provides a way for sugarcane to be transported from the plantations in the north to the processing plants and docks in the south.

1833–1960   British crown colony

St. Vincent is administered as a crown colony from 1833 until 1960. Under the British, plantations primarily grow and process sugar.

1834: August 1   Slavery abolished

Although the British prosper through the use of slave labor in the West Indies, the institution is cruel and highly inefficient. Slaves are unwilling laborers forced to endure fear, suffering, and brutality. Slavery also demoralizes the British colonists so it is abolished in 1834.

Britain has control over most of the Caribbean and the high seas, and is able to enforce the abolition of the slave trade. It compensates slave owners for the loss of their slaves and initiates an apprenticeship system whereby former slaves must work for their former masters without pay.

1838   Apprenticeship system abolished

The system of apprenticeship, implemented as part of the same legislation that had given the slaves their freedom, is abolished. After the apprenticeship system is abolished, most of the former slaves begin farming on small plots of land cleared from the forest.

1861   Indentured workers from India begin arriving

After slavery is abolished, indentured workers from India are brought to St. Vincent to replace the African slave-laborers who have left the plantations.

1877   St. Vincent crown colony government introduced

The local colonial government is organized at a time when the sugar cane industry is in decline. Sugar beets are being grown in Europe, which causes demand for cane sugar to plummet. The island suffers from an economic depression for most of the second half of the nineteenth century.

1897   Local government established in Kingstown

Government is established here because the town overlooks Kingstown Harbour and is protected by Berkshire Hill to the north and Cane Garden Point to the south.

1898   Destructive hurricane

Approximately 300 people are killed after a hurricane sweeps through the islands. Many buildings are heavily damaged.

1900 Arrowroot becomes main commodity

Under the British, sugar production falls throughout the 1800s, and is replaced by arrowroot, a plant once used by the Arawaks as an antidote in wounds caused by poison arrows. St. Vincent is still the world's leading producer of arrowroot. The root is used to make starch and also as a coating on computer paper.

1902: May 7 Soufrière erupts

St. Vincent's volcano erupts, killing over two thousand people. Much of the northern part of the island is obliterated. The cloud of gas and ash rises over six miles and is carried all the way to Barbados, nearly 100 miles to the east. Volcanic bombs the size of coconuts rain down even on Kingstown and the villages at the southern tip of the island. After the eruption, the volcano's new crater is a mile wide and fills with water to become a lake.

1920 Commercial whaling ends

Whaling has been practiced on Bequia (one of the Windward islands located just south of St. Vincent) for many centuries, and commercial whaling dates back to 1830. During the peak of the industry in the 1890s, there were six whaling stations on Bequia and one on Petit St. Vincent (another of the Windward Islands). Due to the cultural significance of whaling on Bequia, the International Whaling Commission still permits islanders a quota of up to three whales per year, although usually only one is killed.

1924 New constitution

Representative government is introduced.

1925 St. Vincent legislative council created

St. Vincent's first legislative elections take place, with limited franchise (relatively few people vote).

1935 Riots by Vincentian workers

In the early twentieth century, many islanders work abroad. A falling demand for labor during the Great Depression in the 1930s causes massive unemployment in many of the Caribbean island colonies and wages plummet. Workers riot and later form trade unions, which initiates the formation of political parties that agitate for the right to self-government.

1950 Trade Union Ordinance

The Trade Union Ordinance legalizes labor organizations in St. Vincent. In the past, labor unions have been illegal and anyone trying to form one could have been charged as a criminal. Ebenezer Theodore Joshua organizes the first labor union, the Federated Industrial and Agricultural Workers Union.

1951 Adult suffrage

Universal adult suffrage to vote for the Executive Council is granted.

1952 First local political party formed

After the first elections with universal adult suffrage, Ebenezer Theodore Joshua helps found the island's first political party, the People's Political Party. Joshua later goes on to be St. Vincent's head of government from 1957 until 1967.

1956 First St. Vincent Music Festival

The Music Festival has become a popular national event. The festival is held every other year and features vocal, choral, and instrumental performances.

1960s Development of tourism

In 1962, the Arnos Vale Airport opens, allowing international air travel. Tourism becomes particularly important in the Grenadines, where yachting is popular. Some of the smaller cays and islands are wholly acquired by private interests and developed into resorts for European and North American visitors. In 1959, Colin Tennant buys Mustique and develops it into an exclusive resort island. In 1966, John and Mary Caldwell purchase Prune Island (renaming it Palm Island after planting hundreds of palm trees), and develop it into a resort.

1960–62 Federation of the West Indies

In 1960, St. Vincent and the Grenadines becomes a separate administrative unit of the Federation of the West Indies. The federation is a short-lived government of several Caribbean island governments that soon breaks apart.

1964 Kingstown deep water harbor inaugurated

A deepened harbor enables large oceangoing vessels to dock at Kingstown.

1969: June St. Vincent granted associate statehood status

As a British Associated State, St. Vincent receives full internal autonomy. Foreign affairs and defense are still Britain's responsibility.

Mid-1970s Rastafarianism comes to St. Vincent

Rastafarianism starts in Jamaica as an Afro-centric movement that responds to growing social inequality. The beliefs are based on the teachings of the Jamaican Marcus Garvey (1887–1940), who in the 1920s advocated African racial pride and the eventual return of blacks to Africa. He believed that an African king would rise up and rescue Africans from oppression. In 1930, Ras Tafari (1892–1975) of Ethiopia is crowned Emperor Haile Selassie I (r. 1930–36, 1941–74). Garvey's adherents believe the Ethiopian emperor is the king that had been foreseen. Emperor Haile Selassie claims to be a direct descendent of King David. The Rastafarians reject the European culture imposed by colonialism and seek a simpler life. By the late 1970s, Rastafarians are known internationally for two of their characteristics: wearing their hair in dreadlocks (long cords of twisted and uncombed hair) and ganja (marijuana). The smoking of marijuana is a sacrament in Rastafarianism.

1977 East Caribbean Flour Mills company begins operating

East Caribbean Flour Mills Ltd. becomes the main industrial estate in the country. The company later expands to become the East Caribbean Group of Companies, which has operations that handle animal feed processing, bag production, and flour and rice milling. It also has subsidiaries in Antigua and Guyana.

1979: April 13 Soufrière erupts

Over the course of several weeks, Soufrière ejects immense quantities of volcanic ash, which covers mountains, forests, and plantation fields. On April 17, a tremendous volcanic blast of ash and gas forces the evacuation of 20,000 people, many of whom go to Bequia. There are no casualties, but much of the banana and arrowroot crops are ruined. The volcano has been quiet since 1971, when a small volcanic upheaval created an island in the middle of the crater lake.

1979: October 27 St. Vincent becomes independent

St. Vincent becomes the last of the Windward Islands to gain independence, although both St. Vincent and the Grenadines keep the British monarch as the nominal head of state, represented by a governor-general. Executive power is in the hands of the prime minister and cabinet. The legislature is a twenty-one-seat House of Assembly.

During the first few months of independence, there is a rebellion on Union Island, the southernmost constituent. A group of Rastafarians wants to secede. The revolt is put down with military assistance from neighboring Barbados. In the end, one person is killed and forty are arrested.

1980s Fresh vegetables become important commodity

During the 1980s, the market for arrowroot declines. Agriculture is the mainstay of the economy, with bananas as the primary cash crop.

1984 Last sugar plant closes

The Mount Bentick factory, near Georgetown, once the center of St. Vincent's sugar industry, closes marking the end of the sugar industry on St. Vincent.

1984 Mitchell appointed prime minister

James F. Mitchell (b. 1931) is appointed prime minister.

1989: July Fire destroys Kingstown

A fire destroys much of Kingstown's center, but reconstruction in 1990 emphasizes increasing tourism.

1989 Number of constituencies increases

Before 1989, St. Vincent and the Grenadines have thirteen political districts. With the population growth, the number increases to fifteen.

1990 Regional Constituent Assembly

Leaders of St. Vincent and the Grenadines, Dominica, Grenada, and St. Lucia start planning for a more limited union of the countries. However, these talks stall in 1995 when several countries change political hands.

1996: December Offshore Finance Authority founded

The government begins targeting offshore financial services as a means of diversifying their economy away from bananas and tourism.

Bibliography

Bobrow, Jill and Dana Jinkins. *St. Vincent and the Grenadines: Gems of the Caribbean*. Waitsfield, Vt.: Concepts Publishing Inc., 1993.

Hamshere, Cyril. *The British in the Caribbean*. Cambridge, Mass.: Harvard University Press, 1972.

Philpott, Don. *Caribbean Sunseekers: St. Vincent & Grenadines*. Lincolnwood, Ill.: Passport Books, 1996.

Suriname

Introduction

Located on the northern coast of South America, Suriname is the continent's smallest independent country, and also a very young one—the former Dutch Guiana won its independence in 1975. Since then it has weathered both political and economic turmoil, but its multiethnic population has continued to live in harmony while retaining an unusual degree of cultural, religious, and linguistic diversity. Bordered by Guyana to the west and French Guiana to the east and located just above the equator, Suriname occupies an area of 63,039 square miles (163,270 square kilometers). To the north lies a swampy coastal plain. The interior consists of a savanna region, mountains, and highlands that are mostly tropical rain forest (four-fifths of Suriname is rain forest). Suriname has an estimated population of 436,000. Its capital, Paramaribo, is located on the coast.

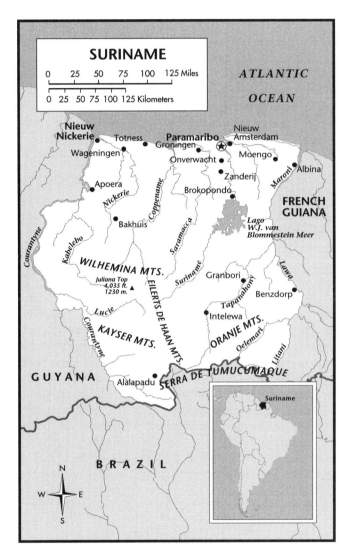

Early History and Explorers

Before it was settled by Europeans, present-day Suriname was inhabited by Surinen, Arawak, and Carib Indians. By about A.D. 1400 the warlike Carib, who, like the Arawak, had migrated from the West Indies, had nearly driven out the other two groups. At the end of the fifteenth century, the region was "discovered" by two Italian explorers, both in the employ of Spain—Christopher Columbus and Amerigo Vespucci. In the sixteenth century, it received considerable attention as the supposed gateway to the fabled El Dorado, a land of gold said to exist in the interior of the continent. The first permanent settlement was founded at the direction of Lord Willoughby, the English governor of Barbados, in the Caribbean. A thriving agricultural colony was established, with plantations producing coffee, cotton, and most importantly, sugar. Its economy was heavily dependent on the labor of African slaves, initially brought from Barbados by the first settlers and later imported directly from Africa to the mainland.

Beginning of Dutch Influence

In 1667 the region was ceded to the Dutch under the Treaty of Breda as part of an exchange through which the British acquired Manhattan Island in North America. For the next 300 years, except for two brief periods when it was retaken by the British during wartime, Dutch Guiana, as it was then called, remained a possession of the Netherlands. The Dutch, with their extensive experience reclaiming land from the sea, were particularly well-suited to develop the area's swampy coastal area for sugar cane production, and agriculture in the colony expanded rapidly, as did the number of African slaves, who came to outnumber the white population by ten to one. The colony was notorious for the harsh treatment its slaves received; their masters were free to discipline them as they saw fit, including such punishments as cutting off arms or legs. Many slaves escaped to the interior, where they remained, resisting capture, and formed their own communities. These escaped slaves came to be known by the Dutch as Bush Negroes, or Maroons, and the culture they developed—in which their native African traditions played a large role—became a distinctive part of the colony's cultural and ethnic heritage. Today the descendants of these slaves account for about ten percent of Suriname's population. (The black Africans who remained on plantations and later migrated to cities and towns, sometimes intermarrying with whites, were called Creoles. Today they make up Suriname's largest single ethnic group.)

Origin of Multiethnic Population

By the late 1700s slave revolts and administrative misman-agement had taken their toll, and Dutch Guiana's agricultural production had declined from its earlier high levels. In the nineteenth century the plantation economy was also faced with the decline of slavery as an institution. Early in the century the slave trade was banned; however, slaves were still smuggled in. By 1863, slavery was outlawed altogether, and a new source of plantation labor had to be found. Beginning around mid-century, contract laborers, who had signed agreements to work for specified periods of time (often about five years), were brought in from Asia. Between 1853 and 1939, roughly 2,500 Chinese, 34,000 East Indians (also called Hin-dustanis), and 33,000 Javanese from Indonesia came to Dutch Guiana to work. Many stayed on and settled in the country, permanently changing its ethnic and religious makeup. Today the majority of Suriname's population is made up of descen-dants of these Asian contract workers, with the single largest group being Hindustanis, who account for about one-third of the population.

Mining of Bauxite

Early in the twentieth century, it was discovered that Dutch Guiana had one of the world's most extensive reserves of bauxite. This mineral, used in the production of aluminum, became the colony's (and later, the country's) single most important source of income, as the sugar industry continued to decline. In 1916 Suralco (the Suriname Aluminum Com-pany) was formed by the U.S. firm Alcoa, and, with demand spurred by World War I, bauxite mining operations began the following year.

Following World War II, the colony moved gradually toward independence. In 1954 it became a territory and received internal autonomy, with a parliament and cabinet overseeing all areas of government except foreign affairs and defense. In 1961 the first nationalist political party was formed. By 1973 Henck Arron, head of the leading national-ist party, was elected prime minister, and he paved the way for independence. Following negotiations with the Dutch, the independent nation of Suriname was officially inaugurated on November 25, 1975, as a parliamentary republic governed by a prime minister and a popularly elected national assembly.

During the preceding decades, the issue of independence had been a source of conflict between the Creoles and the major Asian ethnic groups descended from contract laborers (Chinese, Indians, and Javanese). The Creoles were over-whelmingly in favor of independence, while many Asians were against it, fearing that Suriname would not fare well as an independent country once the Dutch withdrew. When it became apparent that independence was to become a reality, many Asians emigrated to the Netherlands, draining the new country of valuable talent and professional training. Alto-gether, some 150,000 Surinamese left the country.

Rule by Désiré Bouterse

For the first five years of independence, Suriname remained a parliamentary republic as outlined in its constitution, with its major parties representing the interests of the country's larg-est ethnic groups. However, in 1980 the democratically elected government was overthrown in an armed coup that brought Sergeant Désiré Bouterse (b. 1945) to power as head of a military government. Although civilian leaders were appointed over the following years, they were figureheads, and Bouterse and the military retained control of the country until 1987. Bouterse's regime officially declared Suriname a Socialist republic in 1981, consequently straining relations between the United States and the Netherlands, as Suriname became friendly with Cuba. Then international pressure and a declining economy, combined with a guerrilla uprising in the interior, forced him to schedule elections for a new national assembly and president. Since 1988 Suriname has had three democratically elected presidents. (One was removed in a coup in 1990, but democracy was soon restored, and new elections were held the following year). However, Bouterse has never completely relinquished power, playing a role in the government either directly or through the party he controls.

When civilian government returned to Suriname, its lead-ers found themselves confronting serious economic prob-lems, including high unemployment and rising inflation. Discontent with the nation's continuing economic stagnation was largely responsible for bringing Bouterse's New Demo-cratic Party (NDP), to power in 1996, when the new president became Jules Wijdenbosch, a former aide to Bouterse.

Timeline

c. 10,000 B.C.   Hunting and gathering tribes populate Suriname

Ancestors of the Arawak Indians inhabit the region today known as Suriname, hunting animals including mammoths and mastodons. Those on the coast live on fish and shellfish. The Indians in the interior come to rely increasingly on gathering berries, roots, and fruit as prehistoric animals become extinct.

2000 B.C.   Primitive agriculture is introduced

As rudimentary agriculture, especially the farming of cassava and other root crops, is adopted, the peoples of the region settle in temporary villages organized by kinship ties. When the soil in one location is exhausted, they move and rebuild their villages. Their crafts include making clay pottery, boat building, and spinning and weaving cotton cloth.

European Discovery and Conquest

A.D. 1400   Carib Indians are the dominant native group

The warlike Carib Indians have driven out the Surinen Indians (from whom the name *Suriname* comes) and most of the Arawaks, who, like the Caribs, had migrated to the northern coast of South America from the West Indies.

1498   Columbus sails down the Guianan coast

On his third voyage to the New World, Christopher Columbus sails down the coast of Guiana, which includes the land that will become Suriname.

1499   Vespucci lands on the Guianan coast

Amerigo Vespucci, an Italian commissioned by the Spanish crown, lands on the coast of Guiana. Vespucci and Columbus both claim the land for Spain, but no settlement attempt is made.

Late 16th century   Gold is sought on the coast

The Guianan coast attracts adventurers from Spain, as well as Britain, France, and the Netherlands, due to the rumored existence of El Dorado, the city of gold, supposedly in the interior beyond the coast. The search for gold is unsuccessful, but trading stations are set up along the coast.

1602   Dutch settlement begins

The first Dutch settlers arrive in Suriname.

1650   First permanent settlement is established

An expedition launched by Francis, Lord Willoughby, the English governor of Barbados, under the direction of Anthony Rowse, sets up an agricultural settlement on the Surinam River. The fertile land produces sugar, cotton, and coffee. In the following years, the initial site expands into a thriving colony with over 500 sugar plantations whose inhabitants include English colonists, Portuguese Jews fleeing religious persecution in Brazil, and African slaves, who outnumber the white population by a ratio of two to one. By current estimates, a total of between 300,000 and 500,000 African slaves are brought to Suriname before slavery is abolished in the nineteenth century.

1662   Land is awarded to Willoughby and Hide

As a reward for the success of the agricultural settlement in Suriname, King Charles II awards the land to Lord Willoughby and Earl Laurens Hide, another English nobleman.

Dutch Rule

1667   Dutch gain control of Suriname

A Dutch fleet captures Suriname, but the British quickly retake it. However, under the Treaty of Breda, which ends the second Anglo-Dutch war, England cedes the land to the Netherlands as part of a deal through which the British gain Manhattan Island in North America. The region becomes known as Dutch Guiana, and is administered through the Dutch West India Company.

The Dutch increase the land available for agriculture in the coastal swamps of Guiana by reclaiming low-lying lands through the systems used extremely effectively in the Netherlands. They also build dikes, floodgates, and irrigation canals. Production of sugarcane, the major crop, as well as coffee, cocoa, and cotton expands dramatically over the following century and a large population of African slaves is imported to work on the plantations, eventually outnumbering the white settlers ten to one. Dutch Guiana is known for its harsh treatment of slaves whose owners can punish them as they see fit, including chopping off arms or legs.

1686   Indians sign peace treaty

The Indians and white settlers sign a peace treaty, but there are still repeated raids on European plantations by the Indians whose land has been taken from them.

1705   German immigrant artist's book is published

After spending two years collecting and documenting samples of the plant and insect life of Suriname, German artist Maria Sibylla Merian publishes her drawings in the Netherlands in an album entitled *Beschryvinge van de volk-plantige Zuriname* (Description of the settlements of Suriname).

1707–09   Painter Valkenburg travels through Suriname

Dutch painter Dirk Valkenburg travels through Suriname producing drawings and paintings commissioned by plantation owner Jonas Witsen.

1735   Moravian missionaries arrive in Suriname

Missionaries from the Moravian Church begin preaching in Suriname. Eventually, the Moravians become one of the country's chief religious denominations.

1787   Catholicism is introduced

Roman Catholic missionaries begin working with African slaves in Suriname.

Late 1700s   Suriname's economy declines

Government mismanagement and slave revolts weaken the colony's plantation economy, and agricultural production declines from its earlier peak.

1780   Birth of artist Gerrit Schouten

Schouten (date of death unknown), one of Suriname's first native-born white artists, is known for his panoramic paintings of the colony's landscape, as well as still lifes and watercolors.

1796   Stedman publishes account of life in Suriname

John Gabriel Stedman (1744–97), a Scot serving with the British troops assigned to put down uprisings by the Maroons (escaped slaves living in the forests), publishes the book: *Narrative of a five years' expedition against the revolted Negroes of Surinam,* based on a detailed diary he kept while in Suriname. It describes the brutal conditions under which the African slaves live and also documents the fauna and flora of the land. The book's illustrations are taken from Stedman's own drawings, engraved on metal plates by English artist and poet William Blake (1757–1827).

1799–1802   Period of British control

With Britain and the Netherlands at war, Suriname comes under British control.

1804–16   The British take control again

Suriname is retaken by the British.

1807   Britain stops supplying slaves

Suriname's economy declines further when Britain ends its slave trade.

1814   Dutch slave trade is outlawed

The Netherlands bans the sale of slaves in all their territories. However, a brisk slave trade is still continued illegally through smuggling.

1815   The Dutch regain control

Under the Treaty of Vienna, ending the Napoleonic wars, the Netherlands regains control of Suriname, which is renamed Dutch Guiana. The land to the east is taken over by France and renamed French Guiana; the region to the west, now the country of Guyana, is claimed by the British.

1821   Fire destroys much of Paramaribo

A terrible fire in Paramaribo is one of a series of disastrous blazes that occur in Dutch Guiana.

1850–1940   Contract laborers brought in from Asia

To bolster the dwindling labor supply on its plantations, especially after slavery is abolished, Suriname imports contract labor from Asian countries including China, India, and Indonesia. Although these laborers arrive to work only for a specified number of years, so many remain in Suriname permanently that today their descendants make up the country's ethnic majority.

1863   Dutch slavery is abolished

Slavery is abolished in Dutch colonies, but all former slaves are required to continue working on plantations at low wages for ten years. After this period, most depart to live in communities founded by escaped slaves (called by the Dutch "Bush Negroes") in the interior, leaving the plantations with insufficient labor to continue operating, and many close. Within ten years, the number of plantations in Suriname has declined by nearly three-fourths from its level forty years earlier.

1866: January 1   Legislature is established

Through the Surinamese Government Order, Suriname is granted its own representative body, which replaces two Dutch-appointed courts in governing the colony, although a Dutch-appointed governor can still veto parliamentary legislation. The parliament's members are either appointed by the Dutch king or elected by the small minority of the colony's residents who meet voting restrictions based on property ownership and education.

1868   Book of lithographs is published

A book of twenty-five satirical lithographs by Belgium planter Theodore Bray (1818–87) is published in Paramaribo on what is thought to be the first lithographic press in Suriname.

1876   Education becomes mandatory

The Dutch colonial government makes education mandatory in Suriname.

1899   Birth of artist Nola Hatterman

Dutch-born painter Nola Hatterman is one of Suriname's leading postwar artists. She plays a major role in training a new generation of painters, and an art institute is named for her. She dies in 1984.

Early 20th century   Bauxite is discovered

Bauxite is first discovered at Moengo. The forests of Suriname are eventually found to contain one of the most extensive bauxite reserves in the world, and this mineral, from which aluminum is made, becomes the country's major source of income.

c. 1900   Branch of Dutch art association is opened

The Dutch art organization *Arti et Amicitiae* establishes a branch in Suriname. Its activities include selling reproductions of Dutch artworks in annual lotteries.

1916   Aluminum company is formed

The Aluminum Company of America (Alcoa) forms a subsidiary in Suriname called the Suriname Aluminum Company, or Suralco.

1917   Bauxite mining begins

World War I spurs U.S. interest in Suriname's bauxite reserves for purposes of aluminum production.

1922   Shipment of bauxite begins

The first bauxite ore is shipped from Suriname.

1929   Birth of artist Erwin de Vries

De Vries (b. 1929), a Surinamese artist trained in the Netherlands, gains international recognition for his paintings and sculptures, which feature abstract styles. One of the works for which he is best known is the sculpture *Alonso de Ojeda* (1963).

1936: April 25   Birth of political leader Henck Arron

Henck Arron, the first Surinamese prime minister, is born in Paramaribo, and pursues a career in banking, first in the Netherlands and then in Suriname, before becoming involved in politics in 1963, when he is elected to Suriname's national assembly. Becoming a nationalist leader, Arron forms a pro-independence coalition of parties supported by Creoles (Surinamese of African descent), called the National Party Alliance. When Arron's coalition wins the 1973 territorial elections, Arron becomes prime minister and over the next

two years works for his country's independence in negotiations with the Dutch.

When Suriname wins independence in late 1975, Arron is its first prime minister. He is reelected in 1977, but the nation's economic ills lead to political turmoil, and Arron's government is overthrown in 1980 in a military coup led by Désiré Bouterse, who forms a military government. Arron is arrested, and released in 1981, after which he returns to the private sector for a time. In 1988 Arron becomes vice president in the government of Ramsewak Shankar, Suriname's first civilian president since the 1980 coup.

1939–45   World War II

Suriname escapes attack in World War II, one of only two Dutch possessions not captured by Germany and its allies. The war does affect Suriname economically by increasing international demand for aluminum in aircraft and other military equipment, which they supply to the allies.

1945   Agricultural mechanization commission is created

The Commission for the Application of Mechanized Techniques to Agriculture in Suriname is established to aid in developing new plantations and restoring old ones.

1948–51   Constitutional reforms are enacted

Constitutional reforms pave the way for greater autonomy.

1949   Universal suffrage is introduced

Suriname's electorate, which has always totaled less than two percent of the population due to strict voting requirements, is vastly expanded by the granting of universal male and female suffrage.

1952   The Surinaams Museum opens

The Surinaams Museum is launched, with collections in the areas of ethnography (branch of anthropology dealing with nonliterate peoples), history, natural history, and ninteenth- and twentieth-century art.

1954   Suriname receives internal autonomy

Both Suriname and the Netherlands Antilles become territories with internal autonomy (self-rule) within the Kingdom of the Netherlands. Suriname still has a Dutch-appointed governor, and the Dutch also have control of defense and foreign affairs, but Suriname's own parliament and cabinet oversee all other aspects of government.

1957: April   Central Bank of Suriname is founded

The Central Bank of Suriname is established as the government's bank of issue.

1961   Nationalist Republican Party is founded

Suriname's first nationalist party, the Nationalist Republican Party (NRP) is established. In addition to winning independence for Suriname, it also dedicates itself to lessening the inequality between the rich and the poor, reducing the influence of foreign investors, and promoting interethnic unity.

1965   Suralco completes Afobaka dam

After six years, construction of the multimillion-dollar Afobaka Dam, a huge hydroelectric power project on the Suriname River, is completed by the Suriname Aluminum Company to supply power for the transformation of alumina into aluminum. A 870-square-mile (2,253-square-kilometer) lake (Vann Blommestein Lake) is created by the dam, which channels water through six generators. As a result, forty-three villages are flooded, and thousands of Surinamese are forced from their homes.

1966–71   Art academies are launched

The Surinaamse Akademie voor Beeldende Kunsten (1966) and the Nationale Instituut voor Kunst en Kultur (1967) are established, and creation of the Nieuwe School voor Beeldende Kunsten begins (1971).

1968   University of Suriname is founded

The University of Suriname (later renamed Anton de Kom University) is established at Paramaribo.

1972: January 5   Commission is formed to prepare for independence

A special commission is created to prepare a framework for independence.

1973   Nationalist is elected prime minister

Creole leader Henck Arron, elected prime minister, puts Suriname on the road to independence, which has been a goal of the Creole population throughout the postwar period. Independence is opposed by some others in the colony.

1975   Bauxite prices decline

Declining demand for aluminum lowers the price of bauxite on world markets, forcing Suriname to cut production of its major export, and causing the economy of the newly independent country to stagnate.

1975: May   Violence erupts over independence talks

East Indian extremists opposed to independence protest talks between the majority Creole government and the Dutch by setting fire to government buildings in Paramaribo.

1975–80   Emigration problems

In the years following independence, around 150,000 Surinamese, including many of Asian descent, emigrate to the Netherlands, fearing that conditions in the country will deteriorate without Dutch rule. Their emigration adds to the new nation's economic problems, already exacerbated by falling bauxite production, by depriving it of many talented, well-educated people. In addition, the large influx of émigrés contributes to social problems in the Netherlands, including unemployment, urban housing shortages, and interracial tensions.

Independence

1975: November 25   Suriname's independence is declared

In an official ceremony attended by Princess Beatrix of the Netherlands, the colony of Dutch Guiana becomes the independent nation of Suriname, a parliamentary republic governed by a prime minister and a unicameral (single legislative chamber), thirty-nine-member national assembly whose representatives are directly elected by popular vote for four-year terms. The prime minister is advised by a cabinet. The official head of state is the president, who serves largely in a ceremonial capacity.

The first prime minister is Henck Arron, leader of a coalition of parties supported by the nation's Creoles. The opposition leader is Jaggernath Lachmon, whose Progressive Reform Party primarily represents the interests of East Indians, also known as Hindustanis. Johan H. E. Ferrier, colonial governor under the Dutch, becomes president until elections can be held. The Dutch promise $100 million annually in development aid in the first ten years of independence.

1975: December 4   Suriname is admitted to the UN

Suriname becomes a member of the United Nations.

1977   Arron is reelected

Prime Minister Henck Arron is elected to a second term.

1980   Multilingual education is introduced

For the first time, languages other than Dutch are used in the schools to help the children of Indonesian and East Indian Surinamese, who do not speak Dutch at home.

Military Rule

1980: February 25   Military coup ousts civilian government

A dispute between military personnel and the government over promotions, pay levels, and unionization escalates to the level of a coup by junior army sergeants led by Sergeant Désiré "Dési" Bouterse. Following predawn attacks on army headquarters and the central police station in Paramaribo, government troops are overcome and Suriname's elected government is ousted. Bouterse forms an eight-member National Military Council. A civilian council is also created, led by Dr. Henk Chin A. Sen, a former leader of the Nationalist Republican Party (NRP).

1980: August   Military government suspends the constitution

The military government led by Désiré Bouterse suspends the constitution adopted in 1975 (at the time of Suriname's independence) and dissolves the National Assembly.

1981   Socialist republic is formed

The military government of Désiré Bouterse declares itself a socialist republic and forms closer ties with Cuba.

1981   Petroleum reserves are discovered

Petroleum reserves are found in the sand of the Saramacca District. Production begins, but Suriname continues to import petroleum from abroad, largely for electrical power.

1982: February   Military council seizes total power

The National Military Council deprives the civilian government of all power, seizing full control of the country. However, a twelve-member cabinet with a civilian majority is appointed.

1982: December   Dissidents are executed in government crackdown

The military government cracks down on dissent, bombing media and union offices and arresting fifteen leading political opponents of Suriname's military government, including labor leader Cyriel Daal. They are tortured and then executed without a trial, provoking international condemnation. One thousand women demonstrate in the nation's capital to protest the executions. The Netherlands and the United States suspend foreign aid in protest.

1984   Bouterse creates Supreme Council

Bouterse forms a new thirty-three-member legislative body called the Supreme Council, whose members are appointed rather than elected.

1986: July   Insurgents launch a guerrilla war

An insurgent group called the Surinamese Liberation Army led by Bouterse's former bodyguard, Ronnie Brunswijk, begins a series of guerrilla attacks (popularly known as the 'Bush Negro,' or 'Maroon,' insurgency) on economic and military targets in the northeast of the country. It seriously disrupts bauxite mining operations by targeting the Suriname Aluminum Company.

1986: August   Group is appointed to prepare a new constitution

A Council of Ministers, including representatives from the nation's three major political parties, is appointed to prepare a new constitution in conjunction with the Supreme Council.

1987   World bauxite prices collapse

The collapse of bauxite prices on the world market is a serious blow to Suriname's economy.

1987   Government agrees to aid sugar industry

Responding to pressure by labor unions, the government agrees to a program to improve working conditions in the sugar industry and create new jobs.

1987: September   Voters approve a new constitution

Faced with internal and foreign pressure plus a deteriorating economy, Suriname's military rulers authorize elections for a new constitution, which is approved by ninety-three percent of the electorate. It provides for the formation of a new National Assembly, which elects the president (the chief of state) to a five-year term. An executive Council of State, composed of military personnel, can repeal laws passed by the National Assembly.

Return to Civilian Government

1987: November 25   Legislative elections are held

In elections for a new National Assembly, the Front for Democracy and Development, a multiethnic political coalition running against the Bouterse government, wins eighty percent of the vote and forty out of a total of fifty-one seats, while the party of military leader Dési Bouterse wins only two seats. In addition, the first Amerindian representative is elected. Bouterse retains political power, however, by appointing a special council that usurps the responsibilities that should belong to the national assembly.

1988   Dutch aid is resumed

With the return of civilian government, the Netherlands announces an aid package of $721 million to be paid to Suriname over a period of seven to eight years.

1988   Surinamese wins Olympic gold medal in swimming

Anthony Nesty wins an Olympic gold medal in the 100-meter butterfly swim event.

1988: January 25   Civilian president is installed

Ramsewak Shankar, Suriname's first civilian president since the 1980 military coup, takes office. Former president Henck Arron becomes the new vice president. However, the military government of Dési Bouterse maintains its hold on power through a five-member military council.

1990: December 24   Shankar government is overthrown in coup

The government of President Ramsewak Shankar is overthrown, and the military once again assumes control.

1991: May 25   New elections are held

Suriname's military rulers once again bow to pressure and allow new elections to be held. A new coalition, the New Front (NF), wins an overwhelming majority.

1991: September 6   New president is elected

Ronald Venetiaan, head of the New Front, is elected president. His government takes over a country whose considerable economic problems include spiraling inflation, high unemployment, depleted foreign exchange reserves, and declining export earnings.

1992   Bouterse resigns

Military leader Désiré Bouterse is forced to resign his government position but remains politically powerful as head of the New Democratic Party (NDP).

1992   Oil pipeline is completed

Staatsolie, Suriname's national oil company, completes construction of the nation's first cross-country oil pipeline. The thirty mile (sixty kilometer) pipeline will transport crude oil from oil fields to the Tout Lui Faut distribution terminal near the capital city of Paramaribo.

1996: May 23   Bouterse's party wins a majority in new elections

Due in large part to the failure of the previously elected government under Ronald Venetiaan to reverse the nation's economic problems through free-market economic reforms, the New Democratic Party, headed by former military dictator Désiré Bouterse, wins a majority in national elections. But it fails to achieve the two-thirds majority needed to appoint a new president, either in the general elections or in two run-off votes.

1996: September 5   Wijdenbosch is elected president

Jules Wijdenbosch of the New Democratic Party (NDP), a former aide to military leader Dési Bouterse, is elected president, defeating incumbent president Ronald Venetiaan. After general elections in May fail to yield a conclusive vote, the United People's Assembly, combining the nation's elected officials at all levels of government—local, regional, and national—votes Wijdenbosch into office. The election marks the first peaceful transition of power since Suriname gained its independence.

1997: October 25   Coup attempt is thwarted

An attempt to overthrow the government of President Jules Wijdenbosch is foiled in Paramaribo. A gun battle between coup plotters and police is followed by thirteen arrests.

1997: December 2   Human rights probe is announced

President Jules Wijdenbosch announces the formation of a special commission to investigate human rights abuses during the military government of Désiré Bouterse in the 1980s. Questions about the credibility of the commission, however, are raised when it is learned that it will be headed by Ludwig Waaldijk, a lawyer representing Bouterse, who still occupies a post in the current government.

Bibliography

Chin, Henk E. *Surinam: Politics, Economics, and Society.* New York: F. Pinter, 1987.

Cohen, Robert. *Jews in Another Environment: Surinam in the Second Half of the Eighteenth Century.* New York: E.J. Brill, 1991.

Dew, Edward M. *The Trouble in Suriname, 1975–1993.* Westport, Conn.: Praeger, 1994.

Goslinga, Cornelis C. *A Short History of the Netherlands Antilles and Surinam.* Norwell, Mass.: Kluwer Academic Press, 1978.

Hoefte, Rosemarijn. *Suriname.* Santa Barbara, Calif.: Clio Press, 1990.

Hoogbergen, Wim S. M. *The Boni Maroon Wars in Suriname.* New York: Brill, 1990.

Price, Richard, ed. *Alabi's World.* Baltimore: Johns Hopkins University Press, 1990.

———. *Maroon Societies.* 3rd ed. Baltimore: Johns Hopkins University Press, 1996.

Sedoc-Dahlberg, Betty, ed. *The Dutch Caribbean: Prospects for Democracy.* New York: Bordon and Breach, 1990.

Stedman, John Gabriel. *Stedman's Surinam: Life in Eighteenth-Century Slave Society.* Ed. by Richard Price and Sally Price. Baltimore: Johns Hopkins University Press, 1992.

Wooding, Charles J. *Evolving Culture: A Cross-cultural Study of Suriname, West Africa, and the Caribbean.* Washington, D.C.: University Press of America, 1981.

Trinidad and Tobago

Introduction

Trinidad and Tobago are located at the southern edge of the Caribbean chain just seven miles off the coast of Venezuela (Tobago lies another nineteen miles to the northeast). Geologically, Trinidad is an extension of the South American continent, whereas Tobago is part of the Lesser Antilles, the southern-most group of Caribbean Islands. With a total area of 1,980 square miles (5,128 square kilometers), slightly smaller than the state of Delaware, the islands are the second largest of the former British West Indies behind Jamaica. Trinidad is the larger of the two at 4,828 square kilometers. The nation's population is approximately 1,300,000. But the capital city, Port-of-Spain, on the island of Trinidad, is the only major urban center with nearly 500,000 inhabitants. The second largest city has less than 40,000 people.

Geography

The primary physiographic feature of Trinidad is the Northern Range of mountains running east-west on the far northern side of the island. Two lesser hill ranges in the center and south of the island also run east-west. Between the three ranges is a series of rolling flatlands on which Trinidad's agricultural sector is centered. The south of Trinidad is a heavily oil-rich belt that reaches into the Atlantic Ocean. Natural seepage of gas and oil give rise to mud bogs and other curious natural phenomena. The most distinct of these is the world's largest natural asphalt bog, Pitch Lake, at over 110 acres in size.

Tobago's main natural resource is its beaches which make it an attractive tourist destination. The residents of Tobago have historically considered Trinidad to be something of a domineering force that overlooks the needs of their island. The recent push by the national government to exploit Tobago as a tourist haven has elicited further calls for consideration of the needs of the inhabitants along with exploitation of the island's revenue-producing potential. While this issue has not assumed such force as to result in calls for Tobago's independence, it has had political implications.

Ethnic Diversity

Ethnicity is an issue of critical importance throughout the Caribbean, and Trinidad and Tobago are no exceptions. However, whereas most of the Caribbean has a large African underclass and a small white or mixed elite, Trinidad and Tobago have a more heterogeneous ethnic composition. Owing to the historic labor needs of the plantation owners on the two islands, large numbers of East Indians were brought in as laborers in the latter half of the nineteenth century, adding to the existing pool of former African slaves. Neither Africans nor East Indians constitute a majority, with both accounting for around forty percent of the population. Much of the rest of the population is defined as "mixed" (fourteen percent). Europeans make up only one percent of the total.

This ethnic diversity has had serious political implications as political parties tend to run along ethnic lines. Until recently, the African-descended population has dominated government. The first East Indian government did not come to pass until 1995. During the era of British rule, which lasted until 1962, Africans were the best positioned to move into the ranks of power when the British left. Effective labor leaders such as Uriah "Buzz" Butler, helped mold the African work force into a powerful political entity. The secondary posts beneath the British colonial officials tended to be held by Africans. One such figure of great importance was Dr. Eric Williams, a scholar and political activist of international repute. He gained popularity in the 1940s and 1950s with his nationalist and semi-socialist ideologies and went on to dominate politics for nearly three decades until his death in 1981.

The island's ethnic heterogeneity brings with it cultural diversity, particularly in the area of religion. Five religions have large followings. Roman Catholicism is the largest at around thirty-five percent of the population, reflecting the historic presence of the French. Hinduism accounts for another twenty-three percent of the religious affiliates, and other faiths, including Protestants (thirteen percent), Muslims (six percent) and "others"—primarily African-based religions (twenty-two percent)—make up the rest.

The cultural output of Trinidad and Tobago, particularly in the arena of literature, tends to draw upon the nation's ethnic dynamic. Perhaps because of its diverse heritage, Trinidad and Tobago have produced literary and cultural figures of international importance. The novelist V.S. Naipaul, for instance, of East Indian descent, is widely recognized as a major figure in modern literature. Many of his literary themes revolve around the issue of interethnic relations.

Pre-Columbian Peoples and Migrations

The history of Trinidad and Tobago is perhaps best divided into only two periods, before and after the arrival of the Europeans in the late 1400s. By virtue of its geographic position, the islands played a crucial role in the pre-Columbian history of the Caribbean. As the islands closest to the South American mainland, they were the stepping stone for migration into the Caribbean. All major Indian migrations into the Caribbean, with the exception of one, passed through Trinidad and Tobago. The first migration into the Caribbean occurred around 4000 B.C. and originated somewhere in Middle America, probably the Yucatán.

The remaining peoples of the Caribbean came through Trinidad from the south, ultimately from deep in the interior of the South American continent, in a series of migratory cycles starting around 2000 B.C. Each successive wave of migrants arrived with increasingly complex tools and pottery. The migrants' dispersal and settlement patterns resulted in a complex ethnic milieu by the time the Europeans arrived. Tainos, a blanket term refering to a variety of ethnic groups sharing common heritage, occupied the Greater Antilles, the larger northern islands such as Cuba and Hispaniola (modern-day Haiti and the Dominican Republic). The Lesser Antilles, including Trinidad and Tobago, were occupied primarily by two later arriving groups, the Arawaks and the Caribs, who migrated from the South American coast around A.D. 100 and 1200 respectively.

Arrival of Europeans

When the Europeans arrived in the late 1400s, the islands of Trinidad and Tobago did not capture their attention to the same extent as other areas considered to offer more immediate economic reward. The Spanish established possession over Trinidad, but Tobago was left on its own. Britain and France made some initially half-hearted attempts at settling Tobago, but the stout defense of the Carib inhabitants forestalled such endeavors. Tobago changed hands constantly throughout the 1600s and 1700s, primarily between France and England, until the English finally gained it for perpetuity in 1802. The Spanish made some efforts at encouraging the production of exportable agricultural commodities on Trinidad, but the island remained sparsely settled and economically limited until the late 1700s. At that time, anxious to turn Trinidad into a profitable possession, the Spanish encouraged the migration of French planters from other islands in the

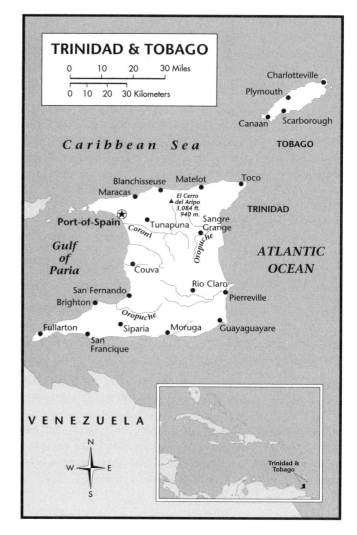

Caribbean to migrate to Trinidad and set up sugar plantations. The large-scale importation of slaves began at this time.

By the time the British abolished slavery in 1833, Africans constituted an overwhelming majority of the population. After abolition, plantation owners looked to India as a source of labor, and starting in 1845 indentured East Indian laborers began arriving. Nearly 150,000 arrived over the next seventy years.

Independence

Until the early 1960s, Trinidad and Tobago remained colonial possessions of Great Britain. As in most of its colonial possession, British officials administered the affairs of the islands without significant input from the inhabitants. A series of labor protests and strikes in the 1930s encouraged the British to enact reforms and allow for greater degrees of self-government. By the time Trinidad and Tobago gained independence in 1962, amid the wave of decolonization occurring globally in the late 1950s and early 1960s, the persons who occupied these secondary administrative posts were in position to assume the reins of command left by the departing British.

The challenge before most governments in the Caribbean is to overcome the limits of their traditional, agricultural-based economies. Trinidad and Tobago have the fortune of sitting atop large deposits of oil and gas that have allowed them to achieve higher levels of economic wealth than many of their island neighbors. The revenues from the petroleum industry have been diverted to social programs, such as education and health care, and to a lesser extent to other economic enterprises, such as small industry. But the nation is still heavily dependent on its oil resources and thus is subject to sudden downturns in the international market. Drops in oil prices in the early 1980s and mid-1990s forced cutbacks in spending and led to changeovers in government, as well as intense debate over the nation's economic structures. Hoping to encourage economic diversification, the government has recently promoted its tourist industry, primarily on the more remote Tobago.

Timeline

c. 7000 B.C.   Oldest Indian site

Trinidad's oldest known peoples settle on the island. They arrive from the South American mainland and will eventually constitute the front end of a steady northward migration into the Caribbean. Evidence of this ancient civilization includes chipped stone remains and bone points and barbs similar to sites in northwestern Guyana and eastern Venezuela. But for now, the migration stops at Trinidad, for no sites of comparable age have been found on northern islands. These first people are defined as Lithic, meaning that the extent of their technological innovation is limited to stone chipping.

c. 4000 B.C.   First major migration into Caribbean

The first inhabitants of the Caribbean begin migrating into the region from the north, probably from the Yucatán peninsula. They are Lithic peoples and are termed "Casimiroid" by archaeologists. They settle on the islands of Cuba, Hispaniola, and Puerto Rico.

c. 2000 B.C.   Second major migration into Caribbean

Distant relatives of Trinidad's initial settlers known as "Ortoiroid" people begin to migrate north from the region of Trinidad and the South American coast. These people are "Archaic," meaning that they have advanced their technological capacities beyond stone chipping to include the production of stone, bone, and shell artifacts.

c. 2000 B.C.   Emergence of pottery civilizations

In the upper reaches of the Orinoco River in modern-day Venezuela, a civilization known as "Saladoid" emerges. Its distinct characteristic is the development of pottery. Migrants carrying this new innovation will follow in the footsteps of their Ortoiroid predecessors and migrate northward into the Caribbean through the island of Trinidad. These peoples are the ancestors of most of the Caribbean Indians that the Europeans will encounter.

c. 1000 B.C.   Frontier with Casimiroids

The northward migrating Ortoiroid peoples arrive on the island of Puerto Rico, come into contact with the Casimiroids, and establish a frontier region between Puerto Rico and Hispaniola.

c. 1000 B.C.   Ceramic peoples arrive at the coast

From their homeland in the interior of South America, the pottery-bearing Saladoid migrants arrive at the coast, near Trinidad, and establish a foothold that will serve as their stepping-off point in the Caribbean. The language of these migrants, Proto-Northern, is the immediate parent language of Carib and Arawak, the two Indian groups who will inhabit the island of Trinidad at the time of the Spanish conquest. It is also the parent language of Taino, the peoples who will inhabit the northern Caribbean at the time of the Spanish conquest.

c. 500 B.C.   Ceramic peoples pass through Trinidad

The Saladoid peoples begin their northward migrations by first embarking for Trinidad from the South American coast. They slowly filter northward in small bands searching for favorable areas in which to settle, selecting uninhabited areas, intermingling with existing populations, or pushing pre-existing peoples out of the way through conquest. Southern Trinidad, with its plains for agricultural production, is one of the first settlement areas. These Saladoid peoples are responsible for introducing the ceramic age—pottery and the techniques of its production—to the Caribbean. As they begin their migration north, they add to their pottery styles by developing new forms, such as jars, bottles, effigy vessels, and incense burners.

c. A.D. 100   Second migration of ceramic peoples

A second, minor northward migration of Saladoid pottery peoples begins at the Trinidad/South American coast nexus. These second-wave migrants are Arawak speakers and are defined by a new type of pottery incision.

c. A.D. 600   Ceramic peoples arrive at Puerto Rico

The pottery-bearing migrants arrive at the Puerto Rico/Hispaniola frontier. During their one thousand-year migration, they have replaced much of the Ortoiroid civilizations that they encountered. At Puerto Rico they divide into at least four new migrant civilizations, collectively called "Ostionoid," which refers to their new cultural styles and adaptations to the local environment. The general difference between the Saladoid pottery style, and the new Ostionoid pottery style is that the former is smaller, finer, and delicately ornamented, while the latter is heavier, bulkier, and more massive. These four groups in turn migrate into the rest of the northern Caribbean, including Cuba, Jamaica, Hispaniola, and the Bahamas. They will become the "Tainos," the Indians that Columbus will encounter in 1492. The origin stories of most all Taino groups uniformly identify their homeland as a cave on the island of Hispaniola, which scholars site as evidence of their common ancestry.

A.D. 1200   Final Indian migration into Caribbean

A final northward migration into the Caribbean begins at the Trinidad/South America coast. The migrants are "Carib" speakers who share their linguistic origin with the Tainos and the Arawaks. As the Carib migrants pass through Trinidad on their way north, a group of them remains on the island. They live alongside, but not necessarily at peace with a group of Arawak speakers, who recently migrated to the island around A.D. 100 from the the continental coast. When Spaniards first arrive at Trinidad in 1498, they find these two language groups on the island. Spaniards portray the Carib peoples as warlike and cannibalistic. (Scholars contend that this portrayal was overstated in order to justify the suppression of the Indians, but they do acknowledge that Caribs had to have been more aggressive to force their way onto already inhabited lands.)

Pre-Twentieth Century

1498: July 31   Columbus lands on Trinidad

During his third voyage to the New World, Christopher Columbus lands on the island of Trinidad. He names the island after the Holy Trinity (hence the name Trinidad) because of the three mountain peaks on its northern side. Columbus sights Tobago, then named Bellaforma, but does not visit it.

1510   Second visit by Spaniards

Spaniards arrive on Trinidad for the first time since Columbus's initial visit in 1498. They proceed to kill or enslave many of the Arawak Indians on the island. The slaves are taken to the islands of Puerto Rico and Santo Domingo. But Spanish settlement on the island is negligible.

1592   Spanish administration begins

Don Antonio de Berrio y Oruna, founds the town of San José de Oruna (now St. Joseph), and officially establishes Spanish authority over the island of Trinidad. The town of San Jose de Oruña will be destroyed by Sir Walter Raleigh in 1595 and not rebuilt until 1606. For nearly 200 years it will be the only European settlement on the island. Immigration to and development of Trinidad occurs very slowly over the coming decades.

1595: March   Sir Walter Raleigh lands

Sir Walter Raleigh arrives at Trinidad with three ships. He is enroute to the South American mainland looking for "El Dorado," a city with a mighty hoard of riches.

1616   First colonists on Tobago

Although the Spaniards knew of Tobago, they chose not to settle it, thereby leaving it to be fought over by other European nations. Perhaps the island had been visited by Europeans, but the first colonists land in 1616. They are English from the island of Barbados. The Carib Indians on the island, however, drive them out. Other colonists follow shortly thereafter, but officially occupation of the island remains in debate for centuries.

1626–1802   Disputed claims to Tobago

Tobago changes hands twenty-six times during this century and a half period. It is claimed and reclaimed by various European powers, primarily Britain and France.

1655   Courlanders ship tobacco

With permission from the English monarch, Charles I (1600–49), people from the principality of Latvia, known as Courlanders, settle on the island of Tobago and export their first shipment of native-grown tobacco. This experiment in settlement and economic enterprise eventually fails.

1702   Slaves for cacao

Some of the first African slaves are brought to Trinidad for the purposes of laboring in the cacao (tropical American tree whose seeds or beans produce cocoa for chocolate) industry. The collapse of the cacao industry in the 1720s temporarily halts the importation of slaves until the late 1780s, when the Spanish administrators encourage slave owners to immigrate to the island to boost the local economy.

1727   Cacao crop fails

The cacao crop fails causing a long-term economic depression on Trinidad that will not be reversed for more than half a century, until the development of new agricultural endeavors, primarily sugar.

1762–63   England gains Tobago

England seizes Tobago from France in 1762 and in 1763 gains official control over the island under the Treaty of Paris, placing it under the jurisdiction of nearby Barbados.

1770: November   Slave revolt on Tobago

A slave revolt occurs at Courland Bay on Tobago. It is led by a slave named Sandy who kills his owner and gathers up to three dozen recruits from nearby plantations. They roam about, surrounding plantations, burning buildings and fields, and threatening owners. Not until army reinforcements arrive from Barbados two weeks after the first incident are the rebels driven into the woods and the rebellion considered to be over. Most of the insurgents avoid capture and flee to Trinidad by canoe.

This revolt at Courland Bay is the first in a series of uprisings and plots on Trinidad and Tobago that will last for the next three or four decades. The rather sudden emergence of public slave opposition reflects not only the islands' growing economy and the increasing number of slaves, but also international political developments associated with revolutions in North America, France, and Haiti. Slaves throughout the Caribbean region will take advantage of European imperial competition and instability to strike out at the slave system during this era.

1771: June   Slave revolt at Bloody Bay

A slave revolt occurs at Bloody Bay on Tobago. Up to eighty slaves are involved. Militia units and army regulars charge the slaves at their hilltop stronghold. Most of the slaves avoid capture by fleeing into the woods, possibly escaping to Trinidad by canoe.

1774: March   Slave revolt at Queen's Bay Estate

A slave revolt occurs at the Queen's Bay Estate on Tobago led by a slave named Sampson. The rebels kill three of the eight whites on the estate before fleeing into the woods with army regulars in close pursuit. The chase lasts one week. Most of the rebels are captured and subjected to torture. The leaders are eventually executed.

1776   Spain encourages settlement by Catholics

The Spanish Crown unofficially begins encouraging Catholics from other Caribbean countries to move to Trinidad.

1782   French take Tobago

The French take the island of Tobago away from the British and initiate policies of settlement and economic development.

1783   Royal Cedula adopted

The Spanish Crown adopts the Royal Cedula of Colonization in an attempt to turn Trinidad into a productive and economically lucrative possession. The cedula, a document which makes official the informal policy of immigration begun in 1776, encourages slavery and the immigration of non-Spanish, Catholic planters from other Caribbean regions who are willing to engage in commercial agricultural production. Many French planters take advantage of the legislation and arrive with their slaves to establish sugar plantations. Although African slaves had been brought to Trinidad early in the century, they begin to arrive in large numbers for the first time. By the end of this year there are only 310 slaves on Trinidad. But this number will increase more than thirty times in the next decade and a half. Africans will soon constitute an overwhelming majority of the population on the island.

1783   Port-of-Spain becomes political center

The *cabildo*, or municipal council, is moved to Port-of-Spain, which remains the political center of Trindidad thereafter.

1787   Sugar production begins

The first sugar plantation goes into production on the island of Trinidad.

1791   Population of Tobago

The total population of Tobago is 15,102, of which only 541 are Europeans and 14,170 are of African descent.

1797   Sugar plantations increase in number

It has been ten years since the first sugar plantation was established, and there are now 159 plantations. The number of slaves on the island of Trinidad is determined to be 10,009 out of a total population of 17,712.

1797   British arrive to challenge Spain

A British armada of twelve ships under the command of Admiral Harvey arrives off the coast of Trinidad. The Spanish admiral on the island chooses to burn his fleet rather than face the British, making way for the uneventful transfer of the island to the British five years later.

1797   Few Indians left

It is estimated that only 1,080 Indians remain of the tens of thousands who inhabited Trinidad at the time of the Spanish arrival.

1802   Peace of Amiens

The peaceful transfer of the island of Trinidad from the Spanish to the British is ratified in the Peace, or Treaty of Amiens. On March 27, the Treaty of Amiens is signed by Britain, France, Spain, and the Netherlands. The "Peace of Amiens," as it is known, brings a temporary fourteen-month peace during the Napoleonic Wars. Importantly, Tobago is returned to British control from the French.

1803   Slave population

A census records that African slaves account for over twenty thousand of the twenty-eight thousand total people on the island of Trinidad.

1805: December   Slave plot uncovered on Trinidad

A rebellion planned for Christmas Day among French-speaking slaves in the suburbs of Port-of-Spain is discovered and quelled.

1807   Slave trade abolished

The British government abolishes the slave trade, thereby cutting off the supply of African-born slaves. The current slave population will remain in bondage for nearly two more decades.

1814   Tobago to the British

The British gain control over Tobago for perpetuity.

c. 1830   Orisha religion gains prominence

Orisha emerges as the primary religion among the African population on Trinidad and Tobago. Orisha is one of a number of African-based religions taking shape in the Caribbean. Others include Voodoo in Haiti and Santería in Cuba. Orisha's origins lay with the Yoruba people in Nigeria. The recent arrival of large numbers of Yoruba laborers accounts for Orisha's rise to prominence.

1833   Slavery is abolished

The British Government abolishes slavery in all of its dominions in the West Indies, including Trinidad and Tobago, over the objections of the many and powerful plantation interests. Plantation owners fear financial ruin because of a shortage of cheap and readily available labor. The British financially compensate many owners for their lost slaves and introduce an apprentice system that continues the slaves' obligations to the plantations another half decade. Eventually the planters look to India for a new source of labor.

1841   Trinidad given archbishopric status

Trinidad is made an archbisophric of the Anglican church in the West Indies.

1844–45   Beginning of East Indian labor

Ten years after the abolition of slavery, the British government allows East Indian workers to migrate to Trinidad as indentured servants under obligation to plantation owners. Indenture is a system whereby a laborer signs a contract, an indenture, with a company to have his transport to Trinidad paid in exchange for an agreement to work for a set number of years. Indenture is often described as being akin to slavery because the contractor and its agents often use their legal bind over the laborer to exploit him. The indenture system begins a fundamental reshaping of the island's ethnic composition. East Indians will continue to arrive in varying concentrations over the next seventy-two years, until indenture is abolished in 1917. Their population will eventually be equal in number to that of the Africans.

1861   First census

The islands' first census records a population of 100,000.

1867   Hurricane

A hurricane strikes the islands causing widespread damage. Because Trinidad and Tobago do not lie in the traditional path of northward-moving hurricanes, this will be only one of two hurricanes to strike the islands in the next century.

1870   East Indian population

Approximately twenty-five percent of Trinidad's population is now East Indian.

1876   Calypso composed

The oldest surviving calypso song, "Juvé Mamicoa," is composed. Calypso is a musical genre that has drawn international attention to Trinidad and Tobago. It has folk origins and shows heavy African rhythmic influences.

1883   Anti-drumming law passed

The Europeans in power on the islands pass an anti-drumming ordinance as a means to limit African-based religious practices.

1884   Sugar economy suffers

The collapse of the Gillespie Brothers' firm in London ruins most of the sugar estates on Tobago, causing an economic depression. The firm had been the main creditor for the estates.

1888   Tobago under Trinidad

After a long history of ambiguity over the jurisdictional control of Tobago, the British government places the island under the control of Trinidad. Tobago retains its own subordinate legislature and taxation policies.

1897   EINA founded

The East Indian National Association (EINA) is founded in Princes Town. It is one of the first Indian nationalist organizations to be founded.

1897   TWA Founded

The Trinidad Workingmen's Association (TWA) is founded. It will become the primary labor organization on the two islands. It will organize heavily among the African-descended laborers and call for an end to Indian indenture.

1897   Butler born

Tubal Uriah Butler (1897–1977), a political activist and labor leader, is born in Grenada. Butler immigrates to Trinidad after WWI and begins working alongside the famed labor leader and nationalist politician, A. A. Cipriani. By the mid 1930s Butler and Cipriani part ways over differences in politics and organizational strategies. But Butler proceeds to expand his base of support, especially among oil workers, and becomes arguably the most recognized and powerful labor leader in the nation. His status as a popular hero is cemented during the strikes of 1937 when he is jailed for two years on charges of sedition (stirring up rebellion against the government in power). After being released, he is jailed again under wartime ordinances and held without trial until 1945. Following his release after the war, he continues to organize and engage in politics. He founds a new union, the British Empire Workers', Peasant's, and Rentpayers' Union in 1946 and proceeds to organize strikes. In the 1950 election, his party receives the most seats in the legislature, but the British government refuses to allow him to come to power. He spends much of the 1950s in England, and by the early 1960s his popular support has been appropriated by the PNM (see 1956).

Twentieth Century

1901   C. L. R. James is born

The Trinidad-born C. L. R. James (1901–89) is born. An author and political figure of international standing, James leaves Trinidad in 1932 for England and the United States. He returns to Trinidad in 1958 to participate in the PNM (see 1956) government. In 1961 he is expelled from the PNM after conflicts with its leader Eric Williams. He joins other émigrés from the PNM to form the Worker's and Farmers' Party, which holds a more radical line and advocates land reform and nationalization of the country's major industries. The government places him under house arrest for a brief period, and in 1966 he leaves again for the U.S. and Britain. James is a committed Marxist whose best known work is the widely read *The Black Jacobins* (1938), a detailed history of the Haitian Revolution of the 1790s.

1903   Water riots

Widespread protests break out in Port-of-Spain that will come to be called the Water Riots. The cause of the disturbances is multifaceted, but it is primarily over the recent decision to meter water usage. Conflicts between the city council of Port-of-Spain and the British-run colonial administration exacerbate the conflict.

1904   Oil exploration

The British show the first sustained interest in exploring the potential for oil reserves on and around Trinidad and Tobago.

1911   First oil export

The first shipment of oil is exported.

1915   Spiritual Baptist religion appears on scene

The Spiritual Baptist religion emerges as prominent on the islands. Its followers and beliefs will intermingle with Orisha.

1917   End of the importation of indentured workers

The migration of indentured East Indian laborers comes to an end. A total of 143,939 Indians have arrived in Trinidad and Tobago since 1845 under the system of indenture. Of these, one in six takes the opportunity to return home after the completion of his contact. Some indenture contracts remain effective for another four years (see 1921).

1917   Anti-Shouters Prohibition

An Anti-Shouter's Prohibition Ordinance is passed. Like its predecessor, the Anti-Drumming Ordinance of 1883, it is designed to hinder African-based religious practices.

1919: November–December   Labor disturbances

Workers' strikes and demonstrations break out on the docks in Port-of-Spain in mid-November and spread throughout the country over the next month. The initial conflict occurs on November 15 after a pay claim submitted by the Trinidad Workingmen's Association on behalf of the dockworkers is rejected. The workers strike and by the end of November, they have been joined by workers throughout Port-of-Spain. A British warship arrives in early December to help restore peace and ensure that the docks remain operational. But the strikes spread to the rural areas of Trinidad and onto the island of Tobago. The work stoppages eventually end in mid-December, but their sheer scale forces the colonial government to consider electoral and economic reforms. One consequence of the strikes is the arrival of the Wood Commission of 1921 to evaluate the social and economic conditions of the islands.

1921   Indenturing comes to an end

All indenture contracts are ended.

1921   Wood Commission

The British colonial government sends the Wood Commission, under the direction of Major Wood, to discover ways to alleviate the social tensions that have resulted in the recent surge in worker discontent. After taking testimony from a variety of local groups and spokespersons, the commission concludes that a modicum of political and economic reform would do much to reduce social conflict. The extent to which the commission's modest proposals are followed fails to end discontent among laborers. Strikes will continue to erupt over the next two decades.

1929   Publication of *Trinidad*

The magazine *Trinidad* begins publication. Although it will last only one year, it is an important outlet for Trinidad's literary community.

1931   Publication of *The Beacon*

A literary magazine named *The Beacon* begins publication. It will circulate for four years (1931–33, 1939) and at its peak of popularity will sell five thousand copies per issue. Like its predecessor, *Trinidad*, *The Beacon* provides an important outlet for Trinidad's nascent writing community. One scholar defines the poetry, short fiction, and essays in these two magazines during their brief existence as the "Trinidad Awakening," a surge in literary expression that lays the foundation for the next generation of internationally recognized authors.

1932   Trade unions legalized

In the wake of a series of labor disturbances, the Trade Union Ordinance is passed that, among other things, legalizes trade unions for the first time.

1932   Naipaul born

V.S. Naipaul (b. 1932) is born on Trinidad. Widely regarded as one of the world's best writers, Naipaul is educated at University College of Oxford University and lives in England. He does not readily refer to himself as a native of Trinidad, but much of his writing is set in Trinidad and explores the island's identity through such central themes as inter-racial relations. Of East Indian heritage himself, Naipaul's main characters are often East Indian and poor. Three of his well-known novels are *Mystic Masseur* (1957), *House for Mr. Biswas* (1961) and *Mimic Men* (1967). He remains an active and highly relevant author.

1934: July   Labor disturbances

Plantation owners lower wages as a response to the Great Depression's (1929–41) negative impact on agricultural prices. Workers, especially those on the sugar estates, respond by striking. At the end of the month, with the strikes spreading, local governments adopt financial relief programs that help bring the strikes to a close. However, worker discontent remains high and will lead to more strikes later in the decade.

1936: April 30–May 10   Women's conference

A conference of the British West Indies and British Guiana Women Social Workers is held in Port-of-Spain. It is led by the charismatic Audrey Jeffers (1898–1968) who emerges as a strong advocate for women's rights. As the first major women's gathering, the conference serves as a rallying point for women's political mobilization.

1937: June   Butler leads strike

With the economic crisis of the Great Depression continuing unabated, rising prices, reduced wages, and high unemployment incite work stoppages. The strikes are initially centered in the southern oil-producing areas and are led by Tubal Uriah Butler (1897–1977), a former oilworker turned labor leader. The strikes spread and are accompanied by acts of violence and property destruction. The disturbances last three weeks and come to a close in early July amid widespread arrests.

1938   Oil as main export

Oil accounts for seventy percent of Trinidad and Tobago's exports.

1945   Citizens win the   right to vote

Universal suffrage is adopted.

1945   Cipriani dies

The labor leader and nationalist political figure A. A. Cipriani dies. A white labor organizer in the 1920s and 1930s, Cipriani is considered one of the island's first nationalist activists for his efforts to unite working people across ethnic lines. Although an adherent of leftist ideas, he adopted a less radical stance than his one-time organizational ally Uriah Butler (see 1937: June), and when Butler split from Cipriani in the early 1930s, Cipriani's influence began to wane. Throughout his career he had organized and led numerous unions and labor movements, and also held numerous elected posts on the Port-of-Spain Municipal Council, reflecting his belief in bringing about change by working within the political system.

1949   League of Women Voters

The League of Women Voters is founded in Port-of-Spain. Its purpose is to encourage women to become aware of and knowledgeable in political affairs.

1956   PNM comes to power

After more than a decade of political activity lacking in cohesiveness since the granting of universal suffrage in 1945, the People's National Movement (PNM) wins the election and establishes the first party-based cabinet government. PNM leader Eric Williams (1911–81) is made chief minister. The PNM will win the next six elections and hold power until 1986.

1958–62 West Indies Federation

Facing the likelihood that it will soon be granting independence to its colonies in the Caribbean, the British government creates the West Indies Federation. Its purpose is to bring the islands together under a common administrative infrastructure that is more viable than the separate individual islands. The breakup of the federation in 1962 is caused by interregional conflict between the wealthier islands which feel the smaller islands are economic drains on them. Jamaica is the first to pull out in 1961, followed by Trinidad and Tobago. The first wave of independence (1962–66) follows.

1960 Statistics on income

Income statistics contained in this year's census are released revealing significant differences in income along ethnic lines. Persons of European descent make over four times as much per year as Africans and approximately six times as much as East Indians.

1962: August 31 Independence

Trinidad and Tobago gain their independence from the British government and become members of the British Commonwealth. Eric Williams (1911–81) is promptly elected prime minister, a position he will hold for the next two decades until his death in 1981. Williams, the foremost political figure of his time, has been an instrumental force in his country's fight for independence. The founder of what will become the largest and most influential party in the nation, the People's National Movement (PNM), he had become Prime Minister before independence in 1961.

1963 Hurricane

The second hurricane in 100 years strikes the islands causing much damage.

1965 Oil leads exports

Oil accounts for eighty percent of Trinidad and Tobago's exports.

1968 Tapia House Group

A political party known as the Tapia House Group is founded under the direction of university professor Lloyd Best. The word Tapia refers to an earthen wall of a traditional hut. The party is the first major leftist movement to mobilize in opposition to the general political conservatism of the post-independence era. The ideas advocated by the Tapia Group will feed into the mobilization and discontent of the "black power" uprising in 1970. Tapia has limited electoral success over the next eighteen years. In 1986 it will dissolve in order to join the National Alliance for Reconstruction, which will

Dr. Eric E. Williams (1911–81) becomes the first prime minister of the newly-independent country in 1962. (EPD Photos/CSU Archives)

win the 1986 election. Best, however, refusing to join the alliance, will reestablish Tapia in 1987.

1970 Carmichael visit

Stokely Carmichael (b. 1941), a black power leader and supporter of Pan-Africanism in the United States, visits Trinidad. Carmichael was born in Trinidad.

1970: February-April 'Black Power' Uprising

A series of protests and strikes break out in Port-of-Spain, events which come to be known as the "Black Power" Uprising. The protests are led by the National Joint Action Committee (NJAC), a radical movement led by Geddes Granger and influenced by the black consciousness movement in the United States. The protests are directed against Prime Minister Eric Williams's perceived compliance with foreign capital. The protesters call Williams an "Afro-Saxon." On April 21, Lt. Raffique Shah (b. 1940), a high ranking military officer, launches a military coup in support of the demonstrators. The government suppresses the coup and ends the protests by arresting and trying large numbers of protesters for treason. The opposition boycotts the next general election in 1971, and a guerrilla group, calling itself the National Union of Freedom Fighters (NUFF) forms.

1973–81 Oil boom

A rapid increase in the price of oil on the world market creates a favorable economic climate for oil-producing nations. Trinidad and Tobago's leaders respond by creating a large number of public jobs and establishing social welfare programs.

1976: August 1 Constitution adopted

The nation's first constitution (adopted in 1962 as a British Order in Council) is replaced by a new one. It establishes the nation as a republic headed by an elected president with a legislature.

1980 Tobago gains a degree of autonomy

New legislation creates a separate House of Assembly for Tobago that consists of twelve members elected at a primary election.

1981: March Eric Williams dies

Prime Minister Eric Williams dies after nearly three decades as the most prominent political figure in the country.

1982 Economic deficit

Faced with a sudden drop in oil revenues, the government finds the nation's balance of trade changing from a surplus to a deficit for the first time since 1973. A long-term recession sets in and will have significant repercussions on politics and financial planning for the next decade. The economy shrinks an average of 4.2 percent per year until 1990.

1983: July Caribbean Basin Initiative

The United States Congress approves the Caribbean Basin Initiative (CBI), which goes into effect in January 1984. CBI is a regional assistance package for the Caribbean designed to isolate socialist-leaning governments by encouraging private enterprise through trade and tax incentives. Some of its main components include free trade with the United States and economic incentives to U.S. business investing in the Caribbean. Supporters herald CBI as an economic windfall; detractors describe it as a masked form of U.S. imperialism. Most nations in Central America and the Caribbean are invited to participate, with the exceptions of those deemed to be communist. Conditions of participation include closer ties to the United States in a variety of legal and economic arenas. Amid growing protectionism in the United States by 1986, CBI is largely considered a failure.

1985 NAR founded

A four-party political coalition known as the National Alliance for Reconstruction (NAR) is founded. It is an extension of a three-party alliance that competed in the election of 1981 representing a rare political alliance between East Indians and Afro-Trinidadians.

1986 Sexual Offences Bill passed

Amid intense interest and debate, the government signs into law a bill known as the Sexual Offences Bill, officially titled, "An Act to Repeal and Replace the Laws of Trinidad and Tobago Relating to Sexual Crimes, to the Procreation, Abduction and Prostitution of Persons and to Kindred Offences." The bill is a highly controversial initiative on the part of the government to regulate and define punishments for certain kinds of sex, ranging from homosexuality to adultery.

1986: December PNM defeated

The PNM is defeated soundly at the polls after three decades in power. Its opponent, the NAR (see 1985) wins thirty-three of thirty-six legislative seats. The PNM's lack of popularity can be attributed to many factors, such as the loss of its charismatic leader, Eric Williams, and its longtime failure to win the support of East Indian voters. But arguably the most important cause of its downfall is the sudden drop in oil prices that depletes state revenues. The NAR's program is more free-market oriented calling for divestment of state-owned utilities and a restructuring of the economy. The new prime minister is the NAR leader, Arthur Robinson (b. 1926).

1987: January Internal self-government for Tobago

Tobago is granted full internal self-government. It is still part of the nation of Trinidad and Tobago, but its leaders now have greater autonomy over certain policymaking areas, such as finance and the environment.

1987 February Police corruption

The new NAR government publishes a report stating that two PNM ministers of the former government, along with five judges and fifty-three police officers, have been involved in drug trafficking.

Late 1987 NAR coalition splits

The NAR coalition falls apart amid charges that the government is not doing enough for East Indians. One of the coalition's main figures, the East Indian Basdeo Panday (b. 1932), accuses Prime Minister Robinson of acting like a dictator.

1988 IMF loan

Confronted with financial shortfalls, the government enters into an agreement with the International Monetary Fund (IMF) for US $231 million.

1988: August   Financial austerity

The government introduces an emergency financial program that includes an increase in prices, fifteen percent devaluation of the currency, and cuts in government spending.

1989   UNC Party founded

Panday and other politicians who either left or were expelled from the NAR during the split of 1987 unite to form the United National Congress (UNC) party. Panday is its leader (See late 1987.)

1990: July   Guerrilla attack

Members of the Jamaat al-Muslimeen, a militant Islamic group led by Imam Yasin Abu Bakr, seize control of the Parliament building in Port-of-Spain. Among the forty-six hostages are Prime Minister Robinson and his cabinet. The rebels demand rapid political changes, including the resignation of the prime minister and new elections in ninety days. Robinson is shot in the leg when he refuses to sign a letter of resignation. Widespread looting and burning occurs throughout Port-of-Spain during the five-day standoff. Twenty-three people are killed and 500 are wounded in the process.

1991: December 16   PNM returns to power

After being weakened by the Jamaat rebellion, the NAR cannot hold off the PNM in the general election. The PNM wins a majority of the legislative seats and its leader Patrick Manning (b. 1946) becomes prime minister. With few alternatives, the new government continues the liberal economic policies of the NAR.

1993: January   Police corruption

A British-led police investigation determines that corruption is widespread throughout the police force.

1993: April   Currency devalued

The government allows the Trinidad and Tobago dollar to be devalued, liberalizing exchange with foreign currency. The currency loses twenty-six percent of its value almost immediately. This is part of a broad liberalization program designed to attract foreign capital and investment in order to help modernize the nation's petroleum industry. A high degree of public discontent follows as purchasing power declines and unemployment increases.

1994   Unemployment on the rise

Unemployment reaches 18.4 percent from 10.3 percent in 1982.

1995   Panday becomes prime minister

The PNM Prime Minister, Patrick Manning, calls a general election one year early in hopes of strengthening his party's control in government, but the plan backfires. Basdeo Panday's (see late 1987 and 1989) UNC party wins seventeen seats in the legislature, the PNM also wins seventeen seats, and the NAR wins two. The stalemate is broken when the NAR and UNC parties form a coalition and elect Panday as prime minister. He is the first East Indian to hold that post. Panday calls for strict observance of non-racial policies, and even goes so far as to meet with representatives of the Jamaat al-Muslimeen.

1998: March   Oil prices fall

Oil prices fall suddenly, forcing the government to scale back proposed spending for the 1998–99 budget year.

1998: April   Minimum wage established

The government establishes a national minimum wage.

1998: June   Death penalty debate

Trinidad and Tobago's Attorney General, Ramesh Majaraj, declares that the government will step-up its efforts to ensure that most of the nation's 100 death-row prisoners are executed. This represents a strong policy reversal from the prior two decades when only one prisoner was executed—with serious debate still being argued about that decision. According to an agreement made with Britain at independence, Britain's Privy Council remains the final court of appeals for Trinidad and Tobago. With support for capital punishment low in Britain, the council has forestalled executions. But with Trinidad and Tobago facing a rise in crime, a public in support of capital punishment, and a closer relationship with the United States's anti-drug campaign, the Panday government believes it can overcome opposition from the human rights supporters to proceed with executions.

Bibliography

Bereton, Bridget. *A History of Modern Trinidad.* Portsmouth, N. H.: Heinemann, 1981.

Craton, Michael. *Testing the Chains: Resistance to Slavery in the British West Indies.* Ithaca: Cornell University Press, 1982.

Fineman, Mark. "Trinidad Returns to the Gallows," *Los Angeles Times,* 17 June 1998, p. A1.

Gunson, Phil, et. al., eds. *The Dictionary of Contemporary Politics of Central America and the Caribbean.* N.Y.: Simon and Schuster, 1991.

Houk, James. *Spirits, Blood and Drums: The Orisha Religion in Trinidad.* Philadelphia: Temple University Press, 1995.

Newson, Linda. *Aboriginal and Spanish Colonial Trinidad: A Study in Culture Contact.* N.Y.: Academic Press, 1976.

Reddock, Rhoda. *Women Labour and Politics in Trinidad and Tobago: A History.* London: Zed Books, 1994.

Rohter, Larry. "For New Trinidad Chief, Race Question Looms Big," *The New York Times*, 1 January 1996, p. A7.

Rouse, Irving. *The Tainos: Rise and Decline of the People Who Greeted Columbus.* New Haven: Yale University Press, 1992.

Sander, Reinhard *The Trinidad Awakening: West Indian Literature of the Nineteen-Thirties.* N.Y.: Greenwood Press, 1988.

Weeks, John, and Peter Ferbel. *Ancient Caribbean.* N.Y.: Garland Publishing, 1994.

United States of America

Introduction

The land that would become the United States of America was devoid of human life until about twelve thousand years ago, long after humans had already settled in Africa, Asia, and Europe. Over 270 million people now populate the United States, and it grows by another million or so every year. Its human history begins with the Native Americans who settled the land and formed hundreds of cultures and political entities as vast and varied as the Apache, Sioux, Iroquois, Algonquin, and Seminole nations. Some of these nations were nomadic, like the Sioux, migrating with giant buffalo herds. Others, like the Seminole, were farmers. Still others lived in longhouses like the Iroquois and required all members of a clan to live under one roof. Some were peaceful while others, like the Iroquois and Huron, warred with one another for centuries. What they all shared in common was defeat and dislocation by European immigrants carrying better weapons, modern diseases, and their own agenda.

Early Immigrants

The newcomers came from diverse backgrounds for a variety of reasons. There were immigrants from Spain in what became Florida and the Southwest who sought riches. Immigrants from France sought trade with Native Americans in the Ohio Valley and Louisiana, especially in furs. Immigrants from England sought farmland and began deforesting and farming the eastern seaboard. Many of them were fleeing persecution for religious beliefs. Others came over as indentured servants, and still others were criminals trying to avoid punishment.

After about 100 years as European colonists, the newcomers in the east would unite to break from their European rulers, expand westward, and evolve into an independent democracy with a capitalist economy. The Revolutionary War (1775–83), as it came to be known, was, in fact, a rebellion. The colonists rebelled against authoritarian and sometimes arbitrary British rule. Particularly loathsome to the colonists were the high taxes imposed on them to support Britain's wars with France and the Native Americans.

The founding fathers of the United States were imperfect men with conflicting goals, but they united to declare independence and empower General George Washington to lead an army against the British. After years of guerrilla warfare, and unable to destroy Washington's Continental Army, the British finally gave the colonies their independence, expecting them to fail on their own.

The Constitution

The founding fathers did not fail, despite their imperfections. Aristocratic slave-owners themselves, some wanted to limit power to the wealthy. They disagreed over whether there should be a central government, how powerful the central government should be, what rights and powers the states had in relation to one another, and how the legislature should be apportioned. The compromises they made in shaping the Constitution allowed them to form a federal political system which has proven strong enough to withstand scandals, assassinations, impeachments, and even a bloody civil war.

The Constitution, based in part on the Iroquois nation's form of government, did not resolve difficult issues like slavery, but it was broad enough and flexible enough to allow most issues to be resolved peacefully through amendments and by providing for a Supreme Court to resolve disputes over its interpretation. The U.S. Constitution, the world's oldest written constitution, was novel for its creation of a balance of power between the states and the central government, and in developing a three-pronged balance of power between the president, Congress, and the Supreme Court.

1783–1861

The period between independence and civil war saw tremendous growth in the country's population, territory, and cultural diversity. The United States had already imported a huge slave population and large waves of poor immigrants from European countries came to the United States hoping to make better lives for themselves. This mix of people steadily expanded west in search of new land, creating both new problems and new opportunities. The vulnerabilities of the Native

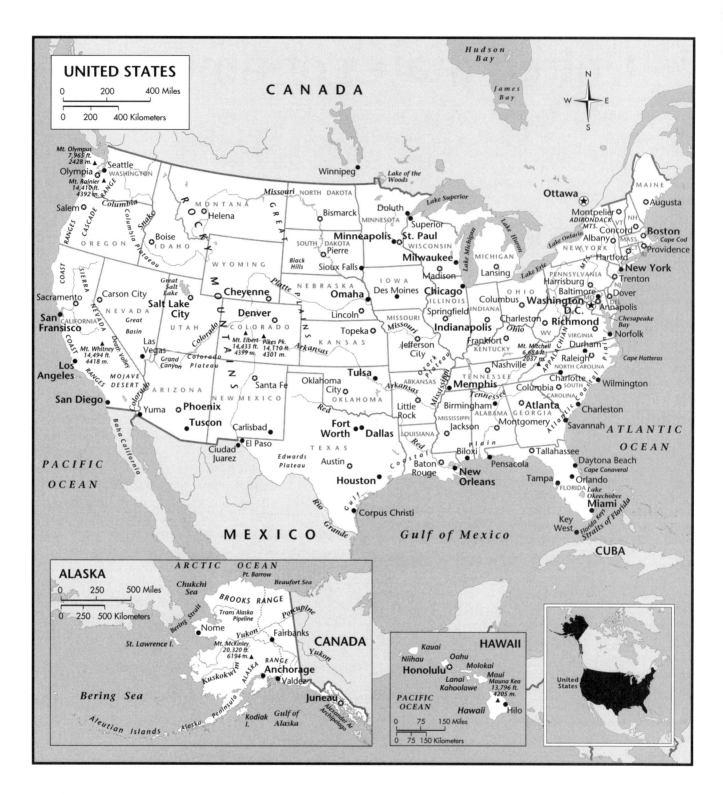

American population and the waning power of the Spanish and French empires aided in the acquisition of new territory.

While the British and their Native American allies were able to stave-off U.S. expansion into Canada, the War of 1812 (1812–14) established real U.S. independence (forcing Britain to recognize the right of the United States to direct its own international commerce and conduct its own foreign affairs) and pushed its boundaries across the continent. New states were carved out of the wilderness as the construction of canals and railroads increased the pace of westward expansion. Religious differences were increasingly tolerated as the mostly Protestant nation welcomed immigrant Catholics and witnessed the birth of Mormonism in the west and Transcendentalism in the east.

If the United States was prevented from incorporating Canadian territory into its own, it found it could acquire land from Mexico. The United States annexed one-third of Mexico's land including the area that would eventually comprise the two most populous states in the union, California and Texas. American settlers in Mexico had already declared Texas independent, but disputes over the common borders between the United States and Mexico led to war in 1848. After the U.S. victory, American settlers in search of gold and land quickly flowed into the Southwest, bringing their slaves with them and cementing a U.S. presence on the Pacific coast.

The Civil War

As the continent was quickly running out of room, so was the nation running out of patience over slavery. A physical brawl among members of Congress occurred when a southern congressman beat an abolitionist congressman with a cane on the floor of the House of Representatives. Americans were also killing each other over slavery in places like Kansas. The Southern states to seceded from the Union in 1861 and established the Confederate States of America. The Civil War was the costliest in American history in terms of money, property, and lives lost. Thousands of young men died in a single day's battle, major cities were burned to the ground, and entire states were laid waste. Slavery was abolished, but emancipation did not end the nation's racial problems. Legal segregation would last another 100 years. The Fourteenth Amendment to the Constitution theoretically gave all citizens "equal protection of the laws," but the protection was often not enforced when it came to African Americans, not only in the South but throughout the country.

Post-war Industrialization

The growth of large corporations run by newly-made millionaires like John D. Rockefeller, Andrew Carnegie, and Cornelius Vanderbilt turned a country of mostly farmers and small shopkeepers into one of mostly city dwellers and factory workers. Immigration policies were expanded and people came from other parts of Europe, including Ireland, Italy, and Poland. Business owners exploited them to keep factory wages low and threatened them to keep strikes at a minimum. Asian immigrants from China and Japan entered the West and were resented and blamed for low wages and poor working conditions. At times, immigration by Asians was banned.

Under the harsh conditions of the early industrial period, the labor movement emerged. Violence over unionization and labor rights was common, although it would take decades for the government to recognize the legitimacy of labor unions. Nevertheless, the government's reduction of imports and the country's rapid population growth helped the growth of American industry surpass that of Britain and Germany by the twentieth century.

Quality of life was improved by technology and information. New transportation and medical breakthroughs combined with the discovery of electricity and the creation of the telephone allowed Americans to live longer and healthier lives, and shortened the distance and time it took for travel and communication. For consumer goods, the seemingly impenetrable wall of physical distance gave way, as Americans were able to purchase meat from the Midwest, produce from the South, and dairy products from New England. Americans began to cultivate leisure time, and professional sports like boxing, baseball and football became increasingly popular.

Expansionism

At the turn of the century, the United States focused its attention on expansion. It sought to extend its territory and influence on the continent by forcing the last of the independent Indian tribes on to reservations through a series of wars that reduced their population to about 250,000. Expansion overseas took the form of intervention as U.S. sugar growers in Hawaii first toppled the royal government, then requested annexation by the United States. In Cuba, U.S. support for the independence movement and the sinking of the battleship *Maine* in Havana harbor provided the pretexts for declaring war on the ailing Spanish empire. At the conclusion of the Spanish-American war (1898), the Philippines, Guam and Puerto Rico became U.S. territories and Cuba became a U.S. protectorate. However, the idea of empire-building was contrary to the American ideals of liberty and self-determination, and the resulting internal conflict over U.S. expansionist policies prevented European-style conquest and colonization of Caribbean and Central American countries by the United States.

Although the United States maintained an isolationist stance in the early part of World War I, its eventual involvement resulted in its emergence as a world power. By the time the United States joined the war effort in 1917, the cost of prolonged trench warfare, both in human lives and capital expenditure, had drained European combatants on both sides sufficiently enough that the infusion of fresh troops and supplies proved decisive in ending the war.

The Great Depression

After the war, harsh financial ups and downs in the economy would continue until the stock market crash of 1929 ushered in more government regulation. The flamboyant "Roaring Twenties" gave way to the Great Depression of the 1930s, an era characterized by worldwide chronic unemployment, bank failures, and industrial and economic stagnation. Seeking to alleviate the crippling effects of the Depression on the country, President Franklin Delano Roosevelt instituted a number of social and legislative reforms and relief programs that collectively became known as the New Deal. The New Deal insured that the federal government would play a large role in regulating the economy and guaranteeing the public a bare minimum standard of living.

The Great Depression saw a rise of radicalism as labor unions fought to guarantee the rights of U.S. workers. The civil unrest of Prohibition, exacerbated by the effects of the Great Depression, gave rise to organized crime, kidnappings, and other illegal activities which, in turn, led to increasing federal involvement in domestic affairs. Agencies like the FBI were created to counter these trends. Passage of a controversial income tax allowed the government to support its domestic programs.

World War II and the 1950s

President George Washington's farewell address advised the nation to avoid foreign entanglements, but Japanese warplanes forever changed that policy on December 7, 1941 at Pearl Harbor. After the Japanese forces attacked the U.S. naval base at Pearl Harbor, Hawaii, entry into World War II was inevitable for the United States. Isolationist policies were abandoned and the United States declared war on Japan the next day. In return, Japan's allies, Italy and Germany, declared war on the United States.

What emerged after the devastaion of World War II was a dual superpower structure, with America and her allies promoting democracy through the North Atlantic Treaty Organization (NATO) and the Soviet Union supporting the doctrines of communism in the Warsaw Pact.

Meanwhile, the wave of renewed industrialization that swept the country to meet the demands of the war lifted the country out of the Great Depression and caused so many demographic changes that U.S. troops returning from abroad found a different America. Women had taken jobs to replace the men who had gone overseas and great migrations had occurred as millions of African Americans from the South migrated to cities in the North, where there were factory jobs and less perceived racial oppression.

After the war, the nation became more affluent, expanding the middle classes. A post-war rise in births created a baby "boom" (adults of this generation are called "baby boomers"), which also added to the growing middle class. Urban populations began to decline as returning soldiers and their families moved into newly-created suburbs. New freeways and airports made this easy. City neighborhoods began to decay through neglect and abandonment. Television became the information and entertainment medium of choice and brought the world into the homes of American families.

The Cold War

The devastation of Europe and Asia and the lessons learned from World War II forced a change in global policy. On the one hand, international agreements were made to increase world trade, support and defend allies, and help countries rebuild themselves. On the other hand, tensions between the major Western powers and the Soviet Union over fundamental ideological differences created a "Cold War" between the two. The Western powers' ideals of capitalism and democracy were in direct conflict with the totalitarian, communistic ideals of the Soviet Union. These differences, coupled with the promotion of communism abroad, began a nuclear arms race that shadowed the world with the fear of global nuclear war.

The United States entered the Korean conflict on the side of the United Nations (UN) who were defending South Korea from a communist takeover by the North Koreans. When the communist Chinese military intervened on behalf of the North Koreans, UN forces were pushed back, resulting in a stalemate between the combatants. Eventually a ceasefire was declared and an uneasy peace was proclaimed, although an official end to the war was never signed. Instead, a demilitarized zone (DMZ) was established between the two nations. Americans remained behind and still have a large contingent of troops on the South Korean side of the DMZ, although their numbers are considerably smaller than the one million troops that are estimated to guard the North Korean side of the DMZ.

The Vietnam War was a prolonged conflict that severely impacted not only the Vietnamese people, but millions of Americans as well. Though there seemed many similarities between the Korean and Vietnamese situations (communism in the north spreading to the south), American intervention was never wholly supported by the Vietnamese people. Communist forces in Vietnam, who were heavily armed by other communist nations, conducted the war with guerilla tactics and a constant pressure that eroded morale among American and South Vietnamese soldiers. As casualties mounted, U.S. public opinion turned against further intervention, and many Americans took to the streets in protest demanding the recall of American troops and end to the war.

The decades-long Cold War never broke out into open warfare between the superpowers, but it greatly contributed to the proliferation of nuclear arms (as each side raced to build more powerful and smarter nuclear weapons), space exploration (the Soviets' launch of Sputnik influenced the U.S. to ante-up for a space program of its own), and fear of annihilation (citizens built bomb shelters in their backyards). The two superpowers came close to global warfare at times, but both powers proved responsible enough to compromise when necessary.

Civil Rights

The glacial progress of civil rights picked up in the 1950s and 1960s. Boycotts, marches, sit-ins and speeches, often accompanied by violence against the protesters, won basic rights for

African Americans and broke down racial barriers. The military was desegregated, Jim Crow laws were abolished and the Supreme Court ordered the desegregation of the schools.

Women's rights groups were not far behind in advancing their cause as well. Equal opportunities in education, sports, and jobs were demanded, and so was equal pay for equal work.

1970s to Present Day

People born in the period following World War II (1945–65) would enter adulthood in a country that was less optimistic than the one their parents knew. Disaster in Vietnam, the resignation of President Nixon, and the energy crisis made the public more cynical than it had been in the past, as did painful economic transitions such as the decline of manufacturing industries. The old industrial Midwest became known as the "rust belt," and people moved west and south into the "sunbelt." California replaced New York as the most populous state, and Los Angeles replaced Chicago as the nation's second most populous city. The middle class continued to move from the cities to the suburbs, taking their money and political power with them, and city neighborhoods spiraled further into decay. Left behind, the mostly African American urban population, frustrated by lack of access to jobs and education and suffering unfair treatment by police and continuing racism, rioted violently in cities across the United States in the late sixties.

Demographic changes continued also as Asian and Latin American immigration grew. In the Southwest and in cities like New York and Chicago, the U.S.-Hispanic population grew to rival that of the African American population.

The dissolution of the Soviet Union in the early 1990s quickly ended the Cold War and placed the United States in a transition period. Military spending was cut, more Federal dollars were targeted towards social programs and fiscal responsibility became the focus of many election campaigns. Without the Cold War focus on external problems, the United States has begun to tackle some of its internal social problems like high unemployment, the lack of health care for all citizens and the growing welfare system. Many social problems remain, however, including the increasing disparity between the rich and poor, and having the highest crime rate among the industrialized nations.

The United States is now an active part of the interconnected global political network and economy that it helped create. Isolationism is no longer feasible for the richest and most powerful nation on earth. However, only the development of a clear sense of moral purpose will help it decide when to answer global and regional conflict with active intervention.

Timeline

c. 50,000–12,00 B.C.   First humans inhabit North America

The first Native Americans (Paleo-Indians) migrate across the Bering Strait Land Bridge to establish themselves in the major river valleys and mountain passes of North America.

c. 25,000 B.C.   Paleo-Indian societies create tools

North American Indian societies make flint, chert, and obsidian tools and spear points.

c. 10,000–6000 B.C.   Paleo-Indian societies begin to diversify

Climate changes occur which kill off the large mammals. Indian societies across North America begin adapting to their particular climate and environment. They begin eating more and various kinds of plants. Fur, cloth, medicine bags, and basketry remains indicate increasingly complex cultures.

c. 6000–1500 B.C.   Some Indian societies become sedentary

Cultivation and processing of native plants begins in some societies, as well as the building of permanent shelters. Opportunities for interaction among neighboring groups occurs. Trade develops.

c. 1500 B.C.–A.D 500   Regional variations emerge

The Indian cultures continue adapting to their environments, giving rise to more advanced tool and weapon technology, increased cultivation and harvesting, and development of complex religious and ceremonial practices. People of the Southwest build irrigation systems to water their fields. Central and Southwest societies begin to grow maize and pottery appears in the Southwest.

Several regions, especially the Plains, use the bow and arrow. Societies in the East and Southeast develop social hierarchies.

c. 500–1000   Agriculture and resource distribution become important

Farming becomes the principal means of subsistence in the East and Southwest. Maize is the major crop. Food storage facilities come into use. Trade and resource distribution become entwined with the emerging social hierarchies.

c. 1000–1492   Indian societies develop complex social structures

The social organization of many Indian societies includes hereditary leadership, multiroom and multidwelling settlements, fortifications, interdependent trade networks, and the concept of wealth.

1492 First European arrives in New World

Setting sail on August 3, 1492 from Spain, Genoese explorer Christopher Columbus (1451–1506) becomes the first European to sight the "New World" (islands of the North American continent). Under the sponsorship of the Spanish King Ferdinand (1452–1516) and Queen Isabella (1451–1504), Columbus leads a small band of 120 men in three ships — the *Niña, Pinta,* and *Santa Maria* — towards what he hopes will be a westward route to India. Instead, he finds several islands in the Caribbean, including Cuba and Hispaniola, where he founds a small colony. Columbus's discovery of land across the Atlantic Ocean creates a boom in trans-Atlantic exploration from the late fifteenth century onwards.

At this time, no more than two million Native Americans inhabit the land that will one day become the United States.

1499 Amerigo Vespucci explores the Americas

Italian-born Spanish explorer Amerigo Vespucci (1451–1512) sails as part of an expedition to the New World and explores the coast of Venezuela. The American continents are named after Vespucci, whose name is Latinized as "Americus." A young German cartographer (map-maker) who reads Vespucci's published letters is the first to dub the continent "America."

1565 Spain establishes the first permanent settlement

The Spanish found the first permanent settlement in what is now the United States at St. Augustine in the future state of Florida.

1607 Jamestown colony founded

The English found the small, ill-fated colony of Jamestown (named after English King James I) in what is now Virginia in 1607. Captain John Smith (1580–1631) is elected the first president of the new colony, but returns to England in 1609.

Pocahontas (1595–1617), a Native American princess, saves the life of John Smith on two occasions when he is at the mercy of her tribe. Pocahontas later becomes a Christian and marries the Englishman John Rolfe (1585–1622). She goes to England in 1616, only to die of smallpox one year later.

1609 Henry Hudson explores North America

Chartered by the Dutch United East India Company, English navigator Henry Hudson (c. 1550–1611) begins the search for the so-called "Northwest Passage," an all-water route to Asia. Hudson's explorations take him to Newfoundland, the New England coast, Cape Cod Bay, and the Chesapeake Bay. He explores the Hudson River (named after him) up to the site of

present-day Albany, and is on friendly terms with the Iroquois people during his journeys.

1612 Tobacco cultivation begins

English colonists in Virginia begin growing tobacco. Tobacco will come to play an essential part in the social and economic life of the colony, especially in the South.

1619 First slaves

The first African slaves are brought to the colony.

1620: November Plymouth Colony founded

The Pilgrims, a group of religious separatists from England, arrive on their ship — the *Mayflower* — at the land which is now Massachusetts and form a settlement at Plymouth. The Pilgrims draw up the "Mayflower Compact," which provides laws and regulations for their new colony. John Carver (c. 1575–1621) is elected the first president, but dies within five months.

The Native American Tisquantum (known as "Squanto," a member of the Wampanoag tribe) is one of the first to help the new colonists, showing them where to hunt, fish, and how to grow native crops such as beans, corn, and squash. Tisquantum had been captured by the English between 1605–17, yet escaped and returned to his home with the ability to speak English.

After a long and difficult winter, the Pilgrims learn quickly from the native peoples, and in the autumn of 1621 host the first Thanksgiving dinner as a celebration of their harvest. The Wampanoag and their chief, Massasoit, are invited to the feast, and when the Pilgrims do not have enough food, the Wampanoag provide food as well.

1624–55 Dutch settle the "New World"

The Dutch establish a settlement at Fort Orange (now Albany, NY) in 1624. In 1626 Dutch colonists purchase Manhattan Island from the Native Americans for twenty-four dollars, and founded the settlement of New Amsterdam (now New York City). The Dutch also conquer the Swedish colony in Delaware and New Jersey (New Sweden) in 1655. An active fur trade is quickly begun.

1630–42 "Great Migration"

During these years, known as the "Great Migration," over 16,000 settlers arrive at the Massachusetts Bay Colony from England.

1636 Founding of Harvard College

The General Colonial Court votes £400 towards the founding of the first college of higher education. Two years later, clergyman John Harvard (1607–38) leaves £780 and 260 books to the college, which is named after him in 1639.

1636: June   Rhode Island Colony founded

Roger Williams (c. 1603–83) is a clergyman who has been banished from the Massachusetts colony for his outspoken criticism of civil authorities and colonization policies as they affected the Native Americans. Williams founds his own colony at Providence, organizing a democratic government with separation of church and state. Rhode Island becomes noted for its religious toleration and democratic, liberal government.

1638   Anne Hutchinson scandal

Religious liberal Anne Hutchinson (1591–1643) is banished from the Massachusetts Bay Colony after she is convicted of slandering the colonial church. Hutchinson, who had preached such principles as salvation through love of God with disregard for the laws of the church, then settles in Rhode Island. She and her family are murdered by Native Americans in 1643. Hutchinson's radical ideas shock the predominantly conservative colonists.

1641   Massachusetts Bay adopts code of laws

A code of 100 laws, known as the "Body of Liberties," is established by the Massachusetts Bay Colony's General Court.

1651–73   British Navigation Laws

Designed to reduce the profitability of Dutch shipping, the Navigation Laws require that all colonial products must first be shipped to Great Britain on English or plantation ships. A 1663 Act further requires that all materials being sent to the colonies must first be shipped to England. The Act of 1673 imposes intercolonial duties (tax) on sugar, tobacco, and several other products.

1663   Grant of Carolina established

The English crown grants the territory between 31° and 36° latitude to eight proprietors. The grant names the land "Carolina."

1664   England monopolizes American settlement

The English seize the Dutch colonies and control all colonization on the eastern seaboard of North America, except Florida where the Spanish remain in power until 1821. Spain also controls the area now comprised of California, Arizona, New Mexico, and Texas until the nineteenth century.

In August, New Amsterdam is officially surrendered by the Dutch, and the English rename it New York.

1670   Founding of Hudson's Bay Company

The Hudson's Bay Company is created and given a monopoly on the trade in the Hudson's Bay Basin.

1673   French explore the Mississippi River

French missionary Jacques Marquette (1637–75) and trader Louis Joliet (1645–1700) follow rivers southward from the French colonies in Canada, eventually reaching the Mississippi River, which they follow as far south as what is now Arkansas. In 1682 explorer Robert de la Salle (1643–87) travels the entire Mississippi, finally reaching the river delta and taking possession of it in the name of the king of France. La Salle attempts to establish a colony at the mouth of the Mississippi (to gain a monopoly on exploration and the fur trade, as well as provide a military base of attack against the Spanish), but is mutinied by his own crew in 1687. New Orleans, however, is finally founded in 1718.

1675–76   King Philip's War

Native Americans begin a series of attacks against white settlers in response to the rapid advance of the colonial frontier. In addition, the Christianization of the Cape Cod peoples has been regarded suspiciously by other tribes including the Wampanoags, who viewed it as an attempt to weaken native power. Philip (son of chief Massasoit of the Wampanoags) forms a league of most of the Native Americans from Maine to Connecticut. Border attacks by this league are answered by a white attack upon a Native American stronghold in Rhode Island. During the war, nearly 500 white men are killed, and twenty villages are destroyed. The Native Americans suffer heavy losses, though exact numbers of casualties are unknown. The war ends with the death of Philip in August, 1676.

1681   Signing of Charter of Pennsylvania

Pennsylvania ("Penn's woods") is named after William Penn (1644–1718), who is granted a charter for the land between 40° and 43° latitude, extending west from the Delaware River. A boundary dispute with Maryland is resolved when two surveyors—Mason and Dixon—finalize the present-day boundary between the two states. Philadelphia is founded in 1682.

1684   Massachusetts Bay Charter annulled

The independent behavior of Massachusetts Bay is an irritation to the English crown, which claims the colony exhibits little respect for the king's authority. The colony's charter is therefore annulled, and Massachusetts as well as the other New England colonies come to be controlled by an English governor. However, when English King James II (1633–1701) flees the throne in 1689, the colonists rise in a revolt and restore the charter government.

1692   Salem witchcraft trials

Nineteen women are hanged and one pressed to death as a result of the paranoia-driven witchcraft trials in Salem, Massachusetts.

1701   Founding of Yale College

Yale, the second-oldest college in the United States, is founded in New Haven, Connecticut.

1702–11   Queen Anne's War

Queen Anne's War is the American extension of the War of the Spanish Succession (1701–14), when changes in the Spanish monarchy threaten to upset the European balance of power). Attacks between the English and French/Native Americans spread over the frontier and even reach the outskirts of Boston. The French province of Acadia is captured in 1710 and becomes the British province of Nova Scotia. The Treaty of Utrecht in 1713 grants Great Britain possession of the Hudson's Bay Company, as well as Newfoundland and Nova Scotia.

1733   The Molasses Act

In response to an economic depression on the sugar-growing islands of the British West Indies, England imposes a tax upon all sugar and molasses imported into the colonies from islands that are not English possessions.

1733   Colony of Georgia founded

Georgia, the last of the thirteen original British colonies in North America, is founded. British, Spanish, and French claims often conflict over portions of Georgia. Georgia is originally created as a buffer between the three nations' colonies and is initially the home to impoverished English debtors. James Oglethorpe (1696–1785)—an English philanthropist who favors a strong policy against the Spanish and wishes to relieve the plight of the jailed debtors by giving them the opportunity to settle new land—obtains a charter for Georgia in 1732, and the first colonists arrive in January, 1733, immediately founding the city of Savannah. Religious liberty is guaranteed to all except Catholics.

1737   Birth of artist John Singleton Copley

Born in Boston to Anglo-Irish parents, John Singleton Copley (1737–1815) becomes renowned for his portraits and historical paintings. In 1755, when Copley is only eighteen, George Washington sits for his portrait. From 1774 onwards, Copley spends his time studying in England and Italy, painting royalty and historical events. Many of these paintings are found in the Tate Gallery in London.

1738   Birth of artist Benjamin West

Benjamin West (1738–1820) is born in Springfield, Pennsylvania. Showing early talent, however, he is sent to Italy to learn the art of painting and is then encouraged to settle in London. The British royalty underwrite him for forty years, during which time he makes several innovations in portrait painting, such as depicting his subjects in modern dress rather than classical costume.

1743–48   King George's War

In King George's War the French and British are again at odds. The most significant event of the war is the capture of Louisburg (one of the strongest fortresses in the New World, located on Cape Breton Island, modern day Canada) by the English after a month-and-a-half-long siege. In the colonies, the Ohio Valley and Cherokee country contine to be points of conflict.

1749   Ohio Company established

The Ohio Company is formed to explore and develop the land around the Ohio River and in the Ohio River Valley. In response the French also send men to claim the Ohio Valley, in a 1753 expedition led by Marquis Duquesne, who establishes Fort Duquesne at the forks of the Ohio. In the same year, the young American surveyor George Washington (1732–99), future president of the United States, is sent to demand the withdrawal of the French from the valley.

1754   Washington leads Virginia troops

In response to the French infiltration of the Ohio Valley, George Washington is second-in-command of Virginia troops who build Fort Necessity in western Pennsylvania. Attacked by the French, however, the soldiers are forced to surrender.

1754: June 19   The Albany Convention

The colonies meet at Albany to develop a common plan of defense against the aggressive French.

1755–63   The French and Indian War

Known in Europe as the Seven Years' War, the French and Native Americans ally against the English in North America. At first, the British attack is very weak; almost all of their 1755 campaigns are defeated by the French. Marquis Louis de Montcalm (1712–59), commander of the French forces, captures many English fortresses aided by his Native American allies. But in 1759 the English launch a massive attack against the northern French territories. In the battle of the Plains of Abraham (Quebec) on September 13, both the British commanding officer James Wolfe (1727–59) and Montcalm lose their lives. Quebec surrenders five days later on September 18, and on September 8, 1760, Montreal surrenders as well. All of Canada passes into the possession of Great Britain.

1761   "Writs of assistance" enacted

Wearied by constant evasion of the 1733 Molasses Act by colonists (who engage in smuggling and illegal trade with the enemy nations during war), the British government enacts the writs of assistance. These writs are general search warrants, allowing British officials to search any place they suspect smuggled goods might be found. Merchants are angered by the writs and believe them to be illegal.

1763–75   Westward expansion beyond the mountains

Settlers begin a gradual westward expansion beyond the Appalachian mountains. Such frontiersmen as Daniel Boone (1734–1820) become famous during this period. Boone himself clears a wilderness road to Kentucky.

1763–65   Tax increases

With the acquisition of large territories of land from the French, the British government decides it needs more funds for defense and Native American administration. In order to raise this money, Britain enforces its navigation laws and begins to tax the colonies directly, using the revenues to maintain an army in America. Colonial governors are told to enforce the trade laws.

1763–75   Series of acts leads to colonial discontent

The 1764 Colonial Currency Act prevents colonies from paying their debts with certain colonial currencies deemed unsound by the British government. The act creates a shortage of money in the colonies at the same moment that the 1764 Sugar Act hampers the colonies' West Indian (Caribbean) trade, where they had obtained much of their money.

The 1765 Stamp Act forces colonists to pay for a stamp on all commercial and legal documents, such as pamphlets, newspapers, almanacs, playing cards, and dice. The 1765 Quartering Act provides that, in the event of a shortage of barracks, British soldiers may be housed in citizens' private residences.

The Stamp Act is repealed in 1766, only to be replaced by the Declaratory Act, which gives the English monarch the right to make laws to bind the colonies in all respects.

The 1767 Townshend Acts impose duties on imported goods including glass, lead, painters' colors, tea and paper. Revenues from these taxes go towards the salaries of royal officials living in the colonies.

In 1774 and 1775 the British continue to clamp down on the colonists' freedom, which clearly sets the stage for a fight for independence.

1763   Algonquin uprising

In response to the British victory, Indian tribes north of the Ohio River—fearing eviction from their land and angered by the dishonesty and bad behavior of British traders—prepare to attack the formerly-French, now British fortifications. Pontiac (d. 1769), chief of the Ottawas, organizes an uprising of Algonquins and some Iroquois. A simultaneous attack captures all but three of the northwestern British posts. By 1765,

however, the British regain control of all the formerly French outposts.

1763: February 10   The Treaty of Paris

According to the terms of the treaty, France gives Britain all of Acadia (Nova Scotia), Canada, Cape Breton, and all of Louisiana situated east of the Mississippi except the Island of Orleans. France retains, however, most of its Caribbean possessions, as well as fishing rights off the coast of Newfoundland. Spain cedes Florida to Britain.

1765: October 7   Stamp Act Congress meets

On October 7 a Stamp Act Congress (consisting of twenty-eight delegates from nine colonies) meets and drafts a "Declaration of Rights and Liberties."

1770: March 5   The Boston Massacre

A popular uprising protesting the presence of British troops in their city becomes a massacre when Bostonians are fired upon. Only three citizens are killed—including Crispus Attucks (c. 1723–70), a free black man and the first to be killed—but the incident deepens Americans' resentment for the Britain's colonial policies.

1773: December 16   The Boston Tea Party

In protest of the taxes on imported goods such as tea—a staple of the colonial diet—Bostonians dressed as Native Americans board ships at night and dump tea into the harbor. These ships from the East India Company had not been permitted to land in Charleston, Philadelphia, or New York in protest of the duties, and finally landed in Boston only to be met by this more unusual form of protest.

1774: September 5   First Continental Congress

Representatives from all colonies except Georgia gather in Philadelphia and draw up a "Declaration of Rights and Grievances" to be sent to the British crown. The delegates pledge that they will disallow English imports after December 1 if American grievances are not addressed.

1775: April 18   Paul Revere's ride

Paul Revere (1735–1818), a gold- and silversmith from Boston and official messenger of the Patriots, sets out on horseback the night of April 18 to warn those living around Lexington—especially leaders John Hancock (1737–93) and Samuel Adams (1722–1803)—that the British are marching towards the town. He had previously arranged a signal for the Patriots, using lantern lights: "one if by land and two if by sea."

1775: April 19   First battles in the War for Independence

The battles of Lexington and Concord are the first armed conflicts between Americans and British troops. A local militia of Americans—mostly farmers known as the Minutemen for their rumored ability to be ready for battle in only a minute—are in training to defend their Massachusetts towns when British troops are sent to destroy stores in Concord. The Minutemen along with local patriots, however, meet the British at Lexington, opening fire with the so-called "shots heard 'round the world." The British are able to advance to Concord and destroy the stores, but after a fight at the bridge between the two towns, the British are forced to retreat, first to Lexington and then to Boston.

Subsequent militias of Minutemen are formed in the colonies of Maryland, New Hampshire, and Connecticut.

1775: June 15   George Washington leads American forces

George Washington (1732–99) is appointed commander-in-chief of the American forces.

1775: June 17   Battle of Bunker Hill

During the Battle of Bunker Hill (actually Breed's Hill) near Boston, British troops launch a frontal attack on the hill manned by 1,600 Americans, stationed there to defend Dorchester Heights. The British are forced to assault the hill three times, finally driving off the Americans who had by then run out of ammunition. The battle becomes a source of pride to Americans: they have been able to hold off one of the most powerful armies in the world, and the British lose about one thousand soldiers, or approximately three times the Patriot casualties.

The Battles of Bunker Hill, Lexington, and Concord motivate the Continental Congress to organize the militias of minutemen into a regular army.

1775: July–March 17, 1776   Siege of Boston

The British become more and more unpopular in the colonies as they refuse to respond to American grievances. The British hire German mercenaries (soldiers), burn the city of Norfolk, Virginia, and lay siege to Boston.

1776: January   Thomas Paine publishes *Common Sense*

Thomas Paine (1737–1809), a political theorist and agitator, does much to shape American public opinion during the Revolutionary War. Perhaps the most influential of all his writings is *Common Sense,* a pamphlet published anonymously which advises Americans to declare their independence immediately from England and make the American cause a symbol of the worldwide democratic struggle for freedom. *Common Sense* is immensely popular, with over half a million copies sold—and Paine is forced to assume responsibility for the work.

1776: July 4   The Declaration of Independence

The Declaration of Independence, the formal act of separation from Great Britain by the thirteen original American colonies, is approved by the Second Continental Congress on July 4, 1776. Thomas Jefferson (1743–1826) is the man chosen to write the document because he is considered by many to have a superior writing style. The title of the document is "The unanimous Declaration of the thirteen United States of America," the first official use of the term "United States of America." The declaration states that all men are created equal and have certain rights that cannot be taken away, and the British Empire has betrayed the Americans' rights to life, liberty, and the pursuit of happiness. The declaration is read out loud to the people on July 8.

1776: December 26   Battle of Trenton

In one of General George Washington's most successful military campaigns, he crosses the Delaware River by night to surprise and capture about a thousand mercenary soldiers.

1777: October 17   Burgoyne surrenders his troops

English general John Burgoyne (1722–92) surrenders his entire force to American general Horatio Gates (c. 1728–1806) after two devastating losses in battle during September and October.

1777: November 15   The Articles of Confederation

In response to the desire for a stronger union, the Articles of Confederation—forerunner to the Constitution and basis of the system of government in place after the Revolutionary War—are approved. It takes many years for the Articles to be written and agreed upon, however, because many Americans are wary of the formation of a central government, which they fear may resemble the oppressive English system of government. Because of this, the revised version of the Articles finally passed in Congress gives very little real power to the federal government. For example, the national Congress could ask the states for money, but not tax them; recommend tariffs, but not enact them; and had no court or army to enforce its decisions. However, the Articles do indeed make contributions to national unity and prepare the way for the Constitutional Convention in 1787.

1777–78: winter   U.S. Army at Valley Forge

Valley Forge, a small town in Pennsylvania, becomes the headquarters of the Continental (United States) Army during the winter of 1777–78. After several defeats, the army (led by George Washington) retreats to this village, about twenty-four miles northwest of Philadelphia, arriving there on December 19, 1777. The winter, however, is extremely severe; men lack proper shelter, warmth, food, clothing (including shoes), and medical supplies, and over three thousand perish. The morale of the troops improves in February, 1778, upon the arrival of Baron von Steuben (1730–94), a hired Prussian soldier who turns Valley Forge into the first training camp of the U.S. armed service. On June 19 the reinvigorated army breaks camp to pursue the British.

1778: February   British attempt peace offer

The British parliament led by Lord North (1732–92) attempts to compromise with the Americans, by among other measures changing the taxation policy. The U.S. Congress rejects the peace offer in June; only complete independence is acceptable for the Americans.

1778: February 6   Treaties of Commerce and Alliance

The defeat and surrender of Burgoyne (see 1777: October 17) inspires the French to openly support the United States, with the aim of re-establishing French prestige (which had been badly damaged during the Seven Years' War). The French have been supplying secret aid to the United States for two years prior to signing the Treaties.

1778: July 8   French fleet arrives

The French fleet arrives in America, giving the United States a chance to battle Britain's powerful navy.

1779: June   Spain enters war

Spain joins the French-American alliance against Britain, with the promise that the French will then aid Spain in gaining back Gibraltar and the Floridas, both lost to Britain.

1780: May   Charleston surrenders

After a long siege, the city of Charleston surrenders to the British.

1780: September 23   Treason uncovered

Benedict Arnold (1741–1801), an American general, is often considered the most famous traitor in American history. He was found guilty of using military supplies for his own private purposes in 1779, but General Washington—his staunchest defender—professed continued faith in Arnold by giving him control of West Point. In 1780 a captured British agent reveals Arnold's plot to surrender West Point to the British army. Arnold escapes capture and prosecution, joining the British army and eventually sailing to England with his family. A highly unpopular figure, however, Arnold dies friendless and alone.

1781: September 30–October 19   Siege of Yorktown and end of war

British General Cornwallis (1738–1805) marches his troops to Yorktown in order to fortify them and organize further attacks on Virginia settlements. While inactive at Yorktown, however, American troops hem Cornwallis in on all sides (including on the Chesapeake Bay, blocked by the French fleet). Cornwallis surrenders with seven thousand men on October 19, 1781, ending the war.

1783: September 3   Peace treaty signed in Paris

The treaty between Great Britian and the United States grants full American independence. However, parts of the treaty dealing with northwestern and southern borders later come to be contested by the British and Spanish, respectively. Yet the United States is given possession of the Northwest Territory (see 1785–87) and other lands east of the Mississippi, as well as fishing rights off the Canadian coast. Navigation of the Mississippi is declared open to both the Americans and the British.

The American negotiators, impatient with the French (who are busy trying to please both the Spanish and Americans), negotiate a separate peace with the British against French wishes.

1784–85   Economic depression

Initial widespread economic difficulties occur in the former British colonies. American ships are not permitted to trade in the British West Indies, an area which had been vitally important to the economy before the Revolution. Though not truly responsible for the economic troubles, the Articles of Confederation are popularly blamed. As early as 1787, however, the economy begins to right itself.

1785–87   Northwest Ordinances

Two Northwest Ordinances encourage settlement of the land in the Northwest Territory (region east of the Mississippi River, north of the Ohio River, south and west of the Great Lakes, and west of Pennsylvania). The first Ordinance, passed in 1785, provides for a survey of the area, while the second (1787) establishes a territorial government. The 1787 Ordinance outlines the method by which the Northwest Territory is to developed: it is to be divided into three to five separate territories, each of these smaller regions gaining self-government once their populations exceed five thousand. When the populations become greater than sixty thousand, the territories can outline a constitution and apply for statehood. The states created from the Northwest Territory are Ohio (1803), Indiana (1816), Illinois (1818), Michigan (1837), and Wisconsin (1848).

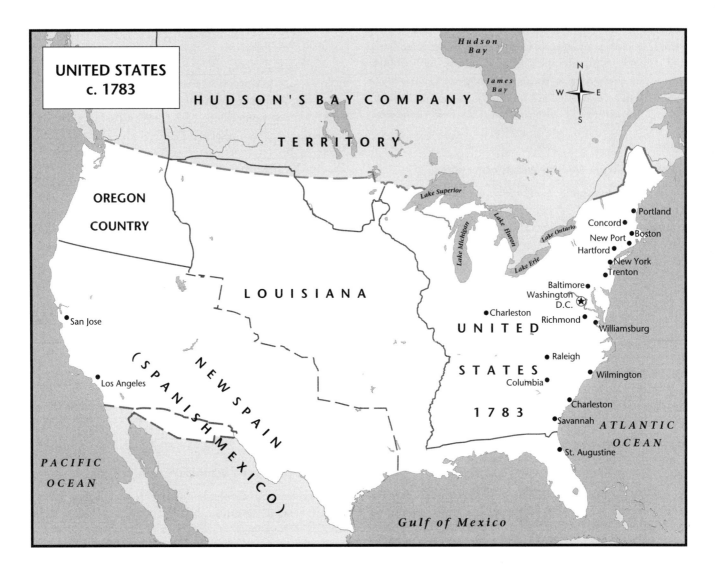

UNITED STATES c. 1783

HUDSON'S BAY COMPANY TERRITORY

Hudson Bay

James Bay

OREGON COUNTRY

LOUISIANA

NEW SPAIN (SPANISH MEXICO)

San Jose

Los Angeles

PACIFIC OCEAN

Lake Superior

Lake Michigan

Lake Huron

Lake Ontario

Lake Erie

Portland
Concord
New Port ● Boston
Hartford
New York
Trenton
Baltimore
Washington D.C. ✪
Charleston
Richmond
Williamsburg

UNITED

STATES

Raleigh

Columbia
Wilmington

1783

Charleston
Savannah
St. Augustine

ATLANTIC OCEAN

Gulf of Mexico

1787: May   The Constitutional Convention

More and more Americans desire an improved central government as frustrations (regarding outbreaks of disorder and obstructions to commerce, as well as the inability of Congress to raise revenue) grow. After four months of writing and negotiating in Philadelphia, with all states except Rhode Island represented, the Constitution of the United States is completed.

1787: September 17   Signing of the Constitution

The United States Constitution is signed by the delegates of the Constitutional Convention, and then sent to states for ratification. Finally in June 1788, New Hampshire—the ninth approving state—ratifies the document, and the Constitution becomes law. The first ten Amendments, known as the Bill of Rights, are adopted on December 5, 1791, and guarantee certain fundamental freedoms.

1789   Birth of writer James Fenimore Cooper

James Fenimore Cooper (1789–1851), whose father is a wealthy Quaker and Federalist member of Congress, spends much of his childhood in Cooperstown, New York, a wild frontier region of great natural beauty. He attends Yale, but is expelled during his third year for a prank and enters the navy. After three years, however, he resigns, marries, and settles down to the life of a country gentleman. Cooper writes over thirty novels in his lifetime, the best of which are stories of the sea and of the Native Americans living in the rugged wilderness of his boyhood. Considered to be one of the first important American novelists, his most famous works include *The Spy* (1821) and *The Last of the Mohicans* (1826).

1789: April 30   George Washington inaugurated as president

George Washington becomes the first president of the United States.

1790–91 New economic policies

Statesman Alexander Hamilton (1757–1804) is at the head of new economic developments. He creates such policies as the Assumption Bill, which allows the federal government to assume the debts of the states and creates the Bank of the United States.

1791 Introduction of power-driven spinning machine

Industrialists Samuel Slater (1768–1835) and Moses Brown apply machinery to the spinning of cotton yarn in Rhode Island. Many consider this accomplishment to be the beginning of the Industrial Revolution in the United States.

1792 Birth of political parties

Due to differences in opinion involving Alexander Hamilton's economic policies (see 1790–91), government leaders divide into political parties. Thomas Jefferson (1743–1826) becomes leader of the Republican (today's Democratic) Party, which is more concerned with how the policies affect the lower-class, farming majority. Hamilton and John Adams (1735–1826) lead the opposing Federalist Party.

1793 Cotton gin introduced

Inventor Eli Whitney (1765–1825) develops the cotton gin, a machine which cleans seeds from cotton fibers. Whitney's invention leads to an enormous increase in the production of cotton in the South, making the fiber an even larger and more significant part of the economy. The cotton gin also has a revolutionizing effect on slavery, for cotton cultivation now requires much less labor.

1794 Neutrality Act

War breaks out between England and France, but despite pressure from Jefferson (favoring France) and Hamilton (favoring Britain), Washington decides to keep the United States neutral. A Neutrality Act is passed in 1794.

1794 Whiskey Rebellion

The so-called "Whiskey Rebellion" occurs in western Pennsylvania, in response to the excise tax on domestic alcohol. Farmers not only dislike the tax, but offenders are forced to make the long and expensive journey to Philadelphia or New York for trial. Various outbreaks of rebellion (including "tarring and feathering") had already broken out among the farmers against tax collectors, and many people begin to talk of seceding from the Federal Union. The federal government asserts its power, however, and brings in a militia of thirteen thousand men to quiet the rebellious counties. Though the rebellion demonstrates the power of the government to enforce its laws, it also creates some resentment among the frontiersmen, which will last well into the nineteenth century.

1794: November 19 Jay's Treaty

Relations between Britain and the United States have been tense since the 1783 treaty. Thus, negotiator John Jay (1745–1829) is sent to London to develop a new, revised treaty. The Americans are angry at Britain's refusal to enter into a commercial treaty, refusal to evacuate their border posts, alleged conspiracy with the Native Americans, and capture of American merchant ships. Jay's Treaty satisfies most people, allowing Americans once again to trade in the West Indies and providing for British abandonment of their border posts.

1795: October 27 Treaty of San Lorenzo (Pinckney's Treaty)

Friction between Spain and the United States involving the border with Florida and navigation of the Mississippi River is resolved with a treaty negotiated by Thomas Pinckney (1750–1828). The treaty establishes a southern border at the thirty-first parallel, and gives Americans the right to navigate the Mississippi to its mouth, with right of deposit at New Orleans for three years.

1796: September 19 Washington's farewell address

Washington, weary of political life, retires. In his farewell address, he advises all Americans to cherish the Union, uphold the Constitution, minimize party differences, and not to take sides with any foreign nation (in response to the British-French conflicts). Later foreign policy-makers have used this statement to encourage America's isolationist (non-involvement) policies.

1797–1800 Troubles during the Adams administration

Jay's Treaty (see 1794: November 19) creates difficulties with France, who sees the treaty as pro-British. This leads to the "XYZ Affair," when American negotiators are sent to France to negotiate, only to be denied an audience (by three men, called X, Y, and Z) with the French foreign minister unless they pay $250,000 in bribes. This leads to an undeclared naval war with France (a U.S. naval department is created) until 1800.

In 1798 Adams responds to the criticism of his opponents—many of whom are French citizens living in the United States—by creating a series of Acts. The Naturalization Act, which extends the time until citizenship may be granted to fourteen years, Alien Act, and Alien Enemies Act are all repressive measures against foreigners living on American soil. The Sedition Act is an attempt to make political opposition a crime.

1797: March 4 John Adams takes the Presidency

Thomas Jefferson becomes the vice-president.

1800 Election of 1800 controversy

Adams is defeated by Jefferson and Aaron Burr (1756–1836) in the election of 1800, but both Jefferson and Burr win the same number of electoral votes. Thus, the election is decided in the House of Representatives in favor of Jefferson.

The difficulty in deciding the election leads to the creation of the Twelfth Amendment, which requires that the president and vice-president be elected by separate ballots.

1802 Birth of reformer Dorothea Dix

Dorothea Dix (1802–87) becomes a humanitarian at a young age, opening her own school for girls in Boston at the age of nineteen. Horrified by the treatment of prisoners (especially those who are mentally handicapped) and the sanitation of the prisons themselves, Dix devotes her life to prison reform. She succeeds in providing for the care of the insane in many state asylums. Dix also serves as superintendent of women nurses in the army throughout the Civil War (1861–65).

1803 Birth of writer Ralph Waldo Emerson

Both writer and pastor, Ralph Waldo Emerson (1803–82) is perhaps best known as one of the great exponents of Transcendentalist philosophy. Transcendentalism—a movement begun in and around Boston which flourishes from the 1830s until the outbreak of Civil War—originates as a rebellion against "old intellectualism," the movement believing instead in the superiority of intuitive knowledge (for example, each man's consciousness is his own supreme judge in spiritual matters). Emerson's essay *Nature* (1836) is considered by many to best explain the philosophy; in *Nature,* mysticism is seen in natural phenomena and "transcends" the material world. Many of the most prominent literary figures of the day adhere to Transcendentalism.

1803: April 30 The Louisiana Purchase

Spain had returned the land west of the Mississippi to French control in 1800, and French emperor Napoleon (1769–1821) now expresses interest in re-establishing a colonial empire in North America. This worries President Jefferson, who does not want to have a strong French power at the mouth of the Mississippi. Jefferson sends foreign ministers to France to attempt to purchase a sufficient area at the mouth of the river to guarantee freedom of navigation and trade. Meanwhile, Napoleon suffers several losses in Europe, and loses interest in the colonial empire. Louisiana (all the land west of the Mississippi and east of the Rockies, with the southern boundary undetermined) is sold to the United States for 80,000,000 francs (about fifteen million dollars) and doubles the size of the country.

Not everyone, however, is pleased with the purchase. Federalists (led by Aarron Burr) fear the new territory will strengthen the power of Jefferson's beloved agrarian class.

1804 Birth of writer Nathaniel Hawthorne

After his father's early death, Nathaniel Hawthorne (1804–64) and his mother move to a farm in the woods of Maine, where he becomes accustomed to solitude. He begins his first (unsuccessful) novel at Bowdoin College, from he graduates in 1825. After moving to Salem, Massachusetts, however, Hawthorne shuts himself away for twelve years "of heavy seclusion," writing poetry and tales. Slow to be recognized in his own country, his works (including *Twice-told Tales,* 1837) gain popularity in England. Hawthorne's works are now considered some of the greatest contributions to American literature. Most notable are *The Scarlet Letter* (1850) and *The House of the Seven Gables* (1851).

1804–06 Lewis and Clark expedition

Jefferson appoints Meriwether Lewis (1774–1809) and William Clark (1770–1838) to explore the Mississippi country and lands west. The expedition, aided by the Shoshone Sacajawea (b. 1784), follows the Mississippi to its source and then follows the Snake and Columbia rivers into the "Oregon country," giving the United States a claim to that land.

1804: July 11 Hamilton-Burr duel

In opposition to each other due to issues provoked by the Louisiana Purchase (see 1803: April 30), Alexander Hamilton and Aaron Burr face each other in a duel. Hamilton is killed.

1807 The Embargo Act

Difficulties in Europe result in restrictions on American trade. Although the United States is neutral and attempts to continue to trade with both France and Britain (who are in conflict), both nations issue decrees refusing to trade with the United States if it trades with their enemy. Therefore, Jefferson issues the Embargo Act, which forbids American ships from traveling to foreign ports. The act is meant to make both Britain and France, for lack of sufficient foreign trade, respect America's neutrality.

1807 Birth of poet Henry Wadsworth Longfellow

A classmate of Nathaniel Hawthorne (see 1804) at Bowdoin College in Maine, Henry Wadsworth Longfellow (1807–82) becomes an extremely popular poet in his lifetime. He spends much of his life as a professor in the United States and abroad. One of his most popular poems is "Evangeline" (1847), a tale in verse of the French exiles in Acadia. Among his best-loved works is "Hiawatha" (1855), based on Indian legends.

1809: March 15 Repeal of Embargo Act

Not having the desired effect, the Embargo Act (see 1807) causes more harm than good to American shipping and is repealed. It is replaced by the Non-Intercourse Act, which allows trade with all countries except Britain and France.

1809 Birth of writer Edgar Allan Poe

Orphaned when only three years old, poet and story writer Edgar Allan Poe (1809–49) is adopted by John Allan (1780–1834), a wealthy and childless merchant in Virginia. Growing up in England and Virginia, Poe spends one year at the University of Virginia before enlisting in the U.S. army, being dismissed in 1830 for 'neglect of duty.' Poe's life has much tragedy: his wife dies in 1847, he attempts suicide in 1848, and has an attack of delirium in 1849. Only a few months later, he becomes ill in Baltimore and dies in the hospital there. Poe's works are known for their dark genius and their original, fantastic, yet horrifying themes. In addition to his many stories, Poe is best known for the poem "The Raven" (1845).

1809 Birth of physician and writer Oliver Wendell Holmes

Oliver Wendell Holmes (1809–94) spends his professional career as a physician, serving as the chair of the anatomy departments of both Dartmouth and Harvard. Although he had begun to write as an undergraduate, it is not until twenty years later—with the publication of *The Autocrat of the Breakfast Table* in 1858—that he becomes famous and well-loved for his writing.

1810–11 Rise of the "War Hawks"

The election of 1810 votes into office many young leaders who are impatient with the neutral policies regarding Britain and France. These men, dubbed the "War Hawks," also support suppression of Native American uprisings, especially those led by Shawnee chief Tecumseh (c. 1768–1813) in the Northwest. The Battle of Tippecanoe in November, 1811, makes William Henry Harrison (1773–1841) a hero to many. The War Hawks believe that British aid and encouragement from Canada are making the Native American opposition stronger.

1812: June 18–December 24, 1814 War of 1812

The United States declares war on Britain in 1812 for impressment (forcing American sailors to work on British ships), for a paper blockade, and for their commerce restrictions, but above all for violating American trading rights as a neutral state.

The British have a great victory in 1814: they capture and burn Washington, D.C., though they fail to capture Baltimore. The burning is in retaliation for the American burning of York (Toronto). The British attack Maine and New Orleans. Andrew Jackson (1767–1845) is successful at defending New Orleans in battle on January 8, 1815, and becomes a national hero.

The Treaty of Ghent on December 24, 1814, surprisingly does not address most of the questions over which the war has been fought. It only provides joint committees for examining boundary issues.

1814: December The Hartford Convention

Many Federalists in New England, dissatisfied since the time of the embargo, had refused to contribute their state militias to the war effort and spoke frequently about state rights. When no federal troops are present to defend the embattled Maine coast, Massachusetts asks her sister states to join in a convention at Hartford. Though secession is discussed, many moderates are present who make it an impossibility. After the treaty of Ghent however, many of New England's grievances are forgotten.

1816 Bank of the United States chartered

The second Bank of the United States is re-chartered after the first bank's charter ended in 1811. This new charter is partially a response to an unprecedented surge of post-war nationalism.

1816 Founding of American Colonization Society

The American Colonization Society is formed to help colonize free blacks in Liberia, Africa.

1817–25 Monroe's presidency and "Era of Good Feelings"

James Monroe's (1758–1831) presidency is marked by an "Era of Good Feelings" in which there is very little party dissension.

1817–18 Seminole War

Under pressure from southern slave owners, Andrew Jackson invades Spanish-controlled Florida to find escaped slaves taking refuge with the Seminole people. Jackson destroys Native American and Spanish settlements, conquers Pensacola, and sets up an American government. The Spanish, realizing they cannot regain Florida except at a very high cost, sell Florida, and it becomes a U.S. possession in 1819.

The Seminole are later ordered to move west of the Mississippi to the new "Indian territory," but most refuse, and a second Seminole War begins in 1835. The war lasts for eight years, during which time fifteen hundred American lives are lost, and the Seminoles are almost completely wiped out. A very few Seminole survivors stay in the Everglades of Florida or move to Oklahoma.

1817 Birth of writer Henry David Thoreau

A Transcendentalist philosopher and writer (see 1803), Henry David Thoreau (1817–62) starts in the late 1830s to consider walks and studies of nature as his most serious occupation.

From 1845–47 he lives in a small shanty in the woods by Walden Pond. During this time, Thoreau writes the great philosophical work *Walden, or Life in the Woods*. He later supports himself by whitewashing, gardening, fence building, and land surveying, but continues to write essays and poems as well as give lectures. Other notable works include *A Yankee in Canada* (1866) and *Civil Disobedience* (1849), an essay inspired by his opposition to the Mexican War (1846–48).

1819   Birth of writer Herman Melville

Herman Melville (1819–91) a bank clerk in search of adventure, joins a whaling ship headed for the South Seas. He deserts the ship in the Marquesas Islands and spends several weeks with a native tribe in the Typee valley (which inspires his first book, *Typee,* 1846) before being picked up by an Australian whaler. He is jailed for mutiny on Tahiti, but escapes and spends some time on the island, a circumstance which sparks the inspiration for his second book, *Omoo* (1847). He writes his masterpiece on the whaling industry, *Moby Dick,* in 1851. Melville's works, however, are not appreciated until thirty years after his death, and he is forced to work as a customs official during the last years of his life.

1820: March 3   The Missouri Compromise

After the Revolutionary War, many northerners as well as southerners agree that the abolition of slavery is desirable.

The statehoods of Louisiana (1812), Indiana (1816), Mississippi (1817), Illinois (1818), and Alabama (1819) had not provoked any debate over the issue of slavery, but when Missouri attempts to enter the Union as a slave state, great opposition arises in the North. Finally, a compromise, known as the Missouri Compromise, is reached: Maine is admitted as a non-slave state, and all other land in the Louisiana Purchase north of 36°30' is to be free as well, while Missouri itself becomes a slave state.

Also, the introduction of the cotton gin (see 1793) and the consequent revolution in the southern economy and culture gradually changes sentiment regarding slavery.

1823: December 2   The Monroe Doctrine

President Monroe's doctrine, given in a speech to Congress, defines American foreign policy for years to come. The doctrine—a response to distrust of the British and unrest in Europe—states that America is never again to be considered a continent open for colonization by European powers. Further, it states that any involvement of European powers in the western hemisphere (America's "sphere of influence") is to be considered a direct affront to U.S. sovereignty.

1825: March 4–March 4, 1829   Presidency of John Quincy Adams

When no majority is garnered in the electoral college, the House of Representatives chooses John Quincy Adams (1767–1848) for president over the popularly-elected Andrew Jackson.

1825   Completion of the Erie Canal

Begun in 1817, the Erie Canal—which connects the Hudson River to Lake Erie—opens to great success. During its construction, the canal had been sarcastically dubbed the "big ditch," as machines and men struggled to build the waterway through wilderness. Yet the 363-mile-long canal, the longest in the world, collects several million dollars in tolls its first year. Buffalo and Rochester become boom towns, and trade with the West increases significantly. By 1860, however, the canal is in decline due to the rapid growth of railroads.

1826   Birth of composer Stephen Foster

The songs of Stephen Foster (1826–64) are considered some of the first truly American compositions. He writes 125 compositions in his lifetime, including "Nelly Bly," "Camptown Races," "Jeannie with the Light Brown Hair," and "My Old Kentucky Home."

1828: July 4   Beginning of the first public railroad

The Baltimore and Ohio Railroad construction begins. It is to be the first public railroad in the nation.

1829: March 4–March 4, 1837   Presidency of Andrew Jackson

One hallmark of the Jackson presidency is his use of the "spoils system," in which government appointments are made based on political party participation and service.

1829   Universal suffrage for white males

For the first time, all white male citizens are permitted to vote, regardless of their religion or whether they own property.

1830   Mormon Church founded

The Church of Jesus Christ of Latter-Day Saints (also called the Mormon Church) is founded by Joseph Smith, Jr. (1805–44). The Book of Mormon—the church's bible, which states that Christ will return to earth and that there will be an earthly paradise for the Mormons—is also printed this year for the first time. A magnetic personality, Smith (who had received "divine inspiration") attracts several hundred followers and founds his church (see also 1847: July 24).

1830–34 Jackson's "Indian policy"

Controversy between the state of Georgia and the Cherokee people leads President Jackson to create an area west of Arkansas as the final home of all southern Native Americans. A commission of Indian affairs is created.

1831 Beginning of abolitionist movement

William Lloyd Garrison (1805–79) establishes the journal *The Liberator* to press for unconditional emancipation of slaves, marking the beginning of the abolitionist movement. Other groups, such as the New England Anti-Slavery Society (1832) and the American Anti-Slavery Society (1833) are soon founded.

1832 Birth of author Louisa May Alcott

Most famous for her children's books, Louisa May Alcott (1832–88) grows up in Concord and Boston, Massachusetts, the center of New England's (Transcendentalist) cultural renaissance. Alcott is educated by her father, and comes to adopt many of his progressive views, including Transcendentalism. She begins to write at the age of sixteen, but is perhaps best known for her novel *Little Women* (1868–69), which is often considered the most popular book of its kind ever written. *Little Women* is autobiographical in nature. Later works (including *Little Men* [1871]) also depict family life in the Victorian era.

1833 College opens its doors to women and African Americans

Oberlin College in Oberlin, Ohio, is the first to admit women and African Americans.

1833 Founding of the Whig Political Party

Though not highly organized and without a platform at first, the newly formed Whig Party is held together by a distrust of Jackson.

1834 Birth of writer Horatio Alger

Horatio Alger (1834–99), writer of "rags to riches" stories for boys, lived in Paris and in Massachusetts as a minister before coming to New York City as a social worker. It is in New York that Alger begins to write the works that are to make him famous, using the popular American theme that success can be achieved through hard work, clean living, and "inevitable" luck. Among the best-known of his highly influential series are *Ragged Dick* (from 1867), *Luck and Pluck,* and *Tattered Tom.*

1836 Texas gains its independence

Colonization of Texas had begun in 1821, when Stephen Austin (1793–1836) was given a land grant with the condition that he settle several families there. Soon substantial numbers of Americans began to settle the area as part of the American westward movement. Differences between the settlers and Mexicans culminate in the Texas Revolution. After defeat by Mexican commander Santa Anna (1795–1876) at the Alamo (the battle that became the rallying cry of the revolution, with "Remember the Alamo!"), Texas commander Sam Houston defeats the Mexicans at the April 21, 1836, Battle of San Jacinto. The Republic of Texas is established the same year.

1837 "Panic of 1837"

The economy crashes due to the waves of speculation (risky investment) and hasty expansion which marked the years 1833–37. The situation is complicated by the ensuing difficulties of many large business houses in Britain (which have invested heavily in the United States), poor crops in the West during 1835 and 1837, and Jackson's Specie Circular pamphlet, which states that the government will only accept gold and silver as payment for federal lands.

1837 Mount Holyoke College founded

Educator Mary Lyon (1797–1849) opens Mt. Holyoke Seminary (now Mt. Holyoke College), the first all-women institution of college rank.

1837: December *Caroline* affair

One American citizen is killed when the Canadian militia seizes the American steamer *Caroline,* in service of Canadian rebels, on the U.S. side of the Niagara River. A Canadian is put on trial for the death of the American and acquitted, thereby averting a potentially serious conflict with Great Britain.

1838: May–March 1839 "Trail of Tears"

Seven thousand federal troops are sent to remove the Cherokee from the state of Georgia, with the intent of relocating them to the "Indian Territory" (present-day Oklahoma). The U.S. government imprisons all Cherokee who refuse to abandon their land, and burns homes and crops. One-fourth of the Cherokee nation, or over 4,000 men, women, and children, die during the long trek.

1838 Underground Railroad organized

As many as 100,000 slaves are thought to have escaped to sanctuary in the North or Canada through this secret network of "safe houses" in the years before the Civil War.

1840s Pioneers follow the Oregon Trail

The Oregon Trail, beginning in Independence, Missouri, and ending in the fertile Willamette Valley of Oregon Territory, is the route taken by American pioneers throughout the 1840s. The difficult, 2,000-mile journey takes five months to complete, with entire families traveling together, usually by covered wagon and oxen. The most dangerous and deadly part of the trail is at the South Pass, the 430-mile stretch through the Rocky Mountains. Nonetheless, from 1842, a thousand or more farmers migrate westward, and by 1846 the population of the Willamette Valley exceeds 5,000.

1841: March 4–March 4, 1845   Two presidents serve

William Henry Harrison (1773–1841), is elected president with the slogan "Tippecanoe and Tyler, too!" (Harrison's fame originates with his victory at the battle of Tippecanoe, see 1810–11). He dies from pneumonia only one month after his inauguration. Vice president John Tyler (1790–1862) assumes the presidency.

1843   Birth of novelist Henry James

Henry James (1843–1916) is considered by many to be one of the great masters of the psychological novel. In the beginning of his career, James writes mostly about the impact of American life on the older European civilization, as in *Portrait of a Lady* (1881). Later works are devoted to English subjects (he lives in England from 1876). In his last period of writing he explores again Anglo-American subjects, as in *The Wings of a Dove* (1902) and *The Ambassadors* (1903).

1845:March 1   Annexation of Texas

Texas is annexed into the United States by a joint resolution of Congress led by President Tyler.

1846–48   War with Mexico

There are many mutual grievances between the United States and Mexico, with Mexico especially angry about the recent annexation of Texas by the United States. When Mexico refuses to negotiate with the United States over the sale of New Mexico, President Polk prepares for war. Polk sends troops into the disputed area between the Rio Nueces and Rio Grande, thereby initiating a small battle after which time he could claim that the Mexicans had "shed American blood on American soil." The U.S. Army under general Zachary Taylor (1784–1850) captures most of the disputed area, while General Winfield Scott (1786–1866) captures Mexico City.

The war ends with the Treaty of Guadeloupe Hidalgo on February 2, 1848. Mexico agrees to the Rio Grande as the border and cedes New Mexico and California to the United States for $15,000,000.

1847: July 24   Arrival of the Mormons in Utah

The Mormons (see 1830) are forced to move several times, first to Ohio, then Mississippi, and finally to Illinois. However, in 1844 a discontented member of the Church and his angry followers (who believe Church founder Joseph Smith is immoral and untruthful) kill Smith and his brother by lynching. Smith's successor, Brigham Young (1801–77), then leads a small party of followers westward, finally arriving in the valley of the Great Salt Lake in the Utah territory. Young declares the valley to be the Promised Land, though the dry land is difficult for farming, and the colony almost perishes due to a plague of locusts. Yet, miraculously, a flock of sea gulls arrives from the Pacific, eating the locusts before they do irreparable damage, and the colony survives. Brigham Young is appointed governor of the Utah Territory in 1850, and begins the practice of polygamy, or the practice of having two or more spouses at the same time. (Young, himself, has about eighteen "formal" wives and numerous other "spiritual" wives).

1848: January 24   Beginning of gold rush

Gold is found on January 24, 1848, near Sutter's Mill, California. This discovery ignites the great gold rush of 1849 (the gold seekers and prospectors come to be called the "49-ers").

1848: July 19   Seneca Falls convention

The first Women's Rights convention in history is held at Seneca Falls, New York.

1849: March 4–March 4 1853   Two presidents serve

Zachary Taylor, of Mexican War fame (see 1846–48), is elected president but dies on July 9, 1850, after which time Vice President Millard Fillmore (1800–74) takes office.

1850   The Compromise of 1850

The Compromise of 1850 is a series of measures passed through Congress in September, 1850. The compromise admits California to the Union as a free state; divides the remainder of the Mexican cession at the 37th parallel, creating the territories of New Mexico and Utah which will ultimately become states with or without slaves, as their constitutions allow; abolishes the slave trade in the District of Columbia; and provides for the enactment of a more effective fugitive slave law.

1851: June   Erie railroad completed

The Erie Railroad becomes the first to make a connection from New York City to the Great Lakes. Later, in 1853, a rail connection between New York and Chicago is finished; thus, American trade becomes a west-east movement, a change that will come to have great economic and political significance.

1852   Harriet Beecher Stowe publishes *Uncle Tom's Cabin*

The first anti-slavery novel to gain popularity, Harriet Beecher Stowe's (1811–96) *Uncle Tom's Cabin* does much to focus the anti-slavery sentiment in the North and create controversy across the United States. *Uncle Tom's Cabin* is the moving story of a slave family, which Stowe is inspired to write after the passage of the Fugitive Slave Law (see 1850: Compromise of 1850). It is said that when President Abraham

Lincoln meets Stowe for the first time (many years later during the Civil War), he exclaims, "So this is the woman who started the war!"

1853: December 30   Gadsden purchase

Mexican leader Santa Anna (c. 1795–1876), badly in need of money, sells the United States land in what is called the Gadsden purchase. The purchase includes what is now the extreme southwest of the United States.

1854–58   "Bleeding Kansas"

Intermittent warfare between pro- and antislavery forces in the Kansas Territory comes to be known as "Bleeding Kansas." Nebraska, farther north and of cooler climate, is almost certain to become a non-slave state; Kansas, on the other hand, is farther south and shares a long border with pro-slavery Missouri. Nonetheless, the settlements of Lawrence and Topeka become home to antislavery populations in Kansas, countered by the pro-slavery towns of Leavenworth and Atchinson. One of the worst outbreaks in violence is the 1855 massacre of the citizens of Lawrence. Eventually, though, Kansas is admitted to the Union (1861) as a free state.

1854: May 30   Kansas-Nebraska Act

The Kansas-Nebraska Act repeals the Missouri Compromise (see 1820). The Act opens the Nebraska Territory to settlement by popular sovereignty, which maintains that the inhabitants of federal lands have the right to decide among themselves whether or not slavery will exist within that territory. The Act also provides for the creation of two territories, Kansas and Nebraska.

The Know-Nothing and Republican (present-day Democratic) political parties are born, as part of opposition to the Kansas-Nebraska Act.

1855   Opening of Soo Canal

The opening of the Soo Canal between Lakes Huron and Superior results in cheap transportation of iron ore, laying the basis for the fast development of the steel industry in the Great Lakes region.

1857: March 7   Dred Scott decision

The Dred Scott Supreme Court decision, in essence, nullifies the principles behind the Missouri Compromise, and cements the status of slaves as "personal property." Dred Scott, a slave, was taken by his owner first into Illinois, a free state, and then into the Missouri Territory north of 36°30', where the Missouri Compromise had forbidden slavery. However, the Supreme Court rules that Scott has not won his freedom by living in the free territory, as he is his owner's "property," and Congress may not enact laws to deprive people of their property in any U.S. territory.

1857: August   Lincoln-Douglas debates

The issue of slavery is openly discussed in a series of seven debates between Abraham Lincoln (1809–65) and Stephen Douglas (1813–61) during a campaign for election to the Senate. These debates are perhaps the best-known in U.S. history, as both men are recognized for their outstanding oration. The Lincoln-Douglas style of debating continues to be practiced.

1859   Discovery of the "Comstock Lode"

The richest silver deposit ever found on American soil is discovered by Henry "Old Pancake" Comstock (1820–70) in western Nevada. His discovery starts such a huge rush on the area that a town, Virginia City, springs up almost overnight. Within a year the town boasts a population of ten thousand, and dance halls, saloons, gambling halls, restaurants, and hotels are built rapidly. In the 1880s, the mines flood, and by 1890 the Comstock Lode is almost deserted. Virginia City becomes a ghost town, its six churches and 100 saloons empty until the mid-1960s, when Virginia City becomes a tourist attraction.

1859: October 19   Raid on Harper's Ferry

Harper's Ferry, a West Virginian town in a strategic location at the juncture of the Shenandoah and Potomac Rivers, is the site of a raid by abolitionist John Brown (1800–59). Brown and his band of twenty-two men successfully capture a U.S. arsenal in Harper's Ferry in order to arm a future slave rebellion. Soon, however, Brown's band is overpowered, and he is convicted of treason and hanged.

1860: December 20   South Carolina threatens secession

After Abraham Lincoln wins the 1860 presidential election, South Carolina adopts an ordinance of secession in protest.

1861   Territories of Nevada and Colorado organized

Later mining rushes lead to the territories of Arizona (1863), Idaho (1863), Montana (1864), and Wyoming (1868) being created.

1861   Last efforts to save the Union

A peace conference in February, 1861, and the Crittenden Compromise resolutions (which promise to extend the Missouri Compromise line all the way to the Pacific) do not succeed in keeping the Union together. From January until May, Mississippi, Florida, Alabama, Georgia, Louisiana, Texas, Virginia, Arkansas, Tennessee, and North Carolina successively secede from the Union, forming the Confederate States of America. Jefferson Davis (1808–89) is elected president and Alexander H. Stephens (1812–83) is elected vice-president of the Confederacy.

1861 Beginning of Civil War: military preparations

After having seized Federal funds and property in the South, the Confederates demand the evacuation of Fort Sumter in Charleston. Meanwhile, President Lincoln calls for 75,000 army volunteers, with everyone in Washington expecting a short conflict. The North has immense advantages over the South: a population of 23 million (versus 5 million free whites in the south), superior financial strength, extensive manufacturing facilities, and better railway communication. The South's cotton-based economy is not as effective for waging war.

1861: July 21 First Battle of Bull Run

The quick defeat of the Federalist (Northern or Union) army by the Confederates opens northerners' eyes to the reality of Southern military might, and leads to more careful and extensive preparation.

1861–62 Naval battles

The North quickly attempts to blockade the Southern shipping ports, leading to the first great sea battle between the Federal ironclad *Monitor* and the Confederate frigate *Merrimac* on March 9, 1862. The Merrimac finally withdraws from battle.

1862: April The Peninsula Campaign

The Federal army led by General George B. McClellan (1826–85) advances on the city of Richmond. The greatly outnumbered Confederates, led by Robert E. Lee (1807–70), fall back, but shortly thereafter receive reinforcements. Heavy fighting around Richmond leads to the Federal withdrawal from the peninsula.

1862 Battles of Antietam and Fredericksburg

After the September 17 Battle of Antietam, Confederate General Lee is pushed back slightly into Virginia. McClellan, however, does not take advantage of the opportunity to pursue him, one of several mistakes in strategy he makes during the early stages of war. Thus, in November, McClellan is replaced by General Ambrose E. Burnside (1824–81). After the devastating Battle of Fredericksburg on December 13—during which time the Federals are badly defeated by the Confederacy—Burnside is replaced with General Joseph Hooker (1814–79).

1862–63 Campaigns in the West

The western theater of the war is marked by a series of Federal victories, led by Brigadier-General Ulysses S. Grant (1822–85). After a short siege on July 4, 1863, Grant succeeds in taking Vicksburg, key to control of the Mississippi River. After the July 8 surrender of Port Hudson, the Federal forces control the entire river and cut off Texas, Arkansas, and Louisiana from the rest of the Confederacy.

1862 Birth of novelist Edith Wharton

Born and raised among the aristocratic society of New York City, Edith Wharton (1862–1937) receives much of her education from private tutors. Wharton's husband becomes ill in 1907, and the couple travels to Europe, settling in Paris. There, Wharton meets author Henry James (see 1843), who inspires her to continue her literary career. She publishes *The Valley of Decision* in 1902, and in 1911 publishes what many consider her finest work, *Ethan Frome*. In 1920 she writes *The Age of Innocence,* which is awarded the Pulitzer Prize the following year. *The Age of Innocence,* as well as many of her other works, is based on the aristocratic New York lifestyle of her youth.

1863: January 1 Emancipation Proclamation

Lincoln issues a proclamation giving freedom to all slaves in the rebel states. Slaves in states loyal to the federal government are freed later.

1863: May 1–4 Battle of Chancellorsville

The Battle of Chancellorsville, Virginia, is a victory for the Confederates, though Confederate General Thomas ("Stonewall") Jackson (1824–63) perishes during the fight. He is accidentally killed by a line of fire from his own troops.

1863: July 1–3 Battle of Gettysburg

In this decisive battle, General Lee is unable to dislodge Federal soldiers and is forced to fall back south of the Potomac River. From this point on, the South launches no major offensives against the North; the Confederacy must stay on the defensive, and the war becomes a test of endurance.

At the dedication of Gettysburg battlefield (in south central Pennsylvania) as a national cemetery (November 19, 1863), President Lincoln delivers the Gettysburg Address, one of the most famous and often quoted orations in United States history. It outlines the importance of the American struggle for liberty, and declares that those who perished at Gettysburg, and in the Civil War in general, did not do so in vain but in order to preserve the American definition of liberty and justice.

1864 "Bessemer steel" comes to U.S.

Steel made in the "Bessemer process" is stronger and more durable than previous metals, which makes it ideal for construction and industry. Henry Bessemer (1813–98) invented this process in England in 1856, and it is used for the first time in America at Wyandotte, Michigan.

1864 Grant and Sherman control forces

Ulysses S. Grant (see 1862–63) is made commander-in-chief of all the armies, while General William Tecumseh Sherman (1820–91) is put in control of the western forces. Beginning in May 1864 Sherman launches a campaign, marching with 100,000 men from Chattanooga, Tennessee through Georgia to Atlanta. He is opposed by one of the ablest Confederate generals, Joseph E. Johnston (1807–91); nonetheless, Sherman succeeds in reaching Atlanta on September 2. He continues on to Savannah, ravaging the countryside as he proceeds. Columbia, S.C., and Charleston, N.C., are taken by Federalists in quick succession.

1865: April 9 Confederacy surrenders at Appomattox; end of war

Robert E. Lee surrenders his forces at Appomattox Court House. Soon after, Johnston surrenders to Sherman, and Jefferson Davis (president of the Confederacy) is captured and imprisoned.

1865: April 14 President Lincoln assassinated

Abraham Lincoln is shot at Ford's Theater in Washington, D.C., by actor John Wilkes Booth (1839–65). Lincoln dies the next morning, and Vice President Andrew Johnson (1808–75) succeeds him.

1865: December 18 Thirteenth Amendment ratified by states

The Thirteenth Amendment prohibits slavery within the United States.

1866 Postwar depression

A rapid decline in prices accompanies a postwar depression.

1866 American Society for Prevention of Cruelty to Animals is founded

Henry Bergh (1811–1888) founds the American Society for Prevention of Cruelty to Animals (ASPCA) in New York.

1866 YWCA founded

The Young Women's Christian Association is founded in Boston.

1867 Howard University is incorporated

Founded by civil war general, Oliver Otis Howard, the university was established for African Americans of all ages, whether male or female, married or single.

1867 Rise of the railroads

This is the beginning of steel rail manufacture in America. The first American elevated railroad is built in New York City on Ninth Avenue. Cornelius Vanderbilt (1794–1877) obtains New York Central Railroad from Albany to Buffalo.

1867 African American suffrage

African American men obtain the right to vote in the South. When some southern states refuse to cooperate, the military is called in to register voters regardless of color. African American voters become a majority in South Carolina, Alabama, Florida, Mississippi, and Louisiana.

1867 First free public schools

An act of legislature creates the first free public schools in New York State. The concept of public education had been gaining in importance before the Civil War, but now it gains even more momentum as public grade schools and high schools are created.

1867: March 30 U.S. buys Alaska

The United States signs a treaty with Russia to purchase Alaska for slightly over $7,200,000. The Senate consents on April 9, and Alaska becomes American territory on October 18.

1867: December 4 Granger movement started

Oliver Hudson organizes the first meeting advocating farmers' interests in Washington, D.C. The Patrons of Husbandry (agriculture) become popularly known as the Grangers, whose purpose is to relieve farmers from the high prices charged by the railroads. The railroads' near monopoly on hauling freight allows them to charge high prices and treat farmers unfairly, so the Grangers become politically active in Minnesota, Kansas, and other prairie states. Laws are passed to control and limit the prices charged by the railroads.

1868 America opens relations with China

The Burlingame Treaty between the U.S. and China allows unrestricted Chinese immigration into the U.S. and gives concessions to the U.S. regarding trade, travel, and diplomatic privileges. Chinese immigration will be restricted, however, in 1880.

1868: February 24 President Johnson impeached

Differences between President Johnson and Congress over reconstruction of the South and Johnson's firing of his Secretary of War lead to an impeachment vote. President Johnson wants to be more lenient on the South than Congress does, and as a result he becomes the first president ever impeached. President Johnson is charged with "high crimes and misdemeanors," but he is acquitted in the Senate 35–19, on May 16, only one vote short of conviction. Votes taken on May 26 also fall one vote short, and President Johnson is acquitted. (A two thirds vote is necessary to convict a president.)

Members of the National Women's Suffrage Association, founded in 1869 by Susan B. Anthony (1820–1906) and Elizabeth Cady Stanton (1815–1902), march in support of voting rights for women in Cleveland, Ohio. (EPD Photos/CSU Archives)

1868: July 28   Fourteenth Amendment ratified

The Fourteenth Amendment gives equal protection and due process rights to citizens and is applicable to states' governments. The equal protection clause is used to eliminate racial discrimination, and the due process clause is used to guarantee fundamental rights, like the right to vote, to most citizens. The scope of this amendment has gradually increased over the years to protect not only African Americans, but also other ethnicities and women.

1868: December 25   Pardon of Confederates

President Johnson unconditionally pardons and amnesties everyone involved in the rebellion. Treason charges against Jefferson Davis, the former president of the Confederacy, are dropped.

1869: May   National Women's Suffrage Association founded

Susan B. Anthony and Elizabeth Cady Stanton found the National Women's Suffrage Association (NWSA). Stanton serves as president of the association dedicated to winning women the right to vote. (See 1920: August 26.)

1869: May 10   Trans-Continental Railroad completed

The Union Pacific and Central Pacific railroads connect at Promontory Point, Utah. A golden spike finishes the rail connection, and the entire country celebrates.

1869: December 30 Knights of Labor organized

Uriah Stephens (1821–82), leading garment workers, organizes a meeting in Philadelphia which leads to a national organization called the Knights of Labor. The organization is open to skilled and unskilled workers of any race and gender. It advocates the eight-hour work day, an end to child labor, the legalization of labor unions, and public ownership of utilities. The organization, headed by Uriah Stephens for ten years, is the forerunner of the AFL (American Federation of Labor) labor union.

1870: January 10 Standard Oil Company created

John D. Rockefeller (1839–1937), who will later become the world's first billionaire, is a principal incorporator of the Standard Oil Company in Cleveland, Ohio. Throughout the years, Rockefeller uses price cutting and other methods to take over or eliminate his business rivals in the oil industry so that, by 1878, Standard Oil controls about ninety percent of America's oil business. Price cutting occurs when a company drops its prices to force a lesser company to drop theirs, eventually bankrupting the other company. After the competition is eliminated, the prices go back up.

1870: March 30 Fifteenth Amendment ratified

The Fifteenth Amendment gives African American men the right to vote.

1870 Rise of the Ku Klux Klan

Former Confederate soldiers create the Ku Klux Klan in Pulaski, Tennessee, in 1865. By the 1870s, they are wearing white hooded robes and attacking African Americans and "carpetbaggers" (Northerners going to live in the South during Reconstruction to make money—they often use carpetbags to carry their possesions—hence the name). Northerners flee the South, African Americans become afraid to vote, and Congress is forced to pass legislation to suppress this terrorist organization. President Grant is given the power to suspend writs of habeas corpus to enforce the anti-terrorist laws. Hundreds of people in South Carolina are arrested, and the terror subsides, but the need for softer reconstruction policies is also recognized.

1870: June 22 Department of Justice is created by Congress

The Department of Justice enforces federal laws. It is headed by the attorney general and assistant attorneys who prosecute civil and criminal issues.

1871: October 8 Great Chicago fire

Legend says that Mrs. O'Leary's cow kicks over a kerosene lamp that causes the raging fire. The fire lasts four days, burns down three square miles of Chicago, and causes $200,000,000 worth of property damages. 250 people die and many people survive by fleeing to the shores of Lake Michigan or the river. Over 100,000 people are left homeless, and it takes two years to rebuild the city center. Architects in Chicago begin building steel-reinforced skyscrapers, and the age of modern business buildings begins.

1873 Financial panic

The over-expansion of industry, railroad speculation, and a drop in European demand for American farm products weakens the economy. On September 18, the powerful banking firm, Jay Cook and Company closes its doors, and America falls into financial panic. The New York stock market closes for ten days, and an economic depression begins that lasts five years.

1875 First American cardinal

John McCloskey (1810–85) is named a cardinal in the Roman Catholic Church, the first American to be so honored.

1875 First Kentucky Derby

The first annual Kentucky Derby horserace is held at Churchill Downs near Louisville, Kentucky.

1876: March 10 Telephone invented

Alexander Graham Bell (1847–1922) demonstrates the telephone in Boston. It is the first time an understandable conversation can be heard over wire. The first sentence is, "Mr. Watson, come here. I want you." Bell goes on to found the Bell Telephone Company in 1877 and begins installing telephones in private homes. Years later he helps invent the wax cylinder record for the first phonograph.

1876: May–October Centennial celebrated in Philadelphia

America's first World's Fair is held to celebrate America's first 100 years. Over 9,000,000 people attend to see exhibits of American progress in democracy, the sciences, and culture. Exhibits from other countries are present, and public morale improves despite an ongoing economic depression.

1876: June 25 Custer's Last Stand

General George A. Custer (1839–76), a cavalry officer during the Civil War, is lured into an ambush by Sioux Chiefs Sitting Bull (1834–90) and Crazy Horse (1849?–77). General Custer and all 264 of his men are killed by 3,000 Sioux and Cheyenne at the Little Big Horn River in southeastern Montana. This is just one of many battles between American Indian tribes and white settlers and miners encroaching onto the western prairies.

Tensions in the West date back at least to the Civil War when the Sioux went to war against the U.S. government in Minnesota in 1862. Reservations created in 1867 achieve only limited success and rebellion is frequent, especially with the Sioux. The warpath that leads to Little Big Horn is the result of over fifteen gold prospectors encroaching on the land near the Black Hills Indian Reservation where the Sioux live.

1877: April   Reconstruction ends

President Rutherford B. Hayes (1822–93) withdraws the last of Federal troops from the South. Northern "carpetbaggers" who had been governing also leave. Their years of rule have been known for high taxes and corruption but also have been helpful in physically restoring the southern states after the Civil War's devastation. Most southern states have laws for free public schools for white and African American students.

1879: October   Thomas Edison invents the interior light bulb

Thomas Alva Edison (1847–1931) expands on the street arch lamp to create a light bulb for practical indoor use. Edison's research facility, in Menlo Park, New Jersey, starts a trend towards joint research among experts and away from the single-inventor approach. This joint research approach is put into practice at Edison's laboratory, resulting in the light bulb breakthrough.

1881: July 2   President Garfield assassinated

Shortly after taking office, President James A. Garfield (1831–81) is shot by Charles J. Guiteau (1840?–82), a Chicago lawyer angry at not being awarded a government job in Washington, D.C. The shooting takes place in a train station and predates the creation of the Presidential Secret Service. Garfield dies of his injuries a few months later, and Guiteau is hanged. Action begins in Washington, D.C. to end the political patronage system in the nation's capital so that fewer people will gain and lose their jobs whenever a new president is elected. The reform occurs with the passage of the Pendleton Civil Reform Act on January 16, 1883.

1886: May   Haymarket Square Riot

The Knights of Labor call a general strike in support of the eight-hour work day in Chicago. On May 3, anarchist speakers on a vacant lot advocate violence and throw stones at the police ordering them to disperse. One person dies when police open fire. On May 4, anarchists speaking at the Haymarket Square again incite violence, and a bomb is thrown at the intervening police who open fire. Eventually, seven police officers die, and many civilians are killed and wounded before the rioting moves on to Milwaukee. The person who throws the bomb is never identified, but eight alleged anarchists are convicted. Four are hanged, one commits suicide, but the remaining three are later freed due to the unfairness of their trials. Many blame the Knights of Labor, fairly or not, for the violence.

1886: October 28   Statue of Liberty dedicated

As a gesture of friendship, the French commission sculptor Frederic Auguste Bartholdi (1834–1904) to design a gigantic statue dedicated to freedom. Gustave Eiffel (1832–1923), designer of the Eiffel Tower in Paris, designs the statue's metal framework, and the statue is shipped from France to America in sections. The statue stands 151 feet tall on Bedloe's Island (now Liberty Island), New York, and is covered in copper.

1886: December 8   American Federation of Labor formed

The AFL (American Federation of Labor) is formed in Columbus, Ohio. The new union is preceded by other labor organizations including the Knights of Labor. Its membership exceeds 150,000, and its first president is Samuel Gompers (1850–1924), who had been president of the Cigarmakers Union.

1889: February 22   Four new states

Congress passes an Omnibus Bill that creates the states of North Dakota, South Dakota, Montana, and Washington. Idaho and Wyoming are admitted a year later. The Oklahoma Territory is opened for settlement and about 100,000 settlers race into the territory on April 22 to claim land. Oklahoma City is founded a day later.

1892: January 1   Ellis Island opens for immigrants

Reception center for immigrants to the United States is opened in New York Harbor.

1892   Chinese Exclusion Act passed

Concerns in the West, especially California, over Chinese immigrants willing to work for subsistence wages result in a ten-year suspension of Chinese laborer immigration in 1882. In 1892 this exclusion policy is made permanent. Chinese laborers already in America are exempted, but anti-Chinese sentiment and violence against Asians in general still runs high.

1892: March 1   Standard Oil broken up

Passage of the Sherman Anti-Trust Act leads to the Ohio Supreme Court's ruling that Standard Oil is guilty of attempting to "establish a virtual monopoly" and orders it broken up. The nine smaller companies created out of the Standard Oil Trust are run by Rockefeller's directors, the Standard Trust is run in the form of a holding company, and Standard Oil eventually recharters in New Jersey.

1893   Hawaii annexed

The Americans who own many sugar plantations in the Hawaiian Islands advocate America's annexation of Hawaii in order to protect their property. Hawaiian Queen Lili-uokalani (1838–1917) attempts to seize the sugar plantations and force their owners to leave Hawaii, but she is dethroned by a revolutionary regime. There is some contention in Washington because the new president, Grover Cleveland (1837–1908), initially wants to restore Liliuokalani to her throne. Cleveland eventually recognizes the Republic of Hawaii, and America continues to use the Pearl Harbor naval base, established in 1875.

1894: June 28   Labor Day established

Congress recognizes the first Monday of every September as Labor Day in honor of American workers. All states and territories recognize the legal holiday, and the plight of the workers gains much popularity with the public.

1895   Professional football begins

The first professional football game is played in Latrobe, Pennsylvania.

1896: May 18   *Plessy v. Ferguson,* segregation

The Supreme Court rules that the principle of "separate but equal" is constitutional, upholding Jim Crow laws in the states (legal segregation or discrimination against blacks). Racial segregation will exist until *Brown v. Board of Education* in 1955, when the Court reverses itself and outlaws segregation in public schools.

1898: April 28–December 10   Spanish American War

The U.S. battleship *Maine* explodes in Spanish-controlled Havana Harbor, Cuba, on February 15. War breaks out in April despite the unlikelihood of Spanish sabotage. The new American steel navy quickly destroys the Spanish fleets in Manila Bay in the Philippines, and later at Santiago harbor, Cuba. Theodore Roosevelt participates in the charge up San Juan Hill, in Cuba, and Manila is captured in the Philippines. Spain sues for peace, and an armistice is signed on August 12.

Under the Treaty of Paris, the United States obtains Puerto Rico, Guam, and some West Indian islands. Spain cedes the entire Philippines for $20,000,000 and also gives up Cuba.

1899: September   U.S. adopts "open door" policy with China

U.S. Secretary of State John Hay (1838–1905) sends letters to Japan and major European countries urging their respect for the port rights China had previously negotiated. The purpose is to protect American economic interests in China and prevent other countries from dominating Chinese trade. In 1900 the Boxer Rebellion occurs when some Chinese attempt to force foreign powers out of China. Hundreds of foreigners are killed before an international force, including Americans,

arrives and defeats the Boxers. America continues its "open door" approach and convinces the other major powers not to colonize China and instead accept monetary compensation for their losses.

1900   Cause of yellow fever discovered

Dr. Walter Reed (1851–1902), an American, is sent to Cuba to research the disease yellow fever, which is causing thousands of deaths per year. He and others conduct experiments with mosquitoes, using human volunteers as guinea pigs, and conclude that mosquitoes carry the disease. Efforts to eradicate the disease begin with the draining of swamps and burning of garbage.

1900   Baseball becomes the American pastime

In 1900 the American league is formed. The National League had been formed in 1876, so the modern baseball system is created. The first World Series will be held in 1903.

1901: September 6   President McKinley assassinated

President William McKinley (1843–1901) is shot and killed in Buffalo, New York, while attending a Pan-American Exposition. Theodore "Teddy" Roosevelt (1858–1919) becomes president (1901–09).

1902   Carnegie contributes to philanthropy

Multimillionaire Andrew Carnegie (1835–1919), a Scotch immigrant who made his fortune in the steel industry, gives tens of millions of dollars towards philanthropic causes. Many trusts and foundations are created for public libraries, educational institutions, and world peace efforts. Carnegie began his career in Pittsburgh and was one of the first Americans to install a Bessemer converter (see 1964). Carnegie's steel empire becomes the United States Steel Corporation.

1903   Roosevelt, the "Trust Buster"

President Roosevelt, and later his successor, President William Howard Taft (1857–1930), counter the growing power of corporations and their holding companies by taking many of them to court for violating the Sherman Anti-Trust Act and later the Clayton Anti-Trust Act. Corporate abuses are curbed.

1903: October 20   Alaska–Canada boundary settled

The discovery of gold in the Alaska Klondike in 1896 leads to a dispute with Canada over how far inland America's boundary with Canada extends. This is the panhandle part of Alaska that extends south along the Pacific coast. President Roosevelt tells the British that America might use military force if an agreement is not reached, and one quickly is. This establishes what will come to be called Teddy Roosevelt's "big stick" policy. He quotes: "I have always been fond of the

West African proverb: 'Speak softly and carry a big stick, you will go far.'" In 1905 Roosevelt expands the Monroe Doctrine to allow American intervention in Latin American affairs to preserve order. This is known as the Roosevelt Corollary to the Monroe Doctrine and more use of the "big stick."

1903: December 17   Wright brothers fly first airplane

Orville Wright (1871–1948) and Wilbur Wright (1867–1912), brothers from Dayton, Ohio, fly the world's first motor-powered airplane at Kitty Hawk, North Carolina. The first flight lasts twelve seconds, and the plane flies 120 feet. In 1906 the brothers patent their invention, and, in 1909 the U.S. Army accepts their airplane.

1906: April 18   Earthquake and fire devastate San Francisco

An earthquake levels buildings, splits streets, and causes a fire that destroys over four square miles of San Francisco, California. The fire rages for three days, 452 people die, and damage exceeds 350 million dollars. The city is rebuilt in four years.

1908   Roosevelt appoints commission on conservation

Roosevelt encourages the conservation of America's natural resources. Over 148 million acres of forests are protected as natural reserves along with 80 million acres of mineral lands. This is the beginning of the National Parks system.

1909   Henry Ford begins mass assembly of automobiles

Henry Ford (1863–1947) applies the standardized system to the automobile manufacturing process, using interchangeable parts and assembly lines to mass produce cars. The "Flivver," otherwise known as the "Model T," is the first car built by this method and is sold by the hundreds of thousands. Henry Ford had begun his career as a machine shop apprentice in Detroit. He founded the Ford Motor Company in 1903.

1909   NAACP founded

William Edward Burghardt DuBois (1868–1963) founds the National Association for the Advancement of Colored People (NAACP) in order to achieve racial equality.

1913: February 25   Sixteenth Amendment

Congress is given the power to tax the income of people and corporations and to decide the rate at which they will be taxed.

1913: May 31   Seventeenth Amendment

This amendment gives the public the right to directly vote for their own senators. Previously state legislatures elected the federal senators, which resulted in much bribery and corruption.

1914: August 15   Panama Canal opens

The $275-million canal shortcuts the long trip around South America and uses a lock system to elevate and lower ships as they travel through the canal. The United States had taken over the project from the French, who were unable to complete it years earlier. Panama becomes a separate country from Columbia through American intervention.

1916   First birth–control clinic

The first birth-control clinic is opened by Margaret Sanger (1883–1966) in Brooklyn, New York.

1917   First woman elected to Congress

Jeannette Rankin (1880–1973) becomes the first woman elected to Congress, serving one term in the House of Representatives representing Montana. Rankin devotes much of her life to women's voting rights and power in government and will be elected to Congress again.

1917: April 2–November 11, 1918   World War I

Germany's unrestricted submarine warfare results in the sinking of American ships, and President Woodrow Wilson (1856–1924; in office 1913–1921), abandoning his isolationist stance, asks Congress for a declaration of war against the Central Powers. The Senate approves the joint resolution with only six dissenting votes, and the House passes it 373–50. The United States enters World War I (1914–1918) with the AEF (American Expeditionary Force) led by General John Joseph Pershing (1860–1948). It fights mostly as a separate unit under its own command. Americans help stop the German offensive towards Paris and begin driving the German armies east. Germany appeals for peace, and an armistice is signed on November 11, 1918.

1918: January 8   Wilson lists his Fourteen Points for peace

President Wilson lists his Fourteen Points to Congress explaining America's goals for peace. Point number fourteen is the creation of a League of Nations to guarantee world peace in the future.

1918   Global influenza epidemic

Almost 500,000 Americans die in this epidemic, which takes about 30,000,000 lives globally. The epidemic begins in Europe, hits America's east coast, and spreads west. People panic, and a plague is feared, but the epidemic suddenly subsides in early 1919.

1919   America refuses to join League of Nations

President Wilson makes two trips to Europe to press his Fourteen Points, including the League of Nations. There is much resistance in Europe and in America. Wilson tours America to build support for the League of Nations, but he falls ill under the stress and never fully recovers. The Senate does not support the League, and the United States never joins.

1919   Post–war depression, red scare, Palmer raids

Four million workers strike during the post-war depression. For two years after the war, America is swept by a "red scare" during which communist activity, both legal and illegal, increases. Attorney General Alexander Palmer (1872–1936) orders the Justice Department to arrest suspects, and many constitutional rights are violated during the "Palmer raids." Two hundred forty-nine people are deported.

1919: January 29   Eighteenth Amendment; Prohibition begins

Beverages containing over one-half of one percent of alcohol are outlawed nationwide.

1920–29   Roaring Twenties

The 1920s become famous for their music, art, literature, and architecture. Authors like F. Scott Fitzgerald (1896–1940), T. S. Eliot (1888–1965), Sinclair Lewis (1885–1951), and Ernest Hemingway (1899–1961) publish great works. African American musicians, including Duke Ellington (1899–1974), play a style of jazz music that combines dance music and improvisation. They perform at clubs like the Cotton Club in Harlem, New York, throughout the 1920s and 1930s. Architect Frank Loyd Wright (1869–1959) builds his famous structures. The art world experiments with modernism, and jazz becomes popular. Mobsters like Al Capone in Chicago become rich dealing in illegal sales of alcohol and gambling, and there is much violence involving organized crime.

1920: August 26   Nineteenth Amendment

Women obtain the constitutional right to vote nationwide with the passage of the Nineteenth Amendment. This is the culmination of the suffragist movement, which arguably began in 1848 with Lucretia Mott (1793–1880) and Elizabeth Cady Stanton's (1815–1902) Women's Rights Convention at Seneca Falls, New York. This year the National League of Women Voters is organized.

1923   Ku Klux Klan returns

Wearing their characteristic white robes and hoods, the new Klan terrorizes and murders African Americans, Catholics, Jews, and immigrants in the South and North. Newspapers report the atrocities to the appalled nation, and the violence subsides.

1924: July 6   Photographs successfully transmitted

The first radiophoto is sent when an image of President Calvin Coolidge (1872–1933) is transmitted from New York to London and back again.

1927: May 20–21   Lindbergh flies across the Atlantic

Charles A. Lindbergh, Jr. (1902–1974) flies solo from New York to Paris in his monoplane, the *Spirit of St. Louis*. The trip takes thirty-three and one-half hours. Throngs of cheering people greet Lindbergh at Le Bourget airport in France and when he returns to America.

1927: October 6   *The Jazz Singer,* the first talking movie

Al Jolson (1886–1950) stars in this first "talky," which is the beginning of the end for silent movies.

1928   Mickey Mouse debut

Walt Disney (1901–66) makes *Steamboat Willie,* a Disney cartoon introducing Mickey Mouse.

1928: June 17–18   Amelia Earhart flies across the Atlantic

Amelia Earhart (1897–1937) becomes the first woman to fly across the Atlantic. She also becomes the first woman to win the Distinguished Flying Cross. In 1937 she and her radioman disappear over the Pacific while attempting to fly around the world.

1929: October 29   Stock market crashes, Great Depression begins

On "Black Tuesday," overspeculation finally causes the New York stock market to drop, with losses of thirty billion dollars by November 13. Waves of bank failures in 1930 wipe out personal savings. On June 17, 1930, President Herbert Hoover (1874–1964) signs the Smoot–Hawley Tariff, raising duties on imports and causing other countries to retaliate. International trade is crippled, and the Great Depression worsens.

1931: May 1   Empire State Building dedicated, tallest skyscraper

Consisting of 102 stories of office space, its art deco design towers 1,250 feet over New York City.

1932   Lindbergh baby kidnapping and trial

On March 1 Charles Lindbergh's infant son is kidnapped and held for ransom. The baby dies, Bruno Hauptmann is convicted for the crime, after a much publicized trial, and is executed. The "Lindbergh Law" is passed making interstate kidnapping a federal offense.

1932: November 8   Franklin D. Roosevelt elected, starts New Deal

Franklin Delano Roosevelt (FDR) wins a landslide victory and begins his New Deal policy to lift the nation out of the Great Depression. All banks are closed the day after his inauguration in order to be inspected, and a series of public works and other government programs are created throughout the

Sandwiches are distributed to unemployed people at City Hall in Cleveland, Ohio during the Great Depression. (EPD Photos/CSU Archives)

1930s to provide jobs for the unemployed. Programs include the Emergency Relief Act, the TVA (Tennessee Valley Authority), the AAA (Agriculture Adjustment Administration) to subsidize farmers, and the NRA (National Recovery Administration) among many others. When the Supreme Court strikes down some of FDR's programs, he threatens to pack (to arrange a court's membership in a way that gets the desired outcome) the Court, causing it to back down. The federal government gains more power to regulate the economy.

1933   Good Neighbor Policy

FDR expands Hoover's friendly policies towards Latin America and pulls American troops out of Central America. The "Roosevelt Corollary" of intervention ends. The Philippines are given more autonomy, and in 1934 plans are made for their eventual independence.

1933: December 5   Twenty–First Amendment ends Prohibition

Prohibition proves impossible to enforce, and the nation grows tired of the mobsters, violence, and general lawlessness that it creates. The Eighteenth Amendment is repealed.

1934: June 6   Securities and Exchange Commission created

The Securities and Exchange Commission (SEC) is created to regulate the stock markets and prevent the fraud and other abuses that contributed to the 1929 crash.

1934   Dust bowl

Drought and poor farming techniques lead to soil erosion and clouds of dust blanket much of America's farmland. The Soil Conservation Act is passed in 1935 as a result.

1935: July   National Labor Relations Act, Social Security

The National Labor Relations Board (NLRB) is created to supervise union elections and hear charges of unfair employer practices. There are over four million union members. Social Security is created to help the elderly, unemployed, and dependent children.

1938: October 30   Invasion from Mars radio broadcast

Orson Welles (1915–85) broadcasts his "Invasion from Mars" radio program. Many listeners panic, not realizing it is fiction.

1939: June 28   Transatlantic passenger flights begin

Pan Am begins the first regular passenger flights from America to Europe. The *Dixie Clipper* flies twenty-two people from Long Island, New York, to Lisbon, Portugal.

1939: August 2   Einstein writes to FDR about atomic energy

Theoretical physicist Albert Einstein (1879–1955), who had immigrated to America from Germany, informs FDR about the potential use and misuse of atomic energy. FDR begins an atomic energy program.

1941: December 7–August 10, 1945   World War II

Japanese fighters bomb the American naval base at Pearl Harbor, Hawaii. The vote to declare war is unanimous in the Senate, and only one representative dissents in the House—Jeannette Rankin. After initial losses, the American navy destroys the Japanese navy during a series of naval actions and island-hopping campaigns in the Pacific Ocean. American air forces bomb Japanese cities into ruin and bring an end to the war in Asia by dropping nuclear bombs on Hiroshima (August 6, 1945) and Nagasaki (August 9, 1945).

The War in Europe includes battles with German armies in North Africa and northern and southern Europe. American air forces also severely bomb German cities. American and Allied forces cross the English Channel and advance into Germany. The War in Europe ends with the occupation of Germany.

1942: February 20   FDR approves Japanese internment camps

During the war, resentment and suspicion of Japan lead to Japanese Americans being taken from their homes and confined to internment camps, even if they are U.S. citizens. Ethnic Japanese men still serve in the American armed forces.

1945: April 12   FDR dies in office

Elected to an unprecedented fourth presidential term, Roosevelt dies three weeks before the Nazis surrender in Europe. Vice-president Harry S. Truman assumes the office of the president.

1946   United Nations is formed

The United Nations (UN), a successor to the League of Nations, is headquartered in New York City. America joins.

1947   African American baseball player breaks color line

Jackie Robinson (1919–72) becomes the first African American baseball player in the major leagues. He plays for the Brooklyn Dodgers.

1947: March 12   Cold War begins

Truman announces the Truman Doctrine, America's policy of containing communism. The Marshall Plan is created to help Europe recover from the war. In 1948 the Soviets blockade West Berlin, and the United States creates the Berlin Airlift to defeat the blockade. North Atlantic Treaty Organization (NATO) is formed in 1949, aimed at safeguarding the Atlantic community, as a western countermeasure against rising Soviet and communist infuence.

1948   Truman orders armed forces desegregated

President Harry S. Truman (1884–1972) orders the end of racial segregation in the armed forces.

1950–54   Red scare, McCarthyism

Senator Joseph R. McCarthy (1908–57) feeds off of the growing fear of communism by making widespread charges of communism in the government, film industry, and elsewhere. He is censured by the Senate in 1954.

1950: June 25–June 27, 1953   Korean War

Communist North Korea invades South Korea. America leads a UN army that drives the North Koreans back into North Korea. Mainland China enters the war, and the war drags on until an armistice is signed in 1953.

1955: February 22   Beginning of highway system

President Eisenhower (1890–1969; in office 1953–61) submits to Congress a proposed $101 billion highway construction program. This eventually results in a massive interstate highway system.

1955: April 12   Polio vaccine

Dr. Jonas E. Salk (b. 1914) creates a vaccine for polio. Congress spends $30 million to vaccinate American school children, and the disease is virtually wiped out.

1955: May 31 Brown v. Board of Education

The Supreme Court unanimously holds that segregated school systems (which separate African American students from white students) violate the Fourteenth Amendment. The school systems are reluctant to integrate their schools.

1955: July Disneyland opens

Walt Disney opens the Disneyland amusement park in Los Angeles, California. Disney World, and other parks will be built later in Orlando, Florida, and elsewhere.

1955: December 1 Montgomery bus boycott

An African American woman named Rosa Parks refuses to give up her seat for a white person and is arrested. The Reverend Dr. Martin Luther King, Jr. (1929–68) helps organize a year-long boycott of the bus system in Montgomery, Alabama, and the company suffers revenue losses. In December, 1956, a federal court injunction forbids bus segregation, and the boycott ends. Southern states still resist integration efforts.

1955: December 5 AFL–CIO super labor union

The AFL (American Federation of Labor) and the CIO (Congress of Industrial Organizations) merge at a convention in New York City creating a 15 million-member union.

1957: June 18 International Atomic Energy Agency

President Eisenhower supports the idea to pool the international community's nuclear resources into the peaceful use of atomic energy.

1957: September 9 Civil Rights Act

Eisenhower asks Congress for a law to protect the voting rights of African Americans, and the Civil Rights Commission is created to curb voting abuses. The governor of Arkansas uses the National Guard to prevent African American students from entering a white high school under a desegregation order. Riots ensue, and on September 25, Eisenhower sends 1,000 paratroopers to escort the African American students to and from school.

1957: October 4 Sputnik, space race begins

The Soviet Union launches *Sputnik,* the world's first manmade satellite. The United States launches one on January 31, 1958, called *Explorer I.* Congress creates the National Aeronautics and Space Administration (NASA).

1959: June 26–27 St. Lawrence Seaway dedicated

The $500 million joint project with Canada opens up the Great Lakes to the Atlantic Ocean for shipping. It is the largest water project since the Panama Canal.

1959 Alaska and Hawaii are admitted as states

Alaska becomes the forty-ninth and Hawaii the fiftieth state.

1960 Nixon–Kennedy T.V. debates

Vice President Richard M. Nixon (1913–94) and Massachusetts Senator John F. Kennedy (1917–63) engage in the first televised presidential debate. Kennedy's election victory is largely credited to his debate performance.

1960: April 1 First weather satellite launched

Tiros I is launched and transmits pictures of Earth from orbit. The age of satellite weather forecasting, hurricane predictions, etc. begins.

1960: May 1 U2 incident

An American U2 spyplane is shot down over the U.S.S.R. (Union of Soviet Socialist Republics) injuring diplomatic relations.

1961: March 1 Kennedy creates Peace Corps

In an executive order President Kennedy creates the Peace Corps, a group of volunteers who assist and educate people in developing countries.

1961: May 5 Space race continues

On April 12 Soviet cosmonaut Yuri Gagarin (1934–68) is the first person to orbit the earth. Alan Shepard (b. 1923), U.S. astronaut, is the first American launched into space on May 5, and astronaut John Glenn (b. 1921) becomes the first American to orbit the Earth on February 20, 1962.

1962: July 10 First communications satellite

The Bell Telephone system's satellite is sent into orbit, improving telephone and telecast capabilities.

1962: September–October Anti–integration riots in Mississippi

President Kennedy sends 15,000 troops to Oxford, Mississippi, to stop the rioting and allow James H. Meredith, an African American, to attend the University of Mississippi.

1962: September–November Cuban Missile Crisis

America discovers Soviet nuclear missiles in Cuba and blockades the island. After much tension, the Soviets remove their missiles, and America promises not to invade Cuba.

1963: August 28 Freedom March

About two hundred thousand protesters, mostly African American, march on Washington, D.C. to demand equal rights and express support for a new civil rights law. Dr. Martin Luther King, Jr., delivers his "I Have a Dream" speech.

1963: November 22   President Kennedy assassinated

President Kennedy is shot and killed in Dallas, Texas. Lee Harvey Oswald (1939–63) is the prime suspect, but he is shot and killed by Jack Ruby two days later. Lyndon Johnson, Kennedy's Vice President, is immediately sworn-in as president upon word of Kennedy's death.

1964: July 2   1964 Civil Rights Act passed; Great Society

This act mandates equal treatment for African Americans and whites in public places, voting, and employment. It opens up many restaurants, hotels, and other areas that were previously off limits to African Americans.

President Johnson (1908–73) initiates his "Great Society" programs designed to reduce poverty and improve quality of life.

1964: August 7   Gulf of Tonkin Resolution

This resolution increases Johnson's military power in Vietnam. America's military presence in Vietnam greatly increases and will eventually peak at half a million troops before public sentiment becomes sharply opposed to American actions, and the troops are slowly withdrawn.

1965: August 11–16   Watts Riots

A race riot in the African American neighborhood of Watts in Los Angeles, California, kills thirty-four and causes over $200 million in damages.

1966 and 1967   Summers of racial violence

Racial violence breaks out in Chicago, Illinios, and Cleveland, Ohio, in the summer 1966. Newark, New Jersey, and Detroit, Michigan, are beset by racial violence in the summer of 1967. Racial violence and riots occur in other cities as well. This racial uprising has roots in the lack of civil rights for minorities and is a rebellion against the overcrowded, poor conditions of minorities in urban America.

1967   African American firsts

On October 2, Thurgood Marshall (1908–93), the attorney who successfully argued the *Brown v. Board of Education* case, becomes the first African American Supreme Court Justice. Carl B. Stokes becomes the first African American mayor of a major city, Cleveland, Ohio.

1968: April 4   Martin Luther King assassinated

Dr. Martin Luther King, Jr. is shot and killed in Memphis, Tennessee, on his hotel room balcony. James Earl Ray is convicted of the murder and spends the rest of his life in prison.

The Black Panther Party is founded in 1966. These Black Panther Party leaders use aggressive slogans to emphasize their demands for social change. (EPD Photos/CSU Archives)

1968: June 5   Robert F. Kennedy assassinated

After wining the California primary in his race to become president, Senator Robert F. Kennedy (1925–68) is shot and killed by Sirhan Bishara Sirhan (b. 1943).

1969   "Vietnamization" of the war in Vietnam

In 1969 President Nixon begins slowly reducing the size of the American military in Vietnam in an effort to "Vietnamize" the war and make the South Vietnamese fight for themselves. The war will drag on until America completely pulls out in 1973, and South Vietnam falls to the communist North on April 30, 1975.

1969: July 20   Man lands on the moon

Neil Armstrong (b. 1930) and Edwin "Buzz" Aldrin, Jr. (b. 1930) leave *Apollo 11,* land their lunar module "Eagle" on the moon, and walk on its surface. Their moonwalk is televised to an estimated 400 million viewers.

1969: August 15–18   Woodstock

The music festival near Bethel, New York, attracts 400,000 young fans.

1969: October 29   School integration hastened

After years of delay, school districts are ordered by the Supreme Court to create and implement school desegregation plans "at once." School districts will wrestle with desegregation throughout the 1970s and 1980s, sometimes resorting to busing to achieve integrated schools.

1970: February 4   Pollution control

President Nixon calls for environmental legislation to clean up America's polluted air, water, and land. Investments are made in water treatment plants, air pollution control, and garbage disposal methods.

1970   Students die in protests

Four students die at Kent State University, Ohio, and two at Jackson State College, Mississippi, when they are fired upon by authorities during demonstrations. The protests are against Nixon sending American forces into Cambodia. Anti-war protests are common throughout America's military involvement in Vietnam.

1970   Mexican–American political movement

Mexican–Americans form community action groups in order to elect candidates to political office. They also work to improve housing and education, and unionize their workers.

1970: August 26   Women's Liberation Movement Day

Women of all economic backgrounds participate in marches across the country to demand equal rights and pay for equal work. The ERA (Equal Rights Amendment) is put forward, but it falls three states short of being ratified in 1982, largely because of disagreement over women serving in the armed forces.

1972: February 21–28   Nixon opens China

President Nixon visits China and begins the opening of diplomatic relations between the United States and mainland China for the first time since the communist takeover. Full diplomatic relations will be resumed on January 1, 1979.

1973   Abortion legalized

The Supreme Court rules in *Roe v. Wade* that a woman has the right to terminate her pregnancy, with some restrictions.

1973: October 17   Oil Embargo

The oil producing countries in the Middle East suspend oil exports to the United States in retaliation for American support of Israel during its Yom Kippur War with its Arab neighbors. The embargo lasts until March 18, 1974, and causes a massive increase in gasoline prices, shocking the American economy.

1974: August 8   President Nixon resigns

President Nixon becomes the first president in American history to resign from office. His resignation comes after a break-in at the Democratic Party headquarters in the Watergate Hotel which is investigated by Special Prosecutor Archibald Cox. Nixon's battle to avoid impeachment includes denying any involvement in the break-in, attempting to make the Central Intelligence Agency (CIA) stop an investigation of the matter by the Federal Bureau of Investigations (FBI), attempting to fire the special prosecutor, and refusing to turn over tapes of his conversations in the White House until ordered to do so by the Supreme Court. Vice President Gerald Ford (b. 1913) assumes the Presidency and, on September 8, pardons Nixon for crimes he may have committed in office.

1976: July 3   Death penalty upheld

The Supreme Court rules that capital punishment (the death penalty) is not "cruel and unusual," and therefore is constitutional. States are permitted to decide for themselves whether to execute convicts.

1977   President Carter pardons draft dodgers

President Jimmy Carter (b. 1924; in office 1977–81) pardons the people who avoided being drafted into the military during the era of the Vietnam War.

1977   Alaska pipeline opens

Oil is piped from oil fields in Alaska through a giant pipeline to the lower forty-eight states.

1977: September 7   Panama Canal Treaty

President James "Jimmy" Earl Carter signs a treaty with Panama that calls for the canal to be eventually handed over to the government of Panama. This resolves issues over Panamanian sovereignty and potential violence against Americans.

1979   More oil inflation

OPEC (Organization of Petroleum Exporting Countries) doubles the price of petroleum, causing inflation in America to reach its highest level ever, around thirteen percent.

1979: March 28   Three Mile Island

The worst American nuclear accident occurs at the Three Mile Island nuclear power plant near Harrisburg, Pennsylvania. Clouds of radioactive steam are released into the open, people are forced to evacuate their homes, and popular support for nuclear power wanes.

1979: November 4   Hostages taken at American Embassy in Iran

Iranian students and other demonstrators storm the American Embassy in Tehran, Iran, and take sixty-six American hostages. This incident follows a revolution in Iran that overthrows the Shah (the Iranian leader) who had received considerable American support. When the Shah is admitted into the United States for medical treatment, outraged Iranians take the hostages who are later held by the new Iranian government. A rescue attempt on April 24 fails, when an American helicopter crashes in the desert, killing eight. The hostages are held 444 days and are released minutes after Ronald Reagan (b. 1911, in office 1981–89) becomes president in 1981.

1980s   Reaganomics

President Reagan's presidency creates many changes in government including "supply side" economics. Federal government programs are given over to states to control, taxes and domestic spending are lowered, while defense spending is increased. There is also much deregulation, including in the airline industry. In 1982 unemployment exceeds ten percent, the highest since the Great Depression, but by 1984 inflation and unemployment both decrease, the value of the American dollar rises, and the economy grows considerably. American budget deficits grow to exceed $100 billion.

1980   U.S. boycotts Moscow Olympics

In protest of the Soviet invasion of Afghanistan, President Carter boycotts the Olympics, which are held in Moscow. No American teams attend.

1980: May 18   Mt. St. Helens erupts

The volcano Mount Saint Helens erupts and explodes in Washington State. The blast is 500 times more powerful than the Hiroshima atomic bomb, and twenty-six people are killed. There are about $2.7 billion in damages, forests are flattened, and lakes and rivers are poisoned.

1981: April 12–14   First Space Shuttle launch

The Space Shuttle *Columbia* successfully lifts off, beginning the American shuttle program, the first with reusable launch vehicles.

1981: September 25   Sandra Day O'Connor

Judge Sandra Day O'Connor (b. 1930) becomes the first female Supreme Court Justice.

1982: January 8   AT&T breaks up

AT&T agrees to let its local "Bell" telephone companies break-off from the parent company in order to end a lengthy antitrust lawsuit and expand its business elsewhere. The "baby bells" are created.

1983: October 23   Beirut bombing

In September President Reagan sends American marines to Beirut, Lebanon, on a multinational peace-keeping mission. A suicide truck bomb causes the deaths of 240 of them, and Reagan withdraws the remaining marines on February 7, 1984.

1986: January 28   Challenger explosion

The Space Shuttle *Challenger* explodes shortly after takeoff, killing six astronauts and one civilian, teacher Christa McAuliffe. NASA is criticized for relaxing safety procedures and speeding up shuttle launch schedules. The shuttle fleet is temporarily grounded.

1987: July 13   Colonel Oliver North testifies about Iran–Contra

On November 19, 1986, the Iran-Contra scandal breaks, and President Reagan denies involvement. The Iran-Contra scandal involves the Reagan Administration selling weapons to Iran and using the profits to fund the contra military forces in Nicaragua, both of which activities are banned by Congress. Colonel Oliver North's (b. 1943) televised testimony before members of Congress is watched by millions. A special prosecutor investigates, and the Tower Commission criticizes the White House staff as well as President Reagan's "management style." After his presidency, Reagan will testify that he has no recollection of the events. Oliver North and national security adviser John Poindexter (b. 1936) will escape punishment by immunity and a pardon by President Bush (b. 1924) in 1991. In 1992 Bush will pardon six other Reagan Administration members.

1987: October 16   Stock market crash

The American stock market crashes a record 508 points in one day, shocking markets around the world.

1989: March 24   Exxon *Valdez* oil spill

The supertanker *Valdez*, owned by the oil company Exxon, crashes into the Alaskan coast. Over 11 million gallons of oil are spilled, causing extensive environmental damage and killing tens of thousands of animals.

1989: October 17   Loma Prieta earthquake

The second worst earthquake in modern American history hits the San Francisco Bay area, causing $6 billion in damages and killing sixty-two people. Freeways, bridges, and buildings collapse.

1989: November 7   First African American governor

Virginia elects the nation's first African American governor since Reconstruction, L. Douglas Wilder.

1989: December 20   U.S. invades Panama

Tensions rise between America and Panama over the treatment of American service people, the drug trade, and other issues. When Panama declares a state of war with America, President Bush sends twenty-four thousand troops to occupy the small country and incarcerate its leader, General Manuel Noriega (b. 1940). The United States helps install a popularly elected government.

1990–91   Gulf War

After Iraq invades and occupies its smaller neighbor, Kuwait, President Bush builds an international coalition to protect neighboring Saudi Arabia. After an intensive bombing campaign and 100 hours of ground fighting, Iraq retreats from Kuwait, and the conflict ends. American casualties number 144, mostly from friendly fire.

1992: January 26   Americans with Disabilities Act

The Americans with Disabilities Act is passed to guarantee disabled people equal access to most businesses.

1992: February 1   Cold War ends

President Bush and Russian President Boris Yeltsin (b. 1931) issue a joint statement that the Cold War has come to an end. As the Soviet Union dissolves, the United States recognizes and opens relations with its succeeding republics.

1992: April 29–May 4   Los Angeles riots

On March 3, 1991, four white Los Angeles police officers are videotaped beating an African American motorist, Rodney King. At trial, an all-white jury acquits the four officers, and rioting erupts minutes later in the African American neighborhood of South Central in Los Angeles, California. The riots last for days and result in fifty deaths and seven thousand arrests. The officers are later convicted on federal charges.

1993   First Lady's office in White House

President Clinton's wife, Hillary Rodham Clinton, becomes the first First Lady to obtain an office in the White House. She attempts to make changes in the nation's health care system but is largely unsuccessful.

1993   AZT helps AIDS victims

A virus known as the HIV (human immunodeficiency) virus becomes widespread during the 1980s. A federal study indicates that the drug AZT (azidothymidine) greatly reduces infection rates from infected mothers to their babies. AZT becomes a widely used drug to treat AIDS, but a cure remains unknown, and the virus continues to spread.

1993: February 26   World Trade Center bombing

Muslim extremists detonate a car bomb in the garage beneath the World Trade Center twin towers in New York City. Seven people are killed, over 1,000 injured, and 50,000 are forced to evacuate the center. Four men are later convicted of the bombing.

1993: November 17   Free trade agreements created

The NAFTA (North American Free Trade Agreement) is narrowly approved by the House of Representatives on November 17. This agreement creates the world's largest free trade zone: Canada, Mexico, and the United States.

The United States and Europe agree to the GATT (General Agreement on Tariffs and Trade). The GATT reduces tariffs on more industries and in more countries than any other trade agreement in history.

1994   Prison population is over one million

The Bureau of Justice Statistics announces that inmates in federal and state prisons exceed one million, giving the United States the world's highest incarceration rate.

1995: March 16   American on Mir

American astronaut Norman E. Thagard becomes the first American to board the Russian (ex-Soviet) space station *Mir* (Russian for "world"). He is the first of a series of American astronauts to visit and work on *Mir* with Russian cosmonauts.

1995: April 19   Oklahoma City bombing

A truck bomb detonates next to a federal building in downtown Oklahoma City. One hundred sixty-nine people are killed, including fifteen children, as part of the building collapses. Timothy McVeigh and Terry L. Nichols are later convicted for what is the deadliest act of domestic terrorism in American history.

1995: October 3   O.J. Simpson is found not guilty of murder

A media event begins in 1994 after former football player O.J. (Orenthal James) Simpson's ex-wife Nicole Brown Simpson and her friend Ronald Goldman are discovered murdered. Police incarcerate O.J. Simpson after he leads them on a sixty-mile low speed televised car chase. The trial is also televised live and watched by millions. O.J. Simpson is found not guilty by a mostly African American jury and released, although a civil jury later finds him liable in civil court and heavily fines him. The trial divides America largely along racial lines.

1995   Federal government "shuts down"

Democratic President Bill Clinton and the Republican Congress refuse to compromise over the federal budget, resulting in nonessential government functions being shut down. About 800,000 federal employs go without work for six days until a compromise is reached.

1995: December 25   Middle East peace accords

President Clinton, Egyptian leader Hosni Mubarak, Jordanian King Hussein, Israeli leader Yitzhak Rabin, and Palestinian leader Yasir Arafat sign an accord beginning Palestinian self–rule.

1996   President Clinton is reelected

Bill Clinton becomes the first Democratic president to be reelected since Franklin Delano Roosevelt.

1997   *Titanic* is most successful movie ever

The movie *Titanic,* based on the 1912 tragic sinking of the ocean liner, *Titanic,* in the North Atlantic, surpasses *Star Wars, ET,* and *Jurassic Park* to become the highest grossing movie of all time.

1997: July 4   Pathfinder lands on Mars

The American spacecraft *Pathfinder* lands on Mars. Because of *Pathfinder's* remote-control robotic car called *Sojourner* equiped with artificial intelligence, NASA engineers can further explore the planet. This is the first of a series of space missions to Mars.

1998   New space station

Initial pieces of the new multinational space station *Freedom* are launched and assembled in orbit.

1998: December   President Clinton impeached

Democratic President Bill Clinton is impeached by the Republican-controlled House of Representatives on charges of perjury and obstruction of justice. The charges stem from a forty million dollar independent counsel investigation of Clinton's financial activities as governor of Arkansas. This leads to further charges that Clinton lied to a federal grand jury when denying a sexual affair with an intern at the White House. Simultaneously Clinton is being impeached as America bombs Iraq. Clinton makes history by being the second president ever impeached.

1999   Acquittal of President Clinton in the Senate

President Clinton is acquitted of all impeachment charges.

1999   NATO strikes against Yugoslavia

The United States leads a NATO military effort against Yugoslavia. The stated goal of the initiative is to stop the "ethnic cleansing" in Kosovo, a region considered a homeland by both ethnic Albanians and Serbs. Yugoslav president Slobodan Milosevic approves Serbian efforts to drive Albanians out of Kosovo (referred to as "ethnic cleansing"), often accomplished by the use of deadly violence. Following the end of the NATO initiative, Serb troups leave the country, and thousands of Albanians return to Kosovo, many exacting violent revenge against Serbs.

Bibliography

Ayres, Stephen M. *Health Care in the Unites States: The Facts and the Choices.* Chicago: American Library Association, 1996.

Bacchi, Carol Lee. *The Politics of Affirmative Action: 'Women', Equality, and Category Politics.* London: Sage, 1996.

Barone, Michael. *The Almanac of American Politics.* Washington, D.C.: National Journal, 1992.

Bennett, Lerone. *Before the Mayflower: A History of Black America.* 6th ed. New York: Penguin, 1993.

Brinkley, Alan. *American History: A Survey.* 9th ed. New York: McGraw-Hill, 1995.

Brinkley, Douglas. *American Heritage History of the United States.* New York: Viking, 1998.

Carnes, Mark C., ed. *A History of American Life.* New York: Scribner, 1996.

Commager, Henry Steele (ed.). *Documents of American History.* Englewood Cliffs, N.J.: Prentice-Hall, 1988.

Davidson, James West. *Nation of Nations: A Narrative History of the American Republic.* 3rd ed. Boston, MA: McGraw-Hill, 1998.

Davies, Philip John. (ed.) *An American Quarter Century: US Politics from Vietnam to Clinton.* New York: Manchester University Press, 1995.

Donaldson, Gary. *America at War since 1945: Politics and Diplomacy in Korea, Vietnam, and the Gulf War.* Westport, Conn.: Praeger, 1996.

Foner, Eric. *The Story of American Freedom.* New York: Norton, 1998.

Garraty, John Arthur. *The American Nation: A History of the United States.* 9th ed. New York: Longman, 1998.

Goldfield, David, ed. *The American Journey: A History of the United States.* Upper Saddle River, NJ: Prentice Hall, 1998.

Hart, James David, ed.. *Oxford Companion to American Literature.* 6th ed. New York: Oxford University Press, 1995.

Hummel, Jeffrey Rogers. *Emancipating Slaves, Enslaving Free Men: A History of the Civil War.* Chicago: Open Court, 1996.

Jenkins, Philip. *A History of the United States.* New York: St. Martin's Press, 1997.

Kaplan, Edward S. *American Trade Policy, 1923–1995.* Westport, Conn.: Praeger, 1996.

Magill, Frank N., ed. *Great Events from History: North American Series.* rev. ed. Pasadena, CA: Salem Press, 1997.

Martis, Kenneth C. *The Historical Atlas of Political Parties in the United States Congress 1789–1989.* New York: Macmillan, 1989.

McNickle, D'Arcy. *Native American Tribalism: Indian Survivals and Renewals.* New York: Oxford University Press, 1993.

Mudd, Roger. *American Heritage Great Minds of History.* New York: Wiley, 1999.

Nash, Gary B. *The American People: Creating a Nation and a Society.* 4th ed., New York: Longman, 1998.

People Who Shaped the Century. Alexandria, VA: Time-Life Books, 1999.

Robinson, Cedric J. *Black Movements in America.* New York: Routledge, 1997.

Tindall, George Brown. *America: A Narrative History.* 5th ed. New York: Norton, 1999.

Tocqueville, Alexis de. *Democracy in America.* New York: Knopf, 1994.

United States Government Manual. Washington, D.C.: US Government Printing Office, 1935-date.

Virga, Vincent. *Eyes of the Nation: A Visual History of the United States.* New York: Knopf, 1997.

Woodward, C. Vann, ed. *The Comparative Approach to American History.* New York: Oxford University Press, 1997.

Uruguay

Introduction

Uruguay is South America's second smallest nation. With an area of 68,039 square miles (176,220 square kilometers), it is about the same size as the state of Oklahoma. Its official name—the República Oriental del Uruguay (the Eastern Republic of the Uruguay)—refers to its location on the eastern bank of the Uruguay River, which forms the nation's western boundary. Uruguay is dwarfed by its larger neighbors, Brazil and Argentina, in population as well as size. Its population of over 3.2 million people compares with 34 million for Argentina and 160 million for Brazil. Nearly half of all Uruguayans live in the capital city of Montevideo, located on the bank of the Río de la Plata.

When the first Europeans arrived in the region that is now known as Uruguay, they found it inhabited by an Amerindian group they called the Charrúa. Believed to have occupied the area for some four thousand years, they were hunters and gatherers. Through warfare, disease, and intermarriage, the Charrúa virtually disappeared as a distinct group by the middle of the nineteenth century.

In the sixteenth century, both the Spanish and Portuguese explored the region of the Río de la Plata river, which runs through present day Uruguay. The Spanish named the entire area the Banda Oriental (Eastern Bank). Settlement did not begin until the late seventeenth and early eighteenth centuries, when the Portuguese founded Colonia (1680) and the Spanish founded the present day capital of Montevideo (1726). The two colonial powers competed for control until 1777, when Spain claimed the Banda Oriental as part of the newly formed Viceroyalty of the Río de la Plata, headquartered in Buenos Aires.

The struggle for independence from Spain began in 1811, led by Uruguay's national hero, José Gervasio Artigas (1964–1850). Artigas's initial efforts were thwarted by the support the Spanish rulers received from Brazilian Portuguese forces as well as the newly independent regime in Buenos Aires. After a two year exile in Argentina, Artigas and his supporters returned and captured Montevideo but ultimately lost it to the

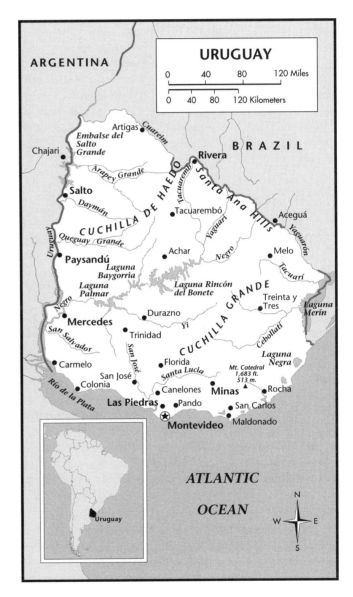

Portuguese by 1920. The region was annexed by Brazil and renamed the Cisplatine Province.

Other heroes of Uruguayan independence were the "Thirty-three Immortals" who, in 1825, entered the region from exile in Argentina under the leadership of Juan Antonio Lavalleja. Hostilities ended in 1828 with a treaty negotiated by Great Britain, which hoped that an independent Uruguay would serve as a buffer zone between Argentina and Brazil, helping to keep peace in the region and thus protecting its own shipping access to the Río de la Plata. Uruguay became an independent republic on July 18, 1830, with José Fructuoso Rivera (1788–1854) as its first president.

Independence did not bring peace to Uruguay. The new nation was torn by competing interests, which gathered around two political parties whose rivalry was to dominate Uruguay's history through the nineteenth and twentieth centuries: the Colorados (Reds) and the Blancos (Whites). Throughout the nineteenth century, political turbulence was the rule. Even though the Colorado party dominated the government (as it was to do in the twentieth century, as well), political allegiances were unstable, and armed uprisings and civil war were common. In 1838 Uruguay's former president, José Fructuoso Rivera, ousted his democratically elected successor, Manuel Oribe (1792–1857), leading to thirteen years of civil war with eventual involvement by Argentina, France, Britain, and Italy. The second half of the nineteenth century was a still unsettled, if less violent, period that included fourteen years of rule by military governments.

The Twentieth Century

Political stability finally came to Uruguay in the twentieth century under the leadership of José Batlle y Ordóñez (1810–87), who served two nonconsecutive terms as president in the first two decades of the new century and dominated social and political policy until his death in 1929. It was the social reforms implemented by Batlle that laid the groundwork for Uruguay's unique social welfare network, initiated at the beginning of the twentieth century, and led to comparisons with Switzerland. (Uruguay is sometimes referred to by the nickname "The Switzerland of South America.") Under Batlle, the government also began to play a more active role in the nation's economy through state-owned enterprises. In the 1930s there was a period of reaction to the progressive policies of the Batlle era, culminating in a government takeover in 1933 by Gabriel Terra (1873–1942), who had been elected president in 1931 but ruled as a dictator until 1938. In the years following World War II (1939–45), Batlle's nephew, Luis Batlle Berres, brought a return of his uncle's liberal policies.

Beginning in the mid 1950s, economic problems began to threaten Uruguay's political stability. The first sign of discontent was the elevation of the Blanco party to power in 1958 for the first time in ninety-three years. However, the new political leadership was unable to quell popular discontent with a faltering economy, and a Marxist terrorist group known as the Tupamaros gained supporters. Increased lawlessness eventually led to more repressive government, and by 1973 Uruguay was under military rule, and remained so for the next twelve years.

Civilian government was restored in 1985, when elected president Julio Sanguinetti took office. Sanguinetti was suc-

ceeded in 1990 by Luis Alberto Lacalle in elections that were notable for the strong showing by the leftist Frente Amplio (Broad Front) party, marking the first time in the nation's history that a third party had become a serious rival to the Colorados and Blancos. Sanguinetti was returned to office in the 1994 elections, in which the Frente Amplio once again recorded a strong showing.

An important economic development of the 1990s was the creation in 1991 of the MERCOSUR common market by Uruguay and three of its South American neighbors, Argentina, Brazil, and Paraguay. (MERCOSUR is an acronym for Mercado Común del Sur—Southern Common Market.) By 1995, the organization had come close to achieving its goal of abolishing all tariffs between member nations.

Timeline

20,000–10,000 B.C.   First inhabitants migrate to Uruguay from Asia

The first inhabitants of present day Uruguay are thought to have come from Asia, crossing the Bering Strait near the end of the Ice Age. Archaeologists have identified 10,000 year old remains of two hunting and gathering groups, the Cuareim and the Catalan.

2000 B.C.   The Charrúa arrive in Uruguay

The Charrúa are one of two groups to appear in the region. (The other is the Tupí Guaraní.) The Charrúa evolve into advanced hunter-gatherers. They hunt with bows and arrows, spears, and slings, and use canoes for transportation. When the first Europeans arrive in present day Uruguay in the sixteenth century, the Charrúa are the most powerful and populous native group. Other groups include the Minuanes, Bohanes, Guenoas, Yaros, and Chanáes.

European Exploration and Colonization

1516   Spanish explorer Juan Díaz di Solís sails down the Río de la Plata river

Díaz di Solís (c. 1470–1516) is credited with the European discovery of present-day Uruguay, which the Spanish call the Banda Oriental (Eastern Bank) because it is located on the east bank of the Uruguay River. Believing he has found a passageway between the Atlantic and Pacific oceans, Díaz di Solís enters the Río de la Plata and lands on its banks. The explorer and all but one member of his party are killed by either the Charrúa or the Guarani.

Evidence indicates that Díaz de Solis and his men are killed by the Guarani, but the so called "Charrúa legend" has existed for centuries; the Uruguayan people—many of whom trace their lineage to the Charrúa people—are proud of their past of bravery and rebellion in the face of oppression. After 1516, the Charrúa will continue to plague the Spanish for over 300 years.

1520   Magellan navigates the Río de la Plata

Portuguese explorer Ferdinand Magellan (c. 1480–1521) enters the bay of the Río de la Plata, landing at the future site of Uruguay's capital, Montevideo. A popular legend claims that the capital receives its name when one of Magellan's sailors called out "monte vide eu" (Portuguese for "I see a mountain").

1603   Spanish colonists introduce livestock to the Banda Oriental

Releasing cattle and horses onto the empty plains of present day Uruguay, the Spaniards create a base for the economy of the future nation.

1624   Arrival of the first missionaries

The first Franciscan and Jesuit missionaries arrive in the region, eventually winning converts among the more peaceful native groups. However, the small size of the Indian population as a whole keeps the number of converts relatively low. The absence of a substantial population of native converts contributes to the eventual secularization of Uruguayan society, compared to those of many of its Latin American neighbors.

1680   The Portuguese found Colonia del Sacramento

Portuguese from Brazil found Colonia del Sacramento on the northern bank of the Río de la Plata, opposite Buenos Aires, Argentina. The settlement becomes the focus of Spanish-Portuguese colonial rivalry over the Banda Oriental for nearly a century.

1695   Church of the Most Holy Sacrament founded

The Church of the Most Holy Sacrament (Iglesia Matriz del Santisimo Sacramento) is Uruguay's oldest church. It has been partly destroyed several times. In 1823, an explosion destroyed part of the church that was being used as an explosives depot.

1726   Montevideo is founded by the Spanish

San Felipe de Montevideo, Uruguay's present day capital, is established by Spain to counter Portuguese claims to the region. It is populated by settlers from Buenos Aires and the Canary Islands, who receive allotments of land from the Spanish government. Montevideo becomes the principal Spanish naval center in the South Atlantic.

1764   Birth of José Gervasio Artigas, Uruguay's first national hero

Artigas (1774–1850) is the foremost hero in the struggle for Uruguayan independence from Spain and Portuguese Brazil. His family is among the founding settlers of Montevideo in the early eighteenth century. For a time, Artigas lives the life of a *gaucho*, or South American cowboy, on his family's ranch. Between 1811 and 1820, he and his followers try unsuccessfully to win independence for their homeland. They capture Montevideo in 1815 and declare independence but are ousted by Brazilian Portuguese forces after a four-year struggle. In 1820, Artigas goes into exile in Paraguay, where he remains until his death thirty years later. Although Uruguay wins its independence under a different revolutionary leader, Artigas is credited with inspiring the people of his region with a sense of national identity.

1777   Banda Oriental becomes part of the Viceroyalty of the Río de la Plata

Spain places the Banda Oriental (present day Uruguay) under the authority of the Viceroyalty of the Río de la Plata. With its capital in Buenos Aires, the viceroyalty is part of a larger system through which Spain administers its territories in the New World. The founding of the viceroyalty increases the existing rivalry between Montevideo and Buenos Aires.

1780–1800   First meat salting plants are built

Saladeros, or meat salting plants, process beef and convert it to *tasajo*, or lean salt beef, which becomes one of Uruguay's leading agricultural export products.

1792   Fort of Saint Teresa is built

Gomez Freire de Andrada, viceroy of Brazil, orders the building of a fort near Punta del Diablo. The Fort of Santa Teresa is captured by Spain soon after it is completed. The Spanish modify the appearance of the fort. The government of Uruguay eventually converts the fort into an historic site and museum, Santa Teresa National Park.

1802   First religious music is written in Uruguay

The first major religious work to be composed in Uruguay is the *Misa para día de difuntos* (Mass for All-Saint's Day; 1802) by Fray Manuel Ubeda (c. 1760–1823).

1806–7   British forces invade and occupy Montevideo

During the Napoleonic Wars (the intermittent wars which are waged by France on other European nations between 1796–1815), the British, who are at war with Spain, invade the Río de la Plata region in 1806. They occupy Montevideo from the beginning of 1807 until July of that year.

The Struggle for Independence

1811: February   Artigas begins his campaign against the Spanish

Under the leadership of José Gervasio Artigas, an army of *orientales* (Uruguayans) rebel against the Spanish colonial regime. Early in 1811, Artigas allies his forces with those of the military rulers who have ousted the Spanish in Buenos Aires. He lays siege to Montevideo between May and October. However, he is defeated by Portuguese forces.

1811–13   Artigas and his supporters go into exile in Argentina

Abandoned by the Buenos Aires regime and overwhelmed by the joint power of the Spanish and Portuguese forces, Artigas leaves the Banda Oriental (eastern bank of the Uruguay river) for voluntary exile across the Uruguay River in Argentina, accompanied by 3,000 troops and 12,000 civilian supporters. From their base in exile, Artigas's troops conduct guerrilla warfare against forces in the Banda Oriental. This historic "exodus" is often regarded as the spiritual birth of the Uruguayan nation.

1813   Artigas rejoins Buenos Aires in the battle for Montevideo

Artigas and his supporters return to the Banda Oriental, and the commander once again allies his forces with those of the independent government that has expelled the Spanish from Buenos Aires. However, he withdraws his support from them when he is denied a guarantee of independent status for the Banda Oriental after the Spanish are removed.

1814: June   Troops from Buenos Aires defeat the Spanish at Montevideo

After Artigas withdraws from their alliance in January of 1814, the Buenos Aires forces continue their siege of Montevideo. In June they drive out the Spanish, after which they are attacked by Artigas and his allies from various provinces of the Banda Oriental.

1815   Artigas captures Montevideo

Montevideo falls to Artigas's forces after their victory at Guayabos. The Uruguayans declare an independent republic including the Banda Oriental and northern provinces of Argentina. They set up a federal system of government modeled on that of the United States.

1816   Uruguay's major library established

The National Library is founded in Montevideo.

1816–20   The Portuguese gain control of the Banda Oriental

A large Portuguese attack force invades the Banda Oriental in 1816 and captures Montevideo in 1817. After three more years of fighting, Artigas is forced into exile in Paraguay, where he remains until his death thirty years later. Brazil annexes the Banda Oriental, renaming it the Cisplatine Province.

1825: April 19   The "Thirty-Three Immortals" enter Uruguay

After five years of Portuguese rule, a band of rebels led by Juan Antonio Lavalleja and José Fructuoso Rivera (c. 1790–1854) crosses the Uruguay River from Argentina and launches an attack to regain freedom for their homeland. They issue a declaration of independence in August. Argentina supports the conflict and is subjected to a naval blockade by Brazil. Ultimately, the British intervene and broker a settlement to protect their shipping interests in the region.

1828: August 27   The Treaty of Montevideo ends hostilities in the Banda Oriental

Brazil and Argentina sign the British mediated agreement that ends the fighting in the Banda Oriental and provides for the creation of an independent nation to serve as a buffer zone between the warring powers. However, both nations retain the right to approve the constitution of the new state.

The Republic of Uruguay

1830s   Cattle industry exports become central to Uruguay's economy

Exports derived from the cattle industry—primarily salted beef and leather—play an increasingly prominent role in Uruguay's economy, and the number of *saladeros* (meat salting plants) multiplies.

1830   Rivera becomes Uruguay's first president

José Fructuoso Rivera (c. 1790–1854), one of the "treinta y tres orientales," known as "Thirty-Three Immortals" or "Thirty-Three Easterners" who fought for Uruguay's freedom in 1825, is designated by the constitution as the new nation's first president and takes office in 1830. During his term, he faces three uprisings led by the hero of the 1825 campaign, Juan Antonio Lavalleja.

1830   Birth of painter Juan Manuel Blanes

Blanes, today considered the foremost nineteenth century Uruguayan painter, is born to a poor family in Montevideo and quits school at the age of eleven to work. After working as a typographer in Montevideo, Blanes moves to Salto, where he paints commissioned portraits and teaches at the

School of Humanities. Returning to Montevideo in 1857, he documents the outbreak of a yellow fever epidemic in the acclaimed painting *Episode of the Yellow Fever*. In 1860, he receives a grant from the Uruguayan government for study abroad and travels to Paris and Florence, where he is influenced by neoclassicism.

Returning to Uruguay after five years, Blanes produces historical paintings and portraits of prominent Latin Americans, including the president of Paraguay. He also paints a series of pictures of *gauchos*. After spending the years 1879–1883 in Florence, Blanes returns to Montevideo where, among other works, he paints a portrait of Uruguayan national hero José Artigas. He also paints *Review of Río Negro by General Roca and His Army* on a commission from the Argentine government. Blanes moves to Pisa, Italy, in 1898 and dies there in 1901.

1830: July 18   Uruguay becomes an independent republic

The constitution of the República Oriental del Uruguay goes into effect, approved by an assembly of elected officials and ratified by both Brazil and Argentina. The new nation is to be a republic with a separation of powers modeled on governments in Europe and North America. The constitution mandates a bicameral legislature, an independent judicial system, and a president elected every four years.

1831   Remaining native population decimated by the military at Salsipuedes

Uruguay's native Charrúa population, which is only about 10,000 by the eighteenth century, is further reduced as the ranks of the *gauchos* (cowboys) are swelled by immigration. The native population declines further due to disease and military attacks, most notably the attack inflicted by soldiers of the newly independent Uruguayan republic at Salsipuedes in 1831. Given the reduction in numbers, together with intermarriage, there are virtually no pureblooded Charrúas left in Uruguay by 1850.

1832   Charles Darwin visits Uruguay

Naturalist Charles Darwin, author of *The Origin of Species,* visits Uruguay. In his journals, he describes the Uruguayan *gauchos* as colorful and courteous but ready to use violence when necessary.

1837   Civil marriage recognized

In a move that foreshadows Uruguay's secularization in the twentieth century, civil marriages become legal.

1830s–1840s   Emergence of the Colorado and Blanco political parties

The two parties that will control Uruguay for most of its existence—and between them hold the majority of political power for most of Uruguay's history—emerge during the first two decades of the nation's history. They grow out of the rivalry between Uruguay's first two elected presidents, Fructuoso Rivera and Manuel Oribe (c. 1796–1857), both heroes of the country's struggle for independence. Rivera, who assumes office in 1830, is designated by the constitution as the nation's first president. Oribe, his elected successor, replaces him in 1835. However, Rivera organizes a revolt, unseats Oribe, and regains the presidency in 1838. The Colorados (Reds) are Rivera's supporters; the Blancos (Whites) are Oribe's. The Blancos traditionally represent rural, conservative interests. They have historically been associated with Argentina, while the Colorados have closer ties to Brazil.

1838: June   Rivera unseats his elected successor, touching off a civil war

With support from France, Uruguay's first president, Fructuoso Rivera, deposes his successor, Manuel Oribe, and assumes office for a second term. This ignites thirteen years of civil war. Oribe goes into exile in Argentina, turning to his ally, Argentine dictator Juan Manuel de Rosas (1793–1877), for aid against Rivera. Rivera declares war on de Rosas and forces his remaining troops out of Uruguay.

1843–52   The Guerra Grande (Great War)

The Great War, part of a lengthy civil war between the Blancos and Colorados, centers on the nine year siege of Montevideo by ousted president Manuel Oribe, aided by the Argentine leader de Rosas. French writer Alexandre Dumas (1803–70) likens Montevideo's plight to the Greek siege of Troy in his book, *Montevideo: A New Troy*. Ultimately, Oribe, de Rosas, and the Blanco party are defeated.

1849   Uruguay's first university is founded

The University of the Republic, Uruguay's only public university, is established in Montevideo. Today it offers programs in engineering, agronomy, medicine, chemistry, and other fields.

1850   Regular stagecoach service begins

Regular stagecoach service is inaugurated. It takes four to five days to cover the distance from Montevideo to San Fructuoso (240 miles/386 kilometers).

1850–70   Sheep farming is introduced

Uruguay's *estancias*, or estates, begin raising sheep as well as cattle. Between 1852 and 1868, Uruguay's sheep flock increases from 0.8 million to 17 million. During the same period, crossbreeding raises the wool yield per head of sheep from fourteen to eighteen ounces to forty ounces. By 1884, wool has replaced hides as Uruguay's principal export. Sheep farming, which requires only one fifth as much land as cattle raising and can be done on lower quality pastures, leads to the creation of a rural middle class.

1850–70 Montevideo is modernized

Uruguay's capital city undergoes expansion and modernization, getting its first gas service in 1853, a bank in 1857, sewers in 1860, a telegraph office in 1866, railroad connections with the rural areas in 1869, and running water in 1871.

1850–1900 Immigration to Uruguay increases

Once the worst of the civil warfare is over, a substantial increase in immigration to Uruguay, mostly from Spain and Italy, occurs. By 1968, over two thirds of Uruguay's population is foreign born. Other immigrant groups include Brazilians, British, and French and Spanish Basques. In the 1870s alone, 100,000 Europeans emigrate to the country.

1852–1863 The failure of political unity

Power passes between the Blanco and Colorado parties in the years that follow the Great War of 1843–52, but factionalism, political rivalries, and uprisings continue to plague the four presidents who hold office in the 1850s and early 1860s. The administration of Bernardo Berro, elected in 1860 and confronted with a rebellion led by Colorado leader Venancio Flores (1809–68) three years later, marks the last time the Blancos control the country until 1958.

1855 Birth of poet Juan Zorrilla de San Martín

Zorrilla de San Martín, Uruguay's most famous nineteenth century poet, is educated in Montevideo and Santa Fe, Uruguay, and Santiago, Chile, where he receives a law degree in 1877. In 1878, he founds a Catholic journal, *El bien público.* He is appointed to the faculty of the National University in Montevideo in 1880. However, he resigns his position five years later and goes into political exile in Argentina because of his opposition to the government of President Máximo Santos (1836–87). There he cooperates with other political exiles in efforts to overthrow Santos. After the downfall of Santos, Zorrilla returns to Uruguay and is elected to the congress in 1886. In 1894, he serves as Uruguayan ambassador to Paris. He dies in 1931.

His masterpiece, *Tabaré,* published in 1905, is an epic poem about life in Uruguay at the beginning of Spanish colonial rule. It describes the country's native Charrúa Indian population, which has mostly disappeared by the middle of the nineteenth century.

1856 Solís Theater is built in Montevideo

Following the introduction of opera in Uruguay in the 1920s, the Teatro Solís, Uruguay's principal theater, is completed. One of South America's largest opera houses, it seats 2,800.

It becomes the venue for performances by Toscanini and Caruso in 1903, and Richard Strauss (1864–1949) conducting his opera *Elektra* in 1923. Today, the Solís still hosts performances by local and international performers. Other opera and concert houses constructed in Uruguay in the nineteenth century include the Teatro San Felipe (1879), the Politeama (1890), the Cibils (1893). The Urquiza, seating 3,000 and later renamed the SODRE, was completed in 1905 and leveled by a fire in 1971.

1856 Birth of statesman and reformer José Batlle y Ordóñez

Batlle is Uruguay's most famous modern political figure. The son of a former Uruguayan president, he serves two terms as president, from 1903 to 1907 and from 1911 to 1915. As a young man he founds the newspaper *El Día* as a platform from which to discuss social issues. During his presidency he institutes many of the social reforms that have made his nation a leader in the area of social welfare. Batlle dies in 1929.

1860s New livestock breeding techniques are introduced

Thanks to new breeding techniques, made possible by the introduction of fencing, Uruguay's sheep flocks expand from three million in 1860 to seventeen million in 1868.

1861 Painter Pedro Figari is born

Figari, one of Uruguay's foremost twentieth century painters, leads a varied and colorful life as a journalist, lawyer, and politician before turning to painting in his late fifties. The first exhibition of his work takes place in Buenos Aires in 1921, when he is sixty. In 1925, Figari moves to Paris for nine years, later returning to Uruguay. He is known for his paintings of Uruguay's people and landscapes, especially his depiction of early nineteenth century scenes. Figari dies in 1938, having produced around 3,000 paintings in his seventeen year career as an artist.

1865 Meat extract plant opens

A meat processing factory is opened at Fray Bentos by a British firm, the Liebig Meat Extract Company. Its products are exported for use by armies in Europe.

1865–68 Flores ousts President Berro and the Colorados retake power

With the aid of both Argentina and Brazil, Colorado leader and former president Venancio Flores (1809–68) launches an armed revolt in 1863 and overthrows the government of Bernardo Berro. Flores rules as a dictator from 1865 until he is assassinated three years later. Ironically, Flores and Berro are assassinated on the same day.

1865–70   The War of the Triple Alliance

In return for Brazilian military aid, Colorado president Venancio Flores is obligated to join Argentina and Brazil in their war against Paraguay's dictator, Francisco Solano López (1827–70). The Triple Alliance is victorious, and Paraguay is defeated.

1870s–1890   Enclosure of livestock ranges

The fencing off of Uruguay's large ranches, or *estancias,* marks a turning point in the country's livestock economy. Fewer workers are needed to tend the herds, and small land-holders are no longer able to farm at the edges of large estates.

1870   First labor organization is formed

The typographers' union in Montevideo becomes the nation's first permanent labor organization.

1871   Formation of the Asociación Rural

Uruguayan landowners form a guild to lobby for their interests. Since the formation of a competing organization, the Federación Rural, the Asociación Rural has become more closely associated with farmers who run small or medium sized operations.

1872   Adoption of coparticipación as a government policy

Recognizing the continuing dominance of the Colorado party, Uruguay's political and military leaders agree to a compromise, coparticipación, that will give the Blancos a share in the government. They are given political control of four of the country's departments (provinces), while the Colorados run the rest of the departments and the federal government. (The Blanco controlled departments are increased to six in 1897.)

1874   Painter Joaquín Torres García is born

García (1874–1949) becomes one of Uruguay's leading twentieth century painters, although he lives outside the country for forty years. At the age of seventeen, he moves to Spain, where he completes early works including murals and the stained glass windows for the cathedral of Palma de Mallorca. He lives briefly in New York, and then moves to Paris, where he is involved in planning the first international abstract art exhibition in 1928. In 1933, Torres García returns to Uruguay, settling in Montevideo, where he later founds a studio, the Taller Torres García, as well as an art school and the Association of Constructivist Art, the school of art with which he is most closely associated. Constructivism is a type of abstract art that incorporates a variety of media such as glass, metal, and wire.

1875   Dramatist Florencio Sánchez is born

Sánchez (1875–1910) becomes Uruguay's most famed dramatist. His realistic plays dramatize the problems of turn of the century Uruguay.

1875   Birth of poet Julio Herrera y Reissig

The works of Herrera y Reissig (1875–1910) are an important part of the flourishing Uruguayan literary output at the turn of the twentieth century, whose creators are collectively referred to as the "Generation of 1900." Born into a distinguished family that includes one of Uruguay's presidents, Reissig has a heart condition that keeps him in fragile health throughout his life, and ends his life at the age of thirty-five. Both his education and his formal employment in civil service jobs are interrupted for health reasons. After 1897 he devotes his life solely to writing, publishing a literary journal, *La Revista*, between 1899 and 1900. Herrera's poetry collections include *Los peregrinos de piedra* (1909), *Los parques abandonados* (published 1919), and *Las lunas de oro* (1924). A proponent of modernism, Reissig wins international acclaim for his poems, which influence the avant garde poetry of the 1920s and 1930s.

1876–1890   Military rule

The continuing weakness of the central government paves the way for military leaders to take power. Uruguay has a succession of four military governments, led by Lorenzo Latorre (1876–80), Franciso Vidal (1880–82), Máximo Santos (1882–86), and Máximo Tajes (1886–90). The incidence of rural uprisings decreases, and civilians begin to resume political control during the administration of Tajes.

1877   Universal education is mandated by law

In the 1870s, social reformer José Pedro Varela works for educational reform. In spite of opposition by the church and the university establishment, he wins the backing of Uruguay's dictator, Lorenzo Latorre, and universal education becomes the law of the land.

1878   First Uruguayan opera is composed

La parisina, the first Uruguayan opera, is written by Tomás Giribaldi (1847–1930).

1879   Government takes over registration of marriages

The marriage registration (Registros de Estado Civil) becomes the province of the state rather than the church, although a religious ceremony commonly precedes the civil one. By 1885, all couples must register, and the civil ceremony precedes the religious one.

1880s–90s   Uruguay leads in technological advances

Montevideo is the site of South America's first electric plant, built in 1886. It also leads the continent in the number of telephones per person in the 1890s. The construction of the nation's first refrigerated meat packing plant in 1904 signals the transition from salted beef to modern meat processing operations.

1882   First soccer club is formed

Soccer has been introduced to Uruguay in the 1870s by British residents. The first soccer club is founded by a professor of English at the University of Montevideo, and more clubs are formed shortly. Regularly scheduled games begin in 1886 at Punta Carretas in Montevideo. Soccer is to become the most popular sport in Uruguay, as in much of South America. Fans of the two major national teams, the Peñarol and the Nacional, carry on a sustained and intense rivalry.

1882: May 18   Birth of composer Eduardo Fabini

Fabini (1882–1951), Uruguay's best known composer, studies violin in Montevideo and later pursues studies in both violin and composition at the Royal Conservatory in Brussels. Upon his return to Uruguay, he embarks on a career that includes both performing and composing. He becomes a major figure in the movement to incorporate nationalistic elements in Uruguayan music, using folk themes and rhythms in his compositions. (Other major composers of this school include Alfonso Broqua and Luis Cluzeau-Mortet.) Fabini's most famous work is *Campo,* a symphonic poem, first performed on April 29, 1922. It is considered the single work most representative of Uruguayan musical nationalism of this period. Famed European composer Richard Strauss (1864–1949) conducts a performance of this work in Buenos Aires in 1923.

1885   Civil marriage is legally required in addition to a religious ceremony

The government requirement that all couples have a civil marriage as well as a church wedding contributes to a decline in the influence of the church.

1886   Reformist newspaper El Día founded

El Día is founded by future president and social reformer José Batlle y Ordóñez (1856-1929). It helps lay the groundwork for the social, political, educational, and labor reforms Batlle will institute in the first two decades of the twentieth century.

1890–97   Civilian government and Blanco rebellions

Political turmoil returns with the reinstatement of civilian government in the 1890s. The governments of Julio Herrera y Obes (1890–94) and Juan Idiarte Borda (1894–97) face two Blanco uprisings led by Aparicio Savaria. Savaria and his followers will carry out one more revolt in 1904, which will be put down by newly elected president José Batlle y Ordóñez, before political stability is finally achieved.

1890s   Uruguay's first trade unions are formed

Uruguay's unions have traditionally had a strong voice in the nation's political life.

1894   Gaucho novel Soledad is published

The novel *Soledad,* about Uruguay's gauchos (cowboys), is published by writer Eduardo Acevedo Díaz (1851–1924). It becomes enormously popular and wins recognition for its author.

1895   Birth of poet Juana de Ibarbourou

Ibarbourou (1895–1979) is one of the more recent members of Uruguay's roster of well known women poets, which also includes early twentieth century poets Delmira Agustini and María Eugenia Vaz Ferreira, as well as Idea Vilariño, a contemporary of Ibarbourou.

Ibarbourou is educated in a convent and in public schools. She marries in 1914 and has a child. Her poems first appear in the publication *La Razón* in Montevideo, and the Argentine magazine *Caras y Caretas* devotes an entire issue to them. Ibarbourou's first collection of poetry, *Las lenguas de diamante* (Tongues of Diamond), is published in 1919. Later collections include *Raíz salvaje* (Wild Root; 1922), *La Rosa de los vientos* (The Rose of the Winds; 1930), and *Perdida* (Lost; 1950). Ibarbourou also publishes a memoir of her childhood, *Chico Carlo* (1944), a children's play (*Los sueños de Natacha*) (1945), and a memoir about her poetic career (*Autobiografía lírica*).

Ibarbourou receives great public acclaim throughout Latin America during the course of her career. In 1929, she is honored in a ceremony presided over by Uruguay's foremost literary figures and attended by representatives of twenty Spanish speaking countries. In 1950, Ibarbourou is named president of the Uruguayan Society of Authors. Seven years later UNESCO (United Nations Educational, Scientific and Cultural Organization) holds a symposium in her honor in Montevideo. However, toward the end of her life, the poet's fame fades, and she dies in poverty.

1896   Bank of Uruguay established

The Banco de la República Oriental del Uruguay (Bank of Uruguay), also known as BROU, is founded as a state enterprise.

1900   Publication of essay "Ariel"

Enrique Rodó (1871–1917) publishes his idealistic and influential essay "Ariel," an analysis of Hispanic culture.

Stability and Reform

1903 Reformer José Batlle y Ordóñez is elected president

Twice elected president (1903–07 and 1911–15), Batlle is Uruguay's most revered politician. He is a member of the Colorado party and a former political journalist who founds his own newspaper, *El Día.* Batlle establishes the nation's social welfare system. Labor reforms enacted during his second term in office include the eight-hour workday, the right to strike, old age pensions, workmen's compensation, and other medical benefits. Batlle also supports state intervention in the economy, and government enterprises become active in industries including utilities, banking, insurance, and transportation. Under Batlle, Uruguay becomes a progressive, modern nation with political stability and a high standard of living. Batlle remains the country's dominant political influence until his death in 1929.

1904: September Final Blanco rebellion defeated

When Batlle assumes the presidency in 1903, his party still does not have full control of the country. In the six departments they control, the Blancos refuse to cooperate with the central government. Civil war breaks out in 1904 and lasts for eight months until the leader of the Blancos, Aparicio Savaria, is killed in the fighting. After hostilities end, the Blancos lose control of their departments. The Colorados emerge with complete control of the country, ending over sixty years of uprisings and civil war.

1905: August 15 Uruguay competes in first South American soccer match

The first international soccer match in South America is held in Buenos Aires between Argentina and Uruguay. The result is a draw, with neither side scoring.

1907–1911 Williman succeeds Batlle in presidency

Claudio Williman, the candidate backed by Batlle, is elected to the presidency, while Batlle, who cannot succeed himself in office, travels to Europe to study political systems abroad. During Williman's administration, a peace treaty is signed with Brazil over a longtime border dispute, and minority parties win increased representation in government. Batlle runs for president again in 1911 and wins.

1907–1919 Batlle's reforms lead to advances for women

In 1907 divorce is legalized, with legal grounds including cruelty by the husband. In 1912 women win the right to sue for divorce without alleging a specific cause. Married women are able to maintain bank accounts separate from their husbands' by 1919.

1909 Birth of acclaimed novelist Juan Carlos Onetti

Onetti (1909–94) is born in Uruguay but lives in Buenos Aires between 1943 and 1955. There he works as a journalist for several publications and for the Reuters News Agency. In 1939, he publishes his first work, the novella *El pozo* (The Pit), which uses interior monologue and depicts an alienated character in an urban environment. Urban alienation will become a prominent theme of his work. Other early books include *Tierra de nadie* (No-man's-land; 1942) and *La vida breve* (A Brief Life; 1950). In the latter, which is probably his best known work, Onetti creates the fictitious city of Santa Maria, which will serve as the setting of later works as well.

Works written after Onetti's return to Montevideo in 1955 include *Para una tumba sin nombre* (A Grave with No Name; 1959); *El astillero* (The Shipyard; 1961) and *Juntacadáveres* (Body Snatcher; 1964), both set in Santa Maria; *La muerte y la niña* (Death and the Little Girl; 1973); and *Cuando ya no importa* (What's the Use?; 1993).

In 1963, Onetti is awarded Uruguay's national literary prize and, in 1980, he wins Spain's Cervantes Prize.

1909 Montevideo port improvements are completed

Montevideo's port facilities are modernized, making the city competitive with Buenos Aires as a center for trade.

1911–1914 Batlle nationalizes key sectors

President Batlle creates public companies to run important sectors of the economy. A savings and loan institution formed in 1911 takes over the printing of money. The State Electric Power Company, created the following year, takes over electric power generation and distribution. In 1914, the railways are nationalized, later coming under the jurisdiction of the State Railways Administration. In addition, the Mortgage Bank of Uruguay is nationalized, and the government creates entities to oversee fisheries and coal, oil, and gas exploration.

1918 New constitution is approved

Uruguay's second constitution is influenced by the policies and reforms of President Batlle y Ordóñez. This document expands civil rights, eliminates restrictions on male suffrage, provides for secret ballot elections and proportional representation, and abolishes the death penalty. Under Batlle's administration, the church has lost much of its political power, including control of education. Its social authority has been eroded by the legalization of divorce and of civil marriage. The 1918 constitution, reflecting these changes, guarantees the separation of church and state.

1920s Compromise and political stability

In the mostly prosperous economic climate of the 1920s, the political stability gained under the Batlle administrations con-

By the early twentieth century, Montevideo becomes one of the leading cities in South America with a population of over half a million. (EPD Photos/CSU Archives)

tinues. The Colorados remain the majority party, but the Blancos have a strong voice in governing the country. In addition, both parties have a variety of factions that are compelled to work out political compromises. Batlle's faction of the Colorados has to make concessions within its own party, weakening, but not halting, the continuation of the former president's reform agenda.

1919–1929   Reforms continue under Batlle's successors

Many of Batlle's planned reforms are implemented by his successors Feliciano Viera (1915–19), Baltasar Brum (1919–23), José Serrato (1923–27), and Juan Campisteguy (1927–31). These include social security for government employees (1919), employee accident compensation (1920), a six-day workweek (1920), a minimum wage for rural workers (1923), and the extension of the social security system to cover all workers, including those in the private sector (1928).

1924   Uruguay wins Olympic soccer medal

At the 1924 Paris Olympics, Uruguay is the first non-European country to become the Olympic soccer champion, defeating Switzerland 3–0 in the final match. In the preliminary matches the Uruguayan team defeats Yugoslavia (7–0), the United States (3–0), France (5–1), and the Netherlands (2–1).

1928   Uruguay wins second Olympic soccer championship

The Uruguayan team captures Olympic gold for the second time in a row in Amsterdam, defeating the Netherlands (2–0), Germany (4–1), Italy (3–2), and, in the finals, Argentina (a 2–1 replay following a 1–1 draw).

1930s   Uruguay's economy suffers during the Great Depression

Like nations throughout the world, Uruguay faces economic setbacks during the global depression of the 1930s. Its generous social welfare program becomes costly for a nation suffering from economic depression, inflation, and declining productivity.

1930   Capitol building centennial is commemorated

A massive celebration marks the centennial anniversary of the construction of Uruguay's capital building in Montevideo.

1930   Uruguay hosts and wins first World Cup soccer competition

In honor of being chosen to host the first World Cup championship, Uruguay plans construction of a new stadium, the Estadio Centenario (which, however, is not completed in time for all the matches). Most European teams balk at traveling to South America to compete, and all except France, Belgium, Yugoslavia, and Romania boycott the games. Uruguay wins the championship after playing only four matches. The most intense competition comes from Uruguay's neighbor and arch rival, Argentina, which it defeats 4–2. Offended by the failure of England and the Central European countries to participate in the games, Uruguay stays away from the 1934 games, the only time a world champion has ever failed to compete in the ensuing World Cup.

1931   Orchestra is established in Montevideo

Montevideo's symphony orchestra is founded with the support of the country's broadcasting network. Its first conductor is Lamberto Baldi.

1933: March   Terra begins period of dictatorial rule

In a *coup d'état,* Gabriel Terra, the conservative leader who has served as the nation's elected president since 1931, disbands the General Assembly and begins to rule by decree.

With the support of conservative factions of both the Colorado and Blanco parties, Terra undertakes a conservative agenda in opposition to the social reforms of the past thirty years. Aiding him is Blanco leader Luis Alberto de Herrera. During Terra's years in power, social and labor legislation is halted and trade unions restricted. Terra is succeeded in 1938 by elected president Alfredo Baldomir. Full civil liberties are restored under Juan José Amézega (1943–47).

1938 Women vote in national elections

For the first time in history, Uruguayan women vote in national elections.

1939–45 World War II

Uruguay declares its neutrality at the outset of World War II (1939–45). However, in 1939, Uruguayans demonstrate their sympathy with the Allied cause by requiring a damaged German battleship, the *Graf Spee,* to leave their waters within the four days mandated by international law. Confronted with three waiting British ships, the Germans are forced on December 17, 1939, to blow up and abandon their vessel and flee to Argentina for safety. As the war continues, Uruguay takes an increasingly pro Allied stance, investigating Nazi sympathizers in 1940 and breaking off relations with Germany and her allies by 1942.

1943–47 Social welfare allowances introduced by Amézega government

During the administration of Colorado president Juan José Amézega, important labor and social welfare measures take effect. In 1943, wage councils are inaugurated, and rural workers are included in the nation's pension system. In the same year, a system of family allowances based on the number of dependent children is first introduced (it is reformed in 1950). The rights of rural workers are established in the 1946 Rural Worker Statute, which protects men and women equally.

1947–51 The return of Batllism

The liberal agenda of President José Batlle y Ordóñez is resumed by his nephew, Luis Batlle Berres, who, as elected vice president, takes over the presidency when Tomás Berreta dies after only six months in office. The social security system is expanded and, in 1948, the state takes over the British owned railroads.

1948 National Land Settlement Institute created

The National Land Settlement Institute is formed to oversee agrarian reform, including redistribution of agricultural land. However, its activities are hampered by a lack of funding.

1949 First hydroelectric plant completed

The construction of Uruguay's first hydroelectric plant, on the Río Negro river, creates South America's largest artificial lake, the Lago Artificial del Río Negro.

1950s Emergence of LFAR, nonpartisan agricultural league

The Liga Federal de Acción Rural (LFAR), or the Federal League for Rural Action, is a nonpartisan agricultural movement that disrupts the traditional ties between farmers and the Blancos. LFAR switches its allegiance between parties in different elections.

1950s Era of economic instability begins

With declines in both livestock and industrial production, the mid-1950s usher in a twenty year period of economic stagnation that contributes to the social unrest of the 1960s and, ultimately, to the advent of military government in the 1970s and 1980s. Exports decline, inflation increases, and a negative balance of payments leads to a decline in the nation's financial reserves.

1950 Uruguay wins its second World Cup championship

In 1950, Uruguay plays in the World Cup for only the third time, having declined to play in the 1934 and 1938 games. Playing in Brazil, the Uruguayan team defeats the favored Brazilian home team 2–1 in a game that sets an unsurpassed world attendance record of 199,860.

1951 Voters approve new constitution

Uruguayan voters approve a constitution replacing the president with a nine member National Council of Government (the Colegiado) composed of six members from the majority party and three from the opposition. Presidential authority rotates among its members on a yearly basis.

1952 First official celebration of Carnaval

Performing groups have held informal Carnaval celebrations since around 1930. Representing five categories of performance, the groups join together to found an organization to oversee Carnaval in Uruguay. The organization, Directores Asociados de Espectaculos Carnavaleros y Populares del Uruguay (DAECPU), conducts competitions and awards prizes (mostly monetary) to performing groups. The performing groups fall into five categories: *parodistas,* singers and dancers who satirize classic plays; *humoristas,* comic actors and singers who perform original material; *revistas,* musical groups led by glamorous female singers; *murgas,* Spanish-influenced groups of white vocalists with percussion accompaniment; and *lubolos,* percussionists of African descent. (Lubolo means "whites with black painted faces," a reference to white Uruguayans who played African *candombe,* a style of percussion music. The music performed by murgas and lubolos is recognized as the typical music of Carnaval.)

1958 Blancos win national elections

For the first time in ninety-three years, the Blancos win a majority of votes in a national election. Ironically, longtime Blanco leader Luis Alberto de Herrera (d. 1959) dies just as his party has finally come to power. Deprived of his leadership, their rule is weakened by factionalism, which increases after the party's second victory in the 1962 election. Altogether, there are eight different Blanco governments between 1959 and 1967.

1960s Emergence of the Tupamaro terrorist group

An economic decline beginning in the 1950s leads to the rise of the Movement of National Liberation, popularly known as the Tupamaros, after an eighteenth century Peruvian Inca leader, Tupac Amarú (c. 1544–71). A socialist urban guerrilla

group, they recruit students, professionals, unionized workers, and other members of the middle class.

The group acquires a sort of "Robin Hood" image by exposing government corruption, robbing banks, and distributing food in poor neighborhoods. Eventually, however, their activities turn more violent and include bombings, kidnappings, and assassinations, including the killing of U.S. government official Dan A. Mifrione. In 1971, they kidnap Great Britain's ambassador to Uruguay, Geoffrey Jackson, and hold him hostage to ensure that scheduled elections will take place. When they stage a one-day assassination of nearly a dozen police and military officials the following year, the government declares all-out war on the group, which is decimated over a period of several months. Their activities are commonly viewed as one of the factors that paves the way for the establishment of a military dictatorship in 1973, even though the group has been destroyed by that time.

1960 Uruguay signs letter of intent with the IMF

Under the Blanco government, Uruguay signs its first letter of intent with the International Monetary Fund (IMF). The letter of intent implies that Uruguay will work more closely towards a mutual relationship with the IMF, and will create a formal relationship with the IMF at a future date. This event is a step towards improving the Uruguayan economy, which has been declining since the early 1950s.

1964 Labor federation is formed by Uruguay's unions

The nation's labor unions cooperate in establishing the Convención Nacional de Trabajadores (National Convention of Workers), or CNT. The CNT is banned when the military takes over the government in 1973, after calling a general strike for which many of its leaders are imprisoned or exiled. There is a virtual halt to all labor activity during most of the twelve-year period of military government. However, another labor federation is formed in 1983. Now known as the Inter-Union Workers Assembly-National Convention of Workers (PIT-CNT), it has a membership of 140,000 in 1993.

1966 The Colorados are returned to power

National elections return a right wing faction of the Colorado party to power. Uruguayans also vote to dismantle the Colegiado, the nine member executive council, and restore the presidency. The Colorados place General Oscar D. Gestido (d. 1967) in office as president.

1967–1971 Government repression under Pacheco

After Gestido's death in 1967, the presidency is assumed by his vice president, Jorge Pacheco Areco, who implements wage and price controls. In response to labor disputes, he institutes repressive measures, declaring states of emergency in 1968 and 1969 and suspending constitutional freedoms. Student demonstrators are shot, and suspected government opponents are jailed. Pacheco defends many of his policies on the grounds that they are necessary to subdue the Tupamaro terrorist movement.

1971: September 9   Tupamaro prisoners escape from jail

Over a hundred Tupamaros escape from prison. President Pacheco enlists the aid of the military to thwart terrorist activity.

1971   Juan Bordaberry is elected president and moves against the Tupamaros

With Uruguay's civilian government making little headway against the Tupamaros, Bordaberry brings the military into the campaign against the terrorists. By 1972, the Tupamaro movement has been crushed.

Military Rule

1973–1985   Bordaberry ushers in twelve years of military rule

Although the goal of putting down the Tupamaro resistance had been met by 1973, Bordaberry continues to expand the role of the military. Congress is dissolved and replaced by a Council of State which rules by the decree of appointed, rather than elected, members. Political restrictions are placed on students, trade unionists, and the press, and members of left wing groups are arrested. By 1976, Uruguay's military regime is considered one of the harshest in Latin America. In that year, Bordaberry resigns, and is succeeded by interim president Alberto Demichelli, followed by Aparicio Méndez (1976–81) and General Gregorio Alvarez (1981–84). In 1977, the military announces a timetable leading to a return to civilian government by 1985.

1980: November 30   Military government's proposed constitution is defeated

Uruguay's military rulers hold a special election for approval of a proposed constitution that would legitimize their control of the government. Among other measures, it calls for creation of a National Security Council under the control of the military that could overturn decisions of the executive branch. Although opposition groups are not allowed to campaign against it, and in spite of an intensive campaign in favor of it by the pro-government media, Uruguayans defeat the new constitution by a vote of fifty-eight to forty-two percent. This defeat signals a turning point that leads to the restoration of democratic rule by 1985.

1984   The Catholic University of Uruguay is founded

Uruguay's first private university is established in Montevideo.

Return to Civilian Government

1985: March   Civilian government is restored; Sanguinetti elected president

In November 1984, Uruguay holds its first national elections in thirteen years. Moderate Colorado leader Julio Sanguinetti (b. 1936) is elected to a five-year term as president and takes office in March 1985. A newly elected legislature is also installed. Sanguinetti, a strong proponent of national reconciliation, releases all political prisoners and backs a policy of amnesty toward former military officials accused of human rights violations. However, Uruguay continues to be plagued by economic problems.

1988   Passenger rail service is discontinued

Because of declining ridership, passenger service is discontinued on Uruguay's government operated rail lines as part of a five-year program to downsize the railroad system.

1988–89   Drought cripples northern Uruguay

A severe drought in the north and northeast causes widespread agricultural damage. Crop growing is badly disrupted, and about 650,000 head of cattle have to be killed in the first four months of 1989. The area experiences shortages of food, electric power, and other necessities.

1989: April   Amnesty law is upheld in special election

In one of only four referendum elections held since World War II, Uruguayans vote on whether to overturn a controversial 1986 law granting amnesty to officials of the former military government. In order to have the election, opponents of the law gather 635,000 signatures, which are subjected to a lengthy verification process. Those opposing the law include labor unions, leftists, and church and human-rights groups. In the voting, the law is upheld, fifty-seven percent to forty-three percent. However, the divisiveness created by the referendum leads to the Sanguinetti government's defeat in the national elections held in November of the same year.

1990   Blancos assume power after national elections

The faction of the Blanco party led by Luis Alberto Lacalle wins the elections held in the fall of 1989 and takes over control of the government the following spring. For the first time since 1971, political power passes from one civilian government to another. Another historic feature of the 1989 election is the impressive showing by a third party: the leftist group Frente Amplio (Broad Front), which has strong results nationally and wins control of Montevideo. This is the first time in Uruguay's history that a third party poses a serious threat to the Blancos and Colorados.

1991   MERCOSUR trade alliance forms

The Southern Common Market, or Mercado Común del Sur (MERCOSUR) is created by Argentina, Brazil, Uruguay, and Paraguay to establish a common market in a region accounting for over half of South America's economic activity. By 1995, almost all tariffs between member nations have been abolished.

1995: March 1   Sanguinetti returned to office in Uruguay's closest national election

Julio Sanguinetti and the Colorado party are returned to office in the 1994 elections, the closest in the nation's history. The Colorados defeat both the Blancos and the Frente Amplio leftist coalition by a narrow margin. The Frente Amplio comes close to capturing the presidency. After regaining office, Sanguinetti embarks on an ambitious economic reform program.

1995   Uruguay exports meat to United States

Uruguay, along with neighboring Argentina, have been granted quotas to begin exporting fresh and frozen meat products to the United States. Both countries must meet U.S. Department of Agriculture standards for certification that they have eliminated hoof-and-mouth disease, the reason they have been prevented from exporting meat to the United States since 1930.

1996   Securities Market Law enacted

A law regulating and promoting the growth of the securities market, the Securities Market Law, is enacted. It establishes a framework for regulation of the market, and is designed to attract private investment in Uruguayan companies.

1999: April   Primary elections

The ruling Colorado Party and the Progressive Encounter Party emerge from the April primary election with strong candidates for president. (See 1999: October.)

1999: October   Elections scheduled

Former Uruguay President Luis Alberto Lacalle of the Blanco Party plans to run for president. His platform is based on a plan to improve trade relations with Paraguay, Argentina, and Brazil. Lacalle was formerly president from 1990–95.

Bibliography

Davis, William Columbus. *Warnings from the Far South: Democracy versus Dictatorship in Uruguay, Argentina, and Chile.* Westport, Conn.: Praeger, 1995.

Finch, M. H. J. *A Political Economy of Uruguay Since 1870.* New York : St. Martin's Press, 1981.

Gilio, Maria Esther. Translated by Anne Edmondson. *The Tupamaro Guerrillas.* New York: Saturday Review Press, 1972.

Gonzalez, Luis E. *Political Structures and Democracy in Uruguay.* South Bend, Ind.: University of Notre Dame Press, 1991.

Hudson, Rex A., and Sandra Meditz. *Uruguay: A Country Study.* Washington, DC : Federal Research Division, Library of Congress, 1992.

Kaufman, Edy. *Uruguay in Transition: From Civilian to Military Rule.* New Brunswick, NJ: Transaction Books, 1979.

Peck, Marie Johnston. *Mythologizing Uruguayan Reality in the Works of Josi Pedro Dmaz.* Tempe,: Center for Latin American Studies, Arizona State University, 1985.

Stone, Kenton V. *Utopia Undone: The Fall of Uruguay in the Novels of Carlos Martinez Moreno.* Lewisburg, Penn.: Bucknell University Press, 1994.

Vanger, Milton I. *The Model Country: Jose Batlle y Ordonez of Uruguay.* Hanover, NH: University Press of New England, 1980.

"Watch it, that's us you're trampling on." *The Economist,* November 22, 1997, vol. 344, no. 8044, p. 36.

Weinstein, Martin. *Uruguay, Democracy at the Crossroads. Nations of Contemporary Latin America.* Boulder, Colo.: Westview Press, 1988.

———. *Uruguay: The Politics of Failure.* Westport, Conn.: Greenwood Press, 1975.

Weschler, Lawrence. *A Miracle, A Universe: Settling Accounts with Torturers.* New York: Pantheon Books, 1990.

Zlotchew, Clark M., and Paul David Seldis, ed. *Voices of the River Plate: Interviews with Writers of Argentina and Uruguay.* San Bernadino, Calif.: Borgo Press, 1993.

Venezuela

Introduction

Located at the northernmost edge of South America, Venezuela was the first part of the continent to be explored by Europeans and the site of the first Spanish settlement. With an area of 352,145 square miles (912,050 square kilometers), it is the continent's fourth-largest country (after Brazil, Argentina, and Colombia) and has coastlines on both the Atlantic Ocean and the Caribbean Sea. Venezuela also claims some 58,000 square miles (150,000 square kilometers) of disputed territory currently controlled by Guyana. Venezuela's diverse terrain can be divided into four main areas: the Andean highlands to the west; the centrally located river plains, or *llano;* the jungles and Guayana highlands to the south; and the coastal zone. In addition, Venezuela includes over 100 Caribbean islands, the largest of which is Margarita. Venezuela had an estimated 1997 population of 21.9 million, of which more than one-sixth lived in the capital city of Caracas.

Initially a focus of the Spanish obsession with the "earthly paradise" of El Dorado, Venezuela has yielded a variety of commodities valuable on world export markets, including coffee, chocolate, and, most recently and dramatically, petroleum, which has dominated the nation's economy since the 1930s. The discovery of oil turned Venezuela from a poor agrarian country into a modern industrial nation with Latin America's highest per capita income. Venezuela currently has the largest known oil reserves outside of the Middle East and Russia and is among the world's major oil producers. It also has significant deposits of high grade iron ore.

Spanish Colonial Rule

Present-day Venezuela was sighted by Christopher Columbus on his third voyage to the New World in 1498. Other explorers followed, and the first permanent settlement was founded in 1523 at Cumaná. With its long coastlines and active trade, Venezuela was a notorious center for piracy and smuggling from the sixteenth to the eighteenth centuries. In 300 years of Spanish colonial rule, various parts of modern day Venezuela belonged to the Audiencia of Santo Domingo, the Viceroyalty of Peru, and the Viceroyalty of New Granada. It was finally united under the Captaincy General of Venezuela in 1777.

Venezuelans initially revolted against their Spanish governor in 1810 and declared independence from Spain in 1811. However, ten years of armed struggle followed before the region was finally freed from Spanish control. This struggle was led by South America's great liberator Simón Bolívar (1783–1830), along with Francisco Miranda. By the time the Spanish were decisively defeated in 1821, Venezuela had joined with New Granada (present-day Colombia) to form the Republic of Gran Colombia. In 1830, Venezuela seceded from this union to form its own republic.

The Struggle for Independence

Like other new South American nations, Venezuela's nineteenth century history was full of feuding and revolts by rival political factions. Until 1859 there were extended periods of political control by the country's two major parties, the conservatives and liberals. In the Federalist Wars (1859–63) tensions erupted between those who favored a strong central government (the Centralists) and those who wanted the individual states to retain more power (the Federalists). The Federalists won the conflict, but ruled only from 1864 to 1868. In 1870, General Antonio Guzmán Blanco (1829–99), a former vice president, seized power, beginning one of the two periods in Venezuela's history when the country was dominated politically by one strong leader for an extended period of time. Blanco ruled as a dictator, but the country made social and economic progress during his regime.

Blanco's resignation from office in 1888 was followed by a period of political instability that lasted until the beginning of Venezuela's next long-term dictatorship, that of General Juan Vicente Gómez (1864–1935), who seized power in a military coup in 1908 and retained it until 1935. For some of this time, he was president; during some periods, other men served as president, but they were under Blanco's control. It was during this era that Venezuela's oil boom began: by 1929, the nation was the world's largest oil exporter. Gómez's regime was only the beginning of a much longer period of nearly exclusive military control that was to last until 1959. However, some of the military leaders who followed Gómez, such as Eleazar López Contreras and Isaías Medina Angarita (1897–1953), ruled in a more liberal manner. A civilian-military *junta* (a group of conrolling authority) ruled from 1945 to

1948, and in 1948 the country's most prominent writer, Rómulo Gallegos (1884–1969), was chosen as president by direct popular election. However, he was deposed by the military after less than one year in office. The final dictator in Venezuela's long era of military rule was Pérez Jiménez, who held power from 1952 to 1958. His repressive regime was known for its large public works projects and widespread government corruption.

A Country for the People

With the 1959 election of Rómulo Betancourt (1908–81) to the presidency, civilian rule returned to Venezuela. Since that time, power has been transferred peacefully from one elected government to another, and from one party to another—the longest period of democratic rule in a South American coun-

try during this period. The major political parties have been Accíon Democratica (AD) and the Christian Democratic Party (COPEI), whose first successful presidential candidate was Rafael Caldera (b. 1916) who served from 1969–74. Since the late 1970s, Venezuela's economy has faltered with declining world oil prices, reducing popular support for the governments that have tried, with limited success, to cope with mounting inflation and foreign debt levels. By 1983, Venezuela, which at one time had paid off its entire foreign debt, was unable to pay the interest on its current debt, and the government was obliged to impose foreign exchange controls and price freezes. As the 1980s progressed, Venezuela faced a new problem—involvement in the Colombian drug trade. In 1988, the country's minister of justice resigned over allegations of connections with drug traffickers, and Venezuela implemented a joint plan with Colombia to monitor their shared border for drug activity.

Under President Jaime Lusinchi (b. 1924), the nation arranged to meet its debt payments, but instituted austere measures that prolonged the recession and brought inflation to twenty percent per year. The high cost of living and chronic shortages of consumer items created widespread public discontent. In February, 1989, news of gasoline price hikes triggered four days of rioting and looting, which had to be halted by the military, and resulted in numerous injuries and deaths. The official death toll was 277, but the actual toll was rumored to be considerably higher. Renewed rioting broke out in 1990, leading to two failed coup attempts against President Carlos Andrés Pérez (b. 1922) in 1992. Discontent with Pérez's government was so severe that he resigned in 1993 before the end of his term and an interim president took over. Former president Rafael Caldera returned to office in 1994 and took strong measures to maintain law and order, including the suspension of constitutional rights, which drew criticism from international human rights groups.

Timeline

20,000–10,000 B.C.   First inhabitants arrive in South America from Asia

Venezuela's earliest inhabitants are thought to have crossed the Bering Strait from Asia, migrated southward through North America, and toward the milder climate of South America.

European Discovery and Colonization

A.D. 15th century   Native groups inhabit present-day Venezuela

Unlike Peru to the south or Mexico to the north, the land that will become Venezuela is not home to an advanced native civilization, such as the Incas or Aztecs, at the time of the European colonization. There are semi-nomadic hunter gatherers living in the river valleys and coastal areas. The men hunt or fish and wage warfare against neighboring communities, while the women gather wild plants. The most advanced groups are settled farmers in the Andean highlands who grow corn and other crops and domesticate animals such as dogs and birds. (They do not, however, raise fowl or livestock.)

Indian groups living in present day Venezuela include the warlike Caribs and the Arawaks, who live in small farming villages. Groups that do not survive European exploration and conquest include the Goyones, Mixtecas, and Taironas. Those who manage to maintain a presence in the region include the Guajiros and the Piaroas.

1498: August   Christopher Columbus sights Venezuela

On his third voyage to the Americas, the Italian explorer Columbus (c. 1446–1506) enters the Gulf of Paria and sights modern-day Venezuela. Sailing up and down the coast, he and his men encounter a small group of Carib Indians and note the gold and pearl ornaments they wear. Columbus draws a map of the area for his patrons, King Ferdinand and Queen Isabella of Spain. He is soon followed by other voyagers to the region.

1499   Alonas de Ojeda and Amerigo Vespucci explore the coast

Soon after Columbus's voyage, the Spanish explorer Alonas de Ojeda (c. 1465–1515), along with the Italian Amerigo Vespucci (1451–1512), voyage to the region and sails along Venezuela's northern coast, from the Orinoco delta to the Gulf of Venezuela, also entering Lake Maracaibo. Ojeda records his fascination at the dwellings of the Indians in these coastal areas—huts raised on stilts, with thatched roofs. Comparing these water borne dwellings, perhaps ironically, to the canals of Venice, Italy, Vespucci names the region "Venezuela" (Little Venice).

1523   First permanent settlement founded at Cumaná

By 1520, some Spaniards have already attempted to settle on the small island of Cubagua near the Paria Peninsula, which is known for its pearls. Although some three hundred houses are constructed, the rocky island, with no source of freshwater, is not suitable for long term habitation, and Cumaná, established on the mainland in 1523, is generally regarded as the first permanent European settlement in the region. Coro is founded four years later.

1527   Coro becomes Venezuela's first provincial capital

Coro is the colonial provincial capital of the Province of Venezuela from 1527 to 1546. It is followed by El Tocuyo (1547–77) and Caracas (1577–present).

1528–46   Charles V grants development rights to German traders

In return for financial and political aid in Europe, Charles V (1500–58, r. 1516–56), king of Spain and emporer of the Holy Roman Empire, gives the German House of Welser the right to explore and develop the western portion of the recently discovered territory. The arrangement proves disastrous. Instead of founding settlements, the Welsers are mostly obsessed with looking for gold. They exploit the Indians and anger the Spaniards in the region. The last commander of the region to be appointed by the Welsers is beheaded by Spanish settlers in 1541. In 1546, the Spanish crown ends the agreement and takes control of the region.

Bartolomé de Las Casas (1474–1566) accompanies Christopher Columbus on his third voyage to the New World in 1498, and later serves as an advocate for non-violence toward natives.
(Library of Congress)

1527–1600 Spanish settlements established, including Caracas

Beginning with the founding of Coro in 1527, the Spanish proceed to found a series of settlements. These include El Tocuyo (1545), Barquisimeto (1552), Valencia (1556), Mérida, Trujillo (1558), and Maracaibo (1567), and the future capital city of Caracas, which is founded in 1567 by Diego de Losada. By the turn of the seventeenth century, there are more than twenty Spanish settlements, both in the Caribbean coastal area and in the Andean highlands.

1530s Bartolomé de Las Casas calls for better treatment for Indians

Bartolomé de Las Casas (1474–1566), a Dominican friar, advocates a tolerant attitude toward Indians in Spanish South America. Appalled by the violent treatment by the Spanish colonial administration, Las Casas petitions for non-violence. He is opposed by those who view the natives as inferior to the Spaniards.

1546 Venezuela divided among colonial jurisdictions

The eastern coastal lands are placed under the control of the audiencia (colonial government) of Santo Domingo, while the western and southern areas are made part of the viceroyalty of Peru.

1561 Killing spree by renegade conquistador Aguirre

Lope de Aguirre (c. 1508–61), a Spanish conquistador from Peru who has gone mad on a voyage up the Amazon, leads a band of men on a rampage through the coastal mainland of Venezuela and to Margarita Island off the coast, killing people from the local populace and his own men as well. He is ultimately killed by Spanish troops. Aguirre's story becomes the basis of a critically acclaimed 1972 film by German director Werner Herzog, *Aguirre, the Wrath of God.*

1580 Smallpox epidemic decimates Indian population

An outbreak of smallpox seriously reduces the size of the Indian population in the Caracas area, possibly by as much as two thirds (from 30,000 to 10,000). Unlike the Indians who have no prior exposure to it, the sturdier immune systems of the Spanish make them much better prepared to ward off the disease. By reducing the Indian population, the epidemic furthers European development of the region.

15th–18th centuries Pirates and smugglers active off the coast of Venezuela

With its long Caribbean coast, Venezuela has become a center for smuggling and piracy by the fifteenth century. English, Dutch, and French pirate ships roam the sea off Venezuela's coast, attacking trading ships. Sometimes they land and attack the towns on the shore. In 1595, the pirate Amias Preston leads 500 men in a raid on Caracas, which leaves the city virtually in ruins.

1620s Cocoa becomes mainstay of Venezuelan economy

Cocoa, grown in coastal valleys, becomes Venezuela's principal export (which it will remain for 200 years). The new cocoa wealth attracts increased Spanish immigration, as well as a need for African slaves to provide the labor to produce the cocoa. Foreign trade flourishes as slaves are traded for cocoa, which is then transported across the Caribbean to Mexico (then known as New Spain).

Venezuelan coffee exports to Mexico grow from 6,960 pounds (3,157 kilograms) in 1622 to 133,400 pounds (60,510 kilograms) by 1650, and to 748,200 pounds (339,383 kilograms) by 1700. The coffee growing area expands to the valleys of the Tuy River and its branches.

1641 Caracas is leveled by earthquake

The city of Caracas is destroyed in a major earthquake, which kills some five hundred inhabitants.

1650–1700 Capuchin missionary activity grows

The Capuchin Franciscans are one of several missionary orders to pursue converts in Venezuela during the colonial period. Others include the Dominicans, Augustinians, and Jesuits. The Franciscans play a large part in the development of Caracas, especially in the area of education. For the Capuchins, Venezuela forms part of a larger network of missionary activity that extends from the Andes Mountains to Paraguay. In 1650, they establish a center south of Cumaná in the Venezuelan interior, and are active in the *llanos* (plains) region between 1650 and 1700. All together, the Capuchins found over 150 settlements in Venezuela.

18th century Life of musician Padre Pedro Palacios y Sojo

This Venezuelan priest founds a music academy, the Chacao Conservatory, in Caracas.

1717 Venezuela comes under jurisdiction of New Granada

Venezuela's western and southern provinces are made part of the newly established Viceroyalty of New Granada.

1721 Central University founded

Venezuela's oldest university, Central University, is established. Other early institutions of learning include the University of Venezuela, founded in 1725, and Universidad de Los Andes in Mérida, established in 1785.

1728–84 Trade controlled by the Caracas Company

A company called the Real Compañia Guipuzcoana de Caracas (also known as the Caracas Company) is formed by Basques in Spain to control trade between Spain and Venezuela, thus eliminating illegal intercolonial trade between Venezuela and Mexico. The Spanish government grants it exclusive trading rights in Venezuela. The Caracas Company is also charged with the tasks of protecting the coast from military and pirate attacks, keeping smuggling under control, and providing slaves for cacao production and other purposes. Due to a variety of problems, Spain ends its agreement with the company in 1784.

1738 Birth of liberator, statesman, and author Simón Bolívar

Renowned as South America's great liberator, Bolívar (1783–1830), who is born to a wealthy Creole family in Caracas, is also counted among Venezuela's great literary figures. He is known for the brilliant writing on international politics and other topics found in his thousands of letters. He also writes Bolivia's first constitution and authors numerous decrees and speeches.

1749 Hundreds protest the Caracas Company's activities

Some eight hundred men, led by an immigrant cocoa grower, Juan Francisco de Léon, march on Caracas to protest the Caracas Company's operating methods, as well as alleged cor-

ruption within the company. Spanish troops put a quick stop to the insurrection and brutally crush its leadership.

1777 Venezuela united under Captaincy General

Divided administratively between the Audiencia of Santo Domingo and the Viceroyalty of New Granada, Venezuela is reunited in 1777 with the establishment by Spain of the Captaincy General of Venezuela.

1778 Birth of artist Juan Lovera

Lovera (1778–1841) is the first Venezuelan artist to use nationalistic, rather than purely religious, themes in his paintings. His two best known works, *10 April 1810* and *5 July 1811*, are displayed at the Capilla de Santa Rosa de Lima, in Caracas.

1781 Birth of poet, journalist, and scholar Andrés Bello

Bello (1781–1865) becomes one of South America's most renowned nineteenth century intellectual figures. A one-time tutor to famed liberator Simón Bolívar, he travels to London with Bolívar in 1810 on a diplomatic mission and stays on, due to political developments in Venezuela. Bello lives in London for eighteen years, working as a journalist for Spanish language newspapers and as a diplomat for the foreign legations of South American countries. Two poems he writes during this period ("Alocución a la poesía" and "A la agricultura de la zona tórrida") are considered milestones of Latin American literature.

In 1929, at the invitation of the Chilean government, Bello moves to Chile, where he will spend the rest of his life. There he edits a government newspaper and produces important works in a wide variety of fields, including linguistics, philosophy, science, and law. These include Chile's Civic Code (1855) and *Gramática de la lengua castellana*, (1847) an acclaimed book on the Spanish language.

1784 Venezuela's first theater is founded

The Teatro del Conde, Venezuela's first theater, is established in Caracas.

The Long Struggle for Independence

1795 Early uprisings

In 1795 there is an unsuccessful revolt against the Spanish by slaves and free workers near the city of Coro. It is harshly suppressed by the government. This revolt is an early uprising, and revolts like this will culminate in the 1890s and 1900s, when the disenfranchised classes will stage minor revolts against wealthy landowners.

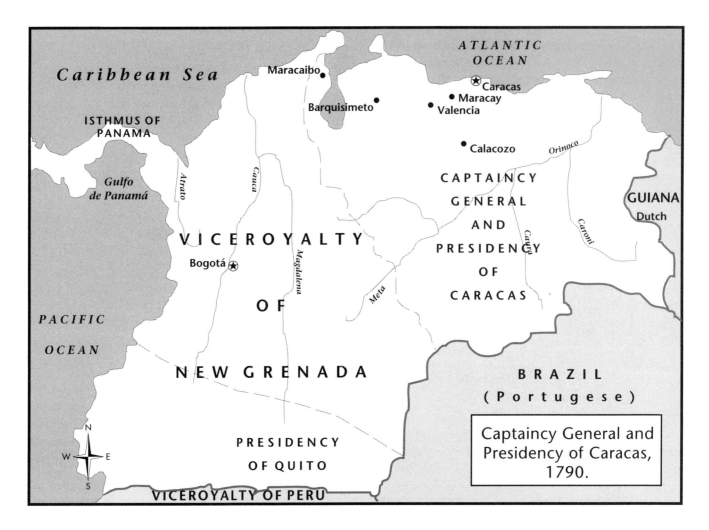

Caribbean Sea

Maracaibo

ATLANTIC
OCEAN

Caracas

Maracay
Valencia

Barquisimeto

ISTHMUS OF
PANAMA

Calacozo

Orinoco

Gulfo
de Panamá

CAPTAINCY

GENERAL

GUIANA
Dutch

AND

PRESIDENCY

OF

CARACAS

Atrato

Cauca

Magdalena

Cayá

Caroni

VICEROYALTY

Bogotá

PACIFIC

OCEAN

Meta

N

OF

W   E

S

NEW GRENADA

BRAZIL
(Portugese)

Captaincy General and
Presidency of Caracas,
1790.

PRESIDENCY

OF QUITO

VICEROYALTY OF PERU

1808   First printing press brought to Caracas

The arrival of the first printing press, which has a major effect on Venezuelan culture, is credited to revolutionary hero Francisco Miranda (c. 1750–1816). Within a year, he begins publishing a newspaper, the *La Gaceta de Caracas* (The Caracas Gazette). Its writers advocate independence from Spain, and the paper discusses current politics and political theories. Its first editor is the distinguished author and poet Andrés Bello (c. 1780–1865).

1810   First Venezuelan book published

The work that many consider to be the first Venezuelan book is the *Manual, Calendar, and Universal Guide for Foreigners* by Andrés Bello. It includes an authoritative account of Venezuelan history.

1810: April 19   Caracas ousts Spanish governor

Like other Spanish colonists in South America, the Venezuelan Creoles (native born descendants of Spanish settlers) begin thinking about independence at the beginning of the nineteenth century. They are inspired by the American and French revolutions, and think they see their opportunity when

the Spanish king is overthrown by Napoleon Bonaparte (1769–1821) in 1808, and Bonaparte's brother is installed in his place. In 1810, the *cabildo,* or town council, of Caracas votes to reject the governor who serves the new French controlled Spanish government, although they initially remain faithful to the deposed king, Ferdinand VII (1784–1833). In response, Venezuela is blockaded by the Spanish, and a decade of armed struggle begins.

1811: July 5   Independence declared

Withdrawing their former support for the deposed king, Ferdinand VII, the Venezuelan Creoles declare full independence from Spain. However, in July of 1812, Venezuela is attacked by 300 Spanish troops and recaptured. In addition, a force of nature works against the independence movement: Venezuelan priests claim that the catastrophic earthquake of March 1812 (see 1812: March 12) is a warning from God about rebelling against established political authority. For years, control of the region alternates between the Spanish and the rebel forces, led first by Francisco Miranda (1750–1816), then by the great liberator of South America, Simón Bolívar.

1812: March 12   Worst earthquake in Venezuela's history rocks Caracas

On Holy Thursday, 1812, the city of Caracas and its environs are hit by the worst earthquake ever to occur in Venezuela. Ninety percent of Caracas is destroyed and 20,000 people perish, either in the earthquake itself or in its aftermath. In addition to other destruction, the quake crushes two churches full of worshippers, all of whom were killed, and swallows up an entire army barracks and all its occupants. Ten thousand people are killed in Caracas, and some two thousand survivors are pulled out of the rubble over the following days. The dead are cremated in huge funeral pyres in an attempt to prevent the spread of disease from rotting corpses. Five thousand are killed in the communities surrounding Caracas. Another 5,000 in the region die of illness from pollution of the water supply.

Because of its timing at the beginning of the movement for independence from Spain, the Caracas earthquake of 1812 is of great historical significance. Many believe that the disaster is a sign that God disapproves of the rebellion against Spain, and consequently the rebellion suffers a major setback. Many troops in the revolutionary forces desert, and the royalists are able to win back control of the region.

1818   Birth of noted musician Felipe Larrazábal

Larrazábal (1818–73) is a leader in establishing a classical music tradition in Venezuela. Larrazábal, a pianist and composer, is also the founder of the Caracas Conservatory.

1821: June 24   Spanish forces defeated at Carabobo

With the decisive Spanish defeat at the Battle of Carabobo, Venezuelan independence is assured. In December 1819 the region had joined with New Granada (now Colombia) in the Republic of Gran Colombia (a federation that is joined by Ecuador in 1922). Liberator Simón Bolívar is declared president. The Republic of Gran Colombia is recognized by the United States in 1822.

1828   Birth of artist Martín Tovar y Tovar

One of Venezuela's more important painters, Tovar y Tovar (1828–1902) is known for his portraits of Venezuelan leaders and his depictions of battle scenes. His best known works are those displayed in the capitol building in Caracas.

The Republic of Venezuela

1830   Venezuela separates from Colombia and forms a republic

After 1821, tensions develop between Venezuela and New Granada, eventually breaking out into open civil warfare. Finally, Venezuela secedes from the Republic of Gran Colombia. An independent Republic of Venezuela is born, under the leadership of revolutionary hero José Antonio Páez (1790–1873), who becomes its first president and remains influential in the nation's political life until 1846.

1830–46   Conservative rule

Under Páez's leadership, political power is wielded by the Conservative Party, which supports the interests of the wealthy landowners and traders. They are firm in maintaining law and order, take financial steps to improve the economy, retain the institution of slavery, and reduce the influence of the church. During this period Venezuela's cacao and coffee industries prosper. However, Venezuela does lose some disputed lands to its neighbor, British Guiana, in 1840.

Aside from Páez, Venezuela's other presidents during this period include José María Vargas (1835–36), Andrés Narvarte (1836–37), and Carlos Soublette (1837–39).

1833   National Library founded

The National Library is the largest of Venezuela's 103 libraries, with over five million volumes. It receives a large private bequest in 1958—100,000 rare books from the estate of Pedro Manuel Arcaya.

1845: March 30   Spain recognizes Venezuelan independence

Spain finally recognizes Venezuela as an independent nation in the Treaty of Madrid. Venezuela had asserted its independence over thirty years earlier. (See 1811: July 5.)

1847   Liberals take over under leadership of Monagas

With the backing of former president Páez, General José Tadeo Monagas (1784–1868) becomes president in 1847. However, he shows more sympathy for the Liberal party than Páez has expected, and, after a showdown between the new president and the Conservatives, which leads to a popular revolt in 1848, the Liberals come to power.

Monagas is succeeded in office in 1851 by his brother, José Gregorio Monagas (1795–1858), who continues his Liberal policies. In 1855, José Tadeo is returned to office. During this decade, slavery is abolished, the vote is extended to more people, and the death penalty is declared illegal. However, economic problems plague the country. José Tadeo Monagas resigns the presidency in 1858 in response to a popular rebellion.

1853   Birth of pianist Maria Teresa Carreño

Venezuelan musician Maria Teresa Carreño (1853–1917) gives her first public concert in New York City at the age of nine. Carreño wins international recognition as a pianist, singer, conductor, and composer. She eventually settles in Germany and becomes one of the first women to manage an opera company.

1859–63   The Federal Wars

The decade of Liberal rule is followed by a period of political instability, centering on the rivalry of two factions, the Centralists, who want a strong central government, and the Federalists, who want the individual states to retain more political power. This factional rivalry erupts into civil war after a constitutional convention approves a new constitution in 1858, designed to mediate between the two sides. Refusing to compromise, the Federalists, led by General Ezequiel Zamorea, rebel against the sitting president, Dr. Manuel Felipe Tovar.

The following years are marked by continued hostilities and rapid changes in government. A succession of leaders holds office briefly, including former president José Antonio Páez, who is allied with the Centralists. Ultimately, the conflict is won by the Federalists and the Treaty of Coche ends the fighting in 1863. Another convention is called, and Federalist leader General Juan C. Falcón (1820–70) is elected president under a new constitution. His vice president, General Antonio Guzmán Blanco (1829–99), will later become president himself.

1864–70   The failure of Federalist government

Under President Falcón, Venezuela is divided into twenty states, part of a republic. Falcón and his Federalists are unable to control civil unrest that breaks out in the states or to formulate an effective economic policy. Falcón is ousted by a coalition of Liberals and Conservatives in 1868, and the country is led by a coalition government. In 1870, Falcón's former vice president, General Antonio Guzmán Blanco, seizes control of Caracas, with support from members of all the leading political parties.

1864   Cult figure José Gregorio Hernández is born

José Gregorio Hernández is a competent and caring physician killed when he is run over by a car in 1919. In the years following his death, a religious cult grows up around Hernández as Venezuelans begin praying to him to cure diseases and other medical conditions.

1870–88   The Blanco regime

Between 1870 and 1908, Venezuelan politics are dominated by one man—Antonio Guzmán Blanco (1829–99), who is both a political liberal and a dictatorial ruler. During these years, he serves a total of three terms as president (1870–77, 1879–84, and 1886–88), with hand picked successors serving the terms in between. Guzmán Blanco rules in an authoritarian manner, ruthlessly suppressing rebellions in 1872 and 1874–75. However, Venezuela makes considerable progress in many areas during the years he is in power.

The country's communication and transportation facilities are modernized, and free, mandatory public education is introduced. Politically, universal suffrage, representative government, and direct elections are instituted. The number of states is reduced and political power is concentrated more heavily in the central government.

1870   Beginning of public education

President Guzmán Blanco issues a decree authorizing compulsory free public education, to be administered by government at the federal, state, and local level. A Ministry of Education is established, and a nationwide school system is organized. The first teacher training schools are also created.

1870   Archbishop exiled

During his years in power, President Guzmán Blanco takes a number of steps to reduce the influence of the Catholic Church. In 1870, he exiles the powerful Archbishop Silvestre Güevara y Lira. Guzmán Blanco also closes the nation's convents and seminaries and turns religious instruction over to the universities. Civil marriage is instituted, and, in 1873, a civil registry of births and death is established.

1872   Construction of capital buildings completed

The capital buildings in Caracas are completed. The complex includes an elegant fountain and several courtyards.

1874–1944   Life of Rufino Blanco Fombona, leading literary figure

In early adulthood, Blanco Fombona pursues a career in politics, serving as Venezuelan consul to Philadelphia at the age of twenty and filling other government positions afterward, both at home and abroad. In 1910, he is exiled from Venezuela by dictator Juan Vicente Gómez (c. 1857–1935), living in France and Spain until 1936. During this time, he remains active in the resistance to Gómez through his political connections and through his writing. His essays, novels, short stories, and poetry receive wide acclaim, and he is nominated several times for the Nobel Prize for Literature. As a poet, he is an important part of the Modernismo literary movement founded by Nicaraguan poet Rubén Darío (1867–1916). This lyrical style of poetry, which emphasizes poetic form, has a major influence on other Venezuelan authors.

1884   Birth of novelist and politician Rómulo Gallegos

Gallegos (1884–1969) makes major contributions to his country both as a writer and as a politician. His best-known novel, *Doña Bárbara* (1929), is about a despotic rancher in Venezuela's cattle country, or *llanos*. The book is generally regarded as a social critique of the tyrannical Gómez regime. After its publication, Gallegos has to go into exile, returning after Gómez's death in 1935. He later serves as minister of education and as a member of congress, as well as organizing the Democratic Action Party in 1941. He also serves briefly as interim president of Venezuela in 1948, spending another period in exile after being deposed in November of that year. In 1958, he returns to Venezuela, to great popular acclaim.

Gallegos's greatest literary achievements are the novels written during his first period in exile. *Doña Bárbara*, the most popular novel in Venezuelan history, is widely adapted for stage and screen. Gallegos himself writes the screenplay for the first film version, released in 1943. (In the 1990s the novel is made into a television series, in a joint Venezuelan Mexican project.) Other well known works by Gallegos include *Cantaclaro* (1931) and *Canaima* (1935).

1888   Blanco resigns as president

After eighteen years in office, Antonio Guzmán Blanco steps down as president and retires to Europe. His resignation is followed by a period of rapid changes of government encompassing the presidencies of General Hermógenes López (1888), whom Guzmán Blanco names to succeed him; Juan Pablo Rójas Paúl (1888–1890), Venezuela's first civilian president since 1834; and Raimundo Andueza Palacio (1890–1892). The longest single presidency after Guzmán Blanco is that of General Joaquín Crespo (1892–1898).

1889   Birth of painter Armando Reverón

Reverón (1889–1954) is born in Valencia. He begins studying art in Caracas in 1908, followed by studies in Madrid. Returning to Caracas in 1915, he becomes part of a group of landscape painters who call themselves the Fine Arts Circle. An exhibit of Reverón's work is mounted at the Academy of Fine Arts in 1919. The painter later moves to the seaside town of Macuto, where he lives in a castle and paints the seascapes for which he is famous. Most of his critical recognition comes after his death.

1890   Birth of novelist Teresa de la Parra

Teresa de la Parra is the pen name of Ana Teresa Parra Sanojo (1890–1936). Born into a wealthy Venezuelan family living in Paris, Parra returns to Venezuela at the age of eighteen, publishes her first story, and wins a literary award that encourages her to pursue her career further. She returns to Paris in 1923 and organizes a literary salon frequented by Latin American expatriates. She spends most of the remainder of her life outside Venezuela, and dies in Madrid.

Parra is Venezuela's best known female novelist. She is famed for her in-depth psychological portraits of women, which are rare in that they portray nineteenth century Venezuelan life through the viewpoint of female protagonists. Her novels include *Ifigenia: diario de una señorita que escribía porque se fastidiaba* (1924) and *Mama Blanca's Souvenirs* (1929), which explores the life of a young girl living on a plantation.

1895   United States intervenes in boundary dispute with British Guiana

A fifty–year–old boundary dispute between Venezuela and British Guiana escalates when the British occupy the Orinoco River in 1886. In 1895, gold deposits are discovered in the disputed territory, and U.S. President Grover Cleveland (1837–1908, president 1885–89 and 1893–97) demands that the disagreement be settled by international arbitration, on grounds that Great Britain is violating the Monroe Doctrine. The arbitration panel decides in Britain's favor, awarding it most of the disputed land, but returning control of the Orinoco Valley to Venezuela.

1897   Birth of poet Andrés Eloy Blanco

Blanco (1897–1955) is one of Venezuela's foremost twentieth century literary figures. Like many other Latin American writers, he is also active in politics and is imprisoned or exiled a number of times for his opposition to various dictatorial regimes, including that of Juan Vicente Gómez. He takes part in drafting Venezuela's 1947 constitution and is appointed foreign minister during the aborted 1948 presidency of fellow writer Rómulo Gallegos (1884–1969).

Blanco becomes the foremost poet of the "Generation of 1928," which rejects the Modernismo movement and stresses the content of poetry over its form. Many of his poems are written in a popular, folkloric style that is easily accessible to the general public. These poems are widely read and some have entered Venezuelan oral tradition. Perhaps the most famous is *Angelitos Negros*. He is also the author of stories, plays, articles, and speeches.

1897: January   First movies are shown

The first public film screenings in Venezuela take place in Maracaibo.

1899–1908   Presidency of Cipriano Castro

The dictatorial regime of General Cipriano Castro (c. 1959–1924) is marked by internal revolts, strained relations with other governments, and bureaucratic mismanagement. During this period, there is a growing awareness of the economic importance of Venezuela's oil resources. An unsuccessful rebellion led by General Manual Antonio Matos between 1901 and 1903 is suppressed. From this point, Venezuela's government will be controlled by military leaders from Castro's home state of Táchira for the next sixty years, except for a brief period between 1945 and 1948.

1902   International debt crisis

In response to Venezuela's failure to pay its large foreign debts, Great Britain, Italy, and Germany impose a naval blockade on the country. Venezuela yields to this international pressure, in what becomes known as the "Venezuela Claims incident." Venezuela's debts are resolved in 1903 by the Washington Protocol, which requires the nation to set aside nearly a third of its customs revenue for debt repayment.

1904 Academy of Medicine founded

Venezuela's National Academy of Medicine is established. Other scientific organizations founded in the first half of the twentieth century include the Academy of Physical, Mathematical, and Natural Sciences (1917), the Institute of Experimental Medicine (1940), and the Venezuelan Association for the Advancement of Science (1950).

1905 Birth of sculptor Francisco Narváez

Narváez (1905–82) is generally regarded as Venezuela's first modern sculptor. The most extensive single collection of his works is found at the art museum in his birthplace, Porlamar on the Isla de Margarita. His sculptures are also found throughout Caracas.

1906 Birth of author Arturo Uslar Pietri

Arturo Uslar Pietri (b. 1906) becomes Venezuela's foremost twentieth century author. For most of his life, he pursues a political career, serving abroad as a diplomat (and a delegate to the League of Nations) and in several prominent government positions at home. He also teaches political science at the university level. In 1945, he is forced into political exile in the United States, where he teaches at Columbia University. Uslar Pietri returns to Venezuela in 1950 and is elected to the Senate in 1959. After the age of sixty, he turns primarily to writing and teaching and also appears as a television commentator.

Uslar Pietri is best known for his novel, *Las lanzas coloradas* (Red Lances; 1931), which is set during Venezuela's war for independence. He has also written novels, essays, plays, short stories, and historical works. A recent work of fiction, *La Visita en El Tiempo* (The Visit in Time), won Spain's Prince of Asturias Prize in 1991.

Gómez Begins Period of Military Rule

1908–1935 Gómez dictatorship

General Juan Vicente Gómez (c. 1857–1935) deposes Cipriano Castro and assumes power in 1908. He retains political control of Venezuela until his death in 1935. During some periods, he rules directly as president; at other times, he appoints other men to the presidency and dictates their policies. Gómez's rule is an early example of a twentieth century authoritarian regime better able to control its citizens through the use of modern technology: a state-run telegraph system can report any rural uprising before it has time to gain momentum, and modern roadways facilitate the movement of government troops. After serving his first official term as president, from 1910 to 1914, Gómez suspends the nation's constitution. During his years in power, he has many of his political adversaries arrested, imprisoned, exiled, or assassi-

nated. He also engages in financial corruption on a large scale, accumulating a large amount of personal wealth.

Gómez presides over the growth of Venezuela's oil export economy, winning favorable terms in agreements with foreign oil companies. During Gómez's regime, Venezuela becomes so wealthy that it completely pays its foreign obligations, becoming the world's only country with no debt. Gómez retains power until his death in 1935.

1909 Divorce becomes legal

Divorce is legalized, eventually becoming as common as in other Western nations.

1909 First Venezuelan movie is produced

A short film is shot in Venezuela. The first Venezuelan feature film, the silent movie *La Dama de las Cayenas*, is produced in 1913. In 1932 the first picture using sound is made: *La venus de Nacar.*

1917 First petroleum produced

Venezuelan petroleum production begins, with 120,000 barrels produced. Venezuela grows to be the world's third largest petroleum producer, after the United States and the former Soviet Union.

1922: December Major oil discovery at Bolívar Coastal Field

Drilling at an oil field near La Rosa, on the coast of Lake Maracaibo, yields quantities of oil far greater than have been discovered in Venezuela thus far. In the first day of drilling, 100,000 barrels of oil are produced. This discovery begins Venezuela's oil boom and the country's transformation from an agrarian society to a modern industrial nation with the highest per capital income in Latin America. Within two years of the discovery, dozens of foreign oil companies begin drilling in the Lake Maracaibo region. Eventually, offshore drilling is started on Lake Maracaibo itself.

1923 Internationally acclaimed artist Jesús Soto is born

Soto is born to a peasant family in Ciudad Bolívar, on the shores of the Orinoco River. He wins a scholarship for study in Caracas and becomes interested in Cubism and other modern artistic movements. He becomes a leading exponent of kinetic art (artworks, particularly sculptures, composed of moving parts). He is appointed director of the School of Fine Arts in Maracaibo in 1947 and has his first solo exhibition two years afterward.

Soto goes to Paris in 1950, forming professional relationships with renowned artists including Alexander Calder and Marcel Duchamp. Soto's work wins a major award at the São Paulo Bienal in 1963 and is exhibited at the Museum of Modern Art of Paris in 1969. In the same year, the Venezuelan government establishes the Jesús Soto Foundation. In 1973, the artist donates his private collection of his own work to his native city. It is housed in the Museum of Modern Art Jesús Soto. Works by Soto are commissioned in leading cities including Paris and Toronto. In 1988, Soto receives a commission from the International Olympic Committee for the Olympics held in Seoul, South Korea.

1925   First labor protest takes place

A labor movement grows up in response to the growth of Venezuela's oil industry in the first part of the twentieth century. In 1925, workers in the Bolívar oilfields protest the high cost of living, but they are silenced by government troops. By 1935, Venezuela has laws that allow for the formation of unions, limit the work day to nine hours, and provide for workmen's compensation and death benefits.

1928   Suppression of student protest has wide ranging consequences

Gómez's repressive measures draw criticism and protest in some segments of society. In 1928, a student protest against government policies takes place, inspired by the Mexican and Russian revolutions of the preceding years. Initially, three students are imprisoned for making anti-government speeches. More students then protest, and 200 are arrested, after which a mass demonstration takes place. The demonstration is violently broken up by armed police, with many participants wounded or killed. Later, aided by youthful military officers, student rebels storm and briefly occupy the presidential palace, from which they are routed by the military. Gómez closes the university, student activists are arrested, and many are sent to work on chain gangs.

The 1928 student revolt proves significant as formative experiences for some of Venezuela's future leaders, including Rómulo Betancourt, Rafael Caldera Rodríguez, and Raúl Leoni, all of whom are participants in the 1928 revolt who escape to exile in other countries. They and their peers become known as the "Generation of 1928."

1928   Workers' Bank is established

The government founds the Workers' Bank as a public housing agency. The bank builds over 20,000 low cost housing units between 1959 and 1967. The majority of these are built in rural areas with the goal of reversing the pattern of rural to urban migration.

1930   Symphony Orchestra of Venezuela is founded

The Symphony Orchestra of Venezuela is established in Caracas. Three other orchestras are subsequently founded in the city.

1935   Sighting and naming of "Angel Falls" are discovered

American bush pilot Jimmie Angel first spots the Angel Falls while looking for gold in Bolívar state, southeastern Venezuela. Though the existence of the falls had been reported as early as 1910 (and it is doubtless that native peoples knew about the falls), Jimmie Angel is the first to name and document them. Angel Falls is the highest waterfall in the world, with a drop of over 979 meters. The drop is so high that in the dry season the water turns into mist before it reaches the bottom of the falls.

1935: December   Aging dictator Gómez dies

Venezuela's longtime dictator, Juan Vicente Gómez, called the "Tyrant of the Andes," dies at the age of 79 of natural causes. Furious crowds in Caracas and Maracaibo go on a rampage following the dictator's death, rioting, looting, and killing relatives and supporters of Gómez.

1936–41   Contreras presidency inaugurates period of progress

General Eleazar López Contreras succeeds Juan Vicente Gómez as president. Governing in a more liberal manner than his predecessor, he welcomes political exiles back into the country and introduces public health and education reforms, and institutes a public works program. In response to an oil workers' strike in 1936, Contreras increases their wages, although he outlaws their union. Social security legislation is passed, and the Central Bank is established in 1939 (see 1939).

1938   Fine Arts Museum is founded

Venezuela's Fine Arts Museum houses paintings and sculpture by artists from Venezuela and other countries.

1939   Central Bank of Venezuela is founded

The Banco Central de Venezuela is established as the federal government's fiscal agent. Its operations include making collections and payments for the government, maintaining the nation's gold and foreign exchange reserves, and setting discount rates. It also issues government notes.

1941–45   Angarita presidency

In a move unusual for military leaders, General Contreras allows his elected successor, Isaías Medina Angarita, to take office on schedule. Angarita turns out to be one of Venezuela's most popular and admired presidents. Under Angarita, Venezuela's first income tax law is passed (1942), as well as oil exploration laws giving the country more leverage against the foreign companies that develop the oil fields. With

A bronze monument of a dead mosquito commemorates the government's success against malaria. In 1945, over a million Venezuelans contracts the disease. Twelve years later, malaria afflicts less than one thousand Venezuelans. (EPD Photos/CSU Archives)

increased political freedom, new political parties are formed, including Accíon Democrática (Democratic Action), or AD, which is to dominate Venezuelan politics from 1958 on. The government even tolerates the formation of a communist party.

1945–1948   The trienio junta seizes power

President Medina Angarita is overthrown on October 18, 1945, by the AD and a faction of the military. The three year period during which their seven member civilian-military council governs the country is known as the *trienio junta.* At the head of the council is AD leader and president Rómulo Betancourt (1908–81). The *junta* begins to implement a number of reforms aimed at lessening the disparity between the nation's rich and its poor. A land reform measure is intended to convert many large estates to small holdings. The government also investigates how a number of the nation's millionaires have amassed their fortunes. In 1946 oil workers are allowed to form a union, called Fedepetrol.

1945–57   War against malaria

Malaria affects over one million Venezuelans in 1945. After a twelve-year campaign against the disease, the number of those afflicted drops to under one thousand. The success of the anti-malarial program is largely due to mass spraying of DDT, a chemical agent.

1947   Women win the right to vote

1947: December   Rómulo Gallegos is elected president

Reformist candidate of the AD party and well-known author Rómulo Gallegos (1884–1969) is elected to succeed the junta that has ruled Venezuela since 1945. He takes over the office in 1948 and immediately takes steps to implement a radical land reform plan and a program to tax the profits of oil companies, so that their wealth can be shared by the whole country. Alarmed by these and other measures, the military ousts Gallegos from office in a coup at the end of 1948, when he has served less than a year in office.

1948–52   Military junta rules Venezuela

After the coup that removes President Rómulo Gallegos from office, the nation is governed by a three member military junta whose members are Carlos Delgado Chalbaud, Marcos Pérez Jiménez (b. 1914), and Luis Felipe Llovera Páez. Gallegos is forced into exile, his AD party is disbanded, classes at the Central University of Venezuela are canceled, and the Communist Party is outlawed. After Chalbaud is assassinated in 1950, true power shifts to Pérez Jiménez.

1951   Caracas hosts "Little Olympics"

Caracas hosts the Bolivarian Games. Known as the "Little Olympics," this event is an international sporting competition featuring athletes from Bolivia, Ecuador, Panama, Peru, and Venezuela.

1951   Venezuelan film wins acclaim at Cannes

La Malandra Isabel becomes the first Venezuelan movie to gain international recognition when it wins an award at the Cannes Film Festival in France.

1951   Source of the Orinoco River is discovered

A joint French Venezuelan expedition discovers the exact source of the Orinoco River at the cliff of a mountain (Cerro Delgado Chalbaud) near the Brazilian border, at an altitude of 3,435 feet (1,047 meters).

1952–58   Presidency of Pérez Jiménez

After voters refuse to cooperate in a rigged election, it is suspended and Pérez Jiménez is simply installed as president, an office he holds for five years, ruling in an authoritarian manner that he calls the "New National Ideal." Political opponents are jailed, tortured, or forced into hiding or into exile. Government corruption is widespread, with Pérez Jiménez amassing great personal wealth. To generate employment,

Over 50,000 sports fans are on hand for the opening ceremonies of the Bolivarian Games Festival at Caracas in 1951. The event, known as the "Little Olympics," features athletes from Bolivia, Columbia, Ecuador, Panama, Peru, and Venezuela. (EPD Photos/CSU Archives)

Pérez Jiménez's government concentrates on large public works projects, including road, railroad, and bridge construction. Other construction projects include public buildings and military installations. In 1956, the government once again begins granting oil concessions to foreign companies.

Opposition to the repressive tactics of Pérez Jiménez grows. The archbishop of Caracas criticizes the working conditions of the country's labor force in a pastoral letter in 1957. In January 1958, students riot in the streets of Caracas. At the height of the unrest, police kill as many as 300 people in two days. Elements in the military begin organizing to overthrow Venezuela's leader, forming the Movement for National Liberation (MLN). Ultimately, the air force leads a rebellion and

ousts Pérez Jiménez on January 23, 1958. The dictator goes into exile in Florida, and Admiral Wolfgang Larrazábal, who has led the coup, takes over leadership of the government as head of a military junta working together with a civilian cabinet.

1953 First television programs are broadcast

Television broadcasts begin on the government run Televisora Nacional. Two privately owned stations, Venevisión and Radio Caracas Television, are formed shortly afterward. Nationwide television networks grow out of both private stations, with additional networks created by the government (Venezuelan Television Network) and the Catholic Church.

1958   Women protesters are attacked by police

Venezuelan women protesting government repression toward the end of the Jimínez regime are attacked by Caracas police wielding machetes.

1958   Rural housing program is initiated

A government program begins construction of low cost rural homes made from cement and blocks to replace traditional huts of mud and thatch. By 1969, over 100,000 homes have been built.

Civilian Government Becomes a Reality

1959: November   Betancourt is chosen as president in free elections

With the ouster of the dictator Pérez Jiménez, civilian government returns to Venezuela for the first time since the assumption of power by Juan Vicente Gómez in 1908. Wolfgang Larrazábal's provisional government wins promises by all the nation's major parties that they will honor the winner of the next presidential election. (This agreement is called the Pact of Punto Fijo.) Rómulo Betancourt (1908–81) and his Democratic Action (AD) party win a plurality of forty-nine percent of the vote.

1959–64   The Betancourt presidency

Betancourt's government undertakes social and economic reforms, expanding education, raising income taxes, encouraging foreign investment, and, in 1960, passing a land reform bill to distribute land to 700,000 farmers. Development projects include a steel mill and hydroelectric plants.

With democratic rule still new to Venezuela, political turmoil plagues Betancourt throughout his administration, with numerous attempts to unseat him, both from the extreme right and left. In many cases, these factions enlist the aid of foreign governments. The best known case is the involvement of the Dominican Republic's strongman, Rafael Trujillo (1891–1961), in a right wing plot to assassinate Betancourt. (Betancourt is wounded but survives, and sanctions are imposed on the Dominican Republic by the Organization of American States (OAS).

Nevertheless, Betancourt does achieve some political stability by granting concessions to traditional sources of political power. To conciliate the military, he continues the draft, improves salaries and living conditions for the troops, and guarantees that no one will be prosecuted for crimes committed during the preceding years of military government. To win the approval of organized labor, Betancourt issues a new labor code recognizing the rights of workers to organize and engage in collective bargaining. Except for continued opposi-

tion by the extreme left, Betancourt succeeds in achieving considerable political stability and conciliation.

1960s   Militant leftists turn to guerrilla tactics

Alienated by President Betancourt's opposition of communism, including his opposition to the Castro government in Cuba, Venezuelan leftists pull out of the president's political coalition and begin forming new and more extreme political groups, including the student-supported Movement of the Revolutionary Left (MIR). In 1962, military officers supportive of the left launch violent revolts against the government at Carúpano and Puerto Cabello. The government then outlaws both MIR and the Venezuelan Communist Party (PCV), which proceeds to go underground, establishing the militant Armed Forces of National Liberation (FALN). Throughout the 1960s, FALN mounts urban and rural guerrilla operations. They are at their most active in the early part of the decade, when they plant bombs in the U.S. embassy in Caracas and at a Sears Roebuck warehouse.

1961   Betancourt pursues anti-Castro policies

After supporting the expulsion of Cuba from the OAS (Organization of American States), President Betancourt breaks diplomatic ties with the communist government of Fidel Castro (b. 1926) in December 1961.

1963: December 1   Raúl Leoni is chosen as president in democratic elections

In spite of a charged political climate that includes terrorist threats, scheduled elections are held in December 1963, and almost ninety percent of Venezuela's registered voters participate. The Democratic Action (AD) party candidate, Raúl Leoni, is elected with thirty-two percent of the vote, running against six other candidates.

1964: March 11   Leoni takes office as president

Newly elected president Raúl Leoni takes office in March 1964, in the first peaceful transfer of power between democratic governments in Venezuela's history. During the Leoni presidency (1964–69), the government continues the land reform efforts begun by its predecessor. Politically, the situation is more stable than before, although there are still efforts to unseat the government by force, including guerrilla activity. Cuba plays a role in these, and the OAS imposes sanctions in response.

Internationally, Venezuela becomes a member of the Latin America Free Trade Association (LAFTA), takes part in the 1967 Punta del Este economic conference, and makes plans to join a regional common market with the other Andean nations of Peru, Chile, Ecuador, and Chile. There is also an agreement in the longtime border dispute with British Guiana. Both sides agree on a common border.

1966 Government stability is threatened from the right and the left

In 1966, the Venezuelan government deals with actual or threatened revolts from both the right and the left. In October, an attempted coup by the military is suppressed. Due to continuing threats from the left, the Central University in Caracas is investigated in December in connection with possible revolutionary activities.

1968: November Major hydroelectric project completed

The Guri Dam across the Caroní River, begins operating. One of the largest hydroelectric plants in the world, it is built by the Corporación Venezolana de Guayana. By 1982, its capacity is 14.5 million kilowatts.

1969: March 16 160 people die in major air disaster

A DC-9 leaving the Maracaibo airport crashes within two minutes of takeoff, killing all eighty-four passengers and crew on board. As it is falling, the plane hits a high tension wire and explodes, setting fire to the suburb of La Coruba, where seventy-six more people die on the ground immediately, or in hospitals later. The crash—the worst in civil aviation history thus far—is blamed on pilot error.

1969–74 Christian Democratic government in office

Rafael Caldera is elected president in 1968, marking the first time a Christian Democrat has won the nation's top office. However, Caldera, elected primarily because of a schism in the AD party, is lacking a strong mandate to govern. There are political kidnappings and assassinations throughout his term in office, and he has problems getting his programs through the legislature. Caldera works to improve education and social welfare programs and create an economy based on a combination of private and state owned enterprise. He also eliminates a previous ban on the Community Party and establishes diplomatic relations with the former Soviet Union.

The government moves toward even greater control of Venezuelan oil, adopting provisions to rescind existing concessions when they expire. It also nationalizes natural gas operations. Venezuela also moves toward closer economic cooperation with its neighbors when it officially joins the Andean pact in 1973.

1970 New agreement signed in dispute with Guyana

Venezuela signs the Port of Spain Protocol, an agreement with Guyana (formerly British Guiana) that provides for a temporary twelve-year truce in the ongoing border dispute between the two nations.

1972 National park created

Los Roques (The Rocks) is declared a national park. It is comprised of a group of islands located 100 miles (160 kilo-

meters) north of mainland Venezuela. The park covers just under 550,000 acres (220,000 hectares) of islands and underwater coral reef with a large, lagoon in the center. Gran Roque, a large protrusion of dark gray and rust-colored rock, is a remnant of the area's volcanic history. Los Roques is the largest national marine park in the Caribbean.

1974–79 AD wins back control of government

In the 1973 elections, Carlos Andrés Pérez of the Democratic Action party wins forty-nine percent of the vote, a larger plurality than that of any president since Betancourt in 1959. His accession to the presidency represents yet another peaceful transfer of power between rival parties. During his term in office, Pérez has the advantages of AD control of both houses of Congress and rising government revenues from high oil prices. His government carries out the nationalization of key industries, including iron mining (1975) and oil (1976). It also invests in ambitious agricultural and industrial programs.

1975: August 29 Oil industry is nationalized

The Oil Industry Nationalization Act is signed by President Pérez. Under this law, exploration concessions to private oil companies are cancelled as of the end of 1975. The government establishes a state owned oil company, Petróleos de Venezuela, or Petrovén, in September 1975. It obtains a fifty year monopoly on Venezuelan oil production.

The Boom Period Ends: Falling Oil Prices and Recession

1978 Falling oil prices signal end of era of prosperity

Venezuela's oil boom starts winding down, taking with it the high income that has fueled government economic programs. Over the next four years, inflation rises to eight percent (very low by South American standards but high for Venezuela) and unemployment to eight percent. Support for Christian Democrat Herrera Campíns, elected president in 1978 with a forty-six percent plurality, declines.

1978 Ministry of the Environment is created

The Venezuelan government creates a Ministerio del Ambiente (Ministry of the Environment), the first such ministry in any South American nation.

1979–1984 Presidency of Herrera Campíns

Christian Democrat Herrera Campíns becomes president just as the country enters a long recession triggered by falling world oil prices. Venezuela's foreign debt and cost of living mount rapidly. Civil unrest, including guerrilla attacks, mounts after 1981. By 1983, when seventy percent of the nation's debt comes due, the government is obliged to stop paying interest, freeze prices for sixty days, and impose controls on foreign exchange.

1982   Guyana border dispute revives

When the twelve year moratorium imposed on the Guyana border dispute by the Port of Spain Protocol ends in 1982, Venezuela takes no steps to extend it. The dispute is brought to the UN in 1985, and has still not been resolved by the early 1990s.

1982   Women's rights are expanded by civil code

The Venezuelan civil code is revised to extend greater rights to women. In addition, Venezuela adopts legislation protecting women from discrimination, either in wages or working conditions.

1983   Caracas Metro opens

The Caracas Metro, or subway system, long under construction, opens in 1983. The *caraqueños,* or people of Caracas, take great pride in this facility, which is punctual, clean, and safe. When using it, they are more careful and courteous than in many other aspects of daily life. One thing that makes the Metro special is the modern sculptures commissioned for the entrance of every station. Additional subway construction continues through 1990, as new lines are opened.

1984–1988   Presidency of Jaime Lusinchi

With the nation's economy faltering, the Christian Democrats are roundly defeated in the 1983 elections, and Democratic Action party (AD) candidate Jaime Lusinchi (b. 1924) is elected president, by a fifty-seven percent margin. Lusinchi's government institutes economic austerity measures that make Venezuela the only Latin American country able to meet its foreign debt payments on time. However, these measures prolong the nation's recession, and the inflation rate reaches twenty percent during Lusinchi's presidency. Economic growth resumes in 1986, and Venezuela reschedules its foreign debt without resorting to the type of economic restructuring advocated by the International Monetary Fund.

1986   Guri hydroelectric plant completed

With the world's fourth largest installed capacity, the newly completed Guri hydroelectric plant becomes Venezuela's single largest source of electrical power. It provides as much energy as 300,000 barrels of oil per year.

1988   Women's group organized to support female candidates for office

Venezuelan women organize Women Leaders United, a movement that supports female political candidates through marches, rallies, and debates. Thanks to their efforts, four women are elected to the Senate, and nineteen to the House of Deputies.

1988: March   Top government official resigns over drug charges

By 1988, Venezuelans have become involved in the Colombian drug trade. In that year, the Minister of Justice resigns over charges of connections with drug traffickers. In 1988–89 Colombia and Venezuela agree to work together to stop the movement of drugs across their shared borders, with both countries increasing border troop deployment.

1988: October 29   Fishermen are mistakenly killed in alleged drug raid

Government troops patrolling Venezuela's western border open fire on local civilians near the town of El Amparo, killing sixteen fishermen. In a cover-up attempt, the incident is initially described as an attack on Colombian guerrillas. Eventually, however, the truth is revealed, resulting in rioting and in political repercussions at the national level.

1989   Former president Carlos Andrés Pérez returns to office

In their disillusionment with the country's current economic situation, Venezuelan voters return former president Carlos Andrés Pérez to office in the 1988 elections. In contrast to his former policies of government spending and nationalization, Pérez imposes austerity measures, including increases in the cost of public transportation and gasoline. Even though these measures provoke widespread protests and nearly topple the government, Pérez continues with his plans to stabilize the economy, including the privatization of businesses and encouragement of foreign investment. In August 1989, Pérez's government is shaken by financial scandal in connection with an $8 billion financial fraud case.

1989   Women legislators form bicameral advocacy group

Venezuela's female members of the Senate and the House of Deputies form the Bicameral Commission for the Rights of Women to advance women's causes.

1989: February 27   "Black Monday" riots protest government economic policies

On the Friday before "Black Monday," the government announces that in response to pressure by the International Monetary Fund, it is raising gasoline and transportation prices. When commuters head for work on Monday morning, they find that the bus companies have already raised fares. This action touches off rioting at bus stops in the Caracas suburb of Guarenas. Upset about shortages caused by merchants holding back goods to wait for higher prices, consumers also begin looting supermarkets. Publicized by television and radio coverage, the actions in Guarenas touches off similar rioting and looting in Caracas and throughout the country.

By the next day, nearly all Venezuelans stay home from work, either participating in, or frightened by, the rampant lawlessness. The government declares a state of emergency and imposes a curfew. The army troops sent to quell the rioting are unused to dealing with civil disturbances and overreact, killing many innocent bystanders and others with stray shots. The unrest continues for four days, causing many injuries, deaths, and large scale destruction of property.

The official death toll is 277, but it is widely believed that the actual figure is much higher, perhaps as high as 1,000. Later the same year, the nation's trade unions call a twenty four hour general strike to protest government policies. In February of 1990, new price hikes, together with news that the government has released officials charged with corruption, brings on renewed riots and looting throughout the country.

1989: August   Government shaken by financial scandal

The Venezuelan government is rocked by the arrest of a former highly placed government official on charges related to the loss of about $8 billion in federal funds through fraud.

1991: August   State run airline is privatized

Venezuela's state owned airline, Venezuelan International Airways (Venezolana Internacional de Aviación—VIASA) is privatized. It is an international airline with service to Europe, North America, the Caribbean, and other South American countries.

1992   Comprehensive environmental law is passed

Venezuela adopts an extensive environmental law governing air, water, and soil pollution, and encompassing issues such as ozone layer damage and forest fires. It includes specific standards for pollution levels and definitions of protected areas.

1992: February 4   Failed coup attempt results in mass killings and arrests

A failed military coup against the government of President Carlos Andrés Pérez results in the killing of 100 soldiers and the arrest of over a thousand soldiers and officers.

1993: May   Pérez resigns before end of term

Although President Pérez's economic policies successfully reduce inflation (it fell from eighty percent in 1989 to about thirty percent by 1994), yet the accompanying unemployment that results produces widespread discontent, especially when coupled with the widespread perception of government corruption extending all the way to the presidency. In 1993, the Congress brings corruption charges against Pérez himself, and the president resigns. Historian Ramón J. Velásquez is

appointed by Congress to serve out the remainder of Pérez's term.

1994   Former president Rafael Caldera returned to office

Caldera, who leaves the Christian Democrats, runs for the presidency in 1993, supported by the Movement Toward Socialism (MAS) and a new political coalition called *Convergencia* (Convergence). He wins by a narrow margin and proceeds to begin reversing the policies of the Pérez government. Caldera uses authoritarian measures to retain control over Congress and over the population.

1994: January 3   Prison riot

Inmates riot at the Sabaneta jail in Venezuela's second largest city, Maracaibo. More than 100 inmates participate in the riot, during which prisoners attacked each other in acts of revenge. Inmates were beaten, stabbed, drowned, and burned alive by other inmates. National guardsmen used bullets and tear gas to quell the disturbance.

1994: June   Caldera suspends constitutional rights

President Caldera suspends six constitutionally guaranteed rights, including freedom of movement and freedom from arbitrary search and arrest. He defends the suspensions, claiming they are necessary for effective enforcement of government, economic, and law enforcement measures. However, his actions draw criticism from international human rights groups. Caldera restores the suspended freedoms in July 1995. However, they remain in force along Venezuela's Colombian and Brazilian borders as part of military operations to control drug trafficking, guerrilla operations, and illegal mining.

1994: June 28   Singer Ricardo Montaner advances international career

Popular singer and songwriter Ricardo Montaner releases an album entitled "Un Manana Y Un Camino" (A Tomorrow and a Road), in the United States, promoted by EMI Music International. Montaner has sold more than seven million albums internationally.

1995   Group of Three free trade agreement goes into effect

A free trade pact signed by Colombia, Mexico, and Venezuela takes effect at the beginning of 1995. The three member nations agree to remove most trade restrictions by 2007.

1996   Oil refinery chemical leak damages coral reef

A chemical leak from an oil refinery in the northwest is thought to have caused the death of some of the coral found in Parque Nacional Morrocoy.

1996   Venezuelan contestants successful in beauty contests

Since 1980, Venezuelan women have won four Miss Universe titles and five Miss World titles, far ahead of any other country in the world. Since 1985, the contestant from Venezuela has been a finalist in every Miss Universe and Miss World pageant. Osmel Sousa, director of the Miss Venezuela contest, is a former dress designer who helps prepare the contestants for international competition.

1996: March 2   MAS party shifts allegiance from Caldera to Christian Democrats

Disillusioned with President Rafael Caldera's economic policies, the Movement Toward Socialism allies itself with the Christian Democrats and the Radical Cause leftist party, abandoning Caldera's Convergence coalition and eliminating the president's majority in both houses of Congress.

1997: July 9   Earthquake kills 82 people in northeastern Venezuela

Venezuela is hit by its worst earthquake in thirty years. The quake, measuring 6.9 on the Richter scale, strikes Sucre province on the northeastern coast. At least eighty-two people are killed, and over five hundred others are injured. Of those who are killed, forty lose their lives when a school building collapses in the town of Cariaco.

1997: November 7–9   Venezuela hosts Ibero–American Summit

The seventh annual Ibero–American Summit, attended by representatives of countries in which Spanish or Portuguese is spoken, is held on Margarita Island. The heads of state of all countries are present except for Ecuador's interim president. Among other actions, the conference affirms its opposition to the Helms–Burton Act passed by the U.S. Congress in 1996, stepping up the U.S. embargo on Cuba. However, the leaders also press Cuban president Fidel Castro, who is present, to implement democratic reforms.

1998: January 3   Electoral reform bill signed

President Caldera signs an election reform law designed to reduce the influence of political parties on the nation's electoral process. Other reforms include limits on the length of campaigns and the amount of television advertising and automated voting systems. The law also stipulates that at least thirty percent of each party's candidates should be women.

1998: December 6   Presidential elections

To accommodate all the candidates, the ballot for the presidential election measures almost 9 inches (22 centimetres) wide and over almost 20 inches (50 centimeters) long. Hugo Chavez Frias, leader of the failed 1992 coup, wins forty percent plurality in the Chamber of Deputies and the Senate, thus becoming the clear winner of the election.

Bibliography

Arroyo Talavera, Eduardo. *Elections and Negotiation: The Limits of Democracy in Venezuela.* New York: Garland Publishing, 1986.

"Beauty? Forget Bangalore." *The Economist,* November 30, 1996, vol. 341, no. 7994, p. 40.

Betancourt, Romulo. *Venezuela, Oil, and Politics.* Boston: Houghton Mifflin, 1979.

Blank, David E. *Venezuela: Politics in a Petroleum Republic.* New York: Praeger, 1984.

Boue, Juan Carlos. *Venezuela: The Political Economy of Oil.* New York: Oxford University Press, 1993.

Coppedge, Michael. *Strong Parties and Lame Ducks: Presidential Partyarchy and Factionalism in Venezuela.* Stanford, Calif.: Stanford University Press, 1994.

Coronil, Fernando. *The Magical State: Nature, Money, and Modernity in Venezuela.* Chicago: University of Chicago Press, 1997.

Ewell, Judith. *Venezuela and the United States: From Monroe's Hemisphere to Petroleum's Empire.* Athens: University of Georgia Press, 1996.

Fox, Geoffrey. *The Land and People of Venezuela.* New York: HarperCollins, 1991.

Gall, Norman. *Oil and Democracy in Venezuela.* Hanover, NH: American Universities Field staff, 1973.

Haggerty, Richard A., ed. *Venezuela: A Country Study,* 4th ed. Washington, D.C.: Library of Congress, 1993.

Hellinger, Daniel. *Venezuela: Tarnished Democracy.* Boulder, Colo.: Westview Press, 1991.

Herman, Donald L. *Democracy in Latin America: Colombia and Venezuela.* New York: Praeger, 1988.

Lombardi, John V. *Venezuela: The Search for Order, the Dream of Progress.* New York: Oxford University Press, 1982.

Smith-Perera, Roberto. *Energy and the Economy in Venezuela.* Cambridge, Mass.: Massachusetts Instute of Technology Press, 1971.

"Venezuela: Pick Your Chavez." *The Economist,* November 21, 1998, p. 40.

Waddell, D. *Venezuela.* Santa Barbara, Calif.: Clio Press, 1990.

Glossary

abdicate: To formally give up a claim to a throne; to give up the right to be king or queen.

aboriginal: The first known inhabitants of a country. A species of animals or plants which originated within a given area.

allies: Groups or persons who are united in a common purpose. Typically used to describe nations that have joined together to fight a common enemy in war.

In World War I, the term Allies described the nations that fought against Germany and its allies. In World War II, Allies described the United Kingdom, United States, the USSR and their allies, who fought against the Axis Powers of Germany, Italy, and Japan.

Altaic language family: A family of languages spoken in portions of northern and eastern Europe, and nearly the whole of northern and central Asia, together with some other regions. The family is divided into five branches: the Ugrian or Finno-Hungarian, Smoyed, Turkish, Mongolian, and Tunguse.

amendment: A change or addition to a document.

Amerindian: A contraction of the two words, American Indian. It describes native peoples of North, South, or Central America.

amnesty: An act of forgiveness or pardon, usually taken by a government, toward persons for crimes they may have committed.

animal husbandry: The branch of agriculture that involves raising animals.

Anglican: Pertaining to or connected with the Church of England.

animism: The belief that natural objects and phenomena have souls or innate spiritual powers.

annex: To incorporate land from one country into another country.

anti-Semitism: Agitation, persecution, or discrimination (physical, emotional, economic, political, or otherwise) directed against the Jews.

apartheid: The past governmental policy in the Republic of South Africa of separating the races in society.

appeasement: To bring to a state of peace.

arable land: Land that can be cultivated by plowing and used for growing crops.

archipelago: Any body of water abounding with islands, or the islands themselves collectively.

archives: A place where records or a collection of important documents are kept.

arctic climate: Cold, frigid weather similar to that experienced at or near the north pole.

aristocracy: A small minority that controls the government of a nation, typically on the basis of inherited wealth.

armistice: An agreement or truce which ends military conflict in anticipation of a peace treaty.

ASEAN *see* Association of Southeast Asian Nations

Association of Southeast Asian Nations: ASEAN was established in 1967 to promote political, economic, and social cooperation among its six member countries: Indonesia, Malaysia, the Philippines, Singapore, Thailand, and Brunei. ASEAN headquarters are in Jakarta, Indonesia. In January 1992, ASEAN agreed to create the ASEAN Free Trade Area (AFTA).

asylum: To give protection, security, or shelter to someone who is threatened by political or religious persecution.

atoll: A coral island, consisting of a strip or ring of coral surrounding a central lagoon.

atomic weapons: Weapons whose extremely violent explosive power comes from the splitting of the nuclei of atoms (usually uranium or plutonium) by neutrons in a rapid chain reaction. These weapons may be referred to as atom bombs, hydrogen bombs, or H-bombs.

austerity measures: Steps taken by a government to conserve money or resources during an economically difficult time, such as cutting back on federally funded programs.

Australoid: Pertains to the type of aborigines, or earliest inhabitants, of Australia.

Austronesian language: A family of languages which includes practically all the languages of the Pacific Islands—Indonesian, Melanesian, Polynesian, and Micronesian sub-families. Does not include Australian or Papuan languages.

authoritarianism: A form of government in which a person or group attempts to rule with absolute authority without the representation of the citizens.

autonomous state: A country which is completely self-governing, as opposed to being a dependency or part of another country.

autonomy: The state of existing as a self-governing entity. For instance, when a country gains its independence from another country, it gains autonomy.

Axis Powers: The countries aligned against the Allied Nations in World War II, originally applied to Nazi Germany and Fascist Italy (Rome-Berlin Axis), and later extended to include Japan.

Baha'i: The follower of a religious sect founded by Mirza Husayn Ali in Iran in 1863.

Baltic states: The three formerly communist countries of Estonia, Latvia, and Lithuania that border on the Baltic Sea.

Bantu language group: A name applied to the languages spoken in central and south Africa.

Baptist: A member of a Protestant denomination that practices adult baptism by complete immersion in water.

barren land: Unproductive land, partly or entirely treeless.

barter: Trade practice where merchandise is exchanged directly for other merchandise or services without use of money.

bicameral legislature: A legislative body consisting of two chambers, such as the U.S. House of Representatives and the U.S. Senate.

bill of rights: A written statement containing the list of privileges and powers to be granted to a body of people, usually introduced when a government or other organization is forming.

black market: A system of trade where goods are sold illegally, often for excessively inflated prices. This type of trade usually develops to avoid paying taxes or tariffs levied by the government, or to get around import or export restrictions on products.

bloodless coup: The sudden takeover of a country's government by hostile means but without killing anyone in the process.

boat people: Used to describe individuals (refugees) who attempt to flee their country by boat.

Bolshevik Revolution: A revolution in 1917 in Russia when a wing of the Russian Social Democratic party seized power. The Bolsheviks advocated the violent overthrow of capitalism.

bonded labor: Workers bound to service without pay; slaves.

border dispute: A disagreement between two countries as to the exact location or length of the dividing line between them.

Brahman: A member (by heredity) of the highest caste among the Hindus, usually assigned to the priesthood.

Buddhism: A religious system common in India and eastern Asia. Founded by and based upon the teachings of Siddhartha Gautama, Buddhism asserts that suffering is an inescapable part of life. Deliverance can only be achieved through the practice of charity, temperance, justice, honesty, and truth.

buffer state: A small country that lies between two larger, possibly hostile countries, considered to be a neutralizing force between them.

bureaucracy: A system of government that is characterized by division into bureaus of administration with their own divisional heads. Also refers to the inflexible procedures of such a system that often result in delay.

Byzantine Empire: An empire centered in the city of Constantinople, now Istanbul in present-day Turkey.

CACM see Central American Common Market.

canton: A territory or small division or state within a country.

capital punishment: The ultimate act of punishment for a crime, the death penalty.

capitalism: An economic system in which goods and services and the means to produce and sell them are privately owned, and prices and wages are determined by market forces.

Caribbean Community and Common Market (CARICOM): Founded in 1973 and with its headquarters in Georgetown, Guyana, CARICOM seeks the establishment of a common trade policy and increased cooperation in the Caribbean region. Includes 13 English-speaking Caribbean nations: Antigua and Barbuda, the Bahamas, Barbados, Belize, Dominica, Grenada, Guyana, Jamaica, Montserrat, Saint Kitts-Nevis, Saint Lucia, St. Vincent/Grenadines, and Trinidad and Tobago.

CARICOM see Caribbean Community and Common Market.

cartel: An organization of independent producers formed to regulate the production, pricing, or marketing practices of its members in order to limit competition and maximize their market power.

cash crop: A crop that is grown to be sold rather than kept for private use.

caste system: One of the artificial divisions or social classes into which the Hindus are rigidly separated according to the religious law of Brahmanism. Membership in a caste is hereditary, and the privileges and disabilities of each caste are transmitted by inheritance.

Caucasian or Caucasoid: The white race of human beings, as determined by genealogy and physical features.

ceasefire: An official declaration of the end to the use of military force or active hostilities, even if only temporary.

censorship: The practice of withholding certain items of news that may cast a country in an unfavorable light or give away secrets to the enemy.

census: An official counting of the inhabitants of a state or country with details of sex and age, family, occupation, possessions, etc.

Central American Common Market (CACM): Established in 1962, a trade alliance of five Central American nations. Participating are Costa Rica, El Salvador, Guatemala, Honduras, and Nicaragua.

Central Powers: In World War I, Germany and Austria-Hungary, and their allies, Turkey and Bulgaria.

centrist position: Refers to opinions held by members of a moderate political group; that is, views that are somewhere in the middle of popular thought between conservative and liberal.

cession: Withdrawal from or yielding to physical force.

chancellor: A high-ranking government official. In some countries it is the prime minister.

Christianity: The religion founded by Jesus Christ, based on the Bible as holy scripture.

Church of England: The national and established church in England. The Church of England claims continuity with the branch of the Catholic Church that existed in England before the Reformation. Under Henry VIII, the spiritual supremacy and jurisdiction of the Pope were abolished, and the sovereign (king or queen) was declared head of the church.

circuit court: A court that convenes in two or more locations within its appointed district.

CIS see Commonwealth of Independent States

city-state: An independent state consisting of a city and its surrounding territory.

civil court: A court whose proceedings include determinations of rights of individual citizens, in contrast to criminal proceedings regarding individuals or the public.

civil jurisdiction: The authority to enforce the laws in civil matters brought before the court.

civil law: The law developed by a nation or state for the conduct of daily life of its own people.

civil rights: The privileges of all individuals to be treated as equals under the laws of their country; specifically, the rights given by certain amendments to the U.S. Constitution.

civil unrest: The feeling of uneasiness due to an unstable political climate, or actions taken as a result of it.

civil war: A war between groups of citizens of the same country who have different opinions or agendas. The Civil War of the United States was the conflict between the states of the North and South from 1861 to 1865.

Club du Sahel: The Club du Sahel is an informal coalition which seeks to reverse the effects of drought and the desertification in the eight Sahelian zone countries: Burkina Faso, Chad, Gambia, Mali, Mauritania, Niger, Senegal, and the Cape Verde Islands. Headquarters are in Ouagadougou, Burkina Faso.

CMEA see Council for Mutual Economic Assistance.

coalition government: A government combining differing factions within a country, usually temporary.

Cold War: Refers to conflict over ideological differences that is carried on by words and diplomatic actions, not by military action. The term is usually used to refer to the tension that existed between the United States and the USSR from the 1950s until the breakup of the USSR in 1991.

collective bargaining: The negotiations between workers who are members of a union and their employer for the purpose of deciding work rules and policies regarding wages, hours, etc.

collective farm: A large farm formed from many small farms and supervised by the government; usually found in communist countries.

collective farming: The system of farming on a collective where all workers share in the income of the farm.

colonial period: The period of time when a country forms colonies in and extends control over a foreign area.

colonist: Any member of a colony or one who helps settle a new colony.

colony: A group of people who settle in a new area far from their original country, but still under the jurisdiction of that country. Also refers to the newly settled area itself.

COMECON see Council for Mutual Economic Assistance.

commerce: The trading of goods (buying and selling), especially on a large scale, between cities, states, and countries.

commission: A group of people designated to collectively do a job, including a government agency with certain law-making powers. Also, the power given to an individual or group to perform certain duties.

common law: A legal system based on custom and decisions and opinions of the law courts. The basic system of law of England and the United States.

common market: An economic union among countries that is formed to remove trade barriers (tariffs) among those countries, increasing economic cooperation. The European Community is a notable example of a common market.

commonwealth: A commonwealth is a free association of sovereign independent states that has no charter, treaty, or constitution. The association promotes cooperation, consultation, and mutual assistance among members.

Commonwealth of Independent States: The CIS was established in December 1991 as an association of 11 republics of the former Soviet Union. The members include: Russia, Ukraine, Belarus (formerly Byelorussia), Moldova (formerly Moldavia), Armenia, Azerbaijan, Uzbekistan, Turkmenistan, Tajikistan, Kazakhstan, and Kyrgyzstan (formerly Kirghiziya). The Baltic states—Estonia, Latvia, and Lithuania—did not join. Georgia maintained observer status before joining the CIS in November 1993.

Commonwealth of Nations: Voluntary association of the United Kingdom and its present dependencies and associated states, as well as certain former dependencies and their dependent territories. The term was first used officially in 1926 and is embodied in the Statute of Westminster (1931). Within the Commonwealth, whose secretariat (established in 1965) is located in London, England, are numerous subgroups devoted to economic and technical cooperation.

commune: An organization of people living together in a community who share the ownership and use of property. Also refers to a small governmental district of a country, especially in Europe.

communism: A form of government whose system requires common ownership of property for the use of all citizens. All profits are to be equally distributed and prices on goods and services are usually set by the state. Also, communism refers directly to the official doctrine of the former U.S.S.R.

compulsory: Required by law or other regulation.

compulsory education: The mandatory requirement for children to attend school until they have reached a certain age or grade level.

conciliation: A process of bringing together opposing sides of a disagreement for the purpose of compromise. Or, a way of settling an international dispute in which the disagreement is submitted to an independent committee that will examine the facts and advise the participants of a possible solution.

concordat: An agreement, compact, or convention, especially between church and state.

confederation: An alliance or league formed for the purpose of promoting the common interests of its members.

Confucianism: The system of ethics and politics taught by the Chinese philosopher Confucius.

conscription: To be required to join the military by law. Also known as the draft. Service personnel who join the military because of the legal requirement are called conscripts or draftees.

conservative party: A political group whose philosophy tends to be based on established traditions and not supportive of rapid change.

constituency: The registered voters in a governmental district, or a group of people that supports a position or a candidate.

constituent assembly: A group of people that has the power to determine the election of a political representative or create a constitution.

constitution: The written laws and basic rights of citizens of a country or members of an organized group.

constitutional monarchy: A system of government in which the hereditary sovereign (king or queen, usually) rules according to a written constitution.

constitutional republic: A system of government with an elected chief of state and elected representation, with a written constitution containing its governing principles. The United States is a constitutional republic.

Coptic Christians: Members of the Coptic Church of Egypt, formerly of Ethiopia.

Council for Mutual Economic Assistance (CMEA): Also known as Comecon, the alliance of socialist economies was established on 25 January 1949 and abolished 1 January 1991. It included Afghanistan*, Albania, Angola*, Bulgaria, Cuba, Czechoslovakia, Ethiopia*, East Germany, Hungary, Laos*, Mongolia, Mozambique*, Nicaragua*, Poland, Romania, USSR, Vietnam, Yemen*, and Yugoslavia. (Nations marked with an asterisk were observers only.)

counterinsurgency operations: Organized military activity designed to stop rebellion against an established government.

county: A territorial division or administrative unit within a state or country.

coup d'ètat or coup: A sudden, violent overthrow of a government or its leader.

criminal law: The branch of law that deals primarily with crimes and their punishments.

crown colony: A colony established by a commonwealth over which the monarch has some control, as in colonies established by the United Kingdom's Commonwealth of Nations.

Crusades: Military expeditions by European Christian armies in the eleventh, twelfth, and thirteenth centuries to win land controlled by the Muslims in the middle east.

cultivable land: Land that can be prepared for the production of crops.

Cultural Revolution: An extreme reform movement in China from 1966 to 1976; its goal was to combat liberalization by restoring the ideas of Mao Zedong.

customs union: An agreement between two or more countries to remove trade barriers with each other and to establish common tariff and nontariff policies with respect to imports from countries outside of the agreement.

cyclone: Any atmospheric movement, general or local, in which the wind blows spirally around and in towards a center. In the northern hemisphere, the cyclonic movement is usually counter-clockwise, and in the southern hemisphere, it is clockwise.

Cyrillic alphabet: An alphabet adopted by the Slavic people and invented by Cyril and Methodius in the ninth century as an alphabet that was easier for the copyist to write. The Russian alphabet is a slight modification of it.

decentralization: The redistribution of power in a government from one large central authority to a wider range of smaller local authorities.

declaration of independence: A formal written document stating the intent of a group of persons to become fully self-governing.

deficit: The amount of money that is in excess between spending and income.

deficit spending: The process in which a government spends money on goods and services in excess of its income.

deforestation: The removal or clearing of a forest.

deity: A being with the attributes, nature, and essence of a god; a divinity.

delta: Triangular-shaped deposits of soil formed at the mouths of large rivers.

demarcate: To mark off from adjoining land or territory; set the limits or boundaries of.

demilitarized zone (DMZ): An area surrounded by a combat zone that has had military troops and weapons removed.

demobilize: To disband or discharge military troops.

democracy: A form of government in which the power lies in the hands of the people, who can govern directly, or can be governed indirectly by representatives elected by its citizens.

denationalize: To remove from government ownership or control.

deportation: To carry away or remove from one country to another, or to a distant place.

depression: A hollow; a surface that has sunken or fallen in.

deregulation: The act of reversing controls and restrictions on prices of goods, bank interest, and the like.

desalinization plant: A facility that produces freshwater by removing the salt from saltwater.

desegregation: The act of removing restrictions on people of a particular race that keep them socially, economically, and, sometimes, physically, separate from other groups.

desertification: The process of becoming a desert as a result of climatic changes, land mismanagement, or both.

détente: The official lessening of tension between countries in conflict.

devaluation: The official lowering of the value of a country's currency in relation to the value of gold or the currencies of other countries.

developed countries: Countries which have a high standard of living and a well-developed industrial base.

dialect: One of a number of regional or related modes of speech regarded as descending from a common origin.

dictatorship: A form of government in which all the power is retained by an absolute leader or tyrant. There are no rights granted to the people to elect their own representatives.

dike: An artificial riverbank built up to control the flow of water.

diplomatic relations: The relationship between countries as conducted by representatives of each government.

direct election: The process of selecting a representative to the government by balloting of the voting public, in contrast to selection by an elected representative of the people.

disarmament: The reduction or depletion of the number of weapons or the size of armed forces.

dissident: A person whose political opinions differ from the majority to the point of rejection.

dogma: A principle, maxim, or tenet held as being firmly established.

dominion: A self-governing nation that recognizes the British monarch as chief of state.

dowry: The sum of the property or money that a bride brings to her groom at their marriage.

draft constitution: The preliminary written plans for the new constitution of a country forming a new government.

Druze: A member of a Muslim sect based in Syria, living chiefly in the mountain regions of Lebanon.

dual nationality: The status of an individual who can claim citizenship in two or more countries.

duchy: Any territory under the rule of a duke or duchess.

due process: In law, the application of the legal process to which every citizen has a right, which cannot be denied.

dynasty: A family line of sovereigns who rule in succession, and the time during which they reign.

Eastern Orthodox: The outgrowth of the original Eastern Church of the Eastern Roman Empire, consisting of eastern Europe, western Asia, and Egypt.

EC *see* European Community

ecclesiastical: Pertaining or relating to the church.

ecology: The branch of science that studies organisms in relationship to other organisms and to their environment.

economic depression: A prolonged period in which there is high unemployment, low production, falling prices, and general business failure.

elected assembly: The persons that comprise a legislative body of a government who received their positions by direct election.

electoral system: A system of choosing government officials by votes cast by qualified citizens.

electoral vote: The votes of the members of the electoral college.

electorate: The people who are qualified to vote in an election.

emancipation: The freeing of persons from any kind of bondage or slavery.

embargo: A legal restriction on commercial ships to enter a country's ports, or any legal restriction of trade.

emigration: Moving from one country or region to another for the purpose of residence.

empire: A group of territories ruled by one sovereign or supreme ruler. Also, the period of time under that rule.

enclave: A territory belonging to one nation that is surrounded by that of another nation.

encroachment: The act of intruding, trespassing, or entering on the rights or possessions of another.

endemic: Anything that is peculiar to and characteristic of a locality or region.

Enlightenment: An intellectual movement of the late seventeenth and eighteenth centuries in which scientific thinking gained a strong foothold and old beliefs were challenged. The idea of absolute monarchy was questioned and people were gradually given more individual rights.

epidemic: As applied to disease, any disease that is temporarily prevalent among people in one place at the same time.

Episcopal: Belonging to or vested in bishops or prelates; characteristic of or pertaining to a bishop or bishops.

ethnolinguistic group: A classification of related languages based on common ethnic origin.

EU *see* European Union

European Community: A regional organization created in 1958. Its purpose is to eliminate customs duties and other trade barriers in Europe. It promotes a common external tariff against other countries, a Common Agricultural Policy (CAP), and guarantees of free movement of labor and capital. The original six members were Belgium, France, West Germany, Italy, Luxembourg, and the Netherlands. Denmark, Ireland, and the United Kingdom became members in 1973; Greece joined in 1981; Spain and Portugal in 1986. Other nations continue to join.

European Union: The EU is an umbrella reference to the European Community (EC) and to two European integration efforts introduced by the Maastricht Treaty: Common Foreign and Security Policy (including defense) and Justice and Home Affairs (principally cooperation between police and other authorities on crime, terrorism, and immigration issues).

exports: Goods sold to foreign buyers.

external migration: The movement of people from their native country to another country, as opposed to internal migration, which is the movement of people from one area of a country to another in the same country.

faction: People with a specific set of interests or goals who form a subgroup within a larger organization.

Fascism: A political philosophy that holds the good of the nation as more important than the needs of the individual. Fascism also stands for a dictatorial leader and strong oppression of opposition or dissent.

federal: Pertaining to a union of states whose governments are subordinate to a central government.

federation: A union of states or other groups under the authority of a central government.

fetishism: The practice of worshipping a material object that is believed to have mysterious powers residing in it, or is the representation of a deity to which worship may be paid and from which supernatural aid is expected.

feudal society: In medieval times, an economic and social structure in which persons could hold land given to them by a lord (nobleman) in return for service to that lord.

final jurisdiction: The final authority in the decision of a legal matter. In the United States, the Supreme Court would have final jurisdiction.

Finno-Ugric language group: A subfamily of languages spoken in northeastern Europe, including Finnish, Hungarian, Estonian, and Lapp.

fiscal year: The twelve months between the settling of financial accounts, not necessarily corresponding to a calendar year beginning on January 1.

fjord: A deep indentation of the land forming a comparatively narrow arm of the sea with more or less steep slopes or cliffs on each side.

folk religion: A religion with origins and traditions among the common people of a nation or region that is relevant to their particular life-style.

foreign exchange: Foreign currency that allows foreign countries to conduct financial transactions or settle debts with one another.

foreign policy: The course of action that one government chooses to adopt in relation to a foreign country.

Former Soviet Union: The FSU is a collective reference to republics comprising the former Soviet Union. The term, which has been used as both including and excluding the Baltic republics (Estonia, Latvia, and Lithuania), includes the other 12 republics: Russia, Ukraine, Belarus, Moldova, Armenia, Azerbaijan, Uzbekistan, Turkmenistan, Tajikistan, Kazakhstan, Kyrgizstan, and Georgia.

free enterprise: The system of economics in which private business may be conducted with minimum interference by the government.

fundamentalist: A person who holds religious beliefs based on the complete acceptance of the words of the Bible or other holy scripture as the truth. For instance, a fundamentalist would believe the story of creation exactly as it is told in the Bible and would reject the idea of evolution.

GDP *see* gross domestic product

genocide: Planned and systematic killing of members of a particular ethnic, religious, or cultural group.

Germanic language group: A large branch of the Indo-European family of languages including German itself, the Scandinavian languages, Dutch, Yiddish, Modern English, Modern Scottish, Afrikaans, and others. The group also includes extinct languages such as Gothic, Old High German, Old Saxon, Old English, Middle English, and the like.

glasnost: President Mikhail Gorbachev's frank revelations in the 1980s about the state of the economy and politics in the Soviet Union; his policy of openness.

global warming: Also called the greenhouse effect. The theorized gradual warming of the earth's climate as a result of the burning of fossil fuels, the use of man-made chemicals, deforestation, etc.

GMT *see* Greenwich Mean Time

GNP *see* gross national product

grand duchy: A territory ruled by a nobleman, called a grand duke, who ranks just below a king.

Greek Catholic: A person who is a member of an Orthodox Eastern Church.

Greek Orthodox: The official church of Greece, a self-governing branch of the Orthodox Eastern Church.

Greenwich (Mean) Time: Mean solar time of the meridian at Greenwich, England, used as the basis for standard time throughout most of the world. The world is divided into 24 time zones, and all are related to the prime, or Greenwich mean, zone.

gross domestic product: A measure of the market value of all goods and services produced within the boundaries of a nation, regardless of asset ownership. Unlike gross national product, GDP excludes receipts from that nation's business operations in foreign countries.

gross national product: A measure of the market value of goods and services produced by the labor and property of a nation. Includes receipts from that nation's business operation in foreign countries

guerrilla: A member of a small radical military organization that uses unconventional tactics to take their enemies by surprise.

gymnasium: A secondary school, primarily in Europe, that prepares students for university.

harem: In a Muslim household, refers to the women (wives, concubines, and servants in ancient times) who live there and also to the area of the home they live in.

harmattan: An intensely dry, dusty wind felt along the coast of Africa between Cape Verde and Cape Lopez. It prevails at intervals during the months of December, January, and February.

heavy industry: Industries that use heavy or large machinery to produce goods, such as automobile manufacturing.

Holocaust: The mass slaughter of European civilians, the vast majority Jews, by the Nazis during World War II.

Holy Roman Empire: A kingdom consisting of a loose union of German and Italian territories that existed from around the ninth century until 1806.

home rule: The governing of a territory by the citizens who inhabit it.

homeland: A region or area set aside to be a state for a people of a particular national, cultural, or racial origin.

homogeneous: Of the same kind or nature, often used in reference to a whole.

Horn of Africa: The Horn of Africa comprises Djibouti, Eritrea, Ethiopia, Somalia, and Sudan.

human rights issues: Any matters involving people's basic rights which are in question or thought to be abused.

humanist: A person who centers on human needs and values, and stresses dignity of the individual.

humanitarian aid: Money or supplies given to a persecuted group or people of a country at war, or those devastated by a natural disaster, to provide for basic human needs.

hydroelectric power plant: A factory that produces electrical power through the application of waterpower.

IBRD *see* World Bank

immigration: The act or process of passing or entering into another country for the purpose of permanent residence.

imports: Goods purchased from foreign suppliers.

indigenous: Born or originating in a particular place or country; native to a particular region or area.

Indo-Aryan language group: The group that includes the languages of India; also called Indo-European language group.

Indo-European language family: The group that includes the languages of India and much of Europe and southwestern Asia.

infanticide: The act of murdering a baby.

infidel: One who is without faith or belief; particularly, one who rejects the distinctive doctrines of a particular religion.

inflation: The general rise of prices, as measured by a consumer price index. Results in a fall in value of currency.

insurgency: The state or condition in which one rises against lawful authority or established government; rebellion.

insurrectionist: One who participates in an unorganized revolt against an authority.

interim government: A temporary or provisional government.

interim president: One who is appointed to perform temporarily the duties of president during a transitional period in a government.

International Date Line: An arbitrary line at about the 180th meridian that designates where one day begins and another ends.

Islam: The religious system of Mohammed, practiced by Moslims and based on a belief in Allah as the supreme being and Mohammed as his prophet. The spelling variations, Muslim and Muhammad, are also used, primarily by Islamic people. Islam also refers to those nations in which it is the primary religion.

isthmus: A narrow strip of land bordered by water and connecting two larger bodies of land, such as two continents, a continent and a peninsula, or two parts of an island.

Judaism: The religious system of the Jews, based on the Old Testament as revealed to Moses and characterized by a belief in one God and adherence to the laws of scripture and rabbinic traditions.

Judeo-Christian: The dominant traditional religious makeup of the United States and other countries based on the worship of the Old and New Testaments of the Bible.

junta: A small military group in power in a country, especially after a coup.

khan: A sovereign, or ruler, in central Asia.

khanate: A kingdom ruled by a khan, or man of rank.

labor movement: A movement in the early to mid-1800s to organize workers in groups according to profession to give them certain rights as a group, including bargaining power for better wages, working conditions, and benefits.

land reforms: Steps taken to create a fair distribution of farmland, especially by governmental action.

landlocked country: A country that does not have direct access to the sea; it is completely surrounded by other countries.

least developed countries: A subgroup of the United Nations designation of "less developed countries;" these countries generally have no significant economic growth, low literacy rates, and per person gross national product of less than $500. Also known as undeveloped countries.

leftist: A person with a liberal or radical political affiliation.

legislative branch: The branch of government which makes or enacts the laws.

less developed countries (LDC): Designated by the United Nations to include countries with low levels of output, living standards, and per person gross national product generally below $5,000.

literacy: The ability to read and write.

Maastricht Treaty: The Maastricht Treaty (named for the Dutch town in which the treaty was signed) is also known as the Treaty of European Union. The treaty creates a European Union by: (a) committing the member states of the European Economic Community to both European Monetary Union (EMU) and political union; (b) introducing a single currency (European Currency Unit, ECU); (c) establishing a European System of Central Banks (ESCB); (d) creating a European Central Bank (ECB); and (e) broadening EC integration by including both a common foreign and security policy (CFSP) and cooperation in justice and home affairs (CJHA). The treaty entered into force on November 1, 1993.

Maghreb states: The Maghreb states include the three nations of Algeria, Morocco, and Tunisia; sometimes includes Libya and Mauritania.

majority party: The party with the largest number of votes and the controlling political party in a government.

Marshall Plan: Formally known as the European Recovery Program, a joint project between the United States and most Western European nations under which $12.5 billion in U.S. loans and grants was expended to aid European recovery after World War II.

Marxism *see* Marxist-Leninist principles

Marxist-Leninist principles: The doctrines of Karl Marx, built upon by Nikolai Lenin, on which communism was founded. They predicted the fall of capitalism, due to its own internal faults and the resulting oppression of workers.

Marxist: A follower of Karl Marx, a German socialist and revolutionary leader of the late 1800s, who contributed to Marxist-Leninist principles.

Mayan language family: The languages of the Central American Indians, further divided into two subgroups: the Maya and the Huastek.

Mecca (Mekkah): A city in Saudi Arabia; a destination of pilgrims in the Islamic world.

Mediterranean climate: A wet-winter, dry-summer climate with a moderate annual temperature range.

mestizo: The offspring of a person of mixed blood; especially, a person of mixed Spanish and American Indian parentage.

migratory workers: Usually agricultural workers who move from place to place for employment depending on the growing and harvesting seasons of various crops.

military coup: A sudden, violent overthrow of a government by military forces.

military junta: The small military group in power in a country, especially after a coup.

military regime: Government conducted by a military force.

militia: The group of citizens of a country who are either serving in the reserve military forces or are eligible to be called up in time of emergency.

minority party: The political group that comprises the smaller part of the large overall group it belongs to; the party that is not in control.

missionary: A person sent by authority of a church or religious organization to spread his religious faith in a community where his church has no self-supporting organization.

monarchy: Government by a sovereign, such as a king or queen.

Mongol: One of an Asiatic race chiefly resident in Mongolia, a region north of China proper and south of Siberia.

Mongoloid: Having physical characteristics like those of the typical Mongols (Chinese, Japanese, Turks, Eskimos, etc.).

Moors: One of the Arab tribes that conquered Spain in the eighth century.

mosque: An Islamic place of worship and the organization with which it is connected.

Muhammad (or Muhammed or Mahomet): An Arabian prophet, known as the "Prophet of Allah" who founded the religion of Islam in 622, and wrote *The Koran,* the scripture of Islam. Also commonly spelled Mohammed.

mujahideen (mujahedin or mujahedeen): Rebel fighters in Islamic countries, especially those supporting the cause of Islam.

mulatto: One who is the offspring of parents one of whom is white and the other is black.

municipality: A district such as a city or town having its own incorporated government.

Muslim: A follower of the prophet Muhammad, the founder of the religion of Islam.

Muslim New Year: A Muslim holiday. Although in some countries 1 Muharram, which is the first month of the Islamic year, is observed as a holiday, in other places the new year is observed on Sha'ban, the eighth month of the year. This practice apparently stems from pagan Arab times. Shab-i-Bharat, a national holiday in Bangladesh on this day, is held by many to be the occasion when God ordains all actions in the coming year.

NAFTA (North American Free Trade Agreement): NAFTA, which entered into force in January 1994, is a free trade agreement between Canada, the United States, and Mexico. The agreement progressively eliminates almost all U.S.-Mexico tariffs over a 10–15 year period.

nationalism: National spirit or aspirations; desire for national unity, independence, or prosperity.

nationalization: To transfer the control or ownership of land or industries to the nation from private owners.

NATO *see* North Atlantic Treaty Organization

naturalize: To confer the rights and privileges of a native-born subject or citizen upon someone who lives in the country by choice.

neutrality: The policy of not taking sides with any countries during a war or dispute among them.

Newly Independent States: The NIS is a collective reference to 12 republics of the former Soviet Union: Russia, Ukraine, Belarus (formerly Byelorussia), Moldova (formerly Moldavia), Armenia, Azerbaijan, Uzbekistan, Turkmenistan, Tajikistan, Kazakhstan, and Kirgizstan (formerly Kirghiziya), and Georgia. Following dissolution of the Soviet Union, the distinction between the NIS and the Commonwealth of Independent States

(CIS) was that Georgia was not a member of the CIS. That distinction dissolved when Georgia joined the CIS in November 1993.

Nonaligned Movement: The NAM is an alliance of third world states that aims to promote the political and economic interests of developing countries. NAM interests have included ending colonialism/neo-colonialism, supporting the integrity of independent countries, and seeking a new international economic order.

Nordic Council: The Nordic Council, established in 1952, is directed toward supporting cooperation among Nordic countries. Members include Denmark, Finland, Iceland, Norway, and Sweden. Headquarters are in Stockholm, Sweden.

North Atlantic Treaty Organization (NATO): A mutual defense organization. Members include Belgium, Canada, Denmark, France (which has only partial membership), Greece, Iceland, Italy, Luxembourg, Netherlands, Norway, Portugal, Spain, Turkey, United Kingdom, United States, and Germany.

nuclear power plant: A factory that produces electrical power through the application of the nuclear reaction known as nuclear fission.

OAPEC (Organization of Arab Petroleum Exporting countries): OAPEC was created in 1968; members include: Algeria, Bahrain, Egypt, Iraq, Kuwait, Libya, Qatar, Saudi Arabia, Syria, and the United Arab Emirates. Headquarters are in Cairo, Egypt.

OAS (Organization of American States): The OAS (Spanish: Organizaciûn de los Estados Americanos, OEA), or the Pan American Union, is a regional organization which promotes Latin American economic and social development. Members include the United States, Mexico, and most Central American, South American, and Caribbean nations.

OAS *see* Organization of American States

oasis: Originally, a fertile spot in the Libyan desert where there is a natural spring or well and vegetation; now refers to any fertile tract in the midst of a wasteland.

occupied territory: A territory that has an enemy's military forces present.

official language: The language in which the business of a country and its government is conducted.

oligarchy: A form of government in which a few people possess the power to rule as opposed to a monarchy which is ruled by one.

OPEC *see* OAPEC

open market: Open market operations are the actions of the central bank to influence or control the money supply by buying or selling government bonds.

opposition party: A minority political party that is opposed to the party in power.

Organization of Arab Petroleum Exporting Countries *see* OAPEC

organized labor: The body of workers who belong to labor unions.

Ottoman Empire: A Turkish empire founded by Osman I in the thirteenth century, that variously controlled large areas of land around the Mediterranean, Black, and Caspian Seas until it was dissolved in 1923.

overseas dependencies: A distant and physically separate territory that belongs to another country and is subject to its laws and government.

Pacific Rim: The Pacific Rim, referring to countries and economies bordering the Pacific Ocean.

pact: An international agreement.

panhandle: A long narrow strip of land projecting like the handle of a frying pan.

papyrus: The paper-reed or -rush which grows on marshy river banks in the southeastern area of the Mediterranean, but more notably in the Nile valley.

paramilitary group: A supplementary organization to the military.

parliamentary republic: A system of government in which a president and prime minister, plus other ministers of departments, constitute the executive branch of the government and the parliament constitutes the legislative branch.

parliamentary rule: Government by a legislative body similar to that of Great Britain, which is composed of two houses—one elected and one hereditary.

partisan politics: Rigid, unquestioning following of a specific party's or leader's goals.

patriarchal system: A social system in which the head of the family or tribe is the father or oldest male. Kinship is determined and traced through the male members of the tribe.

per capita: Literally, per person; for each person counted.

perestroika: The reorganization of the political and economic structures of the Soviet Union by president Mikhail Gorbachev.

periodical: A publication whose issues appear at regular intervals, such as weekly, monthly, or yearly.

political climate: The prevailing political attitude of a particular time or place.

political refugee: A person forced to flee his or her native country for political reasons.

potable water: Water that is safe for drinking.

pound sterling: The monetary unit of Great Britain, otherwise known as the pound.

prime meridian: Zero degrees in longitude that runs through Greenwich, England, site of the Royal Observatory. All other longitudes are measured from this point.

prime minister: The premier or chief administrative official in certain countries.

privatization: To change from public to private control or ownership.

protectorate: A state or territory controlled by a stronger state, or the relationship of the stronger country toward the lesser one it protects.

Protestant: A member or an adherent of one of those Christian bodies which descended from the Reformation of the sixteenth century. Originally applied to those who opposed or protested the Roman Catholic Church.

Protestant Reformation: In 1529, a Christian religious movement begun in Germany to deny the universal authority of the Pope, and to establish the Bible as the only source of truth. (*Also see* Protestant)

province: An administrative territory of a country.

provisional government: A temporary government set up during a time of unrest or transition in a country.

purge: The act of ridding a society of "undesirable" or unloyal persons by banishment or murder.

Rastafarian: A member of a Jamaican cult begun in 1930 as a semi-religious, semi-political movement.

referendum: The practice of submitting legislation directly to the people for a popular vote.

Reformation *see* Protestant Reformation

refugee: One who flees to a refuge or shelter or place of safety. One who in times of persecution or political commotion flees to a foreign country for safety.

revolution: A complete change in a government or society, such as in an overthrow of the government by the people.

right-wing party: The more conservative political party.

Roman alphabet: The alphabet of the ancient Romans from which the alphabets of most modern western European languages, including English, are derived.

Roman Catholic Church: The designation of the church of which the pope or Bishop of Rome is the head, and that holds him as the successor of St. Peter and heir of his spiritual authority, privileges, and gifts.

Roman Empire: A Mediterranean Empire, centered in the Italian peninsula, that was the most powerful state in the region in the first four centuries A.D. The empire helped spread Greek culture throughout its territory. After the fourth century, the empire served as a Christianizing influence. Although the western half of the area fell to barbarian invasions in the fifth century, the eastern half, based in Constantinople, continued until 1453.

romance language: The group of languages derived from Latin: French, Spanish, Italian, Portuguese, and other related languages.

runoff election: A deciding election put to the voters in case of a tie between candidates.

Russian Orthodox: The arm of the Orthodox Eastern Church that was the official church of Russia under the czars.

Sahelian zone: Eight countries make up this dry desert zone in Africa: Burkina Faso, Chad, Gambia, Mali, Mauritania, Niger, Senegal, and the Cape Verde Islands. (*Also see* Club du Sahel.)

savanna: A treeless or near treeless plain of a tropical or subtropical region dominated by drought-resistant grasses.

secession: The act of withdrawal, such as a state withdrawing from the Union in the Civil War in the United States.

sect: A religious denomination or group, often a dissenting one with extreme views.

segregation: The enforced separation of a racial or religious group from other groups, compelling them to live and go to school separately from the rest of society.

self-sufficient: Able to function alone without help.

separatism: The policy of dissenters withdrawing from a larger political or religious group.

serfdom: In the feudal system of the Middle Ages, the condition of being attached to the land owned by a lord and being transferable to a new owner.

Seventh-day Adventist: One who believes in the second coming of Christ to establish a personal reign upon the earth.

shamanism: A religion of some Asians and Amerindians in which shamans, who are priests or medicine men, are believed to influence good and evil spirits.

Shia Muslims: Members of one of two great sects of Islam. Shia Muslims believe that Ali and the Imams are the rightful successors of Mohammed (also commonly spelled Muhammad). They also believe that the last recognized Imam will return as a messiah. Also known as Shiites. (*Also see* Sunni Muslims.)

Shiites *see* Shia Muslims

Shintoism: The system of nature- and hero-worship which forms the indigenous religion of Japan.

Sikh: A member of a politico-religious community of India, founded as a sect around 1500 and based on the principles of monotheism (belief in one god) and human brotherhood.

Sino-Tibetan language family: The family of languages spoken in eastern Asia, including China, Thailand, Tibet, and Burma.

slash-and-burn agriculture: A hasty and sometimes temporary way of clearing land to make it available for agriculture by cutting down trees and burning them.

slave trade: The transportation of black Africans beginning in the 1700s to other countries to be sold as slaves—people owned as property and compelled to work for their owners at no pay.

Slavic languages: A major subgroup of the Indo-European language family. It is further subdivided into West Slavic (including Polish, Czech, Slovak and Serbian), South Slavic (including Bulgarian, Serbo-Croatian, Slovene, and Old Church Slavonic), and East Slavic (including Russian Ukrainian and Byelorussian).

socialism: An economic system in which ownership of land and other property is distributed among the community as a whole, and every member of the community shares in the work and products of the work.

socialist: A person who advocates socialism.

Southeast Asia: The region in Asia that consists of the Malay Archipelago, the Malay Peninsula, and Indochina.

state: The politically organized body of people living under one government or one of the territorial units that make up a federal government, such as in the United States.

subcontinent: A land mass of great size, but smaller than any of the continents; a large subdivision of a continent.

Sudanic language group: A related group of languages spoken in various areas of northern Africa, including Yoruba, Mandingo, and Tshi.

suffrage: The right to vote.

Sufi: A Muslim mystic who believes that God alone exists, there can be no real difference between good and evil, that the soul exists within the body as in a cage, so death should be the chief object of desire, and sufism is the only true philosophy.

sultan: A king of a Muslim state.

Sunni Muslims: Members of one of two major sects of the religion of Islam. Sunni Muslims adhere to strict orthodox traditions, and believe that the four caliphs are the rightful successors to Mohammed, founder of Islam. (Mohammed is commonly spelled Muhammad, especially by Islamic people.) (*Also see* Shia Muslims.)

Taoism: The doctrine of Lao-Tzu, an ancient Chinese philosopher (about 500 B.C.) as laid down by him in the *Tao-te-ching.*

tariff: A tax assessed by a government on goods as they enter (or leave) a country. May be imposed to protect domestic industries from imported goods and/or to generate revenue.

terrorism: Systematic acts of violence designed to frighten or intimidate.

Third World: A term used to describe less developed countries; as of the mid-1990s, it is being replaced by the United Nations designation Less Developed Countries, or LDC.

topography: The physical or natural features of the land.

totalitarian party: The single political party in complete authoritarian control of a government or state.

trade unionism: Labor union activity for workers who practice a specific trade, such as carpentry.

treaty: A negotiated agreement between two governments.

tribal system: A social community in which people are organized into groups or clans descended from common ancestors and sharing customs and languages.

undeveloped countries *see* least developed countries

unemployment rate: The overall unemployment rate is the percentage of the work force (both employed and unemployed) who claim to be unemployed.

UNICEF: An international fund set-up for children's emergency relief: United Nations Children's Fund (formerly United Nations International Children's Emergency Fund).

untouchables: In India, members of the lowest caste in the caste system, a hereditary social class system. They were considered unworthy to touch members of higher castes.

Warsaw Pact: Agreement made May 14, 1955 (and dissolved July 1, 1991) to promote mutual defense between Albania, Bulgaria, Czechoslovakia, East Germany, Hungary, Poland, Romania, and the USSR.

Western nations: Blanket term used to describe mostly democratic, capitalist countries, including the United States, Canada, and western European countries.

workers' compensation: A series of regular payments by an employer to a person injured on the job.

World Bank: The World Bank is a group of international institutions which provides financial and technical assistance to developing countries.

world oil crisis: The severe shortage of oil in the 1970s precipitated by the Arab oil embargo.

Zoroastrianism: The system of religious doctrine taught by Zoroaster and his followers in the Avesta; the religion prevalent in Persia until its overthrow by the Muslims in the seventh century.

Bibliography

Africa

Algeria

Lorcin, Patricia M. E. *Imperial Identities: Stereotyping, Prejudice, and Race in Colonial Algeria.* London: I. B. Tauris Publishers, 1995.

MacMaster, Neil. *Colonial Migrants and Racism: Algerians in France, 1900–62.* New York: St. Martin's Press Inc., 1997.

Metz, Helen Chapin. *Algeria: A Country Study. Area Handbook Studies.* Washington, D.C.: Federal Research Division, Library of Congress. U.S. Government, Department of the Army, 1994.

Sahnouni, Mohamed. *The Lower Paleolithic of the Maghreb: Excavations and Analyses at Ain Hanech, Algeria. Cambridge Monographs in African Archaeology 42.* BAR International Series 689. Oxford: Hadrian Books Ltd., 1998.

Angola

Bollig, M. *When War Came the Cattle Slept: Himba Oral Traditions.* Koln: R. Koppe, 1997.

Ciment, J. *Conflict and Crisis in the Post-Cold War World: Angola and Mozambique: Postcolonial Wars in Southern Africa.* New York: Facts on File, 1997.

Etienne Dostert , P., ed. "The Republic of Angola." *Africa.* Harpers Ferry, WV: Stryker-Post Publications, 1997.

Kaplan, I. *Angola: A Country Study.* Washington, D.C.: American University, Foreign Area Studies, 1979.

Oliver, R. and B. Fagan. *Africa in the Iron Age, c. 500 BC to AD 1400.* New York: Cambridge University Press, 1975.

Benin

Africa on File. New York: Facts on File, Inc., 1995.

Decalo, Samuel. *Historical Dictionary of Benin.* 3rd ed. Lanham, Md.: Scarecrow Press, 1995.

Miller, Susan Katz. "Sermon on the Farm." *International Wildlife.* March/April 1992: 4951.

Botswana

Country Profile of Botswana. McLean, Va.: SAIC, 1998.

Jackson, A. *Botswana, 1939–1945: An African Country at War.* Oxford: Clarendon Press, 1999.

Morton, R. F., and J. Ramsay, eds. *Birth of Botswana: The History of the Bechuanaland Protectorate, 1910–1966.* Gaborone: Longman Botswana, 1988.

Pickford, P. and Pickford, B. *Okavango: The Miracle Rivers.* London: New Holland, 1999.

Ramsay, J.; Morton, B. and Morton, F. *Historical Dictionary of Botswana.* 3rd ed. Lanham, Md.: Scarecrow Press, 1996.

Burkina Faso

McFarland, Daniel Miles, and Lawrence A. Rupley. *Historical Dictionary of Burkina Faso.* Second Ed. *African Historical Dictionaries,* No. 74. Lanham, Md., and London: The Scarecrow Press, 1998.

Sankara, Thomas. *Women's Liberation and the African Freedom Struggle.* London: Pathfinder, 1990.

Wilks, Ivor. *The Mossi and Akan States.* In *History of West Africa.* Third Ed. Ed. J.F.A. Ajayi and Michael Crowder. Harlow, U.K.: Longman, 1985.

Burundi

Lemarchand, René. *Burundi: Ethnic Conflict and Genocide.* Woodrow Wilson Center Press, Cambridge University Press, 1997.

Ramsay, F. Jeffress. *Burundi in Global Studies: Africa.* Connecticut: Dushkin Publishing Group/Brow, Benchmark Publishers, 1995.

Weinstein, Warren, Robert Schire. *Political Conflict and Ethnic Strategies: A Case Study of Burundi.* New York: Maxwell School of Citizenship and Public Affairs, 1976.

Cameroon

Bjornson, Richard. *The African Quest for Freedom and Identity: Cameroonian Writing and the National Experience.* Bloomington: Indiana University Press, 1991.

DeLancey, Mark. *Historical Dictionary of the Republic of Cameroon.* 2nd ed. Metuchen, N.J.: Scarecrow Press, 1990.

"Odd Man in: Cameroon. (African nation to become the 52nd member of the Commonwealth of Nations)." *The Economist (US),* October 7, 1995, vol. 337, no. 7935, p. 51.

Takougang, Joseph. *African State and Society in the 1990s: Cameroon's Political Crossroads.* Boulder, Colo.: Westview Press, 1998.

Cape Verde

Carreira, Antonio. *The People of the Cape Verde Islands: Exploitation and Emigration.* Hamden, Conn.: Archon Books, 1982

Chilcote, Ronald H. *Amilcar Cabral's Revolutionary Theory and Practice: A Critical Guide.* Boulder, Colo.: Lynne Rienner Publishers, 1991.

Foy, Colm. *Cape Verde: Politics, Economics and Society.* London: Pinter Publishers, 1988.

Lobban, Jr. Richard A. *Cape Verde: Crioulo Colony to Independent Nation.* Boulder, Colo.: Westview Press, 1995.

Russell-Wood, A.J.R. *A World on the Move: The Portuguese in Africa, Asia, and America, 1415–1808.* New York: St. Martin's Press, 1993.

Central African Republic

Decalo, Samuel. *Psychoses of Power: African Personal Dictatorships.* Boulder, Colo.: Westview Press, 1989.

Kalck, Pierre. *Central African Republic: A Failure in De-colonisation.* Trans. Barbara Thomson. New York: Praeger, 1971.

———. *Central African Republic.* Santa Barbara, Calif.: Clio Press, 1993.

———. *Historical Dictionary of the Central African Republic.* 2nd ed. Metuchen, N.J.: Scarecrow Press, 1992.

O'Toole, Thomas. *The Central African Republic: The Continent's Hidden Heart.* Boulder, Colo.: Westview Press, 1986.

Titley, Brian. *Dark Age: The Political Odyssey of Emperor Bokassa.* Montreal: McGill-Queen's University Press, 1997.

Chad

Azevedo, Mario Joaquim. *Chad : A Nation in Search of its Future.* Boulder, Colo.: Westview Press, 1998.

Collelo, Thomas, ed. *Chad: A Country Study,* 2nd ed. Washington, DC: Government Printing Office, 1990.

Decalo, Samuel. *Historical Dictionary of Chad.* 2nd ed. Metuchen, NJ: Scarecrow Press, 1987.

Nolutshungu, Sam C. *Limits of Anarchy: Intervention and State Formation in Chad.* Charlottesville: University Press of Virginia, 1996.

Wright, John L. *Libya, Chad, and the Central Sahara.* Totowa, NJ: Barnes & Noble Books, 1989.

Comoros

"Under the Volcano." *Time International,* March 23, 1998, vol. 150, no. 30, p. 35.

The Comoros: Current Economic Situation and Prospects. Washington, D.C.: World Bank, 1983.

Newitt, Malyn. *The Comoro Islands: Struggle Against Dependency in the Indian Ocean.* Boulder, Colo: Westview, 1984.

Ottenheimer, Martin. *Historical Dictionary of the Comoro Islands.* Metuchen, N.J: Scarecrow Press, 1994.

Weinberg, Samantha. *Last of the Pirates: The Search for Bob Denard.* New York: Pantheon Books, 1994.

Congo, Democratic Republic of the

Bobb, F. Scott. *Historical Dictionary of Zaire.* Metuchen, N.J.: Scarecrow Press, 1988.

Kanza, Thomas. *The Rise and Fall of Patrice Lumumba.* Cambridge, Mass.: Schenkman, 1979.

Leslie, Winsome J. *Zaire: Continuity and Political Change in an Oppressive State.* Boulder, Colo.: Westview Press, 1993.

Lumumba, Patrice. *Congo, My Country.* With a foreword and notes by Colin Legum. Transl. by Graham Heath. London: Pall Mall Press, 1962.

Mokoli, Mondonga M. *State Against Development: The Experience of Post-1965 Zaire.* Westport, Conn.: Greenwood Press, 1992.

Congo, Republic of the

Clark, John F. "Elections, Leadership, and Democracy in Congo." *Africa Today* 41, no. 3 (1994): 41–60.

Decalo, Samuel, Virginia Thompson, and Richard Adloff. *Historical Dictionary of Congo.* Lanham, Md.: The Scarecrow Press, Inc., 1996.

Fegley, Randall. *The Congo.* World Bibliographical Series, vol. 162, Oxford: Clio Press, 1993.

Hilton, Anne. *The Kingdom of the Kongo.* Oxford: Clarendon Press, 1985.

Vansina, Jan. *The Tio Kingdom of the Middle Congo: 1880–1892.* London: Oxford University Press, 1973.

Cote d'Ivoire

Ajayi, J.F. Ade, and Michael Crowder, eds. *History of West Africa, 2.* New York: Colombia University Press, 1974.

Clark, John F. and David E. Gardinier, eds. *Political Reform in Francophone Africa.* Boulder, Colo.: Westview Press, 1997.

EIU Country Reports. London: Economist Intelligence Unit, April 1999.

Handloff, Robert E. *Côte d'Ivoire: A Country Study. Area Handbook Series.* Third Edition. Washington, DC: Federal Research Division, Library of Congress, 1991.

Mundt, Robert. *Historical Dictionary of the Ivory Coast.* Metuchen, N.J.: Scarecrow Press, 1995.

Djibouti

Darch, Colin. *A Soviet View of Africa: An Annotated Bibliography on Ethiopia, Somalia, and Djibouti.* Boston: G. K. Hall, 1980.

Koburger, Charles W. *Naval Strategy East of Suez: The Role of Djibouti.* New York: Praeger, 1992.

Schrader, Peter J. *Djibouti.* Santa Barbara, Calif.: Clio Press, 1991.

Tholomier, Robert. *Djibouti: Pawn of the Horn of Africa.* Metuchen, N.J.: Scarecrow, 1981.

Woodward, Peter. *The Horn of Africa: State Politics and International Relations.* London: I. B. Tauris, 1996.

Egypt

Daly, M.W., ed. *The Cambridge History of Egypt*. New York: Cambridge University Press, 1998.

Metz, Helen Chapin. *Egypt, a Country Study,* 5th ed. Washington, D.C.: Library of Congress, 1991.

Rubin, Barry M. *Islamic Fundamentalism in Egyptian Politics*. New York: St. Martin's Press, 1990.

Shamir, Shimon, ed. *Egypt from Monarchy to Republic: A Reassessment of Revolution and Change*. Boulder, Colo.: Westview Press, 1995.

Equatorial Guinea

Fegley, Randall. *Equatorial Guinea: An African Tragedy*. New York: P. Lang, 1989.

Klitgaard, Robert E. *Tropical Gangsters*. New York: Basic Books, 1990.

Liniger-Goumaz, Max. *Historical Dictionary of Equatorial Guinea*. 2nd ed. Metuchen, NJ: Scarecrow Press, 1988.

———. *Small is Not Always Beautiful: The Story of Equatorial Guinea*. Translated from the French by John Wood. Totowa, N.J.: Barnes & Noble Books, 1989.

Sundiata, I.K. *Equatorial Guinea: Colonialism, State Terror, and the Search for Stability*. Boulder, Colo.: Westview Press, 1990.

Eritrea

Connell, Dan. *Against All Odds*. Trenton, N.J.: Red Sea Press, 1993.

Doombos, Martin, et al., eds. *Beyond the Conflict in the Horn*. Trenton, NJ: Red Sea Press, 1992.

Gebremedhin, Tesfa G. *Beyond Survival: The Economic Challenges of Agriculture and Development in Post-Independence Eritrea*. Trenton, NJ: Red Sea Press, 1997.

Okbazghi, Yohannes. *Eritrea: A Pawn in World Politics*. Gainesville: University of Florida Press, 1991.

Ethiopia

Bahru, Z. *A History of Modern Ethiopia, 1855–1974*. Athens: Ohio University Press, 1991.

Hassen, M. *The Oromo of Ethiopia: A History 1570–1860*. Cambridge: Cambridge University Press, 1990.

Kaplan, S. *The Beta Israel (Falasha) in Ethiopia: From Earliest Times to the Twentieth Century*. New York: New York University Press, 1992.

Marcus, H. G. *A History of Ethiopia*. Berkeley: University of California Press, 1994.

Prouty, C. and Rosenfeld, E. *Historical Dictionary of Ethiopia*. Metuchen, NJ: The Scarecrow Press, Inc., 1994.

Gabon

Alexander, Caroline. *One Dry Season*. New York: Alfred A. Knopf, 1989.

Gall, Timothy L., ed. *Worldmark Encyclopedia of the Nations*. 9th ed. Detroit: Gale Research, 1998.

Iliffe, John. *Africans: The History of a Continent*. Cambridge University Press, 1995.

Ungar, Sanford J. *Africa: The People and Politics of an Emerging Continent*. New York: Simon & Schuster, 1989.

The Gambia

Else, David. *The Gambia and Senegal*. Oakland, Calif.: Lonely Planet Publications, 1999.

Gailey, Harry A. *Historical Dictionary of The Gambia*. 2nd ed. African Historical Dictionaries, No. 4. Metuchen, N.J.: Scarecrow Press, 1987.

Gall, Timothy L., ed. *Worldmark Encyclopedia of Cultures and Daily Life*. vol. 1: Africa 1998.

Vollmer, Jurgen. *Black Genesis, African Roots: A voyage from Juffure, The Gambia, to Mandingo country to the slave port of Dakar, Senegal*. New York: St. Martin's Press, 1980.

Wright, Donald R. "The world and a very small place in Africa (history of Niumi)." *Sources and Studies in World History*. New York: M.E. Sharpe, Armonk, 1997.

Ghana

Ardayfio-Schandorf, Elizabeth and Kate Kwafo-Akoto. *Women in Ghana: An Annotated Bibliography*. Accra: Woeli Publishing Services, 1990.

Berry, LaVerle. *Ghana: A Country Study. Area Handbook Series*. Third Edition. Washington, DC: Federal Research Division, Library of Congress, 1995.

Davidson, Basil. *Black Star: A View of the Life and Times of Kwame Nkrumah*. Boulder: Westview Press, 1989.

Glickman, Harvey, ed. *Political Leaders of Contemporary Africa South of the Sahara: A Biographical Dictionary*. New York: Greenwood Press, 1992.

Guinea

Africa on File. New York: Facts on File, 1995.

Gall, Timothy L., ed. *Worldmark Encyclopedia of Cultures and Daily Life*. vol. 1: Africa. Detroit: Gale Research, 1998.

Nelson, Harold D. et al, eds. *Area Handbook for Guinea. Foreign Area Studies*. Washington, DC.: American University, 1975.

O'Toole, Thomas. *Historical Dictionary of Guinea*. Third Edition. Lanham, MD: London: The Scarecrow Press, 1994.

Guinea-Bissau

Africa on File. New York: Facts on File, 1997.

EIU Country Reports. London: Economist Intelligence Unit, Ltd., April 8, 1999.

Forrest, Joshua. *Guinea-Bissau: Power, Conflict, and Renewal in a West African Nation*. Boulder: Westview Press, 1992.

Lopes, Carlos. *Guinea-Bissau: From Liberation Struggle to Independent Statehood*. Boulder, Colo.: Westview Press, 1987.

Pedlar, Frederick. *Main Currents of West African History, 1940–1978*. London: The Macmillan Press, Ltd., 1979.

Kenya

Cohen, W. David, and E. S. *Atieno Odhiambo. Burying SM: The Politics of Knowledge and the Sociology of Power in Africa.* Portsmouth, NH: Heinemann, 1992.

Miller, Norman, and Roger Yeager. *Kenya: The Quest for Prosperity* (2nd edition). Boulder, Colo.: Westview Press, 1994.

Mwaniki, Nyaga. "The Consequences of Land Subdivision in Northern Embu, Kenya." *The Journal of African Policy Studies,* 2(1), 1996.

Ogot, A. Bethwell. *Historical Dictionary of Kenya.* Metuchen, NJ: The Scarecrow Press, Inc., 1981.

Lesotho

Eldredge, E. A. *A South African Kingdom: The Pursuit of Security in Nineteenth-Century Lesotho.* Cambridge: Cambridge University Press.

Haliburton, G. *Historical Dictionary of Lesotho.* African Historical Dictionaries, No. 10, Metuchen, N.J.: The Scarecrow Press, Inc., 1977.

Khaketla, B. M. *Lesotho 1970: An African Coup Under the Microscope.* London: Hurst, 1971.

Liberia

Africa on File. New York: 1995 Facts on File, Inc., 1995.

Africa South of the Sahara. London: Europa Publishers, 1998.

Dunn, D. Elwood, and Svend E. Holsoe. *Historical Dictionary of Liberia.* African Historical Dictionaries, No. 38. Metuchen, N.J., and London: The Scarecrow Press, Inc., 1985.

Dunn, D. Elwood and S. Byron Tarr. *Liberia: A National Polity in Transition.* Metuchen, N.J., and London: The Scarecrow Press, Inc., 1988.

Nelson, Harold D., ed. *Liberia: A Country Study.* Third Edition. Washington, DC: The American University, 1985.

Libya

Gall, Timothy, L., ed. *Worldmark Encyclopedia of Cultures and Daily Life.* vol. 1: Africa. Detroit: Gale Research, 1998.

Haley, P. Edward. *Qadhafi and the United States Since 1969.* New York: Praeger, 1984.

Simonis, Damien, *et al. North Africa.* Hawthorn, Aus.: Lonely Planet Publications, 1995.

Simons, Geoff. *Libya: The Struggle for Survival.* New York: St. Martin's Press, 1996.

Vanderwalle, Dirk ed. *Qadhafi's Libya: 19691994.* New York: St. Martin's Press, 1995.

Madagascar

Allen, Philip M. *Madagascar: Conflicts of Authority in the Great Island.* Boulder, Colo.: Westview Press, 1994.

Covell, Maureen. *Historical Dictionary of Madagascar.* Lanham, Md.: Scarecrow Press, 1995.

Kent, Raymond K. *Early Kingdoms in Madagascar 1500–1700.* New York: Holt, Rinehart and Winston, 1970.

Stratton, Arthur. *The Great Red Island.* New York: Charles Scribner's Sons, 1964.

Verin, Pierre. *The History of Civilisation in North Madagascar.* Rotterdam: A.A. Balkema, 1986.

Malawi

Baker, Colin. *State of Emergency: Crisis in Central Africa, Nyasaland 1959–1960.* London: I. B. Tauris Publishers, 1997.

Crosby, Cynthia A. *Historical Dictionary of Malawi.* Metuchen, N.J.: The Scarecrow Press, 2nd edition, 1993.

Phiri, D. D. *From Nguni to Ngoni: A History of the Ngoni Exodus from Zululand and Swaziland to Malawi, Tanzania and Zambia.* Limbe, Malawi: Popular Publications, 1982.

Mali

Economist Intelligence Unit. *Country Profile: Mali, 1998.* London: The Economist, 1998.

Historical Dictionary of Mali. Second edition. Metuchen, New Jersey: Scarecrow Press, 1986.

Mann, Kenny. *Ghana, Mali, Songhay: The Western Sudan.* Parsippany, NJ: Dillon Press, 1996.

McIntosh, Roderick J. *The Peoples of the Middle Niger: The Island of Gold.* Malden, MA: Blackwell Publishers, 1998.

Mauritania

Africa. Hawthorn, Aus.: Lonely Planet Publications, 1995.

Goodsmith, Lauren. *The Children of Mauritania.* Minneapolis: Carolrhoda Books, 1993.

McLachlan, Anne and Keith. *Morocco Handbook.* Bath, Eng.: Footprint Handbooks. 1993.

Pazzanita, Anthony G. *Historical Dictionary of Mauritania.* Lanham, Md.: Scarecrow Press, Inc., 1966.

Thompson, Virginia, and Adloff, Richard. *The Western Saharans.* Totowa, N.J.: Barnes and Noble Books, 1980.

Mauritius

Bunge, Frederica M., ed. *Indian Ocean. Five Island Countries.* Washington, D.C.: Dept of the Army, 1983.

Bunwaree, Sheila S. *Mauritian Education in a Global Economy.* Stanley, Rose Hill, Mauritius: Editions de l'Océan Indien, 1994.

Butlin, Ron (ed). *Mauritian Voices. New Writing in English.* Newcastle Upon Tyne: Flambard Press, 1997.

Selvon, Sydney. *Historical Dictionary of Mauritius.* Metuchen, N.J.: Scarecrow Press, 1991.

Morocco

Cook, Weston, F. *The Hundred Year War for Morocco: Gunpowder and the Military Revolution in the Early Modern Muslim World.* Boulder, Colo.: Westview Press, 1985.

Jereb, James F. *Arts and Crafts of Morocco.* New York: Chronicle Books, 1996.

Hermes, Jules M. *The Children of Morocco.* Minneapolis: Carolrhoda Books, 1995.

Hoisington, William A., Jr. *Lyautey and the French Conquest of Morocco.* New York: St. Martins Press, 1995.

Simonis, Damien, et al. *North Africa.* Hawthorn, Aus.: Lonely Planet Publications, 1995.

Mozambique

Azevedo, Mario Joaquim. *Historical Dictionary of Mozambique.* Metuchen, N.J.: Scarecrow Press, 1991.

Davidson, Basil. *Africa in History.* New York: Simon and Schuster, 1991.

Magnin, Andre; and Jacques Soulillou, eds., *Contemporary Art of Africa.* New York: Harry N. Abrams, Inc. Publishers, 1996.

Newitt, M. D. D. *A History of Mozambique.* Bloomington: Indiana University Press, 1995.

Slater, Mike. *Mozambique.* London: New Holland Publishers Ltd, 1997.

Namibia

Breytenbach, Cloete. *Namibia: Birth of a Nation.* South Africa: LUGA Publishers, 1989.

Cliffe, Lionel. *The Transition to Independence in Namibia.* Boulder, Colo.: Lynne Rienner Publishers, 1994.

Namibia: A Nation Is Born. Washington, D.C.: U.S. Dept. of State, 1990.

Swaney, Deanna. *Zimbabwe, Botswana, and Namibia: A Lonely Planet Travel Guide.* Hawthorn, Australia, Lonely Planet Publications, 1995.

Niger

Beckwith, Carol. *Nomads of Niger.* New York: Henry N. Abrams, 1983.

Charlick, Robert. *Niger. Personal Rule and Survival in the Sahel.* Boulder, Colo.: Westview, 1991.

Cooper, Barbara. *Marriage in Maradi: Gender and Culture in a Hausa Society in Niger 1900-1989.* Portsmouth, N.H.: Heinemann, 1997.

Decalo, Samuel. *Historical Dictionary of Niger.* Metuchen, N.J.: Scarecrow Press, 1979.

Gall, Timothy L., ed. *Worldmark Encyclopedia of Cultures and Daily Life.* vol. 1: Africa. Detroit: Gale Research, 1998.

Nigeria

Diamond, Larry, Anthony Kirk-Greene, and Oyeleye Oyediran, eds. *Transition Without End: Nigerian Politics and Civil Society under Babangida.* Boulder, Colo.: Lynne Rienner Publishers, 1997.

Graf, William D. *The Nigerian State: Political Economy, State Class, and Political System in the Post-Colonial Era.* Portsmouth, N.H.: Heineman, 1988.

Ihonvbere, Julius O., and Timothy Shaw. *Illusions of Power: Nigeria in Transition.* Trenton, NJ: Africa World Press, 1998.

Osaghae, Eghosa E. *Crippled Giant: Nigeria Since Independence.* Bloomington: Indiana University Press, 1998.

Wesler, Kit W. *Historical Archaeology in Nigeria.* Trenton, N.J.: Africa World Press, 1998.

Rwanda

Hodd, M. *East African Handbook,* Chicago: Passport Books, 1994.

Nyrop, Richard., *et al. Rwanda: A Country Study.* Washington, D.C.: U.S. Government Printing Office, 1984.

Pierce, Julian R. *Speak Rwanda.* New York: Picador USA, 1999.

Webster, J.B., B.A. Ogot, and J.P. Chretien. "The Great Lakes Region: 1500–1800." In *The General History of Africa,* Volume V, Calif.: Heinemann, UNESCO, 1998

Sao Tome and Principe

Carreira, Antonio. *The People of the Cape Verde Islands: Exploitation and Emigration.* Connecticut: Archon Books, 1982.

Davidson, Basil. *Black Mother: The Years of the African Slave Trade.* Boston: Little, Brown, and Co., 1961.

Denny, L.M. and Donald I. Ray. *São Tomé and Príncipe: Economics, Politics and Society.* London and New York: Pinter Publishers, 1989.

Garfield, Robert. *A History of São Tomé Island, 1470–1655: The Key to Guinea.* San Francisco: Mellen Research University Press, 1992.

Hodges, Tony and Malyn Hewitt. *São Tomé and Príncipe: From Plantation Colony to Microstate.* Boulder and London: Westview Press, 1988.

Thomas, Hugh. *The Slave Trade.* New York: Touchstone Books-Simon and Schuster, 1997.

Senegal

Clark, Andrew Francis, ed. *Historical Dictionary of Senegal.* 2nd ed. African Historical Dictionaries, No. 23. Metuchen, N.J.: The Scarecrow Press, Inc., 1994.

Else, David. *The Gambia and Senegal.* Oakland, Calif.: Lonely Planet Publications, 1999.

Gellar, Sheldon. *Senegal: An African Nation between Islam and the West.* Second Edition. Boulder, Colo.: Westview Press, 1995.

Sharp, Robin. *Senegal: A State of Change.* Oxford, UK: Oxfam, 1994.

Seychelles

Bennett, George. *Seychelles.* Oxford, England; Santa Barbara, CA: Clio Press, 1993.

Franda, Marcus. *The Seychelles, Unquiet Islands.* Boulder, Colo.: Westview 1982.

McAteer, William. *Rivals in Eden: A History of the French Settlement and British Conquest of the Seychelles Islands, 1724–1818.* Sussex, Eng.: Book Guild, 1990.

Vine, Peter. *Seychelles.* London, Eng.: Immel Pub. Co., 1989.

Sierra Leone

Africa South of the Sahara. "Sierra Leone." London: Europa Publishers, 1997.

Alie, Joe. *A New History of Sierra Leone.* New York: St. Martin's Press, 1990.

Foray, Cyril P. *Historical Dictionary of Sierra Leone.* African Historical Dictionaries, No. 12. Metuchen, N.J.: The Scarecrow Press, 1977.

Gall, Timothy L., ed. *Worldmark Encyclopedia of Cultures and Daily Life.* vol. 1: Africa. Detroit: Gale Research, 1998.

Somalia

Barnes, Virginia Lee. *Aman: The Story of a Somali Girl.* New York: Pantheon, 1994.

Clarke, Walter S., and Jeffrey Ira Herbst. *Learning from Somalia: The Lessons of Armed Humanitarian Intervention.* Boulder, Colo.: Westview Press, 1997.

DeLancey, Mark. *Blood and Bone: The Call of Kinship in Somali Society.* Lawrenceville, N.J.: Red Sea, 1994.

———, et al, eds. *Somalia.* Santa Barbara, Calif.: Clio, 1988.

DeLancey, Mark. *Blood and Bone: The Call of Kinship in Somali Society.* Lawrenceville, N.J.: Red Sea, 1994.

Metz, Helen Chapin, ed. *Somalia: A Country Study.* 4th ed. Washington, DC: Library of Congress, 1993.

South Africa

Brynes, Rita M. ed. *South Africa: A Country Study.* Washington D.C.: U.S. Government Printing Office, 1997.

Mostert, N. *Frontiers: The Epic of South Africa's Creation and the Tragedy of the Xhosa People.* New York: Knopf, 1992.

Riley, E. *Major Political Events in South Africa, 1948–1990.* New York: Facts on File, 1991.

Thompson, L. M. *A History of South Africa.* London: Yale University Press, 1990.

Sudan

Alier, Abel. *Southern Sudan.* Reading:Ithaca Press,1990.

Bovill, Edward William. *The Golden Trade of the Moors.*Princeton, New Jersey: Marcus Wiener Publishers,1995.

Metz, Helen Chapin, ed. *Sudan: A Country Study.* Federal Research Division/Library of Congress,1991.

Stewart, Judy. *A Family in Sudan.* Minneapolis: Lerner Publishing Co., 1988.

Voll, John Obert and Sarah Potts Voll. *The Sudan: Unity and Diversity in a Multicultural State.* Boulder, Col.: Westview Press, 1985.

Swaziland

Bonner, P. M. "Swati II, c. 1826–1865." In *Black Leaders in Southern African History.* Edited by Christopher Saunders. London: Heinemann, 1979: 61–74.

Davies, R. H.; O'Meara, D. and Dlamini, S. *The Kingdom of Swaziland: A Profile.* London: Zed Books, 1985.

Grotpeter, J. J. *Historical Dictionary of Swaziland.* Metuchen, N.J.: The Scarecrow Press, 1975.

Kuper, H. *Sobhuza II: Ngwenyama and King of Swaziland.* London: Duckworth, 1978.

Williams, G. and B. Hackland. *The Dictionary of Contemporary Politics of Southern Africa.* New York: Macmillan, 1988.

Tanzania

Hodd, M. *East African Handbook.* Chicago: Passport Books, 1994.

Hughes, A. J. *East Africa.* Baltimore: Penguin Books, 1969.

Iliffe, J. *A Modern History of Tanganyika.* Cambridge, Eng.: Cambridge University Press, 1979.

Maddox, Gregory, et al., eds. *Custodians of the Land: Ecology and Culture in the History of Tanzania.* Athens: Ohio University Press, 1996.

Togo

Ajayi, J.F.A. and Michael Crowder, eds. *History of West Africa,* Vol. 1. London,Eng.: Congman Group Limited, 1971.

———. *History of West Africa,* Vol. 2. Londond, Eng.: Congman Group Limited, 1971.

Decalo, Samuel. *Historical Dictionary of Togo.* 3rd ed. Metuchen, NJ: Scarccrow Press, 1996.

Packer, George, *The Village of Waiting.* New York: Vintage Books, 1988.

Tunisia

Ali, Wijdan. *Modern Islamic Art: Development and Continuity.* Gainesville, Fla.: University Press of Florida, 1997.

Lancel, Serge. Nevill, Antonia (translator). *Hannibal.* Oxford, Eng.: Blackwell Publications, 1998.

Ling, Dwight L. *Morocco and Tunisia: A Comparative History.* Washington DC: University Press of America, 1979.

Memmi, Albert. *Pillar of Salt.* Boston: Beacon Press, 1992.

Uganda

Byrnes, Rita. ed. *Uganda: A Country Study.* Washington, D.C.: U.S. Government Printing Office, 1992.

Gall, Timothy L., ed. *Worldmark Encyclopedia of Cultures and Daily Life.* vol. 1: Africa. Detroit: Gale Research, 1998.

Hodd, M. *East African Handbook.* Chicago: Passport Books, 1994.

Jorgenson, Jan Jelmett. *Uganda: A Modern History.* New York: St. Martin's Press, 1981.

Nzita, Richard. *Peoples and Cultures of Uganda.* Kampala: Fountain Publishers, 1993.

Zambia

Burdette, M. M. *Zambia Between Two Worlds.* Boulder, Colo.: Westview Press, 1988.

Dresang, E. *The Land and People of Zambia.* Philadelphia: Lippincott, 1975.

Grotpeter, J. J.; Siegel, B. V. and Pletcher, J. R. *Historical Dictionary of Zambia.* Lanham, Maryland: The Scarecrow Press, Inc., 1998.

Zimbabwe

Dewey, William Joseph. *Legacies of Stone: Zimbabwe Past and Present.* Tervuren: Royal Museum for Central Africa, 1997.

Nelson, H. D. *Zimbabwe: A Country Study.* Washington, D. C.: U.S. Government Printing Office, 1983.

Rasmussen, R. Kent and Rubert, Steven, C. *Historical Dictionary of Zimbabwe.* 2nd ed. Metuchen, N.J.: The Scarecrow Press, 1990.

Sheehan, Sean. *Zimbabwe.* New York: M. Cavendish, 1996.

Americas

Antigua and Barbuda

Coram, Robert. *Caribbean Time Bomb: the United States' Complicity in the Corruption of Antigua.* New York: William Morrow and Company, Inc., 1993.

Dyde, Brian. *Antigua and Barbuda: the Heart of the Caribbean.* London: Macmillan Publishers, 1990.

Kurlansky, Mark. *A Continent of Islands: Searching for the Caribbean Destiny.* Reading, Mass.: Addison-Wesley Publishing Co., 1992.

Argentina

American University. *Argentina: A Country Study,* 3rd ed. Washington, DC: Government Printing Office, 1985.

Sarlo Sabajanes, Beatriz. *Jorge Luis Borges: A Writer on the Edge.* New York: Verso, 1993.

Timerman, Jacobo. *Prisoner Without a Name, Cell Without a Number.* New York: Knopf, 1981.

Tulchin, Joseph S. *Argentina: The Challenges of Modernization.* Wilmington, Del.: Scholarly Resources, 1998.

Worldmark Press, Ltd. *Worldmark Encyclopedia of Cultures and Daily Life.* vol. 2: Americas 1998.

Bahamas

Hamshere, Cyril. *The British in the Caribbean.* Cambridge, Mass.: Harvard University Press, 1972.

Kurlansky, Mark. *A Continent of Islands: Searching for the Caribbean Destiny.* Reading, Mass.: Addison-Wesley Publishing Co., 1992.

Marx, Jenifer. *Pirates and Privateers of the Caribbean.* Malabar, Fla.: Krieger Publishing Company, 1992.

Barbados

Beckles, Hilary. *A History of Barbados: From Amerindian Settlement to Nation-State.* New York: Cambridge University Press, 1990.

Pariser, Harry S. *Adventure Guide to Barbados.* Edison, N.J.: Hunter Publishing, 1995.

Wilder, Rachel, ed. *Barbados.* Boston: Houghton Mifflin Co., 1993

Belize

Bolland, O. Nigel. *Belize: A New Nation in Central America.* Boulder, Colo.: Westview, 1986.

Edgell, Zee. *Beka Lamb.* London: Heinemann, 1982.

Fernandez, Julio A. *Belize: Case Study for Democracy in Central America.* Brookfield, Vt.: Avebury, 1989.

Gall, Timothy L., ed. *Worldmark Encyclopedia of Cultures and Daily Life.* vol. 2: Americas. Detroit: Gale Research, 1998.

Mallan, Chicki. *Belize Handbook.* Chico, Calif.: Moon Publications, 1991.

Bolivia

Blair, David Nelson. *The Land and People of Bolivia.* New York: J.B. Lippincott, 1990.

Hudson, Rex A. and Dennis M. Hanratty. *Bolivia, a Country Study.* 3rd ed. Washington, DC: Government Printing Office, 1991.

Morales, Waltrand Q. *Bolivia: Land of Struggle.* Boulder, Colo.: Westview Press, 1992.

Lindert, P. van. *Bolivia : A Guide to the People, Politics and Culture.* New York: Monthly Review Press, 1994.

Parker, Edward. *Ecuador, Peru, Bolivia. Country fact files.* Austin, TX: Raintree Steck-Vaughn, 1998.

Brazil

Carpenter, Mark L. *Brazil, an Awakening Giant.* Minneapolis, MN: Dillon Press, 1987.

Levine, Robert M. *Historical Dictionary of Brazil.* Metuchen, N.J.: Scarecrow Press, 1979.

——— and John J. Crocitti. *The Brazil Reader: History, Culture, Politics.* Durham, N.C.: Duke University Press, 1999.

Poppino, Rollie E. *Brazil: The Land and People.* New York: Oxford University Press, 1973.

Roop, Peter, and Connie Roop. *Brazil.* Des Plaines, Ill.: Heinemann Interactive Library, 1998.

Canada

Bothwell, Robert. *A Short History of Ontario.* Edmonton: Hurtig Publishers Ltd., 1986.

Dickinson, John A. and Brian Young. *A Short History of Quebec.* Toronto: Copp Clark Pitman Ltd., 1993.

McNaught, Kenneth. *The Penguin History of Canada.* London: Penguin, 1988.

Morton, Desmond. *A Short History of Canada.* 2nd revised ed. Toronto: McClelland & Stewart Inc., 1994.

Woodcock, George. *A Social History of Canada.* Markham, Ont.: Penguin Books Canada, 1989.

Chile

Arriagada Herrera, Genaro. Trans. Nancy Morris. *Pinochet: The Politics of Power.* Boston : Allen & Unwin, 1988.

Blakemore, Harold. *Chile.* Santa Barbara, Calif.: Clio Press, 1988.

Collier, Simon. *A History of Chile.* Cambridge: Cambridge University Press, 1996.

Falcoff, Mark. *Modern Chile, 1970–89: A Critical History.* New Jersey: Transaction Books, 1989.

Hudson, Rex A., ed. *Chile, a Country Study.* 3rd ed. Federal Research Division, Library of Congress. Washington, D.C., 1994.

Colombia

Bushnell, David. The Making of Modern Colombia: A Nation in Spite of Itself. Berkeley: University of California Press, 1993.

Davis, Robert H. *Colombia.* Santa Barbara, Calif.: Clio Press, 1990.

———. *Historical Dictionary of Colombia.* Metuchen, N.J.: Scarecrow Press, 1977.

Hanratty, Dennis M., and Sandra W. Meditz. *Colombia: A Country Study.* 4th ed. Washington, DC: Federal Research Division, Library of Congress, 1990.

Pearce, Jenny. *Colombia: The Drugs War.* New York: Gloucester Press, 1990.

Posada Carbs, Eduardo. *The Colombian Caribbean: A Regional History, 1870–1950.* New York: Clarendon Press, 1996.

Costa Rica

Biesanz, Richard; Biesanz, Karen Zubris; Biezanz, Mavis Hiltunen. *The Costa Ricans.* New Jersey: Prentice Hall, 1982.

Gall, Timothy L., ed. *Worldmark Encyclopedia of Cultures and Daily Life.* vol. 2: Americas. Detroit: Gale Research, 1998.

Stone, Doris. *Pre-Columbian Man in Costa Rica.* Cambridge: Peabody Museum Press, 1977.

Todorov, Tzvetan. *The Conquest of America.* New York: Harper & Row, 1984.

Cuba

Balfour, Sebastian. *Castro.* New York: Longman, 1990.

Brune, Lester H. *The Cuba-Caribbean Missile Crisis of October 1962.* Claremont, Calif.: Regina Books, 1996.

Gall, Timothy L., ed. *Worldmark Encyclopedia of Cultures and Daily Life.* vol. 2: Americas. Detroit: Gale Research, 1998.

Rudolph, James D., ed. *Cuba: A Country Study,* 3rd ed. Washington, D.C.: U.S. Government Printing Office, 1985.

Wyden, Peter. *Bay of Pigs: The Untold Story.* New York: Simon & Schuster, 1979.

Dominica

Baker, Patrick L. *Centering the Periphery: Chaos, Order, and the Ethnohistory of Dominica.* Montreal: McGill-Queen's University Press, 1994.

Philpott, Don. *Caribbean Sunseekers: Dominica.* Lincolnwood, Ill.: Passport Books, 1996.

Whitford, Gwenith. "Mining on 'Nature Island': the Dominican Government's Resource Extraction Plans Anger Conservationists." *Alternatives Journal,* Winter 1998, vol. 24, no. 1, p. 9+.

Dominican Republic

Cambeira, Alan. *Quisqueya la bella: the Dominican Republic in Historical and Cultural Pperspective.* Armonk, N.Y.: M.E. Sharpe, 1997.

Horowitz, Michael M. *Peoples and Cultures of the Caribbean: An Anthropological Reader.* New York: Natural History Press, 1971.

Logan, Rayford W. *Haiti and the Dominican Republic,* New York: Oxford University Press, 1968.

Plant, Roger. *Sugar and Modern Slavery: Haitian Migrant Labor and the Dominican Republic.* Totowa, N.J.: Biblio Dist., 1986.

Moya Pons, Frank. *The Dominican Republic: A National History.* New Rochelle, N.Y.: Hispaniola Books, 1995.

Ecuador

Bork, Albert William. *Historical Dictionary of Ecuador.* Metuchen, N.J.: Scarecrow Press, 1973.

Hemming, John. *The Conquest of the Incas.* San Diego: Harcourt Brace Jovanovich, 1970.

Rathbone, John Paul. *Ecuador, the Galápagos, and Colombia.* London: Cadogan Books, 1991.

Roos, Wilma, and Omer van Renterghem. *Ecuador in Focus: a Guide to the People, Politics and Culture.* New York: Interlink Books, 1997.

El Salvador

Browning, David, *El Salvador: Landscape and Society.* London: Clarendon Press, 1971.

Flemion, Philip, *Historical Dictionary of El Salvador* Metuchen, N.J.: The Scarecrow Press, 1972.

Haggerty, Richard, *El Salvador: A Country Study.* Washington, D.C.: Department of the Army, 1990.

Grenada

Gall, Timothy L., ed. *Worldmark Encyclopedia of the Nations.* 9th ed. Detroit: Gale Research, 1998.

Gunson, Phil, et. al., eds. *The Dictionary of Contemporary Politics of Central America and the Caribbean.* New York: Simon and Schuster, 1991.

Rouse, Irving. *The Tainos: Rise and Decline of the People Who Greeted Columbus.* New Haven: Yale University Press, 1992.

Weeks, John, and Ferbel, Peter. *Ancient Caribbean.* New York: Garland Publishing, 1994.

Guatemala

Gall, Timothy L., ed. *Worldmark Encyclopedia of the Nations.* 9th ed. Detroit: Gale Research, 1998.

Handy, Jim. *Gift of the Devil: A History of Guatemala.* Toronto: Between the Lines Press, 1984.

Nyrop, Richard, (ed.). *Guatemala: A Country Study.* (Washington DC: Department of the Army, 1983.

Schele, Linda. *A Forest of Kings: the Untold Story of the Ancient Maya.* New York: Morrow, 1990.

South America, *Central America and the Caribbean,* 6th ed. London: Europa Publication, 1997.

Guyana

Daly, Vere T. *A Short History of the Guyanese People*. London: Macmillan Education, 1975.

Gall, Timothy L., *Worldmark Encyclopedia of the Nations*. 9th ed. Detroit: Gale Research, 1998.

Mecklenburg, Kurt K. *Guyana Gold*. Carlton Press, 1990.

Singh, Chaitram. Guyana: *Politics in a Plantation Society*. New York: Praeger Publishers, 1988.

Haiti

Abbott, Elizabeth. *Haiti: The Duvaliers and their Legacy*. New York: Simon and Schuster. 1991.

Aristide, Jean-Bertrand. *Dignity*. Charlottesville and London: University Press of Virginia, 1996.

Gall, Timothy L., ed. *Worldmark Encyclopedia of the Nations*. 9th ed. Detroit: Gale Research, 1998.

McFadyen, Deidre; LaRamee, Pierre (editors). *Haiti: Dangerous Crossroads*. Boston: South End Press, 1995.

Honduras

Gall, Timothy L., ed. *Worldmark Encyclopedia of the Nations*. 9th ed. Detroit: Gale Research, 1998.

Schulz, Donald E. and Deborah S. Schulz. *The United States, Honduras and the Crisis in Central America*. Boulder, Colo.: Westview Press, 1994.

Todorov, Tzvetan (translated by Richard Howard) *The Conquest of America*. New York: Harper & Row, 1984.

Jamaica

Bayer, Marcel. *Jamaica: A Guide to the People, Politics, and Culture*. Trans. John Smith. London, Eng.: Latin American Bureau, 1993.

Davis, Stephen. *Reggae Bloodlines: In Search of the Music and Culture of Jamaica*. New York: Da Capo Press, 1992.

Gall, Timothy L., ed. *Worldmark Encyclopedia of the Nations*. 9th ed. Detroit: Gale Research, 1998.

Sherlock, Philip, and Hazel Bennett. *The Story of the Jamaican People*. Princeton, NJ: Markus Wiener Publishers, 1998.

Stone, Carl. *Class, State, and Democracy in Jamaica*. New York: Praeger, 1986.

Mexico

Burke, Michael E. *Mexico: An Illustrated History*. New York: Hippocrene Books, 1999.

Briggs, Donald C. *The Historical Dictionary of Mexico*. Metuchen, N.J.: Scarecrow Press, 1981.

Gall, Timothy L., ed. *Worldmark Encyclopedia of the Nations*. 9th ed. Detroit: Gale Research, 1998.

National Geographic. 190:2 (Aug. 1996) "Emerging Mexico". Entire isssue devoted to Mexico.

Randall, Laura, ed. *Changing Structure of Mexico*. Armonk, N.Y.: M.E. Sharpe, 1966.

Williamson, Edwin. *The Penguin History of Latin America*. New York: Penguin Books, 1992.

Nicaragua

Gall, Timothy L, ed. *Worldmark Encyclopedia of the Nations*. 9th ed. Detroit: Gale Research, 1998.

Kagan, Robert. *A Twilight Struggle: American Power and Nicaragua. 19771990*. New York: Free Press, 1996.

Rudolph, James D., ed. *Nicaragua: A Country Study*. 2nd ed. Washington, D.C.: Government Printing Office, 1994.

Walker, Thomas W., ed. *Revolution & Counterrevolution in Nicaragua*. Boulder, Colo.: Westview Press, 1991.

Panama

Flanagan, E. M. *Battle for Panama: Inside Operation Just Cause*. Washington, D.C.: Brassey's, Inc., 1993.

Guevara Mann, Carlos. *Panamanian Militarism: A Historical Interpretation*. Athens, Oh.: Ohio University Center for International Studies, 1996.

Hedrick, Basil C. and Anne K. *Historical Dictionary of Panama*. Metuchen, N.J.: Scarecrow Press, 1970.

Major, John. *Prize possession: The United States and the Panama Canal, 1903–1979*. New York: Cambridge University Press, 1993.

Noriega, Manuel Antonio. *America's Prisoner: The Memoirs of Manuel Noriega*. 1st ed. New York : Random House, 1997.

Pearcy, Thomas L. *We Answer Only to God: Politics and the Military in Panama, 1903–1947*. Albuquerque: University of New Mexico Press, 1998.

Paraguay

Gall, Timothy L., ed. *Worldmark Encyclopedia of the Nations*. 9th ed. Detroit: Gale Research, 1998.

Paraguay: A Country Study. Washington: U.S. Government Printing Office, 1990.

Wiarda, Howard J. and Harvey F. Kline, eds. *Latin American Politics and Development*. Boulder: Westview Press, 1996.

Peru

Gall, Timothy L., ed. *Worldmark Encyclopedia of the Nations*. 9th ed. Detroit: Gale Research, 1998.

Hemming, John. *The Conquest of the Incas*. New York: Harcourt Brace Jovanovich, 1970.

Holligan de Dmaz-Lmmaco, Jane. Peru: A Guide to the People, Politics and Culture. New York: Interlink Books, 1998.

Hudson, Rex A., ed. *Peru in Pictures*. Minneapolis: Lerner, 1987.

———. *Peru: A Country Study*. 4th ed. Washington, D.C.: Library of Congress, Federal Research Division, 1993.

Strong, Simon. *Shining Path : Terror and Revolution in Peru*. New York : Times Books, 1992.

Puerto Rico

Fernandez, Ronald. *The Disenchanted Island: Puerto Rico and the United States in the Twentieth Century*. 2d ed. Westport, Conn.: Praeger, 1996.

Figueroa, Loida. *History of Puerto Rico*. New York: Anaya, 1974.

Gall, Timothy L., ed. *Worldmark Encyclopedia of Cultures and Daily Life*. vol. 2: Americas. Detroit: Gale Research, 1998.

Morales, Carrion, Arturo. *Puerto Rico: A Political and Cultural History*. New York: Norton, 1983.

Morris, Nancy. *Puerto Rico: Culture, Politics, Indentity*. Westport, Conn.: Praeger, 1995.

St. Kitts and Nevis

Gall, Timothy L., ed. *Worldmark Encyclopedia of the Nations*. 9th ed. Detroit: Gale Research, 1998.

Hamshere, Cyril. *The British in the Caribbean*. Cambridge, MA: Harvard University Press, 1972.

Moll, V.P. *St. Kitts-Nevis*. Santa Barbara, CA: Clio, 1994.

Olwig, karen Fog. *Global Culture, Island Identity: Continuity and Change in the Afro-Caribbean Community of Nevis*. Philadelphia: Harwood, 1993.

St. Lucia

Claypole, William, and Robottom, John, *Caribbean Story, Book Two: The Inheritors*. Essex: Longman, 1986.

Craton, Michael, *Testing the Chains: Resistance to Slavery in the British West Indies*. Ithaca: Cornell University Press, 1982.

Gall, Timothy L., ed. *Worldmark Encyclopedia of the Nations*. 9th ed. Detroit: Gale Research, 1998.

Gunson, Phil, et. al., (eds.), *The Dictionary of Contemporary Politics of Central America and the Caribbean*. New York: Simon and Schuster, 1991.

Weeks, John, and Ferbel, Peter, *Ancient Caribbean*. New York: Garland Publishing, 1994.

St. Vincent and the Grenadines

Bobrow, Jill and Dana Jinkins. *St. Vincent and the Grenadines: Gems of the Caribbean*. Waitsfield, Vt.: Concepts Publishing Inc., 1993.

Hamshere, Cyril. *The British in the Caribbean*. Cambridge, Mass.: Harvard University Press, 1972.

Philpott, Don. *Caribbean Sunseekers: St. Vincent & Grenadines*. Lincolnwood, Ill.: Passport Books, 1996.

Suriname

Chin, Henk E. *Surinam: Politics, Economics, and Society*. New York: F. Pinter, 1987.

Cohen, Robert. *Jews in Another Environment: Surinam in the Second Half of the Eighteenth Century*. New York: E.J. Brill, 1991.

Dew, Edward M. *The Trouble in Suriname, 1975–1993*. Westport, CT: Praeger, 1994.

Hoogbergen, Wim S. M. *The Boni Maroon Wars in Suriname*. New York: Brill, 1990.

Sedoc-Dahlberg, Betty, ed. *The Dutch Caribbean: Prospects for Democracy*. New York: Bordon and Breach, 1990.

Trinidad and Tobago

Bereton, Bridget, *A History of Modern Trinidad* (Portsmouth: Heinemann, 1981).

Black, Jan, et. al., *Area Handbook for Trinidad and Tobago* (Washington D.C.: U.S. Government Printing Office, 1976).

Gunson, Phil, et. al., (eds.), *The Dictionary of Contemporary Politics of Central America and the Caribbean* (NY: Simon and Schuster, 1991).

Reddock, *Women Labour and Politics in Trinidad and Tobago: A History* (London: Zed Books 1994).

Weeks, John, and Ferbel, Peter, *Ancient Caribbean* (NY: Garland Publishing, 1994).

United States

Ayres, Stephen M. *Health Care in the Unites States: The Facts and the Choices*. Chicago: American Library Association, 1996.

Bacchi, Carol Lee. *The Politics of Affirmative Action: 'Women', Equality, and Category Politics*. London: Sage, 1996.

Barone, Michael. *The Almanac of American Politics*. Washington, D.C.: National Journal, 1992.

Bennett, Lerone. *Before the Mayflower: A History of Black America*. 6th ed. New York: Penguin, 1993.

Brinkley, Alan. *American History: A Survey*. 9th ed. New York: McGraw-Hill, 1995.

Brinkley, Douglas. *American Heritage History of the United States*. New York: Viking, 1998.

Carnes, Mark C., ed. *A History of American Life*. New York: Scribner, 1996.

Commager, Henry Steele (ed.). *Documents of American History*. Englewood Cliffs, N.J.: Prentice-Hall, 1988.

Davidson, James West. *Nation of Nations: A Narrative History of the American Republic*. 3rd ed. Boston, MA: McGraw-Hill, 1998.

Davies, Philip John. (ed.) *An American Quarter Century: US Politics from Vietnam to Clinton*. New York: Manchester University Press, 1995.

Donaldson, Gary. *America at War since 1945: Politics and Diplomacy in Korea, Vietnam, and the Gulf War*. Westport, Conn.: Praeger, 1996.

Foner, Eric. *The Story of American Freedom*. New York: Norton, 1998.

Garraty, John Arthur. *The American Nation: A History of the United States*. 9th ed. New York: Longman, 1998.

Goldfield, David, ed. *The American Journey: A History of the United States*. Upper Saddle River, NJ: Prentice Hall, 1998.

Hart, James David, ed.. *Oxford Companion to American Literature*. 6th ed. New York: Oxford University Press, 1995.

Hummel, Jeffrey Rogers. *Emancipating Slaves, Enslaving Free Men: A History of the Civil War*. Chicago: Open Court, 1996.

Jenkins, Philip. *A History of the United States*. New York: St. Martin's Press, 1997.

Kaplan, Edward S. *American Trade Policy, 1923–1995*. Westport, Conn.: Praeger, 1996.

Magill, Frank N., ed. *Great Events from History: North American Series.* rev. ed. Pasadena, CA: Salem Press, 1997.

Martis, Kenneth C. *The Historical Atlas of Political Parties in the United States Congress 1789–1989.* New York: Macmillan, 1989.

McNickle, D'Arcy. *Native American Tribalism: Indian Survivals and Renewals.* New York: Oxford University Press, 1993.

Mudd, Roger. *American Heritage Great Minds of History.* New York: Wiley, 1999.

Nash, Gary B. *The American People: Creating a Nation and a Society.* 4th ed., New York : Longman, 1998.

People Who Shaped the Century. Alexandria, VA: Time-Life Books, 1999.

Robinson, Cedric J. *Black Movements in America.* New York: Routledge, 1997.

Tindall, George Brown. *America: A Narrative History.* 5th ed. New York: Norton, 1999.

Virga, Vincent. *Eyes of the Nation: A Visual History of the United States.* New York: Knopf, 1997.

Woodward, C. Vann, ed. *The Comparative Approach to American History.* New York: Oxford University Press, 1997.

Uruguay

Gall, Timothy L., ed. *Worldmark Encyclopedia of the Nations.* 9th ed. Detroit: Gale Research, 1998.

Hudson, Rex A., and Sandra Meditz. *Uruguay: A Country Study.* Washington, DC : Federal Research Division, Library of Congress, 1992.

Weinstein, Martin. *Uruguay, Democracy at the Crossroads. Nations of Contemporary Latin America.* Boulder, Colo.: Westview Press, 1988.

Zlotchew, Clark M. Paul David Seldis, ed. *Voices of the River Plate : Interviews with Writers of Argentina and Uruguay.* San Bernadino, CA: Borgo Press, 1993.

Venezuela

Fox, Geoffrey. *The Land of People of Venezuela.* New York: HarperCollins, 1991.

Gall, Timothy L., ed. *Worldmark Encyclopedia of Cultures and Daily Life.* vol. 2: Americas. Detroit: Gale Research, 1998.

Haggerty, Richard A., ed. *Venezuela: A Country Study,* 4th ed. Washington, D.C.: Library of Congress, 1993.

Hellinger, Daniel. *Venezuela: Tarnished Democracy.* Boulder, Colo.: Westview Press, 1991.

Asia

Afghanistan

Adamec, Ludwig W. *Historical Dictionary of Afghanistan.* Metuchen, NJ: Scarecrow Press, 1991.

Gall, Timothy L., ed. *Worldmark Encyclopedia of the Nations.* 9th ed. Detroit: Gale Research, 1998.

Giradet, Edward. *Afghanistan: The Soviet War.* London: Croom Helm, 1985

Nyrop, Richard F. and Donald M. Seekins, eds. *Afghanistan: A Country Study.* 5th ed. Washington, DC: U.S. Government Printing Office, 1986.

Australia

Bassett, Jan. *The Oxford Illustrated Dictionary of Australian History.* New York: Oxford University Press, 1993.

Bolton, Geoffrey, ed. *The Oxford History of Australia.* New York: Oxford University Press, 1986–90.

Gunther, John. *John Gunther's Inside Australia.* New York: Harper & Row, 1972.

Heathcote, R.L. *Australia.* London: Longman, Scientific & Technical, 1994.

Rickard, John. *Australia, A Cultural History.* London: Longman, 1996.

Azerbaijan

Altstadt, Audrey. *The Azerbaijani Turks.* Stanford, Calif.: Hoover Institution Press, 1992.

Gall, Timothy L., ed. *Worldmark Encyclopedia of the Nations.* 9th ed. Detroit: Gale Research, 1998.

Nichol, James. "Azerbaijan." in *Armenia, Azerbaijan and Georgia.* Area Handbook Series, Washington, DC: Government Printing Office, 1995, pp. 81148.

Bahrain

Crawford, Harriet. *Dilmun and its Gulf Neighbors.* Cambridge: Cambridge University Press, 1998.

Gall, Timothy L., ed. *Worldmark Encyclopedia of the Nations.* 9th ed. Detroit: Gale Research, 1998.

Nugent, Jeffrey B., and Theodore Thomas, eds. *Bahrain and the Gulf: Past Perspectives and Alternate Futures.* New York: St. Martin's Press, 1985.

Robison, Gordon. *Arab Gulf States.* Hawthorn, Aus.: Lonely Planet Publications, 1996.

Bangladesh

Baxter, Craig. *Bangladesh: A New Nation in an Old Setting.* Boulder and London: Westview Press, 1984.

Bigelow, Elaine. *Bangladesh: the Guide.* Dhaka: AB Publishers, 1995.

O'Donnell, Charles Peter. *Bangladesh: Biography of a Muslim Nation.* Boulder and London: Westview Press, 1984.

Newton, Alex. *Bangladesh: a Lonely Planet Travel Survival Kit.* Hawthorne, Victoria, Australia: Lonely Planet, 1996.

Republic of Bangladesh. U.S. Department of State Background Notes, November 1997.

Bhutan

Aris, Michael. *Bhutan: The Early History of a Himalayan Kingdom.* Warminster, England: Aris & Phillips, 1979.

Crossette, Barbara. *So Close to Heaven: The Vanishing Buddhist Kingdoms of the Himalayas.* New York: A. A. Knopf, 1995.

Matles, Andrea, ed. *Nepal and Bhutan: Country Studies.* Washington, DC: Federal Research Division, Library of Congress, 1993.

Pommaret-Imaeda, Françoise. Lincolnwood, Ill.: Passport Books, 1991.

Strydonck, Guy van. *Bhutan: A Kingdom in the Eastern Himalayas.* Boston: Shambala, 1985.

Brunei Darussalam

Bartholomew, James. *The Richest Man in the World: The Sultan of Brunei.* London: Penguin Group, 1990.

Major, John S. *The Land and People of Malaysia & Brunei.* New York: HarperCollins 1991.

Pigafetta, Antonio. *Magellan's Voyage.* Trans. and ed. by R. A. Skelton. New Haven: Yale University Press, 1969.

Singh, D.S. Ranjit. *Brunei 1839–1983: The Problems of Poltical Survival.* London: Oxford University Press, 1984.

Vreeland, N. et al. *Malaysia: A Country Study.* Area Handbook Series. Fourth edition. Washington, D.C.: Department of the Army, 1984.

Cambodia

Barron, John and Anthony Paul. *Murder of a Gentle Land. The Untold Story of Communist Genocide in Cambodia.* New York: Reader's Digest Press. 1977.

Becker, Elizabeth. *When the War Was Over: The Voices of Cambodia's Revolution and Its People.* New York: Simon and Schuster. 1986.

Chandler, David P. *A History of Cambodia.* Boulder, Colo.: Westview Press. 1983.

Gall, Timothy L., ed. *Worldmark Encyclopedia of Cultures and Daily Life.* vol. 3: Asia. Detroit: Gale Research, 1998.

Ponchaud, Francois. *Cambodia Year Zero.* London: Allen Lane. 1978.

China

Bailey, Paul. *China in the Twentieth Century.* New York: B. Blackwell, 1988.

Cotterell, Arthur. *China: A Cultural History.* New York: New American Library, 1988.

Ebrey, Patricia B. *The Cambridge Illustrated History of China.* New York: Cambridge University Press, 1996.

Hsu, Immanuel Chung-yueh. *The Rise of Modern China.* 5th ed. New York: Oxford University Press, 1995.

Worden, Robert L., Andrea Matles Savada, and Ronald E. Dolan, eds. *China, a Country Study.* 4th ed. Washington, D.C.: Library of Congress, 1988.

Cyprus (Greek and Turkish zones)

Crawshaw, Nancy. *The Cyprus Revolt: The Origins, Development, and Aftermath of an International Dispute.* Winchester, Mass.: Allen & Unwin, 1978.

Gall, Timothy L., ed. *Worldmark Encyclopedia of Cultures and Daily Life.* vol. 3: Asia. Detroit: Gale Research, 1998.

Salem, Norma, ed. *Cyprus: A Regional Conflict and its Resolution.* New York: St. Martin's Press, 1992.

Solsten, Eric, ed. *Cyprus, a Country Study.* 4th ed. Washington, D.C. Government Printing Office, 1993.

Tatton-Brown, Veronica. *Ancient Cyprus.* Cambridge, Mass.: Harvard University Press, 1988.

Federated States of Micronesia

Denoon, Donald, ed. *The Cambridge History of the Pacific Islanders.* Cambridge: Cambridge University Press, 1997.

Gall, Timothy L., ed. *Worldmark Encyclopedia of the Nations.* 9th ed. Detroit: Gale Research, 1998.

Levesque, Rodrigu, ed. *History of Micronesia.* Honolulu: University of Hawaii Press, 1994.

Spate, Oskar. *The Pacific Since Magellan, vol. 1: The Spanish Lake.* Canberra: Australian National University Press, 1979.

Fiji

Gall, Timothy L., ed. *Worldmark Encyclopedia of the Nations.* 9th ed. Detroit: Gale Research, 1998.

Ogden, Michael R. "Republic of Fiji," *World Encyclopedia of Political Systems,* 3rd edition, New York: Facts on File, 1999.

Scarr, Deryck. *Fiji: A Short History.* Honolulu: The Institute for Polynesian Studies, 1984.

"Vaughn, Roger. "The Two Worlds of Fiji," *National Geographic,* October 1995, vol. 88, no. 4, p. 114.

India

Heitzen, James and Robert L. Worden, eds. *India: A Country Study.* Washington, DC: Federal Research Division, Library of Congress, 1996.

Robinson, Francis, ed. *The Cambridge Encyclopedia of India, Pakistan, Bangladesh, Sri Lanka, Nepal, Bhutan and the Maldives.* Cambridge: Cambridge University Press. 1989.

Schwartzberg, Joseph E., ed. *A Historical Atlas of South Asia.* 2nd impression. New York and Oxford: Oxford University Press, 1992.

Wolpert, Stanley. *India.* Berkeley: University of California Press, 1991.

Indonesia

Bellwood, Peter S. *Prehistory of the Indo-Malaysian Archipelago.* Honolulu, Hawaii: University of Hawaii Press, 1997.

Broughton, Simon, ed. *World Music: The Rough Guide.* London: The Rough Guides Ltd., 1994.

Cribb, Robert. *Historical Dictionary of Indonesia.* Metuchen, N.J.: Scarecrow Press, 1992.

Frederick, William H., ed. *Indonesia: A Country Study.* Washington, D.C.: Library of Congress, 1993.

Schwarz, Adam. *A Nation in Waiting: Indonesia in the 1990s.* Boulder, Colo.: Westview Press, 1994.

Iran

Albert, David H. *Tell the American People: Perspectives on the Iranian Revolution.* Philadelphia: Movement for a New Society, 1980.

Bacharach, Jere L. *A Near East Studies Handbook, 570–1974.* Seattle: University of Washington Press, 1974.

Famighetti, Robert (ed.). *The World Almanac and Book of Facts.* New York: St. Martin's Press, 1998.

Nyrop, Richard F. (ed.). *Area Handbook of Iran: A Country Study.* Washington, D.C.: The American University, 1978.

Salinger, Pierre. *America Held Hostage: The Secret Negotiations.* Garden City, N.Y.: Doubleday & Company, 1981.

Iraq

Bulloch, John. *Saddam's War: The Origins of the Kuwait Conflict and the International Response.* Boston: Faber and Faber, 1991.

Chaliand, Gerard, ed. *A People without a Country: The Kurds and Kurdistan.* New York: Olive Branch Press, 1993.

Mansfield, Peter. *A History of the Middle East.* New York: Viking, 1991.

Metz, Helen Chapin, ed. *Iraq: A Country Study.* 4th ed. Washington, D.C.: Library of Congress, Federal Research Division, 1990.

Simons, Geoff L. *Iraq: From Sumer to Saddam.* New York: St. Martin's Press, 1994.

Israel

Blumberg, Arnold. *The History of Israel.* Westport, Conn.: Greenwood Press, 1998.

———. *Zion Before Zionism.* Syracuse, N.Y.: Syracuse University Press, 1985.

Grant, Michael. *The History of Ancient Israel.* New York: Charles Scribner's Sons, 1984.

Metz, Helen Chapin. *Israel: A Country Study.* Washington, DC : Library of Congress, 1990

Sachar, Abram Leon. *A History of the Jews.* New York: Alfred A. Knopf, 1955.

Shanks, Hershel. *Ancient Israel.* Englewood Cliffs, N.J.: Prentice-Hall, 1988.

Japan

Demente, Boye Lafayette. *Japan Encyclopedia.* Lincolnwood, Ill.: NTC Publishing, 1995.

Dolan, Ronald E., and Robert L. Dolan, eds. *Japan, a Country Study.* 5th ed. Washington, D.C.: Library of Congress, 1992.

Perkins, Dorothy. *Encyclopedia of Japan: Japanese History and Culture, from Abacus to Zori.* New York: Facts on File, 1991.

Richardson, Bradley M. *Japanese Democracy: Power, Coordination, and Performance.* New Haven, Conn.: Yale University Press, 1997.

Thomas, J. E. *Modern Japan: A Social History Since 1868.* New York: Longman, 1996.

Jordan

Ali, Wijdan. *Modern Islamic Art: Development and Continuity.* University Press of Florida, Gainesville, Fla. 1997.

Contreras, Joseph and Christopher Dickey. "The Day After: King Hussein's Second Bout of Cancer Raises Questions About Jordan's Political Future." *Newsweek,* Vol. 132, No. 6, p. 38, August 10, 1998.

Shahin, Mariam. "Cracks Become Chasms," *The Middle East.* December 1997, No. 273, p. 13-14.

Kazakstan

Conflict in the Soviet Union: The Untold Story of the Clashes in Kazakhstan. New York: Human Rights Watch, 1990.

Edwards-Jones, Imogen. *The Taming of Eagles: Exploring the New Russia.* London: Weidenfeld & Nicolson, 1992.

Kalyuzhnova, Yelena. *Kazakstani Economy: Independence and Transition.* New York: St. Martin's Press, 1998.

Olcott, Martha Brill. *The Kazakhs.* Stanford, CA: Hoover Institution Press, Stanford University, 1987.

World Bank. *Kazakhstan: The Transition to a Market Economy.* Washington, DC: World Bank, 1993.

Kiribati

"Bones Found in '40 May Have Been Hers," *Honolulu Star-Bulletin,* December 3, 1998.

Krauss, Bob. "Heroes Heed Call From Sea," *Honolulu Advertiser*, March 10, 1999.

MacDonald, Barrie. *Cinderellas of the Empire: Towards a History of Kiribati and Tuvalu.* Canberra, Australia: Australian National University Press, 1982.

"New Meaning to the Term Down Under. The Tiny Islands of Kiribati, Tuvalu and Nauru Are Pressuring Australia to Reduce Greenhouse Gas Emissions.)" *The Economist (US),* September 27, 1997, vol. 344, no. 8036, p. 41.

"South Pacific Group Cuts French Ties," *Honolulu Star-Bulletin*, October 3, 1995.

Thompson, Rod. "Hilo Plans Gift for Christmas Isle Neighbors, " *Honolulu Star-Bulletin,* December 18, 1998.

Korea, Democratic People's Republic of

Gills, Barry K. *Korea Versus Korea: A Case of Contested Legitimacy.* New York: Routledge, Inc., 1996.

Oliver, Robert Tarbell. *A History of the Korean People in Modern Times: 1800 to the Present.* Newark,: University of Delaware Press, 1993.

Smith, Hazel. *North Korea in the New World Order.* New York: St. Martin's Press, 1996.

Soh, Chung Hee. *Women in Korean Politics.* 2nd ed. Boulder, Colo.: Westview Press, 1993.

Tennant, Roger. *A History of Korea.* London: Kegan Paul International, 1996.

Korea, Republic of (ROK)

Gills, Barry K. *Korea Versus Korea: A Case of Contested Legitimacy.* New York: Routledge, Inc., 1996.

Lone, Stewart. *Korea Since 1850.* New York: St. Martin's Press, 1993.

Oberdorfer, Don. *The Two Koreas: A Contemporary History.* Reading, Mass.: Addison-Wesley Pub. Co., 1997.

Oliver, Robert Tarbell. *A History of the Korean People in Modern Times: 1800 to the Present.* Newark: University of Delaware Press, 1993.

Tennant, Roger. *A History of Korea.* London: Kegan Paul International, 1996.

Kuwait

Ali, Wijdan. *Modern Islamic Art: Development and Continuity.* Gainesville, FL:

Anscombe, Frederick F. *The Ottoman Gulf: The Creation of Kuwait, Saudi Arabia and Qatar.* New York: Columbia University Press, 1997.

Cordesman, Anthony H. *Kuwait: Recovery and Security after the Gulf War.* Boulder, Colo.: Westview Press, 1997.

Crystal, Jill. *Oil and Politics in the Gulf: Rulers and Merchants in Kuwait and Qatar.* New York: Cambridge University Press, 1990.

Khadduri, Majid. *War in the Gulf, 1990–1991: The Iraq-Kuwait Conflict and its Implications.* New York: Oxford University Press, 1997.

Robison, Gordon. *Arab Gulf States.* Hawthorn, Aus.: Lonely Planet Publications, 1996.

Kyrgyzstan

Attokurov, S. *Kïrgïz Sanjïrasï.* Bishkek: Kyrgyzstan, 1995.

Bennigsen, Alexandre & S. Enders Wimbush. *Muslims of the Soviet Empire: A Guide.* Bloomington & Indianapolis: Indiana Universiity Press, 1986.

Huskey, Eugene. "Kyrgyzstan: The politics of demographic and economic frustration." in: *New States, New Politics: Building the Post-Soviet Nations,* edited by Ian Bremmer and Ray Taras, Cambridge & New York: Cambridge University Press, 1997, 655-680.

Krader, Lawrence. *The Peoples of Central Asia.* Bloomington, Indiana: Uralic and Altaic Series, vol. 26, 1966.

Olcott, Martha B. "Kyrgyzstan." in: *Kazakstan, Kyrgyzstan, Tajikistan, Turkmenistan and Uzbekistan* Area Handbook Series, Washington D.C.: Government Printing Office, 1997, 99-193.

Laos

Cummings, Joe. *A Golden Souvenir of Laos.* New York: Asia Books. 1996.

Kremmer, Chistopher. *Stalking the Elephant King. In Search of Laos.* Honolulu, University of Hawai'i Press. 1997.

Scott, Joanna C. *Indochina's Refugees: Oral Histories from Laos, Cambodia, and Vietnam.* Jeferson, NC: McFarland. 1989.

Stieglitz, P. *In a Little Kingdom.* New York: M. E. Sharpe. 1990.

Stuart-Fox, Martin. *A History of Laos.* New York: Cambridge University Press. 1997.

Lebanon

Abul-Husn, Latif. *The Lebanese Conflict: Looking Inward.* Boulder, Colo.: Lynne Rienner Publishers, 1998.

Bleaney. C.H. *Lebanon: Revised Edition. World Bibliographical Series,* Volume 2, Oxford, Eng.: Clio Press, Ltd., 1991.

King-Irani, Laurie. "War Gods Roar Again, Appear Unstoppable: Jet Streaks So Close She Could See Pilot," *National Catholic Reporter.* Vol. 34, No. 17, p. 9. February 27, 1998.

Norton, Augustus Richard. "Hizballah: From Radicalism to Pragmatism?" *Middle East Policy.* Vol. 5, No. 4, pp. 174-186. January 1998.

Tuttle, Robert. "Americans return to AUB," *The Middle East.* No. 272, p. 42. November 1997.

Malaysia

Bellwood, Peter S. *Prehistory of the Indo-Malaysian Archipelago.* Honolulu,: University of Hawaii Press, 1997.

Bevis, William W. *Borneo Log: The Struggle for Sarawak's Forests.* Seattle,: University of Washington Press, 1995.

Broughton, Simon, ed. *World Music: The Rough Guide.* London: The Rough Guides Ltd., 1994.

Kahn, Joel S. and Loh Kok Wah, Francis, eds. *Fragmented Vision: Culture and Politics in Contemporary Malaysia.* Honolulu,: University of Hawaii Press, 1992.

Kaur, Amarjit. *Historical Dictionary of Malaysia.* Metuchen, NJ: Scarecrow Press, 1993.

Maldives

Ellis, Kirsten. *The Maldives.* Hong Kong: Odyssey, 1993.

Republic of Maldives, Office for Women's Affairs. *Status of Women, Maldives.* Bangkok: UNESCO Principal Regional Office for Asia and the Pacific, 1989.

Marshall Islands

Johnson, Giff. *Collision Course at Kwajalein: Marshall Islanders in the Shadow of the Bomb.* Honolulu: Pacific Concerns Resource Center, 1984.

Langley, Jonathan, and Wanda Langley. "The Marshall Islands." *Skipping Stones,* Winter 1995, vol. 7, no. 5, p. 17+.

Levesque, Rodrigu, ed. *History of Micronesia.* Honolulu: University of Hawaii Press, 1994.

Scarr, Deryck. *History of the Pacific Islands: Kingdom of the Reefs.* Sydney: Macmillan Australia, 1990.

Mongolia

Bergholz, Fred W. *The Partition of the Steppe: The Struggle of the Russians, Manchus, and the Zunghar Mongols for Empire in Central Asia, 1619–1758.* New York: Peter Lang Publishing, Inc., 1993.

Bruun, Ole and Ole Odgaard, eds. *Mongolia in Transition.* Richmond, England: Curzon Press Ltd., 1996.

de Hartog, Leo. *Russia and the Mongol Yoke: The History of the Russian Principalities and the Golden Horde, 1221–1502.* London: British Academic Press, 1996.

Greenway, Paul. *Mongolia.* Hawthorn, Australia: Lonely Planet Publications, 1997.

Sanders, Alan J. K. *Historical Dictionary of Mongolia.* Lanham, Md: Scarecrow Press, Inc., 1996.

Myanmar (Burma)

Diran, Richard K. *The Vanishing Tribes of Burma.* New York: Amphoto Art, 1997.

Parenteau, John. *Prisoner for Peace: Aung San Suu Kyi and Burma's Struggle for Democracy.* Greensboro, N.C. : Morgan Reynolds, 1994.

Renard, Ronald D. *The Burmese Connection: Illegal Drugs and the Making of the Golden Triangle.* Boulder, Colo.: L. Rienner Publishers, 1996.

Rotberg, Robert I., ed. *Burma: Prospects for a Democratic Future.* Brookings Institute Press, 1998.

Silverstein, Josef. *The Political Legacy of Aung San.* Ithaca, N.Y.: Cornell University Press, 1993.

Nauru

Bunge, Frederica and Melinda Cooke, eds. *Oceania: a regional study.* Washington, D.C.: U.S. Government Printing Office, 1984.

Hanlon, David. *Remaking Micronesia.* Honolulu: University of Hawai'i Press, 1998.

McKnight, Tom. *Oceania: the geography of Australia, New Zealand, and the Pacific Islands.* Englewood Cliffs: Prentice Hall, 1995.

"New Meaning to the Term Down Under. The Tiny Islands of Kiribati, Tuvalu and Nauru Are Pressuring Australia to Reduce Greenhouse Gas Emissions.)" *The Economist (US),* September 27, 1997, vol. 344, no. 8036, p. 41.

Viviani, Nancy. *Nauru: phosphate and political progress.* Honolulu: University of Hawaii Press, 1970.

Nepal

Bista, Dor Bahadur. *People of Nepal.* Kathmandu: Ratna Pustak Bhandar, 1987.

Chauhan, R. S. *Society and State Building in Nepal: From Ancient Times to Mid-Twentieth Century.* New Delhi: Sterling, 1989.

Matles, Andrea, ed. *Nepal and Bhutan: Country Studies.* Washington, D.C.: Federal Research Division, Library of Congress, 1993.

Sanday, John. *The Kathmandu Valley: Jewel of the Kingdom of Nepal.* Lincolnwood, IL.: Passport Books, 1995.

Seddon, David. *Nepal, a State of Poverty.* New Delhi: Vikas, 1987.

New Zealand

Belich, James. *Making Peoples: A History of the New Zealanders from Polynesian Settlement to the End of the Nineteenth Century.* Honolulu: University of Hawaii Press, 1996.

Mascarenhas, R.C. *Government and the Economy in Australia and New Zealand: The Politics of Economic Policy Making.* San Francisco: Austin & Winfield, 1996.

McKinnon, Malcolm. *Independence and Foreign Policy: New Zealand in the World since 1935.* Auckland: Oxford University Press, 1993.

Oddie, Graham, and Roy W. Perrett, ed. *Justice, Ethics, and New Zealand Society.* New York: Oxford University Press, 1992.

Rice, Geoffrey W., ed. *The Oxford History of New Zealand.* 2nd ed. New York: Oxford University Press, 1992.

Oman

Arab Gulf States: A Travel Survival Kit. 2d edition. Melbourne Aus.: Lonely Planet Publications, 1996.

Miller, Judith. "Creating Modern Oman: An Interview With Sultan Qabus." *Foreign Affairs,* May–June 1997, Vol. 76, No. 3, pp. 1318.

Molavi, Afshin. "Oman's Economy: Back on Track." *Middle East Policy,* January 1998, Vol. 5, No. 4, pp. 110.

Osborne, Christine. "Omani Forts win Heritage Award." *The Middle East,* November 1995, No. 250, pp. 4344.

Riphenburg, Carol. J. *Oman: Political Development in a Changing World.* Westport, CT: Praeger Publishers, 1998.

Pakistan

Blood, Peter R., ed. *Pakistan, a Country Study.* 6th ed. Washington, D.C.: Federal Research Division, Library of Congress, 1995.

Burki, Shahid Javed. *Historical Dictionary of Pakistan.* Metuchen, NJ: Scarecrow Press, 1991.

Mahmud, S. F. *A Concise History of Indo-Pakistan.* Karachi: Oxford University Press, 1988.

Schwartzberg, Joseph E., ed. *A Historical Atlas of South Asia.* 2nd impression. Oxford and New York: Oxford University Press, 1992.

Taylor, David (revised by Asad Sayeed). "Pakistan: Economy," in *The Far East and Australasia 1997.* London: Europa Publications, 1996, pp. 873-79.

Palau

Hanlon, David. *Remaking Micronesia.* Honolulu: University of Hawai'i Press, 1996.

Hijikata, Hisakatsu. *Society and Life in Palau.* Tokyo: Sasakawa Peace Foundation, 1993.

Liebowitz, Arnold. *Embattled Island: Palau's Struggle for Independence.* Westport, CT: Praeger, 1996.

Morgan, William. *Prehistoric Architecture in Micronesia.* Austin: The University of Texas Press, 1988.

Roff, Sue Rabbitt. *Overreaching in Paradise: United States Policy in Palau Since 1945.* Juneau, Alaska: Denali Press, 1991.

Papua New Guinea

Campbell, I.C. *A History of the Pacific Islands.* Berkeley: University of California Press, 1989.

"Death of a Peacemaker." *The Economist,* October 19, 1996, vol. 341, no. 7988, p. 41.

Grattan, C. Hartley. *The Southwest Pacific since 1900.* Ann Arbor: University of Michigan Press, 1963.

Sinclair, James. *Papua New Guinea: the First 100 Years.* Bathurst: Robert Brown and Associates, 1985.

Spriggs, Matthew. *The Island Melanesians.* Cambridge, Mass.: Blackwell Publishers, 1997.

Philippines

Brands, H.W. *Bound to Empire: The United States and the Philippines.* New York: Oxford Univ. Press, 1992.

Dolan, Ronald E., ed. *Philippines: A Country Study.* 4th ed. Washington, D.C.: Library of Congress, 1993.

Karnow, Stanley. *In Our Image: America's Empire in the Philippines.* New York: Random House, 1989.

Steinberg, David Joel. *The Philippines, a Singular and a Plural Place.* 3rd ed. Boulder, Colo.: Westview, 1994.

Thompson, W. Scott. *The Philippines in Crisis: Development and Security in the Aquino Era 198692.* New York: St. Martin's 1992.

Qatar

Abu Saud, Abeer. *Qatari Women Past and Present.* Essex, Eng.: Longman Group Limited, 1984.

Anscombe, Frederick F. *The Ottoman Gulf: The Creation of Kuwait, Sa'udi Arabia, and Qatar.* New York: Columbia Press, 1997.

Crystal, Jill. *Oil and Politics in the Gulf: Rulers and Merchants in Kuwait and Qatar.* New York: Cambridge University Press, 1995.

Samoa

Fialka, John J. "From Dots in the Pacific, Envoys Bring Fear, Fury to Global-Warming Talks," *Wall Street Journal,* September 31, 1997, A24.

Hallowell, Christopher. "Rainforest Pharmacist," *Audubon* 101:1 (January 1999), 28.

Holmes, Lowell D., and Ellen Rhoads Holmes. *Samoan Village Then and Now.* Orlando, Fla.: Holt, Rinehart and Winston Inc., 1974, repr. 1992.

Meleisea, Malama. *Change and Adaptation in Western Samoa.* Christchurch, New Zealand: MacMillan Brown Centre for Pacific Studies, 1992.

Samoa: A Travel Survival Kit. Sydney, Aus.: Lonely Planet Publications, 1996.

Sa'udi Arabia

Abir, Mordechai. *Saudi Arabia: Government, Society, and the Gulf Crisis.* London and New York: Routledge, 1993.

Caesar, Judith. *Crossing Borders: An American Woman in the Middle East.* Syracuse, NY: Syracuse University Press, 1997.

Long, David E. *The Kingdom of Saudi Arabia.* Gainesville, Fla.: University Press of Florida, 1997.

Peterson, J. J. *Historical Dictionary of Saudi Arabia.* Metuchen, N.J.: Scarecrow Press, 1993.

Wilson, Peter W. and Douglas F. Graham. *Saudi Arabia: The Coming Storm.* Armonk, N.Y.: M. E. Sharpe, 1994.

Singapore

Chew, Ernest and Edwin Chew, ed. *A History of Singapore.* New York: Oxford University Press, 1991.

Chiu, Stephen Wing-Kai. *City States in the Global Economy: Industrial Restructuring in Hong Kong and Singapore.* Boulder, Colo.: Westview Press, 1997.

Lee, W.O. *Social Change and Educational Problems in Japan, Singapore, and Hong Kong.* New York: St. Martin's Press, 1991.

LePoer, Barbara Leitch, ed. *Singapore: A Country Study.* 2nd ed. Washington, D.C.: Library of Congress, 1991.

Trocki, Carl A. *Opium and Empire: Chinese Society in Colonial Singapore.* Ithaca, N.Y.: Cornell University Press, 1990.

Solomon Islands

Bennett, Judith. *Wealth of the Solomons: A History of a Pacific Archipelago, 1800–1978.* Honolulu: University of Hawai'i Press, 1987.

Denoon, Donald, ed. *The Cambridge History of the Pacific Islanders.* Cambridge: Cambridge University Press, 1997.

White, Geoffrey and Lindstrom, Lamont, eds. *The Pacific Theater: island representations of World War II.* Honolulu: University of Hawai'i Press, 1989.

Sri Lanka

Anderson, John Gottberg, and Ravindral Anthonis, editors. *Sri Lanka.* Hong Kong: Apa Productions, 1993.

Baker, Victoria J. *A Sinhalese Village in Sri Lanka: Coping with Uncertainty.* Ft. Worth: Harcourt Brace College Publishers, 1998.

Robinson, Francis, editor. *The Cambridge Encyclopedia of Pakistan, Bangladesh, Sri Lanka, Nepal, Bhutan and the Maldives.* Cambridge: Cambridge University Press., 1989.

Ross, Russell R., et al. *Sri Lanka: A Country Study* (Area Handbook Series). Washington, D.C.: Library of Congress, 1990.

Vesilind, Pritt J. "Sri Lanka." *National Geographic,* January 1997, vol. 191, no. 1, pp. 110+.

Syria

Dourian, Kate. "City of Apamea, Once Lost in Sand, Partially Restored." *Washington Times.* September 3, 1994.

Katler, Johannes. *The Arts and Crafts of Syria.* Thames and Hudson: London, 1992.

Kayal, Michele. "Ruins to Riches." *Washington Post.* February 13, 1994.

LaFranchl, Howard. "Ancient Syria's History Rivals That of Egypt, Mesopotamia." *The Christian Science Monitor.* February 17, 1994.

Parmelee, Jennifer. "Tracking Agatha Christie." *The Washington Post.* August 25, 1991.

"Syria." *Background Notes.* Washington, D.C.: Central Intelligence Agency, November 1994.

Syria: A Country Report. Washinton, D.C.: Library of Congress Research Division, 1999.

Taiwan

Hood, Steven J. *The Kuomintang and the Democratization of Taiwan*. Boulder, Colo.: Westview, 1997.

Long, Simon. *Taiwan: China's Last Frontier*. New York: St. Martin's Press, 1990.

Marsh, Robert. *The Great Transformation: Social Change in Taipei, Taiwan Since the 1960s*. Armonk, N.Y.: M.E. Sharpe, 1996.

Shepherd, John Robert. *Statecraft and Political Economy on the Taiwan Frontier, 1600–1800*. Stanford, Calif.: Stanford University Press, 1993.

Tajikistan

Rakhimov, Rashid, et al. *Republic of Tajikistan: Human Development Report 1995*, Istanbul, 1995.

Rashid, Ahmed. *The Resurgence of Central Asia or Nationalism?*, Zed Books, 1995.

Thailand

Dixon, C. J. *The Thai Economy: Uneven Development and Internationalisation*. London: Routledge, 1999.

Pattison, Gavin and John Villiers. *Thailand*. New York: Norton, 1997.

West, Richard. *Thailand, the Last Domino: Cultural and Political Travels*. London: Michael Joseph, 1991.

Tonga

Aswani, Shankae and Michael Graves. "The Tongan Maritime Expansion." *Asian -Perspectives* 1998. 37: 135201.

———. *Island Kingdom: Tonga Ancient and Modern*. Christchurch, NZ: Canterbury University Press, 1992.

Denoon, Donald, ed. *The Cambridge History of the Pacific Islanders*. Cambridge: Cambridge University Press, 1997.

Perminow, Arne. *The Long Way Home: Dilemmas of Everyday Life in a Tongan Village*. Oslo: Scandinavian University Press, 1993.

Turkmenistan

Edwards-Jones, Imogen. *The Taming of Eagles: Exploring the New Russia*. London: Weidengeld and Nicolson, 1993.

Hunter, Shireen T. *Central Asia Since Independence*. Westport, CN: Praeger, 1996.

Kazakhstan, Kyrgyzstan, Tajikistan, Turkmenistan, and Uzbekistan, country studies. Washington: US Government Printing Office, 1997.

Maslow, Jonathan Evan. *Sacred Horses: The Memoirs of a Turkmen Cowboy*. New York: Random House, 1994.

Olcott, Martha Brill. *Central Asia's New States: Independence, Foreign Policy, and Regional Security*. Washington: United States Institute of Peace Press, 1996.

Tuvalu

Adams, Wanda. "'Double Ghosts' remembers early traders," *Honolulu Advertiser*, March 29, 1998.

Chappell, David. *Double Ghosts: Oceanic Voyagers on Euroamerican Ships*. Armonk, N.Y.: M.E. Sharpe Press, 1997.

Friendship and Territorial Sovereignty: Treaty Between the United States of America and Tuvalu, signed at Funafuti February 7, 1979. Washington, D.C.: U.S. Government Printing Office, 1985.

"New Meaning to the Term Down Under. The Tiny Islands of Kiribati, Tuvalu and Nauru Are Pressuring Australia to Reduce Greenhouse Gas Emissions.)" *The Economist (US)*, September 27, 1997, vol. 344, no. 8036, p. 41.

Kristoff, Nicholas D. "In Pacific, Growing Fear of Paradise Engulfed," *New York Times*, March 2, 1997.

United Arab Emirates

Ali Rashid, Noor. *The UAE Visions of Change*. Dubai: Motivate Publishing, 1997.

Forman, Werner. *Phoenix Rising: the United Arab Emirates, Past, Present & Future*. London: Harvill Press, 1996.

Peck, Malcolm C. *The United Arab Emirates: A Venture in Unity*. Boulder, Colorado: Westview Press, 1986.

Taryam, Abdullah Omran. *The Establishment of the United Arab Emirates 1950–85*. London: Croom-Helm, 1987.

Zahlan, Rosemarie Said. *The Making of the Modern Gulf States*. London: Unwin Hyman Ltd., 1989.

Uzbekistan

Alworth, Edward A. *The Modern Uzbeks: From the Fourteenth Century to the Present*. Stanford, Calif.: Hoover Institution Press, 1990.

Alworth, Edward A., ed. *Central Asia: 130 Years of Russian Dominance, A Historical Overview*. Durham, N.C.: Duke University Press, 1994.

MacLeod, Calum and Bradley Mayhew. *Uzbekistan*. Lincolnwood, Ill.: Passport Books, 1997.

Rashid, Ahmed. *The Resurgence of Central Asia: Islam or Nationalism?* Atlantic Highlands, N.J.: Zed Books, 1994.

Undeland, Charles and Nicholas Platt. *The Central Asian Republics: Fragments of Empire*. New York: The Asia Society, 1994.

Vanuatu

Allen, Michael, ed. *Vanuatu: Politics, Economics, and Ritual in Island Melanesia*. Sydney, Aus.: Academic Press, 1981.

Jennings, Jesse, ed. *The Prehistory of Polynesia*. Cambridge, MA: Harvard University Press, 1979.

Speiser, F. *Ethnology of Vanuatu*. translated by D. Stephenson. Hawaii: University of Hawaii Press, 1996 [1923].

Stanley, D. *South Pacific Handbook*, 3d ed. Chico, Calif.: Moon Publications, 1986.

Viet Nam

Hickey, Gerald Cannon. *Free in the Forest: Ethnohistory of the Vietnamese Central Highlands, 1954–1976*. New Haven: Yale University Press, 1982.

Kahin, George McT. *Intervention: How America Became Involved in Vietnam.* Garden City, N.Y.: Anchor Books. 1987.

Karnow, Stanley. *Vietnam. A History.* New York: The Viking Press. 1983.

Sheehan, Neil. *A Bright Shining Lie.* New York: Random House. 1988.

Thuy, Vuong G. *Getting to the Know the Vietnamese and Their Culture.* New York: Frederick Ungar Publishing. 1975.

Yemen

Al-Suwaidi, Jamal. *The Yemeni War of 1994: Causes and Consequences.* London: Saqi Books, 1995.

Carapico, Sheila. *Civil Society in Yemen : The Political Economy of Activism in Modern Arabia.* New York: Cambridge University Press, 1998.

Chaudhry, Kiren Aziz. *The Price of Wealth: Economies and Institutions in the Middle East.* Ithaca, NY: Cornell University Press, 1997.

Crouch, Michael. *An Element of Luck: To South Arabia and Beyond.* New York: Radcliffe Press, 1993.

Halliday, Fred. *Revolution and Foreign Policy: The Case of South Yemen, 1967–87.* New York: Cambridge University Press, 1990.

Europe

Albania

Battiata, Mary. "Albania's Post-Communist Anarchy," *Washington Post,* March 21, 1998, A1–A18.

Biberaj, Ekiz. *Albania: A Sicuakust Maverick.* Boulder, Colo.: Westview Press, 1990.

Durham, M.E. *High Albania.* Boston: Beacon Press, l985.

Pipa, Arshi. *The Politics of Language in Socialist Albania.* New York: Columbia University Press for Eastern European Monographs, 1989.

Andorra

Carter, Youngman. *On to Andorra.* New York: W. W. Norton, 1964.

Deane, Shirley. *The Road to Andorra.* New York: William Morrow, 1961.

Duursma, Jorri. *Fragmentation and the International Relations of Micro-States: Self-determination and Statehood.* Cambridge: Cambridge University Press, 1996.

Armenia

Batalden, Stephen K. and Sandra L. Batalden. *The Newly Independent States of Eurasia: Handbook of Former Soviet Republics.* Phoenix, AZ: The Oryx Press, 1993.

Croissant, Michael P. *The Armenia-Azerbaijani Conflict: Causes and Implications.* Westport, Conn: Praeger, 1998.

Curtis, Glenn E., ed. *Armenia, Azerbaijan, and Georgia: Country Studies.* Washington, D.C.: Federal Research Division, Library of Congress, 1995.

Economist Intelligence Unit, The. *Country Profile: Georgia, Armenia, 1998–1999.* London: The Economist Intelligence Unit, 1999.

Goldberg, Suzanne. *Pride of Small Nations: The Caucasus and Post-Soviet Disorder.* Atlantic Heights, N.J.: Zed, 1994.

Kaeger, Walter Emil. *Byzantium and the Early Islamic Conquests.* Cambridge: University Press, 1992.

Lang, David Marshall. *Armenia: Cradle of Civilization.* London, Boston: Allen and Unwin, 1978.

McEcedy, Colin. *The New Penguin Atlas of Medieval History.* London: Penguin, 1992.

Austria

Barkey, Karen and Mark von Hagen. *After Empire: Multiethnic Societies and Nation-Building: the Soviet Union and Russian, Ottoman, and Habsburg Empires.* Boulder, Colo.: Westview Press, 1997.

Brook-Shepherd, Gordon. *The Austrians: A Thousand-Year Odyssey.* London: Harper Collins, 1996.

Johnson, Lonnie. *Introducing Austria: A Short History.* Riverside, Calif.: Ariadne, 1989.

Steininger, Rolf, and Michael Gehler, eds. *Österreich im 20. Jahrhundert.* 2 vols. Vienna: Böhlau, 1997.

Belarus

Gross, Jan Tomasz. *Revolution from Abr*oad: *The Soviet Conquest of Poland's Western Ukraine and Western Belorussia.* Princeton, N.J: Princeton University Press, 1988.

Marples, David R. *Belarus: From Soviet Rule to Nuclear Catastrophe.* London: Macmillan, 1996.

Sword, Keith, ed. *The Soviet Takeover of the Polish Eastern Provinces, 1939–41.* New York: St. Martin's Press, 1991.

Zaprudnik, I.A. *Belarus: At a Crossroads in History.* Boulder, Colo.: Westview Press, 1993.

Belgium

Fitzmaurice, John. *The Politics of Belgium: A Unique Feudalism.* London: Hurst, 1996.

Files, Yvonne. *The Quest for Freedom: The Life of a Belgian Resistance Fighter.* Santa Barbara, Calif.: Fithian Press, 1991.

Hilden, Patricia. *Women, Work, and Politics: Belgium 1830–1914.* Oxford: Clarendon Press, 1993.

Hooghe, Liesbet. *A Leap in the Dark: Nationalist Conflict ad Federal Reform in Belgium.* Ithaca: Cornell University Press, 1991.

Warmbrunn, Werner. *The German Occupation of Belgium: 1940–1944.* New York: P. Lang, 1993.

Wee, Herman van der. *The Low Countries in Early Modern Times.* Brookfield, Vt.: Variorum, 1993.

Bosnia and Herzegovina

Burg, Steven L., and Paul S. Shoup. *The War in Bosnia-Herzegovina: Ethnic Conflict and International Intervention.* Armonk, N.Y.: M.E. Sharpe, 1999.

Filipovic, Zlata. *Zlata's Diary: A Child's Life in Sarajevo.* New York: Viking, 1994.

Glenny, Misha. *The Fall of Yugoslavia: The Third Balkan War.* New York: Penguin, 1996.

Lampe, John R. *Yugoslavia as History: Twice There Was a Country.* Cambridge: Cambridge University Press, 1996.

Pinson, Mark, ed. *The Muslims of Bosnia-Herzegovina: Their Historic Development from the Middle Ages to the Dissolution of Yugoslavia.* 2nd ed. Cambridge, Mass.: Harvard University Press, 1996.

Prstojevic, Miroslav. *Sarajevo Survival Guide.* Trans. Aleksandra Wagner with Ellen Elias-Bursac. New York: Workman Publishing, 1993.

Rogel, Carole. *The Breakup of Yugoslavia and the War in Bosnia.* Westport, Conn.: Greenwood, 1998.

West, Rebecca. *Black Lamb and Grey Falcon.* Reprint. New York: Penguin Books, 1982

Bulgaria

Crampton, R.J. *A Concise History of Bulgaria.* Cambridge, New York: Cambridge University Press, 1997.

Curtis, Glenn E., ed. *Bulgaria, a Country Study.* 2nd ed. Federal Research Division, Library of Congress. Washington, D.C., 1993.

Melone, Albert P. *Creating Parliamentary Government: The Transition to Democracy in Bulgaria.* Columbus: Ohio State University Press, 1998.

Minaeva, Oksana. *From Paganism to Christianity: Formation of Medieval Bulgarian Art (681–972).* Frankfurt am Main, New York: P. Lang, 1996.

Paskaleva, Krassira, ed. *Bulgaria in Transition: Environmental Consequences of Political and Economic Transformation.* Brookfield, VT: Ashgate Publications, 1998.

Sedlar, Jean W. *East Central Europe in the Middle Ages, 1000–1500. A History of East Central Europe,* 3. Seattle and London: University of Washington Press, 1994.

Croatia

Cuvalo, Ante. *The Croatian National Movement, 1966–72.* New York: Columbia University Press, 1990.

Glenny, Michael. *The Fall of Yugoslavia: The Third Balkan War.* New York: Penguin, 1992.

Irvine, Jill A. *The Croat Question: Partisan Politics in the Formation of the Yugoslav Socialist State.* Boulder, Colo.: Westview Press, 1993.

Tanner, Marcus. *Croatia: A Nation Forged in War.* New Haven, Conn.: Yale University Press, 1997.

Czech Republic

Bradley, J. F. N. *Czechoslovakia's Velvet Revolution: A Political Analysis.* New York: Columbia University Press, 1992.

Kalvoda, Josef. *The Genesis of Czechoslovakia.* New York: Columbia University Press, 1986.

Kriseova, Eda. *Vaclav Havel: The Authorized Biography.* New York: St. Martin's Press, 1978.

Leff, Carol Skalnik. *The Czech and Slovak Republics: Nation Versus State.* Boulder, Colo.: Westview Press, 1997.

Denmark

Kjrgaard, Thorkild. *The Danish Revolution, 1500–1800: An Ecohistorical Interpretation.* Cambridge: Cambridge University Press, 1994.

Monrad, Kasper. *The Golden Age of Danish Painting.* New York: Hudson Hills Press, 1993.

Pundik, Herbert. *In Denmark It Could Not Happen: The Flight of the Jews to Sweden in 1943.* New York: Gefen Publishing House, 1998.

Estonia

Gerner, Kristian, and Stefan Hedlund. *The Baltic States and the End of the Soviet Empire.* New York: Routledge, 1993.

Hiden, John and Patrick Salmon. *The Baltic Nations and Europe.* London and New York: Longman, 1994.

Iwaskiw, Walter R. *Estonia, Latvia, and Lithuania: Country Studies.* Washington, DC: Federal Research Division, Library of Congress, 1996.

Lieven, Anatol. *The Baltic Revolution.* New Haven and London: Yale University Press, 1993.

Taagepera, Rein. *Estonia: Return to Independence.* Boulder, Colo.: Westview Press, 1993.

Finland

Lander, Patricia Slade. *The Land and People of Finland.* New York: HarperCollins, 1990.

Maude, George. Historical Dictionary of Finland. Metuchen, N.J.: Scarecrow Press, 1994.

.Rajanen, Aini. *Of Finnish Ways.* New York: Barnes & Noble Books, 1981.

Schoolfield, George C. *Helsinki of the Czars: Finland's Capital, 1808–1918.* Columbia, SC: Camden House, 1996.

Singleton, Fred. *A Short History of Finland,* 2nd ed. Cambridge, Eng.: Cambridge University Press, 1998.

France

Agulhon, Maurice. *The French Republic, 1879–1992.* Cambridge, Mass.: B. Blackwell, 1993.

Corbett, James. *Through French Windows: An Introduction to France in the Nineties.* Ann Arbor, Mich.: University of Michigan Press, 1994.

Gildea, Robert. *France Since 1945.* Oxford: Oxford University Press, 1996.

Gough, Hugh, and John Horne. *De Gaulle and Twentieth-Century France.* New York: Edward Arnold, 1994.

Hollifield, James F., and George Ross, eds. *Searching for the New France.* New York: Routledge, 1991.

Noiriel, Gérard. *The French Melting Pot: Immigration, Citizenship, and National Identity.* Minneapolis, Minn.: University of Minnesota Press, 1996.

Northcutt, Wayne. *The Regions of France: A Reference Guide to History and Culture.* Westport, Conn.: Greenwood Press, 1996.

Young, Robert J. *France and the Origins of the Second World War.* Basingstoke, England: Macmillan, 1996.

Georgia

Braund, David. *Georgia in Antiquity: A History of Colchis and Transcaucasian Iberia, 550 BC–AD 562.* New York: Oxford University Press, 1994.

Goldstein, Darra. *The Georgian Feast: The Vibrant Culture and Savory Food of the Republic of Georgia.* New York : HarperCollins, 1993.

Schwartz, Donald V., and Razmik Panossian. *Nationalism and History: The Politics of Nation Building in Post-Soviet Armenia, Azerbaijan and Georgia.* Toronto, Canada : University of Toronto Centre for Russian and East European Studies, 1994.

Suny, Ronald Grigor. *The Making of the Georgian Nation.* 2nd ed. Bloomington, Ind.: Indiana University Press, 1994.

———, ed. *Transcaucasia, Nationalism, and Social Change: Essays in the History of Armenia, Azerbaijan, and Georgia.* Ann Arbor: University of Michigan Press, 1996.

Germany

Davies, Norman. *A History of Europe.* New York: Oxford University Press, 1996.

Dülffer, Jost. *Nazi Germany 1933–1945: Faith and Annihilation.* London: Arnold, 1996.

Eley, Geoff, ed. *Society, Culture, and the State in Germany: 1870–1930.* Ann Arbor: The University of Michigan Press, 1996.

Friedländer, Saul. *Nazi Germany and the Jews.* Volume I - "The Years of Persecution, 1933–1939." New York: HarperCollins, 1997.

Fulbrook, Mary. *A Concise History of Germany.* Updated edition. New York: Cambridge University Press, 1994.

Gies, Frances and Joseph. *Cathedral, Forge, and Waterwheel: Technology and Invention in the Middle Ages.* New York: HarperCollins, 1994.

Kramer, Jane. *The Politics of Memory: Looking for Germany in the New Germany.* New York: Random House, 1996.

Greece

Costas, Dimitris, ed. *The Greek-Turkish Conflict in the 1990s.* New York: St. Martin's Press, 1991.

Jouganatos, George A. *The Development of the Greek Economy, 1950–1991.* Westport, Conn.: Greenwood Press, 1992.

Laisné, Claude. *Art of Ancient Greece: Sculpture, Painting, Architecture.* Paris: Terrail, 1995.

Lawrence, A.W. *Greek Architecture.* New Haven, Conn.: Yale University Press, 1996.

Legg, Kenneth R. *Modern Greece: A Civilization on the Periphery.* Boulder, Colo.: Westview Press, 1997.

Pettifer, James. *The Greeks: The Land and People Since the War.* New York: Viking, 1993.

Hungary

Bartlett, David L. *The Political Economy of Dual Transformations: Market Reform and Democratization in Hungary.* Ann Arbor, MI: University of Michigan Press, 1997.

Corrin, Chris. *Magyar Women: Hungarian Women's Lives, 1960s–1990s.* New York: St. Martin's, 1994.

Hoensch, Jorg K. *A History of Modern Hungary, 1867–1994.* 2nd ed. New York: Longman, 1996.

Litvan, Gyorgy, ed. *The Hungarian Revolution of 1956: Reform, Revolt, and Repression, 1953–1956.* Trans. Janos M. Bak and Lyman H. Legters. New York: Longman, 1996.

Szekely, Istvan P. and David M.G. Newberry, eds. *Hungary: An Economy in Transition.* Cambridge: Cambridge University Press, 1993.

Iceland

Durrenberger, E. Paul. *The Dynamics of Medieval Iceland: Political Economy and Literature.* Iowa City: University of Iowa Press, 1992.

Jochens, Jenny. *Women in Old Norse Society.* Ithaca: Cornell University Press, 1995.

Lacy, Terry G. *Ring of Seasons: Iceland: Its Culture and History.* Ann Arbor: University of Michigan Press, 1998.

Roberts. David. *Iceland.* New York: H.N. Abrams, 1990.

Ireland

Breen, Richard. *Understanding Contemporary Ireland: State, Class, and Development in the Republic of Ireland.* New York: St. Martin's Press, 1990.

Daly, Mary E. *Industrial Development and Irish National Identity, 1922–1939.* Syracuse, N.Y.: Syracuse University Press, 1992.

Hachey, Thomas E. *The Irish Experience: A Concise History.* Armonk, N.Y.: M. E. Sharpe, 1996.

Harkness, D. W. *Ireland in the Twentieth Century: Divided Island.* Hampshire, England: Macmillan Press, 1996.

MacDonagh, Oliver, et al. *Irish Culture and Nationalism, 1750–1950.* New York: St. Martin's Press, 1984.

Sawyer, Roger. *"We Are But Women": Women in Ireland's History.* New York: Routledge, 1993.

Italy

Baranski, Zygmunt G., and Robert Lumley, ed. *Culture and Conflict in Postwar Italy: Essays on Mass and Popular Culture.* New York: St. Martin Press, 1990.

Duggan, Christopher. *A Concise History of Italy.* New York: Cambridge University Press, 1994.

Furlong, Paul. *Modern Italy: Representation and Reform.* New York: Routledge, 1994.

Ginsberg, Paul. *A History of Contemporary Italy: Society and Politics, 1943–1988.* London: Penguin, 1990.

Hearder, Harry. *Italy: A Short History.* New York: Cambridge University Press, 1990.

Holmes, George. *The Oxford History of Italy.* New York: Oxford University Press, 1997.

Latvia

Gerner, Kristian, and Stefan Hedlund. *The Baltic States and the End of the Soviet Empire.* London and New York: Routledge, 1993.

Iwaskiw, Walter R. *Estonia, Latvia, and Lithuania: Country Studies.* Washington, D.C.: Federal Research Division, Library of Congress, 1996.

Lieven, Anatol. *The Baltic Revolution.* New Haven and London: Yale University Press, 1993.

Plakans, Andrejs. *The Latvians: A Short History.* Stanford, Calif: Hoover Institution Press, Stanford University, 1995.

Liechtenstein

Background Notes: Liechtenstein. Washington, D.C.: U.S. Department of State, Bureau of Public Affairs, Office of Public Communication, Editorial Division, USGPO, 1989.

Duursma, Jorri C. *Fragmentation and the International Relations of Micro-States: Self-determination and Statehood.* Cambridge: Cambridge University Press, 1996.

The Principality of Liechtenstein: A Documentary Handbook. Vaduz: Press and Information Office of the Government of the Principality of Liechtenstein, 1967.

Raton, Pierre. *Liechtenstein: History and Institutions of the Principality.* Vaduz: Liechtenstein-Verlag AG, 1970.

Lithuania

Gerner, Kristian and Stefan Hedlund. *The Baltic States and the End of the Soviet Empire.* London/New York: Routledge, 1993.

Hiden, John and Patrick Salmon. *The Baltic Nations and Europe.* London and New York: Longman, 1994.

Hiden, John. *The Baltic States and Weimar Ostpolitik.* Cambridge: Cambridge University Press, l987.

———. *The Baltic States: Years of Dependence, 1940–1980.* Berkeley: University of California Press, 1983.

Vardys, Vytas Stanley. *Lithuania: The Rebel Nation.* Boulder, Colo.: Westview Press, 1997.

Luxembourg

Barteau, Harry C. *Historical Dictionary of Luxembourg.* Lanham, Md.: Scarecrow Press, 1996.

Clark, Peter. *Luxembourg.* New York: Routledge, 1994.

Dolibois, John. *Pattern of Circles: An Ambassador's Story.* Kent, Oh.: Kent State University Press, 1989.

Hury, Carlo. *Luxembourg.* Oxford, England: Clio Press, 1981.

Newcomer, James. *The Grand Duchy of Luxembourg: The Evolution of Nationhood.* Luxembourg: Editions Emile Borschette, 1995.

Macedonia

Billows, Richard A. *Kings and Colonists: Aspects of Macedonian Imperialism.* New York: E.J. Brill, 1995.

Danforth, Loring M. *The Macedonian Conflict: Ethnic Nationalism in a Transnational World.* Princeton, NJ: Princeton University Press, 1995.

Kofos, Evangelos. *Nationalism and Communism in Macedonia: Civil Conflict, Politics of Mutation, National Identity.* New Rochelle, N.Y.: A.D. Caratzas, 1993.

Poulton, Hugh. *Who Are the Macedonians?* Bloomington: Indiana University Press, 1995.

Shea, John. *Macedonia and Greece: The Struggle to Define a New Balkan Nation.* Jefferson, NC: McFarland, 1997.

Malta

Berg, Warren G., *Historical Dictionary of Malta.* Lanham, Md.: Scarecrow, 1995.

Caruana, Carmen M., *Education's Role in the Socioeconomic Development of Malta.* Westport Connecticut: Praeger, 1992.

Europa World Yearbook, Vol. 2, 39th Edition, 1998.

Evans, J.D., *The Prehistoric Antiquities of the Maltese Islands.* New York: Oxford University Press, 1971.

Moldova

Belarus and Moldova: Country studies. Washington: Department of the Army, 1996.

Bruchis, Michael. *The Republic of Moldavia: From the Collapse of the Soviet Empire to the Restoration of the Russian Empire.* Transl. by Laura Treptow. Boulder, Col.: East European Monographs, 1996.

Hitchins, Keith. *Rumania, 1866–1947.* Oxford: Clarendon Press, 1994.

Papacostea, Serban. *Stephen the Great: Prince of Moldavia, 1457–1504.* Transl. by Seriu Celac. Bucharest: Editura Enciclopedica, 1996.

Treptow, Kurt W. *Historical Dictionary of Romania.* Lanham, Md.: Scarecrow, 1996.

Monaco

Duursma, Jorri. *Self-determination, Statehood, and International Relations of Micro-states: The Cases of Liechtenstein, San Marino, Monaco, Andorra, and the Vatican City.* New York: Cambridge University Press, 1996.

Edwards, Anne. *The Grimaldis of Monaco.* New York: William Morrow, 1992.

Sakol, Jeannie and Caroline Latham. *About Grace: An Intimate Notebook.* Chicago: Contemporary Books, 1993.

Netherlands

Andeweg, R.B. *Dutch Government and Politics.* New York: St. Martin's, 1993.

Fuykschot, Cornelia. *Hunger in Holland: Life During the Nazi Occupation.* Amherst, N.Y.: Prometheus Books, 1995.

Israel, Jonathan Irvine. *Dutch Primacy in World Trade, 1585–1740.* New York: Oxford University Press, 1990.

Schilling, Heinz. *Religion, Political Culture, and the Emergence of Early Modern Society: Essays in German and Dutch History.* New York: E.J. Brill, 1992.

Slive, Seymour. *Dutch Painting 1600–1800.* New Haven, Conn.: Yale University Press, 1995.

Norway

Berdal, Mats R. *The United States, Norway and the Cold War 1954–60.* New York: St. Martin's, 1997.

Poland

Blazyca, George and Ryszard Rapacki, eds. *Poland into the 1990s: Economy and Society in Transition.* New York: St. Martin's, 1991.

Engel, David. *Facing a Holocaust: The Polish Government-in-Exile and the Jews, 1943–1945.* Chapel Hill: University of North Carolina Press, 1993.

Staar, Richard F., ed. *Transition to Democracy in Poland.* New York: St. Martin's, 1993.

Steinlauf, Michael. *Bondage to the Dead: Poland and the Memory of the Holocaust.* Syracuse, N.Y.: Syracuse University Press, 1997.

Tworzecki, Hubert. *Parties and Politics in Post-1989 Poland.* Boulder, Colo.: Westview, 1996.

Portugal

Birmingham, David, *A Concise History of Portugal*, Cambridge: Cambridge University Press, 1993.

Hermano Saraiva, Jóse, *Portugal: A Companion History.* Manchester: Carcanet Press, 1995.

Herr, Richard ed. *The New Portugal: Democracy and Europe.* Berkeley: University of California at Berkeley, 1992.

Russell-Wood, A.J.R. *A World on the Move: The Portuguese in Africa, Asia, and America, 1415–1808*, New York: St. Martin's Press, 1993.

Winius, George D. ed., *Portugal, The Pathfinder: Journeys from the Medieval toward the Modern World, 1300–ca.1600*, Madison: The Hispanic Seminary of Medieval Studies, 1995.

Romania

Bachman, Ronald D., ed. *Romania: A Country Study.* 2d ed. Washington, D.C.: Library of Congress, 1991.

Commission on Security and Cooperation in Europe. *Human Rights and Democratization in Romania.* Washington, D.C.: Commission on Security and Cooperation in Europe, 1994.

Economist Intelligence Unit. *Country Report: Romania.* London: The Economist Intelligence Unit, 1999.

Treptow, Kurt W. and Marcel Popa, eds. *Historical Dictionary of Romania.* Lanham, Md.: Scarecrow, 1996.

Russian Federation

Barner-Barry, Carol, and Cynthia A. Hody. *The Politics of Change: The Transformation of the Former Soviet Union.* New York: St. Martin's Press, 1995.

Channon, John, and Robert Hudson. *The Penguin Historical Atlas of Russia.* London: Viking, 1995.

Curtis, Glenn E., ed. *Russia: A Country Study.* Washington: Library of Congress, 1998.

Daniels, Robert V., ed. *The Stalin Revolution: Foundations of the Totalitarian Era.* 4th ed. Boston: Houghton Mifflin, 1997.

Fitzpatrick, Sheila. *The Russian Revolution.* New York: Oxford University Press, 1994.

MacKenzie, David, and Michael W. Curran. *A History of Russia, the Soviet Union, and Beyond.* 5th ed. Belmont, CA: West/Wadsworth, 1999.

Raymond, Boris, and Paul Duffy. *Historical Dictionary of Russia.* Lanham, Md.: Scarecrow, 1998.

San Marino

Catling, Christopher. *Umbria, the Marches and San Marino.* Lincolnwood, Ill.: Passport Books, 1994.

Duursma, Jorri C. *Fragmentation and the International Relations of Micro-States.* Cambridge: Cambridge University Press, 1996.

United States Bureau of Public Affairs; Background Notes: *San Marino.* Washington

World Reference Atlas: *San Marino.* London: Darling Kindersley Publishing, Inc., 1998.

Slovakia

Goldman, Minton F. *Slovakia Since Independence: A Struggle for Democracy.* Westport, Conn.: Praeger, 1999.

Jelinek, Yeshayahu A. *The Parish Republic: Hlinka's Slovak People's Party: 1939–1945.* New York: Columbia University Press, 1976.

Kirshbaum, Stanislav J. *A History of Slovakia: The Struggle for Survival.* New York: St. Martin's Press, 1995.

Leff, Carol Skalnik. *The Czech and Slovak Republics: Nation Versus State.* Boulder, Colo.:Westview Press, 1997.

Slovenia

Cohen, Lenard. Broken Bonds: *The Disintegration of Yugoslavia.* Boulder, Colo.: Westview Press, 1993.

Fink-Hafner, Danica, and John R. Robbins, eds. *Making a New Nation: The Formation of Slovenia.* Brookfield, Vt. Aldershot: Dartmouth Publishing, 1997.

Owen, David. *Balkan Odyssey.* New York: Harcourt Brace, 1995.

Rogel, Carole. *The Breakup of Yugoslavia and the War in Bosnia.* Westport, Conn.: Greenwood Press, 1998.

Silber, Laura, and Allan Little. *Yugoslavia: Death of a Nation.* New York: TV Books, 1996.

Zimmerman, Warren. *Origins of a Catastrophe: Yugoslavia and its destroyers—America's last ambassador tells what happened and why.* New York: Times Books, 1996.

Spain

Cantarino, Vicente. *Civilización y Cultura de España,* 3rd. ed., Englewood Cliffs, N.J.: Prentice Hall, l995.

Hopper, John. *The New Spaniards.* Suffolk, Eng.: Penguin, l995.

Jordan, Barry. *Writings and Politics in the Franco's Spain.* London, Eng.: Routledge, 1990.

Heide, Sigrid. *In the Hands of My Enemy: A Woman's Personal Story of World War II.* Trans. Norma Johansen. Middletown, Conn.: Southfarm Press, 1996.

Jochens, Jenny. *Women in Old Norse Society.* Ithaca, N.Y.: Cornell University Press, 1995.

Leahy, Philippa. *Discovering Spain*. New York: Crestwood House, 1993.

Wernick, Robert. "For Whom the Bell Tolled. (the Spanish Civil War)." *Smithsonian*. April 1998, vol. 28, no. 1, pp. 110+.

Sweden

Palmer, Alan. *Bernadotte: Napoleon's Marshal, Sweden's King*. London: Murray, 1990.

Roberts, Michael. *From Oxenstierna to Charles XII: Four Studies*. New York: Cambridge University Press, 1991.

Rothstein, Bo. *The Social Democratic State: The Swedish Model and the Bureaucratic Problem of Social Reforms*. Pittsburgh: University of Pittsburgh, 1996.

Scobbie, Irene. *Historical Dictionary of Sweden*. Metuchen, N.J.: Scarecrow Press, 1995.

Switzerland

Bacchetta, Philippe, and Walter Wasserfallen, ed. *Economic Policy in Switzerland*. New York: St. Martin's Press, 1997.

Eu-Wong, Shirley. *Culture Shock!: Switzerland*. Portland, Or.: Graphic Arts Center Pub. Co., 1996.

Hilowitz, Janet Eve, ed. *Switzerland in Perspective*. New York: Greenwood Press, 1990.

New, Mitya. *Switzerland Unwrapped: Exposing the Myths*. New York: I.B. Tauris, 1997.

Steinberg, Jonathan. *Why Switzerland?* 2nd ed. Cambridge: Cambridge University Press, 1996.

Turkey

Fodor's Turkey. New York: Fodor's Travel Publications, 1999.

Mitchell, Stephen. *Anatolia: Land, Men, and Gods in Asia Minor*. New York: Oxford University Press, 1993.

Stoneman, Richard. *A Traveller's History of Turkey*. New York: Interlink Books, 1998.

Time-Life Books. *Anatolia: Cauldron of Cultures*. Alexandria, Va.: Time-Life Books, 1995.

Ukraine

Hosking, Geoffrey A., ed. *Church, Nation and State in Russia and Ukraine*. New York: St. Martin's Press, 1991.

Kuzio, Taras. *Ukraine under Kuchma: Political Reform, Economic Transformation and Security Policy in Independent Ukraine*. New York: St. Martin's Press, 1997.

Shen, Raphael. *Ukraine's Economic Reform: Obstacles, Errors, Lessons*. Westport, Conn: Praeger, 1996.

Subtelny, Orest. *Ukraine: A History*. Toronto: University of Toronto Press, 1994.

Wilson, Andrew. *Ukrainian Nationalism in the 1990s: A Minority Faith*. New York: Cambridge University Press, 1997.

United Kingdom

Abrams, M. H., ed. *The Norton Anthology of English Literature*. 2 vols. 6th ed. New York : Norton, 1993.

Cook, Chris. *The Longman Handbook of Modern British History, 1714–1995*. 3rd ed. New York: Longman, 1996.

Delderfield, Eric F. *Kings & Queens of England & Great Britain*. New York: Facts on File, 1990.

Figes, Kate. *Because of Her Sex: The Myth of Equality for Women in Britain*. London: Macmillan, 1994.

Foster, R.F., ed. *The Oxford History of Ireland*. New York: Oxford University Press, 1992.

Glynn, Sean. *Modern Britain: An Economic and Social History*. New York: Routledge, 1996.

Jenkins, Philip. *A History of Modern Wales, 1536–1990*. New York: Longman, 1992.

Judd, Denis. *Empire: The British Imperial Experience from 1765 to the Present*. London : HarperCollins, 1996.

The Oxford History of Britain. New York: Oxford University Press, 1992.

Powell, David. *British Politics and the Labour Question, 1868–1990*. New York: St. Martin's, 1992.

Vatican

Accattoli, Luigi. *Life in the Vatican with John Paul II*. Trans. Marguerite Shore. New York : Universe, 1998.

Reese, Thomas J. *Inside the Vatican: The Politics and Organization of the Catholic Church*. Cambridge, Mass.: Harvard University Press, 1996.

Rosa, Peter de. *Vicars of Christ: The Dark Side of the Papacy*. New York: Crown Publishers, 1988.

Volpini, Valerio. *Vatican City: Art, Architecture, and History*. New York, Portland House: Distributed by Crown Publishers, 1986.

Yugoslavia

Cohen, Lenard J. *Broken Bonds: Yugoslavia's Disintegration and Balkan Politics in Transition*. Boulder, CO: Westview, 1995.

Denitch, Bogdan Denis. *Ethnic Nationalism: The Tragic Death of Yugoslavia*. Minneapolis: University of Minnesota Press, 1996.

Dyker, David A., and Ivan Vejvoda, ed. *Yugoslavia and After: A Study in Fragmentation, Despair and Rebirth*. New York: Longman, 1996.

Lampe, John R. *Yugoslavia as History: Twice There Was a Country*. New York: Cambridge University Press, 1996.

Pavkovic, Aleksandr. *The Fragmentation of Yugoslavia: Nationalism in a Multinational State*. New York: St. Martin's Press, 1997.

Ramet, Sbrina P. *Balkan Babel: The Disintegration of Yugoslavia from the Death of Tito to Ethnic War*. Boulder, Colo.: Westview, 1996.

Index

Cyprus v3:126
Democratic Republic of the Congo
    v1:121
Estonia v4:153
France v4:192
Gabon v1:207, v1:208
Germany v4:218, v4:223, v4:226
Ghana v1:230
Guinea v1:243
Guinea-Bissau v1:253, v1:255
India v3:149
Indonesia v3:172
Iraq v3:217, v3:222
Ireland v4:299
Israel v3:244
Japan v3:255, v3:260, v3:261
Jordan v3:278
Kazakstan v3:281, v3:284, v3:285,
    v3:289, v3:291
Kenya v1:260, v1:266, v1:267, v1:270,
    v1:274, v1:275
Kyrgyzstan v3:343
Laos v3:361
Latvia v4:329
Libya v1:309
Lithuania v4:346
Madagascar v1:311
Malawi v1:325
Mali v1:336, v1:343
Mauritius v1:360
Mayan v2:60
Moldova v4:385
Mongolia v3:418
Mozambique v1:381
Nauru v3:436
Nepal v3:441
Niger v1:400, v1:404
Nigeria v1:412, v1:419
North Korea v3:314
Norway v4:412
Pakistan v3:472
Palau v3:488
Papua New Guinea v3:498
Peru v2:391
Philippines v3:505, v3:510
Romania v4:461
Rwanda v1:426
São Tomé and Príncipe v1:439, v1:440
Seychelles v1:455
Sierra Leone v1:466, v1:467
Somalia v1:470, v1:481
Sri Lanka v3:559, v3:560, v3:561,
    v3:565, v3:566
Sudan v1:501
Suriname v2:439, v2:441
Swaziland v1:507, v1:508
Syria v3:577
Tanzania v1:516
Thailand v3:612

Togo v1:528, v1:530, v1:531
Tunisia v1:534
Turkmenistan v3:634
Uganda v1:548
United States v2:461, v2:477
Uzbekistan v3:662
Vietnam v3:672
Yemen v3:691
Zambia v1:556
Zimbabwe v1:571
Aguilar, Ernesto Paz v2:308
Aguilar, Nicolás v2:234
Aguirre Cerda, Pedro v2:137
Aguirre, Lope de v2:510
Agung v3:173
Ahearn, Bertie v4:301, v4:637
Ahidjo, Ahmadou v1:77
Ah-Kakaw (Double-Comb) v2:255, v2:265
Ahmad v3:690
Ahmad, Mirza Gulam v3:159
Ahmad, Muhammad v1:500
Ahmad Shah v3:198
Ahmad Shah Abdali v3:7, v3:157, v3:476
Ahmad, Sultan of Brunei v3:77
Ahmed Bin Said v3:466
Ahmed Ghozali, Sid v1:21
Ahmed, Kifleh v1:155
Ahmed, Sayyid Abdallah ibn-Ahmad el-
    Wazir. Emir Seif al-Islam
    v3:690
Ahmida, Ali Abdullatif v1:309
Ahmose (r. 1570–46 B.C.) v1:164
Ahmose I v1:501
Aho, Esko v4:174
Aho, Juhani v4:167
Ahomadégbé, Justin v1:42, v1:43
Ahtisaari, Martii v4:174
Ahura Mazdak v3:192
Aidid, Farah v1:471, v1:475, v1:479,
    v1:480, v1:481
Aidid, Hussein v1:481
Aidoo, Ama Ata v1:235
AIDS v2:103, v2:490
Airlangga v3:172
Aitmatov, Chingiz v3:292
Ajasin, Adekunle v1:423
Akaev, Askar v3:347
Akatdamkoeng v3:612
Akayev, Aidar v3:349
Akbar v3:155, v3:476
Ake, Claude v1:423
Akihito v3:263, v3:269
Akintola, Samuel Ladoke v1:418
Akira Kurosawa v3:263
Akuffo v1:224
Alacalufe Indians v2:129
Alak Betatar. *See* Muhammad Shah, Sultan
    of Brunei

Alam, Muhammad Sultan of Brunei v3:78
Alarcón, Fabián v2:215, v2:228
Alas, Leopoldo ("Clarín", 1852–1901)
    v4:548
Alaska
    Russia v2:477
    U.S. statehood v2:486
Ala-ud-Din (r. 1296–1316) v3:154, v3:475
Albania v4:7–v4:18
    agriculture v4:10
    Christianity v4:7, v4:10, v4:11, v4:17
    environment v4:7
    human rights v4:15, v4:16, v4:18
    Islam v4:7, v4:11, v4:17
    literature v4:9, v4:14
    mining v4:7
    poetry v4:16
    religion v4:7, v4:10, v4:11, v4:17
    Roman Catholic Church v4:8
    strikes v4:17
    women v4:15, v4:16
Albéniz, Isaac v4:548
Alberdi, Juan Bautista v2:22
Alberoni, Cardinal Giulio v4:504, v4:507
Albert I of Belgium v4:68, v4:70
Albert I, Prince v4:388, v4:390, v4:392
Albert II of Belgium v4:73
Albert, prince of England v4:627
Albuquerque, Afonso de v3:608
Alcorta, Amancio v2:20
Alcott, Louisa May v2:473
Aldrin, Edwin "Buzz" Jr. v2:487
Alea, Tomas Gutierrez v2:189
Aleandro, Norma v2:33
Alegria, Ricardo v2:414
Aleijadinho (Antônio Francisco Lisboa)
    v2:90
Aleixandre, Vicente v4:554
Alem, Leandro N. v2:25
Alemán Lacayo, José Arnoldo v2:345,
    v2:359
Alessandri Palma, Arturo v2:136, v2:138
Alessandri Rodríguez, Jorge v2:138
Alexander v4:363
Alexander, King v4:430, v4:527
Alexander, King of Greece v4:249
Alexander, King of Macedonia v4:368
Alexander I, Tsar v4:208
Alexander I of Yugoslavia v4:77, v4:82,
    v4:115, v4:658
Alexander II, Tsar v3:42, v3:237, v3:244,
    v4:97, v4:592, v4:601
Alexander III v3:219, v4:476, v4:487,
    v4:588
Alexander III, Pope v4:163, v4:640
Alexander III, Tsar v3:237, v3:244
Alexander VI (Rodrigo Lanzol y Borgia),
    Pope, v4:641
Alexander VI, Pope v4:544

Bamba, Ahmadou v1:446
Ban Kulin v4:75, v4:78, v4:79
Banana plantation(s)
    Guatemala v2:254
    Honduras v2:297, v2:304, v2:308
    St. Lucia v2:422, v2:426
    St. Vincent and the Grenadines v2:432
Banda Oriental. *See* Uruguay
Banda, Hastings Kamuzu v1:329, v1:330,
    v1:332, v1:333
Bandaranaike, Sirimavo v3:567
Bangladesh v3:54–v3:65
    environment v3:54
    Islam v3:54, v3:55
    literature v3:55
    music v3:55
    religion v3:54
Banharn Silpa-archa v3:620
Bani-Sadr v3:202, v3:205, v3:206
Banking
    Argentina v2:28
    Australia v3:24
    Bahamas v2:42
    Bolivia v2:77
    Bulgaria v4:106
    Chile v2:141
    China v3:117
    Croatia v4:120
    Cyprus v3:127, v3:131
    Democratic Republic of the Congo
        v1:122
    Djibouti v1:158
    Ecuador v2:222
    Equatorial Guinea v1:185, v1:187
    Ethiopia v1:200
    Iran v3:204
    Ireland v4:298
    Jamaica v2:315
    Kenya v1:272
    Kyrgyzstan v3:349
    Lebanon v3:371
    Luxembourg v4:352
    Malawi v1:331
    Malaysia v3:392
    Malta v4:377
    Mexico v2:340
    Myanmar v3:430
    Nauru v3:439
    Nigeria v1:422
    Norway v4:424
    Pakistan v3:482
    Panama v2:371
    Peru v2:399
    San Marino v4:509
    Singapore v3:548
    Somalia v1:476
    Suriname v2:441
    Switzerland v4:579, v4:582
    Syria v3:577

    Thailand v3:614
    United Arab Emirates v3:650, v3:651
    Venezuela v2:517
    Yemen v3:693
Bannerman, Charles v1:229
Banzer Suárez, Colonel Hugo v2:69, v2:80
Bao Dai v3:671, v3:678, v3:680, v3:682
Baptista, Pedro v1:559
Barak, Ehud v3:252
Baranauskas, Antanas v4:344
Baranca Velásquez, Efrain v2:267
Barbados v2:9, v2:47–v2:56
    agriculture v2:49
    cricket v2:52, v2:53
    epidemics v2:48
    Grantley Adams Airport v2:55
    indigo cultivation v2:49
    influence in St. Lucia v2:425
    influence in Suriname v2:437, v2:439
    journalism v2:51
    library v2:52
    literature v2:53
    mounted police (photo) v2:54
    museums v2:53
    Panama Canal v2:52
    Rastafarianism v2:54
    religion v2:53
    science v2:53
    Slave Trade Abolition Act v2:51
    slavery v2:48, v2:49, v2:50
    suffrage v2:51, v2:53
    sugarcane v2:49
    Telecommunications v2:54
    University of the West Indies v2:54
Barbie, Klaus v2:81
Barbosa, Duarte v3:49
Barbosa, Jorge v1:85, v1:88
Barbosa, Jose v2:411
Barbuda. *See* Antigua and Barbuda
Barclay, Arthur v1:291
Barclay, Edwin J. v1:292
Bardhylus of Illyria v4:10
Barillas, Manuel v2:262
Barletta, Nicolas Ardito v2:372
Barnard, Christian v1:493
Barons, Krisjanis v4:325
Baroque culture v4:40
Baroque Era
    Spain v4:535
Baré Mainassara, Ibrahim v1:409
Barre, Mohammed Siad. *See* Siad Barre,
    Mohammed
Barrera, Angel v1:180, v1:183
Barreto, Bruno v2:105
Barreto, Francisco v1:573
Barrientos Ortuño, General René v2:80
Barrios de Chamorro, Pedro Joaquín v2:356
Barrios de Chamorro, Violeta v2:345,
    v2:353, v2:354, v2:358

Barrios, Gerardo v2:235
Barrios, Justo Rufino v2:303
Barrow, Errol v2:48, v2:53, v2:55
Barry, Diawadou v1:245
Barry, Ibrahima v1:245, v1:246
Barry, Mamdou v1:248
Barth, Heinrich v1:403
Barthel, Josy v4:358
Bartholdi, Frederic Auguste v2:480
Bartok, Bela v4:154, v4:264
Barzani, Massoud v3:216, v3:230
Barzani, Mustafa v3:227, v3:230
Basanavicius, Dr. Jonas v4:344
Basarab v4:463
Basarab-Laiota v4:464
Baseball
    Cuba v2:179, v2:180, v2:185
    Dominican Republic v2:212
    Mexico v2:338
    Nicaragua v2:350, v2:352,v2:354
    Panama v2:365, v2:368, v2:369,
        v2:371
    Peru v2:402
    Puerto Rico v2:414
    United States v2:481, v2:485
Bashir, Omar Hassan Ahmed al- v1:501
Basho, Matsuo v3:260
al-Biruni (973-1048) v3:11, v3:397
Basil II of Byzantium v4:89, v4:93
Basilio Acuna, Jose v2:172
Basque separatists
    France v4:202
Basta, General George v4:465
Bastille v4:178, v4:188
Bastos, Augusto Roa v2:381
al-Battani v3:220
Bathory, Prince Gabriel v4:465
Bathsheba v3:239
Batista y Zaldívar, Fulgencio v2:184
Batista, Fulgencio v2:174, v2:186
Batmunkh, Jambyn v3:412, v3:420, v3:421
Bator, Sukhe v3:418
Battle Ax culture v4:160, v4:162
Batu Khan v3:286, v3:287, v3:413, v4:479,
    v4:603
Baudouin of Belgium v4:71, v4:72
Bautista Saavedra, Juan v2:77
Bauxite
    Jamaica v2:319
    Suriname v2:438, v2:441, v2:443
Bavadra, Dr. Timoci v3:143
Baxter v3:457
Baxter, James v3:459
Bayar, Celâl v4:595
Bayazid v3:195
Bayezid I (r. 1389-1402) v4:95
Bayezid I, Sultan v4:463
Baymen
    Belize v2:61

H

Murphy, Joseph. *See* Collins, Tom
Musa, Mansa v1:57, v1:145, v1:217,
v1:241, v1:445
Musaddiq, Muhammad v3:197, v3:199
Museums
Botanical Gardens of Dominica v2:195
Brazil v2:99
Denmark v4:143, v4:145, v4:147
France v4:184, v4:201, v4:202
Georgia v4:210
Greece v4:248, v4:255
Guatemala v2:260
Hungary v4:262
Macedonia v4:367
Mexico v2:329, v2:330
Monaco v4:393
Museo Nacional (Cuba) v2:183
Netherlands v4:402, v4:408
Nicaragua v2:351
Norway v4:416, v4:423
Panama v2:368, v2:371
Peru v2:398
Poland v4:433
Portugal v4:451
Prado (Spain) v4:547, v4:555
St. Vincent and the Grenadines v2:433
Slovak Republic v4:514, v4:516
Suriname v2:441
Switzerland v4:579
Vatican Archives v4:641
Venezuela v2:517
Museveni, Yoweri v1:431, v1:547, v1:548,
v1:550, v1:553
Musgrave, Sir Anthony v2:317
Mushanov, Nikola v4:100
Music
Algeria v1:19
Argentina v2:27, v2:28
Armenia v4:31
Austria v4:41
Azerbaijan v3:44
Bangladesh v3:55
Belarus v4:56
Belgium v4:59, v4:62, v4:63, v4:66,
v4:69, v4:70, v4:71
Bolivia v2:72, v2:77, v2:78, v2:79
Bosnia and Herzegovina v4:85
Brazil v2:90, v2:93
Cameroon v1:81, v1:82, v1:83
Canada v2:116
Cape Verde v1:84, v1:90
Chile v2:131, v2:135, v2:136, v2:159
Congo, Democratic Republic of the
v1:120, v1:121
Côte d'Ivoire v1:144, v1:149
Cuba v2:179, v2:180, v2:182, v2:185,
v2:187
Czech Republic v4:127, v4:128
Denmark v4:143, v4:144, v4:146

Dominican Republic v2:211
Equatorial Guinea v1:182
Estonia v4:153, v4:154, v4:157
Finland v4:166, v4:167, v4:168,
v4:173, v4:174
France v4:178, v4:185, v4:189, v4:191,
v4:193, v4:194
Gambia, The v1:222, v1:223
Georgia v4:209
Germany v4:218, v4:223, v4:224,
v4:227, v4:229
Greece v4:250
Hungary v4:263, v4:264, v4:265
Iceland v4:276, v4:278
India v3:160, v3:163, v3:168
Indonesia v3:172, v3:182
Ireland v4:287, v4:288
Italy v4:304, v4:311, v4:312, v4:313
Jamaica v2:318
Japan v3:267
Liechtenstein v4:336, v4:337
Mali v1:346
Mauritius v1:363
Mexico v2:332
Monaco v4:391
Netherlands v4:398, v4:400, v4:404,
v4:407
New Zealand v3:455, v3:456, v3:458
Nicaragua v2:350
Nigeria v1:424
Norway v4:415, v4:417, v4:419,
v4:420, v4:423
Panama v2:370
Peru v2:394, v2:395, v2:399, v2:400
Philippines v3:504
Poland v4:432, v4:433, v4:435, v4:436
Portugal v4:452
Puerto Rico v2:408, v2:414
reggae v2:318
Republic of the Congo v1:134
Russia v4:486
Rwanda v1:428, v1:430
Singapore v3:546
Slovak Republic v4:518, v4:519
South Korea v3:324, v3:328
Spain v4:548, v4:549
St. Lucia v2:428
St. Vincent and the Grenadines v2:435
Sweden v4:563, v4:565
Switzerland v4:578
Tanzania v1:517, v1:518, v1:521,
v1:522
Trinidad and Tobago v2:450
Uganda v1:548
Ukraine v4:604
United Kingdom v4:623, v4:629,
v4:631, v4:633
United States v2:472, v2:483

Venezuela v2:511, v2:513, v2:517,
v2:523
Yugoslavia v4:655, v4:657
Musicians
Anikulapo-Kuti, Fela v1:424
Arrau, Claudio v2:135
Blondy, Alpha v1:149
Campos, Juan Morel v2:408
Carreño, Maria Teresa v2:513
Diaghilev, Sergei v4:486
Feliciano, Jose v2:414
Field, John v4:287
Galway, James v4:298
Guerra, Juan Luis v2:211
Hajibeyov, Uzeyir v3:42
Kamba, Paul v1:134
Kiwele, Joseph v1:121
Laredo, Jaime v2:79
Lecuona, Ernesto v2:187
Lutaaya, Philly v1:548
Mali v1:346
Marley, Bob v2:318–9
Mercury, Freddy v1:522
Olimide, Koffi v1:83
Oryema, Geoffrey v1:548
Saidi, Sitti bit v1:522
Samite v1:548
Sculthorpe, Peter v3:29
Segovia, Andres v4:550
Shankar, Anoushka v3:168
Shankar, Rabendra (Ravi) v3:160
Stubbs, Thomas v3:24
Tavarez, Manuel G. v2:408
Zadeh, Vagif Mustafa v3:44
Musinga, Mwami v1:429
Muskie, Edmund v3:205
Muslims. *See* Islam
Mussolini, Benito v1:189, v1:191, v1:306,
v1:307, v1:475, v4:14,
v4:99, v4:116, v4:315,
v4:504, v4:509
Mutanabbi, Abut at Fayyil Ahmad bin al-
Husayn al v3:574
Mutara, Charles Rudahigwa v1:429, v1:430
Mutawwakil, Mohammed v1:370, v4:450
Mutesa, Kabaka v1:551
Mutota v1:573
Muyongo, Mishake v1:399
Muzaffar al-Din Shah v3:197
Muzorewa, Abel Tendekayi v1:571,
v1:578, v1:579, v1:580
Mwambutsa IV v1:66, v1:68, v1:70
Mwendwa, Kitili v1:272
Mwinyi, Ali Hassan v1:525
Myanmar v3:423–v3:434
banking v3:430
Buddhism v3:423, v3:427, v3:430
currency v3:431
dance v3:425, v3:426

Bolivia v2:68, v2:77
Brunei v3:79
Brunei Darussalam v3:76
Cameroon v1:76
Canada v2:122
Chad v1:105, v1:106, v1:107
China v3:114
Côte d'Ivoire v1:143
Democratic Republic of the Congo
 v1:123, v1:124
Ecuador v2:226
Equatorial Guinea v1:188
Gabon v1:211, v1:213
Ghana v1:236
Indonesia v3:171, v3:176, v3:182
Iran v3:197, v3:201, v3:206, v3:211,
 v3:212
Iraq v3:216, v3:223, v3:224, v3:227,
 v3:229, v3:231, v3:232
Japan v3:268
Kazakstan v3:282, v3:294
Kuwait v3:231, v3:333, v3:334,
 v3:336, v3:337, v3:338
Kyrgyzstan v3:349
Libya v1:302, v1:303, v1:307, v1:309
Madagascar v1:323
Malaysia v3:380, v3:386
Maldives v3:401
Mali v1:343
Malta v4:376
Mauritania v1:354, v1:355
Netherlands v4:405
Nigeria v1:412, v1:417, v1:419,
 v1:420, v1:423, v1:424
Norway v4:412, v4:423
Oman v3:464
Pakistan v3:484
Panama v2:371, v2:372
Peru v2:398
Qatar v3:513, v3:518, v3:520
Republic of the Congo v1:137, v1:138,
 v1:139
Saudi Arabia v3:532, v3:539, v3:540,
 v3:542
Sierra Leone v1:467
Somalia v1:478
South Africa v1:496
Suriname v2:444
Syria v3:577, v3:579, v3:580
Trinidad and Tobago v2:447, v2:451,
 v2:452, v2:454
Tunisia v1:542, v1:543, v1:545
Turkmenistan v3:630, v3:637
United Arab Emirates v3:645, v3:650,
 v3:653
United States v2:488
Venezuela v2:507, v2:516, v2:521
Yemen v3:691, v3:692, v3:693
Zimbabwe v1:580

Oiterong, Alfonso v3:489
Ojeda, Alonas de v2:509
Ojiambo, Julia Auma v1:273
Ojibwa Indians v2:109, v2:111
Ok-sang, Im v3:329
Okar, Gideon v1:421
Okeke, Uche v1:417
Okello, Basilio v1:553
Okello, John v1:519, v1:524
Okello, Tito v1:553
Olaf I v4:410, v4:413
Olaf II v4:410, v4:413
Olaf V, King v4:412, v4:418, v4:423
Olcott, Henry S. v3:159
Ole Bull v4:410, v4:415, v4:416, v4:417,
 v4:419
Oleg the Wise v4:479, v4:603
Oleksy, Jozef v4:440
Olimide, Koffi v1:83
Oliveira, António de v4:453
Oliver, Dr. Borg v4:375
Oller, Francisco v2:412
Olter, Bailey v3:137, v3:138
Olympiad. *See* Olympic Games
Olympic Games v4:256, v4:569
 Finland v4:170
 France (1968) v4:200
 Germany v4:234, v4:237
 Greece, first v4:248
 McKenley, Herb (Jamaica) v2:319
 Mexico v2:325, v2:338
 United Kingdom v4:632
 United States v2:459
Olympio, Sylvanus v1:527, v1:530, v1:531
O'Mahoney, John v4:290
Oman v3:463–v3:469
 education v3:468
 Islam v3:463, v3:465
 medicine v3:464
 mining v3:465
 oil v3:464
 religion v3:463, v3:465
Omar Ali Saifadeen, Sultan of Brunei v3:76,
 v3:78, v3:80, v3:81
Ona Indians v2:18
Onabolu, Aina v1:413
Ondaatje, Michael v3:560
O'Neale, Charles Duncan v2:53
Oneko, Achieng' v1:270
Ong Teng Cheong v3:549
Onganía, General Juan Carlos v2:29, v2:30
Ono Indians v2:129
Onobrakpeya, Bruce v1:417
Opangault, Jacques v1:135
Opera
 Georgia v4:210
 Germany v4:229
 Hungary v4:265
 Latvia v4:327

Pasta, Carlo Enrico v2:397
Peru v2:395
Russia v4:486
Slovak Republic v4:515
Spain v4:552, v4:553
Sweden v4:563
Opium War v4:628
Oppenheimer, J. Robert v3:266
Orchestras
 Macedonia v4:368
 Sweden v4:565
 Venezuela v2:517
Oregon Trail v2:473
Orejuela, Miguel Rodriguez v2:159
Organization of American States v2:13,
 v2:520
Organization of East Caribbean States
 (OECS) v2:14, v2:427
Organization of Indigenous People of
 Pastaza (Ecuador) v2:227
Organization of Petroleum Exporting
 Countries v2:226
Ori v4:31
Orisha v2:450
Orodes I v3:192
Orontes v4:28
Orozco, José Clemente v2:331
Orpen, William v4:293
Ortega Saavedra, Daniel v2:354, v2:346,
 v2:359
Ortega, General Humberto v2:358
Ortiz, Roberto v2:27
Oryema, Geoffrey v1:548
Osario, Oscar v2:237
Oscar I, King of Sweden v4:564
Osei Agyeman Otumfo, Prempe II v1:232
O'Shea, Katherine v4:292
Osman v1:13, v3:125, v3:243, v4:12, v4:30,
 v4:80, v4:246, v4:365,
 v4:591, v4:652
Otakar II v4:125
Othon, King v4:247
Ottawa Indians v2:465
Otto I, Roman emperor v4:220
Otto III, Holy Roman Emperor v4:426,
 v4:428
Otto of Saxony (Otto the Great) v4:214,
 v4:220, v4:260
Otto, King of Bavaria
 influence in Greece v4:247
Ottoman Empire
 Greece v4:245
 Hungary v4:261
 Turkey v4:584, v4:586, v4:587, v4:590
Oueddei, Goukouni v1:105, v1:98, v1:308
Ouédraogo, Jean Baptiste v1:62
Ouedraogo, Kadre Desiré v1:64
Ouedraogo, Macaire v1:62

education v1:521
environment v1:516
Hinduism v1:516
human rights v1:525
Islam v1:516
labor movements v1:522
labor unrest v1:522
music v1:517, v1:518, v1:521, v1:522
sports v1:517, v1:521, v1:522, v1:525
strikes v1:522
Taoism. *See* Daoism
Tapes Indians v2:378
Taraki v3:17
Tariq v1:368
Tartesians v4:532, v4:538
Tarzi, Mahmud v3:14
Tasman, Abel Jansen v3:139, v3:141, v4:401, v3:450, v3:452
Tatarchev, Ivan v4:105
Taufa'ahua Tupou IV v3:624, v3:626
Tavares, Eugenio v1:90
San Marino v4:510
Taylor, Charles v1:287, v1:296, v1:298, v1:299, v1:467
Taylor, Zachary v2:330, v2:474
Te Kanawa, Kiri v3:457
Teatro Colón v2:25
Tecumseh v2:471
Tecúm-Umán v2:256
Tegnér, Esaias v4:563
Tehuelche Indians v2:129
Tejada, Lydia Gueiler v2:81
Telecommunications
Barbados v2:54
Nicaragua v2:350
United States v2:486
Venezuela v2:519
Tell, William v4:573
Tembo, Christon v1:567
Tembo, John v1:333
Temple of the Masks (Guatemala) v2:255
Tenochca. *See* Aztec Indians
Tenreiro, Francisco v1:438
Tenzin Drukda v3:69
Teodorov, Teodor v4:99
Tepelena, Ali Pasha v4:13
Tepes, Vlad v4:463
Tepexpan man v2:326
Teresa de Jesus v2:141
Tereshchenko, Sergei v3:293, v3:296
Tereshkova, Valentina Vladimirovna v4:492
Ter-Petrosian, Levon v4:33, v4:34
Terrorism
France v4:202, v4:204
Italy v4:319, v4:320
Oklahoma City bombing v2:490
Sweden v4:569

Turkey v4:599
United Kingdom v4:634, v4:636
United States v2:490
Teuta of Illyria v4:10
Teutonic Order of Knights v4:342
Tew, Thomas v1:316, v1:317
Tewodros II v1:194, v1:197, v1:198
Thagard, Norman E. v2:490
Thailand v3:604–v3:620
agriculture v3:612
architecture v3:609
banking v3:614
Buddhism v3:604, v3:606, v3:607, v3:608, v3:609, v3:610
education v3:611, v3:612, v3:613, v3:616
environment v3:617, v3:618
literature v3:607, v3:612, v3:617
medicine v3:615, v3:616
women v3:619
Thakurufaanu, Mohammed v3:398
Than Schwe v3:432
Thango, François v1:121
Thangton Gyalpo v3:68
Thani, Abdullah Al- v3:517, v3:518
Thani, Hamad bin Khalifah Al- v3:513, v3:519, v3:520, v3:521
Thani, Qasim bin Mohammed Al-v3:517
Thanom Kittikachorn v3:605, v3:616, v3:617
Thapa, Bhimsen v3:445
Thapa, Surya Bahadur v3:449
Thatcher, Margaret v4:631, v4:634
The Gambia. *See* Gambia
Theater
Austria v4:43
Bosnia and Herzegovina v4:86
Brecht, Bertolt v4:233
Canada v2:123
Cantinflas v2:334
China v3:112, v3:113
Costa Rica v2:172
Czech Republic v4:128
Estonia v4:153
France v4:184, v4:185
Georgia v4:208
Germany v4:229, v4:233
Greece v4:240, v4:250
Guatemala v2:261
Hungary v4:263
Ireland v4:287, v4:290, v4:294
Jamaica v2:318
Japan v3:254, v3:258, v3:260
Kazakstan v3:292
Liberia v1:294
Mexico v2:334, v2:335
Myanmar v3:426, v3:431
Nicaragua v2:346, v2:357
Panama v2:364, v2:366

Peru v2:397
Russia v4:486
Senegal v1:451
Singapore v3:546
Somalia v1:475
Spain v4:535, v4:545, v4:548
Sweden v4:563, v4:564
United Kingdom v4:619, v4:620
Venezuela v2:511
Weill, Kurt v4:233
Theato, Michael v4:356
Theodorakis, Mikis v4:250
Theodore of Epirus v4:94
Theodorico v4:507
Theodosius I, the Great v4:78, v4:539
Theotocopoulos, Domenico. *See* El Greco
Theroux, Paul v1:548
Thieu Tri v3:676, v3:684
Thomas à Becket v4:615
Thomson, Joseph John v1:260, v4:630
Thoreau, David v2:471
Thorn, Gaston v4:357
Thorvaldsen, Bertel v4:141
Thucydides v4:461
Thuku, Harry v1:268
Thurston, Sir John v3:140
Thutmose III, Pharaoh of Egypt v1:161, v1:163, v3:122
Tiahaha, Martha Christina v3:174
Tiahuanaco Indians v2:68
Tian Zhuangzhuang v3:117
Ticos. *See* Costa Rica
Tiglath-pileser I v3:218
Tigran the Great v4:28
Tihuanaco Indians v2:70, v2:82, v2:391
Tilak, Bal Gangadhar v3:160
Timerman, Jacobo v2:31
Timman, Abu v3:224
Timur v3:12, v3:154, v3:195, v3:221, v3:287, v3:345, v3:476, v3:594, v3:629, v3:631, v3:655, v3:658, v4:29, v4:208
Timur-i-Lang. *See* Timur
Ting Ling v3:113
Tinoco, Federico v2:168
Tipu Sultan v3:158
Tiridates I v4:28
Tiridates III of Armenia v4:26, v4:28
Tisquantum v2:462
Titanic v4:620
Tito, Josip Broz v1:174, v4:9, v4:15, v4:77, v4:83, v4:85, v4:110, v4:114, v4:117, v4:119, v4:363, v4:371, v4:649, v4:660, v4:662
influence in Macedonia v4:368
influence in Slovenia v4:528
To Rot, Peter v3:498